BIOLOGICAL SCIENCE

WILLIAM T. KEETON

JAMES L. GOULD

WITH CAROL GRANT GOULD

BIOLOGICAL SCIENCE

VOLUME ONE | FIFTH EDITION

W · W · NORTON & COMPANY · NEW YORK · LONDON

PRINTED IN THE UNITED STATES OF AMERICA

This book is composed in Aster. Composition by New England Typographic Service, Inc. Manufacturing by R. R. Donnelley & Sons, Company. Book design by Antonina Krass.

FIFTH EDITION

Library of Congress Cataloging-in-Publication Data

Keeton, William T.
Biological science/William T. Keeton, James L. Gould, with Carol Grant Gould. —5th ed.
p. cm.
ISBN 0-393-96223-7
1. Biology. I. Gould, James L., 1945-
II. Gould, Carol
Grant. III. Title.
QH308.2.K44 1993
574—dc20 92-19326
ISBN 0-393-96223-7 (cl)
ISBN 0-393-96224-5 (V. I pa)
ISBN 0-393-96225-3 (V. II pa)

W. W. Norton & Company, Inc., 500 Fifth Avenue, New York, N.Y. 10110
W. W. Norton & Company Ltd., 10 Coptic Street, London WC1A 1PU
2 3 4 5 6 7 8 9 0

ABOUT THE COVER

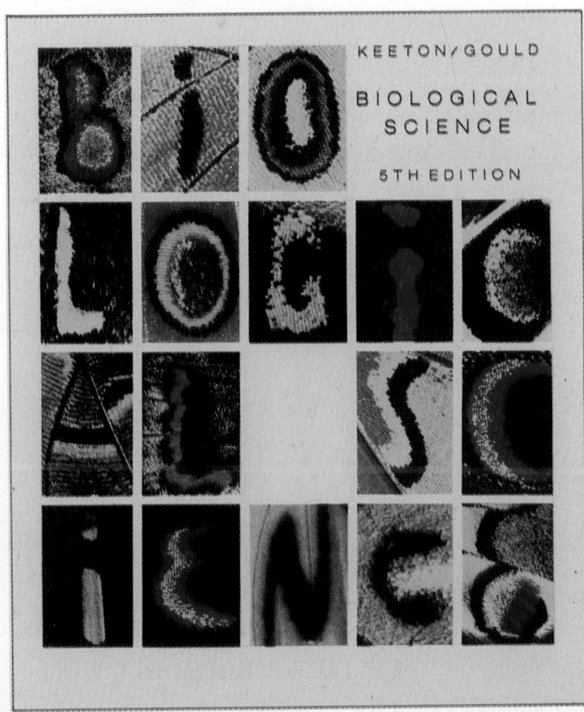

IN ONE VOLUME OR TWO

The colorful letters on the cover, spelling the title of this book, are enlarged photographs of patterns found on the wings of butterflies and moths. They were photographed from all over the world by Kjell B. Sandved of the Smithsonian Institution's National Museum of Natural History in Washington, D.C.

Butterfly-wing patterns may discourage predators and identify potential mates; their developmental histories are under intense scrutiny. As a striking example of the diversity of forms in just one group of species, they demonstrate the vast richness of the variation on which evolution acts, a central theme of modern biology.

CONTENTS IN BRIEF

VOLUME ONE

VOLUME TWO

PART IV THE GENESIS AND DIVERSITY OF ORGANISMS

PART V THE BIOLOGY OF ORGANISMS

PART VI ECOLOGY

CONTENTS

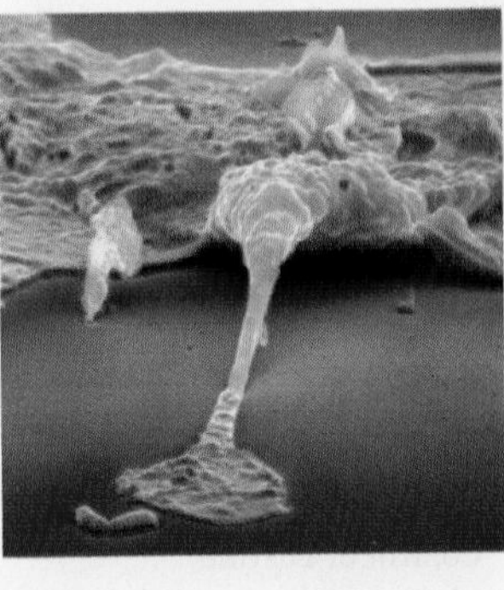

PART III EVOLUTIONARY BIOLOGY

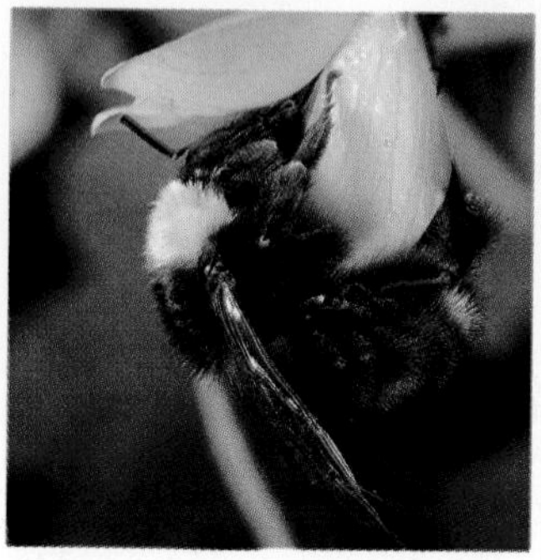
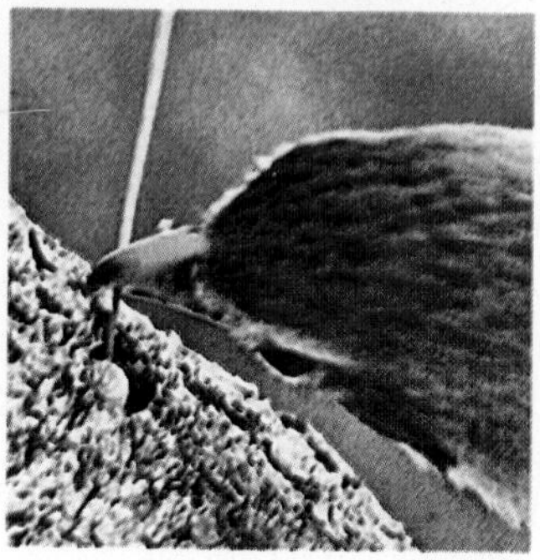

PREFACE

When Bill Keeton's first edition of *Biological Science* appeared in 1967, it started a revolution so complete that we now take his vision for granted. Instead of separate introductory courses (and texts) in microbiology, botany, and zoology, Bill saw that all of biology could (and should) be united. Today it seems obvious that the cells and molecules of plants, animals, and unicellular organisms are very similar—indeed, so nearly identical in most respects that the differences are full of evolutionary significance. Today we begin with the assumption that natural selection operates on organisms of all three groups in similar ways, and that studying the interactions of different groups with one another is essential to any real understanding of ecology. Moreover, as Bill showed most convincingly from the outset, the basic physiological challenges faced by animals, plants, and microorganisms—gas exchange, nutrient procurement, internal circulation, coordination of function, and so on—can be best understood by juxtaposition and contrast. For him, and now for nearly everyone, biology is a unified subject. It is this view of biological science that is his enduring legacy to us all, and the guiding principle upon which this book continues to be based.

We had three main objectives in preparing the Fifth Edition: (1) to bring the book up to date in both depth and scope, so that it continues to reflect new discoveries and to anticipate shifting emphases in the advanced courses for which it may be the student's only preparation; (2) to continue to improve the clarity of the presentation wherever possible, adding more intuitive explanations and more functional examples, and thus making even the most complex subject matter accessible to a wider range of students; (3) to keep the book manageably brief, which sometimes required the abbreviation or deletion of less important topics. Above all, we wanted to reinforce the evolutionary theme in all parts of the text, and to provide more satisfactory molecular explanations of the mechanisms of biology in all chapters. From our own experience and the comments of other teachers of introductory biology, it was clear that the content of every chapter had to be scrutinized once again for accuracy, emphasis, and effectiveness. In the end, every chapter benefited from this process.

CHANGES IN ORGANIZATION

Several major changes in the new edition will be immediately apparent. The sequence of chapters originally laid out by Bill Keeton alternated cellular and organismal topics, so that each semester of his course had some of each. More and more, however, the two semesters of introductory biology are taught by different instructors (often from different departments); one contingent usually focuses on cellular and molecular topics, often from an evolutionary perspective, while the other deals principally with evolution, diversity, physiology, and ecology. It is a continuing tribute to the quality and flexibility of Bill's writing that many schools have taught from *Biological Science* in this order even though the book had a different sequence. Although the order of chapters now follows (roughly) a more common levels-of-organization approach, we have taken special care that the new edition still works with Bill's original order of teaching, and that it continues to provide the coherence of presentation and development that is essential for mastering biological principles.

The reorganization of the book has several advantages. Part II, THE PERPETUATION OF LIFE (formerly Part III), now follows THE CHEMICAL AND CELLULAR BASIS OF LIFE, Part I. Within Part II the order of chapters has been rearranged to allow the student to move directly from the subcellular emphasis of Part I into the molecular basis of information flow. To accomplish this logical transition, we have postponed the material on cell division and classical genetics: cell division now immediately precedes embryology and development, and inheritance comes at the end, leading directly into the discussion of the genetic basis of evolution in Part III, EVOLUTIONARY BIOLOGY. Part III now ends with a section on phylogeny, which lays the groundwork for Part IV, THE GENESIS AND DIVERSITY OF ORGANISMS. The chronicle of diversity in Part IV in turn provides the background for the comparative physiology of organisms in Part V, THE BIOLOGY OF ORGANISMS. And Part V culminates in the study of the mechanisms and evolution of behavior, which leads naturally into ECOLOGY, Part VI.

Some teachers will be pleased to find that we have restored the Selected Readings to the ends of chapters in the Fifth Edition, and that we have added Concepts for Review and Study Questions at the ends of chapters. We have not followed what we feel is the ineffectual practice of including multiple-choice questions at the ends of chapters, but have instead focused on questions that encourage students to review the basic concepts and big ideas, and that provoke thought about the material. These questions are a helpful complement to the more thorough and disciplined study regime provided by Carol H. McFadden's excellent *Study Guide*.

Other obvious changes include the availability of the book in a more portable two-volume format in addition to the conventional single-volume version, and the inclusion of many new four-color illustrations throughout the text. Users of the previous edition will see immediately that we have added many new photographs and line drawings to this edition to summarize, dramatize, and reinforce points made in the text. We have avoided the all-too-common shortcut of developing illustrations independently of the text. Readers of *Biological Science* have come to expect an unfailing harmony of text and illustration, and we have worked hard to make sure that difficult concepts are reinforced visually, and that all terms and ideas presented in illustrations are fully documented in the text.

A GUIDE TO SPECIFIC CHANGES

In the interest of brevity we list only major changes in this edition. The *Instructor's Manual* provides a more thorough description of the revisions and their rationale.

Chapter 1 (Introduction) has added new topics in modern biology, forecasting new material in each of the six parts.

PART I: THE CHEMICAL AND CELLULAR BASIS OF LIFE

Chapter 2 (Some Simple Chemistry) further explains the polarity of water molecules with a discussion of how soaps and detergents work.

Chapter 3 (The Chemistry of Life) introduces "designer" enzymes and new information on the chemical composition of the cell.

Chapter 4 (At the Boundary of the Cell) has new material on clathrin and the formation of vesicles.

Chapter 5 (Inside the Cell) has more information on the mailing-label strategy of targeting proteins, and more about peroxisomes, microfilaments, microtubules, intermediate filaments, mechanisms of cell movement (including cilia and flagella, previously in the muscle chapter), and the intracellular transport of organelles and vesicles, especially the roles of kinesin and dynamin.

Chapter 6 (Energy Tranformations: Respiration) places added emphasis on the role of electronegativity of oxygen versus other atoms, and has a clearer and simpler summary diagram for the anatomy of respiration.

Chapter 7 (Energy Transformation: Photosynthesis) now explores the physiological ecology of granal versus stromal thylakoids, and also contains a new *Exploring Further* section on the structure of the photosynthetic reaction center.

PART II: THE PERPETUATION OF LIFE

Chapter 8 (The Structure and Replication of DNA) now includes a discussion of the replication of DNA in organelles.

Chapter 9 (Transcription and Translation) includes new material on how transcription is terminated, the mechanism of exon splicing, the process of translation, and how ribosomes bind to the endoplasmic reticulum.

Chapter 10 (Mobile Genes and Genetic Engineering) is new. It deals with mechanisms of genetic mobility including transduction, transformation, plasmids, lytic versus lysogenic viruses, retroviruses, and transposons. It also explores the evolutionary significance and interrelationships for these processes or entities, and the practical use of each for genetic engineering. There are also new discussions of the polymerase chain reaction and gene therapy.

Chapter 11 (Control of Gene Expression) contains new material on the structure of DNA-binding proteins, the structure and function of the CAP-activator system, transcription factors, inducers, and enhancers. New sections have also been added on telomeres, genetic imprinting, alternative splicing, translation inhibitors (including anti-sense RNA), and mRNA and protein-digesting enzymes that control molecular lifespans. There is now a discussion of mutations of control versus structural regions and an expanded section on cancer, including the mechanisms of metastasis, oncogene formation, and oncogene operation.

Chapter 12 (Cellular Reproduction) has new sections on cyclins and control of the cell cycle, the evolutionary logic of meiosis, the timing of meiosis in plants, and the investment in diploid and haploid phases in the life history of a species.

Chapter 13 (The Course of Animal Development) no longer incorporates a discussion of plant development, which has been integrated into a later chapter on plant hormones. This change more fully recognizes that development patterns in plants are fundamentally different from those in animals. In plants, since rigid cell walls make most morphogenesis impossible, differential growth is critical; there is no need for many different organs since plants are autotrophic, and growth continues in select tissues throughout the life of the plant rather than being turned off, as in animals. These important developmental strategies are compared and contrasted in their respective chapters.

Chapter 14 (Mechanisms of Animal Development) more clearly distinguishes induction and differentiation, and provides a fuller discussion of the role of CAMs in cell migration and morphogenesis, forecasting their role in immunology. There is now a discussion of somites and the strategy of iteration of subunits in bilaterally symmetric animals, a treatment of morphogen action and pattern formation in *Drosophila*, including homeotic genes and homeobox sequences, and a description of the likely role of retinoic acid in vertebrate development.

Chapter 15 (Immunology) was largely rewritten and reillustrated to update and simplify the treatment. Most of the material on gene evolution is now incorporated in subsequent evolution chapters in the context of how new alleles arise; this chapter retains a discussion of hypermutation and the hypothesis that immune-system molecules evolved from CAMs. New material has been added on the lymphatic system, and the structure, life history, and effects of the AIDS virus.

Chapter 16 (Inheritance) now covers both Mendelian and non-Mendelian patterns of inheritance in one chapter. This combined treatment eases the transition from the preceding molecular discussions to the sections on the allelic distributions in populations that follow in Chapter 17.

PART III: EVOLUTIONARY BIOLOGY

Chapter 17 (Variation, Selection, and Adaptation) incorporates new material on the genetic bases of variation, frequency-dependent selection, and sexual selection. An *Exploring Further* section questions the evolutionary value of sexual reproduction and introduces the gene-repair, red-queen, and tangled-bank hypotheses.

Chapter 18 (Speciation and Phylogeny) has more on punctuated equilibrium, the Burgess shale fauna, the relationship between development and evolutionary change, the quantification of relatedness in different classification schemes, cladistics, and molecular taxonomy.

PART IV: THE GENESIS AND DIVERSITY OF ORGANISMS

Chapter 19 (The Origin of Life) has a new section on the possible role of comets and asteroids in contributing water, organic molecules, and other conditions favorable for life to the earth, together with a discussion of ribozymes as possibly the first enzymes and information-storage molecules. Material on the endosymbiotic hypothesis has been updated, and there is now an improved discussion of the ambiguities inherent in kingdom classifications. The kingdom-classification scheme adopted for the Fifth Edition has been thoroughly modernized, based on the latest and most reliable sequence comparisons.

Chapter 20 (Viruses and Bacteria) has more on viroids, prions, and aquatic viruses; it also has a more ecological, functional, and evolutionary treatment of bacteria. Along with a modern, sequence-based phylogeny of bacteria, it presents an explicit comparison of Gram-positive and Gram-negative cell walls, more on mycoplasmas and myxobacteria, and a more comprehensive treatment of archaebacteria.

Chapter 21 (Archaezoans and Protists) covers organisms of both groups together, omitting the chlorophyll *c* algae (now in Chapter 22) and some single-celled groups that belong with the fungi. The new sequence-based phylogeny is correlated with the latest ultrastructural findings.

Chapter 22 (Chromistans and Plants) encompasses both Chromista and Plantae; it includes a new functional/evolutionary comparison of the algae and higher plants, a discussion of sexual selection in plants, and a summary of plant tissues.

Chapter 23 (Fungi) has an overview of fungal niches, including the one occupied by the pneumonia-causing species that kills many AIDS patients.

Chapter 24 (Invertebrate Animals) places greater emphasis on the progression of developmental patterns and the many parallels with plant evolution, specifically in regard to surface-to-volume ratios and the transition from water to land.

Chapter 25 (Chordate Animals) includes an overview of animal tissues, as well as more on mass extinctions and the use of sequence analysis to trace human evolution.

PART V: THE BIOLOGY OF ORGANISMS

Chapter 26 (Nutrient Procurement and Processing by Plants and Other Autotrophs) has an expanded explanation of water movement, recast in terms of water potential (to allow for the effects of turgor pressure). There is also more on nitrogen fixation.

Chapter 27 (Nutrient Procurement and Processing by Animals and Other Heterotrophs) now mentions avian fermentors, compares the waste-disposal problem in plants with that in animals, and has a new section on plant poisons.

Chapter 28 (Gas Exchange) has an expanded discussion of the control of stomatal opening and closing, and the implications for water use by plants. There is also a new section on water conservation by animals during breathing, with particular reference to the role of countercurrent exchange in water recovery.

Chapter 29 (Internal Transport in Unicellular Organisms and Plants) now discusses plant circulation in terms of water potential, to allow for consideration of temperature gradients, as opposed to just osmotic gradients. There is also a comprehensive presentation of the TATC (transpiration-adhesion-tension-cohesion) theory, as well as new sections on the water cycle in plants and adaptations for water conservation.

Chapter 30 (Internal Transport in Animals) has a major new section on temperature regulation, including the costs and benefits of homeothermy, fevers, and temporal heterothermy. A revised discussion of heat conservation and cooling focuses on the role of countercurrent exchange. Also included are a discussion of the evolution of hemoglobin, an examination of O_2-CO_2 exchange in corpuscles, and an overview of the many functions of the circulatory system, by way of introduction to the next few chapters.

Chapter 31 (Regulation of Body Fluids) has a revised description of kidney function.

Chapter 32 (Development and Chemical Control in Plants) now includes discussions of flower development and homeobox control. There is also more on root growth, tropisms, spacing, and root-to-shoot ratios, as well as an expanded section on photoperiodism and flowering.

Chapter 33 (Chemical Control in Animals) includes a new discussion of atrial natriuretic factor (ANF), nitric oxide, somatostatin, and the hormonal regulation of blood composition and volume. There is also an updated discussion of insulin and diabetes, an explanation of the G-protein transduction system, and more on the evolution of hormones.

Chapter 34 (Hormones and Vertebrate Reproduction) has a description of how the abortion-inducing drug RU 486 works.

Chapter 35 (Nervous Control) discusses the role of chloride channels in cystic fibrosis and of nitric oxide as a transmitter. A shortened discussion of presynaptic phenomena (habituation, sensitization, and conditioning) is now incorporated into the text rather than being set off in a box.

Chapter 36 (Sensory Reception and Processing) has a discussion of the recent discovery of olfactory-receptor genes and an updated description of the molecular basis of visual and auditory transduction. There is also a new discussion of frequency-tuning within the ear and lateral line organs. The discussion of visual processing has been brought up to date and related to effects of neural anomalies, including the possible biological cause of dyslexia.

Chapter 37 (Muscles) now takes into account the use of hydrostatic movement by vertebrates.

Chapter 38 (Animal Behavior) combines and shortens what

had been two chapters, with some behavioral topics moved to more appropriate places in Parts III and VI. There is also a new section on risks and deception, and the discussion of programmed learning has been expanded.

PART VI: ECOLOGY

Chapter 39 (Ecology of Populations and Communities) incorporates completely rewritten sections on population regulation and different forms of density-dependent limitation. The definition of a niche has been revised, and a new section has been added on social organization, including a discussion of the costs and benefits of sociality, the mechanisms of resource control, and the nature and role of altruism. The discussion of human ecology is now here, and the projections of human population growth have been updated.

Chapter 40 (Ecosystems and Biogeography) includes a revised discussion of trophic levels and food webs. The costs and benefits of livestock ranching are also considered, and there is more about human effects on the cycling of materials, including the probable roles of CO_2, methane, and pollution in altering climate. This theme is complemented by an enlarged discussion of the fluorocarbons in ozone destruction and an updated discussion of the chemical basis of plant loss to acid rain and ozone. The section on island biogeography has been updated.

ESSENCE OF BIOLOGY: HYPERCARD AND WINDOWS$_{TM}$ SUPPLEMENTS

This edition of *Biological Science* is accompanied by two new supplements: a HyperCard review and a Windows$_{TM}$ review, developed in collaboration with Grant F. Gould. Suitable for use with Apple Macintosh and DOS 386 computers, respectively, these reviews consist of "stacks" of "cards" summarizing each part, chapter by chapter, together with a set of multiple-choice questions for each chapter, as well as a glossary. Each review card provides a succinct discussion of a concept and allows direct access to the review questions, the glossary, and a table of contents. Most cards include illustrations, some of which are animated to help demonstrate dynamic processes. The first use of any term in the glossary is in boldface, and clicking a mouse-driven cursor on it takes the student directly to its definition. The definitions themselves incorporate cross references to other glossary terms, so that clicking on a boldfaced word in a definition takes the user to that card. The review questions include responses to each choice that explain why the answer selected is correct or incorrect. Other useful features will become clear as the student explores this novel interactive learning aid. We wish to thank the students in the introductory biology course at Princeton for serving as guinea pigs for the preliminary version of the review and for providing valuable feedback; Jessica Avery at W.W. Norton suggested many thoughtful improvements.

ACKNOWLEDGMENTS

A revision of this magnitude of a book with such high standards to maintain would have been impossible without the help of many reviewers. In particular, we would like to thank Wayne M. Becker, University of Wisconsin; Robert A. Bender, University of Michigan; Dennis Bogyo, Valdosta State College; Carole Brown, Wake Forest University; Christine L. Case, Skyline College; Thomas Cavalier-Smith, University of British Columbia; Anne M. Cusic, University of Alabama; Peter J. Davies, Cornell University; Michael Foote, Wake Forest University; Joseph Frankel, University of Iowa; Florence Gleason, University of Minnesota; Lane Graham, University of Manitoba; Barbara Hilyer, University of Alabama; Carl Hopkins, Cornell University; John B. Jenkins, Swarthmore College; Dan Jones, University of Alabama; Alan R. Kabat, Harvard University; Glenn Klassen, University of Manitoba; Ken Marion, University of Alabama; Robert M. May, Oxford University ; Scott Orcutt, University of Akron; Maggie T. Pennington, College of Charleston; Wiltraud Pfeiffer, University of California, Davis; Thomas L. Poulson, University of Illinois; Thomas B. Roos, Dartmouth College; Steve Strand, University of California, Los Angeles; W. Edward Sullivan, Princeton University; Heinz Valtin, Dartmouth Medical School; Joseph W. Vanable, Purdue University; Liz Van Volkenburgh, University of Washington, Charles F. Westoff, Princeton University, and D. Reid Wiseman, College of Charleston.

The many new and uniformly excellent pictures and drawings in the Fifth Edition speak more eloquently than we can of the contributions of Ruth Mandel, our photo editor, and of Michael Reingold and Michael Goodman, the artists. Clark Carroll and John McAusland helped in the preparation of the artwork. The copy editing was expertly handled by Emily Arulpragasam, while Lee Marcott, as project editor, managed somehow to coordinate everything and keep everyone on track. The greatest contribution to both the rigor and the aesthetic appeal of the text was made by our tireless editors, James D. Jordan and Joseph Wisnovsky. To all of these individuals, our heartfelt thanks.

J.L.G.
C.G.G.

Princeton, New Jersey
March 1992

BIOLOGICAL SCIENCE

Chapter **1**

INTRODUCTION

LIFE

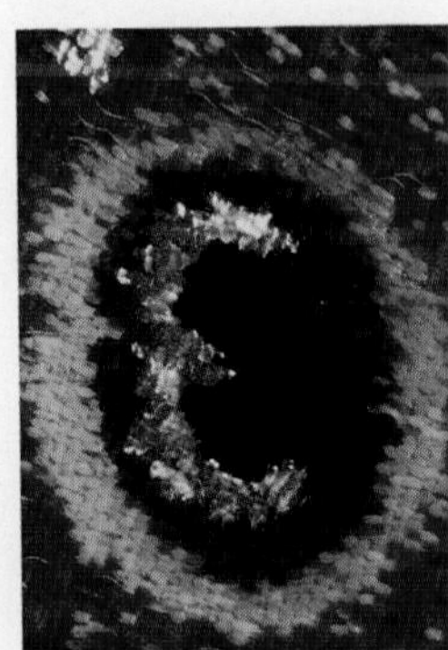

very day, the sun radiates vast amounts of energy to the earth and our moon, and to all the other planets in our solar system (Fig. 1.1). And every day, the planets reradiate that energy into space. Only on earth is there a unique delay in the re-emission of a tiny fraction of that energy: for a brief moment, a minute portion of it is trapped and stored. This minor delay in the flow of cosmic energy powers life. Plants capture sunlight and use its energy to build and maintain stems and leaves and seeds, while animals secure the energy of sunlight by eating plants, or by eating other animals that have eaten plants. At each stage in the many processes of living and dying, waste heat is produced, which joins the pool of energy being relayed back into space.

Biology is the study of this very special category of energy utilization—it is the science of living things. The world is teeming with life: millions of species of organisms of every description inhabit the earth, feeding directly or indirectly off lifeless sources of energy like the sun. What is there in this maze of diversity that unites all biology into one field? What, for instance, do amoebae, redwoods, and people have in common that lends unity to biological science? What differentiates the living from the nonliving? What, in short, *is* life?

Most dictionaries define life as the property that distinguishes the living from the dead, and define dead as deprived of life. These singularly circular and unsatisfactory definitions give us no clue to what we have in common

1.1 The earth as seen from space

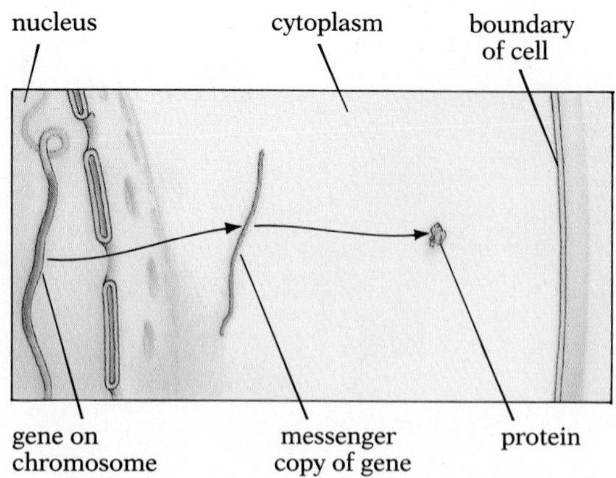

1.2 A simple model of information flow in a cell The internal chemistry of every cell is controlled ultimately by the cell's genes. The genes, located on chromosomes in the cell nucleus, consist of long sequences of four different chemical compounds. A copy of a gene is transcribed into a messenger molecule when needed, and is then transported into the cytoplasm (the portion of the cell that lies outside of the nucleus). There, the messenger is used to direct the construction of a specific protein, which in turn can act as a structural component of the cell or can facilitate and regulate chemical reactions.

with protozoans and plants. The difficulty for the scientist as well as for the writer of dictionaries is that life is not a separable, definable entity or property; it can't be isolated on a microscope slide or distilled into a test tube. To early "mechanistic" philosophers like Aristotle and Descartes, life was wholly explicable in terms of the natural laws of chemistry and physics. The "vitalists," philosophers of the opposing school, were convinced that there was a special property, a "vital force," absent in inanimate objects, that was unique to life. Though some scientists continue to think in terms of an unnamed and intangible special property, vitalism has been essentially dead in biology for at least half a century. The more we learn about living things, the clearer it becomes that life's processes are based on the same chemical and physical laws we see at work in a stone or a glass of water.

If life is not a special property, what is it? One answer may be found by comparing living and nonliving things. Organisms from bacteria to humans seem to have several attributes in common. For instance, all are chemically complex and highly organized. All use energy (metabolize), organize themselves (develop), and reproduce. All change (evolve) over generations. So far as we know, no nonliving thing possesses all these attributes. In addition, and perhaps most important, only the living organism has a set of instructions, or "program," resident in its genes, that directs its metabolism, organization, and reproduction, and is the raw material upon which evolution acts (Fig. 1.2).

THE SCIENTIFIC METHOD

We live in a science-conscious age. Almost every day the news media report some fresh development in agriculture, medicine, or space technology. Commercials use "scientists" in starched white lab coats to woo us with the latest products of "scientific" research. Yet even among well-educated people there is a surprising lack of understanding of what science really is. Some view it as a sort of magic, as indeed it was in the days when many scientists were at least part-time alchemists or astrologers. The sinister scientist seeking forbidden knowledge whatever the cost to society pervades literature in such characters as Faustus, Dr. Jekyll, and, of course, Frankenstein. This image of science persists in some circles, manifesting itself as an uninformed fear of, for instance, research on recombinant DNA or on nuclear fusion. Equally simplistic is the image of scientists as preoccupied, absentminded recluses shut contentedly in academic ivory towers, divorced from—and not greatly interested in—reality.

For the most part, though, scientists are neither sinister nor reclusive; rather, they are people whose native problem-solving mentality and curiosity about the natural world have simply deepened as they have grown older. It is this active curiosity about the living world, springing in most cases from a profound respect for nature, that binds together those who do research in the biological sciences. The unknown and perhaps unknowable instills in us humility and awe. This sense of mystery, accompanied by a belief in the existence of an underlying order, motivates virtually all scientists.

But motivation is only the first requirement for being a scientist. The second is the rigorous and creative application of the ***scientific method***. Like so many truly great ideas, the scientific method is a basically simple concept and is used to some extent by almost everyone every day. As the English biologist T. H. Huxley (1825–1895) put it, the scientific method is "nothing but trained and organized common sense." Its power in the hands of a scientist stems from the rigor and ingenuity of its application.

Formulating hypotheses Science is concerned with the material universe, seeking to discover facts about it and to fit those facts into conceptual schemes, called theories or laws, that will clarify the relations between them. Science must therefore begin with observations of objects or events in the physical universe. The objects or events may occur naturally, or they may be the products of planned experiments; the important point is that they must be *observed*, either directly or indirectly. Science cannot deal with anything that cannot be observed.

Science rests on the philosophical assumption (well justified by its past successes) that virtually all events of the universe can be described by physical theories and laws, and that we get the data with which to formulate those theories and laws through our senses. Needless to say, natural laws are descriptive rather than prescriptive; they do not say how things *should* be, but instead how things are and probably will be. Scientists readily acknowledge the imperfection of human sensory perception: the major alterations our neural processing imposes upon our picture of the world around us are themselves a subject of scientific study. In addition, experience has shown that there is often an interaction between phenomenon and observer: however careful we may be, our preconceived notions and even our physical presence may affect our observations and experiments. But to recognize the imperfection of sensory perception and observation is not to suggest that we may get scientific information from any other source (scriptural revelation, for example). No other means are open to us as scientists.

The first step in the scientific method is to formulate the question to be asked. This is not as simple as it sounds: scientists must decide which of the endless series of questions our escalating knowledge inspires are important and worth answering, and which are trivial. The next step is to make careful observations in an attempt to answer the question. Here too there are difficulties: the researcher must decide what to observe and, since measuring everything is impossible, what to ignore. The scientist must also decide how to make the measurements and how to record the data. This is no trivial matter: an oversight or a mistake can render years of work useless. Next, something must be done with the observations. Simply to amass data is not enough; the data must be analyzed and fitted into some sort of coherent pattern or generalization. A formal generalization, or ***hypothesis***, is a tentative causal explanation for a group of observations. The step from isolated bits of data to generalization can be taken with confidence only if enough observations have been made to give a firm basis for the generalization, and then only if the individual observations have been reliably made. But even when data have been carefully collected, a hypothesis does not automatically follow. Often data can be interpreted in several ways, or may appear to make no sense whatsoever.

Testing hypotheses The making of a general statement, or hypothesis, is not the end of the process. Scientists must devise ways of testing their hypotheses by formulating predictions based on them and checking to see if these predictions are accurate. Again, this is often not easy: a hypothesis may supply the basis for many predictions, but probably only one can be tested at a time. A scientist must decide which predictions can be most readily tested, and of those, which one provides the toughest challenge for the hypothesis (Fig. 1.3).

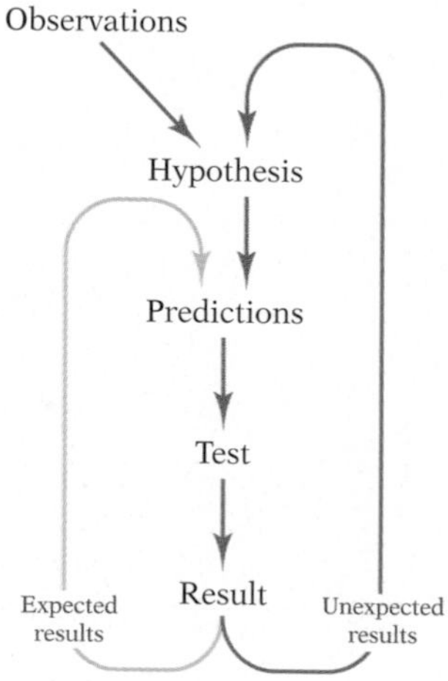

1.3 The scientific method
As this diagram suggests, experimental results are used, when necessary, to modify a hypothesis and to test predictions, but the cycle of testing never really ends.

Perhaps most difficult of all, when researchers find a discrepancy between a hypothesis and the results of their tests, they must be ready to change their generalization. If, however, all evidence continues to support the new idea, it may become widely accepted as probably true and be dignified by the appellation "theory." It is important to realize that scientists do not use the term "theory" as does the general public. To many people a theory is a highly tentative statement, a poor makeshift for fact. But when scientists dignify a statement by the name of theory, they imply that it has a very high degree of probability and that they have great confidence in it.

A ***theory*** is a hypothesis that has been repeatedly and extensively tested. It is supported by all the data that have been gathered, and helps order and explain those data. Many scientific theories, like the cell theory, are so well supported by essentially all the known facts that they themselves are "facts" in the nonscientific application of that term. But the testing of a theory never stops. No theory in science is ever absolutely and finally proven. Good scientists must be ready to alter or even abandon their most cherished generalizations when new evidence contradicts them. They must remember that all their theories, including the physical laws, are dependent on observable phenomena, and not vice versa. Even incorrect theories, however, can be enormously valuable in science. We usually think of mistaken hypotheses as just so much intellectual rubbish to be cleared away before science can progress, but tightly drawn, explicitly testable hypotheses, whether right or wrong, catalyze progress by focusing thought and experimentation.

The controlled experiment In its simplest form the scientific method begins with careful observations, which must be shaped into a hypothesis. The hypothesis suggests predictions, which must be tested. The tests, furthermore, must be *controlled*, a matter of vital importance, since only a controlled test has any hope of illuminating anything. Controlling a test, or experiment, means making sure that the effects observed result from the phenomenon being tested, not from some other source. The most familiar way of controlling an experiment is to perform the same process again and again, varying only one minute part of it each time, so that if a difference appears in the result, we can easily trace the cause. When Louis Pasteur, for instance, took his memorable stand against the prevailing theory of spontaneous generation (the idea that life can arise spontaneously from nutrients), he designed his experiment so that there could be no doubt about the outcome, whatever it might be. He took identical flasks, filled them with identical nutrient solutions, subjected them to identical processes, but left some of them open to the air, while the others were effectively sealed. In time bacteria and mold developed in the open flasks, but the liquid in the sealed flasks remained clear. Since exposure to the outside air was the only variable in the procedure, it was then obvious that the bacteria and the mold-producing organisms, rather than being spontaneously generated from the broth itself, must have come from the air.

Intuition We can see, then, that in actual practice science is neither easy nor mechanical. Most scientists will readily admit that every stage in the scientific process requires not just careful thought but a large measure of intuition and good fortune. At the end of the last century, for instance, physicists saw many small problems, or "anomalies," in Newtonian physics. Experience justifies living with small anomalies when a theory works well, and most minor problems prove in time to be irrelevant or mistaken—the result of observations that were faulty or interpretations that missed a point. Sometimes, though, anomalies in a theory may signal a conceptual error. Albert Einstein (Fig. 1.4) sensed that two of the many irritating anomalies in Newtonian physics were crucial difficulties, and rearranged the pieces of the puzzle to create a revolutionary new theory that accounted for those two anomalies. Like most new hypotheses, Einstein's theory of relativity did not win over the world of science at once. For one thing, it seemed far more complex than the Newtonian mechanics with which everyone was familiar, and it left at least as many (though different) loose ends dangling as had the theory it sought to replace. The potential appeal of Einstein's theory, however, was great: it made some very unlikely but testable predictions—that the light from stars, for instance, should bend as it passed near the sun. When these were subsequently proven to be correct, the theory of relativity was accepted rather quickly, loose ends and all, where a far more plausible but less dramatic hypothesis might well have been ignored.

1.4 Albert Einstein

Scientific investigation, then, depends on a combination of subjective judgments and objective tests, a delicate mixture of intuition and logic. Done well, scientific research is truly an art: the ability to make insightful guesses and imagine clever and critical ways in which to test them is usually the distinguishing characteristic of great scientists. But in the final analysis, the basic rules are the same for all: observations must be accurate and hypotheses testable. And as testing proceeds, hypotheses must be altered when necessary to conform to the evidence.

Limitations of the scientific method The insistence on testability in science severely limits the range of its applications. For example, the idea —widely held by scientists and nonscientists alike—that there is a God working through the natural laws of the universe is simply not testable, and hence cannot be evaluated by science. Science seeks neither to confirm nor to refute it.

Another limitation of science is that it cannot make value judgments: it cannot say, for example, that a painting or a sunset is beautiful. And science cannot make moral judgments: it cannot say that war is immoral. It cannot even say that a river should not be polluted. Science can, however, analyze responses to a painting; it can analyze the biological, social, and cultural implications of war; and it can demonstrate the consequences of pollution. It can, in short, try to predict what people will consider beautiful or moral, and it can provide them with information that may help them make value judgments or moral judgments about war or pollution. But the act of making the judgments is not itself science.

1.5 The Greek philosopher Aristotle
The central portion of *The School of Athens,* a mural by the Renaissance painter Raphael depicting philosophers and scholars of all ages. Aristotle is at right; the Greek philosopher Plato is at left.

THE RISE OF MODERN BIOLOGICAL SCIENCE

EARLY SCIENCE

Science, as we have seen, is the endeavor to understand the natural world. Science must have originated with early humans as they realized that their subjective observations could generate rules of practical utility—when to plant crops, for example, or how to recognize the approach of rain. Though a far cry from modern science and the scientific method, this intellectual observation of cause and effect marked the beginning of scientific thought.

The most important advance in early science came with the Greeks: instead of seeing the universe as ruled arbitrarily by a collection of gods who intervened frequently and capriciously in human events, they began to view the world as operating in a consistent, rule-governed fashion with a minimum of supernatural intervention. Philosophers could therefore set about attempting to discover the natural laws—philosophical principles, as they called them—that the universe obeyed.

The Greeks, particularly Aristotle (384–322 B.C.; Fig. 1.5), made systematic observations and from them formed generalizations, or hypotheses, largely divorced from utilitarian goals. They developed and elaborated formal logic as a powerful intellectual tool, and employed it in their pioneering practice of making deductions from their hypotheses—what we would call making predictions, except that no effort was spent on experimental verification. The emphasis of Greek science was philosophical, its goal the creation of a unifying world picture rather than the working out of details. In addition to the grand philosophical system of Aristotle, there were others, notably that of Anaximander, who believed that everything in the universe was composed of fire, earth, air, and water; that of Pythagoras, who was convinced that all the secrets of the universe were contained in numerical ratios; and that of Democritus, for whom all matter was composed of invisible atoms. Not all Greek science was quite so metaphysical, however, and one particularly fortuitous combination of observation and deduction that is still with us today is the method of medical diagnosis originated by Hippocrates.

Roman culture, though marked by excellence in literature, history, and the arts, added little to the scientific knowledge acquired by the Greeks, and progress of every sort greatly declined after barbarians from northern Europe sacked Rome in the fifth century and ushered in the Dark Ages. Though the Arabs to the east continued to practice science according to the Greek texts they had maintained, what little of Greek science survived in Europe was preserved in remote or fortified monasteries until the slow process of religious, cultural, and military conversion of the barbarians in Europe by Charlemagne and his successors (beginning about A.D. 800) was complete.

At first, scholars had all they could do simply to reabsorb Greek knowledge. The works of Aristotle in particular were enormously influential. His logical, unifying, and aesthetically pleasing world view was especially attractive to a civilization emerging from centuries ruled by the irrational, divisive, and uglier elements of human nature. Aristotle's universe seemed replete with perfection, beauty, and harmony. The earth lay at its center,

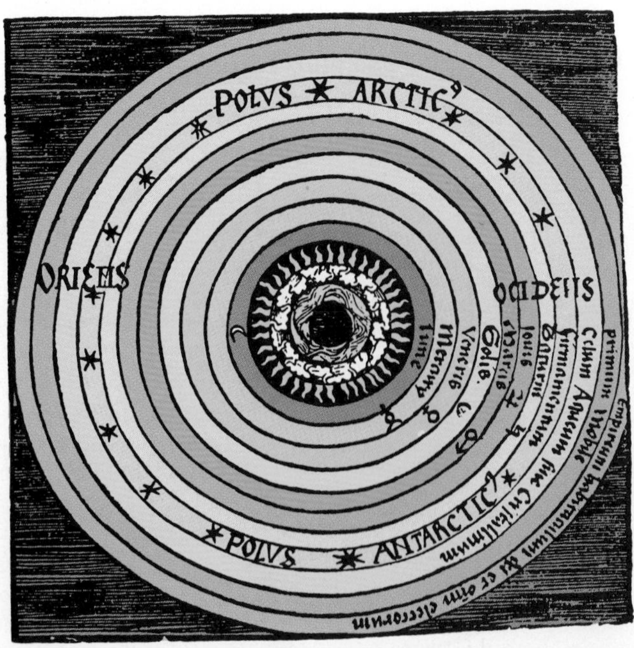

1.6 Aristotle's vision of the universe
The Aristotelian scheme of the universe had the earth in the center, surrounded by a series of concentric spheres containing water (the oceans), air, fire, the moon, Mercury, Venus, the sun, Mars, Jupiter, Saturn, the stars, and the region inhabited by the divine beings who move the spheres. Ptolemaic astronomy sought to explain the observed motion of the planets and stars in terms of Aristotle's system of concentric spheres.

and the moon, sun, planets, and stars moved about the earth on a set of perfect, transparent, concentric spheres (Fig. 1.6). This idealized conception became the basis of what we call Ptolemaic astronomy.

EARLY DISCOVERIES IN ASTRONOMY AND PHYSICS

Perhaps the first modern scientist was a Polish canon, Nicolaus Copernicus (1473–1543). He proposed that the earth orbited the sun, rather than the reverse. This hypothesis had the intuitively satisfying consequence of explaining why the more distant planets, such as Jupiter, reverse their direction of travel against the background of stars roughly once a year: the reversal is a simple consequence of the earth's "passing" them on its inside track around the sun. It also explained why Venus and Mercury never appear very far from the sun: they are circling the sun rather than the earth, and their orbits lie inside that of our planet. Nevertheless, Copernicus' idea was not immediately accepted; the Aristotelian model had become too thoroughly integrated into Western thought.

The fatal blow to the Aristotelian model came from Galileo Galilei (1564–1642; Fig. 1.7), of the University of Padua. Galileo was probably the first person to apply the scientific method rigorously. He was able to show by careful measurement that much of Aristotelian physics was incorrect. For instance, Aristotle had reasoned that if an object weighed twice as much as another, it would fall twice as fast, but when Galileo actually measured the rates of movement (down an inclined plane, as he had no way to time free falls), he found them virtually identical. This finding meant that there was an enormous mistake in the accepted picture of the physical world. Galileo also discovered the principle of inertia: once moving, an object will con-

1.7 Galileo Galilei

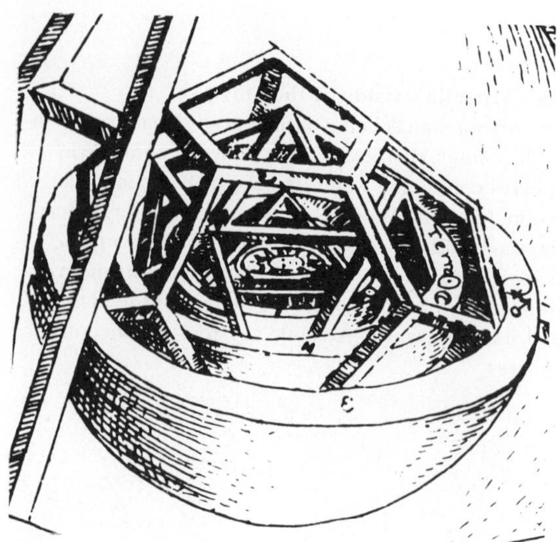

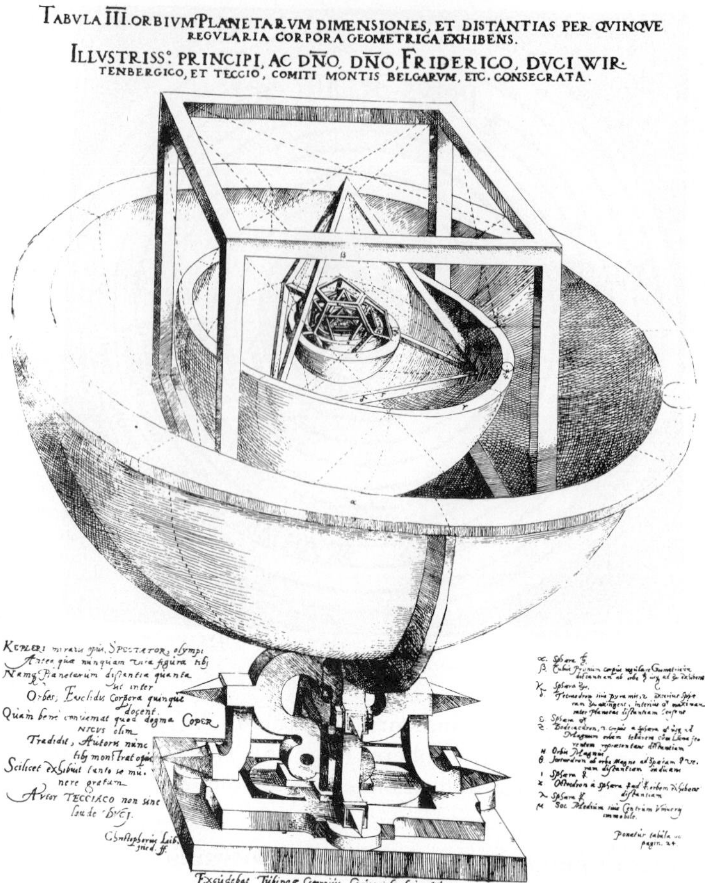

1.8 Kepler's view of the universe
Right: A model of the universe from Kepler's *Mystery of the Cosmos.* The outermost sphere is Saturn's. Above: The detail shows the spheres of Mars, the earth, Venus, and Mercury, with the sun in the center.

1.9 Isaac Newton

tinue at the same speed in the same direction unless acted upon by some other force (such as friction). The discovery of inertia ultimately had a profound effect upon the Western world view. Aristotle had thought that continued motion required continued force, and so someone—God, presumably—had to be continuously at work to keep the planets moving. Now, however, it appeared that God needed merely to have put the planets into place and set them in motion; hence in time the idea emerged of a "God of Secondary Causes" who had created the world, defined the natural laws, and then for the most part left things to take care of themselves. This concept became an underlying tenet for much of Western science in the eighteenth and nineteenth centuries.

Galileo seems also to have been the first scientist to point a telescope at the sky, and wherever he did so he saw that Aristotle's guesses were wrong. The moon was not a perfect sphere of some special celestial substance, but had mountains and craters; the sun was not perfect and immutable, but had spots that appeared, moved, and vanished; Venus had phases, which meant that it was a reflective rather than a luminous object, and it did not follow the path required by the earth-centered system; Jupiter had four tiny moons circling it, in the same way that the planets, small compared to the sun, had been said by Copernicus to circle our star; and Saturn was not a sphere but, seen through Galileo's crude lenses, appeared to have horns.

After Galileo's work, who could any longer doubt the value and wisdom of looking directly at nature? Perhaps the first and certainly the most interesting individual to be inspired by Copernicus and Galileo was the astrologer/astronomer Johannes Kepler (1571–1630), who, convinced both of the Copernican theory and of a God-given order and harmony in the universe, set out to find the divine plan. He began by trying to fit the orbits of the planets into a series of the five perfect geometric solids—the faces of a cube drawn inside the sphere of Saturn's orbit would just contain the sphere of Jupiter's, and a tetrahedron inside it would then just hold the Martian orbital sphere, and so on (Fig. 1.8). Later Kepler even tried to represent the planetary motions as musical chords. Along the way, he discovered the true order of the solar system: the planets move in elliptical orbits at speeds that vary according to their distance from the sun.

Kepler's work more than anyone else's demonstrated the value of a faith that order exists in nature. If there were no underlying order, science would be a waste of time. Faith that the physical world can be understood in terms of orderly relationships and universal laws, that physical events have comprehensible, impersonal causes, and that the whims of gods, magicians, or evil spirits need no longer be invoked to explain physical events, motivated that archetypal scientist Isaac Newton (1642–1727; Fig. 1.9). His discovery of the Law of Gravitation, the principles of optics, the so-called Newtonian mechanics, and the calculus fully justified that faith, and marked the birth of modern physics.

1.10 Cutting the vocal nerve
This is the initial letter in Vesalius' *Fabric of the Human Body* (1542).

THE BEGINNINGS OF MODERN BIOLOGY

The forerunners of modern biological investigation appeared at about the same time as Copernicus and Galileo. Three individuals set the basic course the life sciences were to follow. The earliest was Andreas Vesalius (1514–1564), who made the first serious studies of human anatomy by dissecting corpses. He discovered that the body is composed of numerous complex subsystems, each with its own function, and he pioneered the comparative approach, using other animals to work out the purpose and organization of these anatomical units. A typical (if rather grisly) example is his demonstration that the nerve from the brain to the throat, common to so many animals, is responsible for controlling vocalization. When he took a squealing pig, dissected out the nerve, and cut it, the struggling animal instantly became mute even though its vocal apparatus remained intact (Fig. 1.10).

This powerful style of comparative and experimental study was carried forward by the English physician William Harvey (1578–1657), who showed conclusively that the heart pumps the blood, and the blood circulates. The heart, in short, is not in some metaphysical sense the seat of emotions, but a mechanical device with a clear function. As a result of these studies and the anatomical work that followed, an increasingly mechanistic point of view toward life began to develop.

The third of the pioneers was Antony van Leeuwenhoek (1632–1723; Fig. 1.11). Just as Galileo had the brilliant idea of pointing the newly invented

1.11 Antony van Leeuwenhoek

1.12 **Louis Pasteur**

1.13 **Charles Darwin**

telescope at the heavens, so Leeuwenhoek had the idea of using the microscope—with which he inspected cloth, as a draper's assistant—to look at living things. The most important of his many discoveries were microorganisms (including bacteria), sperm and the eggs they fertilized, and the cells of which all living things seemed to him to be composed.

For biology, unlike physics, centuries of painstaking observation were required to establish the science's fundamental generalizations. The cell theory, for instance, was not given its essentially modern form until 1858, and only about 130 years ago, in 1862, did Louis Pasteur (Fig. 1.12) disprove the theory of spontaneous generation. With the realization that Leeuwenhoek's microorganisms might be responsible for disease, the English surgeon Joseph Lister proved the effectiveness of antiseptics (from *anti-*, "against," and *sepsis*, "decay") in 1865, and Pasteur greatly expanded the use of vaccination.

By far the most important figure in the history of biology, however, is Charles Darwin (1809–1882; Fig. 1.13). The publication in 1859 of his *The Origin of Species*, presenting the theory of evolution by natural selection, suddenly provided a coherent, organizing framework for the whole of biology. His work sparked the explosive growth of biological knowledge that continues today. As the most important unifying principle in biology, the theory of evolution underlies the logic of every chapter of this book.

DARWIN'S THEORY

The theory of evolution by natural selection, as modified since Darwin, will be treated in detail in Chapter 17. But since we shall be referring to it in earlier chapters, we must examine the essential concepts of the theory at the outset. It consists of two major parts: the concept of evolutionary change, for which Darwin presented a great deal of evidence, and the quite independent concept of ***natural selection*** as the agent of that change.

THE CONCEPT OF EVOLUTIONARY CHANGE

Until only two hundred years ago, it seemed self-evident that the world and the animals that fill it do not change: robins look like robins and mice like mice year after year, generation after generation, at least within the short period of written history. This commonsense view is very like our untutored impression that the earth stands still and is circled by the sun, moon, planets, and stars: it accords well with day-to-day experience, and until evidence to the contrary appeared, it provided a satisfying picture of the living world. The idea of an unchanging world also corresponded to a literal reading of the powerfully poetic opening of the Book of Genesis, in which God is said to have created each species independently, simultaneously, and relatively recently—a little over six thousand years ago by traditional scriptural reckonings.

But problems with the commonly held scriptural theory of creation arose from many sources; scientists attempted first, quite naturally, to discount the evidence as ambiguous, and then, when that proved impossible, to construct a new explanation. Let's look at the evidence for evolution that confronted Darwin and his contemporaries.

A

B

1.14 Challenges to traditional ideas of the origin of species
The discovery of fossils of now-extinct species brought into question the static view of life. Shown here are the remains of a baby mammoth (A) that had been preserved in the permafrost in Siberia. The discovery in the New World of organisms unfamiliar to Europeans, such as the anteater (B), also required a reinterpretation of traditional ideas of the origin of species.

The most dramatic findings came from geology. In the eighteenth century a picture of a changing earth had begun to emerge. Extinct volcanoes and their lava flows had been discovered; most geological strata were found to represent sedimentary deposits, laid down layer upon layer a millimeter at a time in columns three thousand meters or more deep; the gradual erosive action of wind and water were seen to have leveled entire mountains and carved out valleys; unknown forces had caused mountains to rise where ocean floors had once been. This latter fact in particular was impressed upon Darwin when he discovered fossilized seashells high in the Andes. Each of these phenomena implied continuous change during vast periods of time.

Another problem for the static view of life was presented by the New World fossils themselves. Many represented plants and animals wholly unknown in Europe, and though theologians had argued that the organisms these fossils represented were alive in the New World, increasingly intensive exploration of the Americas indicated that the hundreds of species of dinosaurs, for instance, were really extinct. In addition, many previously unknown and often bizarre animals inhabited the Americas (Fig. 1.14). As the realization grew that the number of animal species for which evidence was accumulating ran at least into the hundreds of thousands, Noah's ark began to seem very small indeed. In fact, it appeared that the extinct species greatly outnumbered the living, and that new constellations of species had come and gone several times in the past. Moreover, the lowest, oldest rocks contained only the most primitive fossils—seashells, for instance—and these were followed in order by the more modern forms: fish appeared later, for example; reptiles still later; then birds and mammals. The hypothesis of a young earth populated almost overnight by a single bout of creation began to seem very unlikely.

Jean Baptiste de Lamarck (1744–1829; Fig. 1.15) was the first to offer the major alternative explanation of the fossil record: evolution. Lamarck had arranged fossils of various marine molluscs in order of increasing age; he

1.15 Jean Baptiste de Lamarck

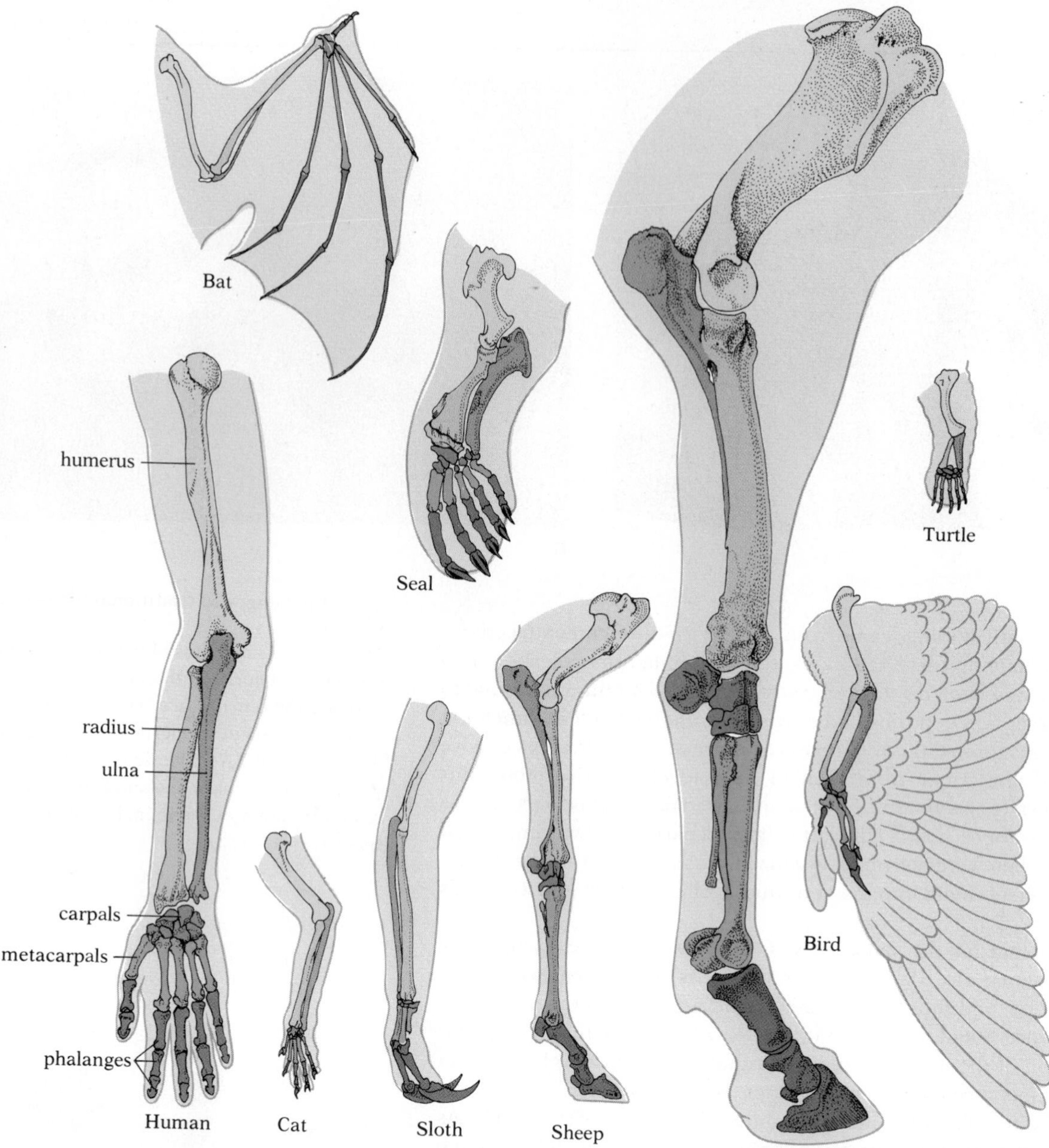

1.16 A comparison of the bones in some vertebrate forelimbs

The labeled and color-coded bones of the human arm at left permit identification of the same bones in the other forelimbs depicted. In the bat the metacarpals (hand bones) and phalanges (finger bones) are elongated as supports for the membranous wing. In the seal the bones are shortened and thickened in the flipper. The cat walks on its phalanges (toes), the metacarpals having come to form a part of the leg. The sloth normally hangs upside down from tree limbs—hence its recurved claws. The horse walks on the tip of one toe, which is covered by a hoof (specialized claw), and the sheep walks on the hoofed tips of two toes (it is therefore cloven-hoofed, though only one hoof can be seen in this side view). The carpals (wristbones) of both the horse and the sheep are elevated far off the ground, because the much-elongated metacarpals (hand bones) have become a section of the leg. Small splintlike bones that are vestiges of other ancestral metacarpals can be seen on the back of the upper portion of the functional metacarpals of both horse and sheep. All the animals mentioned so far—human, bat, seal, cat, sloth, horse, and sheep—are mammals, but the same bones can also be seen in the leg of a turtle and the wing of a bird. (All limbs are drawn to the same scale except that of the turtle, which is enlarged.)

saw clearly that certain species had slowly changed into others, and concluded that this process of slow change had continued right to the present day. As Lamarck put it in 1809, "it is no longer possible to doubt that nature has done everything little by little and successively," over a nearly infinite period of time. In Lamarck's view, the living world had begun with simple organisms in the sea, which eventually moved onto the land, and evolution had culminated in the appearance of our species, the inevitable result of the gradual trend toward change and "increasing perfection."

Lamarck was basically on the right track, though as we will see his mechanism for evolutionary change was incorrect. But he was ignored for the very understandable reason that he could not offer sufficient evidence for the *fact* of evolution. Darwin, only fifty years later, was in a far better position: there was much more evidence of the sort Lamarck had pointed to, and Darwin, a respected geologist, was well acquainted with it. Furthermore, he had the ability to spot important data in the midst of apparent chaos. He could find powerful support for the idea of evolution where Lamarck and others—if they looked at all—saw only irrelevancies.

One of the most important lines of evidence put forward by Darwin was the existence of morphological resemblances among living species (the findings of what we today call comparative anatomy). If, for example, we observe the forelimbs of a variety of different mammals, we see essentially the same bones arranged in the same order (Fig. 1.16). The basic bone structure of a human arm, a cat's front leg, and a seal's flipper is the same; the same bones are present even in a bird's wing. True, the size and shape of the individual bones vary from species to species, and some bones may be missing entirely in one species or another, but the basic construction is unmistakably the same. To Darwin the resemblance suggested that each of these species had descended from a common ancestor from which each had inherited the basic plan of its forelimb, modified to suit its present function. The observation that structures with important functions in some species appear in vestigial nonfunctional form in others further convinced Darwin of the reality of evolutionary change. Why otherwise would pigs, which walk on only two toes per foot, have two other toes that dangle uselessly well above the ground? Why would certain snakes, such as the boa constrictor, and many species of aquatic mammals, such as whales, have pelvic bones and small, internal hind-limb bones (Fig. 1.17)? Why would

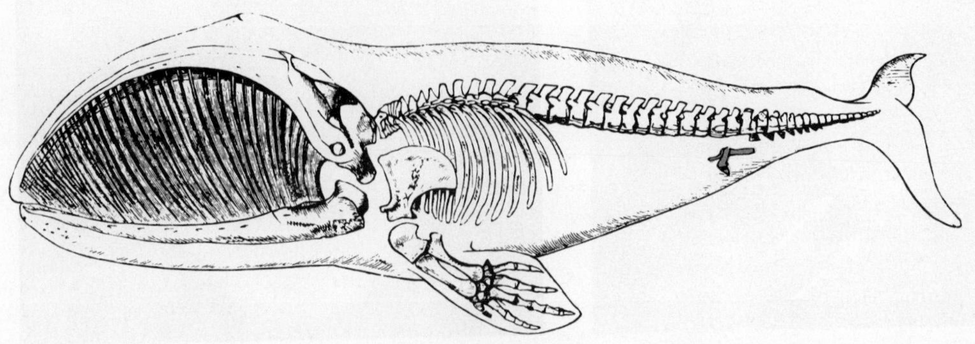

1.17 Rudimentary hind limb of a whale Whales lost their hind limbs long ago, when they returned to the sea from the land, but they retain rudimentary bones that correspond to the pelvic girdle and the thighbone (color).

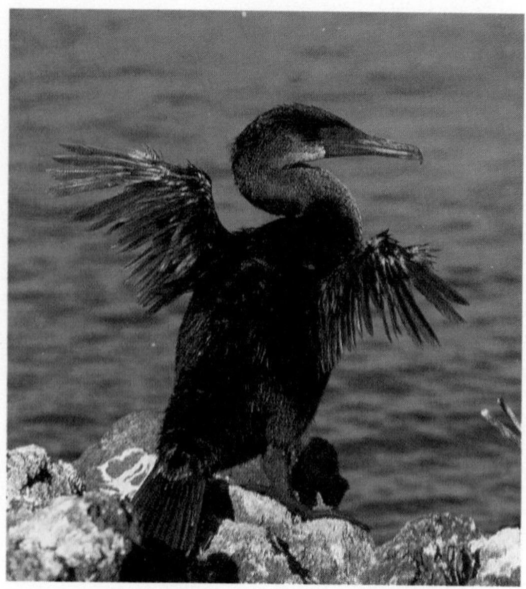

1.18 Flightless cormorant on the Galápagos Islands

flightless birds such as penguins, ostriches, kiwis, and the cormorants of the Galápagos Islands still have rudimentary wings—or feathers, for that matter (Fig. 1.18)? Why would so many subterranean and cave-dwelling species have useless eyes buried under their skin?

Embryology—the study of how living things develop from eggs or seeds to their adult forms—also provided powerfully suggestive evidence. Darwin pointed out that in marine crustaceans as different as barnacles and lobsters the young larvae are virtually identical, implying a common descent. Telltale traces of their genealogy are obvious in vertebrates as well. Human embryos, for instance, have gill pouches and well-developed tails that disappear before the time of birth (Fig. 1.19). It seemed clear to Darwin that such inappropriate structures are inherited vestiges of structures that functioned in ancestral forms, and that may still function in other species descended from the same ancestors.

Another particularly convincing line of evidence offered by Darwin was the well-known ability of breeders to produce dramatic changes in both plants and animals. How could anyone contemplating the historical evidence of the alterations in domesticated species produced by artificial selection doubt that vast changes are possible, given sufficient time? Great Danes, sheep dogs, Irish setters, Yorkshire terriers, poodles, bulldogs, and dachshunds, for instance, are all members of the same species, bred from tamed wolves to look like almost anything breeders have fancied. Similarly, cabbage, brussels sprouts, cauliflower, broccoli, kohlrabi, rutabaga, collard greens, and savoy have all been bred from the same species, the wild form of which looks nothing like its domesticated progeny (Fig. 1.20). So, too, the many varieties of chickens, cattle, horses, flowers, grains, and so on have been bred over the years. Who could compare the colors and shapes of the wild rose or jonquil to the many colors and shapes of the far larger domesticated roses or daffodils, with new varieties bred each year, and doubt that a

1.19 Embryological evidence of evolution
Left: Pharyngeal ("gill") pouches (arrows) in a 4-week human embryo. Right: Tail in a 5-week human embryo.

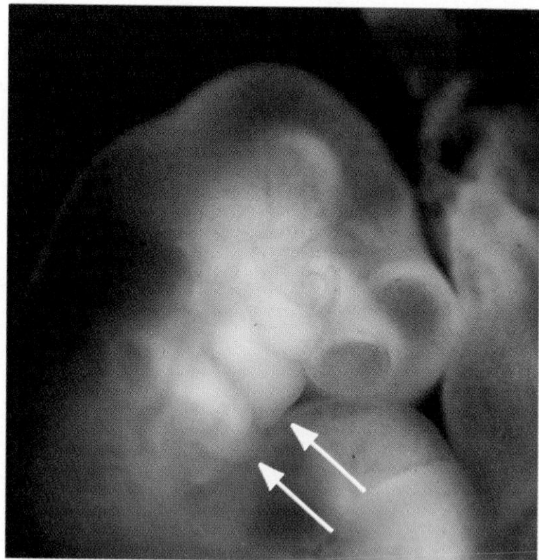

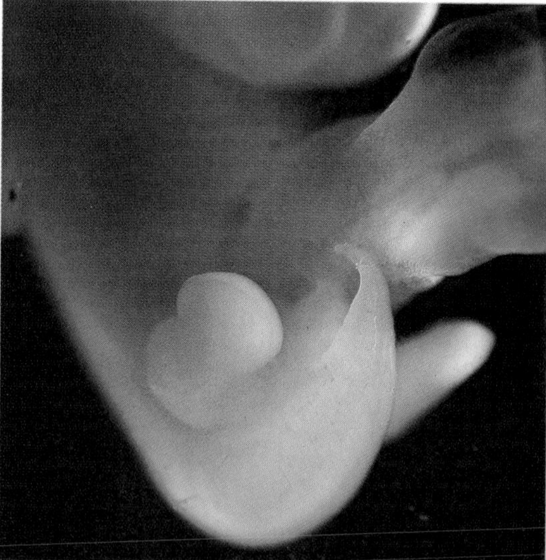

species has the capacity to change enormously even in a hundred years? In everything Darwin looked at—fossils, anatomy, embryology, and breeding—he saw the same message: species can and do change (Fig. 1.21).

THE CONCEPT OF NATURAL SELECTION

But what mechanism accounts for the changes? Lamarck's now-discredited hypothesis was one of the first attempts at a plausible explanation. Lamarck was impressed by how well suited each animal was to its particular position in the web of life, even though the environment had changed enormously again and again over countless millions of years. To account for this ability to adapt, he imagined that God had given each species a tendency toward perfection which allowed for small alterations in morphology, physiology, and behavior to accommodate changes in the environment, and that these alterations could be inherited by the offspring.

Belief in Lamarck's idea of a natural tendency toward perfection and the inheritance of acquired characteristics required no more faith in his day than did belief in other invisible everyday forces, such as gravity and magnetism. His hypothesis was a perfectly logical extension of the prevalent Western view that God had set things going by creating nature and nature's laws, and had then left things for the most part to run themselves. But where in the vestigial legs of whales or the dangling toes of pigs was there evidence of perfection? Instead the clear mark of compromise was everywhere. Plants and animals were well adapted to their places in the environment, but they were by no means perfect.

Darwin proposed a different mechanism—natural selection—requiring no internal tendency other than the one toward variation so obvious in nature. Darwin had conceived the idea of natural selection two years after his return from his voyage to the Americas on the *Beagle*, but was only goaded into publishing, twenty years later, by the receipt in 1858 of A. R. Wallace's manuscript proposing essentially the same theory. (We normally associate Darwin's name with the theory because of the impressive evidence he presented—he had been collecting it for two decades—and because of his thorough exploration of the theory's many ramifications.)

1.20 Selective breeding of the wild cabbage The wild species *Brassica oleracea* (left) has been bred to create cauliflower (upper right), brussels sprouts (middle right), and cabbage (lower right). Each represents a selective exaggeration of one part of the wild plant—the flower heads for cauliflower, the side buds for brussels sprouts, and the leaves for cabbage. Despite the extreme morphological differences, however, the three domestic vegetables can be interbred.

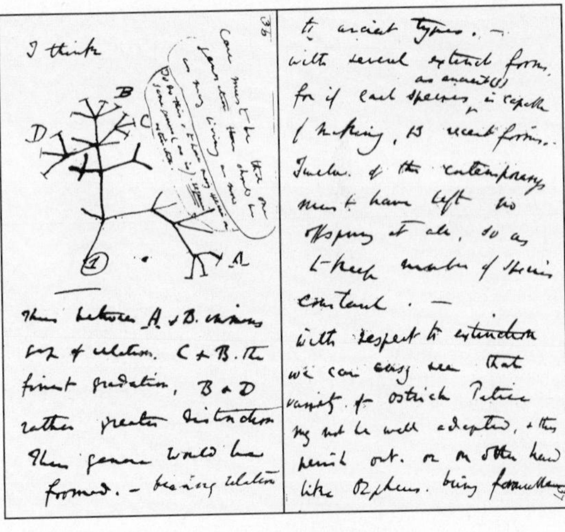

I think

Thus between A & B immense gap of relation. C & B the finest gradation, B & D rather greater distinction. Thus genera would be formed. — bearing relation

to ancient types. — with several extinct forms, for if each species an ancestral is capable of making, 13 recent forms. — Series of the contemporary must have left no offspring at all, so as to keep number of species constant. — With respect to extinction we can easily see that variety of ostrich Petise may not be well adapted, & thus perish out, or on other hand like Orpheus. being favourable

1.21 Pages from one of Darwin's notebooks Scientists of the nineteenth century and earlier, among them Charles Darwin, often recorded in notebooks and journals their experiments, observations drawn from field trips, or simply their thoughts about hypotheses in the making. In addition to the many books he wrote, Darwin produced notebooks on a wide variety of subjects, including his trip to South America and the Galápagos (1831–1836), later described in the *Journal of Researches during the Voyage of the Beagle* (1840). The notebook from which this excerpt is taken was written in 1837, one year before Darwin saw that natural selection was the likely mechanism for evolution. These pages show Darwin's first recorded drawing of an evolutionary tree, a metaphor for the diversity and interrelatedness of species that has continued in the scientific literature to the present time. The trunk represents a common ancestor, the limbs major groups, and the twigs particular species, either extinct (indicated by a crossbar) or living.

1.22 Thomas Robert Malthus

In essence, Darwin put together two ideas. The first was that numerous variations exist within species, and that variations are largely heritable. Immersed as he was in the Victorian preoccupation with plant and animal breeding, Darwin knew that while cuttings produce plants identical to the parent, sexual reproduction produces individual offspring that differ both from their parents and from each other. Variation is a fact of life: breeders, as we know, are able to select for desirable traits and create new, morphologically distinct lines of plants and animals.

Darwin's second inspiration came on 28 September 1838, when he reread the *Essay on the Principle of Population* by the economist Thomas Robert Malthus (Fig. 1.22). Malthus pointed out that humans produce far more offspring than can possibly survive; population growth always outruns any increase in the food supply and is held in check largely by war, disease, and famine. Vast numbers of people thus live perpetually on the edge of starvation. Both Darwin and Wallace were struck at once by the consequence of applying the gloomy Malthusian logic to plants and animals: like humans, the creatures of each overpopulated generation must compete for the limited resources of their environment, and some—indeed most—must die. Each female frog, for example, produces thousands of eggs per year, and a fern produces tens of millions of spores, yet neither population is growing noticeably. Any organism with naturally occurring heritable variations that increase its chances in this life-or-death contest will be more likely than others to survive long enough to have offspring, some of which will inherit these variations. They in turn will have an above-average chance to survive the struggle, and so will form an increasingly large part of the population. As a result of this "selection," the population as a whole will become better adapted—that is, it will evolve—and the never-ending struggle for existence will then turn on the possession of still better adaptations. To distinguish this process from the sort of directed, artificial selection practiced by agriculturalists, Darwin called it *natural* selection.

The contrast between artificial and natural selection that served Darwin so well provides an instructive summary of the evolutionary process. In both, far more offspring are born than will reproduce; in both, differential reproduction, or selection, occurs, causing some inherited characteristics to become more frequent and prominent in the population and others to become less so as the generations pass. But in the breeding of domesticated plants and animals, selection results from the deliberate choice by the breeder of which individuals to propagate. In nature, it takes place simply because individuals with different sets of inherited characteristics have unequal chances of surviving and reproducing. Notice, by the way, that selection does not change individuals. An individual cannot evolve. The changes are in the makeup of populations.

Artificial and natural selection also differ significantly in the *degree* of selection, and its effect on the rate of change. Breeders can practice rigorous selection, eliminating all unwanted individuals in every generation and allowing only a few of the most desirable to reproduce. They can thus bring about very rapid change (Fig. 1.23). Natural selection, which involves a large measure of chance, is usually much less rigorous: some poorly adapted individuals in each generation will be lucky enough to survive and reproduce, while some well-adapted members of the population will not.

1.23 Selective breeding of pigeons By practicing rigorous selection, breeders can achieve major changes in relatively few generations. The ancestral rock dove is shown in the center. The domestic breeds (clockwise from upper left) are fantail, double-crested Saxon, Schmalkalden Moorhead, English carrier, pigmy pouter, Norwich cropper, and frillback.

Hence evolutionary change is usually rather slow; major changes may take thousands or even millions of years, depending on the degree of selection pressure imposed by the environment and by other species.

Darwin's evidence for evolutionary change and the common descent of at least the major groups of organisms was widely accepted in his time, but the idea of natural selection by small steps remained controversial until the 1930s. Some biologists had difficulty seeing how an elaborate and specialized structure like an eye, for instance, could evolve, since the first rudimentary but necessary steps might lack obvious survival value. As we will see in later chapters, an expanded understanding of the nature and organization of genes and their role in development has now made it clear that natural selection does explain most evolutionary change. Darwin extolled the beauty and simplicity of such a system in the final sentence of later editions of *The Origin of Species:* "There is grandeur in this view of life, with its several powers, having been originally breathed by the Creator into a few forms or into one; and that, whilst this planet has gone cycling on according to the fixed law of gravity, from so simple a beginning endless forms most beautiful and most wonderful have been, and are being evolved."

In summary, then, Darwin's explanation of evolution in terms of natural selection depends upon five basic assumptions:

1. Many more individuals are born in each generation than will survive and reproduce.
2. There is variation among individuals; they are not identical in all their characteristics.
3. Individuals with certain characteristics have a better chance of surviving and reproducing than individuals with other characteristics.
4. Some of the characteristics resulting in differential survival and reproduction are heritable.
5. Vast spans of time have been available for change.

All the known evidence supports the validity of these five assumptions.

EVOLUTIONARY RELATIONSHIPS

Darwin's insights into evolution and its mechanisms, together with a variety of techniques (including, most recently, molecular analyses of genes themselves), have permitted a fairly reliable reconstruction of the course of evolution of life on earth, and of the relationships between the species alive today. A diagram of the relationships between the eight major groupings ("kingdoms") of living species recognized at present is shown in Fig. 1.24. The organisms with the longest histories are the two kingdoms of bacteria, which are fundamentally different from the other groups because they lack nuclei; their chromosomes are mixed with the rest of the cellular contents instead of being segregated into a separate compartment. As a result, the bacteria are called ***procaryotes*** ("before nuclei"), while the other six kingdoms comprise the ***eucaryotes*** (having "true nuclei"). (The many other important differences between these two groups of kingdoms will be summarized in Chapter 5.)

The eucaryotes include the two kingdoms that now dominate the earth—

fly agaric, a fungus

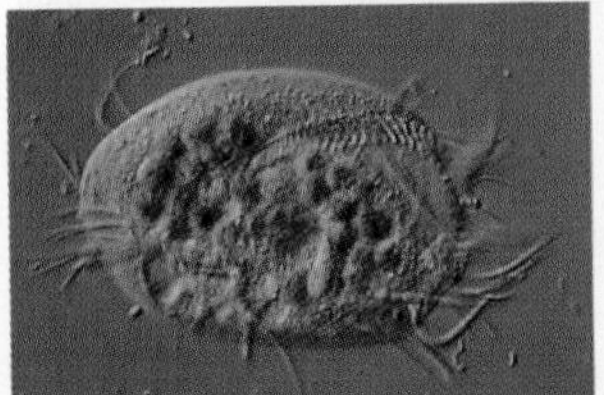
a protist

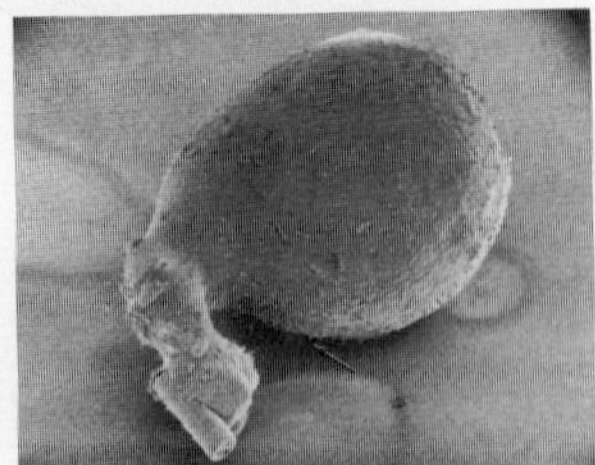
an archaezoan

archaebacteria

great egret

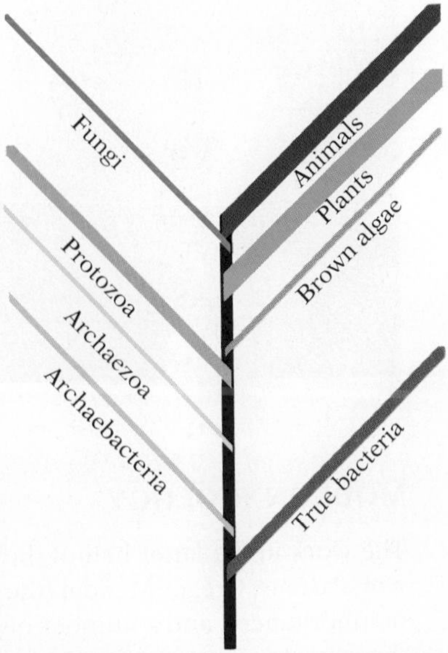

dahlias

kelp, a brown alga

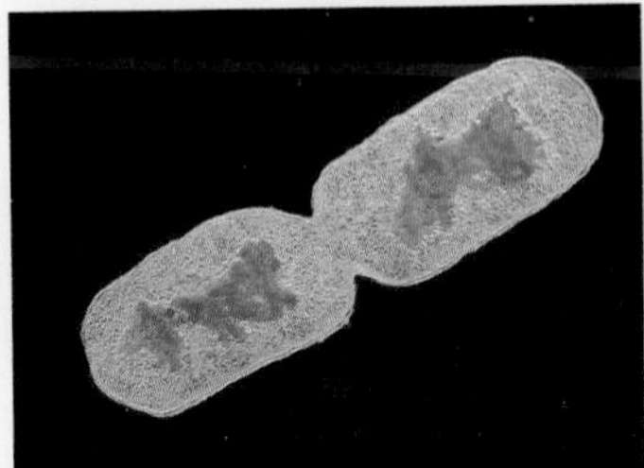
a true bacterium

1.24 Relationships among the eight kingdoms of life
In this representation, kingdoms are shown in color, and the width of the line indicates the relative number of species in each kingdom. The distance from the bottom of the figure at which a group's line departs from the central "trunk" of this evolutionary tree indicates the chronological order in which the kingdoms are thought to have arisen. Common names are used for most groups; technically, animals are in Kingdom Animalia, plants are in Plantae, brown algae are in Chromista, and "true" bacteria are in Eubacteria.

the plants and the animals—as well as three smaller but significant kingdoms, the protozoans (most unicellular organisms), the fungi (mushrooms, for example), and the brown algae (like kelp); they also include one tiny, little-known group, the archaezoans, which are by far the most primitive of the eucaryotes. As we will see, the organisms in these kingdoms are marvelously diverse, having evolved to exploit a bewildering variety of habitats, and yet all share a common set of genetic and biochemical processes that unify the study of life.

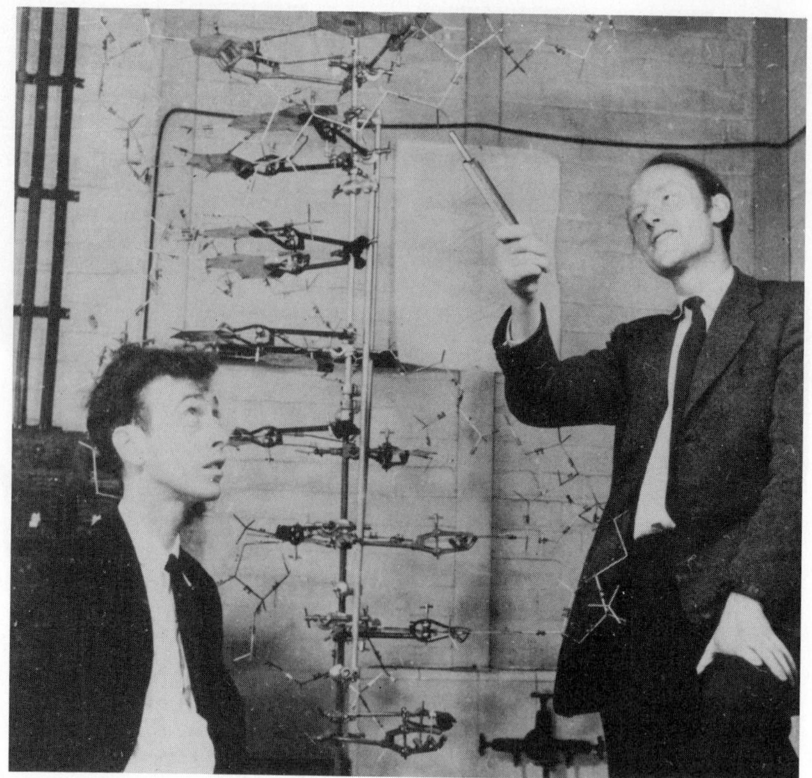

1.25 Watson and Crick with their model of the DNA double helix

MODERN BIOLOGY

The work in the latter half of the nineteenth century by scientists like Darwin, Pasteur, Gregor Mendel (the monk who discovered the basic principles of inheritance), and a number of developmental biologists combined to set the stage for the emergence of modern biology.

Although no one event formally marks the beginning of modern biology, the discovery of the structure of DNA in 1953 was an especially important foundation for later work. The double-helix model proposed by James Watson and Francis Crick (Fig. 1.25) provided a physical basis for Mendel's genes, which are the basic units of Darwinian selection. As we will see in later chapters, the structure immediately suggested how DNA could act as a template for its own reproduction and as a means for issuing instructions for building and operating cells.

Watson and Crick's model of DNA has helped fuel four decades of research on the messages encoded in this remarkable molecule; understanding the code is proving critical to every biological discipline. In relation to cell biology (our focus in Part I), research on DNA has led to a detailed appreciation of how cells are organized, both physically and chemically. One of the more impressive feats of investigators has been the deciphering of the chemical codes used as "mailing labels" on DNA products—codes that tell the elaborate transport system in the cell where to deliver each molecule.

In relation to genetics (Part II), we will see how research on DNA has been used to construct a nearly complete picture of how cells process information to regulate their metabolism, growth, and reproduction. This knowledge is critical in modern treatments of sickness and inborn defects; in the manufacture of chemicals of potentially enormous clinical and commercial importance like synthetic hormones and vaccines; and in the endowment of some plants with new genes that confer benefits like disease resistance. An equally impressive by-product of DNA research is progress toward understanding the body's main line of defense against pathogens (the immune system), as well as the precisely orchestrated development of a complex, trillion-celled organism like a human being from a single, unspecialized cell. Another is the invention, since 1953, of many techniques for manipulating the genetic material and its protein products. Together, the development of these techniques and the investigation of biological processes at the level of molecules are usually referred to as molecular biology.

The benefits of the new molecular biology for whole-organism studies are even more remarkable because they are less obvious. In Part III we will have a glimpse of the enormous impact molecular techniques are having on the study of evolution. Biologists now understand the chemical nature of the variation that is the basis of natural selection, and so are coming to grips with the question of how different species arise in nature and how they adapt to novel conditions. Small changes in genes controlling development, for example, or chromosomal rearrangements that do not actually change the genetic instructions, can produce new species almost instantly. We will see in Part IV that an understanding of how DNA and its chemical collaborators work is providing valuable insights into the origin and early evolution of life, as well as a reliable and often surprising reconstruction of the evolution of the major groups of organisms that populate the earth today. Molecular techniques now allow precise determination of species relationships; they reveal, for instance, that superficial similarities notwithstanding, pandas are not bears, but overgrown raccoons. In Part V, we will see how the molecular approach has exposed the workings of many physiological systems, including those responsible for circulation, gas exchange, movement, and internal coordination. For example, we can now analyze at the molecular level how neurons communicate, and how hormones and other chemical signals (like insulin) regulate the activity of cells, tissues, and organs. So, too, the failures of organs to operate properly—breakdowns that lead to degenerative diseases—are beginning to be understood (and corrected) at the molecular level. Finally, in Part VI, we will see how molecular techniques are aiding our understanding of ecological relationships—how, for example, the pesticide contaminant dioxin exerts its devastating ecological effects by being inappropriately altered by the detoxification enzymes animals use to dispose of unfamiliar (and thus potentially dangerous) chemicals. And, of course, the potential of molecular engineering in solving or at least ameliorating environmental problems, like water pollution and even greenhouse-effect warming, is one avenue of hope for the future of our planet.

Now more than ever, as we stand able to read out the genetic programs of organisms word for word, biology is the most exciting, intellectually stimulating, and promising discipline among the sciences.

STUDY QUESTIONS

1. What roles do anomalies and intuition play in the scientific method? (p. 5)
2. How might evolution occur in the absence of natural selection? (pp. 15–18)
3. What are three important requirements in a population for selection to take place? Can evolution occur in the absence of any one of these? (pp. 15–18)
4. How did Darwin's intimate knowledge of plant and animal breeding help him formulate his theory? (pp. 14, 16)

CONCEPTS FOR REVIEW

- Distinction between mechanists and vitalists
- Scientific method
- Early evidence for evolution
- Distinction between evolution and natural selection
- Requirements for evolution by natural selection
  - Heritable variation
  - Surplus offspring
  - Differential survival

SUGGESTED READING

COMROE, J. H., 1977. *Retrospectroscope.* Von Gehr Press, Menlo Park, Calif. *A fascinating study of how important scientific discoveries are made. The author concludes that great advances usually arise out of research directed at wholly unrelated problems.**

DARWIN, C., 1859. *The Origin of Species. Of the many reprints of this classic work, the edition by R. E. Leaky (Hill and Wang, New York, 1979) provides perhaps the best introduction and illustrations.**

GINGERICH, O., 1982. The Galileo affair, *Scientific American* 247 (2). *The complex interactions between ecclesiastical politics and Galileo's difficult personality are traced in illuminating detail.*

HERBERT, S., 1986. Darwin as a geologist, *Scientific American* 254 (5). *Readable account of Darwin's important contributions to nineteenth-century geology.*

KOESTLER, A., 1959. *The Sleepwalkers.* Macmillan, New York. *This informal and gossipy account of the Copernican revolution focuses on the personalities and motivations of Copernicus, Galileo (to whom Koestler is unsympathetic), and Kepler. It is a good supplement to Kuhn's book on Copernicus, listed next.**

KUHN, T. S., 1959. *The Copernican Revolution.* Random House, New York.*

KUHN, T. S., 1962. *The Structure of Scientific Revolutions.* University of Chicago Press, Chicago. *In this influential book Kuhn argues that science works in two ways—that of Normal Science (in which experiments are designed to investigate the dominant theory, or "paradigm") and that of Revolutionary Science (which arises when the dominant theory has accumulated so many anomalies that the field becomes unstable and a replacement is needed).**

TAYLOR, F. S., 1949. *A Short History of Science and Scientific Thought.* W. W. Norton, New York. *An excellent brief history with numerous excerpts from the major writings of important scientists.**

* Available in paperback.

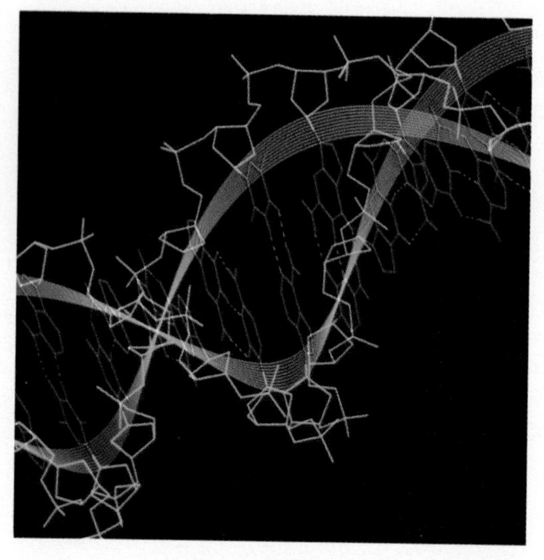

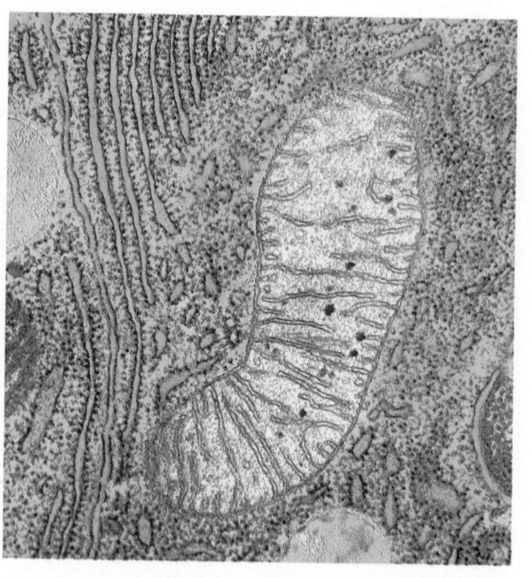

PART I

THE CHEMICAL AND CELLULAR BASIS OF LIFE

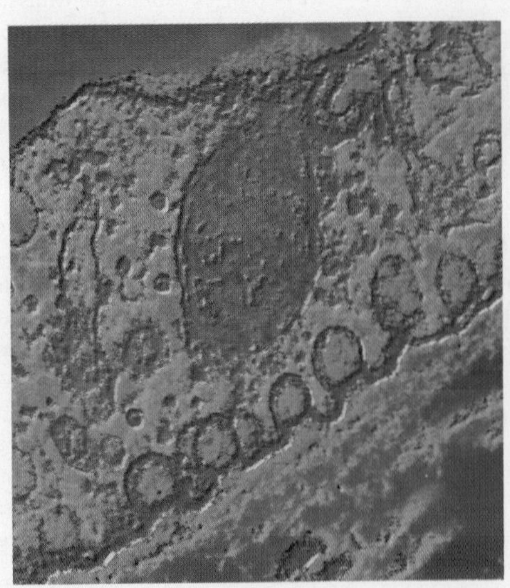

Part I

Part Opening Photographs:

(1) A computer model of a DNA molecule; nucleotides are shown in red. Instructions encoded by genes help build each cell and control its internal chemistry. These cellular blueprints are passed on from generation to generation as a sequence of nucleotides in a gene's DNA.

(2) Color-enhanced transmission electron micrograph (TEM) of a mitochondrion (yellow) surrounded by membranous endoplasmic reticulum. Cells maintain highly ordered chemical environments, manufacture specific molecules on demand, and transport them through a variety of membranous channels. Subcellular organelles play specialized roles in the cell; the mitochondrion shown here provides energy to the cell by completing the digestion of glucose.

(3) A field of sunflowers. For most organisms, the energy for life comes initially from the sun, and is stored by green plants like these sunflowers through the chemical process of photosynthesis. Energy stored in the molecular by-products of photosynthesis is then harvested by plants (and all organisms that feed directly or indirectly on plants) to do work.

(4) Color-enhanced TEM of a blood capillary cell. Many cells are adapted for specialized roles. In this case a cell uses energy to transport serum in vesicles (purple) from the capillary lumen (blue) at the top of the cell to the outside of the capillary (bottom) by a process that is still not fully understood. Similar cells in the lining of the lungs pick up oxygen from the lungs for circulation to the tissues and discard carbon dioxide.

Chapter **2**

SOME SIMPLE CHEMISTRY

he chemistry of life is both simple and complex. It is simple in the sense that the laws of biochemistry can predict with great precision many biochemical interactions. But, though often predictable, the chemistry of life is essentially intangible, depending on atoms we cannot see interacting by means of forces we cannot feel. Yet out of this invisible network of atomic interactions arises all that we see, hear, feel, taste, and smell of the world around us.

All life processes obey the laws of chemistry and physics. The same forces that assembled matter eons ago from the dust of the Big Bang regulate life today (Fig. 2.1). In the behavior of molecules, atoms, and subatomic particles lies the key to such complex biological phenomena as the trapping and storing of solar energy by green plants, the extraction of usable energy from organic nutrients, the growth and development of organisms, the patterns of genetic inheritance, and the regulation of the activities of living cells. The study of biology, then, begins with—and continually returns to—the basic laws of chemistry and physics.

2.1 The Cone Nebula
The Cone Nebula, like most nebulae, consists of enormous amounts of gas left over from the Big Bang. Stars continue to condense out of the gas.

THE ELEMENTS

All the matter of the universe is composed of a limited number of basic substances called elements. There are 92 naturally occurring elements, and in addition many synthetic elements have been manufactured in the laboratory; the current total for both natural and artificial elements is well over a hundred.

TABLE 2.1 *Elements important to life*

Symbol	*Element*	**Atomic number**/ *Typical mass number*	*Approximate percentage of earth's crust by weight*	*Approximate percentage of human body by weight*
H	Hydrogen	**1**/1	0.14	9.5
B	Boron	**5**/11	Trace	Trace
C	Carbon	**6**/12	0.03	18.5
N	Nitrogen	**7**/14	Trace	3.3
O	Oxygen	**8**/16	46.6	65.0
F	Fluorine	**9**/19	0.07	Trace
Na	Sodium	**11**/23	2.8	0.2
Mg	Magnesium	**12**/24	2.1	0.1
Si	Silicon	**14**/28	27.7	Trace
P	Phosphorus	**15**/31	0.07	1.0
S	Sulfur	**16**/32	0.03	0.3
Cl	Chlorine	**17**/35	0.01	0.2
K	Potassium	**19**/39	2.6	0.4
Ca	Calcium	**20**/40	3.6	1.5
V	Vanadium	**23**/51	0.01	Trace
Cr	Chromium	**24**/52	0.01	Trace
Mn	Manganese	**25**/55	0.1	Trace
Fe	Iron	**26**/56	5.0	Trace
Co	Cobalt	**27**/59	Trace	Trace
Ni	Nickel	**28**/59	Trace	Trace
Cu	Copper	**29**/64	0.01	Trace
Zn	Zinc	**30**/65	Trace	Trace
Se	Selenium	**34**/79	Trace	Trace
Mo	Molybdenum	**42**/96	Trace	Trace
Sn	Tin	**50**/119	Trace	Trace
I	Iodine	**53**/127	Trace	Trace

Each element is designated by one or two letters that stand for its English or Latin name. Thus H is the symbol for hydrogen, O for oxygen, C for carbon, Cl for chlorine, Mg for magnesium, K for potassium (Latin, *kalium*), Na for sodium (Latin, *natrium*), and so on. Only a few of the 92 naturally occurring elements are important in life processes (Table 2.1).

Matter cannot be subdivided without limit. Progressive subdivision ultimately leads to units indivisible by ordinary chemical means. These units are called ***atoms***. The atoms of a particular element are alike in many essential characteristics and differ in many measurable ways from the atoms of all other elements. A single atom is customarily represented by the chemical symbol for the element. N, for example, can represent either a single atom of nitrogen or the element itself.

ATOMIC STRUCTURE

Though atoms can be considered the basic chemical units of matter, they are themselves composed of still smaller particles. Many of these particles belong to the world of subatomic physics and are of little immediate concern to biologists. But three of them—the proton, the neutron, and the electron—play a central role in determining the activity of elements. In their interactions lie the power and the cohesion that make life possible.

The atomic nucleus All the positive charge and almost all the mass of an atom are concentrated in its nucleus, or center, which contains two kinds of so-called primary particles, the ***proton*** and the ***neutron***. Each proton carries an electric charge of +1. The neutron, as its name implies, has no charge. The proton and the neutron have roughly the same mass, though strictly speaking the neutron is slightly heavier (Table 2.2).

The number of protons in the nucleus is unique for each element. This number, called the ***atomic number***, is sometimes written as a subscript immediately before the chemical symbol. Thus ${}_{1}H$ indicates that the atomic number of hydrogen is 1; that is, its nucleus contains only one proton. Similarly, ${}_{8}O$ indicates that each oxygen nucleus contains eight protons.

It is often desirable to indicate the total number of protons and neutrons in a nucleus; this number is called the ***mass number***, because it approximates the total mass (commonly called the atomic weight) of the nucleus. The mass number is usually written as a superscript immediately preceding the chemical symbol. For example, most atoms of oxygen contain eight neutrons; the mass number is therefore 16, and the nucleus can be symbolized as ${}^{16}O$ or, if we wish to show both the atomic number and the mass number, as ${}^{16}_{8}O$.

Though the number of protons is the same for all atoms of the same element, the number of neutrons is not always the same, and neither, consequently, is the mass number. For example, most oxygen atoms, as we have seen, contain eight protons and eight neutrons and have a mass number of 16; some, however, contain nine neutrons and therefore have a mass number of 17 (symbolized as ${}^{17}O$), and still others have ten neutrons and a mass number of 18 (symbolized as ${}^{18}O$). Atoms of the same element that differ in mass, because they contain different numbers of neutrons, are called ***isotopes***; ${}^{16}O$, ${}^{17}O$, and ${}^{18}O$ are three isotopes of oxygen. Some elements have as many as 20 naturally occurring isotopes; others have as few as two.

Figure 2.2 illustrates three different isotopes of hydrogen: ${}^{1}_{1}H$, which is the usual form of the element; ${}^{2}_{1}H$, a stable isotope generally called deuterium; and ${}^{3}_{1}H$, an unstable isotope called tritium. Because both deuterium and tritium can be detected easily in living tissue, they have been used extensively

TABLE 2.2 *Fundamental particles*

Particle	*Mass (daltons[1])*	*Electric charge*
Electron	0.001	−1
Proton	1.672	+1
Neutron	1.674	0

[1] One dalton equals 10^{-24} gram. For definitions of units of measurement, see Glossary.

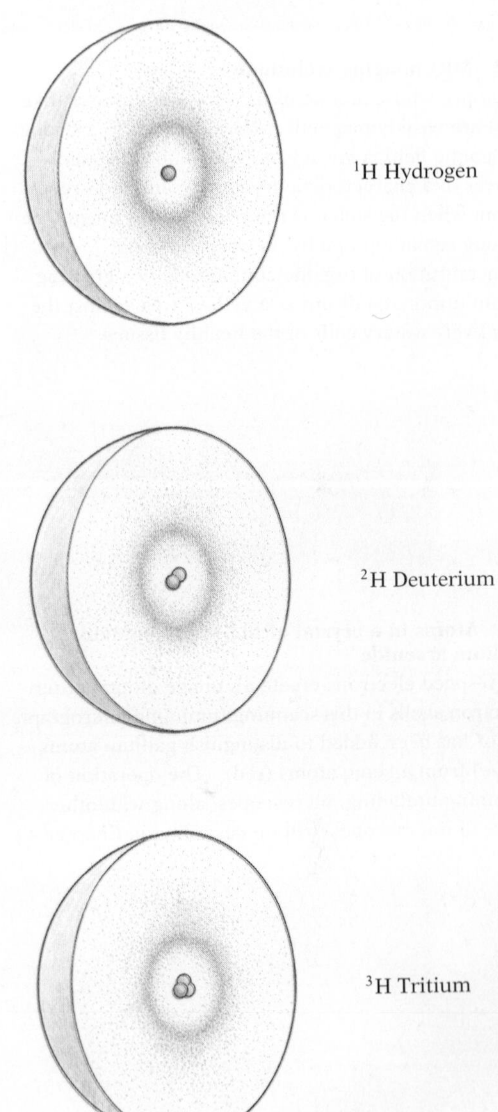

2.2 The three principal isotopes of hydrogen
Each of the three isotopes has one proton (blue) in its nucleus and one electron orbiting the nucleus. The isotopes differ in that ordinary hydrogen (${}^{1}H$) has no neutrons in its nucleus, deuterium (${}^{2}H$) has one, and tritium (${}^{3}H$), which is unstable, has two. The volume within which the single electron can be found 90 percent of the time (a sphere) is indicated by stippling; the denser the stippling in this cross-section view, the greater the likelihood that at any given moment the electron will be found in that portion of the sphere.

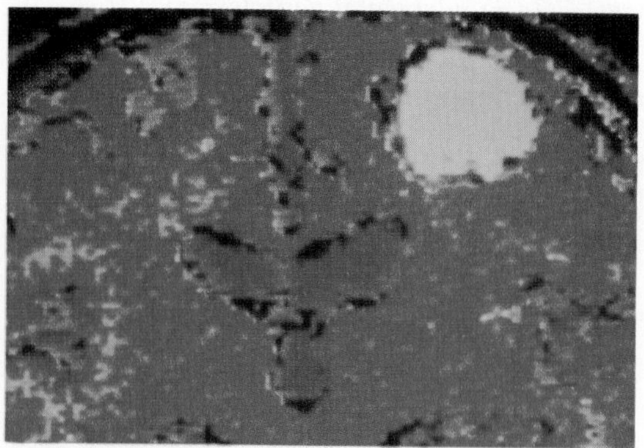

2.3 MRI imaging technique
Isotopes whose atomic mass is an odd number (like ^{31}P) are weakly magnetic. When exposed to a strong magnetic field, a given isotope will absorb radio waves of a characteristic frequency, and then reemit them when the signal is removed. In this image, the strong signal emitted by the relatively high concentration of organic compounds in a growing brain tumor stands out as a yellow area against the relatively watery cells of the healthy tissues.

in tracing the movements of hydrogen in biochemical reactions. As we will note again and again, isotopes of various elements are invaluable research tools for biologists (Fig. 2.3).

The electrons The portion of the atom outside the nucleus contains the third kind of primary particle—the ***electron***. Though electrons have very little mass (see Table 2.2), their behavior is the single most crucial factor in the chemistry of life. Each electron carries a charge of −1: one negative electronic charge unit—exactly the opposite of a proton's charge.

In a neutral atom, the number of electrons around the nucleus is exactly the same as the number of protons in the nucleus. The positive charges of the protons and the negative charges of the electrons cancel each other, making the total atom neutral. Consequently, in a neutral atom, the atomic number represents both the number of protons inside the nucleus and the number of electrons outside the nucleus. If, then, we see the symbol $^{35}_{17}Cl$, we can tell that a neutral atom of this isotope of chlorine has 17 protons, 18 neutrons, and 17 electrons. Similarly, the symbol $^{39}_{19}K$ means that this isotope of potassium contains 19 protons, 20 neutrons, and 19 electrons.

The electrons are not in fixed positions outside the nucleus. Each is in constant motion, making 10^{15}–10^{16} orbits of the nucleus each second. Hence, it is impossible to know exactly where a given electron is at any particular moment; photographs made with the latest imaging technology record only a wispy shell boundary (Fig. 2.4). For this reason some illustrations of atoms, such as Fig. 2.2, do not show the electron itself, but indicate the region where the electron is likely to be. All illustrations of atomic structure exaggerate the size of the nucleus. If a proton were the size indicated in Fig. 2.2, the outer edge of the electron cloud would extend 150 meters in all directions.

2.4 Atoms in a crystal of the semiconductor gallium arsenide
High-speed electrons create an image of hazy outer electron shells in this scanning tunneling micrograph. Color has been added to distinguish gallium atoms (blue) from arsenic atoms (red). (The operation of scanning tunneling microscopes, along with other sorts of microscopes, will be described in Chapter 4.)

The average distance of an electron from the nucleus is a function of its energy; the higher its energy, the greater its probable distance from the nucleus. But in any particular atom, only certain discrete amounts, or "levels," of energy—like steps of a staircase—are possible. To occupy a certain step, or energy level, an electron must possess a specific amount of energy. To achieve a higher energy level an electron must absorb additional energy from some outside source. Conversely, when an electron falls into the next lower level, it emits the same amount of energy it previously took to move up from that level (Fig. 2.5). We refer to an electron occupying the lowest step available to it in the atom as being in the "ground state." Once it has absorbed enough energy to move up to the next energy level, it is said to be in an "excited state."

An electron in the excited state has a strong tendency to return to its ground state by emitting, in some form, the additional energy just acquired. Most often the energy is released as light. The decay of excited electrons in the lining of fluorescent tubes, for instance, helps us light our artificial habitats. As we will see, this fleeting moment of excitation, lasting 10^{-8} sec, is critical to living things: life on earth is based on the ability of specialized molecules, found in plants and photosynthetic bacteria, to capture and make use of the potential energy of excited electrons during that brief moment before they drop back down the energy staircase.

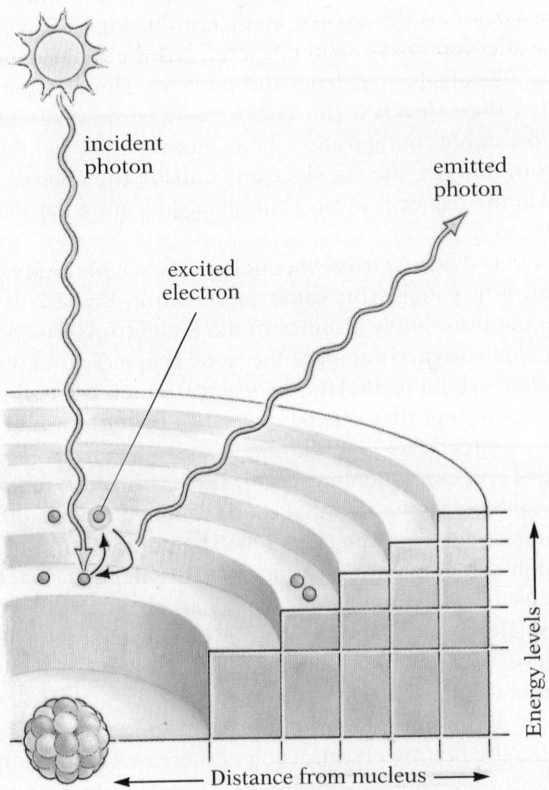

2.5 Energy levels of electrons
The electrons in an atom occupy discrete energy levels. If an electron absorbs the right amount of energy (shown as a photon, a discrete particle of light) and there is a vacancy in a higher energy level, it can move up the energy staircase to this level. Normally this "excited" electron quickly reemits the absorbed energy (here again shown as a photon) and returns to its original energy level.

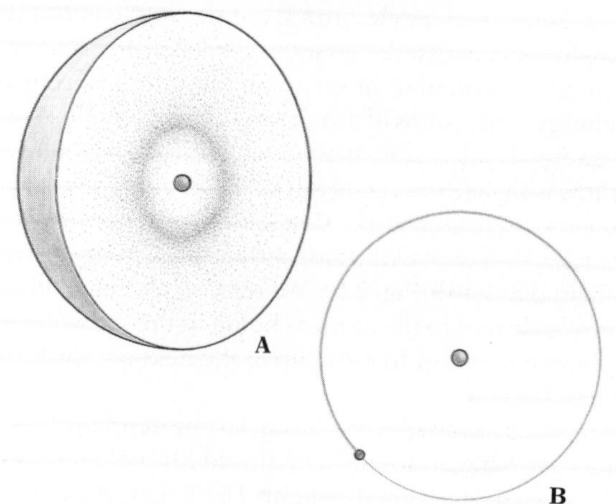

2.6 Two ways of representing the hydrogen atom Since no one has ever seen the particles that make up an atom, all our knowledge of what atoms look like is indirect, and we can only picture them as models that fit the data. (A) The nucleus is shown here as a central blue area, with the "cloud" around it in cross section representing the region where the electron is likely to be. The circle encloses the orbital of the electron—the volume, a sphere, within which the electron will be found 90 percent of the time (see also Fig. 2.2). (B) Sometimes, for convenience, only the circle indicating the circumference of the orbital is shown; the electron may be represented by a small ball on the circle.

The volume within which an electron can be found 90 percent of the time is known as its ***orbital***. In illustrations of atoms the orbital is often represented by a circle, the electrons sometimes being shown as round spots on the circle (Fig. 2.6).

The energy level of the hydrogen electron, which is the level nearest the nucleus, is often referred to as the *K* level (or *K* shell), and the orbital of the electron, which is spherical, is designated the 1*s* orbital. The *K* level can contain only two electrons. What happens, then, in an atom with more than two electrons, such as the oxygen atom, with its eight? Two of these electrons can be accommodated at the *K* level, but the other six must move at higher energy levels, farther from the nucleus. The next possible energy level is called the *L* level; it can contain a maximum of eight electrons. Since the most stable configuration for an atom is one in which its electrons have minimum energy, the six electrons outside the *K* level in an oxygen atom are all at the *L* energy level. Thus an oxygen atom has two *K* electrons and six *L* electrons.

The most likely distance from the nucleus of each of the six *L* electrons of an oxygen atom is roughly the same; as shown in Fig. 2.7, it is somewhat greater than the most likely distance of the *K* electrons. However, the orbitals of the *L* electrons are not all of the same shape. Two of these electrons have a spherical orbital (called the 2*s* orbital), which is like the 1*s* orbital of the *K* electrons except that it extends farther from the nucleus (Fig. 2.7). But the other *L* electrons have dumbbell-shaped orbitals (designated 2*p*). The *L* energy level can contain three of these *p* orbitals, each oriented at right angles to the other two, so that each is aligned along a different one of the three dimensions of space (Fig. 2.8). The combination of the three 2*p* orbitals, which can contain a maximum of two electrons each, and the 2*s* orbital, which can also contain two electrons, accounts for the overall maximum of eight electrons for the *L* energy level. Obviously, not all these possible orbitals are filled in an oxygen atom.

When elements have more than 10 electrons (two at the *K* level and eight at the *L*), the additional electrons are accommodated at energy levels beyond *L*. Like the first two levels, each of these can contain only a limited number of electrons. Thus the third level (*M*) can contain a maximum of 18

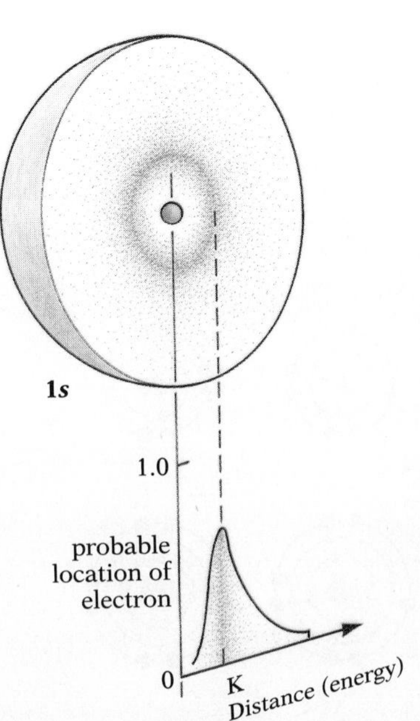

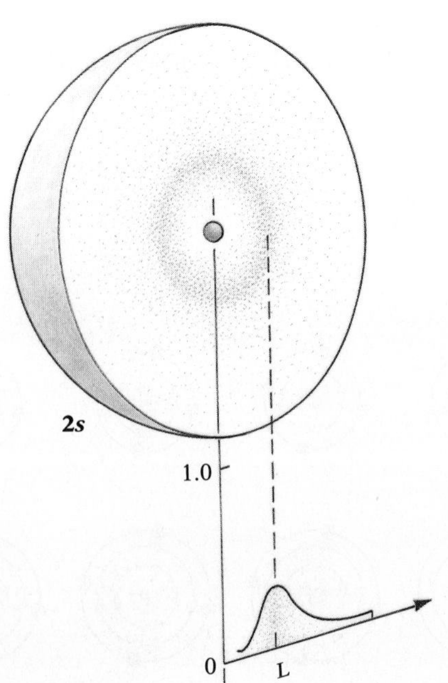

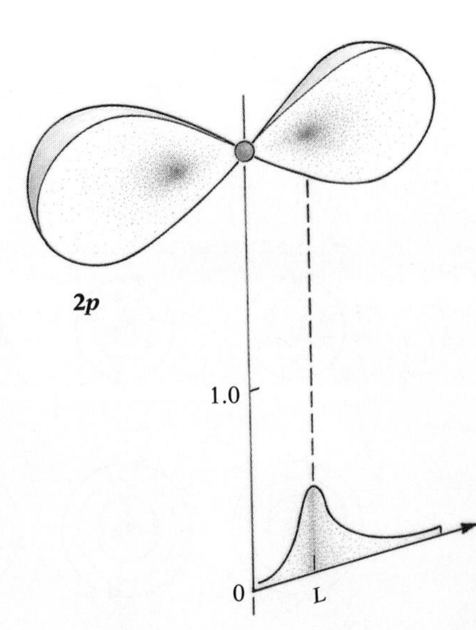

2.7 Representations of electron orbitals
Top: The orbitals of *s* electrons are approximately spherical, those of *p* electrons roughly dumbbell-shaped. The numerals before *s* and *p* indicate the energy level. Thus the 1*s* electron is at the first energy level (*K*), nearest the nucleus; the 2*s* and 2*p* electrons are at the *L* level, a higher energy level, and hence are at a greater average distance from the nucleus than the 1*s* electron. Note that, despite the very different shapes of their orbitals, the 2*s* and 2*p* electrons are at the same energy level—their most probable distances from the nucleus are the same, as shown in the graphs (bottom).

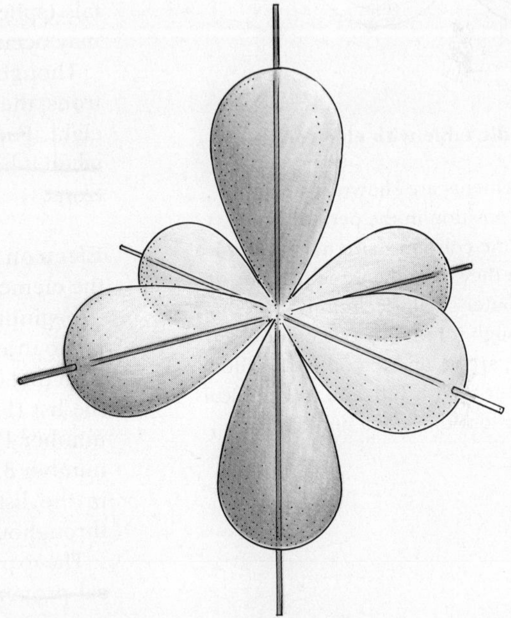

2.8 The three 2*p* electron orbitals
Each of the dumbbell-shaped orbitals is oriented in a different dimension of space, at right angles to the other two.

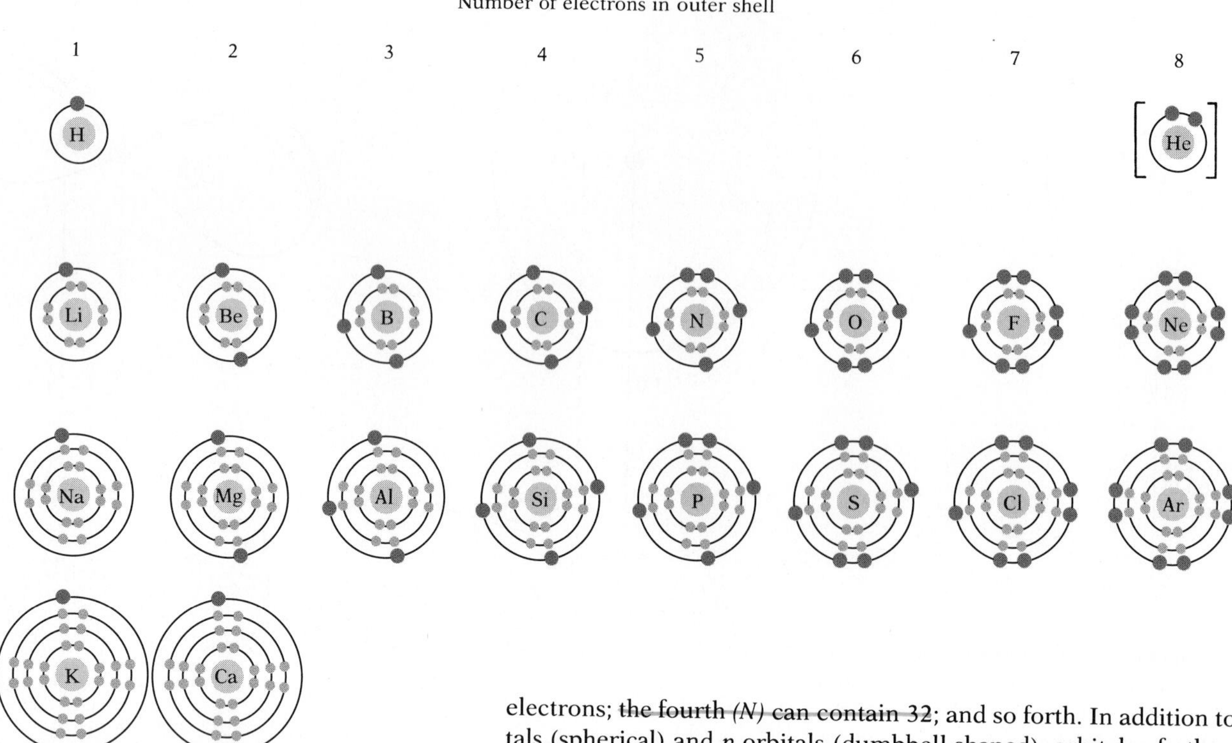

2.9 Partial periodic table with electron distributions
The first twenty elements are shown arranged according to their position in the periodic table. Elements in the same column share many chemical properties because they have the same number of electrons in their outer shell. (Helium is placed in column 8 even though it has only two outer electrons because, like neon, argon, and the other so-called noble gases, its outer shell is full, and its chemical properties are therefore those of a noble gas.)

electrons; the fourth *(N)* can contain 32; and so forth. In addition to *s* orbitals (spherical) and *p* orbitals (dumbbell-shaped), orbitals of other shapes may occur at these outer energy levels.

Though the third and successive levels can hold more than eight electrons, they are in a particularly stable configuration when they contain only eight. For our purposes, then, the first level can be considered complete when it holds two electrons, and every other level when it holds eight electrons.

Electron distribution and the chemical properties of elements When the elements are arranged in sequence according to their atomic numbers—beginning with hydrogen, which has the atomic number 1, and proceeding to uranium, the last of the natural elements, with number 92—it can be seen that elements with very similar properties occur at regular intervals in the list (Fig. 2.9). For example, fluorine, number 9, is more like chlorine, number 17, bromine, number 35, and iodine, number 53, than like oxygen, number 8, or neon, number 10, the two elements immediately adjacent to it in the list. This tendency for chemical properties to recur periodically throughout the sequence of elements is called the Periodic Law.

The explanation for this periodicity is that the reactivity and other chemical properties of elements are largely determined by the number of electrons in their outermost shell (i.e., at their outermost energy level). If that shell is complete, as in helium (atomic number, 2), neon (10), or argon (18), the element has very little tendency to react chemically with other atoms (Fig. 2.9). If the outermost shell has one electron fewer than the full complement as for fluorine, chlorine, bromine, and iodine, the element has certain characteristic chemical properties; if, like oxygen, it lacks two electrons, the element has somewhat different properties; if it lacks three electrons, like nitrogen, the element has very different properties.

A convenient way to represent the electron configuration of the outer shell is to symbolize each electron by a dot placed near the chemical symbol for the element under consideration. Thus fluorine and chlorine, which have seven electrons at their outer energy level, would have identical electron symbols:

$$:\ddot{\underset{\cdot\cdot}{\mathrm{F}}}\cdot \qquad :\ddot{\underset{\cdot\cdot}{\mathrm{Cl}}}\cdot$$

Similarly, hydrogen with one electron in its shell, carbon with four in its outer shell, nitrogen with five, and oxygen with six would be represented as follows:

$$\mathrm{H}\cdot \qquad \cdot\dot{\underset{\cdot}{\mathrm{C}}}\cdot \qquad \cdot\ddot{\underset{\cdot}{\mathrm{N}}}\cdot \qquad \cdot\ddot{\underset{\cdot}{\mathrm{O}}}:$$

Radioactive decay Atomic structure, as we will see, is predictable. Protons, neutrons, and electrons all seek stability as atoms join together to complete their outer shells, or as unstable isotopes give off parts of themselves and reach a more stable state. Though the various isotopes of an element carry different numbers of neutrons, their identical electron distributions give them the same chemical properties. Their physical properties, however, differ in two ways that are important to biological research and to human health.

Unusual isotopes are taken up by tissues just as well as the more common forms of their respective elements, but since isotopes differ significantly in atomic weight, they can be distinguished from each other by weight-sensitive techniques such as centrifugation. For example, there was at one time considerable controversy over whether the oxygen gas (O_2) released by plants comes from carbon dioxide (CO_2) or water (H_2O), the two raw materials of photosynthesis. The issue was settled by supplying plants with water containing a heavy isotope of oxygen (^{18}O rather than ^{16}O), while providing normal CO_2. The mass of the O_2 released by the plants was then compared with normal O_2 and found to be about 12 percent heavier, thus proving that water is the ultimate source for the oxygen in the air.

The other biologically significant physical property of some isotopes is their tendency to decay into a more stable form, giving off various particles on their way to physical stability. The stability of an isotope is measured by the ***half-life*** of the isotope: the time it takes half the atoms in a sample to decay. Tritium (^3H), for instance, has a half-life of about 12 years; ^{32}P, roughly 14 days; ^{14}C, 5700 years; ^{40}K, 1.3×10^9 years; and so on.

Radioactive isotopes are extraordinarily useful in biology, since an isotope added to a sample emits radiation that scientists can track (Fig. 2.10). With a "labeled" isotope of carbon dioxide in CO_2, for instance, we can trace how plants use carbon to build sugars. Because they are taken up by tissue as readily as their more stable counterparts, radioactive isotopes can also be used as "tracers" to help doctors locate circulatory blockages, pinpoint tumors, or predict potential problem areas. Naturally occurring isotopes make possible the dating of many rocks and fossils. The ratio of the radioactive isotope ^{14}C to the stable ^{12}C, for instance, is relatively constant in

2.10 Autoradiography of a soybean plant
A radioactive isotope, ^{32}P, was taken up by the plant's roots and transported to sites of nucleic acid synthesis. The plant was then laid on a photographic emulsion, which recorded the sites of radioactive decay.

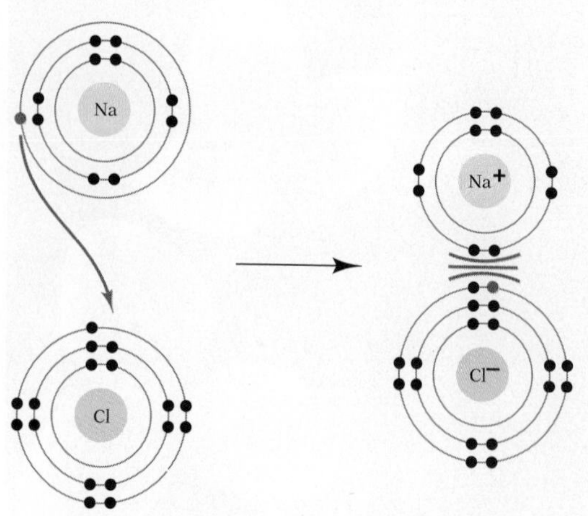

2.11 Ionic bonding of sodium and chlorine Sodium has only one electron in its outer shell, while chlorine has seven. Sodium acts as an electron donor, giving up the one electron in its outer shell, whereupon the complete second shell functions as its new outer shell. Chlorine acts as an electron acceptor, picking up an additional electron to complete its outer shell. But after sodium has donated an electron to chlorine, the sodium, left with one more proton than it has electrons, has a positive charge. Conversely, the chlorine, with one more electron than it has protons, has a negative charge. The two charged atoms, called ions, are attracted to each other electrostatically by their unlike charges. The result is sodium chloride (NaCl).

the CO_2 of the atmosphere, but once a plant has captured a CO_2 molecule and built it into a product like cellulose, the decay of ^{14}C atoms causes the ratio of ^{14}C to ^{12}C to decline steadily with time. Hence, the $^{14}C/^{12}C$ ratio in a sample provides a moderately accurate measure of age. With long-lived isotopes this "radioactive clock" technique can be extended far into the past: the ratio of uranium to the lead it decays into can reveal the age of rocks as much as 4 billion years old.

Isotopes, like most things, have their dark sides as well. A radioactive atom in a living cell poses two potential threats. First, it can, like uranium, decay into another element by losing protons, thereby altering the chemistry of its molecule completely. More often, radioactive isotopes produce highly reactive molecules that have too many or too few electrons to balance the electrical charge of their protons, and thus have a net charge. Beta decay, in which atoms throw off electrons, can produce this result directly. Gamma radiation, by bombarding nearby atoms with photons, can energize adjacent electrons so that they escape from their atoms. Since the behavior of electrons, as we will see, determines the chemistry of life, such unpredictable and uncontrolled movement of electrons can disrupt the precisely ordered and carefully regulated workings of the cell. For instance, a change in a critical part of a cell's DNA, resulting from the redistribution of electrons in beta decay or gamma radiation, can trigger the complicated chain of events that leads to cancer. Occurring in the cells of the reproductive system, such a change in the DNA can cause defects in subsequent offspring. Changes of this sort arise in all of us every day from exposure to the sun's radiation and the natural decay of radioactive elements in the earth's atmosphere and crust, and each cell has a battery of defense mechanisms to counteract them.

CHEMICAL BONDS

The arrangement of electrons in the outer shell of the atoms of most elements gives those atoms an ability to bind to others to form new and more complex aggregates. When two or more atoms are bound together in this fashion, the force of attraction that holds them together is called a chemical bond. The atoms of each particular element can form only a limited number of such bonds; the arrangement of its electrons and the nature of the various charges they exhibit ensure that each element has its own characteristic bonding capacity.

IONIC BONDS

We have said that atoms are in a particularly stable configuration when the outer electron shell is complete—that is, in most cases, when it contains eight electrons. There is consequently a general tendency for atoms to form complete outer shells by reacting with other atoms. The tendency of atoms to gain complete outer shells forms the basis upon which all chemistry is built.

Consider, for example, an atom of sodium (atomic number, 11). This atom has two electrons in its first shell, eight in the second, and only one in the third. One way sodium might gain a complete outer shell would be to acquire seven more electrons from some other atom or atoms. But the sodium atom would then have an enormous excess of negative charge, and since like charges repel each other, the electrons would tend to push each other away from the sodium. In point of fact, sodium cannot obtain a full outer shell by appropriating seven additional electrons. In nature it gives up the lone electron in its third shell to some electron acceptor, leaving the complete second shell as the new outer shell (Fig. 2.11).

Next, consider an atom of chlorine (atomic number, 17). This atom has two electrons in its first shell, eight in its second shell, and seven in its third shell. In other words, its outer shell is almost complete, lacking only a single electron. It cannot lose the seven electrons in its outer shell for reasons similar to those preventing sodium from gaining seven electrons. Only by gaining an extra electron from some electron donor can chlorine acquire a complete outer shell.

If a strong electron donor like sodium (an atom with a strong tendency to get rid of an electron) and a strong electron acceptor like chlorine (an atom with a strong tendency to acquire an extra electron) come into contact, an electron may be completely transferred from the donor to the acceptor. The result, in the present example, is a sodium atom with one electron fewer than normal and a chlorine atom with one more electron than normal. Once it has lost an electron, the sodium is left with one more proton than it has electrons, and it therefore has a net charge of +1. Similarly, the chlorine atom that gained an electron has one more electron than it has protons and has a net charge of −1. Such charged atoms (or charged aggregates of atoms) are called ***ions***, and are symbolized by the appropriate chemical symbol followed by a superscript indicating the charge. Sodium and chlorine ions are written Na^+ and Cl^-.

A sodium ion with its positive charge and a chlorine ion (usually called a chloride ion) with its negative charge tend to attract each other, since opposite charges attract. This important kind of electrical interaction is known as ***electrostatic attraction***. In this instance electrostatic attraction holds the two ions together to form the compound we know as table salt, or sodium chloride, NaCl (Fig. 2.11). Such a bond, involving the complete transfer of an electron and the mutual electrostatic attraction of the two ions thus formed, is termed an ***ionic bond***.

Ionic bonding may entail the transfer of more than one electron, as in calcium chloride, another common salt. Calcium (atomic number, 20) has two electrons in its outermost shell, and it loses both to form the calcium ion, Ca^{++} (Fig. 2.12). Chlorine, however, requires only one electron to complete an octet in its outer shell, as we have already seen. Hence it takes two chlorine atoms to act as acceptors for the two electrons from a single calcium atom, and a total of three ions bond together to form calcium chloride, symbolized as $CaCl_2$ (the subscript 2 indicates two chlorine atoms for each calcium atom). We can say, then, that calcium has a bonding capacity, or ***valence***, of +2, while sodium has a valence of +1 and chlorine a valence of −1.

Ionic bonding occurs between strong electron donors and strong electron acceptors. It is not common between configurations that have intermediate numbers of electrons in the outer shells, or between two strong electron donors or electron acceptors.

Under some conditions the product of the binding of a calcium atom to

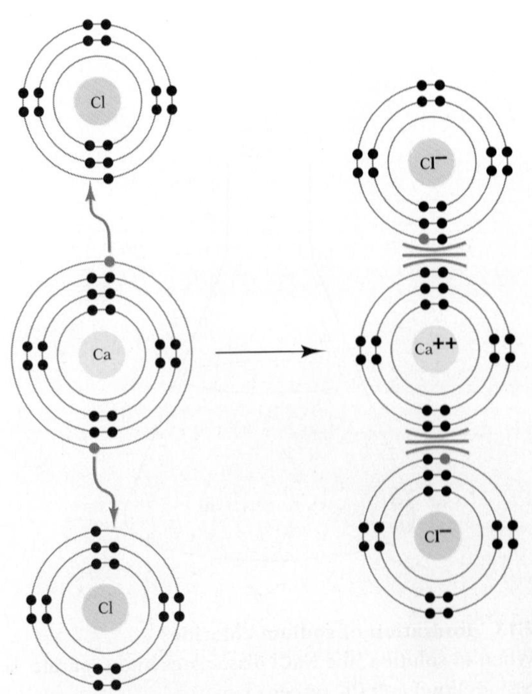

2.12 Ionic bonding of calcium and chlorine
Calcium has two electrons in its outer shell. It donates one to each of two chlorine atoms, and the two negatively charged chloride ions thus formed are attracted to the positively charged calcium ion to form calcium chloride ($CaCl_2$).

two atoms of chlorine is one molecule of calcium chloride. A ***molecule*** is generally defined as an electrically neutral aggregate of atoms bonded together strongly enough to be regarded as a single entity. In many instances, however, ***ionization*** (the transfer of one or more electrons from one atom to another to form ions) occurs without true molecular formation. Substances like sodium chloride (NaCl) and calcium chloride ($CaCl_2$), for example, in which the bonds are almost exclusively ionic, have a pronounced tendency to dissociate into separate ions when in solution. (We will look at why this happens shortly.) When they are ionized in solution, then, they do not exist as molecules; NaCl forms two separate entities, an Na^+ ion and a Cl^- ion (Fig. 2.13). Similarly, $CaCl_2$ in solution forms three separate entities, a Ca^{++} ion and two Cl^- ions. In nature, ionic compounds even in the solid state often do not form discrete molecules in the usual sense. In solid sodium chloride, many sodium and chlorine atoms are bound together into a large crystal (Fig. 2.14). There are no separate molecules composed of one sodium atom bonded to one chlorine atom, as the molecular symbol NaCl might seem to indicate. In a sense, the entire crystal can be conceived of as a single molecule.

Since ions are charged particles, they behave differently in living systems from neutral atoms or molecules, and substances wholly or partly ionized in water play many important roles in the functioning of biological systems. In later chapters we will see the effects of charge on the movements of materials through the membranes of living cells, and the partitioning of positive and negative ions that gives rise to the differences in electrical potential essential for nerve and muscle activity.

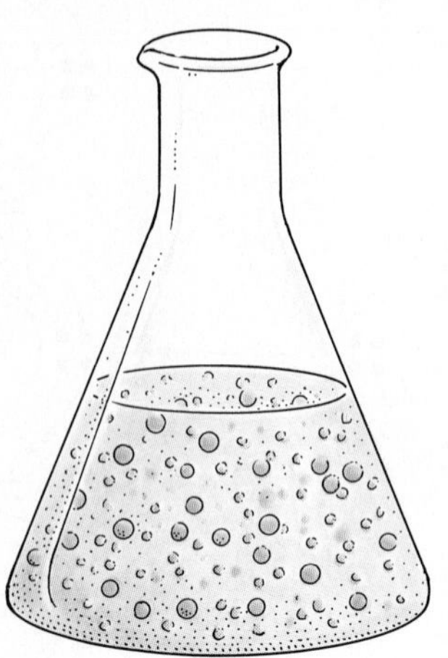

2.13 Ionization of sodium chloride
When in solution, the NaCl dissociates into separate Na^+ (yellow) and Cl^- (green) ions.

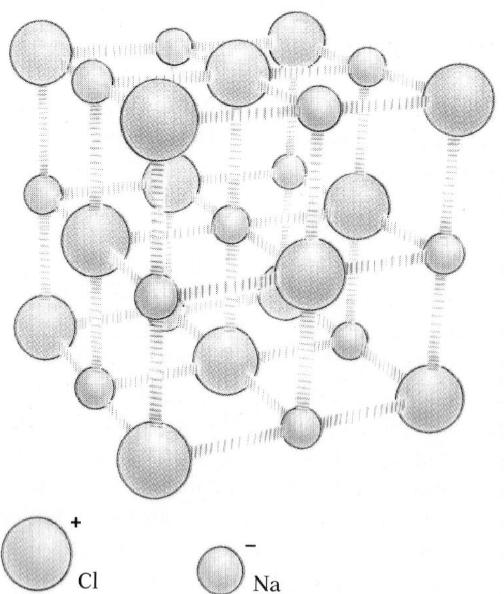

2.14 The arrangement of ions in crystalline table salt (sodium chloride)
(A) The imaginary lattice, representing the lines of electrostatic attraction, indicates the spatial arrangement of the Na^+ and Cl^- ions.
(B) Salt crystals—potassium chloride (KCl) and sodium chloride (NaCl).

Acids and bases Living cells are extemely sensitive to the chemistry of the fluids that bathe them. The delicate balance of acids and bases, in particular, is crucial to most tissues. An ***acid*** is a substance that **increases** the concentration of hydrogen ions (H^+) in water; a ***base*** is a substance that **decreases** the concentration of hydrogen ions, which in water is equivalent to increasing the concentration of hydroxyl ions (OH^-).

The degree of acidity or basicity (usually called alkalinity) of a solution is commonly measured in terms of a value known as ***pH***, which is a measure of the concentration of hydrogen ions. The pH scale generally ranges from 0 on the acidic end to 14 on the alkaline end. A solution is neutral, neither acidic nor alkaline (that is, it contains equal concentrations of H^+ ions and OH^- ions), if its pH is exactly 7. Solutions with a pH of less than 7 are acidic (with a higher concentration of H^+ ions than of OH^- ions); the lower the pH, the more acidic the solution. Conversely, solutions with a pH higher than 7 are alkaline (with a higher concentration of OH^- ions than of H^+ ions); the higher the pH, the more alkaline the solution. You should realize that a change of one pH unit means a tenfold change in the concentration of hydrogen ions. Thus the concentration of H^+ ions in the solution of a very strong acid may be as much as 100,000,000,000,000 (10^{14}) times greater than in the solution of a very strong base. Figure 2.15 illustrates the range of pHs we normally encounter.

Except in parts of the animal digestive tract and a few other isolated areas, most cells function best when conditions are nearly neutral. Most of the interior material of living cells has a pH of about 6.8. The blood plasma and other fluids that bathe the cells in our own bodies have a pH of 7.2–7.3. Numerous special mechanisms aid in stabilizing these fluids so that cells will not be subject to appreciable fluctuations in pH. Foremost among these mechanisms are certain chemical substances known as ***buffers***, which have the capacity to bond to H^+ ions, thereby removing them from solution whenever their concentration begins to rise, and conversely, to release H^+ ions into solution whenever their concentration begins to fall. Buffers thus help minimize fluctuations in pH, which would otherwise be considerable since many of the biochemical reactions normally occurring in living organisms either release or use up H^+ ions. The most important biological buffer in vertebrates is carbonic acid, which stabilizes the pH of the blood as it circulates through the tissues.

pH is the negative logarithm of the hydrogen ion concentration:

$$\text{pH} = \log\left(\frac{1}{[H^+]}\right) = -\log[H^+]$$

	pH	Concentrations of ions (moles/liter) H^+	OH^-	
Caustic soda (NaOH)	14	10^{-14}	10^{0}	ALKALINE
	13			
	12	10^{-12}	10^{-2}	
Detergent				
	11			
	10	10^{-10}	10^{-4}	
Baking soda	9			
Seawater	8	10^{-8}	10^{-6}	
Pure water	7	10^{-7}	10^{-7}	NEUTRAL
Saliva				
	6	10^{-6}	10^{-8}	ACIDIC
Unpolluted rainwater				
Coffee	5			
Typical acid rain				
Beer	4	10^{-4}	10^{-10}	
Orange juice	3			
Carbonated soft drink	2	10^{-2}	10^{-12}	
Stomach acid	1			
Hydrochloric acid (HCl)	0	10^{0}	10^{-14}	

Increasing (OH^-) ↑

Increasing [H^+] ↓

2.15 The pH scale
The concentration of hydrogen ions in a solution is measured by pH. At pH 7, the concentration of hydrogen ions (H^+) exactly balances the concentration of hydroxyl ions (OH^-), and so the solution is neutral. At lower pHs (corresponding to higher H^+ concentrations) solutions are acidic; at higher pHs (corresponding to lower H^+ concentrations) solutions are alkaline, or basic. Notice that the pH number matches the concentration of H^+ in moles/liter—for example, pH 8 corresponds to an H^+ concentration of 10^{-8}. A mole of any compound contains the same number of atoms—Avogadro's number, 6.02×10^{23}—as a mole of any other substance, and its weight in grams corresponds to the molecular weight (the summed atomic masses) of the substance.

COVALENT BONDS

Ionic bonds, as we have seen, involve the complete transfer of electrons from one atom to another. But in most cases bonding occurs not by complete transfer, but by a sharing of electrons between the atoms involved. Bonds of this sort, based on shared electrons, are called ***covalent bonds***. These may be nonpolar or polar.

Nonpolar covalent bonds To see how covalent bonds are formed, consider an atom of hydrogen. A complete first shell for hydrogen would contain two electrons, one more than each atom has normally. If the hydrogen gained an electron from some other atom it would have a full shell, but twice as much negative charge as positive charge (one proton and two electrons). Hydrogen does not, in fact, ionize in this manner. It tends to do the reverse: it loses its single electron, forming H^+ ions, which are simply isolated protons since the hydrogen nucleus contains no neutrons. But suppose there is no strong electron acceptor available and the hydrogen cannot ionize. One possible reaction then is for two atoms of hydrogen to bond to each other and form what is called molecular hydrogen (H_2):

$$H\cdot + H\cdot \longrightarrow H:H$$

In this molecule, each atom shares its electron equally with the other atom, so that each hydrogen has, in a sense, two electrons (Fig. 2.16).

Covalent bonds are not limited to the sharing of one electron pair between two atoms. Sometimes two atoms share two or three electron pairs and form double or triple bonds. When two atoms of oxygen bond together, they form a double bond (remember that an oxygen atom needs two electrons to complete its outer shell), and when two atoms of nitrogen (atomic number, 7) bond together, they form a triple bond, because each nitrogen atom needs three additional electrons to fill its outer shell:

$$:\,:\ddot{O}::\ddot{O}:\,: \qquad\qquad :N:::N:$$

A covalent bond may be represented simply by a line between two atoms, instead of a pair of dots; the other electrons in the outer shells are then ignored. Shown in this manner, H_2, O_2, and N_2 appear as follows:

$$H{-}H \qquad O{=}O \qquad N{\equiv}N$$

Since hydrogen atoms tend to form only one bond, oxygen two bonds, and nitrogen three bonds, we say that hydrogen has a covalent bonding capacity of 1; oxygen, a capacity of 2; and nitrogen, a capacity of 3. Covalent bonding capacity is equivalent to valence: it corresponds to the number of vacancies in the outer shell of an atom with one to three vacancies, and to

A

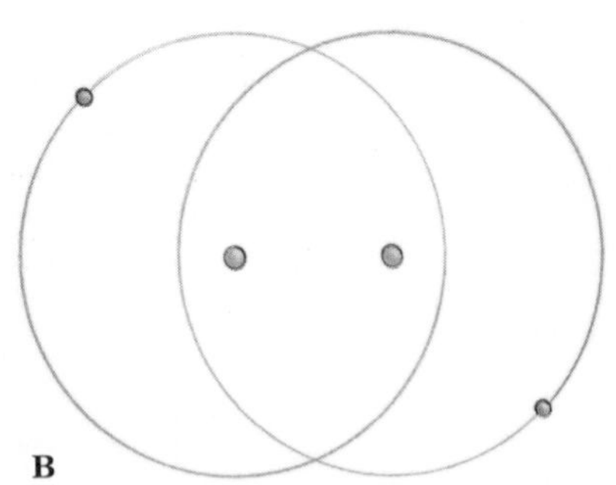

B

2.16 Covalent bonding of two hydrogen atoms (A) The sharing of electrons is indicated by overlapping electron clouds. (B) Alternatively, the sharing may be indicated by interlocking orbital rings. The latter system of representation parallels the textual representation of H_2 as H:H.

the number of sharable electrons in an atom with one to three outer electrons. The maximum covalent bonding capacity is 4: in an atom with four outer electrons, this is the number both of sharable electrons *and* of vacancies. Carbon is the most important example of an atom with a bonding capacity of 4; the ability of carbon to form so many bonds is in part what makes the diversity of life's chemistry possible.

Polar covalent bonds Suppose that instead of being bonded to each other, two hydrogen atoms are covalently bonded to an oxygen atom, forming water (H_2O):

$$\mathrm{H}\cdot + \mathrm{H}\cdot + \cdot\ddot{\underset{\cdot}{\mathrm{O}}}: \longrightarrow \mathrm{H}:\ddot{\underset{\ddot{\mathrm{H}}}{\mathrm{O}}}:$$

Oxygen, with six electrons in its outer shell (Fig. 2.9), needs two more. By sharing electrons with two hydrogen atoms, the oxygen atom can obtain a full outer octet, while at the same time each hydrogen obtains a complete first shell of two electrons. A covalent bond between a hydrogen atom and an oxygen atom is somewhat different from one between two hydrogen atoms or between two oxygen atoms, however. No two elements have exactly the same affinity for electrons. Consequently, when a covalent bond forms between two different elements, the shared electrons tend to be pulled closer to the more attractive element. Such a bond is called a ***polar*** covalent bond because the charge is distributed asymmetrically.

The formal measure of an atom's attraction for free electrons is its ***electronegativity;*** this depends upon the number of vacancies in the outer shell —an atom like oxygen with only one or two electron openings is generally more electronegative than one with three, and so on—and upon the distance of the outer shell from the nucleus (Fig. 2.17).

Since many covalent bonds are polar and since the degree of polarity of a bond varies over a wide range—from situations in which the shared electron is much closer to one of the atoms to situations in which it is only slightly closer—there is actually no sharp distinction between ionic bonds and covalent bonds. Ionic bonds represent one extreme, with the electrons pulled completely from one atom to the other, and nonpolar covalent bonds represent the other extreme, with the electrons pulled with equal force, and hence shared, by two atoms. Polar covalent bonds represent the usual case, a middle ground between these two extremes: the electrons are pulled closer to one atom than to the other, but not all the way.

The phenomenon of polarity helps explain many of the properties of various molecules in living systems. Whole molecules can be polar as a result of the polarity of bonds within them. One example is the water molecule. We can imagine how, even though the two hydrogen–oxygen bonds are polar, the atoms in the water molecule might be aligned in a straight line so that the charge would be distributed symmetrically within the molecule, which would then be nonpolar:

$$\mathrm{H}:\ddot{\underset{\cdot\cdot}{\mathrm{O}}}:\mathrm{H}$$

But this is not the actual arrangement. When the second energy level of oxygen has been filled by shared electrons, the resulting slight polarity of the covalent bonds induces the four pairs of electrons to adopt an arrange-

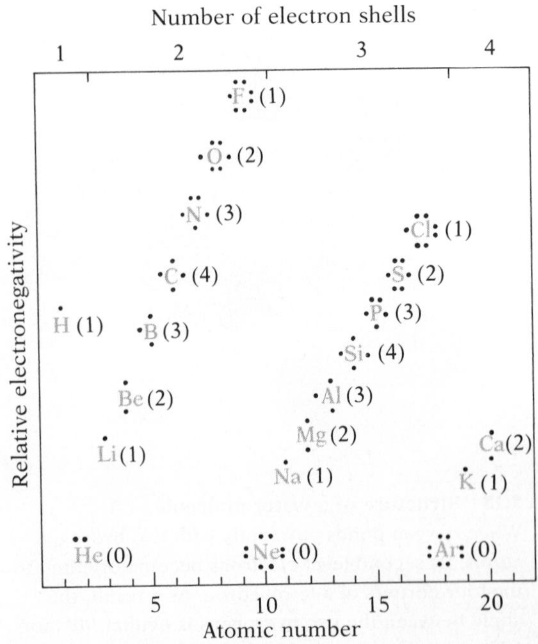

2.17 Electronegativity
The relative tendency of an atom to attract electrons depends on the number of spaces in the outer shell left to be filled and on the distance of the outer shell from the nucleus. Hence, lithium (Li) with seven vacancies is less attractive to electrons (that is, less electronegative) than carbon (C), which has four. Oxygen (O), with only two missing electrons, is yet more electronegative. This graph also helps explain why in methane (CH_4) the shared electrons will be nearer the carbon atom, while in carbon dioxide (CO_2) they will be nearer the oxygens: carbon is more electronegative than hydrogen, but less electronegative than oxygen. The electronegativity of the noble gases (which have filled outer shells) cannot be measured, and so is simply estimated. This knowledge of relative electronegativity permits us to make important predictions about many biochemical reactions. (The covalent bonding capacity of each atom is shown in parentheses.)

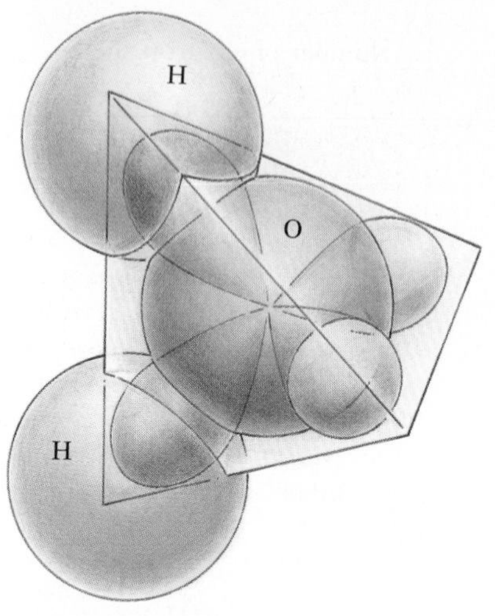

2.18 Structure of a water molecule
When oxygen bonds covalently with two hydrogen atoms, its second-level electrons become oriented to the four corners of a tetrahedron. As a result, the angle between the two hydrogens is neither 90° nor 180°, as might be expected from the perpendicular arrangement of the $2p$ orbitals, but rather is 104.5°.

ment in which the $2s$ orbital and the three $2p$ orbitals, their shapes highly modified, are oriented to the four corners of a tetrahedron (Fig. 2.18). Instead of the hypothetical linear arrangement, the three atoms form a bent-chain, or V-shaped, structure, with the oxygen at the apex of the V and the two hydrogen atoms as the arms:

$$\begin{array}{c} \quad\;\; \cdot\cdot\;(-) \\ (+)\;\mathrm{H} : \mathrm{O} : \\ \quad\;\; \cdot\cdot \\ \quad\;\; \mathrm{H} \\ \quad\;\; (+) \end{array}$$

Since the electrons are drawn closer to the oxygen atom, there is a concentration of negative charge near the oxygen end of the molecule. Therefore, the molecule is polar (Fig. 2.19, right).

The carbon dioxide molecule, on the other hand, exhibits no polarity: its double bonds hold its atoms in rigid linear alignment (Fig. 2.19, left). Hence CO_2 is nonpolar.

As we will see shortly when we look at the properties of water in detail, the polarity of certain molecules often has crucially important biological implications.

BIOLOGICALLY IMPORTANT WEAK BONDS

Strong versus weak bonds To maintain internal stability, or ***homeostasis***, living organisms must be able to change to meet the constant fluctuations of their environments. The changes all begin at the molecular level, and are powered by the liberated energy derived from the sorts of strong, energy-rich covalent bonds we have just discussed. Covalent bonds are called strong because breaking them is hard, usually requiring between 50 and 110 kilocalories of energy (most often supplied as heat) per mole.[2] Bond breakage usually results from collisions with rapidly moving molecules. Since the energy from even the most rapidly moving molecules at physiological temperatures is almost never above 10 kcal/mole, covalent bonds are stable and show little tendency to rupture spontaneously.

But life, as we have said, depends on a capacity for change, as well as on stability. The crucial sources of this ability to change are weak noncovalent bonds, which can readily be broken and re-formed. Ionic bonds in aqueous solutions, for instance, are relatively weak, averaging about 10 kcal/mole. The average duration of an ionic bond (the interval between formation of the bond and a collision with a molecule moving rapidly enough to break it) is quite short. The consequence, as we will see, is that several weak bonds must act in concert to produce molecular stability sufficient for most of life's metabolic processes. The weak bonds (or interactions, as they are sometimes called) of biological significance include ionic bonds, hydrogen bonds, and van der Waals interactions.

[2] A mole is the amount of a substance, in grams, that equals the combined atomic mass of all the constituent atoms in a molecule of that substance; there are aproximately 6×10^{23} molecules in a mole. A calorie (spelled with a small *c*) is defined as the quantity of energy, in the form of heat, required to raise the temperature of one gram of pure water one degree, from 14.5°C to 15.5°C. One kilocalorie (kcal) is 1000 calories. Nutritionists use a different scale to measure energy; their Calorie (spelled with a capital *C*) is equal to one kilocalorie on the standard scale. The number of calories required to break the bonds in a mole of any particular compound tells us something about the strength of those bonds.

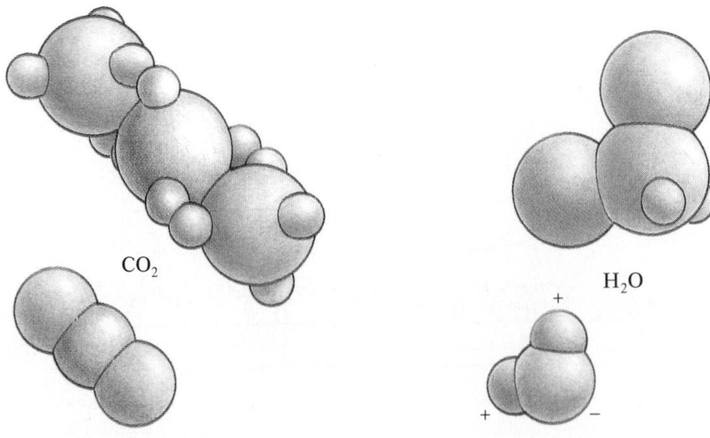

2.19 Polarities of two biologically important molecules
Left: In carbon dioxide, each of the four electrons of the carbon's outer orbital is shared between the carbon and an oxygen. The result is a linear molecule with no electrical polarity. Right: In water, two hydrogen atoms are bonded covalently to one oxygen atom, but the shared electrons are pulled closer to the oxygen because of its higher electronegativity. Because the atoms in water are arranged at an angle of 104.5°, the charge distribution is asymmetrical, with negative charge concentrated at the oxygen end; as a result the molecule as a whole is polar.

Hydrogen bonds The electrostatic attraction between oppositely charged portions of neighboring polar molecules results in ***hydrogen bonds***. Water molecules provide an excellent example. The hydrogen atoms in each water molecule are covalently bonded to the oxygen atom, but because of the polarity of the bond—the electrons being closer to the oxygen end than to the hydrogen ends—each hydrogen has a net positive charge. It is therefore attracted by the oxygen atoms, with their net negative charge, in other nearby water molecules. Since each of the hydrogens, while remaining covalently bonded to the oxygen atom of its own molecule, can form a weak attachment with the oxygen of another water molecule, and the oxygen can form a weak attachment with two external hydrogens, each water molecule has the potential for being simultaneously linked by hydrogen bonds to four other water molecules (Fig. 2.20). In a sense, then, a volume of water is a continuous chemical entity, because of the hydrogen bonding between the individual water molecules.

The distinction between hydrogen bonds and ionic bonds is clear: hydrogen bonds result from the electrostatic attraction between polar but electrically neutral molecules like water, while ionic bonds result from the electrostatic attraction between oppositely charged atoms (ions). However, an important bond also results from the electrostatic attraction between ions and polar molecules, as we will see when we discuss the hydration sphere—the shell of polar water molecules drawn around an ion in solution. In aqueous solutions, pure hydrogen bonds usually have a bonding energy of about 4–5 kcal/mole; ionic bonds, about 10 kcal/mole; and polar/ionic bonds such as those of the hydration sphere, about 7–8 kcal/mole.

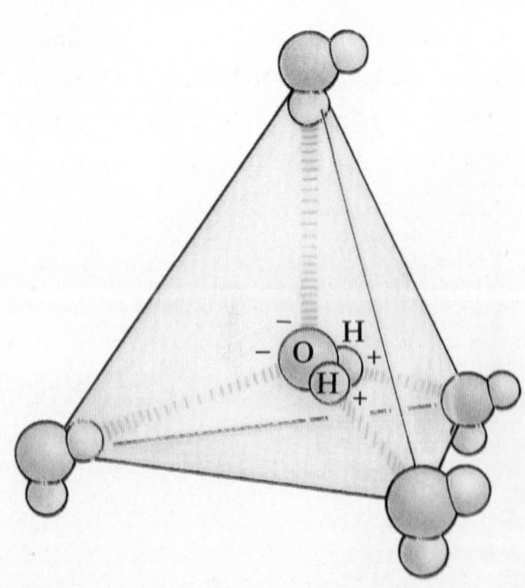

2.20 Hydrogen bonding between water molecules
Like the central H_2O molecule shown here, each water molecule can form hydrogen bonds (red bands) with four other water molecules. The array then assumes the shape of a tetrahedron. Water molecules near the edge of this imaginary tetrahedron can simultaneously form hydrogen bonds with two or three other water molecules, creating an interlocking array of tetrahedrons.

Van der Waals interactions Much weaker than ionic or hydrogen bonds are the linkages known as ***van der Waals interactions***, which have bonding energies of only 1–2 kcal/mole. These linkages occur between electrically neutral molecules (or parts of molecules) when they are so close to each other that the electrons in their outer orbitals are set in synchronous, mutually avoiding motion. The result of this momentary synchrony is that at one very precise distance the normal repulsion between the two sets of outer

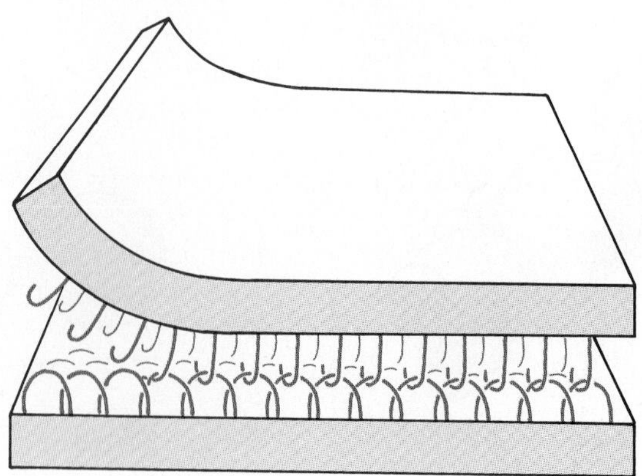

2.21 Stability from weak bonds
An array of many individually weak bonds—here represented as the many hooks and eyes of a Velcro fastener—can be surprisingly strong as a unit.

electrons is lessened and the atoms are able to bond weakly to each other. As the next chapter will show, van der Waals interactions play a crucial role in the enzymatic reactions that control virtually all the processes of life. They are also important in stabilizing aggregations of nonpolar molecules in cells.

The role of weak bonds Weak bonds play a crucial role in stabilizing the shape of many of the large molecules found in living matter—DNA and proteins in particular—and they often hold together groups of such molecules in orderly arrays. Like the minute hooks and eyes of a Velcro fastening, the bonds are individually quite weak (their average duration is only 10^{-11} sec), but can be strong and stable when many act together (Fig. 2.21). However, just as a Velcro patch is easy to unfasten when we begin at one end, attacking a few "bonds" at a time, so too an array of molecules held together by weak bonds (as opposed to the more gripperlike covalent bonds) can be disassembled and rearranged with relative ease by forces that dislodge the fastening bonds one by one. These bonds are important because, as we will see, an enormous number of life processes depend on just these sorts of changes.

SOME IMPORTANT INORGANIC MOLECULES

Chemists have traditionally referred to complex molecules containing the element carbon as ***organic*** compounds. All other compounds are called ***inorganic***, a designation that should not mislead you into assuming that these compounds play no role in life processes. Many inorganic (non-carbon-based) substances are, in fact, basic to the chemistry of life.

WATER

Life on earth is totally dependent on water. Between 70 and 90 percent of all living tissue is water, and the chemical reactions that characterize life all take place in a water-containing medium. Life based on some substance other than water could conceivably exist elsewhere in the universe, but such life would be vastly different from anything in our experience—so different that we might not recognize it as life even if we should stumble upon it.

Water as a solvent One of the main reasons water is so well adapted as the medium for life is that it is a superb solvent for many important classes of chemicals. It is a better solvent than most common liquids because of the marked polarity of the water molecule and the corresponding great propensity of water to form hydrogen bonds. Thanks to this polarity, both ionic substances and substances that are nonionic but polar are soluble in water (Fig. 2.22). Let us consider the effect of water on each.

We have mentioned that the ionic bonds linking the atoms of a salt such as NaCl are relatively weak when the salt is in an aqueous medium; but within a dry crystal of the same salt the bonds are comparatively strong. Why the difference? When the crystal is put into water, the attraction of the negatively charged oxygen ends of the water molecules for the positively charged sodium ions, and the similar attraction of the positively charged hydrogen ends of the water molecules for the negatively charged chloride ions, overcome by force of numbers the mutual attraction between the Na^+ and Cl^- ions. In water, then, the ionic bonds are broken with extreme ease because of the competitive attraction of the water; the Na^+ and Cl^- ions dissociate, and each of the ions becomes surrounded by a sphere of regularly arranged water molecules that are electrostatically attracted to it—a process called ***hydration*** (Fig. 2.23). The ionic bonds between Na^+ and Cl^- atoms are now weaker simply because the ions are kept far apart, and the strength of electrostatic attraction decreases exponentially with distance.

From the point of view of some biological processes, an ion and its hydration sphere must be regarded as a single entity—as though the whole were a true molecule. For example, if the question is whether or not a given type of ion can move through tiny pores in cell membranes, then it is the size of the hydrated ion that must be compared with the dimensions of the pores.

Water is also an excellent solvent for nonionic molecules if they are polar.

2.22 Dead Sea
Because water is polar and NaCl is ionic in solution, relatively large amounts of salt can be dissolved in water. In the Dead Sea, nearly 30 percent of the fluid is dissolved salt, and the density is so great that humans float high in the water.

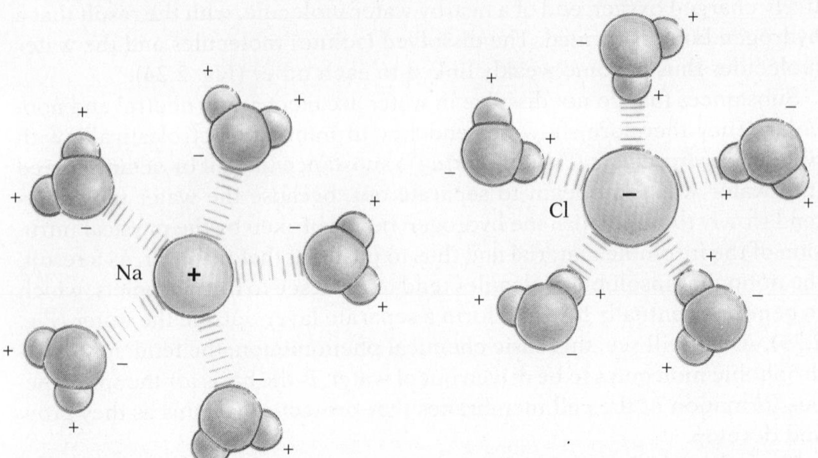

2.23 Hydration spheres of Na^+ and Cl^-
When dissolved in water, each of the Na^+ and Cl^- ions is hydrated—that is, surrounded by water molecules electrostatically attracted to it. Note that the oxygen of the water molecules is attracted to the positively charged Na^+, while the hydrogen of the water molecules is attracted to the negatively charged Cl^-. Water molecules in a hydration sphere are called bound water. This bonding between ion and polar molecules (red bands) makes evident the common electrostatic basis of ionic bonds and polar (hydrogen) bonds.

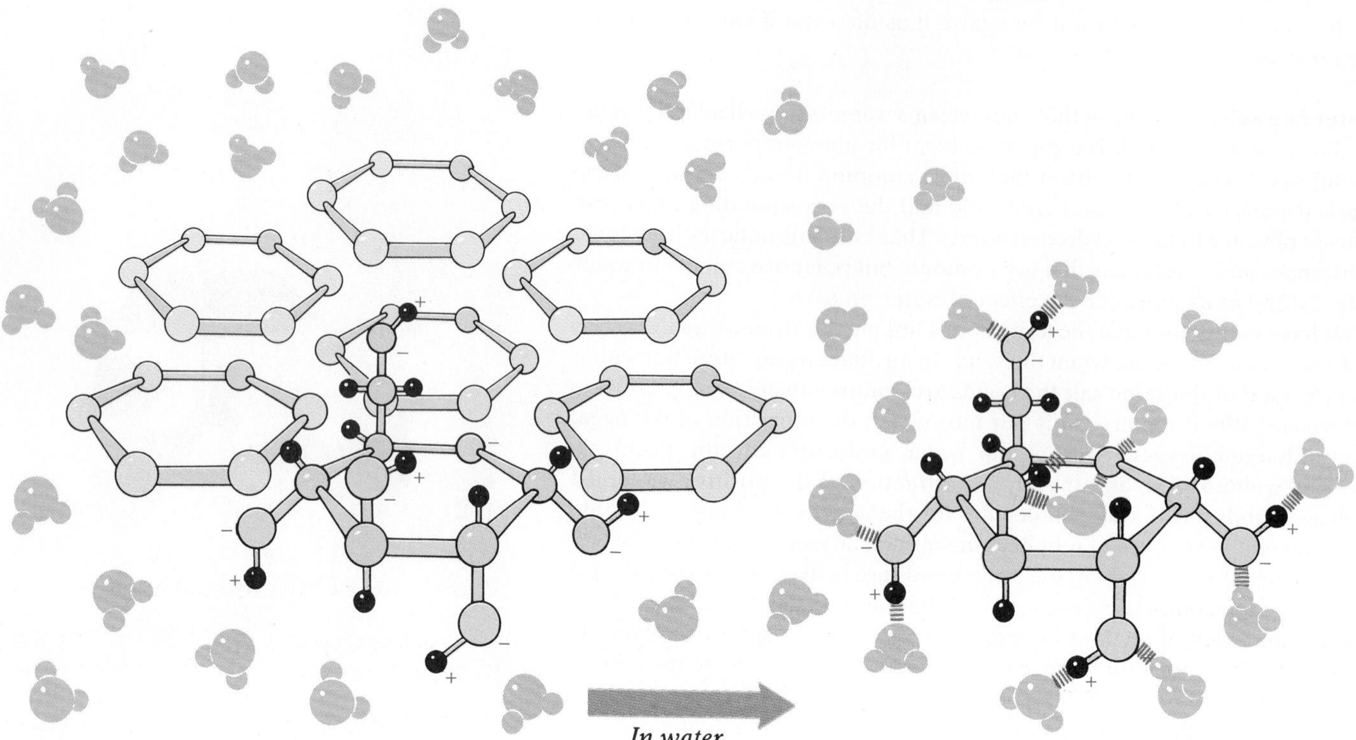

2.24 Polar basis of solubility
When a polar substance such as glucose, an energy-rich sugar (left), is placed in contact with water, the water molecules are attracted to the polar atoms of the sugar. (For clarity the polar —OH groups are shown for only two of the sugar molecules.) The water forms hydrogen bonds with the substance, surrounding it with water molecules, and so dissolves it (right).

Such molecules—ethyl alcohol, for instance—are called ***hydrophilic*** ("water-loving"). They dissolve in water as a result of electrostatic attraction between the charged parts of the solute molecules and the oppositely charged parts of the water molecules. This occurs especially when, as in many biologically important compounds, the molecule has an oxygen with a hydrogen attached to it (—OH). As in water molecules, the hydrogen in such a group has a net positive charge and is therefore attracted by the negatively charged oxygen end of a nearby water molecule, with the result that a hydrogen bond is formed. The dissolved (solute) molecules and the water molecules thus become weakly linked to each other (Fig. 2.24).

Substances that do not dissolve in water are electrically neutral and nonpolar. They therefore show no tendency to interact electrostatically with water. A ***hydrophobic*** ("water-fearing") substance like oil or octane stirred into water will soon begin to separate out, because the water molecules tend slowly to reestablish the hydrogen bonds broken by the physical intrusion of the insoluble material and thus to push out that material. As a result, the nonpolar, insoluble molecules tend to coalesce to form droplets, which in general eventually fuse and form a separate layer outside the water (Fig. 2.25). As we will see, this basic chemical phenomenon, the tendency of hydrophobic molecules to be driven out of water, is the basis for the spontaneous formation of the cell membranes that protect organisms as they grow and develop.

The polar nature of water also helps explain the mundane but important

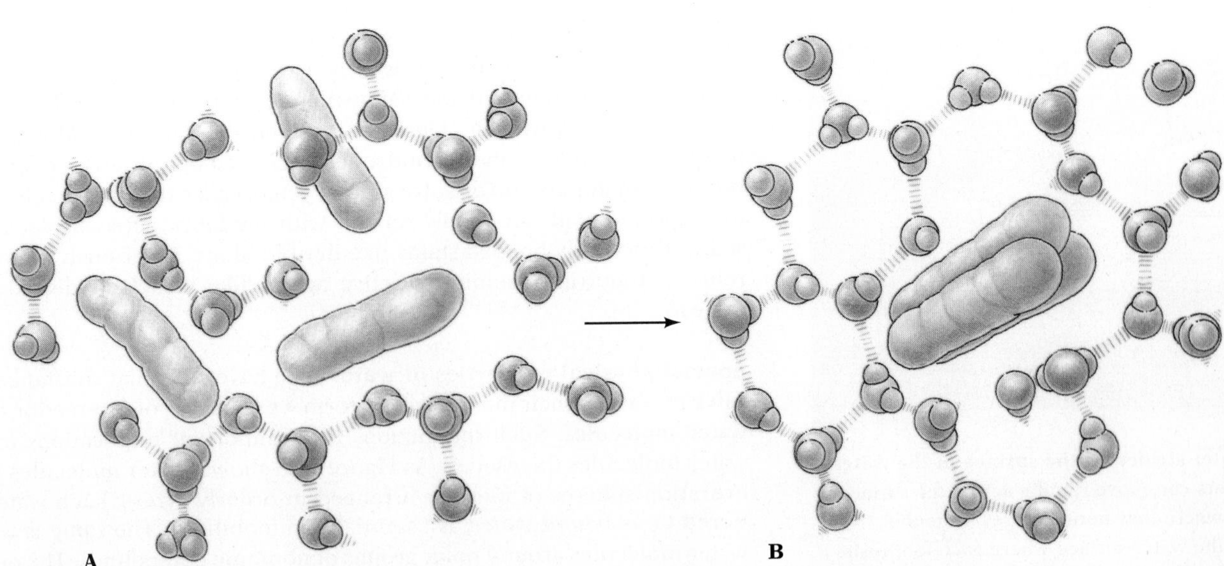

2.25 Water-induced clumping of hydrophobic molecules

Dispersed hydrophobic molecules disrupt the polar bonding pattern of pure water, so that few hydrogen bonds can form in the solution (A). As hydrophobic molecules (schematically represented as brown ovals) encounter one another randomly in a solution of water, they tend to become trapped in clumps by polar bonding of water molecules to one another (B). Because there is more polar bonding when hydrophobic molecules are clumped, the solution becomes stabilized in this form.

workings of soaps and other detergents. Stains left by water-soluble chemicals readily dissolve away when washed in water, but grease and other hydrophobic materials leave stains that cannot be lifted so easily. Detergents (Fig. 2.26) are able to remove grease from a natural fabric because they have an ionic head (which makes them soluble in water) and a hydrophobic tail (which will tend to be driven into tight-packed company with the grease molecules and dissolve them by incorporating these unwanted substances

O^- — O=S=O — O — CH_2 — CH_2 — CH_2 — CH_2 — CH_2 — CH_2 — CH_2 — CH_2 — CH_2 — CH_2 — CH_2 — CH_3

A B

Na^+ grease molecules ionic head hydrophobic tail

2.26 Detergents

Sodium dodecyl sulfate is a strong detergent often used to dissolve cell membranes and other hydrophobic molecules in experiments requiring the separation of these components for further analysis. It has a long, straight hydrophobic tail and, because it ionizes in water, a charged head (A). Water molecules dissolve the heads and drive the tails into tightly packed clumps that dissolve hydrophobic grease molecules (B). During washing, whether in a laboratory preparation or a home washing machine, the entire assembly is rinsed out.

2.27 A water strider on the surface of the water Water striders can move rapidly across the surface of still water, where they hunt for prey. Note the dimples in the water surface where each foot rests.

2.28 Water beading The polar bonding between water molecules causes droplets to form on hydrophobic surfaces like this feather.

into hydrophobic detergent droplets). Nonpolar fabrics (like most synthetics) pose a special problem because water cannot easily wet them (which it must if the detergent is to be carried into the fibers); even if the water does penetrate well, the water will often herd the stain molecules into close association with the fabric rather than the detergent. Molecules of fluids used by dry cleaners stand a better chance because they readily wet nonpolar materials and dissolve grease, yet, because they are small enough to evaporate, will not simply remain with the fabric after displacing the stain. (Covalently bonded stains like dry blood are notoriously difficult to remove, requiring chemically active agents like bleach or digestive enzymes.)

Special physical properties of water We have seen that substances dissolve in water if their molecules can form weak bonds, or interactions, with water molecules. Such interactions have important implications for the water molecules themselves. As Figure 2.23 shows, water molecules in the hydration spheres of ions are arranged in orderly arrays; such water, referred to as ***bound water***, is essentially immobilized. The same is true of water molecules around polar groups of nonionic compounds. The orderly arrays of bound water are very different from those of pure water (Fig. 2.20), and the physical properties of bound water are consequently different from those of free water; the greater the proportion of bound water in a given volume, the lower the freezing point and the higher the boiling point of that volume of water. Since much of the water inside living cells is bound water, the physical properties of the cell contents are very different from those of pure water, even though water is the principal constituent of cells.

The strong ordering of water molecules by hydrogen bonding has important implications for life processes. For example, water has a high ***surface tension:*** the surface of a volume of water is not easily broken. The effects of surface tension are evident when a water strider or other insect walks on the surface of a pond without breaking the surface (Fig. 2.27) or when the water in a glass that is filled slightly above the rim does not spill. Water has a high surface tension because hydrogen bonds link the molecules at the surface to each other and to the molecules below them. Before the legs of the water strider (or any other object, for that matter) can penetrate the water's surface, they must break some of these hydrogen bonds and deform the orderly array of water molecules. Similarly, in an overfilled glass the hydrogen bonds that bind the extra water molecules to the water molecules below them prevent water from spilling.

Just as water molecules are attracted electrostatically to areas of charge on dissolved molecules, so also are they attracted to the charged groups that characterize hydrophilic surfaces. Consequently such surfaces are ***wettable*** —that is, water spreads over them and binds loosely to them. By contrast, hydrophobic surfaces—those of most plastics and waxes, for example—lack surface charge and are not wettable; water on them will form isolated droplets, but will not spread out (Fig. 2.28).

The propensity of water to bind to hydrophilic surfaces explains the phenomenon of ***capillarity***—the tendency of aqueous liquids to rise in narrow tubes. If the end of a narrow glass tube is inserted below the surface of a volume of water, water will rise in the tube to a level well above that of the water outside (Fig. 2.29). Because glass is very hydrophilic, the water molecules, electrostatically attracted to the numerous charged groups on its surface, tend to creep along the inside of the tube. As the ring of water molecules in contact with the inner surface creeps upward, it pulls along other water molecules to which it is linked by hydrogen bonds. The water level stops rising when the pull of gravity just counteracts the electrostatic forces that contribute to capillarity. The larger the diameter of the tube, however, the smaller the percentage of water molecules in direct contact with the glass and, correspondingly, the smaller the rise in the water. Even though the relatively few molecules in contact with the glass have a tendency to creep upward, they are held back by their cohesion with the rest of the water in the tube.

Capillarity is by no means restricted to glass tubes. Water will climb any charged surface. We are all familiar with the way it climbs up the fibers of paper towels and spreads through the fibers of many sorts of cloth.

We have said that each water molecule has the potential for forming hydrogen bonds with four other water molecules (Fig. 2.20). In the liquid state this potential is not fully realized because molecular motion prevents stabilization, but as water is cooled the extent of hydrogen bonding increases. Hydrogen bonding reaches its full potential in water that has frozen into ice. When all four possible bonds have formed, each is oriented in space at the greatest possible distance from the other three. Consequently the bonds are directed toward the four corners of the tetrahedron. The resulting three-dimensional lattice of water molecules in ice is an open one (Fig. 2.30A); the

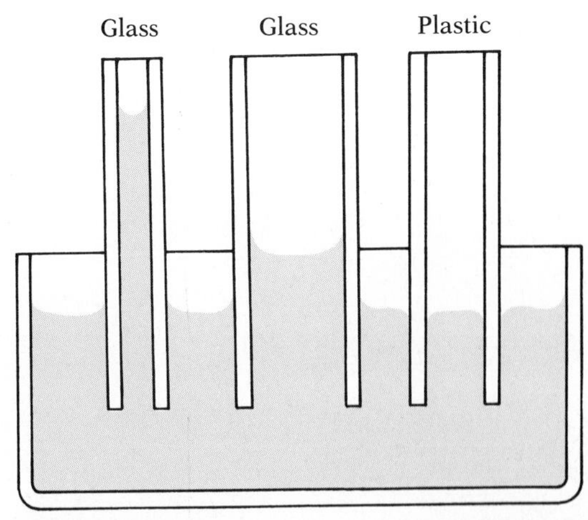

2.29 Capillarity
Water rises higher in a glass tube of small bore (left) than in one of large bore (center) because in the smaller tube a higher percentage of the water molecules are in direct contact with the glass and can form hydrogen bonds with charged groups on the glass. By contrast, water cannot "stick" to the surface of a plastic tube (right) because plastic is uncharged.

2.30 The molecular structure of ice
Because of the tetrahedral arrangement around each water molecule, the lattice is an open one, with considerable space between molecules (A). In liquid water the arrangement is not quite so rigid, and the packing of molecules is therefore slightly denser, but the general lattice arrangement is nonetheless largely preserved. (Planes have been added to help show the three-dimensional disposition of the molecules.) Note the hexagons created by the hydrogen bonding in this bit of ice. This conformation is the basis of the hexagonal shape of most snow crystals (B). Each snowflake contains about 10^{16} water molecules.

rigidly tetrahedral structure maintains space between the molecules, so they can be only loosely packed.

When ice is warmed to the melting point, a few of the hydrogen bonds rupture, and those that remain become less rigidly oriented. The resulting deformation of the lattice and tighter packing of the molecules makes the water denser. The density reaches its maximum when the water is at 4°C; above this temperature, the further packing that might be expected as a consequence of the increasing disruption of the lattice is more than offset by the expansion, shown by virtually all substances, that results from the increased molecular motion of heat energy. In summary, unlike most other substances, which become increasingly dense as the temperature falls, water first becomes denser and then begins to expand again below 4°C. This means that ice, being less dense than cold water, floats and—further—that ponds and streams freeze from the top down rather than from the bottom up. The crust of ice that forms at the surface insulates the water below it from the cold air above and thereby often prevents the pond or stream from freezing solid, even in very cold weather. This special property of water makes life possible in the many ponds and streams that would otherwise freeze solid in the winter.

The role of water in regulating environmental temperature The hydrogen bonds in water give it a high internal cohesion, which enables it to absorb much heat energy without undergoing a very large increase in temperature and to release much heat energy without undergoing a great drop in temperature. When most substances absorb heat, their molecules move more rapidly in relation to one another; temperature is an indication of the amount of such molecular motion. In water, by contrast, much of the absorbed heat energy is dissipated in increased vibration of the hydrogens, each of which is shared between the oxygen to which it is covalently bound and the oxygen of another water molecule, to which it is electrostatically bound. As a result, relatively little of the added heat energy is expressed as movement of whole water molecules, and the temperature increase is therefore modest. The high ***heat capacity*** of water (the amount of heat energy that must be added or subtracted to change the temperature by one degree), together with its high ***heat of vaporization***—the amount of heat energy required for turning water from liquid to vapor (evaporation)—permits it to act as an effective buffer against extreme temperature fluctuations in the environment (Fig. 2.31). In this way water helps stabilize the earth's temperatures within the range favorable to life.

Besides damping fluctuations in temperature, water plays an important role in determining the absolute temperature at the earth's surface, because the water vapor in the atmosphere exerts what has been called a greenhouse effect. The vapor absorbs much of the sunlight striking it from above and also much of the radiation reemitted by the earth. The absorbed radiation warms the atmosphere, which, in turn, warms the earth's surface.

2.31 Water vapor
The three states of water are visible in this volcanic pool in Yellowstone Park during the winter.

CARBON DIOXIDE

As we have seen, carbon has only four electrons in its outer electron shell, and as a result it has a covalent bonding capacity of 4. Carbon dioxide (CO_2) is the compound formed when two atoms of oxygen bond to one atom of carbon. Though this substance contains carbon, it is generally thought of as inorganic, because it is simpler than all but a few of the compounds classified as organic.

Only a very small fraction of the atmosphere, roughly 0.033 percent, is CO_2; yet atmospheric carbon dioxide is the principal inorganic source of carbon, and carbon is the principal structural element of living tissue. Before CO_2 can take part in chemical reactions, it must usually first dissolve in water, which it does very readily on the thin aqueous films that coat most cells. CO_2 then reacts with the water to form carbonic acid (H_2CO_3):

$$CO_2 + H_2O \longrightarrow H_2CO_3$$

This reaction involves so little energy change that it is easily reversible, and CO_2 can readily be released from water solution when conditions are appropriate:

$$H_2CO_3 \longrightarrow CO_2 + H_2O$$

Carbon dioxide and water are the raw materials from which green plants manufacture many complex organic compounds essential to life, as we will see in detail in Chapter 7. When these complex compounds have run their course in the life system, they are broken down again to carbon dioxide and water, and the carbon dioxide is eventually released into the atmosphere. The simple compound carbon dioxide, then, is the beginning and the end of the immensely complex carbon cycle in nature.

OXYGEN

Molecular oxygen (O_2) constitutes approximately 21 percent of the atmosphere. It is necessary for the maintenance of life in most organisms, though a few can live without it. It can be utilized directly, without change, by both plants and animals in the process of extracting usable energy from nutrient molecules. Its role, as we will see, is to serve as the ultimate acceptor of electrons. This is a crucial task: without oxygen to accept electrons, most cells can run at only 5 percent of their normal efficiency. Oxygen is not very soluble in water, but enough dissolves to supply the needs of aquatic organisms, provided (1) that the water is not too hot and (2) that the water's surface is exposed to the air or, alternatively, that green plants are growing in it, thus constantly releasing oxygen into it by the process of photosynthesis. Indeed, it is the production of oxygen by green plants that is the source of virtually all atmospheric oxygen.

Though water, oxygen, and carbon dioxide are truly basic to life as we know it, still other compounds are used to capture, store, transport, and utilize the energy that fuels life. In the next chapter we will examine these complex, carbon-based compounds in an effort to understand the chemical reactions that make life possible.

STUDY QUESTIONS

1. Compare and contrast polar bonds, ionic bonds, and covalent bonds. (pp. 32–34, 36–37)
2. Why are electrons so important to the chemistry of life? (pp. 30–40)
3. What is electronegativity and why is it important? (pp. 37–38)
4. What sorts of chemicals dissolve well in water? Explain their solubility. (pp. 41–44)
5. Do hydrogen bonds have to involve hydrogen? (p. 39)

CONCEPTS FOR REVIEW

- Electrons
    - Electron energy levels
    - Electron number and valence
    - Switching between energy levels: energy absorption and reemission
- Electronegativity
    - Role in ionization
    - Role in creating polar molecules
- Electrostatic force
    - Role in ionic bonds
    - Role in hydrogen bonds
    - Role in dissolving ionic substances
- pH
- Relative strength of covalent, ionic, and hydrogen bonds
- Distinctive characteristics of water important for life
    - Cohesiveness
    - Polarity
    - Ionization
    - Heating and cooling
- Hydrophobic versus hydrophilic molecules
- Behavior of hydrophobic substances in water

SUGGESTED READING

Dickerson, R. E., and I. Geis, 1976. *Chemistry, Matter, and the Universe.* W. A. Benjamin, Menlo Park, Calif. *An excellent introduction to chemistry from a biological perspective.*

Frieden, E., 1972. The chemical elements of life, *Scientific American,* 227 (1). *On procedures for determining whether an element is essential to life, with particular emphasis on four elements (fluorine, silicon, tin, and vanadium).*

Chapter 3

THE CHEMISTRY OF LIFE

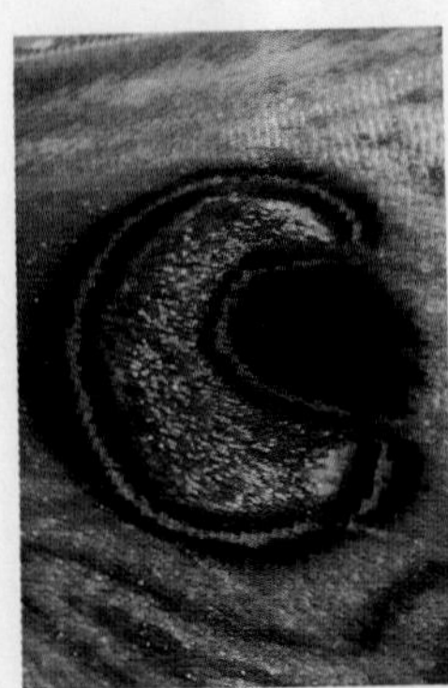

Chemistry does what all good science does: it makes complex phenomena, if no less remarkable, at least easier to predict. Chemistry explains why certain reactions among molecules will take place, and why particular molecular combinations will be stable. Since the diversity of molecules in living organisms is great, and the possibilities for combining them are numerous, an understanding of chemistry is an indispensable predictive tool. Chemistry enables us to make molecular sense of the diversity around us; it lets us see how relatively few elements can combine to constitute living matter in all its varied forms.

The source of the vast molecular diversity in living things is the bonding capacity of just one of the 92 elements—carbon. Carbon's power lies in its versatile structure: four unpaired electrons in its outer shell, which allow it to form covalent bonds with up to four other atoms, make possible enough different molecular connections to generate an almost endless variety of carbon-based—organic—molecules. In this chapter we will look first at the important kinds of organic compounds—carbohydrates, lipids, proteins, and nucleic acids—and then at how some of the crucial well-ordered chemical changes inside cells are orchestrated and controlled.

SOME SIMPLE ORGANIC CHEMISTRY

Though carbon can and does bond to a variety of elements, its four unpaired electrons are most commonly bonded to hydrogen, oxygen, nitrogen, or

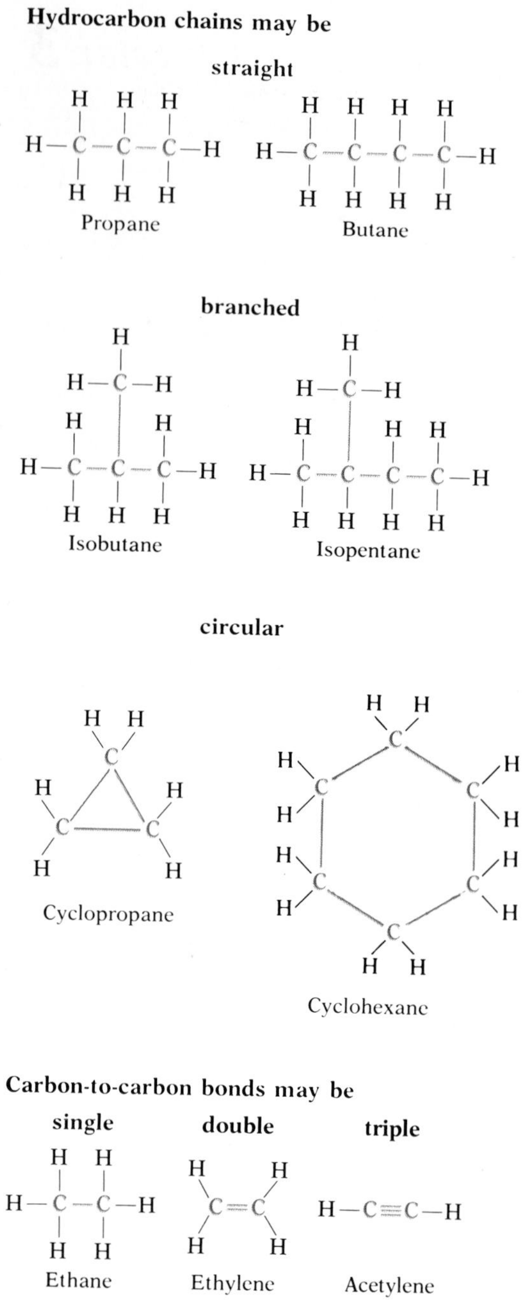

Hydrocarbon chains may be

straight

Propane

Butane

branched

Isobutane

Isopentane

circular

Cyclopropane

Cyclohexane

Carbon-to-carbon bonds may be

single **double** **triple**

Ethane Ethylene Acetylene

3.1 Examples of hydrocarbons
The molecules appear flat in these conventionally drawn structural diagrams, though they are, in fact, three-dimensional. The bonds around a carbon atom that forms only single bonds (all the carbons seen here except those in the last two molecules) are oriented toward the four corners of a tetrahedron.

more carbon. Compounds containing only carbon and hydrogen, the ***hydrocarbons***, are of central importance in organic chemistry; the number of different compounds of this kind is immense. The readiness with which carbon-to-carbon bonds can form and produce chains of varying lengths and shapes has generated a great variety of hydrocarbons. Hydrocarbon chains may be simple, branched, or may form circles of varying numbers of carbons (Fig. 3.1). Obviously, the more atoms a molecule contains, the more different arrangements of those atoms will be possible. Compounds with the same atomic content and molecular formula but different atomic arrangements are called ***isomers*** (Fig. 3.2). Very large organic molecules—ordinary hydrocarbons as well as the other classes of carbon-based molecules—may have hundreds of isomers, with many differing physical properties.

Another source of variety in hydrocarbons is the capacity of adjacent carbon atoms to form single, double, or triple bonds (Fig. 3.1). And, of course, substitution of other elements or groups of elements for hydrogen atoms in hydrocarbons makes possible an almost infinite number of ***derivative hydrocarbons***. The total number of hydrocarbons and derivatives that form in nature has been conservatively estimated at more than half a million. This great capacity for diverse atomic organization makes the hydrocarbon group ideal for building chemicals with unique properties, each precisely suited to the job at hand. For example, life-sustaining compounds like the sugars can be tailored by metabolic processes to suit the varying needs of the cell. From the sugars, cells can extract energy, derive building materials, or construct molecules that help direct cellular processes. On a very different scale, the organic carbon base allows one organism to use another, plant or animal, as food, and thus serves as a versatile common denominator connecting all organisms in an interdependent chain.

The four major classes of complex organic compounds are carbohydrates, lipids (both derivative hydrocarbons), proteins, and nucleic acids. Molecules in each of these classes are often identified on the basis of the subgroups they contain. Each subgroup, or ***functional group***, has its own characteristic properties, which help determine solubility, reactivity, and other traits of the chemical "personality" of the whole molecule. We will refer to many of the groups listed in Table 3.1 in this and later chapters.

CARBOHYDRATES

Carbohydrates are derivative hydrocarbons composed of carbon, hydrogen, and oxygen. In simple carbohydrates the hydrogen and oxygen are characteristically present in the same proportions as in water: there are two hydrogen atoms and one oxygen atom for each carbon atom. Consequently the group $—CH_2O$ recurs frequently in carbohydrate molecules; it is diagrammed

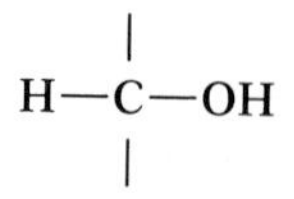

$$\text{H—C—OH}$$

Some carbohydrates, such as starch and cellulose, are very large, complex molecules. But like most very large organic molecules, they are com-

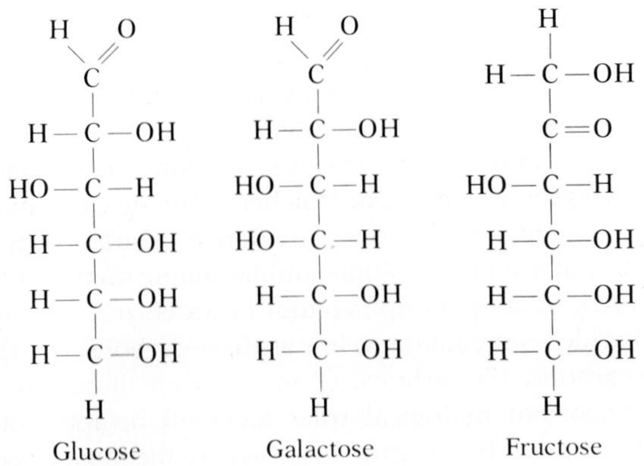

Glucose Galactose Fructose

3.2 Three isomeric hexoses
Each of these six-carbon sugars has the same molecular formula, $C_6H_{12}O_6$; hence each is an isomer of the others.

TABLE 3.1 *Important functional groups*

Group	*Name*	*Properties*	
—OH	**Hydroxyl**	Polar (soluble, because it is able to form hydrogen bonds)	
—C(H)(H)—OH	**Alcohol**	Polar (soluble)	
—C(=O)OH	**Carboxyl**	Polar (soluble); often loses its hydrogen, becoming negatively charged (an acid)	—C(=O)O$^-$
—N(H)H	**Amino**	Polar (soluble); often gains a hydrogen, becoming positively charged (a base)	—N(H)(H$^+$)H
—C(=O)H	**Aldehyde**	Polar (soluble)	
>C=O	**Ketone**	Polar (soluble)	
—C(H)(H)H	**Methyl**	Hydrophobic (insoluble); least reactive of the side groups	
—P(=O)(OH)OH	**Phosphate**	Polar (soluble); usually loses its hydrogens, becoming negatively charged (an acid)	—P(=O)(O$^-$)O$^-$

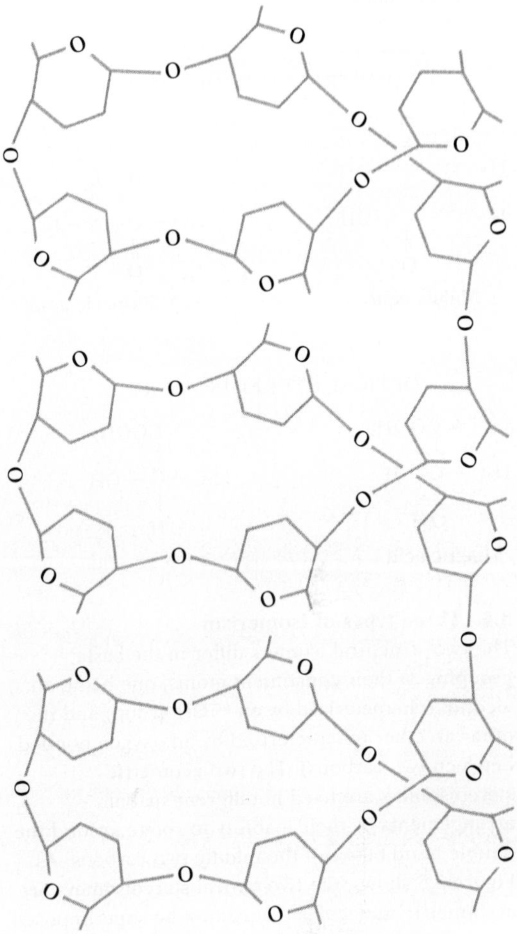

posed of many simpler "building-block" compounds bonded together (Fig. 3.3). Understanding the constituent, or building-block, compounds is the first step toward understanding the more complex substances.

3.3 A moderately complex organic compound
One of the largest of the moderately complex organic compounds is starch, the main energy-storage molecule in plants. Despite its apparent complexity, the starch molecule is actually composed of a repetitive string of glucose units, each represented here as a hexagon. Only a small part of one starch chain is shown. A branched molecule composed of units of several different compounds would be far more complex.

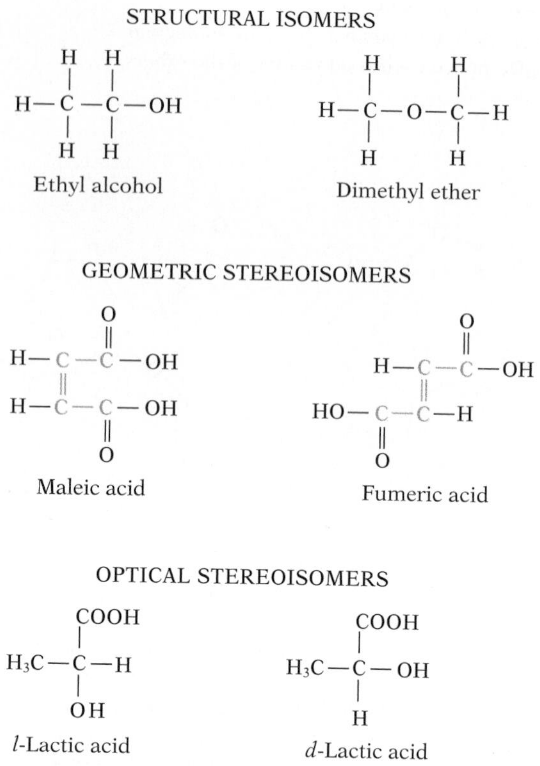

STRUCTURAL ISOMERS

```
   H  H              H         H
   |  |              |         |
H— C —C—OH        H—C—O—C—H
   |  |              |         |
   H  H              H         H
```

Ethyl alcohol   Dimethyl ether

GEOMETRIC STEREOISOMERS

```
          O                         O
          ||                        ||
H— C — C—OH              H— C — C—OH
   ||                          ||
H— C — C— OH         HO— C — C—H
          ||                 ||
          O                  O
```

Maleic acid   Fumeric acid

OPTICAL STEREOISOMERS

```
     COOH                 COOH
      |                    |
H3C—C—H           H3C—C— OH
      |                    |
     OH                    H
```

l-Lactic acid   *d*-Lactic acid

3.4 Three types of isomerism
The two structural isomers differ in the basic grouping of their constituent atoms, one being an alcohol (characterized by an —OH group) and the other an ether (characterized by an oxygen bonded between two carbons). The two geometric stereoisomers are fixed in different spatial arrangements by their inability to rotate around the double bond between the middle two carbons. As Figure 3.5 shows, the two optical stereoisomers are asymmetric molecules that cannot be superimposed on each other.

Simple sugars The basic carbohydrate molecules are simple sugars, or ***monosaccharides***. All sugars when in straight-chain form contain a —C=O group (Fig. 3.2). If the double-bonded O is attached to the terminal C of a chain, the combination is called an aldehyde group; if it is attached to a nonterminal C, the combination is called a ketone group (Table 3.1). The —OH (hydroxyl) groups, which are attached to all the carbons except those with a double-bonded oxygen, are polar. Hence sugars readily form hydrogen bonds with water and are soluble, unlike simple nonpolar hydrocarbon molecules, which tend to clump together in water.

Monosaccharides can contain as few as three carbons, but they may also contain five carbons, six carbons, or more. Both three- and five-carbon sugars play important biological roles and will be mentioned in later chapters, but the six-carbon sugars (hexoses) are the most important building blocks for more complex carbohydrates.

There are many six-carbon sugars, glucose and fructose being two of the most important. Since they all have the proportions of oxygen and hydrogen typical of carbohydrates, all have the same molecular formula, $C_6H_{12}O_6$, and are therefore isomers of one another. Glucose and fructose are structural isomers (Fig. 3.2); the basic groupings of their constituent atoms are different, making one an aldehyde and the other a ketone sugar.

In addition to structural isomerism, there is another, more subtle, kind of isomerism called stereoisomerism. In a given pair of stereoisomers, identical groups are attached to the carbon atoms, but the spatial arrangements of the attached groups are different. The two middle compounds shown in Figure 3.4 are geometric stereoisomers; if the carbon-to-carbon bonds in these two molecules were single, the two would in fact be the same compound, because free rotation is possible around a single bond. A double bond, however, holds its atoms in a rigid configuration.

When considering optical stereoisomers (Fig. 3.4, bottom), keep in mind that molecules are not flat, even though they are often drawn that way. In a carbon atom the four unpaired electrons form the corners of a tetrahedron; if in two molecules the groups attached to corresponding electrons are different, the resulting molecules will be different (Fig. 3.5). Glucose and galactose are optical stereoisomers because the position of one —OH group is different in each (Fig. 3.2). Though the optical stereoisomers of a compound may appear very similar, they are usually quite different in their biological properties and behavior; as Figure 3.5 illustrates, for instance, subtle differences in shape may determine which isomer can bind to or react with a particular molecule and which cannot. Hence, different isomers can play very different roles in the chemistry of cells.

Glucose usually exists in a ring form—most often, as a ring composed of five carbons and one oxygen (Fig. 3.6). Glucose plays a unique role in the chemistry of life. As the primary product of photosynthesis in plants, glucose becomes the ultimate source of all the carbon atoms in both plant and animal tissue. Moreover, the energy stored in its covalent bonds is usually, directly or indirectly, the source of the energy that powers cells. Other six-carbon monosaccharides, among them fructose and galactose, are constantly being converted into glucose or synthesized from glucose. And even such classes of compounds as fats and proteins can be converted into glucose or synthesized from glucose in the living body.

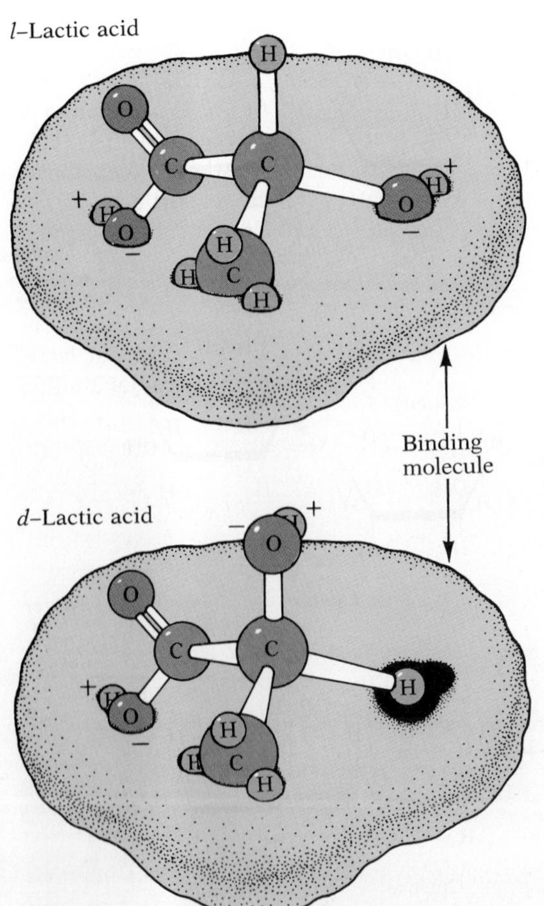

3.5 Optical stereoisomers and chemical specificity Of these two stereoisomers of lactic acid, only one (top) can fit all the holes in the hypothetical schematic molecule to which it binds. The other will not fit the holes no matter which way it is turned. Subtle distinctions of this kind are crucially important: they enable molecules to recognize and bind certain other particular molecules as part of the network of critical chemical reactions necessary for life.

3.6 Two forms of glucose Glucose may exist in the straight-chain aldehyde form shown at left or as a ring structure, as shown in the center. The ring structure is the most common. (By convention, the unmarked corners of the hexagon signify carbon atoms.) Both representations fail to convey the true shape of the molecules, since the four bonds of each carbon atom are directed to the four corners of a tetrahedron. The illustration at right is a more realistic representation of the ring form, but such realism is impossible to preserve in representing any but the simplest organic molecules.

3.7 Two examples of derivative monosaccharides Glucosamine, a chemical used in the synthesis of some protein building blocks as well as in insect exoskeletons, is a glucose molecule with an amino group ($—NH_2$) substituted for an —OH group. Similarly, glucose-6-phosphate, which is used in an important step in the harvesting of energy from glucose, is glucose with a phosphate group added.

Glucosamine

Glucose -6-phosphate

In addition to ordinary monosaccharides composed only of carbon, oxygen, and hydrogen, there are a variety of derivative monosaccharides containing other elements. For example, some have a phosphate group attached to one of the carbons, and others an ***amino*** group: a nitrogen with two hydrogens, $—NH_2$ (Fig. 3.7).

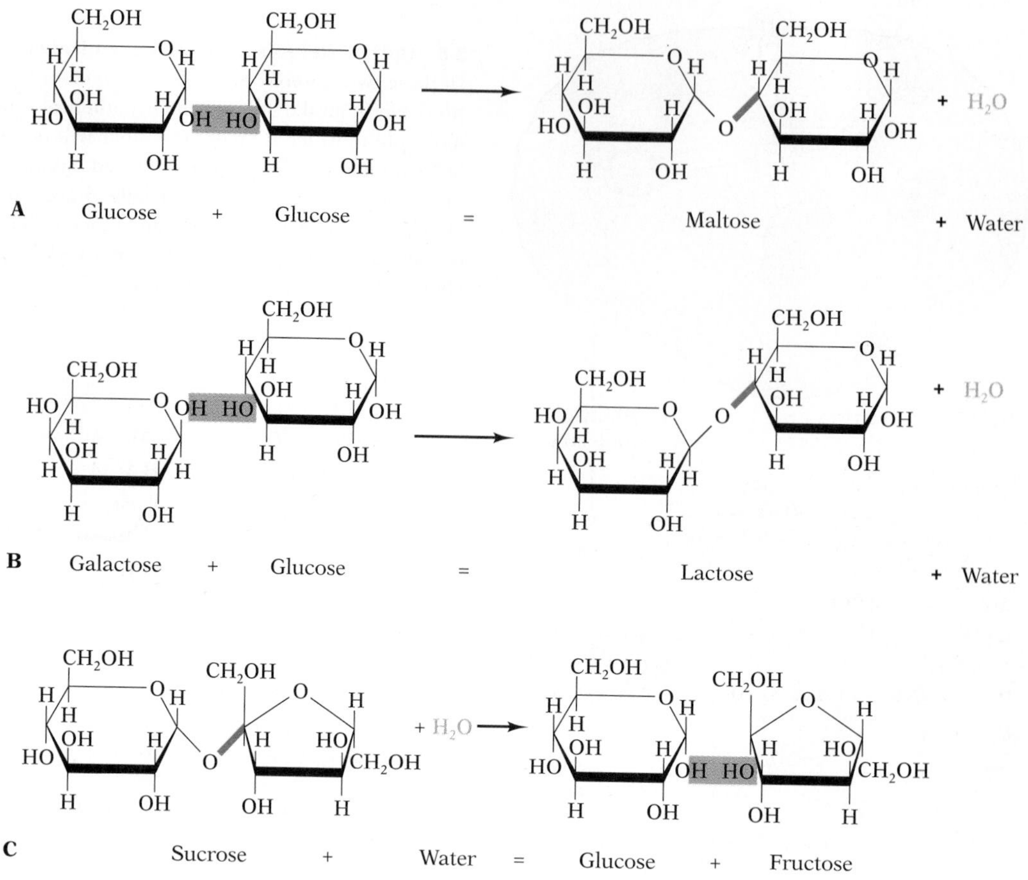

A Glucose + Glucose = Maltose + Water

B Galactose + Glucose = Lactose + Water

C Sucrose + Water = Glucose + Fructose

3.8 Synthesis and digestion of disaccharides Removal of a molecule of water (blue) between two molecules of sugar (a condensation reaction) results in formation of a bond (red) between the two. In the two examples shown here (A, B), intermediate steps are omitted. In the top example, a bottom-to-bottom (α) linkage forms between two glucose molecules, yielding maltose. In the bottom example, a top-to-bottom (β) linkage forms between galactose and glucose, yielding the milk sugar lactose. Some adults suffer from milk intolerance because they cannot digest this linkage. The hydrolysis reaction leading to the breakdown of sucrose involves adding back a water molecule (C).

Disaccharides The ***disaccharides*** are compound sugars composed of two simple sugars bonded together through a series of reactions that involves the removal of a molecule of water. This kind of reaction series is consequently called a ***condensation*** or ***dehydration reaction***.

Let us first examine the disaccharide ***maltose***, or malt sugar. This compound is synthesized by a condensation reaction between two molecules of glucose. The reaction can be described by the following equation, which summarizes several intermediate steps:

$$2C_6H_{12}O_6 \longrightarrow C_{12}H_{22}O_{11} + H_2O$$

Figure 3.8 illustrates how this happens: the hydrogen atom from a hydroxyl group (—OH) of one glucose combines with a complete hydroxyl group from another glucose molecule to form water. The oxygen valence freed by the removal of hydrogen and the carbon valence freed by the removal of —OH are filled by the bonding together of the oxygen of one glucose molecule with the carbon of the other glucose molecule. As a result, the two glucose units are connected by an oxygen atom shared between them, producing the disaccharide maltose.

Sucrose, our common table sugar, is also a disaccharide. It is synthesized by a condensation reaction between a molecule of glucose and a molecule of fructose. ***Lactose***, or milk sugar, is a disaccharide composed of glucose and galactose joined by a condensation reaction (Fig. 3.8A,B). In fact, it is a general (and important) rule that the synthesis of complex molecules from simpler units almost always produces water.

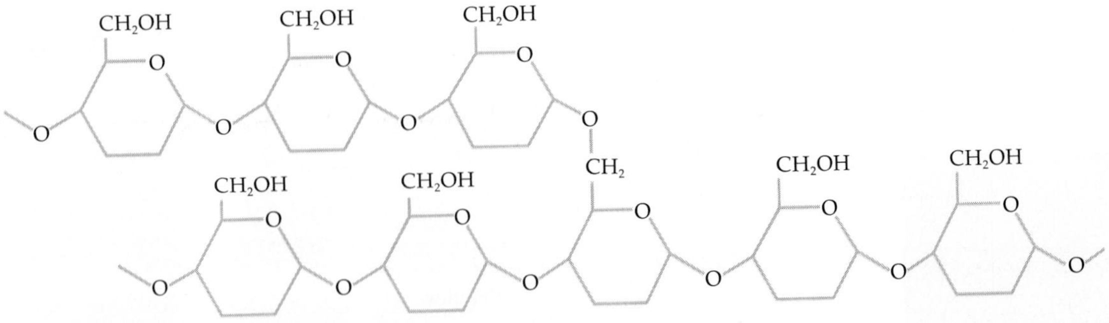

Synthesized by condensation reactions, disaccharides can be broken down to their constituent simple sugars by the reverse process: adding back a water molecule. This reaction, called ***hydrolysis***, involves splitting a water molecule into a hydrogen atom and a hydroxyl group, and then adding them to the subunits through a series of steps. The reactions can be summarized as

$$C_{12}H_{22}O_{11} + H_2O \longrightarrow 2C_6H_{12}O_6$$

Hydrolysis reactions are of particular importance in digestion, as we will see in a later chapter, because digestion breaks down complex molecules into simple building blocks, ready for subsequent use (Fig. 3.8C).

We can now define a monosaccharide more precisely than was hitherto possible. By contrast with compound sugars, a monosaccharide is a sugar produced after the complete hydrolysis of compound sugars.

Polysaccharides The prefix *poly-* means "many," and ***polysaccharides*** are complex carbohydrates composed of many simple-sugar building blocks bonded together in long chains (Fig. 3.9). They are synthesized by exactly the same kind of condensation reaction as the disaccharides and, like them, can be broken down into their constituent sugars by hydrolysis.

A number of complex polysaccharides are of great importance in biology. ***Starches***, for example, are the principal carbohydrate storage products of higher plants. They are composed of many hundreds of glucose units bonded together. In some forms of starch the chain of sugars is unbranched, and in others it is branched; both types are common in plant material. ***Glycogen*** is the principal carbohydrate storage product in animals and is sometimes called animal starch. Its molecules are much like those of starch; they have the same type of bond between adjacent glucose units, but the chains are more extensively branched. ***Cellulose*** is the most common carbohydrate on earth. It is a highly insoluble, unbranched polysaccharide used by plants as their major supporting material. The bonds between its

3.9 Branched starch
Shown here is a small segment of a molecule of starch. This starch is branched, but some forms (as in Fig. 3.3) are unbranched. Like the cellulose of plants, starch is a polymer of glucose, but in cellulose the glucose is connected by β linkages (see Fig. 3.10), whereas in starch the glucose is connected by α linkages.

3.10 Cellulose and chitin
Cellulose is composed of long chains of β-linked glucoses. Chitin is composed of β-linked acetylglucosamines (glucoses with a combined acetyl-amino side group).

CH_2OH CH_2OH CH_2OH CH_2OH

Cellulose

CH_2OH CH_2OH CH_2OH

$NHCOCH_3$ $NHCOCH_3$

Chitin

A

B

glucose units are β-linkages rather than α-linkages (Fig. 3.10); animals can digest the bonds of starch and glycogen, but most animals are unable to hydrolyze those of cellulose. ***Chitin***, the polysaccharide that serves as the major structural component of insect exoskeletons and fungal cell walls, is functionally equivalent to cellulose (Fig. 3.11).

All reactions in which small molecules (called ***monomers***) bond together to form long chains are called polymerization reactions; polymerization of monosaccharides, for example, creates a polysaccharide. The products of polymerization are called ***polymers***. Polymers play a critical role in biology, as we will see.

3.11 (A) Mantis exoskeleton and (B) SEM of paper fibers
The chitin of insect exoskeletons (visible here in the shed skin of a mantis) is chemically similar and functionally equivalent to the cellulose of plants, seen here as fibers in uncoated paper. (Books like this one have a coating of clay over the fibers; this coating, which gives the paper a shine, prevents absorption and spreading of the ink, and thus permits the printing of high-resolution photographs.)

LIPIDS

Like carbohydrates, lipids, a second major group of derivative hydrocarbons, are composed principally of carbon, hydrogen, and oxygen, but they may also contain other elements, particularly phosphorus and nitrogen. In their simplest form, lipids are hydrocarbons with a carboxyl group (—COOH) at one end (Fig. 3.12). Such lipids are primarily nonpolar, by virtue of their long hydrocarbon "tails." They are therefore relatively insoluble in water, but soluble in organic solvents such as ether. Most lipids, as we will see, are more complex, having an ionic group attached to the carboxyl end; the long hydrophobic tails are universal features.

```
HO     O
  \   //
    C
    |
H — C — H
    |
H — C — H
    |
H — C — H
    |
H — C — H
    |
H — C — H
    |
    H
```

3.12 A simple lipid

Fats Among the best-known lipids are the neutral fats. Important as energy-storage molecules in living organisms, the fats also provide insulation, cushioning, and protection for various parts of the body (Fig. 3.13). Each molecule of fat is composed of fatty acids joined together by glycerol.

Glycerol (also sometimes called glycerin) has a backbone of three carbon

3.13 Fat-storage cells
Lipids are stored in spherical fat cells called adipocytes, seen here. Small blood vessels (capillaries) and support fibers (collagen) hold these cells in place. In the same manner as the styrofoam beads they resemble, these human adipocytes cushion and insulate underlying parts of the body. From *Tissues and Organs: A Text-Atlas of Scanning Electron Microscopy* by Richard G. Kessel and Randy H. Kardon. Copyright 1979 W. H. Freeman and Company. Used with permission.

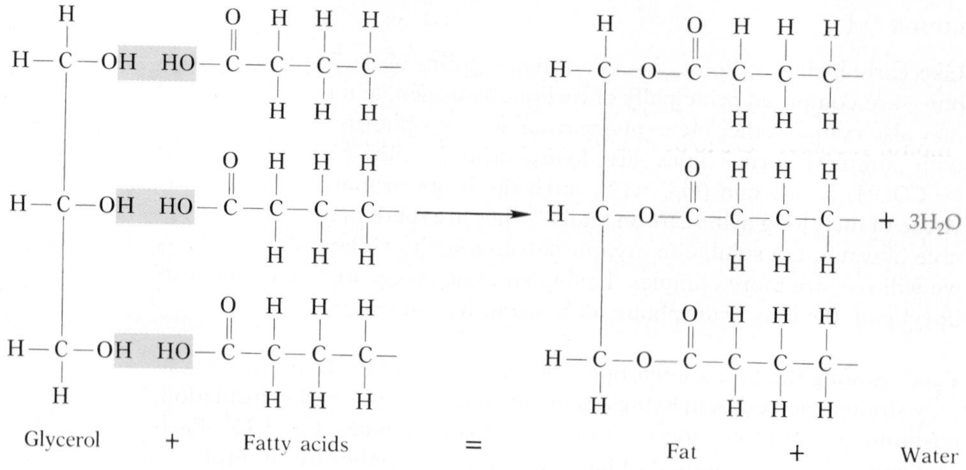

```
      H              O   H   H   H                 H      O   H   H   H
      |              ‖   |   |   |                 |      ‖   |   |   |
  H — C — OH   HO — C — C — C — C —           H — C — O — C — C — C — C —
      |                  |   |   |                 |          |   |   |
      |                  H   H   H                 |          H   H   H
      |              O   H   H   H                 |      O   H   H   H
      |              ‖   |   |   |                 |      ‖   |   |   |
  H — C — OH   HO — C — C — C — C —  ———→     H — C — O — C — C — C — C —  + 3H2O
      |                  |   |   |                 |          |   |   |
      |                  H   H   H                 |          H   H   H
      |              O   H   H   H                 |      O   H   H   H
      |              ‖   |   |   |                 |      ‖   |   |   |
  H — C — OH   HO — C — C — C — C —           H — C — O — C — C — C — C —
      |                  |   |   |                 |          |   |   |
      H                  H   H   H                 H          H   H   H

   Glycerol    +     Fatty acids        =          Fat         +        Water
```

3.14 Synthesis of a fat
Removal of three molecules of water by condensation reactions results in the bonding of three molecules of fatty acid to a single molecule of glycerol. (Intermediate steps are omitted in this example.) Conversely, three molecules of water will be added by hydrolysis when this molecule of fat is digested. The carbon chains of the fatty acids are usually longer than shown here.

atoms, each carrying a hydroxyl (—OH) group (Fig. 3.14). Since, by definition, all alcohols have at least one $—CH_2OH$ group, glycerol is an alcohol.

Fatty acids, like all organic acids, contain a carboxyl group. When both a double-bonded oxygen and an —OH group are attached to the same carbon atom, the double-bonded oxygen tends to cause the —OH part of this carboxyl to lose its hydrogen, making the group ionic and causing the compound to behave as an acid (see Table 3.1).

There are many different fatty acids, varying in carbon-chain length, in the number of single or double carbon-to-carbon bonds, and in other characteristics. The fatty acids in edible fats and oils contain an even number of carbon atoms, and most of them have relatively long carbon backbones, usually from 4 to 24 carbons, or more; three of the most common are stearic acid (18 carbons), palmitic acid (16 carbons) (Fig. 3.15), and linoleic acid (18 carbons).

Organic acids and alcohols have a tendency to combine through condensation reactions. Since glycerol has three hydroxyls, it can combine with three molecules of fatty acid to form a molecule of fat (Fig. 3.14). Hence fats are sometimes also called triglycerides.

The various fats differ in the specific fatty acids, or types of fatty acids, composing them. You have doubtless read of the controversy in medical and nutritional circles concerning saturated and unsaturated fats. ***Saturated fats*** are simply those incorporating fatty acids with the maximum possible number of hydrogen atoms attached to each carbon, and hence no carbon-to-carbon double bonds (Fig. 3.15). The fatty acids in ***unsaturated fats*** (or perhaps we should say oils, since they are usually liquid at room temperature) have at least one carbon-to-carbon double bond—that is, they are not completely saturated with hydrogen. (The double bond induces a "kink" in the otherwise linear structure, which prevents solidification.) There is now good evidence that an elevated intake of saturated fats is one of many factors that predispose human beings to atherosclerosis—a disease of the arteries in which fatty deposits in the arterial walls cause partial obstruction of blood flow, which can lead to strokes and heart failure.

Since fats are synthesized by condensation reactions (removal of water), they, like complex carbohydrates, can be broken down into their building-block compounds by hydrolysis, as happens in digestion. And because fat, though slow to be metabolized, contains 2.5 times as much usable energy per gram as monosaccharides, it is a good substance for long-term energy storage. In fact, if all the energy the average individual stores in fat (about a month's supply) were to be maintained in the form of sugar, we would each be 25–30 kg heavier.

Phospholipids Various lipids contain a phosphate group at the carboxyl end of the chain. Among the most common of these phospholipids are those composed of one unit of glycerol, two units of fatty acid, and a phosphate group often linked with a nitrogen-containing group (Fig. 3.16). The phos-

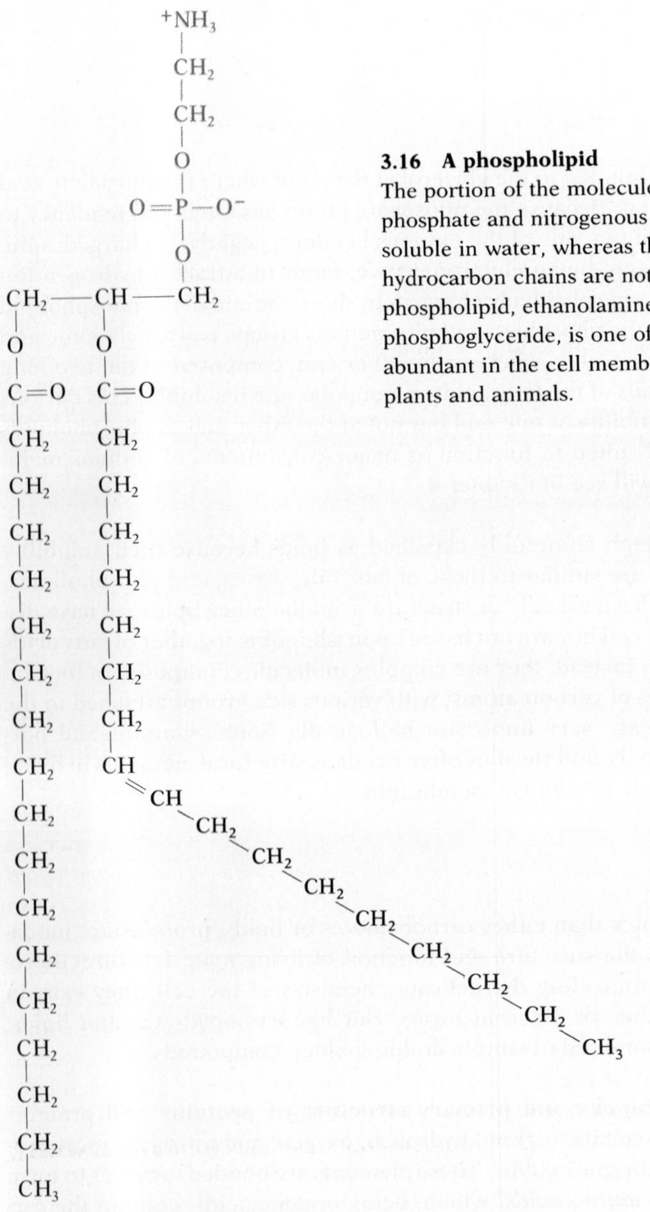

3.16 A phospholipid
The portion of the molecule with the phosphate and nitrogenous groups (blue) is soluble in water, whereas the two hydrocarbon chains are not. This particular phospholipid, ethanolamine phosphoglyceride, is one of the two most abundant in the cell membranes of higher plants and animals.

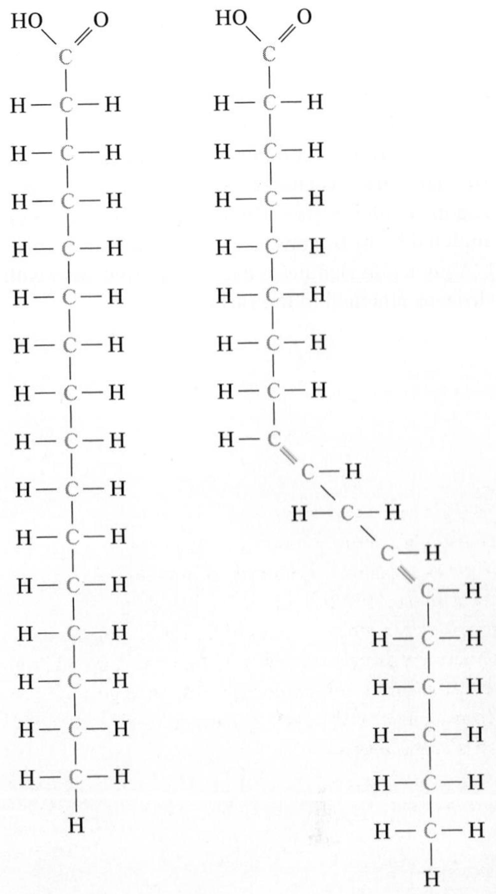

Palmitic acid

Linoleic acid

3.15 Examples of saturated and unsaturated fatty acids
Palmitic acid is saturated with hydrogen—that is, it contains the maximum number of hydrogens possible. By contrast, linoleic acid, with its two inflexible carbon-to-carbon double bonds, accommodates four fewer than the maximum number of hydrogens.

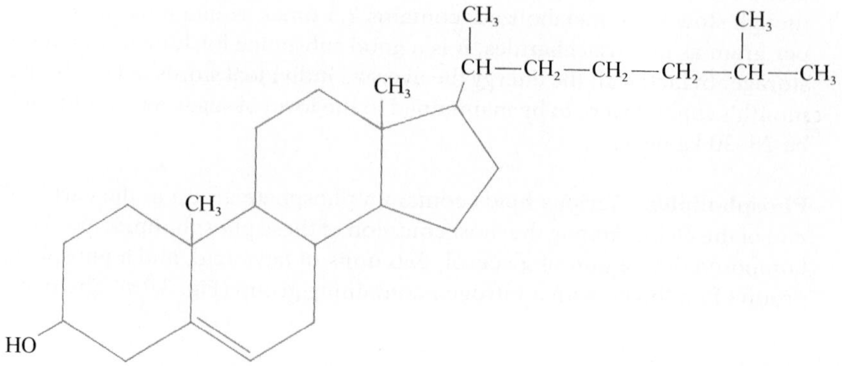

3.17 A steroid
All steroids have the same basic unit of four interlocking rings, but differ in their side groups. This particular steroid is cholesterol. (By convention, a hexagon signifies a six-carbon ring with its valences completed by hydrogens; see cyclohexane in Figure 3.1. A pentagon signifies a five-carbon ring, also with hydrogens attached to the carbons.)

phate group is bonded to the glycerol at the point where the third fatty acid would be in a fat. Because the phosphate group has a marked tendency to lose a hydrogen ion, one of the oxygens becomes negatively charged; similarly, the nitrogen, being electronegative, tends to attract a hydrogen ion and thus to become positively charged. In short, the end of the phospholipid molecule with the phosphate and nitrogenous groups is strongly ionic and hence soluble in water, whereas the other end, composed of the two long hydrocarbon tails of the fatty acids, is nonpolar and insoluble. This curious property of solubility at one end but not at the other makes phospholipids especially well suited to function as major constituents of cellular membranes, as we will see in Chapter 4.

Steroids Though commonly classified as lipids because their solubility characteristics are similar to those of fats, oils, waxes, and phospholipids, the steroids differ markedly in structure from the other lipids we have discussed (Fig. 3.17). They are not based upon a bonding together of fatty acids and an alcohol. Instead, they are complex molecules composed of four interlocking rings of carbon atoms, with various side groups attached to the rings. Steroids are very important biologically. Some vitamins and hormones are steroids, and steroids often occur as structural elements in living cells, particularly in cellular membranes.

PROTEINS

Far more complex than either carbohydrates or lipids, proteins are fundamental to both the structure and function of living material. Directly responsible for controlling the delicate chemistry of the cell, they exist in literally thousands of different forms. But like carbohydrates and lipids, proteins are composed of simple building-block compounds.

The building blocks and primary structure of proteins All proteins contain four elements: carbon, hydrogen, oxygen, and nitrogen; most proteins also contain some sulfur. These elements are bonded together to form subunits called ***amino acids***, which, being organic acids, contain the carboxyl (—COOH) group (Fig. 3.18). In addition, they each have an amino

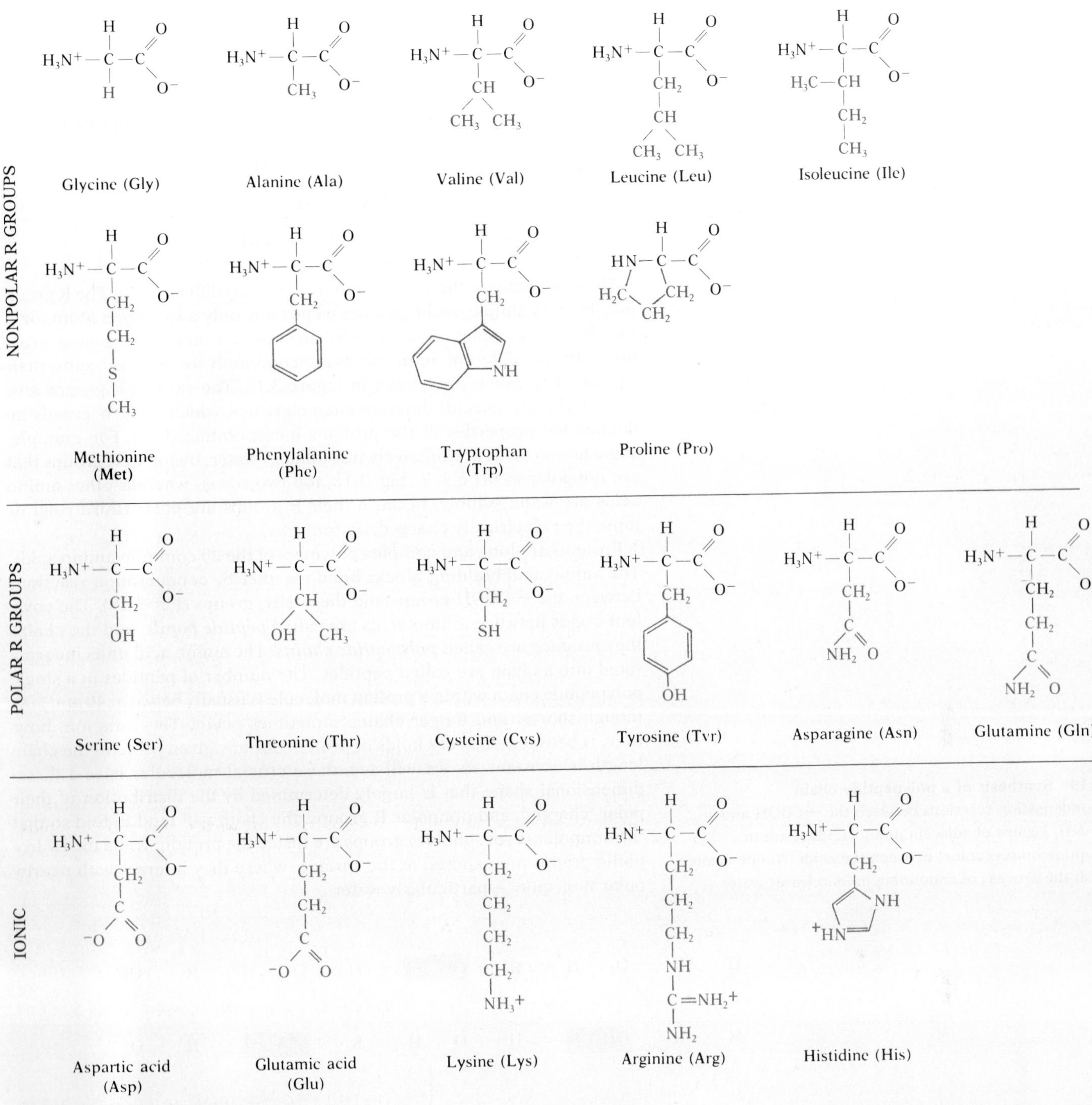

3.18 The 20 amino acids common in proteins classified by R group
The amino acids are shown in their ionized form. All have the same arrangement of a carboxyl group and an amino group attached to the same carbon; they differ in their R groups (brown). Top two rows: These nine amino acids have nonpolar R groups and are relatively insoluble in water. (Glycine is an exception. Its R group, a single hydrogen atom, is nonpolar, but is too small to outweigh the charge of the amino and carboxyl groups. The molecule therefore behaves more like a polar amino acid and is water-soluble. Proline is also unusual. It is technically not an amino acid, because the nitrogen is bonded to part of the R group. However, it is included because it is regularly incorporated into proteins along with the true amino acids.) Third row: These six amino acids have polar R groups and are soluble. Bottom row: These five amino acids, with R groups ionized at intracellular pH levels, are electrically charged and thus water-soluble; the first two, being negatively charged, are acidic, whereas the last three, with a positive charge, are alkaline.

(—NH_2) group. The —COOH and the —NH_2 group are attached to the same carbon atom. Finally, each amino acid has a side chain, designated R:

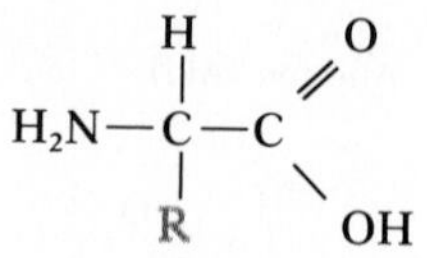

$$H_2N-\underset{R}{\overset{H}{C}}-C\begin{matrix}\nearrow O\\ \searrow OH\end{matrix}$$

The side chains of the various amino acids may differ greatly. The R group may be very simple, as in glycine, where it is only a hydrogen atom, or it may be very complex, as in tryptophan, where it includes two ring structures. Twenty different amino acids are commonly found in proteins; their structural formulas are shown in Figure 3.18. The various R groups give each of the amino acids different characteristics, which, in turn, greatly influence the properties of the proteins incorporating them. For example, some amino acids are relatively insoluble in water, owing to R groups that are nonpolar at pH 6.5–7 (Fig. 3.18, top two rows), whereas other amino acids are water-soluble, because their R groups are polar (third row) or ionic (i.e., electrically charged; bottom row).

Proteins are long and complex polymers of the 20 common amino acids. The amino acid building blocks bond together by condensation reactions between the —COOH groups and the —NH_2 groups (Fig. 3.19). The covalent bonds between amino acids are called ***peptide bonds***, and the chains they produce are called ***polypeptide chains***. The amino acid units incorporated into a chain are called peptides. The number of peptides in a single polypeptide chain within a protein molecule is usually between 40 and 500, though shorter and longer chains sometimes occur. The variation, however, is between different kinds of protein; for any given protein, the chain length is constant. As we will see, these crucial molecules have a three-dimensional shape that is largely determined by the distribution of their polar, charged, and nonpolar R groups: the chain will tend to fold so that the nonpolar (hydrophobic) groups are inside the protein, while the hydrophilic groups are exposed at the surface, where they interact with nearby polar molecules—particularly water.

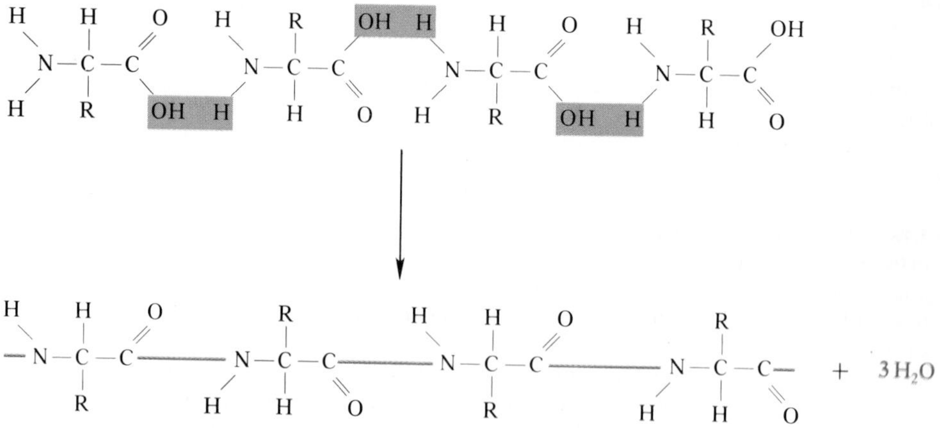

3.19 Synthesis of a polypeptide chain
Condensation reactions between the —COOH and —NH_2 groups of adjacent amino acids result in peptide bonds (color) between the acids. Notice again that the process of combining units releases water.

3.20 The structural formula of a cystine bridge A cystine bridge is formed when two cysteine peptides (red) from different parts of a protein are linked by a disulfide bond.

Protein molecules often consist of more than one polypeptide chain. The chains may be held together by numerous weak bonds, especially hydrogen bonds; for example, a single molecule of hemoglobin, the red oxygen-carrying protein in blood, is composed of four polypeptide chains linked by hydrogen bonds. Insulin, an important hormone secreted by the pancreas in vertebrates, exemplifies a protein with polypeptide chains held together by both hydrogen and covalent bonds. The covalent bonds, called ***disulfide bonds***, are between the sulfur atoms of two units of the amino acid cysteine (Fig. 3.18); two cysteines readily react with each other to form a symmetrical linkage called a cystine bridge (Fig. 3.20). As Figure 3.21 shows, disulfide bonds can also link two parts of a single polypeptide chain, maintaining it in a bent or folded shape.

We have so far discussed the so-called ***primary structure*** of protein molecules—the number of polypeptide chains, the number and sequence of amino acids in each, and the location of disulfide bonds. Since the primary structure is different for every kind of protein, the potential number of different proteins is enormous. For a relatively short polypeptide chain of 100 amino acids, for instance, 20^{100} sequences are possible. Even though most organisms have only 1000–30,000 different proteins, each of the many millions of different species of organisms could in theory have its own peculiar proteins. In fact, though, the common evolutionary heritage of organisms, combined with the basic similarity of the chemical tasks to be accomplished, has constrained protein diversity.

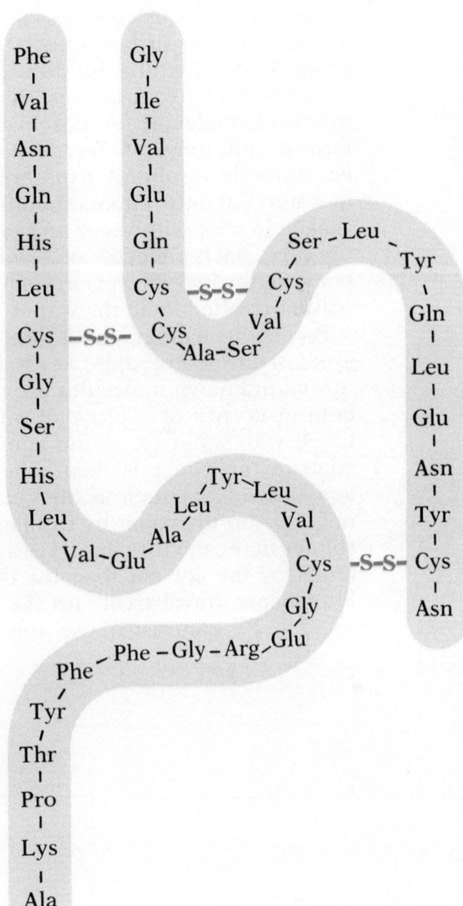

3.21 The structure of bovine insulin The molecule consists of two polypeptide chains joined by two disulfide bonds. There is also one disulfide bond within the shorter chain (right). Hydrogen bonds (not shown) between the chains and between segments of the same chain are also present.

Exploring Further

CHROMATOGRAPHY

Chromatography is one of the most valuable techniques available for separating the substances found in blood or cells, or separating a single substance into its constituent parts. It is chromatography that first demonstrated that bacteria have more than 1000 different proteins, and allowed most of them to be individually isolated. After isolating a single protein using chromatography, we can hydrolyze it and then use chromatography again to separate the amino acids from one another. The many kinds of chromatography all share the same fundamental mechanism—the simultaneous exposure of the mixture being studied to two different substances, such as two solvents that will not mix, or a solvent and an adsorbent solid. (An adsorbent is so named because molecules of a gas, dissolved solute, or liquid adhere to its surface.) Each molecule of each solute in the mixture will diffuse back and forth between the two substances; the relative affinity of a solute molecule for those two solutes will determine how long, on average, it remains in each. For example, if solute A, a material highly soluble in water, but only minimally soluble in phenol, is shaken in a jar with water and phenol (which do not mix), each molecule of A will divide its time between the two solvents in such a fashion that it will more often be in the water.

Paper chromatography is one of the simplest kinds of chromatography. Several drops of a mixture of unknown molecules are placed near one bottom corner of a piece of filter paper moistened with water (A, in the figure). The bottom edge of the paper is then dipped into a nonaqueous solvent such as phenol. As the solvent migrates up the paper by capillary action, those solutes in the mixture that have a much higher affinity for the solvent than for the water in the filter paper travel freely up the paper with the solvent. By contrast, those solutes that have a much higher affinity for the moisture in the filter paper do not travel far, but instead quickly transfer from the flowing solvent and bind to the stationary film of water on the filter paper. Materials with intermediate affinities for the solvent as compared to the water travel intermediate distances. At the end of a measured time interval, the various solutes in the original mixture have come to rest at different places along the filter paper (B). Frequently, further separation is achieved by using a second solvent and allowing the solutes to travel across the paper in a new direction (C and D). The technique is then known as two-axis chromatography. An example of the results obtainable by this method can be seen in the photograph.

Another popular way to separate substances is ***electrophoresis***. After chromatographic separation along one axis, an electric field is set up along the other; molecules with different ionic strengths and polarities migrate at different speeds toward one end or the other.

Column chromatography works like simple paper chromatography, except that instead of filter paper, a glass column packed with some hydrated adsorbent material, such as starch or silica gel, is used. The mixture to be tested, in a nonaqueous solvent, is poured into the top of the column and allowed to filter downward (or is pulled down by a pump). Each component in the mixture tends to move at its own rate, which depends on its relative affinity for the flowing solvent and the stationary beads of adsorbent material. That is, substances with a much higher affinity for the solvent will flow straight through, while those with some affinity for the beads will be delayed. When enough solvent is poured into the column to keep all the materials moving, each of them emerges from the bottom of the column at a different time, and they are collected in separate containers, ready for further analysis.

The spatial conformation of proteins Proteins are not laid out simply as straight chains of amino acids. Instead, they coil and fold into very complex spatial conformations, which play a crucial role in determining the distinctive biological properties of each protein. Much of this three-dimensional

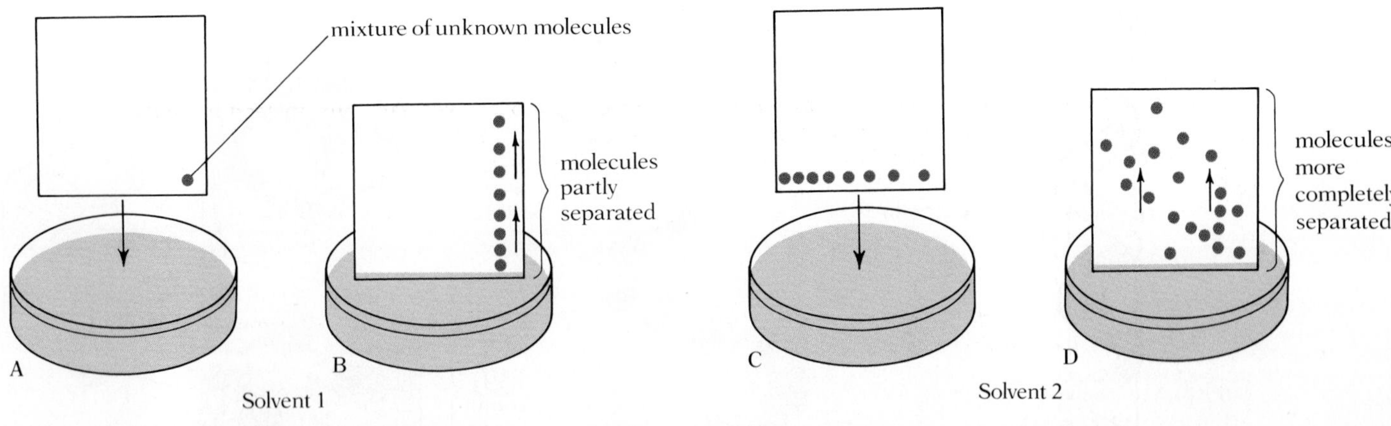

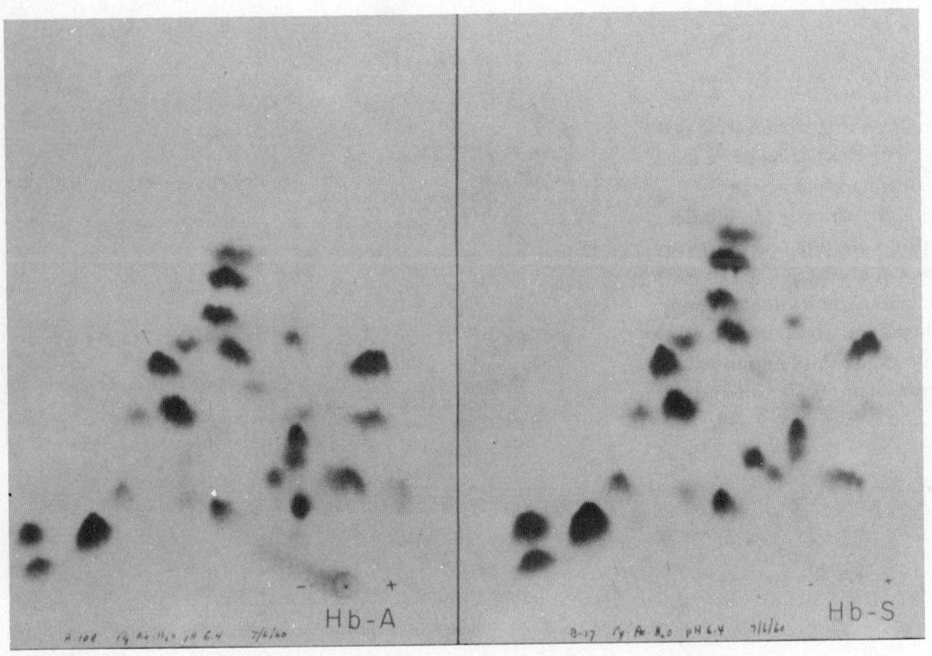

character is a consequence of weak interactions between peptides in the protein. As Linus Pauling and Robert B. Corey of the California Institute of Technology showed in 1951, certain precise degrees of coiling allow internal hydrogen bonds to form and stabilize what is called an ***alpha* (α) *helix***

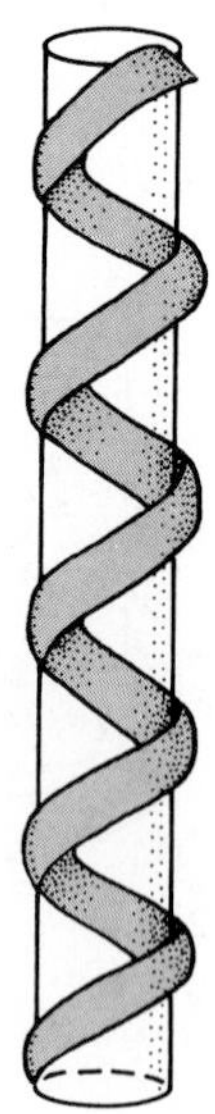

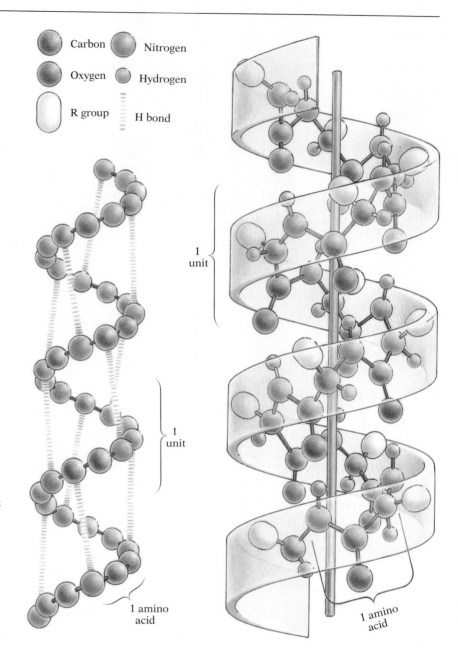

3.22 Alpha-helical secondary structure of some proteins
(A) The helix may be visualized as a ribbon wrapped around a regular cylinder. (B) The backbone of a polypeptide chain (the repeating sequence of N–C–C–N–C–C–N–C–C–) is shown coiled in a helix (all other atoms and R groups are omitted). It takes approximately 3.6 amino acid units (N–C–C) to form one complete turn of the helix. The hydrogen bonds shown extend between the amino group of one amino acid and the oxygen of the third amino acid beyond it in the polypeptide chain. (C) A ball-and-stick model of an α-helical section of a protein.

(Fig. 3.22). A helix can be visualized as a ribbon wound around a regular cylinder (Fig. 3.22A). In a protein, each complete turn of the helix takes up approximately 3.6 amino acid units of the polypeptide chain (Fig. 3.22B). The chain is held in this helical shape by hydrogen bonds formed between the amino group of one amino acid and the oxygen of the third amino acid beyond it in the polypeptide chain—which is the amino acid next to it in the axial direction of the helix (Fig. 3.22C). The ***conformation***, or arrangement in space, of amino acids in a peptide chain is called ***secondary structure***. The α helix was the first example of secondary structure to be discovered.

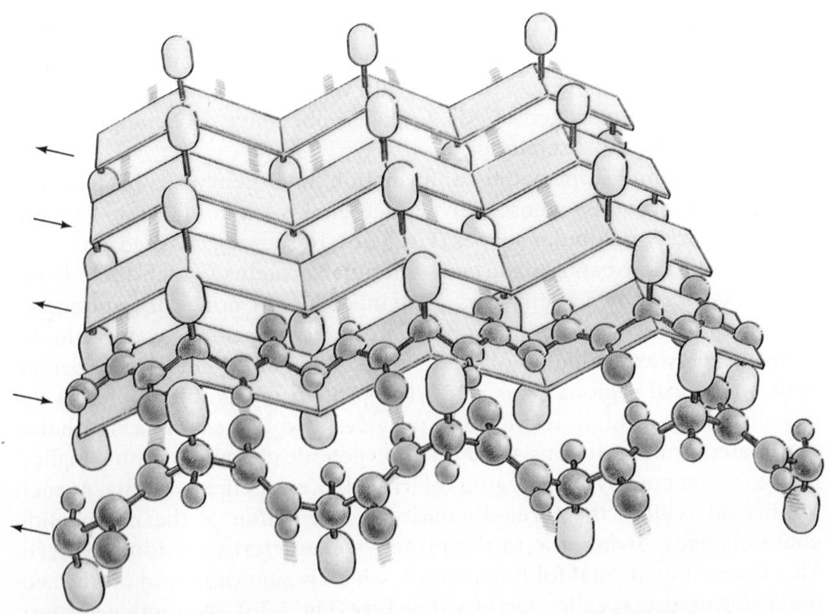

3.23 Pleated-sheet secondary structure of some proteins
Diagrammatic representation of five parallel polypeptide chains in β conformation, with the imaginary pleated sheet shown for four.

The helical pattern is seen at its simplest in some ***fibrous proteins***. One category of these insoluble proteins, extensively used in studies of protein structure, includes the ***keratins***, which provide the structural elements for many of the specialized derivatives of skin cells. Keratins with extensive α-helical secondary structure, such as nails, hooves, and horns, are hard and brittle. The hardness results from an extraordinarily large number of covalent cystine bridges: up to one amino acid in four is cysteine. Others, such as hair and wool, are soft and flexible and can easily be stretched (especially when moistened and warmed). The stretching is possible because there are many fewer cystine bridges. The intrachain hydrogen bonds are easily broken, and the polypeptide chains can then be pulled out of their compact helical shapes into a more extended form. The chains tend to contract to their normal length, with re-formation of the hydrogen-bonded α helix, when the tension on them is released (or when they are dried and cooled).

Another stable arrangement of peptides within a polypeptide chain gives rise to the second major type of secondary structure. This type, designated beta (β) structure, is often called the ***pleated sheet***. In this conformation, also seen at its simplest among some of the keratins, many side-by-side polypeptide chains are cross-linked by interchain hydrogen bonds (Fig. 3.23). The resulting arrangement is flexible and strong, but resists stretching because the polypeptide chains are already almost fully extended. Probably the best-studied β-keratin is silk; other examples include spider webs, feathers (Fig. 3.24), and the scales, claws, and beaks of reptiles and birds.

3.24 The feather of a bird, a β-keratin
This scanning electron micrograph shows the base of a parakeet's tail feather.

Another kind of fibrous protein, with its own distinctive secondary structure, is ***collagen***, the most abundant protein in higher vertebrates. Collagen can constitute one-third or more of all the body protein, and is especially abundant in skin, tendons, ligaments, and bones, and in the cornea of the eye. A molecule of collagen is composed of three polypeptide chains, each first helically coiled and then wound around the other two to form a triple

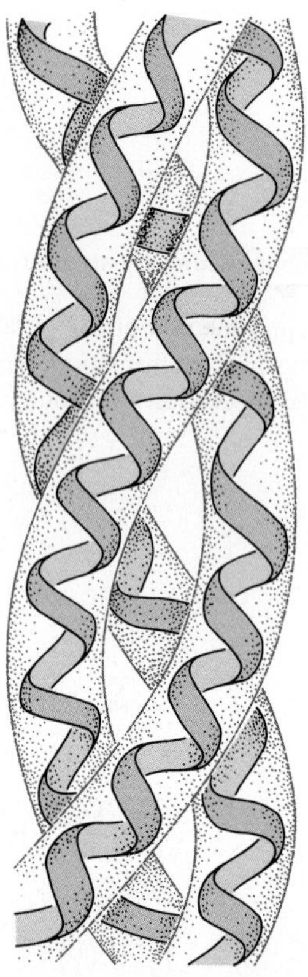

3.25 Model of a portion of a molecule of collagen Three polypeptide chains, each helically coiled, are wound around one another to form a triple helix. The "sheaths" here and in Figures 3.26 and 3.29 are intended as a reminder that each molecule consists not merely of a backbone, but also of R groups, which give it volume.

helix (Fig. 3.25). What facilitates the intertwining of the three chains is that every third amino acid in the chains is glycine, whose R group, being only a single hydrogen atom (Fig. 3.18), takes up very little room. The chains are held together by hydrogen bonds. Collagen fibers are exceedingly strong and very resistant to stretching.

Far more complex in spatial conformation than the fibrous proteins are the globular proteins, whose polypeptide chains are folded into complicated spherical or globular shapes (Fig. 3.26). Because of charged and polar R groups on their exposed surfaces, globular proteins, which include organic catalysts known as enzymes, proteinaceous hormones, the antibodies of the immune system, and most blood proteins, are usually water-soluble. Typically, they are made up of sections of α helix and β sheet interspersed with nonhelical regions (Fig. 3.27). The protein myoglobin, which is the oxygen-storage protein in muscles, provides a typical example of a helix-dominated protein. It consists of one polypeptide chain containing eight α helices connected by short regions of irregular (nonhelical) coiling. At each nonhelical region, the three-dimensional orientation of the polypeptide chain changes, giving rise to the protein's characteristic folding pattern. This three-dimensional folding pattern, which is superimposed on the secondary structure, is called ***tertiary structure*** (Fig. 3.26). In practice, tertiary structure is difficult to determine. The protein must first be crystallized. Then X rays are beamed through the crystals; deflected by the electrons of the thousands of atoms, they form a pattern that is then deciphered by a computer. This process is called X-ray crystallography (Fig. 3.28).

When a globular protein is composed of two or more independently folded polypeptide chains loosely held together, usually by weak bonds, the manner in which the already folded subunits fit together is called ***quaternary structure*** (Fig. 3.29).

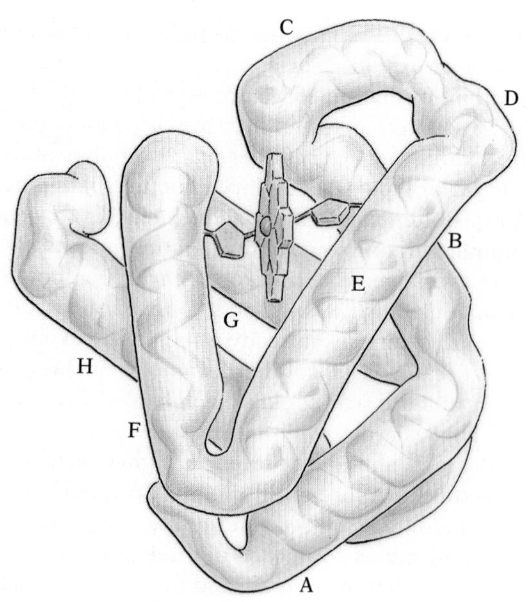

3.26 Tertiary structure of myoglobin Myoglobin, a globular protein related to hemoglobin and, like hemoglobin, characterized by a strong affinity for molecular oxygen, is a single complexly folded polypeptide chain of 151 amino acid units; attached to the chain is a nonproteinaceous group called heme (represented by the disk). The polypeptide chain consists of eight sections of α helix (labeled A through H), with nonhelical regions between them. These nonhelical regions are a major factor in determining the tertiary structure of the molecule—that is, the way the helical sections are folded together. Section D is oriented perpendicular to the plane of the page.

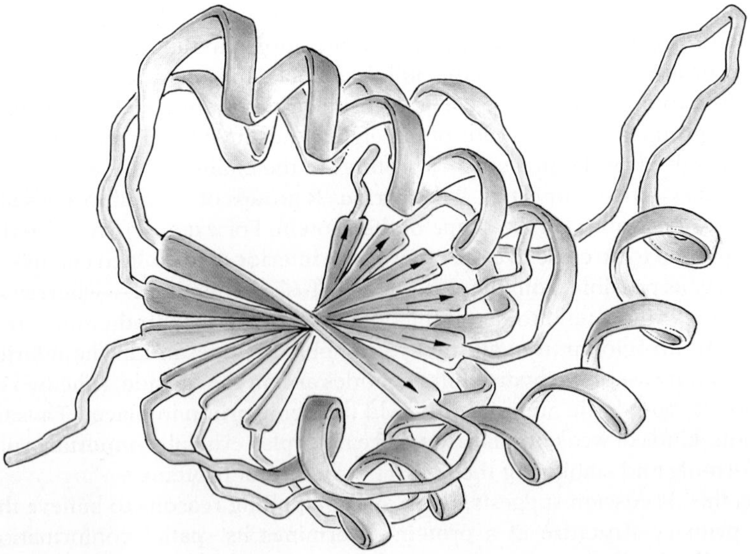

3.27 Mixture of helices and sheets in an enzyme The enzyme lactate dehydrogenase incorporates five α helices and six β sheets, indicated schematically here as coils (red) and ribbons (blue).

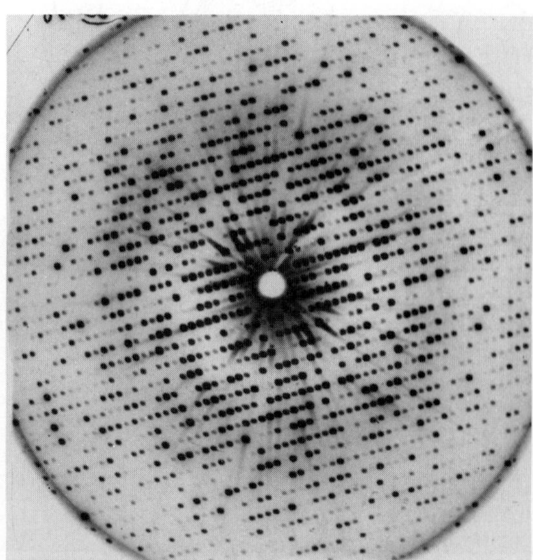

3.28 X-ray diffraction pattern of myoglobin The intensity of each dot in this photograph of an X-ray diffraction pattern of sperm whale myoglobin provides information about the location of atoms in the molecule.

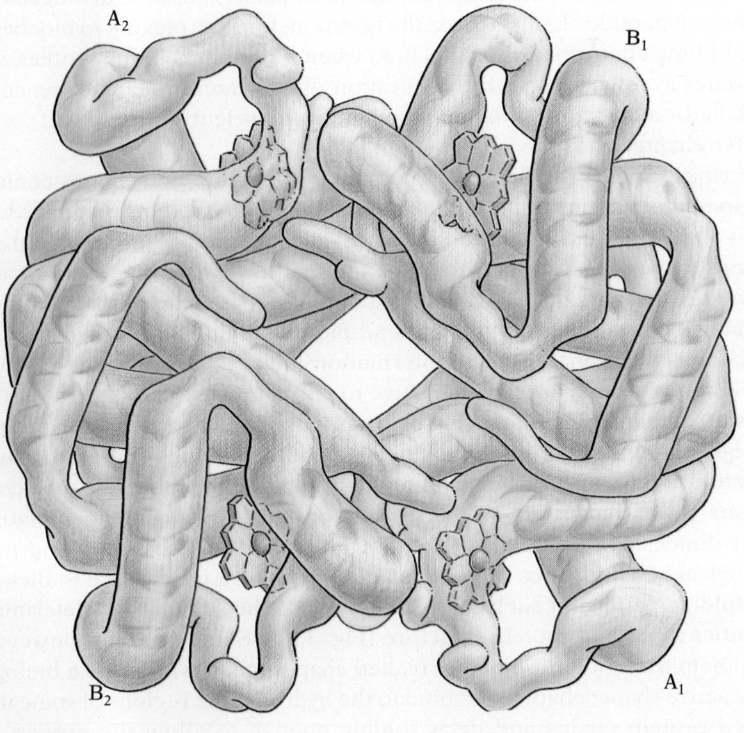

3.29 Quaternary structure of hemoglobin A single molecule of hemoglobin is composed of four independent polypeptide chains, each of which has a globular conformation and its own prosthetic group. The spatial relationship between these four—the way they fit together—is called the quaternary structure of the protein. Chains A_1 and A_2 are identical, as are B_1 and B_2.

Several aspects of a protein's primary structure (that is, its amino acid sequence) contribute to producing its tertiary and quaternary structure. If, for example, a polypeptide chain contains two cysteine units, the intrachain disulfide bond joining them may introduce a fold in the chain or stabilize one created in other ways (Fig. 3.21). The most common cause of folding is

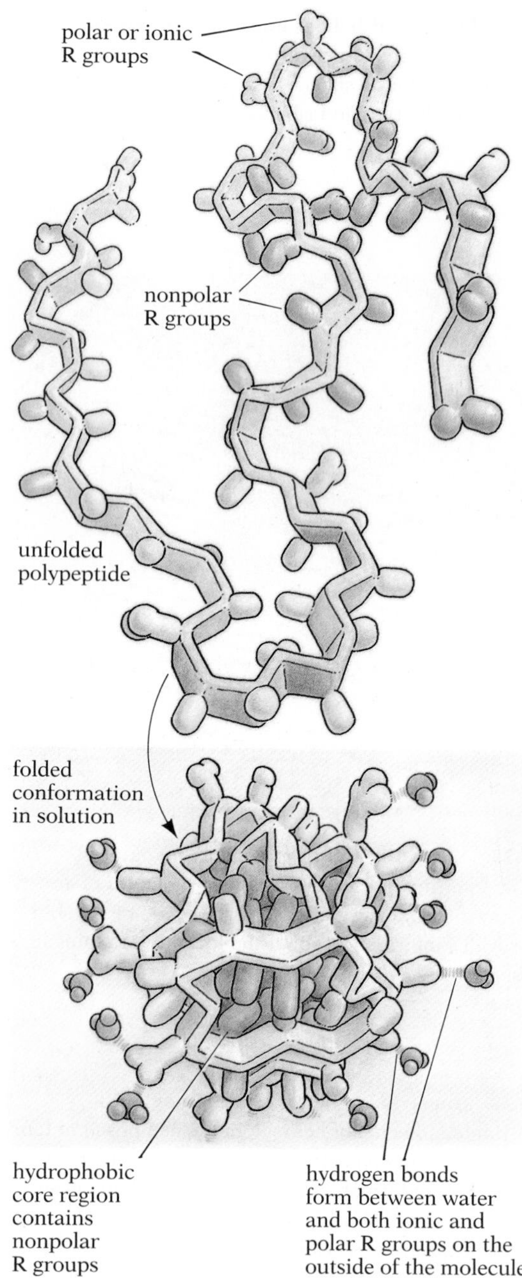

3.30 Folding of a globular protein in solution
As water molecules form polar bonds with polar and ionic side groups, polypeptide chains can fold spontaneously so that the nonpolar groups are herded into the middle.

proline. Wherever there is a proline, a kink or bend occurs, because the structure of proline is such that it cannot conform to the geometry of an α helix; proline, though one of the building-block units of protein, is not technically a true amino acid, since its R group circles around and links with its amino group (Fig. 3.18). Four of the eight bends in globular myoglobin, in fact, result from the presence of prolines in the chain.

The distinctive properties of the various R groups of the amino acids also impose constraints on the shape of the protein. For example, hydrophobic groups tend to be close to each other in the interior of the folded chains—as far away as possible from the water that suffuses living tissue—whereas hydrophilic groups tend to be on the outside, in contact with the water (Fig. 3.30). In myoglobin, too, all the hydrophobic peptides are in the interior, and all but two of the hydrophilic peptides are on the outside. (The two exceptions, both ionic amino acids, hold the heme group in place.) Thus the various kinds of weak bonds discussed earlier play crucially important roles in forming and stabilizing the tertiary structure of proteins.

As this discussion suggests, there are compelling reasons to believe that the primary structure of a protein determines its spatial conformation. More specifically, it appears that the primary structure determines the energetically most favorable, and therefore most stable, possible arrangement of the polypeptide chains. Hence the question, long perplexing to biochemists, of how conformation is specified when a protein is being synthesized becomes synonymous with the question of how amino acid sequence is specified—a question no longer so perplexing to scientists, as we will see in a later chapter.

Further support for the idea that primary structure determines conformation comes from studies of ***denatured*** proteins—proteins that have lost most of their secondary, tertiary, and quaternary structure, and with it their normal biological activity, through exposure to high temperature or extreme pH. That a denatured protein should lack the characteristic biological activity of the natural protein is an indication that its conformation is functionally essential. Since conformation is dependent in large part on weak bonds (which are very sensitive to temperature and pH), it is easily disrupted by anything that breaks or alters those bonds. Even brief exposure to high temperatures (usually above 60°C) or to extremes of pH will cause denaturation of most globular proteins. But under favorable test-tube conditions some denatured proteins can spontaneously regain their native three-dimensional conformation; they can refold, and recover their normal biological activity. Since only the primary structure is available to dictate the folding pattern in such cases, it alone must be sufficient to determine all other aspects of protein structure (Fig. 3.31). Some proteins, however, need help from other molecules (called chaperones) to fold into a biologically active shape; chaperones bind to the hydrophobic regions of some unfolded protein chains and delay folding enough to allow the protein to "discover" its most stable tertiary structure.

Conjugated proteins Attached to some proteins are nonproteinaceous groups called ***prosthetic groups;*** an example is the heme group of myoglobin, a disklike structure with an iron atom in its center (Fig. 3.26). Prosthetic groups may be as simple as a single metal ion bonded to the

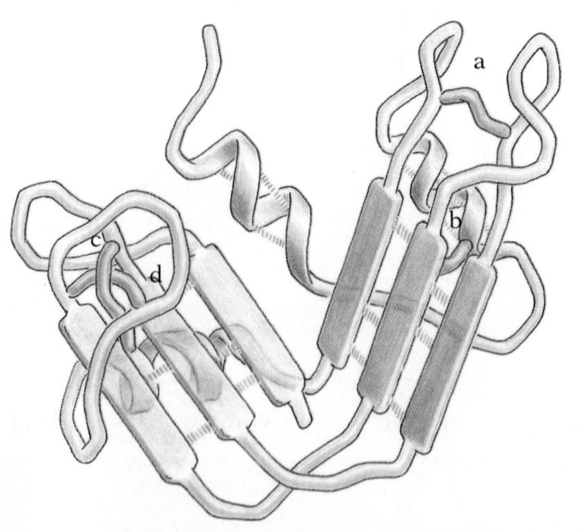

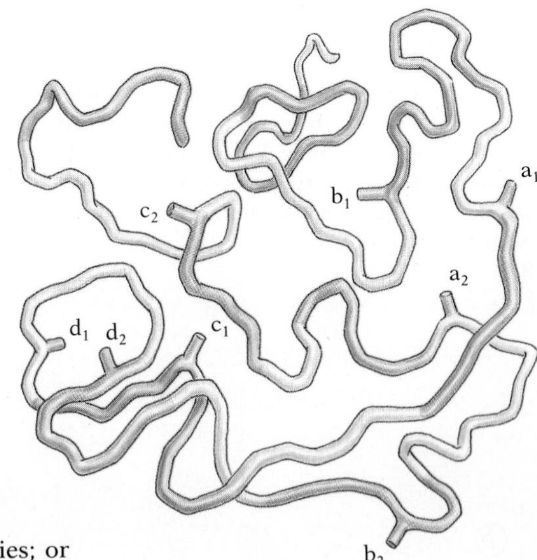

polypeptide chain; they may be sugars or other carbohydrate entities; or they may be of lipid form. Whatever the nature of the prosthetic group, its presence alters the properties of the protein in important ways. Without their heme groups, for example, myoglobin and hemoglobin lose their high affinity for molecular oxygen. All proteins that contain nonproteinaceous substances are called ***conjugated proteins***.

NUCLEIC ACIDS

Nucleic acids constitute a fourth major class of organic compounds crucial to all life. They are the materials of which genes, the units of heredity, are composed. They are also the messenger substances that convey information from the genes in the nucleus to the rest of the cell, information that not only determines the structural attributes of the cell but also regulates its ongoing functional activities.

Like polysaccharides and proteins, nucleic acid molecules are long polymers of smaller building blocks. In this case the building blocks are called ***nucleotides;*** they are themselves composed of still smaller constituent parts: a five-carbon sugar, a phosphate group, and an organic nitrogen-containing base. Both the phosphate group and the nitrogenous base are covalently bonded to the sugar (Fig. 3.32).

3.31 Denaturation and renaturation of ribonuclease
When ribonuclease, a normally globular protein (left), is denatured, with both its weak bonds and its four intrachain disulfide bonds (green) broken, it unfolds into an irregularly coiled state (right). In this denatured condition the ribonuclease lacks its usual ability to digest RNA. When the denaturing agents are removed and favorable conditions restored, the protein spontaneously refolds into its native conformation (right) and regains its capacity for biological activity as an enzyme. Even the four disulfide bonds re-form correctly. Since there are 105 possible ways to join the cysteines but the enzyme folds only to bring the correct pairs together, we must conclude that the conformation does not result just from the positions of the cysteines; most disulfide bonds serve to stabilize, rather than to determine structure.

phosphate group
P
O
nitrogenous base
sugar
HO
OH

3.32 Diagram of a nucleotide
A phosphate group and a nitrogenous base are attached to a five-carbon sugar. The sugar shown is a ribose, part of RNA; the sugar in DNA—deoxyribose—lacks the oxygen (red) at the bottom.

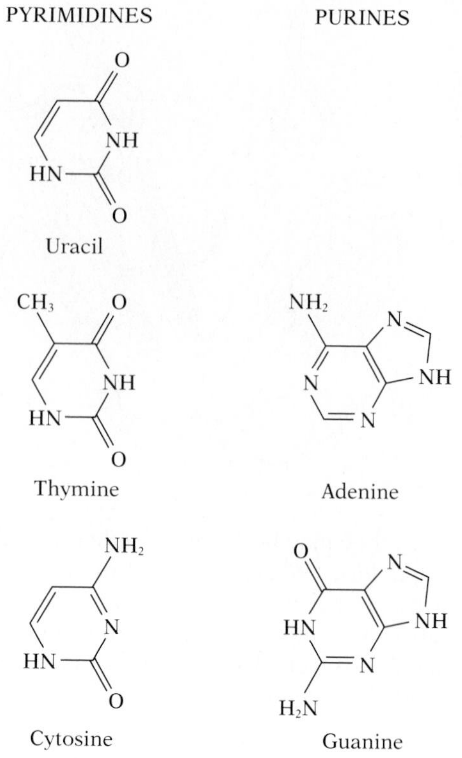

3.33 The nitrogenous bases in DNA and RNA The three single-ring bases, uracil, thymine, and cytosine, are pyrimidines; the two double-ring bases, adenine and guanine, are purines. Adenine, guanine, and cytosine are used in both DNA and RNA; thymine is restricted to DNA, while uracil takes its place in RNA.

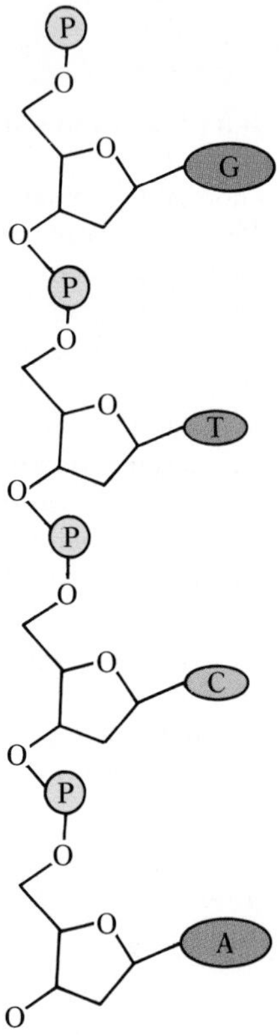

3.34 Portion of a single chain of DNA Nucleotides are linked by bonds between their sugar and phosphate groups. The nitrogenous bases (G, guanine; T, thymine; C, cytosine; A, adenine) are side groups.

Deoxyribonucleic acid This nucleic acid, commonly called ***DNA***, is the one genes are made of. Four different kinds of nucleotide building blocks occur in DNA. All have, as the name suggests, deoxyribose as their sugar component, but they differ in their nitrogenous bases, which may be one of four different substances. Two bases, ***adenine*** and ***guanine***, are double-ring structures of a class known as purines; the other two, ***cytosine*** and ***thymine***, are single-ring structures known as pyrimidines (Fig. 3.33).

The nucleotides within a DNA molecule are bonded together in such a way that the sugar of one nucleotide is always attached to the phosphate group of the next nucleotide in the sequence (Fig. 3.34). Thus a long chain of alternating sugar and phosphate groups is established, with the nitrogenous bases oriented as side groups off this chain. The sequence in which the four different nucleotides occur is essentially constant in DNA molecules of the same species, but differs between species. It is this sequence that determines the specificity of each type of DNA. It is, in fact, the sequence of the nucleotides in DNA that encodes hereditary information, which is expressed through control of protein synthesis. More particularly, the sequence of nucleotides in DNA determines the sequencing of amino acids (the primary structure) in proteins; we will examine this process in considerable detail in Chapter 9.

DNA molecules do not ordinarily exist in the single-chain form shown in Figure 3.34. Instead, two such chains, oriented in opposite directions, are

3.35 Portion of a DNA molecule uncoiled
The molecule has a ladderlike structure, with the two uprights composed of alternating sugar and phosphate groups and the cross rungs composed of paired nitrogenous bases. Each cross rung has one purine base and one pyrimidine base. When the purine is guanine (G), the pyrimidine with which it is paired is always cytosine (C); when the purine is adenine (A), the pyrimidine is thymine (T). Adenine and thymine are linked by two hydrogen bonds (dashed lines), guanine and cytosine by three. Note that the two chains run in opposite directions—that is, the free phosphate is at the upper end of the left chain and at the lower end of the right chain.

arranged side by side like the uprights of a ladder, with their nitrogenous bases constituting the cross rungs of the ladder (Fig. 3.35). The two chains are held together by hydrogen bonds between adjacent bases. Finally, the entire double-chain molecule is coiled into a double helix (Fig. 3.36).

The regular helical coiling and the hydrogen bonding between bases impose two extremely important constraints on how the cross rungs of the ladderlike DNA molecule can be constructed. First, each rung must be composed of a purine (double ring) and a pyrimidine (single ring); only in this way will all cross rungs be of the same length, allowing the formation of a regular helix. Second, if the purine is adenine, the pyrimidine must be thymine, and, similarly, if the purine is guanine, the pyramidine must be cytosine; only these two pairs are capable of forming the required hydrogen bonds (Fig. 3.35). Since it does not matter, however, in which order the members of a pair appear (A–T or T–A; G–C or C–G), the double-chain molecule can have four different kinds of cross rungs, as shown in Figure 3.35. The biological significance of this arrangement is that the base sequence of one chain uniquely specifies the base sequence of the other, so that the two strands can be separated and exact copies made every time a cell divides.

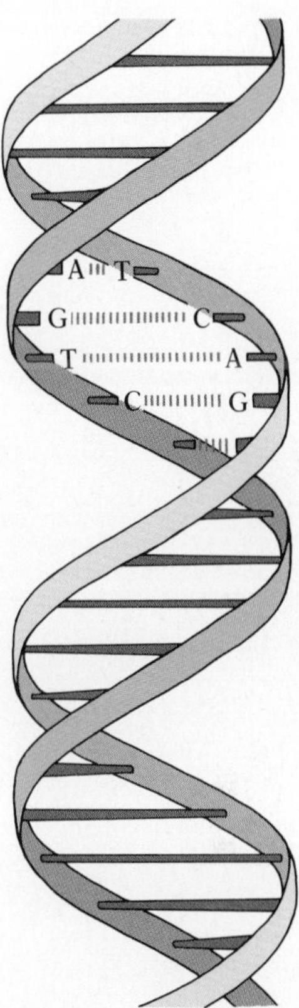

3.36 A model of the DNA molecule
The double-chained structure is coiled in a helix. As shown in detail in the second segment, it consists of two polynucleotide chains held together by hydrogen bonds (red bands) between their adjacent bases.

TABLE 3.2 *Chemical composition of a typical mammalian cell*

Chemical	*Percent of weight*
DNA	0.25
RNA	1
Protein	18
Lipids	5
Polysaccharides	2
Inorganic ions	1
Water	70
Other	3

Source: Bruce Alberts, Dennis Bray, Julian Lewis, Martin Raff, Keith Roberts, and James D. Watson, *Molecular Biology of the Cell,* 2nd ed. Garland Publishing, New York, 1989.

Ribonucleic acid A second important category of nucleic acids comprises the ribonucleic acids, or ***RNA***. There are several types of RNA, each with a different role in protein synthesis. Some act as messengers carrying instructions from the DNA of the genes to the sites of protein synthesis in the cell. Others are structural components of subcellular structures, called ribosomes, on which the process of protein synthesis takes place. Still others transport amino acids to the ribosomes, so that they may be incorporated into proteins. A few even direct chemical reactions. We will discuss each of these types of RNA in much more detail in Chapter 9. Here let us note only that all types of RNA differ from DNA in three principal ways: (1) The sugar in RNA is ribose, whereas that in DNA is deoxyribose (Fig. 3.32). (2) Instead of thymine, one of the four nitrogenous bases of DNA, RNA contains a very similar base called ***uracil*** (Fig. 3.33). (3) RNA is ordinarily single-stranded, whereas DNA is usually double-stranded.

The relative contributions of DNA, RNA, proteins, and lipids to the volume of a typical cell are compared in Table 3.2.

CHEMICAL REACTIONS

In previous sections we mentioned several types of chemical reactions that take place within organisms: condensation reactions between simple sugars to form polysaccharides, and hydrolysis of polysaccharides back to simple sugars; condensation reactions of fatty acids and glycerol to form fats, and the reverse hydrolysis; condensation reactions of amino acids to form polypeptide chains and proteins, and the reverse hydrolysis. But we have said nothing about the conditions under which these reactions will take place. It is now time for a brief examination of those conditions.

All the processes of life depend on the ordered flow of energy. As we said in the preceding chapter, the behavior of electrons is the single most crucial factor in the chemistry of life. Virtually all the energy for living things comes as light from the sun and is captured by electrons, which are thereby excited into higher orbitals. The energy released by such electrons as they move to more highly electronegative atoms in precisely ordered chemical reactions is harvested to fuel all the processes of life. To understand biology, then, it is essential to understand how the transfer of energy in chemical reactions takes place as one set of covalent bonds is replaced by another, an important subject in the field known as thermodynamics.

FREE ENERGY IN LIVING SYSTEMS

Instead of talking at this point about condensation or hydrolytic reactions or any other specific type of reaction, let's consider a generalized one. Suppose, for example, that two substances, A and B, can react with each other in solution to produce two new compounds, C and D:

$$\underset{\text{reactants}}{A + B} \longrightarrow \underset{\text{products}}{C + D}$$

What determines whether a reaction, like one of the condensation or hydrolysis reactions discussed earlier, will tend to take place spontaneously? The answer to this question turns on a concept of physics—that ***energy,*** which is defined as *the capacity to do work*, must be available. ***Free energy*** (as the term is used in chemistry and biology) is the energy in a system available for doing work under conditions of constant temperature and pressure. Where there is energy to be tapped, whether it be in the weight of the water stored behind a dam, or in the covalent bonds of sugars like glucose, or in an electron that has been excited into a higher orbital by sunlight, or in the tightly bound nuclei of the atoms in a nuclear reactor, the potential for work is present (Fig. 3.37).

The ***First Law of Thermodynamics***—the Law of Conservation of Energy—tells us that *the total energy in the universe is constant:* energy needed to do work in a particular system—in a cell, for example—cannot be generated from nothing; any energy not already available internally must be obtained from a source outside that system, which thereby loses a corresponding amount of energy to balance the books. The ***Second Law of Thermodynamics*** states that *in the universe as a whole the total amount of free energy*—that is, the energy actually available for doing work—*is declining*. This is because practically every energy transfer generates heat that is then no longer available for doing work. The magnitude of this waste is enormous, as the need for cooling towers in power plants and radiators in car engines to dissipate unused (and unusable) thermal energy makes evident. Even in our own bodies waste heat can be a serious problem, and highly specialized mechanisms have evolved for releasing this waste energy to the environment.

3.37 Bald eagle swooping down on prey
Many birds of prey use gravitational potential energy to power a rapid dive toward the animals they hunt. This eagle is braking from its dive, talons extended to grasp its prey.

One simple and universal law of cellular chemistry holds that whether or not a reaction can proceed spontaneously depends on the net change in free energy that accompanies the reaction. In other words, the course of the reaction depends on whether the free energy of the set of covalent bonds in the reactants is greater or less than the free energy of the new set of covalent bonds in the products. In order to quantify free-energy changes, we use the symbol ΔG to denote changes in free energy under a defined set of conditions: temperature 25°C, pressure of 1 atmosphere, pH 7.0, with both the reactants and products at a concentration of 1 mole/liter.[1] With this in mind, let's write a hypothetical reaction from a thermodynamic perspective, looking at the free energy of the initial reactants (G_i), the free energy of the final products (G_f), and the change in free energy engendered by the reaction (ΔG):

	initial state		final state	
(reactants)	A + B	⟶	C + D	*(products)*
	G_i	⟶	G_f	$(\Delta G = G_f - G_i)$

The term ΔG, the change in free energy as a result of the reaction, is the crucial variable. *If the reaction results in products with less free energy in the covalent bonds than the reactants possessed, the reaction is "downhill" and can proceed spontaneously.* Since the free energy liberated from the covalent bonds of the reactants is usually released as heat, such reactions are said to be exothermic (heat-producing) or, more generally, ***exergonic*** (energy-releasing). If, on the other hand, the reaction requires an input of ex-

[1] In thermodynamics texts the formal symbol for changes in free energy under these conditions is $\Delta G'^{\circ}$. Since this is the only version of ΔG we will discuss, we can dispense with the superscripts.

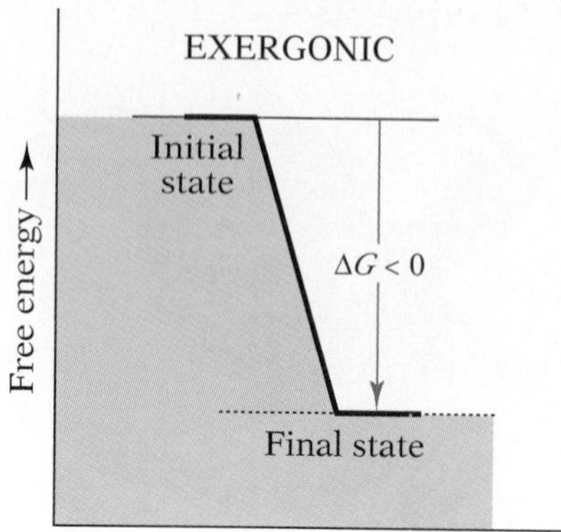

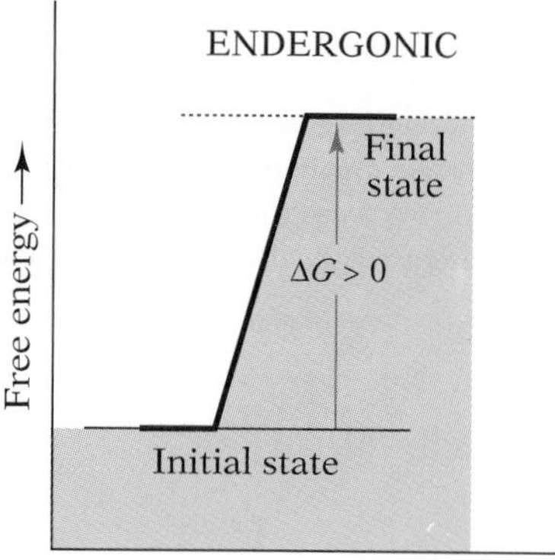

3.38 Exergonic versus endergonic reactions Reactions either liberate some of the free energy of the bonds within the reactants, or consume free energy from outside the system. Those that release energy (in the amount represented by ΔG) to go downhill from the initial state to the final state (top) are said to be exergonic, while those that require energy to go uphill (bottom) are said to be endergonic. (As we shall see, these diagrams describe ideal conditions and are therefore oversimplified.)

ternal energy, it cannot proceed spontaneously. An "uphill" reaction of this kind, in which the covalent bonds of the products have more free energy than those of the reactants, is said to be ***endergonic***. These two alternatives are illustrated schematically in Figure 3.38.

In summary, then, we can say that all spontaneous reactions are downhill—exergonic. This fact is responsible for a general property of matter, both living and nonliving: all systems tend to lose free energy, and ultimately reach a state in which their free energy is as low as possible, just as rocks on a hill tend to move down to the bottom rather than up to the top.[2] Throughout this book we will indicate the free-energy change, ΔG, resulting from biological reactions to the right of the reaction formula. You will need to remember that when ΔG is negative (meaning that the covalent bonds of the reactants had more energy than those of the products), the reaction is exergonic; when ΔG is positive, the covalent bonds of the products have gained energy, so the reaction is endergonic. This relationship holds even when factors like temperature do not correspond to the standard conditions listed above: though the exact magnitude of ΔG will shift as conditions change, the *relative* magnitude of the ΔGs of different reactions in the same cell normally will not. In other words, if reaction I liberates twice the free energy of reaction II under standard conditions, the 2 to 1 ratio will persist under nonstandard conditions of temperature or pressure, provided they are the same for both reactions. Hence the ΔGs of different reactions under the same conditions can be directly compared, and many aspects of cell chemistry can then be predicted.

THE EQUILIBRIUM CONSTANT

Our discussion so far has implied that two reactant molecules can combine spontaneously to create products only if free energy is liberated. By extension, this suggests that if a reaction is exergonic, all the reactants in a mixture will ultimately be turned into products. However, we have ignored the possibility of a ***back reaction***, with the products C and D combining to regenerate the reactants A and B:

$$C + D \longrightarrow A + B$$

Though this reaction is endergonic (uphill), it can take place—very slowly—because there is a source of energy within every system. This source, which we have ignored so far, is the energy of motion: the ***kinetic***, or ***thermal, energy*** of the molecules. Every molecule has both a certain characteristic amount of stored energy in its bonds *and* energy of motion, which

[2] Physicists and chemists frequently relate free energy and "order," which is defined as an arrangement of components in a system that is unlikely to occur by chance. Consider the unlikely circumstance in which all the molecules of air in a room are on one side, with a vacuum on the other. This orderly arrangement disintegrates rapidly as the molecules of air spread throughout the room, with a corresponding loss of free energy. The more orderly a system is, the more free energy it possesses; but since increasing disorder is inevitable, so also is the spontaneous loss of free energy. The amount of disorder in a system—the energy unavailable for doing work—is called entropy.

depends on its speed. In any solution, some molecules move very fast, others more slowly. In our example, when two fast-moving product molecules, C and D, collide, their energy of motion will sometimes be converted into covalent bond energy to produce two (slow-moving) reactants, A and B. Though this back reaction is rare, it is very important when the forward reaction is near completion. At this point the reactants A and B are so scarce that the rare, kinetic-energy-dependent back reaction may be just as likely as the forward reaction.

When the forward reaction, slowed in consequence of the increasing scarcity of reactants, is just counterbalanced by the rare back reaction, the two processes are in equilibrium and no further net change in the concentration of substances takes place. Chemists define this ultimate stable ratio of products to reactants as the equilibrium constant (K_{eq}):

$$K_{eq} = \frac{\text{product concentration}}{\text{reactant concentration}}$$

This relationship succinctly summarizes the results of any reaction, and so complements ΔG in describing chemical reactions. For instance, an equilibrium constant of 10 means that when equilibrium has been reached under conditions of stable temperature and pressure, the products of the reaction outnumber the reactants by a factor of 10:

$$\text{A} + \text{B} \xrightarrow{K_{eq} = 10} \text{C} + \text{D} \qquad (\Delta G = -1.4 \text{ kcal/mole})$$

Such a reaction is downhill: the products are more abundant than the reactants, and the reaction liberates energy. The predominance of the forward reaction is indicated by having the arrow pointing from the reactants to the products longer than the one pointing back. In a reaction with an equilibrium constant of 0.1, on the other hand, the reactants outnumber the products by 10 to 1:

$$\text{E} + \text{F} \overset{K_{eq} = 0.1}{\leftrightharpoons} \text{G} + \text{H} \qquad (\Delta G = +1.4 \text{ kcal/mole})$$

Such a reaction is uphill; it consumes energy, and so relatively few of the initial reactant molecules will be converted into products. And finally, an equilibrium constant of 1.0 indicates that the products and the reactants are equally abundant:

$$\text{I} + \text{J} \overset{K_{eq} = 1.0}{\rightleftharpoons} \text{K} + \text{L} \qquad (\Delta G = 0)$$

As we would expect, there is no free-energy change in this reaction. The numerical relationship between K_{eq} and ΔG at 25°C is summarized in Table 3.3.

Special note must be taken of one consequence of the relationship in cellular chemistry between the equilibrium constant and the concentration of products and reactants. Remember that the equilibrium constant of a reaction is a ratio—a simple empirical description of the outcome of the reaction—and depends, ultimately, on the ΔG of the reaction. Hence, the ultimate ratio of products to reactants in no way depends on the starting

TABLE 3.3 *Relationship between K_{eq} and ΔG at 25°C*

ΔG (kcal/mole)	K_{eq}	
4.1	0.001	Endergonic reactions
2.7	0.01	
1.4	0.1	
0	1.0	
−1.4	10.0	Exergonic reactions
−2.7	100.0	
−4.1	1000.0	

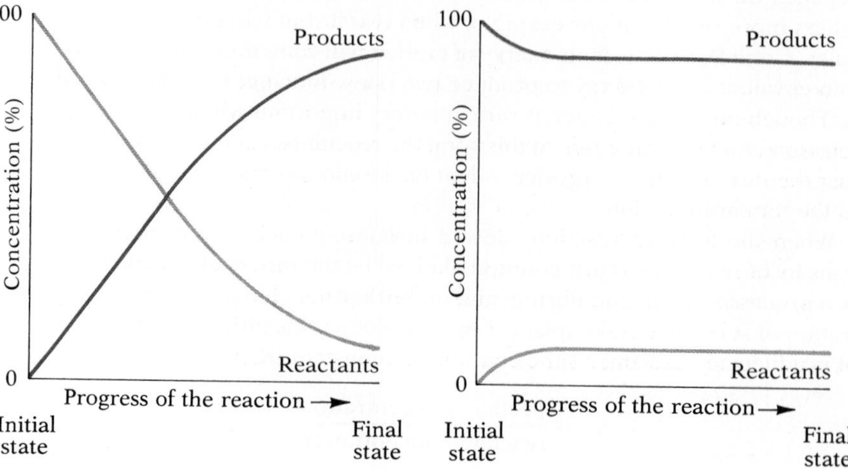

3.39 The relationship between the equilibrium constant and concentration
The equilibrium constant, specifying the final ratio of products to reactants, is independent of the starting concentration. In the example shown here, the equilibrium constant is 10, which means that products (red curves) will be 10 times more common than reactants (blue) when the reaction ends. This is the outcome whether we begin with pure reactants (left) or pure products (right).

conditions. This point is illustrated in Figure 3.39, in which, whether we start with a great deal of reactant or none at all, we wind up with the ratio specified by the equilibrium constant.

ACTIVATION ENERGY AND REACTION PATHWAYS

The question next arises, How do particular compounds follow particular reaction pathways? For example, suppose the cell manufactures compounds D and Z, as follows:

$$A + B \leftrightharpoons C + D \qquad W + X \leftrightharpoons Y + Z$$

What is to prevent B from reacting with W, X, Y, or Z instead of with A? Some of these combinations may well have negative ΔGs, and so be energetically favorable. What serves to prevent all but the "correct" reactants from combining?

For two molecules to combine, they must be brought unusually close to each other in a particular orientation and, frequently, one or more pre-existing bonds must be broken. This requires energy—specifically known as ***activation energy*** (E_a)—so even an exergonic reaction has an endergonic first step. The barrier that must be overcome by activation energy is illustrated schematically in Figure 3.40. The only source of energy for this "priming" is the kinetic energy of colliding molecules.

Because of the strength of covalent bonds, a substantial amount of energy may be necessary to break the pre-existing bonds of reactants—often far more than the amounts listed for the ΔGs in Table 3.3. Consider the following enormously exergonic reaction:

$$2H_2 + O_2 \longrightarrow 2H_2O$$

Even though this combination of oxygen and hydrogen can be explosive—for example, it provides much of the energy that pushes the space shuttle into orbit—the two reactants can coexist as a stable mixture almost indefinitely. A single spark, however, will initiate an explosive reaction. The same stability is a property of most reactants in living systems: the energy necessary to bring the reactants together and break their covalent bonds is far greater than the energy of all but the very few most rapidly moving molecules in a solution. The activation-energy barrier therefore prevents most reactions from taking place at a significant rate (Fig. 3.40). Without such a barrier the complex high-energy molecules (such as carbohydrates, lipids, proteins, and nucleic acids) on which life depends would be unstable, and would break down.

Once a reaction does get started, the combination of one pair of reactants may release enough energy (usually in the form of heat) to activate the next pair, and so on, in a chain reaction. This is precisely what happens in a rocket engine, and in the combustion of a dry piece of firewood. The wood, as we all know, can lie in a woodpile for years without bursting into flames spontaneously, but once set on fire, it literally consumes itself as the free energy liberated by the combining of carbon and oxygen into CO_2 supplies the activation energy to continue the burning. Put quite simply, heating a mixture will increase the *rate* of reaction (though it cannot affect the ultimate ratio of products to reactants, since that is a constant—K_{eq}). But cells will literally cook at temperatures much above the 37°C of most mammals and birds, so they can use heat as a way of overcoming the activation-energy barrier only to a very limited extent. If the reactions necessary for life are to take place, cells must use some other method, and it must be one that lowers the barrier selectively, so that some exergonic reactions run, while others do not. Cellular chemistry, then, is essentially controlled by the selective lowering of particular activation-energy barriers. How is this crucial task managed?

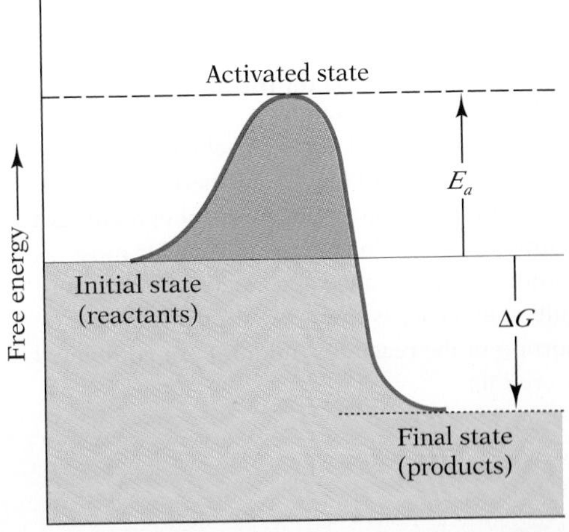

3.40 The energy changes in an exergonic reaction Though the reactants are at a higher energy level than the products, the reaction cannot begin until the reactants have been raised from their initial energy state to an activated state by the addition of activation energy (E_a). It is the need for activation energy that ordinarily prevents high-energy substances like glucose from breaking down, and hence makes them stable; the higher the activation-energy barrier, the slower the reaction and hence the more stable the substance. When activation energy is available, the reactants form a temporary and unstable activated complex, which breaks down to yield the end products of the reaction; in the process both activation energy and free energy (in the amount represented by ΔG) are released.

The effect of catalysts Chemists discovered years ago that certain chemicals speed up reactions between other chemicals. As we have seen, a simple mixture of hydrogen and oxygen does not react, but if we provide the initial activation energy (a spark) the mixture will explode. The same explosion will take place if we add instead a small quantity of platinum. After the reaction is over, the platinum will still be present, unchanged.

A substance that, like the platinum, speeds up a reaction but is itself unchanged when the reaction is over (even though it may have been temporarily changed during the reaction) is known as a ***catalyst***. A catalyst affects only the *rate* of reaction; it simply speeds up reactions that are thermodynamically possible to begin with. Like heat, a catalyst cannot alter the direction of a reaction, its final equilibrium, or the reaction energy involved.

In terms of our discussion, a catalyst decreases the activation energy needed for the reaction to take place (Fig. 3.41), thereby increasing the pro-

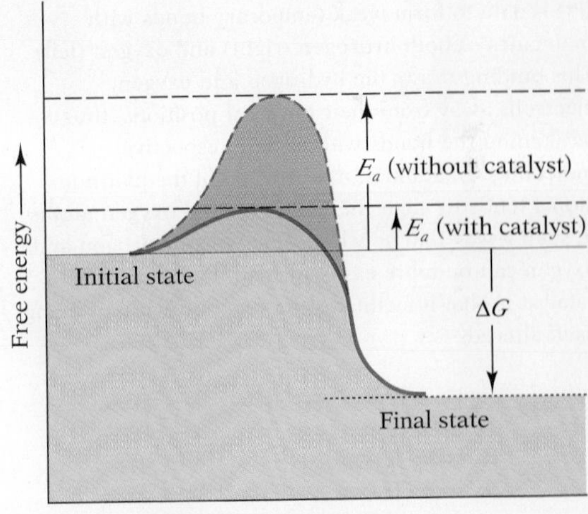

3.41 Reduction of necessary activation energy by catalysts The activation energy (E_a) necessary to initiate the reaction is much less in the presence of a catalyst than in its absence. It is this lowering of the activation-energy barrier by enzyme catalysts that makes possible most of the chemical reactions of life. Note that the amount of free energy liberated by the reaction (ΔG) is unchanged by the catalyst—it is the same for both the catalyzed and the uncatalyzed reaction—and only the activation energy is changed.

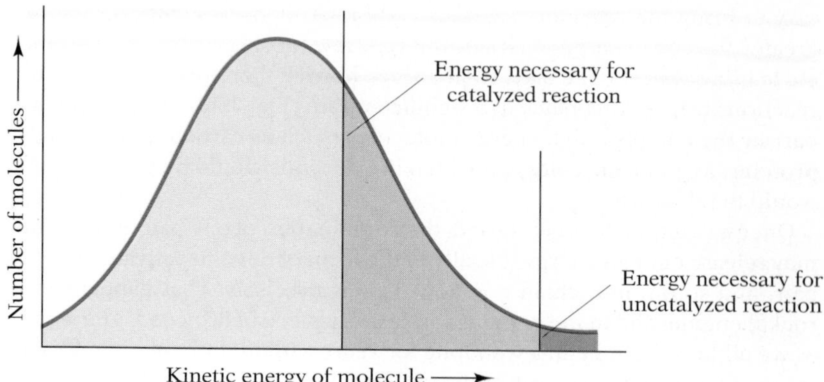

3.42 Effect of a catalyst on the ability of kinetic energy to activate a reaction
As indicated by this bell curve, reactant molecules exhibit a wide range of kinetic, or thermal, energy. Only a minute fraction (right) have enough energy to overcome the activation-energy barrier for most reactions. In the presence of a catalyst, such as an enzyme, the barrier is lower, so a much larger proportion of the reactants (middle) can combine to form products.

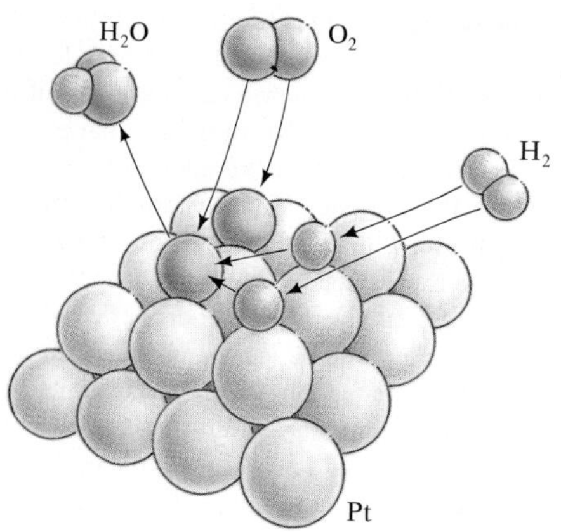

3.43 Action of platinum as a catalyst
Because of its loosely bound outer electron, platinum (Pt) is able to form weak temporary bonds with molecules of both hydrogen (right) and oxygen (left). This binding draws the hydrogen and oxygen electrons away from their covalent positions, thus weakening the bonds within their respective molecules. In addition, the spacing of the platinum atoms tends to align the hydrogen and oxygen atoms in such a way that new bonds between hydrogen and oxygen can be more easily formed. Platinum is a catalyst in that it facilitates the reaction without being itself altered.

portion of reactants energetic enough to react (Fig. 3.42). A catalyst does this by binding the reactants in an intermediate state in which the reactants are correctly oriented to each other and important internal bonds are weakened (Fig. 3.43). As a consequence of binding, then, conditions are highly favorable for the reaction.

An inorganic catalyst such as platinum is relatively unselective about the reactants it "helps"; the spacing of its atoms on a crystal face happens to match the distance between pairs of small covalently bonded atoms, and its loosely bound outer electron readily interacts with the outer shells of other atoms. But cells need catalysts that only promote specific reactions; all other combinations of potential reactants are undesirable. Lacking the necessary activation energy, and without specific catalysts to lower this thermodynamic barrier, these reactants are unable to combine with one another. Over the ages, cells have evolved an enormous variety of highly specialized organic catalysts, called ***enzymes***, which direct cellular chemistry along useful pathways. Virtually all enzymes are globular proteins, though as we will see in Chapter 9, a few are constructed out of RNA.

ENZYMES

Before we look at the way enzymes catalyze the selective chemistry of life, let's summarize what we know about the thermodynamics of biological reactions: (1) A chemical reaction can proceed spontaneously if it releases free energy. (2) Because the activation energy necessary for biochemical reactions is relatively high, they occur only very slowly without the intervention of catalysts. (3) Catalysts, including enzymes, alter neither the equilibrium constant of a reaction nor the net change in free energy—they cannot, by themselves, cause a reaction to run uphill, but they can make specific downhill (exergonic) reactions occur quickly.

Enzyme specificity and the active site Unlike inorganic catalysts, enzymes are highly selective. A particular enzyme generally interacts with only one type of reactant or pair of reactants, customarily called the ***substrates***. The enzyme thrombin, for instance, acts only on certain proteins, and only at very specific sites. It "recognizes" the bond between the amino acids arginine and glycine, which it then hydrolyzes (an important step in the formation of blood clots). Like all catalysts, enzymes lower the activation energy required (Fig. 3.41), so that the kinetic energy of many of the substrate molecules becomes sufficient to cause the reaction to take place (Fig. 3.42). As a result, enzymes vastly speed up the reactions they catalyze; a single molecule of enzyme may cause thousands or even hundreds of thousands of molecules of reactant to combine into product each second.

Because of their efficiency and specificity, enzymes both steer specific substrates into particular reaction pathways and block them from others, thus guiding the chemistry of life with great precision. Biochemists have long held that the key to this specificity is surface activity. Enzymes, as we have said, are globular proteins, extremely complex molecules with intricate three-dimensional contours and distinctive surface geometries. A given enzyme reacts only with substrates whose molecular configurations —conformation and location of charged groups—"fit" its surface. Thus, an enzyme's specificity depends on its three-dimensional molecular conformation.

The conclusion that the action of enzymes depends on their three-dimensional shape is consistent with the observation that when proteins are denatured—when their three-dimensional conformation is disrupted—they lose their characteristic biological activity; their enzymatic properties vanish (Fig. 3.44). Also consistent with this point of view is the observation that most enzymes are highly sensitive to changes in pH, and are active only within a limited pH range (Fig. 3.45). Apparently, change in pH results in the breakage of many of the weak bonds that help stabilize the conformation of proteins and, at the same time, leads to formation of new bonds, with consequent changes in the shape of the protein.

Enzyme and substrate can be visualized as fitting together like a lock and key, or like pieces of a puzzle. And it is true that the two must be roughly complementary if they are to combine. The reactive portion of the substrate molecule and the portion of the enzyme known as the ***active site*** must fit together in space intimately enough to become temporarily bonded, like the platinum and the hydrogen and oxygen in Figure 3.43. In this way they form a transient enzyme-substrate complex:

$$E + S \longrightarrow ES \longrightarrow E + P$$

(E stands for enzyme, S for substrate, and P for product.) But enzymes and their substrates probably do not always have to fit together exactly before

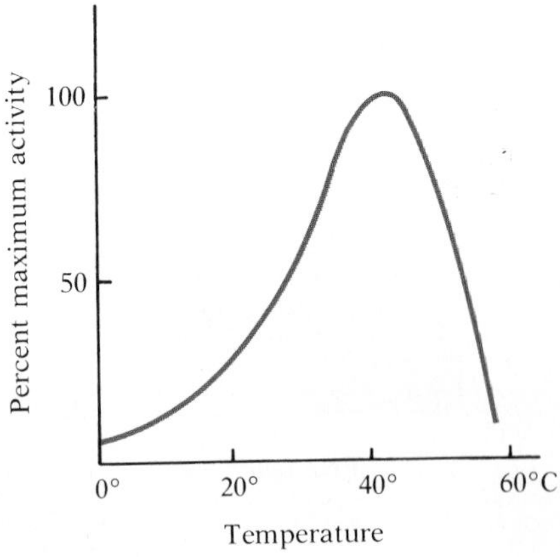

3.44 Enzyme activity as a function of temperature
Though temperature sensitivity varies somewhat from one enzyme to another, the curve shown here may be taken as applying to an "average" enzyme. Its activity rises steadily with temperature (approximately doubling for each 10°C increase) until thermal denaturation causes a sudden sharp decline, beginning between 40° and 45°. The enzyme becomes completely inactivated at temperatures above 60°, presumably because its three-dimensional conformation has been severely disrupted. Bacteria living in volcanic pools, by contrast, have well-stabilized enzymes with activity optima near 100°; below 60°, however, they have almost no activity.

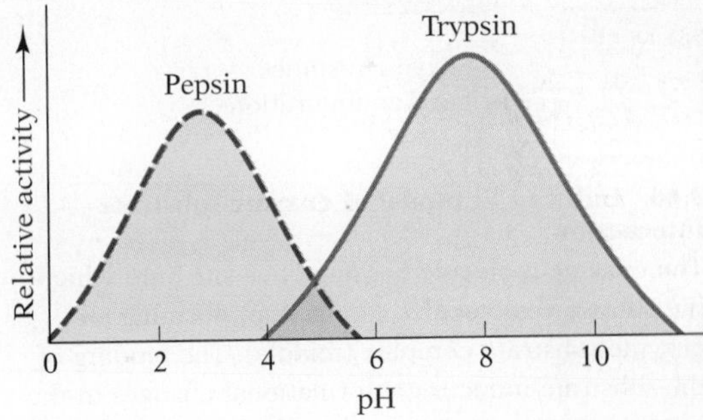

3.45 Enzyme activity as a function of pH
Most enzymes are very sensitive to pH, but they differ markedly in their pH optima. Pepsin and trypsin are both enzymes that digest protein, but the pH ranges within which they are active overlap only slightly. Pepsin, a stomach enzyme, is most active under strongly acidic conditions, while trypsin, an enzyme secreted into the small intestine, is most active under neutral and slightly alkaline conditions.

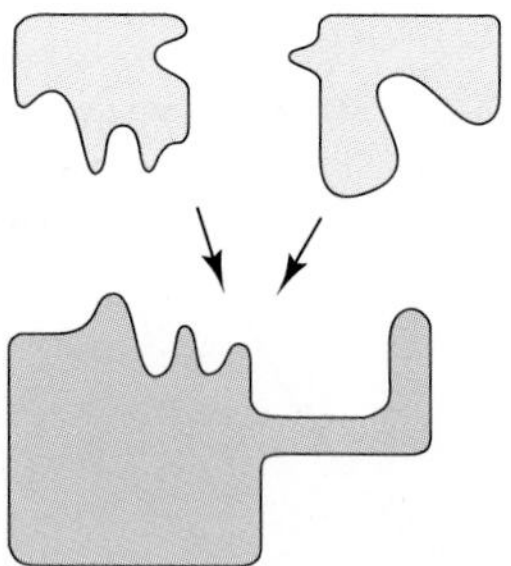

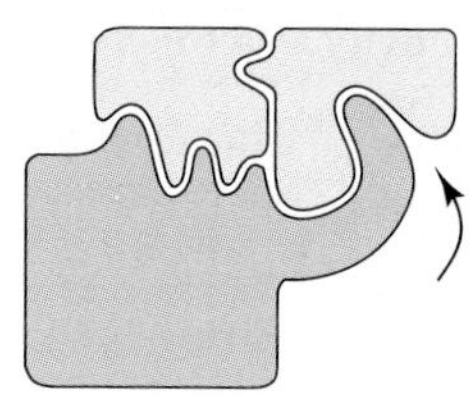

Enzyme-substrate complex

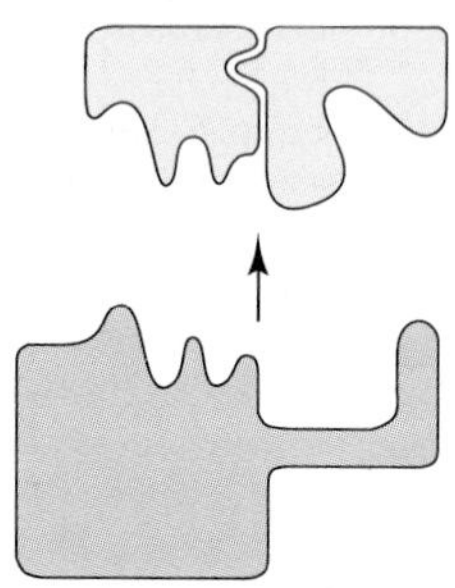

Enzyme resumes original configuration

3.46 Induced-fit model of enzyme-substrate interaction

The enzyme molecule has an active site onto which the substrate molecules can fit (top), forming an enzyme-substrate complex (middle). The binding of the substrate induces conformational changes in the enzyme that maximize the fit and force the complex into a more reactive state. The enzyme molecule reverts to its original conformation when the product is released (bottom).

ES (the enzyme-substrate complex) can form; according to the widely accepted ***induced-fit hypothesis***, many (perhaps most) enzymes undergo conformational changes in the course of bonding, which improve the fit and make ES more reactive (Fig. 3.46).

Spatial complementarity is only one of the prerequisites for enzyme-substrate interaction. Another is that E and S be chemically compatible and capable of forming numerous and precise weak bonds with each other. For though enzyme and substrate are sometimes held together by covalent bonds, the bonds are far more often the same types of weak bonds—ionic, hydrogen, and van der Waals—that stabilize protein conformation. They are bonds that can be made and broken rapidly in the collisions that result from random thermal motion at normal temperatures. The type of substrate to which a given enzyme molecule can become bonded depends on the amino acids constituting its active site—more specifically, on the exposed R groups of these amino acids and the details of their arrangement relative to one another. Suppose the active site of a particular enzyme is a curving groove into which the reactive portion of the substrate must fit. Suppose further that most of the exposed R groups in this groove are electrically charged. It is obvious that the reactive portion of the substrate must be complementarily charged or polar; an electrically neutral nonpolar substrate molecule, or one with the same charge as the active site, could not react with the active site of this enzyme even if it could, by chance, fit into the groove. Conversely, only a hydrophobic substrate could interact with an active site made up largely of hydrophobic R groups; both electrically charged and polar molecules would be incompatible with such a site.

Figure 3.47 offers a model of what is currently known about the active site of one actual digestive enzyme (carboxypeptidase[3], which catalyzes the removal of the terminal amino acid from one end of a polypeptide chain during digestion in the small intestine). The enzyme's active site is visualized as a cleft into which the end of the substrate molecule can fit. The substrate is thought to form several weak bonds (five are shown here) with the R groups of amino acids that constitute part of the active site. In addition, the substrate binds to a zinc held in place by the R groups of three other amino acids. Note that the critical amino acids in the active site (nos. 71, 196, 72, 69, 145, 248, and 270) are not adjacent to each other in the polypeptide chain, which means that the complex folding of the protein—its tertiary structure—has brought amino acids from several regions of the protein into close spatial proximity to form the active site (Fig. 3.48). This is the usual pattern; active sites nearly always include some nonadjacent amino acids. We can now understand more fully why elevated temperature or a major pH change may greatly reduce an enzyme's activity: anything that changes the precise folding pattern of the polypeptide chain is likely to alter the critically important arrangement of amino acids in the active site.

Carboxypeptidase is also typical of enzymes in that the entrance to the

[3] The suffix "-ase" designates an enzyme, and is derived from the name "diastase" coined in 1833 by two French chemists from the Greek *diastasis*, to separate. The function of most enzymes can be guessed from their name. In the case of carboxypeptidase, the "peptid" is from the Greek *pepsis*, to digest; "carboxy" indicates that the digestion is from the carboxyl end of the protein.

3.47 Model of an active site of an enzyme Shown here in schematic form is the base of the cleft where the active site of carboxypeptidase is located. (The hydrophobic entrance to the cleft is out of the drawing, to the bottom. The entire enzyme, with the active site highlighted, is depicted in Figure 3.48). Part of a substrate molecule is shown in the cleft, linked to the enzyme by five weak bonds (striped bands). Seven of the amino acids of the active site are indicated by their abbreviated names; the numbers beside the names refer to the positions of the amino acids in the enzyme polypeptide chain. The function of this enzyme is to separate the terminal amino acid (top) from the amino acid chain (extending down out of the figure) at the covalent bond indicated by the arrow. The highly electronegative zinc and the charged ozygen draw away the electrons of this bond and so initiate its rupture.

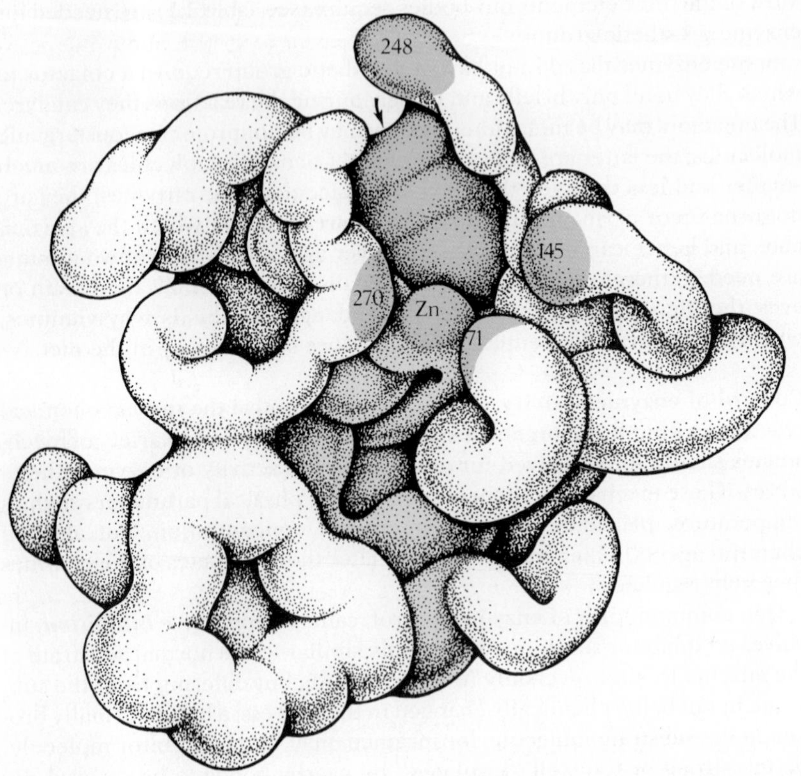

3.48 Location of the active site in an enzyme The folding of the long chain of amino acids of carboxypeptidase brings together the zinc atom and three of the four amino acids that bind to the substrate at the active site, even though—as their numbers indicate—they are located at different places in the chain. The zinc atom itself is held in place by three different amino acids—nos. 69, 72, and 196 (not shown). The fifth part of the site folds into place (arrow) when the substrate binds to the enzyme. The region at the top center in dark color is the active site, while the area in light color is the hydrophobic entrance to the cleft. Most of the length of the polypeptide chain serves to (1) create an appropriately shaped cleft, (2) precisely position the R groups at the active site, (3) provide for the hingelike action of the induced-fit arm, and (4) supply a second location for enzyme regulation (discussed later).

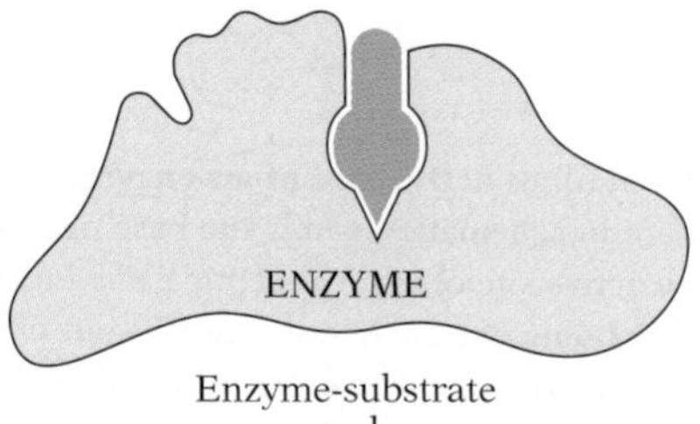

Enzyme-substrate complex

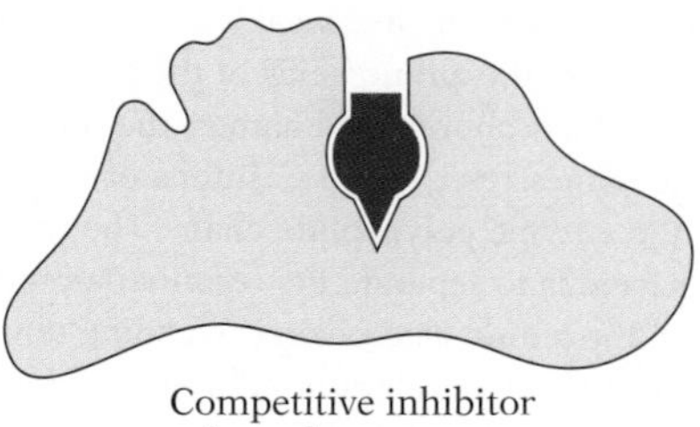

Competitive inhibitor bound to enzyme

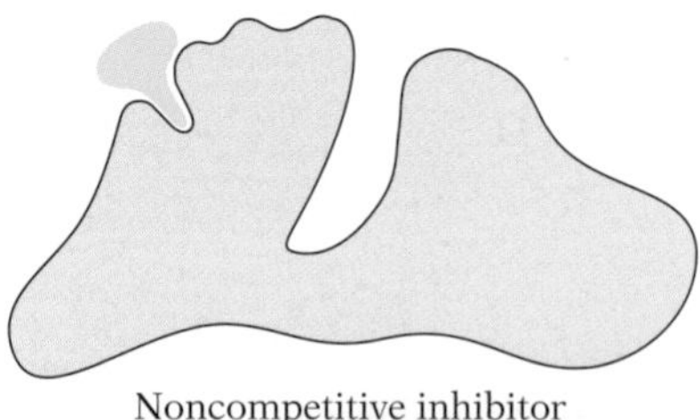

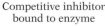

Noncompetitive inhibitor bound to enzyme

3.49 Competitive and noncompetitive inhibition of an enzyme
Top: The substrate is bound to the catalytic site of the enzyme. Middle: The binding of a competitive-inhibitor molecule to the catalytic site prevents the substrate from binding. Bottom: A noncompetitive inhibitor bound to a different site on the enzyme induces an allosteric change that prevents the active site from catalyzing reactions.

cleft in which the active site is located is hydrophobic, so that the water molecules that surround the substrate are stripped away as the substrate enters the groove. The binding energy of the five simultaneous weak bonds is just strong enough to stabilize the enzyme-substrate complex—a less perfect fit with fewer bonds would be unstable. The ability of carboxypeptidase to form these particular five bonds accounts in part for its specificity. The functioning of this enzyme also illustrates the induced-fit strategy: as the binding begins at the other four parts of the active site, the part of the chain containing tyrosine (no. 248) moves in from the periphery to trap the substrate (the polypeptide chain to be broken). The highly electronegative zinc atom and the charged oxygen of glutamate (no. 270) help activate the substrate by drawing away the electrons of the N—C bond that holds the terminal amino acid to the rest of the chain (Fig. 3.47). The resulting cascade of electron shifts between atoms of differing degrees of electronegativity makes the outcome—elimination of the terminal bond—inevitable. Finally, it should be clear why the two products—the terminal amino acid and the remainder of the chain—separate from the enzyme after the reaction: the new set of bonds in the products have redistributed their electrons, so the tenuous array of hydrogen bonds and van der Waals interactions that depended on precise electron interactions with the enzyme at the active site has been disturbed, and the products drift free.

Figure 3.47 illustrates another important point. Many enzymes contain a prosthetic group essential to their activity. A metal atom is often part of the prosthetic group; in carboxypeptidase, as we have seen, the metal is zinc. Most of the trace elements our bodies require (see Table 2.1) are needed for enzyme prosthetic groups.

Some enzymes that do not have a prosthetic group require a cofactor to which they bond only briefly and loosely during the reactions they catalyze. The cofactors may be metal ions, or they may be nonproteinaceous organic molecules; the latter are called ***coenzymes***. Coenzyme molecules are much smaller and less complex than protein molecules. Like enzymes, they are not used up or permanently altered by the reactions in which they participate, and hence can be used over and over again. Only very tiny amounts are needed, therefore, but if the supply falls below normal, the health or even the life of the organism may be endangered. This is why vitamins, which act as parts of essential coenzymes, are so necessary in the diet.

Control of enzyme activity Since enzymes control the myriad chemical reactions within living organisms, it is not surprising that a variety of mechanisms should have evolved for controlling the activity of enzymes themselves. These mechanisms depend not only on physical parameters such as temperature, pH, and substrate or enzyme concentration, but also on chemical agents, which mask, block, or alter the active sites of the enzymes they help regulate.

One common form of enzyme control, called ***competitive inhibition***, involves an inhibitor substance sufficiently similar to the normal substrate of the enzyme to bind reversibly to its active site, but differing from the substrate in not being chemically changed in the process; a bond normally broken in the substrate molecule, for instance, may, in the inhibitor molecule, be too strong or too well "insulated" by nearby bonds to be severed. By binding to the active site, the inhibitor (I) masks the site and prevents the normal substrate molecules from gaining access to it (Fig. 3.49). Thus the reaction

$$E + I \longrightarrow EI$$

competes with the reaction

$$E + S \longrightarrow ES \longrightarrow E + P$$

because both involve the same enzyme, which is present in only very small quantities. Which of the two reactions will predominate depends on their relative energetics, and, even more, on the relative concentrations of I and S. If there is much inhibitor and a low concentration of substrate molecules, a high percentage of the enzyme will be bound as EI and therefore unavailable; on the other hand, if there is much S and only a small concentration of I, then most of the enzyme molecules will be free to catalyze the reaction of substrate molecules to form the product. Carbon monoxide poisoning is an example of competitive inhibition. The carbon monoxide competes with oxygen for the active sites in hemoglobin (Fig. 3.29), an enzymelike substance in the blood of vertebrates that carries oxygen to the body's cells. Carbon monoxide binds so strongly to the active sites that oxygen is effectively excluded. Moreover, the inhibition is essentially irreversible: once bound, carbon monoxide remains in place. As a result of the oxygen deprivation that ensues, living tissue, particularly brain tissue, can be damaged or destroyed even if the concentration of carbon monoxide is relatively low. Most competitive inhibitors, however, bind their target only briefly; this reversible inhibition means that inhibitor and substrate molecules are always competing: a change in the concentration of either is quickly reflected by an alteration in enzyme activity.

A second category of reversible inhibition, called ***noncompetitive inhibition***, depends on the operation of two kinds of binding sites in the same enzyme molecules—the usual active sites to which substrate can bind, and other sites to which inhibitors can bind. The most common kind of noncompetitive inhibition is ***allosteric inhibition***. An allosteric enzyme is one that can exist in two distinct spatial conformations, which usually reflect alterations of tertiary structure. Often when the molecule is in one conformation the enzyme is active, and when it is in the other conformation it is inactive (or less active) because the substrate-binding site has been disrupted. Allosteric inhibition, the binding of inhibitor molecules—usually called negative ***modulators*** (or sometimes negative effectors)—stabilizes the enzyme in its inactive conformation (Fig. 3.49). Quite often the product itself (or the product of a later reaction in the same biochemical pathway) assumes the modulator role: when present in such high concentrations that no more is needed, it turns off the process responsible for its own synthesis. This self-limiting strategy is known as ***feedback inhibition***. Other types of allosteric enzymes have binding sites for positive modulators, which induce conformational changes enhancing enzyme reactivity.

Instead of having different kinds of binding sites, some allosteric enzymes have two or more sites of a single kind, and so can bind substrate at two or more locations simultaneously. The binding of substrate at one active site causes conformational changes that make the remaining sites more reac-

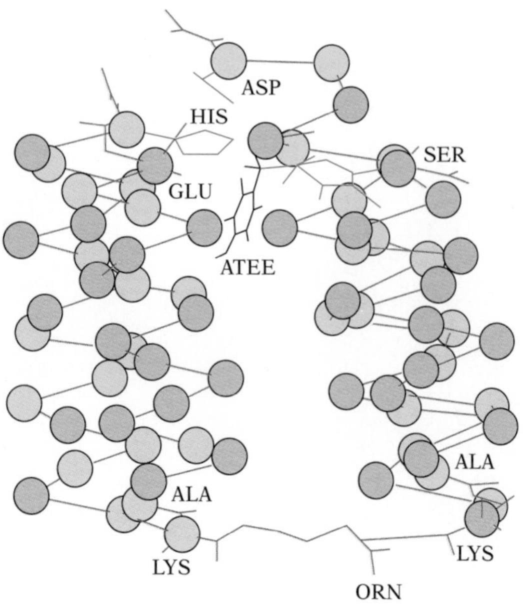

3.50 A synthetic enzyme
This 73-peptide designer enzyme mimics the action of chymotrypsin, a protein-digesting enzyme 245 amino acids long that is secreted by the pancreas. The active site with the R groups shown in blue is just as specific as the natural enzyme, is blocked by the same competitive inhibitors, and increases the reaction rate 100,000 times over that of an enzyme-free solution. Nevertheless, this abbreviated version of chymotrypsin is 1000 times slower than its natural counterpart, indicating that the complex globular structure imparted by the other 172 amino acids probably serves to optimize catalytic activity. (The part of the substrate actually digested is shown in black.)

tive. This phenomenon, called ***cooperativity***, is exemplified in hemoglobin. A single molecule of hemoglobin is capable of carrying four oxygen molecules. The binding of the first oxygen molecule induces changes in the quaternary structure of the hemoglobin molecule that give the other three binding sites a higher affinity for oxygen. Thus in cooperativity, the first substrate molecule functions as the modulator. It stabilizes the allosteric protein in one of its possible conformations—in the case of hemoglobin, its most reactive conformation.

Because of the different effects that competitive inhibitors, noncompetitive inhibitors, and allosteric modulators have on the active sites of enzymes, biochemists find them exceedingly valuable in probing the nature of enzymes. For example, to gain a clearer idea of the active site—its physical shape and its reactive R groups—they may synthesize a series of slight variants of a competitive inhibitor and study the effectiveness of each in binding to the active site. Or, to investigate the roles of the various parts of an enzyme molecule, they may make use of irreversible inhibitors—chemicals that act as enzyme poisons by forming permanent covalent bonds with the functional groups necessary for catalysis. More recently, the techniques of molecular biology have begun to permit selective substitution or deletion of individual peptides in enzymes. As a result, the action of any amino acid in a protein can be explored experimentally. One consequence of the knowledge this technique has brought is that biochemists can now plan and synthesize artificial enzymes (Fig. 3.50). These "designer enzymes" can be made with specificities and activities unknown in nature. Such enzymes could have nearly limitless potential in treating disease or even solving problems of environmental pollution.

STUDY QUESTIONS

1. How might a change in pH affect the "personality" of polar and ionic amino acids, and the activity or solubility of an enzyme? (pp. 81–82)

2. Why is an α helix more elastic than a pleated sheet? (pp. 66–67)

3. Calculate the approximate bonding energy holding the two 1,000,000-nucleotide-long strands of *E. coli* DNA together. How do you suppose the strands are separated to allow duplication? (pp. 39–40, 72–73, 77–79)

4. What effect does raising the temperature or increasing the concentration of the relevant enzyme have on an equilibrium constant? (pp. 76–79)
5. Why is activation energy a blessing for organisms? (pp. 78–79)
6. Why is it essential for a cell to be able to control enzyme activity? What are some ways this task is accomplished, and why are some more appropriate for certain sorts of chemical pathways? (pp. 84–86)

CONCEPTS FOR REVIEW

- Important functional groups
- Carbohydrates
  - Molecular definition
  - Basic types and their roles
- Condensation versus hydrolysis reactions
- Lipids
  - Molecular definition
  - Saturated versus unsaturated
  - Phospholipids
- Proteins
  - Basic structure of amino acids
  - Amino acid "personalities" and their importance
  - Primary versus secondary, tertiary, and quaternary structure
  - Alpha helices versus beta sheets
- Nucleic acids
  - Molecular definition
  - Double-chain binding
- Thermodynamics
  - Free energy
  - Laws of thermodynamics
  - Equilibrium constant: definition and use
  - Activation energy
  - Role of catalysts
- Enzymes
  - Basis of specificity
  - Basis of activity
  - Mechanisms of regulation

SUGGESTED READING

ATKINS, P. W., 1987. *Molecules.* Scientific American Library, New York. *A beautifully produced "molecular glossary" illustrating the chemical formula, three-dimensional structure, and biological action of many common or unusually interesting organic molecules.*

DOOLITTLE, R. F., 1985. Proteins, *Scientific American* 253 (4). *Reviews the properties of amino acids and the structure of proteins, and discusses the evolution of different modern proteins from common ancestral enzymes.*

DRESSLER, D., and H. POTTER, 1991. *Discovering Enzymes.* W. H. Freeman, New York. *A well-written and illustrated history of the study of enzymes, with particular emphasis on how the digestive enzyme chymotrypsin works.*

KARPLUS, M., and MCCAMMON, J. A., 1986. The dynamics of proteins, *Scientific American* 254 (4).

KENDREW, J. C., 1961. The three-dimensional structure of a protein molecule, *Scientific American* 205 (6). (Offprint 121) *How the complete folding pattern of myoglobin—the first protein whose conformation was determined—was worked out.*

KOSHLAND, D. E., 1973. Protein shape and biological control, *Scientific American* 229 (4). (Offprint 1280) *On the importance of protein conformation in determining enzymatic activity; how substances that cause changes in the shape of a protein can regulate its activity.*

LEHNINGER, A. L., 1965. *Bioenergetics.* W. A. Benjamin, New York. *Excellent discussion of thermodynamics from a biological perspective.*

STROUD, R. M., 1974. A family of protein-cutting proteins, *Scientific American* 231 (1). (Offprint 1301) *A good discussion of how enzymes like chymotrypsin work.*

STRYER, L., 1988. *Biochemistry,* 3rd ed. W. H. Freeman, San Francisco. *Beautifully produced, clearly written, but highly technical exposition of biochemistry.*

THOMPSON, E. O. P., 1955. The insulin molecule, *Scientific American* 182 (5). (Offprint 42) *On the first determination of the primary structure of a protein—by Frederick Sanger, who labored for ten years before he worked out the amino acid sequence of insulin in 1954.*

Chapter **4**

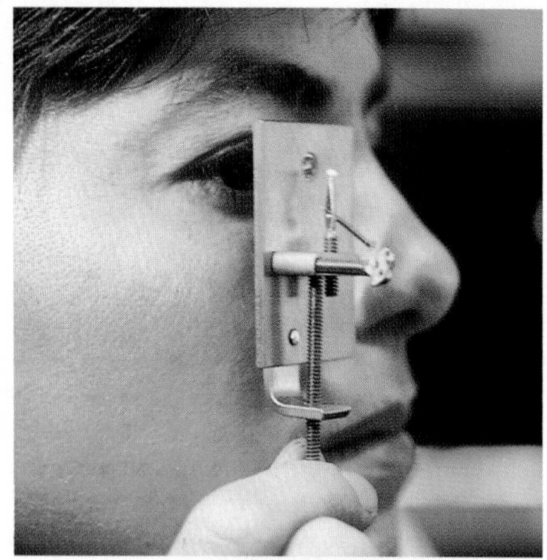

A

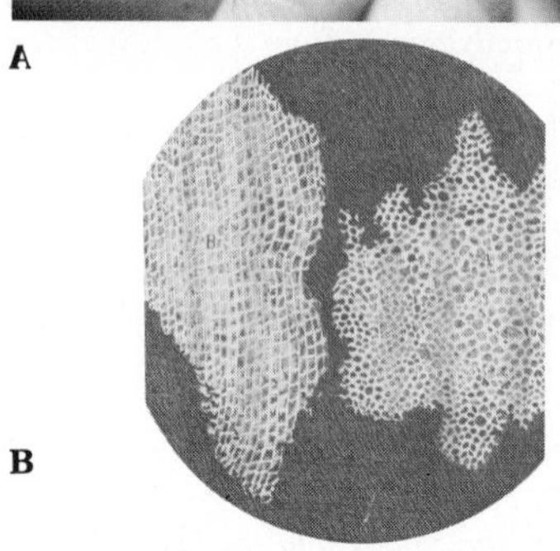

B

4.1 Van Leeuwenhoek's microscope
(A) This magnifier was first developed to inspect samples of cloth but was soon adapted for viewing living things. (B) Drawings of cork which appeared in Robert Hooke's *Micrographia*, published in 1665.

AT THE BOUNDARY OF THE CELL

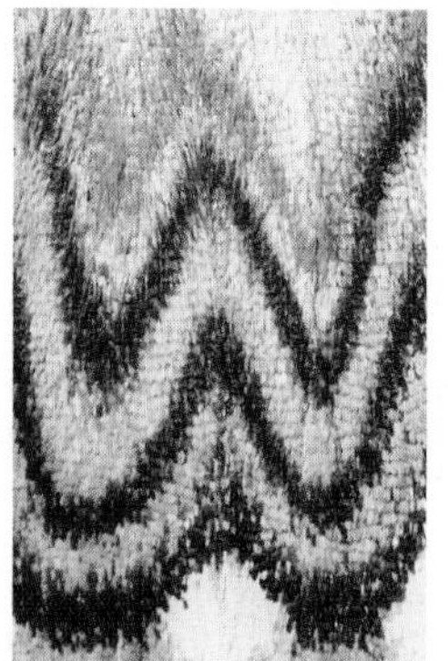

e saw in the last chapter that organisms are composed of a great variety of chemicals, some simple and some complex. But these chemicals do not of themselves possess the properties we recognize as life. We cannot, for example, put a population of amoebae through a blender and expect the resulting mixture of organic molecules to reorganize itself spontaneously into living entities.

Instead, life depends on a precise compartmentalization and organization of organic molecules. An intricate membrane must protect the interior of the cell—the nucleus, with its DNA, and the ***cytoplasm***, which is composed of organic fluids (the ***cytosol***), internal membranes, and a variety of specialized, self-contained entities known as organelles. The cell membrane separates the cell's delicate internal chemistry from the vagaries and dangers of the external environment, holding some chemicals in, passing some through, barring others from entering. And just as the cell membrane protects a favorable chemical atmosphere from the world outside, so structures within the cell separate themselves from the rest of the cell by means of their *own* membranes. In this way the cell's chemical processes are partitioned off—so that, for example, food is digested in membrane-enclosed compartments that prevent digestion of the contents of the rest of the cell.

In addition to being the cornerstone of all living things, the cell is a microcosm of the richness and complexity of life itself. Multicellular organisms often consist of legions of cells which perform highly specialized tasks. In this chapter we begin the study of the physiology and diversity of cells by examining the outer membrane (also known as the ***plasma membrane***) and its function in the life of the cell. Later chapters will probe the machinery of the cell's interior and the role each part plays in the elaborate chemistry of life.

THE CELL THEORY

The discovery of cells and of their structure is linked to the development of magnifying lenses, particularly the microscope. In the seventeenth century Antony van Leeuwenhoek and his contemporaries refined the production of lenses sufficiently to construct microscopes satisfactory for simple observations (Fig 4.1). Thus in 1665 Robert Hooke was able to report to the Royal Society of London on "the first microscopical pores I ever saw, and perhaps, that were ever seen," in a piece of cork. Hooke's microscopic examination of cork marks the beginning of the study of cells. Intensive work on cells was not pursued, however, until the early nineteenth century.

The idea that all living things are composed of cells—the ***cell theory***—is commonly credited to two German investigators, the botanist Matthias Jakob Schleiden and the zoologist Theodor Schwann, who published their conclusions in 1838 and 1839 respectively. An important extension of the cell theory, proposed in 1858 by the German physician Rudolf Virchow, was that all living cells arise from pre-existing living cells (*"omnis cellula e cellula"*), and that there is therefore no spontaneous creation of cells from nonliving matter. The theory of ***biogenesis***, life from life, contradicted the prevailing belief in spontaneous generation, then widely held not only by the general public but by scientists as well. It was Louis Pasteur in France who, a few years later (1862), supplied proof for Virchow's theory in a series of now classic experiments.

Pasteur's first step was to place various nutrient broths in long-necked flasks and then bend the necks of the flasks into curves (Fig. 4.2). Next, he boiled the broths in the flasks to kill any microorganisms that might be in them. While the flasks were left standing, microbe-laden dust particles in the air moving into the flasks were trapped in the films of moisture on the humid curves of the necks; the curved necks acted as filters. Though the broths might be left standing in their swan-neck containers for months or even a year or more, no life appeared in them. Identical broths boiled in flasks with straight necks—the control solutions—did not remain free of microorganisms. They were soon teeming with life. Similarly, if the swan neck was broken off, the experimental broth rapidly developed colonies of molds and bacteria. The control solutions required by rigorous scientific procedure were crucial to the proof of Pasteur's theory: since handling of the control solutions differed in only one respect—exposure to airborne microbes—the changed outcome had to be attributed to that difference. Thus Pasteur showed that the source of the microorganisms that fermented or putrefied such substances as milk, wine, and sugar-beet juice was the air. The organisms did not arise spontaneously from the nutrient media.

The two components of the cell theory—that all living things are composed of cells and that all cells arise from other cells—give us the basis for a working definition of living things: living things are chemical organizations composed of cells and capable of reproducing themselves.

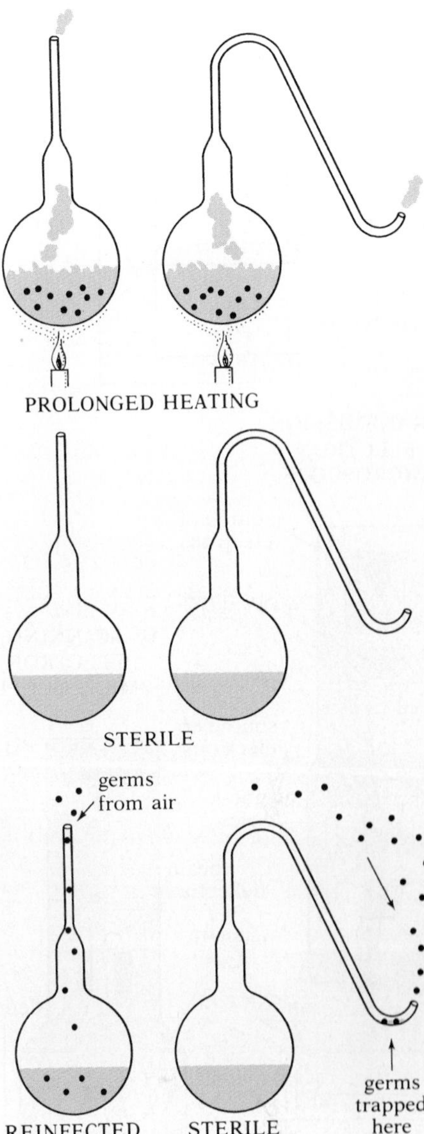

4.2 Pasteur's experiment
Nutrient broths in two kinds of flasks, one with a straight neck, the other with a bent neck, were boiled to kill any microbes they might contain (top). The sterile broths were then allowed to sit in their open-mouthed containers for several weeks (middle). Microorganisms entering the straight-necked flask contaminated the broth, but those entering the bent neck of the other flask were trapped in films of moisture in the curves of the neck (bottom).

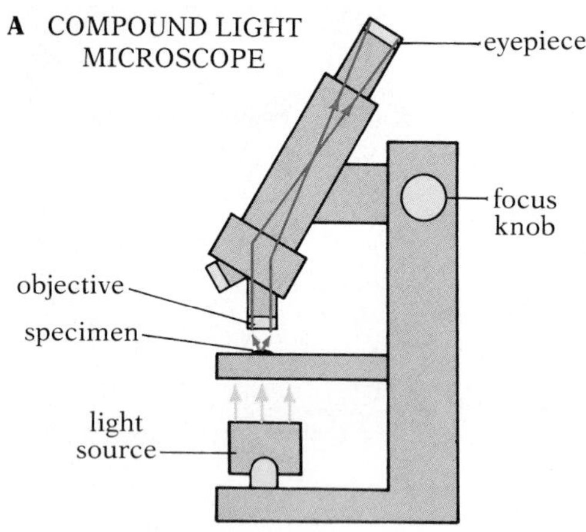

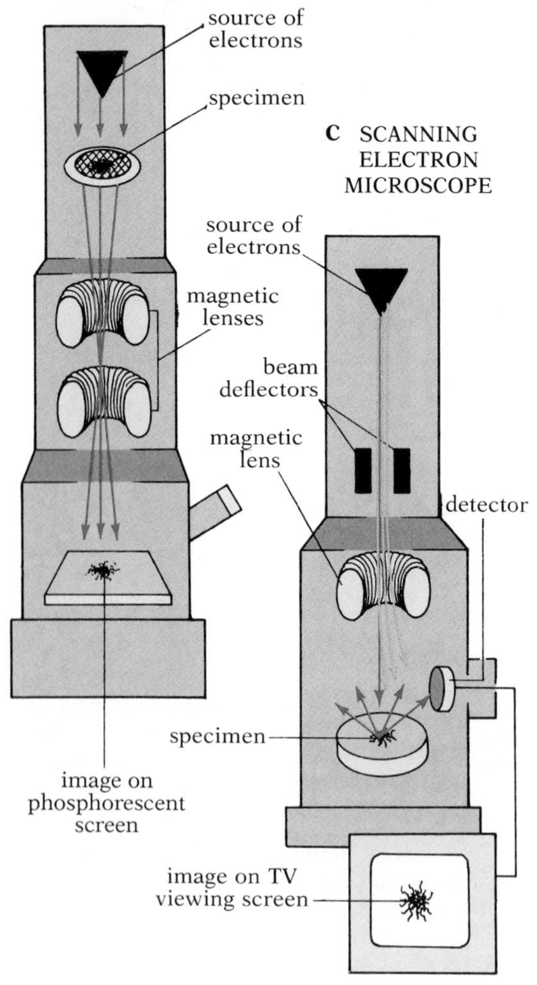

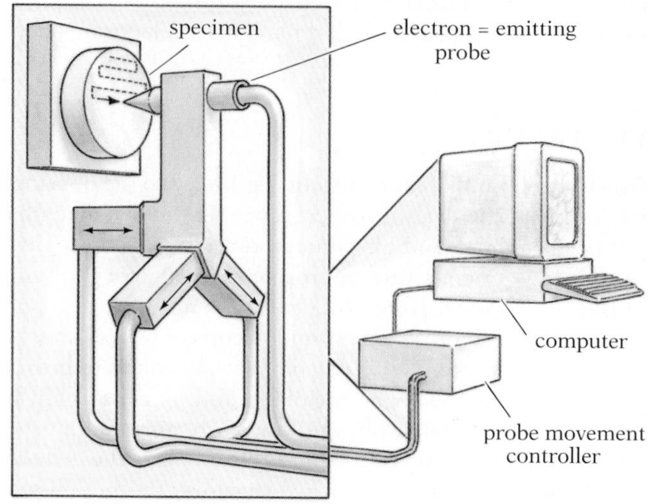

4.3 Microscopes
(A) In a compound light microscope, light passes through a specimen to the objective lens. Here the light is refracted and directed to the eyepiece, where it is focused for a camera or for the eye. (B) In a transmission EM, electrons provide the illumination, which passes through the specimen and is then focused by magnets to form an image for photographic film or on a phosphorescent screen. (C) In a scanning EM, a focused beam of electrons moves back and forth across the specimen, while a detector monitors the consequent emission of secondary electrons and reconstructs an image. (D) The scanning tunneling microscope moves an electron-emitting tip over the sample, constantly adjusting its height to maintain a fixed distance above the sample; these variations in tip elevation allow reconstruction of the specimen's contours. The minute movements necessary are accomplished by a mounting arm attached to piezoelectric crystals. Since the dimensions of such crystals change as different amounts of voltage are applied, it is possible to control the position of the tip exactly.

VIEWING THE CELL

Much of our knowledge of subcellular organization has been made possible by the development of better and more powerful microscopes. In the detailed analysis of subcellular structure, three attributes of microscopes are of particular importance: magnification, resolution, and contrast. Magnification is a means of increasing the apparent size of the object being viewed until it provides an adequate stimulus to our eyes. Resolution is the capacity to show adjacent forms or objects as distinct. Contrast is important in distinguishing one part of a cell from another.

The ordinary compound light microscope has many features basic to the operation of all microscopes. Light passes through a specimen and is then captured, bent, and brought into focus by lenses (Fig. 4.3A). Depending on

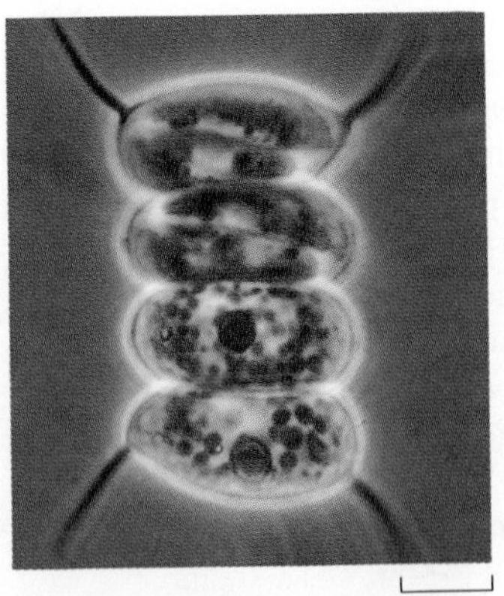

A

B

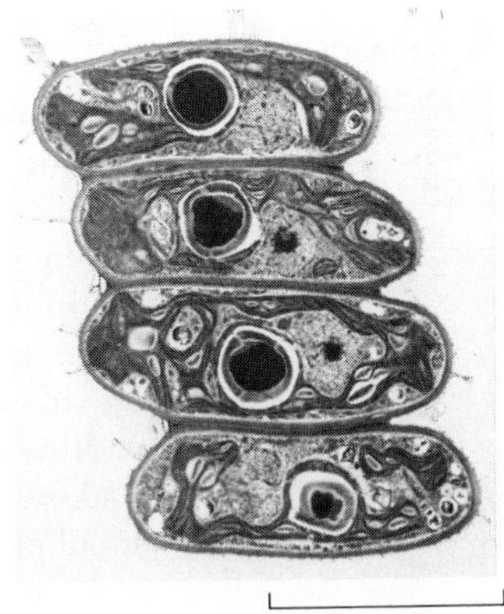

C

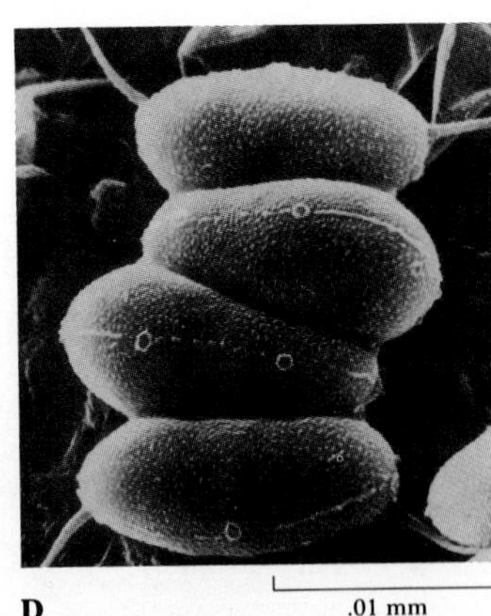

D

4.4 Views of the green alga *Scenedesmus* obtained with microscopes of various types.
(A) Photograph of an unstained specimen as seen with a phase-contrast light microscope. (B) Photograph taken by the Nomarski process. (C) Transmission electron micrograph. Many of the membranous and particulate intracellular organelles show up much more clearly here than under the light microscope. They are made visible by a stain containing heavy-metal atoms, which combines differentially with various structures. (D) Scanning electron micrograph, providing a three-dimensional view of surface features. The scale bars below the photographs in this and later figures indicate the dimensions of the organisms shown.

the magnification and the size of the specimen, a whole cell or only a tiny part may be in the field of view at any one time.

We vary the magnification by using lenses with different shapes that accept light from larger or smaller sections of the specimen: the higher the magnification, the smaller the amount of light that reaches the eyepiece. The limit on useful magnification in light microscopes is not a matter of exhausting the illumination, however. Instead it arises from the tendency of light to bend as it passes near an edge. This phenomenon, known as diffraction, spoils images by bending light out of the straight-line path as it moves from the source, through the specimen, to the objective. The result is a degraded picture with decreased resolution. The amount of useful magnification possible is limited by the wavelengths of ordinary light to about 1000 times the actual size of the object in focus. Though magnification of 1000 is an enormous improvement over the unaided eye, it is still not enough to let us see many of the smaller subcellular structures.

In the light microscope, contrast can be as important as magnification. Contrast is necessary if we are to distinguish structures from their backgrounds. Most cellular components are colorless and have essentially the same texture. But different parts of the cell often differ in their affinities for various dyes, so these areas can be stained with different color dyes or with different intensities of dye to make them stand out from each other. Unfortunately, staining usually kills the cell, and may thereby change its internal structure. New techniques such as phase-contrast and Nomarski optics, both of which depend on elaborate optical manipulation, have greatly increased the value of the light microscope because they create contrast in cells optically without staining (Fig. 4.4A–B).

The electron microscope (EM) opened new vistas in the study of cells by using a beam of electrons instead of light as its source of illumination. Because resolution improves (that is, diffraction decreases) as the wavelength of the illumination becomes shorter, and because electron beams have much shorter wavelengths than visible light, electron microscopes can resolve objects about 10,000 times better than light microscopes (Fig. 4.4C–D). Many details of cellular structure would not be known but for the EM.

In a transmission EM (Fig. 4.3B) the electrons pass through the thinly sliced ("sectioned") specimen, are focused by magnets rather than by

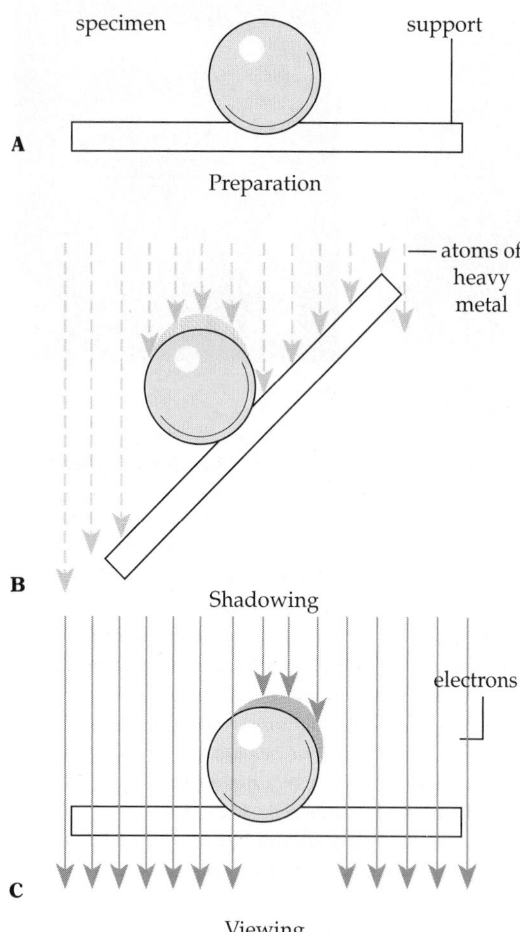

lenses, and fall on a photographic plate or a phosphorescent screen, where they produce an image of the specimen. Because cells are essentially transparent to electrons, a specimen being prepared for the transmission EM must be differentially stained with an electron-dense chemical that binds to specific cell structures. Needless to say, the picture can be only as good as the staining technique used to create it, and for this reason there are many different techniques. One is to use stains containing heavy-metal atoms. These stains bind to different internal cellular structures, blocking the passage of electrons in these places (Fig. 4.4C). Another is to tilt the specimen to allow atoms of the electron-dense substance to fall onto it (Fig. 4.5). The shadows and highlights in the resulting EM picture create a three-dimensional effect—a kind of topographical map of the specimen which can reveal important surface details.

A scanning EM can also produce a three-dimensional view. A specimen coated with atoms of metal is scanned from above by a moving beam of electrons (Fig. 4.3C). This focused probe does not penetrate the specimen, but instead causes so-called secondary electrons to be emitted from the surface. The intensity of the emission of secondary electrons depends on the angle at which the probe beam strikes the surface, and therefore varies with the contours of the specimen. Hence a point-by-point recording of the emission produces a three-dimensional picture (Fig. 4.4D). Though the resolution of the scanning EM does not approach that of the transmission EM, its ability to create a better three-dimensional effect is an advantage for many applications. And since it provides information by scanning surface features, a specimen can frequently be studied whole and intact. Moreover, since "shadowing" is not required, the same specimen can be turned repeatedly and observed from various perspectives.

The most recent elaboration of the electron microscope, the scanning tunneling microscope (STM), uses an ultrafine electron-emitting tip, which is moved back and forth over the sample, systematically scanning it from top to botton (Fig. 4.3D). The tip almost touches the specimen, and so must be moved up and down to accommodate the contours of the sample. Biological specimens are coated with metal and an electrical potential is set up between the tip and the coating. The tip's elevation is constantly adjusted to maintain a set current flow, which keeps it at a fixed distance from the bit of the specimen just below the tip. These up/down movements of the tip are used by a computer to reconstruct the topography of the sample, and a picture is created on a video monitor (Fig. 4.6).

4.5 Shadow staining for the transmission EM
The specimen (A) is tilted and dusted, or "shadowed," with atoms of a heavy metal such as platinum (B). (Another technique is to "shoot" the metal atoms from an angled source onto a horizontal specimen.) When an electron beam is aimed at the coated specimen (C), the unevenly distributed metal plating prevents most of the electrons from penetrating the shadowed areas, and produces a three-dimensional view of the surface of the specimen, as in a photograph of polio virus (D).

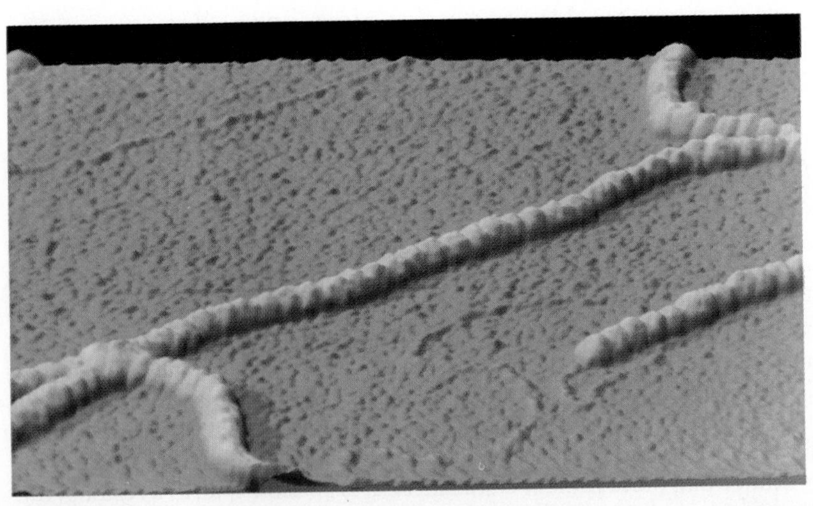

4.6 STM image of DNA
This computer-enhanced picture shows both ordinary DNA (thin coils) and DNA complexed with a protein —SSB (discussed in detail in Chapter 8)—which holds the double helix in a much more open configuration during the process of chromosomal duplication (thick coils).

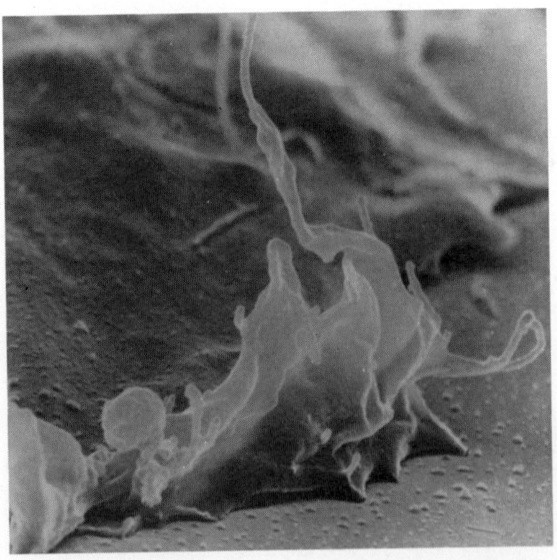

4.7 Ruffling cell
The flat, thin protrusions extending from this mouse cell enable it to sample its environment and determine the right direction to move. (The mechanisms underlying this "ruffling" and the cellular movement it controls will be described in the next chapter.)

FUNCTIONS OF THE CELL MEMBRANE

At one time, the cell membrane was considered little more than a bag to hold in all the organic chemicals that somehow combine to produce life. As we will see, however, the cell membrane is far more than a passive envelope giving mechanical strength and shape to the cell. It is critical to cell movement, and to the cell's sampling of the extracellular environment (Fig. 4.7); moreover, it bears the primary responsibility for regulating the chemical traffic between the precisely ordered interior of the cell and the essentially unfavorable and potentially disruptive outer environment. All substances moving into or out of a cell must pass through a membrane barrier, and the membrane of each cell can be quite specific about what is to pass through, and at what rate, and in which direction. The cell membrane exercises this control in two ways: by utilizing natural processes such as diffusion, and by transporting specific substances in and out.

DIFFUSION

As we have seen, temperature affects the rates of chemical reactions by increasing the kinetic energy (the average velocity) of the molecules involved.

Imagine a small stationary box containing 17 marbles in a tight cluster near one end (Fig. 4.8A). When we shake the box, the marbles disperse al-

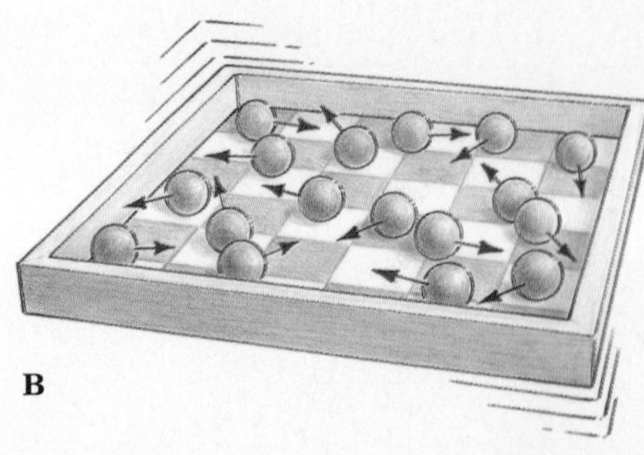

4.8 Mechanical model for diffusion
(A) All 17 marbles are placed in a cluster at one end of a rectangular box. (B) When the box is shaken to make the marbles move randomly, they become distributed throughout the box in nearly uniform density.

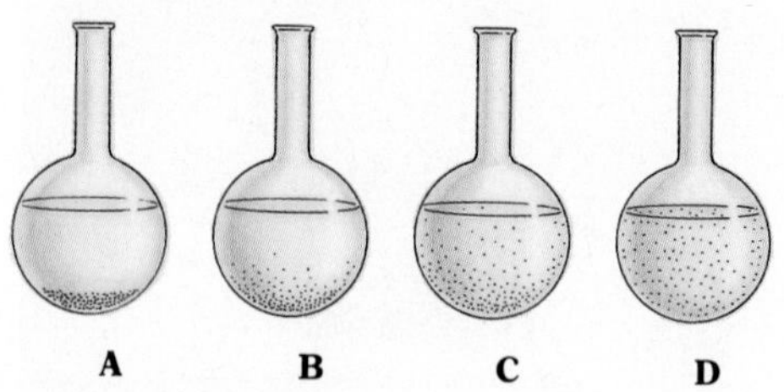

4.9 Diffusion in a liquid
Particles of solute are at the bottom of a flask of water (A). The particles slowly diffuse away from the cluster until (D) they are distributed with nearly uniform density through the water. If the water is cold and there are no convection currents to help move the particles, it may take a considerable period of time to reach the uniform distribution shown in D.

most evenly over the bottom (Fig. 4.8B). Obvious as this result may seem, it is worth a closer look, because the marbles can be thought of as molecules, and the shaking as adding kinetic, or thermal, energy to the system.

First, it is immediately apparent that of all the possible directions in which a given marble might move, more lead *away* from the center of the cluster than toward it. Hence *random movement will tend to disrupt the cluster rather than to maintain it.* As we indicated in Chapter 3, in the absence of any counteracting external influence, a dynamic system will tend to move toward the more probable disorderly state rather than toward the less probable orderly state. This is precisely what happens, for instance, when a lump of sugar dissolves in a cup of warm coffee: the sugar molecules move from the region of high concentration (the sugar crystal) to regions of lower concentration; eventually the sugar molecules disperse throughout the liquid. The warmer the liquid, the more kinetic energy the molecules in solution will have on average, and the faster diffusion will take place.

Notice that the argument is statistical. It is possible that, as a result of random motion, all 17 scattered marbles will form a tight cluster at one end of the box. This result has a definite possibility of occurring, but one so slight that it can justifiably be disregarded. The kind of reasoning used here is typical of most scientific reasoning: the facts and laws of science are statistical rather than absolute. They describe natural phenomena in terms of *probable* outcomes; they do not assert that a certain outcome will occur 100 percent of the time.

We can now make a generalization based on our example of the marbles in the box and on others like it: all other factors being equal, *the net movement of the particles of a particular substance is from regions of higher to regions of lower concentration of that substance.* Note that we speak of *net* movement. There will always be some particles moving in the opposite direction, but, overall, the movement will be away from the centers of concentration. An obvious result is that the particles of a given substance tend to become relatively equidistant from one another within the available space. When this uniform density has been reached, the system is in equilibrium; the particles continue to move, but there is little net change in the system.

Movement of particles the size of molecules from one place to another in the manner we have been discussing is called ***diffusion***. Diffusion is fastest by far in gases, where there is much space between the particles and hence relatively little chance of collisions that retard movement. Diffusion in liquids is much slower (Fig. 4.9); in the absence of convection currents a substance can take a very long time—years, in fact—to move in appreciable quantity only a meter through cold water. Diffusion in solids is, of course, much slower still: there is very little space between the molecules of a solid, and collisions occur almost before the molecules get going. In all these instances, however, regardless of the rate of diffusion, the net effect is movement away from regions of higher concentration, as long as all regions are at the same temperature and pressure. In living organisms, where molecules are generally in a warm aqueous solution and the distances involved are measured in fractions of a millimeter, diffusion is a highly significant process; an amino acid or nucleotide in an aqueous medium will typically diffuse about one cell diameter (10–50 microns) in less than 0.5 seconds.

So far, we have discussed diffusion in terms of movement from a higher to a lower concentration along a gradient. In the living world, however, diffusion is not strictly a function of concentration, since conditions are seldom constant where life processes are at work. It is therefore more useful for us to look at diffusion in terms of the free energy of the particles involved. A local concentration of a substance is a relatively orderly and unlikely arrangement. We can see, for instance, that energy (among other things) would be needed to change a mixture of sugar molecules and coffee back into the original lump of sugar and unsweetened coffee. A random, disorderly arrangement of molecules has necessarily less potential for doing work than an orderly one, and has concomitantly less free energy. As you may recall from Chapter 3, the amount of disorder in a system is known as ***entropy***. Since, as we know from the Second Law of Thermodynamics, the amount of free (useful) energy in the universe is always decreasing, entropy is always increasing.

Diffusion, then, is spontaneous because orderly molecules, concentrated together, have greater free energy than dispersed molecules: it is a downhill reaction from order to disorder. The mixture (or product) has less free energy than the separate original substances (the reactants). Like the chemical reactions discussed in Chapter 3, the rate of diffusion if we begin with two pure substances is fastest at the outset, and slows as the equilibrium of complete mixture is approached and the ratio of available reactants to product decreases. If we could observe diffusion at the molecular level we would see the sugar molecules speeding away from the lump during the early part of the reaction. Later, however, as the substances became more evenly mixed, the frequency of the back reaction returning sugar molecules to the location of the dissolving lump would rise, until equilibrium was reached. In fact, diffusion is a chemical reaction with its own free energy, which depends on the characteristics of the substances involved:

$$\text{sugar} + \text{coffee} \rightleftharpoons \text{sweetened coffee} \qquad (\Delta G = -x)$$

Free energy is a more broadly applicable basis than concentration gradients for understanding diffusion. Consider a situation in which there is a slight concentration gradient in one direction, and a pronounced temperature gradient in the reverse direction (Fig. 4.10). The opposing effects of these two gradients produce a net movement of molecules that depends entirely on the relative free energies of the two gradients—from the region of higher temperature to the area of higher concentration in this case.

The importance of diffusion and its basis in free energy with respect to cells is clear: the concentration of organic molecules and a select group of ions inside a cell is a very unlikely arrangement. Without the cell membrane, the free energy of the cellular chemistry would be lost as the contents diffused into the environment. Two conclusions follow. First, there must be a barrier between the inside and the outside of the cell to maintain the integrity of the cellular chemistry. And second, the free-energy gradient across the cell membrane is available to do work.

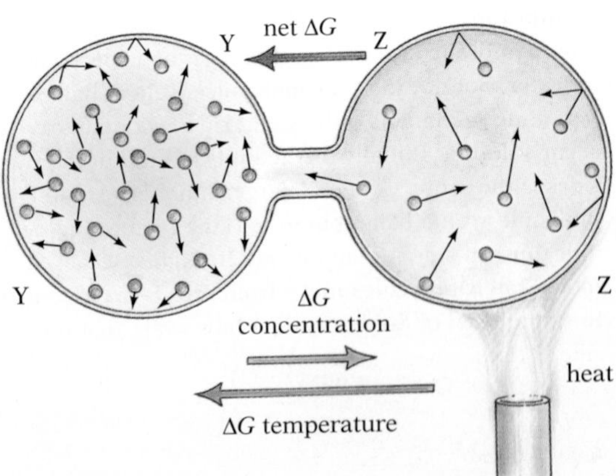

4.10 Multiple gradients and free energy
This two-chambered vessel has one side (Y) with a slightly higher pressure than the other (Z), but the temperature in Z is much higher than that in Y. If concentration alone were important, the net diffusion would be from Y, the region of higher concentration, to Z, the region of lower concentration. But the higher the temperature in a given system, the greater the thermal motion of the particles in that system; and the greater the thermal motion, the greater the free-energy content. Because the difference in free energy associated with the temperature gradient from Z to Y outweighs the difference in free energy associated with the concentration gradient from Y to Z, net diffusion will be from Z to Y.

OSMOSIS

To envision the way cell membranes can function, let's consider another model: a chamber divided into halves by a membrane partition. Let us as-

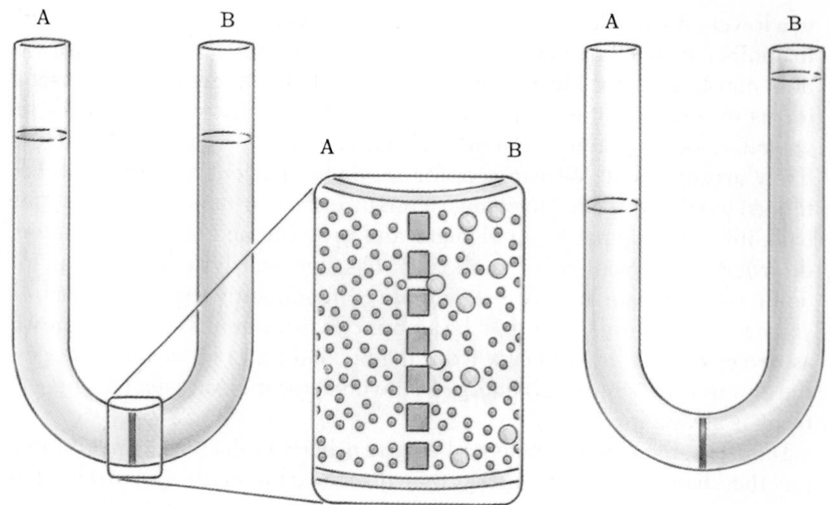

4.11 U-tube divided by a selectively permeable membrane
The membrane at the base of the U-tube is permeable to water, but not to sugar molecules (yellow balls). Left: Side A contains only water; side B contains a sugar solution. Initially the quantity of fluid in the two sides is the same. Center: A larger number of water molecules (blue balls) bump into the membrane per unit time on side A than on side B. Right: Because more water molecules move from A to B than from B to A, the level of fluid on side A falls while that on side B rises.

sume, further, that particles of some substances can pass through the membrane while particles of other substances cannot. Such a membrane is said to be ***differentially permeable*** (or ***selectively permeable***). How will the membrane affect the diffusion of materials between the two halves of the chamber? Suppose the chamber is a U-tube divided in half by a selectively permeable membrane (Fig. 4.11). Suppose side A contains pure water and side B an equal quantity of sugar solution (sugar dissolved in water), both sides being subject to the same initial temperature and pressure. If the membrane is permeable to water but not to sugar, water molecules will be able to pass in both directions, from A to B and from B to A.

This movement of a solvent (usually water) through a selectively permeable membrane is called ***osmosis***. Biological membranes are selectively permeable, and the movement of water through them can be predicted on the basis of osmosis. As we will see, some solutes, such as small lipid-soluble molecules, also pass through biological membranes freely.

Since water is already present on both sides of the membrane in the U-tube in Figure 4.11, it might at first seem that the movement of water molecules across the membrane would have no net effect. But consider the differences between the pure water and the sugar solution more carefully. We have seen that substances diffuse from regions of higher concentration to regions of lower concentration. Water, being more concentrated on side A, will tend to diffuse to side B; sugar molecules, trapped by the membrane on side B, will not cross.

We can also see why water molecules move from A to B by picturing to ourselves the events at the membrane itself. On side A, all the molecules that bump into the membrane during a given interval are water molecules, and because the membrane is permeable to water, many of these will pass through the membrane from A to B. By contrast, on side B, some of the molecules bumping into the membrane during the same interval will be water molecules, which may pass through, and some will be sugar molecules, which cannot pass through because the membrane is impermeable to them. At any given instant, then, some of the membrane channels on side B are in contact with sugar molecules and some are in contact with water, whereas on side A all the molecules entering channels are water molecules. Hence more water molecules will move across the membrane from side A to side B per unit time than in the opposite direction; the net osmosis will be from A to B.

Or we can think of the matter more abstractly, in terms of entropy. The arrangement of water molecules in pure water is orderly, in that every molecular location is occupied by a water molecule, whereas the arrangement in the sugar solution is disorderly, in the sense that any given molecular location may be occupied by either a water molecule or a sugar molecule. Since, as we have said, an orderly system possesses more free energy than a disorderly one, it follows that the orderly water molecules in the pure water (side A) have more free energy than the disorderly water molecules in the sugar solution (side B). There is a free-energy gradient for water from side A to side B, and, according to our generalization concerning diffusion, there will be a net movement of water down this gradient, from A to B. Perhaps the easiest way to remember what happens in the U-tube example is to recall the general rule of diffusion and apply it to this case: as we said at the outset, water moves from a region of high water concentration (i.e., pure water) to one of lower water concentration.

We are now in a position to make some additional generalizations. *The free energy of water molecules is always decreased if osmotically active substances* (dissolved or colloidally suspended particles) *are present in the water.* (Colloidal particles are generally larger than the separated individual molecules of a dissolved substance, yet small enough so that—unlike the still larger particles of a true suspension—they do not settle out at an appreciable rate but remain dispersed within the fluid medium.) The ***osmotic concentration*** of a fluid—the number of osmotically active particles per unit volume—thus bears a direct relationship to that fluid's free energy. In the U-tube example, for instance, *the decrease in the free energy of water molecules is proportional to the osmotic concentration.* The reason for the diminution in free energy is that the osmotically active particles to some degree disrupt the orderly three-dimensional array of the water molecules (see Fig. 2.30, p. 45).

Each solution, then, has a certain free energy, depending on its osmotic concentration. Under conditions of constant temperature and pressure, this free energy can be calculated; it is called ***osmotic potential***. (Pure water is arbitrarily assigned an osmotic-potential value of zero. Since osmotic potential decreases as osmotic concentration increases, all solutions have values of less than zero.) If two different solutions are separated by a membrane permeable only to water, and temperature and pressure are constant, *the net movement of water will be from the solution with the lower osmotic concentration to the solution with the higher osmotic concentration.* The steeper the osmotic-concentration gradient, the more rapid the movement. That is to say, the water flows from regions of higher osmotic potential to regions of lower potential at a rate proportional to the degree of difference in osmotic potential. The principles we have been discussing are summarized in Table 4.1.

If the net movement of water in the U-tube is from side A to side B, the volume of liquid will increase on side B and and decrease on side A. Does

TABLE 4.1 *Some rules of osmotic diffusion*

I. The free energy of water molecules is always decreased if osmotically active substances (dissolved or colloidally suspended particles) are present in the water.
II. The decrease in the free energy of water molecules is proportional to the osmotic concentration.
III. The net movement of water will be from the solution with the lower osmotic concentration to the solution with the higher osmotic concentration.

Exploring Further

OSMOTIC POTENTIAL, OSMOTIC PRESSURE, AND WATER POTENTIAL

As we have seen, osmotic potential is useful for thinking about how two solutions with differing osmotic concentrations interact. It is easily related to the underlying difference between them in free energy, which results in the movement of the solvent. Many researchers, however, prefer to think in terms of the pressure that must be exerted on a solution to keep it in equilibrium with pure water when the two are separated by a selectively permeable membrane. In our U-tube example, this pressure corresponds to the hydrostatic pressure exerted by the sugar solution at equilibrium; it is known as ***osmotic pressure***. Clearly, *the osmotic pressure of a solution is a measure of the tendency of water to move by osmosis into it.* The more dissolved particles in a solution, the greater the tendency of water to move into it, and the higher the osmotic pressure of the solution. Thus, under constant temperature and pressure, water will move from the solution with the lower osmotic pressure to the solution with the higher osmotic pressure when the two solutions are separated by a selectively permeable membrane.

While the terms "osmotic pressure" and "osmotic potential" are regularly used by physiologists studying animals, plant physiologists more often refer to ***water potential***, which is essentially the same as the free energy of water. At a pressure of one atmosphere, pure water is assigned a water potential of zero. Since the water potential decreases as the osmotic concentration increases, all solutions have values of less than zero. In this sense, water potential is like osmotic potential. But unlike osmotic potential, which is a function of solute concentration alone, water potential (like free energy) is also a function of temperature and pressure. When two solutions are separated by a selectively permeable membrane, water will move from the solution with the higher water potential to the solution with the lower water potential.

Familiarity with all of these terms is useful, because they are all common in the biological literature. Except in our discussions of plant physiology, however, we will ordinarily use the term "osmotic potential."

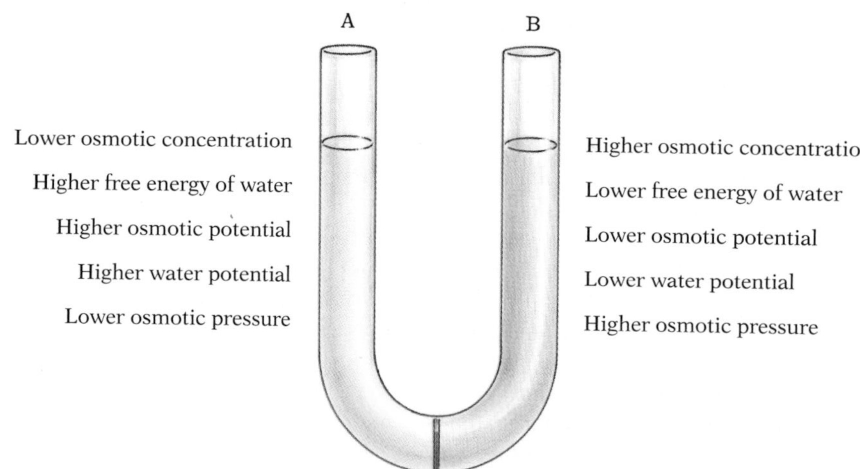

the property of selective permeability cause this process to continue indefinitely, or will an equilibrium point be reached? Clearly, if the membrane is completely impermeable to sugar molecules, conditions on the two sides will never be equal, no matter how many water molecules move from A to B. The fluid in B will remain a sugar solution, though an increasingly weak one, and the fluid in A will remain pure water. Nevertheless, under normal conditions, the fluid level in B will rise to a certain point and then cease to rise further. Why? The column of fluid is, of course, being pulled downward by gravity. As the column rises, therefore, its weight exerts increasing downward hydrostatic pressure. As the pressure increases, the free energy of the water in the sugar solution rises, because pressure too is a form of free (useful) energy. Eventually the column of sugar solution becomes so high, and its pressure and free energy so great, that water molecules are pushed across the membrane from B to A as fast as they move into B from A.

When water is passing through the membrane in opposite directions at the same rate, the system is in dynamic equilibrium, with the free energy—the osmotic potential—of the pure water on one side of the membrane just matching the free energy—the osmotic potential and hydrostatic pressure—of the column of solution on the other side. Obviously, the greater the concentration difference across the membrane, the greater the difference in osmotic potential between the two sides and the higher the column of solution will rise before this difference is counterbalanced by the difference in hydrostatic pressure.

It is important to understand that osmotic concentration is not concentration by weight, but rather molecular or ionic concentration—the total number of solute particles per unit volume. If there are several kinds of solutes in the same solution, then the osmotic concentration of that solution is determined by the total (per unit volume) of *all* the particles of all kinds. If a dissolved substance ionizes, then each ion functions osmotically as a separate particle: one mole of sodium chloride (NaCl) dissolved in water produces two moles of particles—one of Na^+ ions and one of Cl^- ions. Colloidal particles may also contribute to the total osmotic concentration.

OSMOSIS AND THE CELL MEMBRANE

By now you probably realize that we have discussed diffusion and osmosis at such length because the cell membrane is selectively permeable, and the processes of diffusion and osmosis are fundamental to cell life. Though the membranes of different types of cells vary widely in their permeability characteristics—the membrane of a human red blood corpuscle,[1] for instance, is over a hundred times more permeable to water than the membrane of *Amoeba*, a single-celled organism—a few rough generalizations can be made: Cell membranes are relatively permeable to water and to certain simple sugars, amino acids, and lipid-soluble substances. They are relatively impermeable to polysaccharides, proteins, and other very large molecules. In short, cell membranes let pass only the building blocks of complex organic compounds, not the compounds themselves. The permeability of cell membranes to small inorganic ions varies greatly, depending on the particular ion, but in general negatively charged ions can cross more rapidly than positively charged ions, though neither can do so as readily as uncharged particles.[2]

[1] A red blood corpuscle begins as a cell (in mammals) but loses its nucleus as it matures and becomes specialized to transport oxygen. However, it remains sufficiently like true cells to be used as an example of many cellular properties.

[2] The process in which, in addition to solvent, some solutes selectively cross the membrane is often called dialysis.

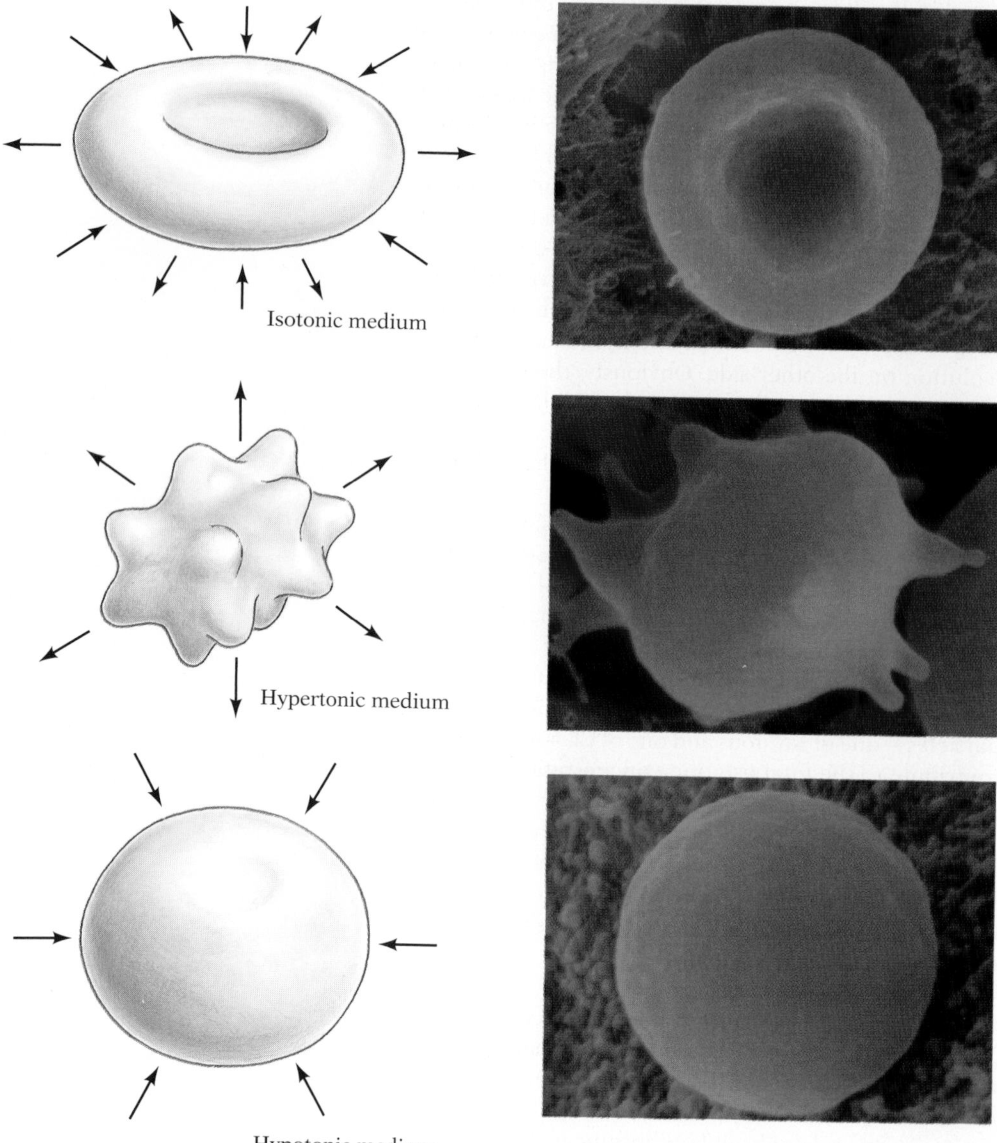

4.12 Osmotic relationships of a cell
In an isotonic medium, water gain and water loss are equal, and so the cell neither shrinks nor grows. In a hypertonic medium, there is a net loss of water from the cell, and the cell shrinks. In a hypotonic medium, the cell swells as water moves from the medium inside. The accompanying photographs are of human red blood cells.

What implications do these generalizations hold for life? On the one hand, selective permeability enables cells to retain the large organic molecules they synthesize. On the other, the tendency of water to pass through selectively permeable membranes into regions of higher osmotic concentration can be harmful or even fatal. When a cell is in a medium that is ***hypertonic*** relative to it (a medium to which it loses water by osmosis, usually because the medium contains a higher concentration of osmotically active particles), the cell tends to shrink (Fig. 4.12); if the process goes too far, it may die. Conversely, when a cell is in a medium ***hypotonic*** to it (a medium from which it gains water, usually because the medium contains a lower concentration of osmotically active particles), the cell tends to swell; unless it has special mechanisms for expelling the excess water, or special structures that prevent excessive swelling (as most plant cells do), it may burst. A cell in an ***isotonic*** medium (one with which the cell is in osmotic balance, usually because it contains the same concentration of osmotically active particles) neither loses nor gains appreciable quantities of water by osmosis.

Obviously, the osmotic relationship between the cell and the medium surrounding it is a critical factor in the life of the cell. Some cells are normally bathed by an isotonic fluid and therefore have no serious osmotic problems. Human red blood corpuscles are an example; they are normally bathed by blood plasma, with which they are in relatively close osmotic balance. Most simpler oceanic plants and animals also exemplify cells in an isotonic medium; their cellular contents have an osmotic concentration close to that of sea water. All cells, however, have a higher osmotic concentration than fresh water. Freshwater organisms therefore live in a hypotonic medium and face the problem of accumulating excessive water within their cells by osmosis. Their very existence has depended on the evolution of ways of preventing their cells from becoming so turgid—so distended by their fluid content—that they would burst. (Solutions to this problem will be discussed in Chapter 31.)

But controlling the flow of water is only one problem. Though the selective permeability of the membrane effectively traps large molecules inside, it does not provide any mechanism for concentrating the organic building blocks necessary for constructing substances like DNA, proteins, and polysaccharides in the first place. So while the cell membrane acts in many respects like an inert osmotic partition, it must also do more. For nutrients to be captured and retained, wastes expelled, and cell volume controlled, the cell membrane must have the capacity to pass many chemicals in only one direction. The secret of this critical ability lies beyond mere selective permeability, in the structure of the cell membrane itself.

STRUCTURE OF THE CELL MEMBRANE

Researchers have struggled for decades to explain the remarkable and apparently contradictory abilities of the cell membrane. The key to membrane function lies in understanding its structure which, in turn, must be inferred from its various properties. For instance, permeability studies had long shown that lipids and many substances soluble in lipids move with relative ease between the cell and the surrounding medium. Early researchers thus deduced that the cell membrane must contain lipids to allow fat-soluble substances to move across the membranes by being dissolved in it. The first major step toward our present understanding was made in the late 1930s by J. F. Danielli of Princeton University and H. Davson of University College, London. They formulated the idea that the membrane might be composed of two layers of phospholipids oriented with their polar (hydrophilic) ends exposed at the two surfaces of the membrane, and their nonpolar (hydrophobic) hydrocarbon chains buried in the interior, hidden from

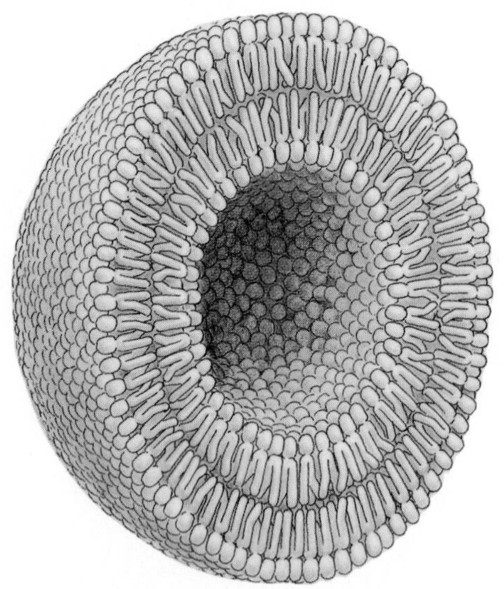

4.13 A liposome
Mixing phospholipids with water produces spherical phospholipid bilayers, each enclosing a droplet of water. The spontaneous formation of these spheres, called liposomes, is a result of the energetically favorable interaction of the hydrophilic ends of the phospholipids with the water molecules. Cell membranes are structured in exactly this way. Hence, they are basically stable, forming almost automatically and requiring no energy to maintain.

the surrounding water. A structure based on hydrophobic/hydrophilic interactions would be at once very stable and elastic. Indeed, as we now know, spheres—called ***liposomes***—composed of phospholipid bilayers will spontaneously form when phospholipids are mixed with water (Figs. 4.13, 4.14). Micrographs of cell membranes reveal a structure nearly identical to that of liposomes (Fig. 4.15), and it seems highly likely that the tendency of lipids to form bilayer cavities made possible the evolution of the first living things.

Though the phospholipid-bilayer model accounted for the stability, flexibility, and lipid-passing characteristics of the membrane with ease, it did not explain the selective permeability of the membrane to certain ions and chemicals. Davson and Danielli suggested that both sides of the membrane might be coated with protein, with charged, protein-lined pores that allowed small molecules and certain ions to pass through the membrane. This part of the model proved wrong.

THE FLUID-MOSAIC MODEL

In 1972 S. J. Singer of the University of California at San Diego and G. L. Nicolson of the Salk Institute proposed the ***fluid-mosaic model***, a hypothesis that is now almost universally accepted. The model incorporates the Davson-Danielli conception of a bilayer of phospholipids oriented with their hydrophobic tails directed toward the interior and their hydrophilic heads exposed to the aqueous environment on both surfaces. In the fluid-mosaic model, however, the arrangement of the proteins is dramatically different. Instead of coating the membrane, the various specialized proteins are inserted in the membrane to mediate a wide range of critical functions, which we will examine presently (Fig. 4.16).

Of the proteins confined to the surfaces (***peripheral proteins***), those on the inner surface usually differ markedly from those on the outer surface.

4.14 Single bilayer liposomes
Liposomes can be manufactured to deliver concentrated dosages of drugs to sites of infection, inflammation, or cancerous growth. Electron micrograph shows a freeze-fracture of artificial liposomes in a cell membrane.

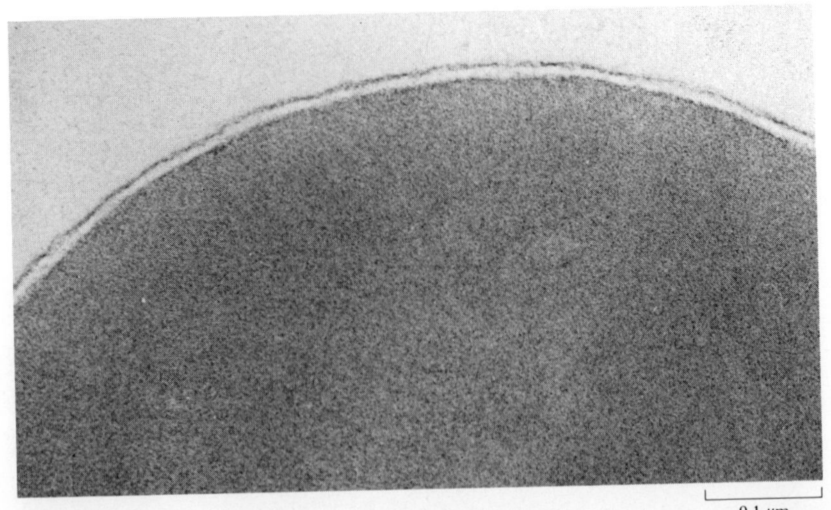

4.15 Electron micrograph showing membrane of sectioned human red blood corpuscle
The cytoplasm of the corpuscle is in the bottom half of the picture. The membrane consists of two dark lines (the phosphate-group "heads") separated by a lighter area (the hydrocarbon tails).

This asymmetry is just one of the many in membranes. For instance, phospholipids differ between the inner and outer half of the bilayer. Carbohydrate-group attachments are found only on the outside; proteins located wholly or largely within the lipid bilayer (***integral proteins***) are almost always oriented with one particular end in and the other out. Integral proteins may exhibit one of several arrangements: some are entirely buried within the bilayer, whereas others have parts that project through the surface; some are confined to the outer half of the lipid core, and others to the inner half; some extend entirely through the bilayer, projecting into the watery medium on both sides. As would be expected, hydrophilic amino acids (those with polar or electrically charged R groups) predominate in the portions of the protein molecules that project out of the lipid bilayer into the water, whereas hydrophobic (nonpolar) amino acids are abundant in the

4.16 The fluid-mosaic model of the cell membrane
A double layer of lipids forms the main continuous part of the membrane; the lipids are mostly phospholipids, but in plasma membranes of higher organisms cholesterol (brown) is also present. Proteins occur in various arrangements. Some, called peripheral proteins, are entirely on the surface of the membrane, to which they are anchored by a covalent bond with a membrane lipid. Others, called integral proteins, are wholly or partly embedded in the lipid layers; some of these may penetrate all the way through the membrane. Three protein units are joined by covalent bonds (not shown) to form part of a single protein molecule bounding a membrane-spanning pore. Proteins make up about half of the weight of membranes. The hexagons represent carbohydrate groups.

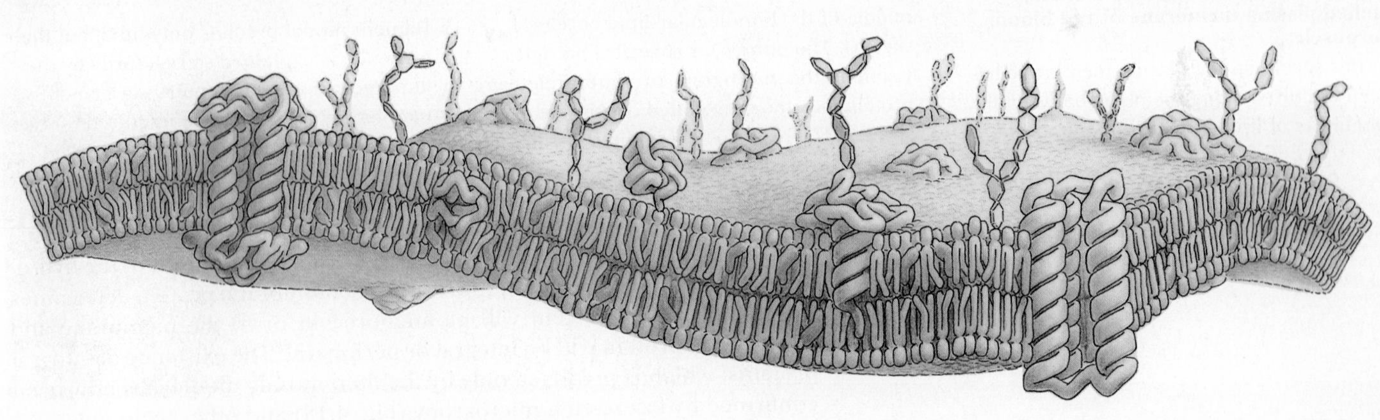

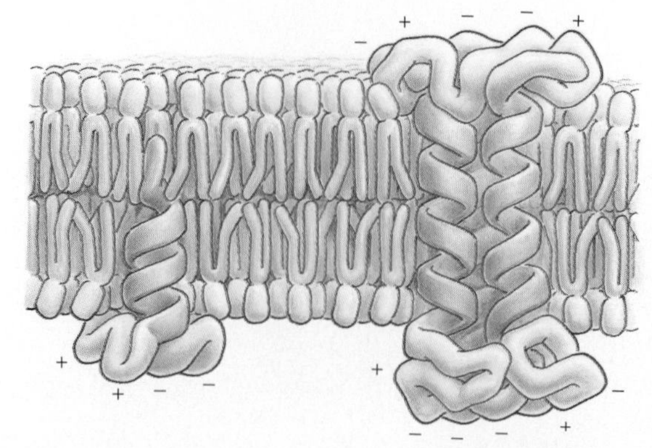

4.17 Orientation of proteins within membranes
The parts of the polypeptide chain containing most of the hydrophilic amino acids (polar or charged R groups; blue) tend to project into the watery medium outside the lipid layers, whereas the parts of the chain with hydrophobic amino acids (brown) tend to be folded into the inner, lipid portion of the membrane. The diameter of the protein strands has been reduced for clarity.

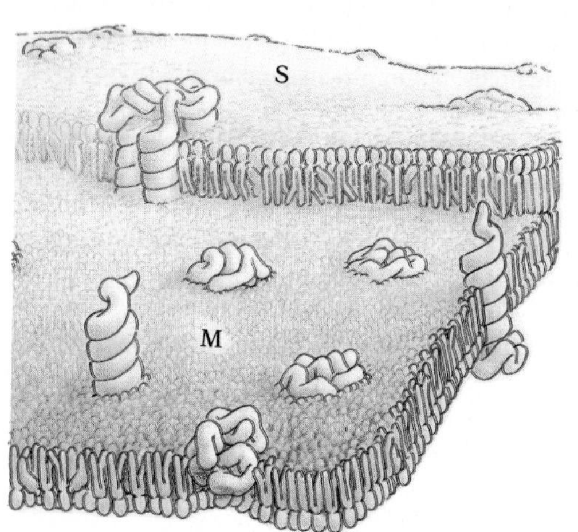

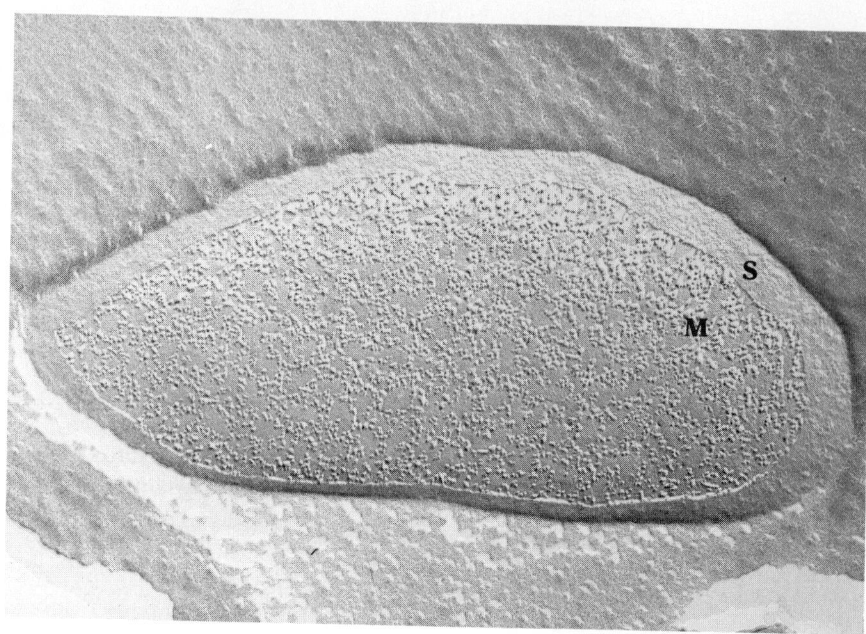

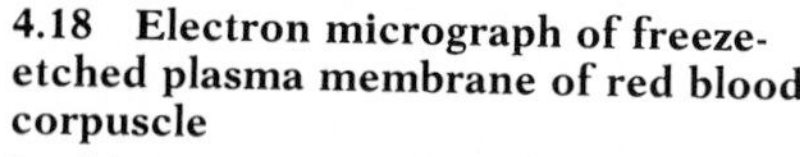

4.18 Electron micrograph of freeze-etched plasma membrane of red blood corpuscle
In this specimen the plasma membrane has been fractured along the plane between the two layers of lipids—that is, along the middle of the bimolecular lipid core (see sketch). The numerous spherical particles visible in the micrograph are interpreted as protein (see the gray coiled entities in sketch). They appear where the Davson-Danielli model predicts only lipid, but their presence is explained satisfactorily by the fluid-mosaic model. S: outer surface of membrane; M: interior of fractured membrane.

portions buried in the bilayer (Fig. 4.17). Indeed, the location of the hydrophilic and the hydrophobic amino acids in a membrane protein determines which part of the protein will be anchored in or to the membrane and whether the protein will be integral or peripheral. The existence of integral proteins, which is predicted only by the fluid-mosaic model, has now been confirmed by freeze-etch microscopy (Fig. 4.18) and other techniques.

Exploring Further

FREEZE-FRACTURE AND FREEZE-ETCHING

Freeze-fracture-etching, a technique for preparing specimens for electron microscopy, has become an indispensable tool for providing detailed confirmation of membrane structure. The specimen is first rapidly frozen and then fractured along the plane of its bilipid membrane (A–B). Some of the ice is then removed from the specimen by sublimation (conversion directly to vapor), which exposes the inside surface of the membrane and gives the specimen an etched appearance (C). Carbon and a metal, usually platinum, are then applied at an angle to the specimen (D) so as to shadow any irregularities in the membrane. Next the original specimen is removed from the platinum cast or surface replica thus formed (E). The replica can now be examined by microscopy. EMs of freeze-etched cell membranes or other structures in cells usually have a three-dimensional appearance (Fig. 4.18).

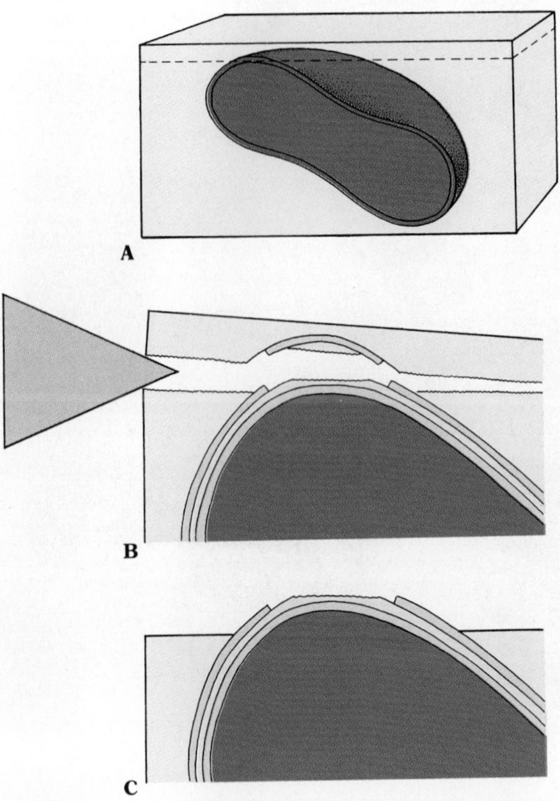

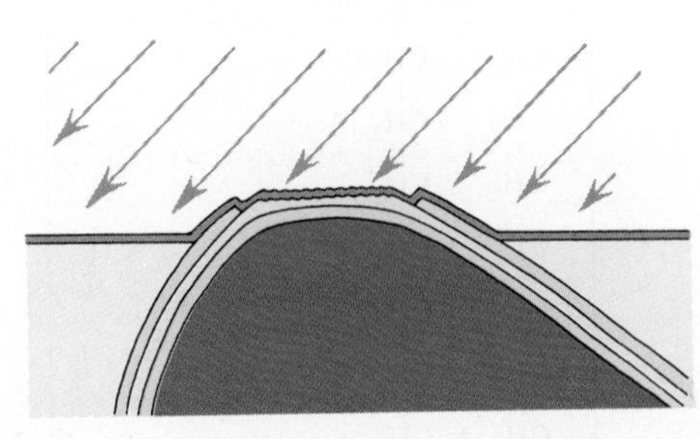

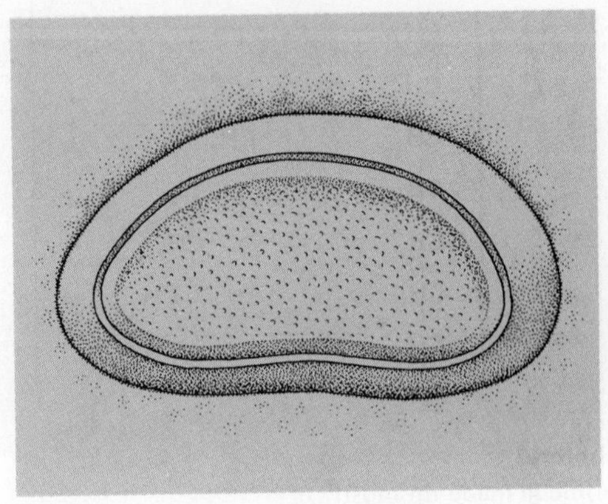

According to the fluid-mosaic model, the structure of the membrane is not static. The individual lipid molecules can move laterally, in the plane of the membrane, so that a particular molecule found in one position at a given moment may be in an entirely different position only seconds later. Mobility of the lipids is greatest in membranes that are high in unsaturated

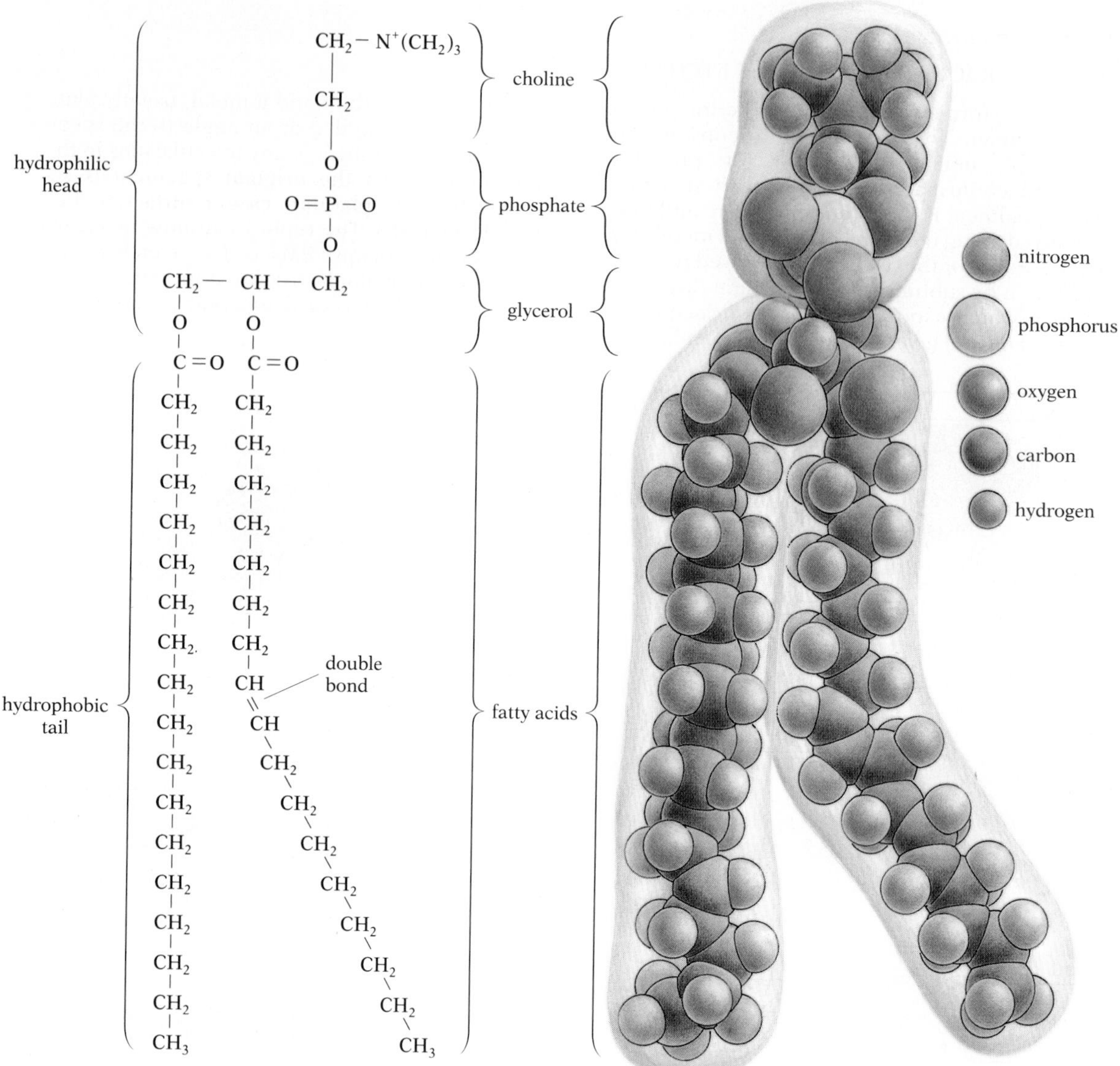

4.19 A phospholipid
The cell membrane is made up mostly of phospholipids. Phosphatidylcholine, a common membrane phospholipid found almost exclusively in the outer half of bilayer membranes, consists of a polar head (a positively charged choline, a negatively charged phosphate, and an uncharged glycerol) joined to two hydrophobic fatty acid chains. The "kink" in the right tail is created by a double carbon bond. Because this tail is unsaturated—that is, not every carbon has its full complement of hydrogen atoms—the phospholipid will be less tightly packed in the membrane, and hence will be more mobile.

phospholipids (Fig. 4.19) and contain no cholesterol. Speeds of 2μm (micrometers) per second are possible in such membranes—an astonishing mobility when we consider that many organisms (the bacterium *Escherichia coli*, for example) are only about 2μm long. When cholesterol is present, it binds weakly to adjacent phospholipids and can have one of two effects. If the membrane phospholipids are mostly saturated, cholesterol prevents them from packing so closely and regularly that they literally crystallize into a stiff and solid layer. When, on the other hand, the phospholipids are mostly unsaturated (so that the "kinks" in their hydrophobic tails keep them loosely packed), cholesterol can fill these gaps, bind to adjacent phospholipids, and thus join them together (Fig. 4.20). Plant membranes lack cholesterol, and instead depend on the cell wall for stability.

The cholesterol concentration and the degree of saturation vary enormously between species, and even among tissues in the same organism, depending on the amount of flexibility needed. Though medical opinion is far from unanimous, there is great concern that too much cholesterol and saturated fat in the human diet can lead to hardening of cell membranes, particularly those lining the arteries. Hardening of the arteries (atherosclerosis) is a major cause of stroke and heart disease.

Proteins in the cell membrane can move laterally to some extent, but much less than lipids. Complete freedom of movement would be incompatible with the specific functional demands placed on the membrane proteins. Certain proteins in the membrane of nerve cells, for example, are essential to the transmission of nerve impulses from one cell to another; they are found only at points where one nerve cell is close to another—not in other positions, where they could not fulfill their function. Similarly, proteins responsible for pumping sodium ions out of the cells that line the intestine are located in the membrane on only one side of the cells (the side away from the intestinal cavity). In short, at least some of the membrane proteins are anchored in place, thereby limiting the fluidity of the membrane. In some cases the anchoring probably results when tight associations between two or more intrinsic proteins give rise to structural and functional complexes too large to move easily. In other cases peripheral and integral proteins may be bound weakly to each other. Even the lipid molecules may not all be free to move; there is evidence that the lipids immediately adjacent to intrinsic proteins may be loosely bound to the proteins and thus immobilized.

In the fluid-mosaic model, the pores in the membrane are depicted as channels through one or a group of protein molecules (Fig. 4.16). The ability of unanchored proteins to drift laterally in the lipid bilayer explains the observed mobility of many membrane pores. The distinctive properties of the various R groups of the amino acids in the proteins give the pores some selectivity; not all ions or molecules small enough to fit in the pores can actually move through them.

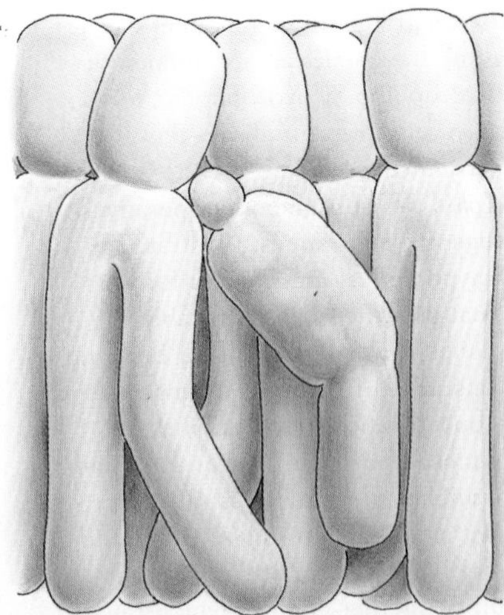

4.20 Cholesterol in the membrane
Cholesterol (brown) binds weakly but effectively to two adjacent phospholipids, thereby partially immobilizing them. The result is a less fluid and mechanically stronger membrane. The amount of cholesterol varies widely according to cell type, with the membranes of some cells possessing nearly as many cholesterol molecules as phospholipids while others lack cholesterol entirely. For the structural formula of cholesterol, see Figure 3.17, p. 60.

MEMBRANE CHANNELS AND PUMPS

The lipid bilayer that makes the cell membrane forms spontaneously from phospholipids manufactured by the cell, creating a flexible but effective barrier between the inside of the cell and the world outside. Moreover, the bilayer provides an anchoring plane for a variety of membrane proteins. The largely lipid composition of the membrane explains why small lipid-soluble molecules can diffuse into and out of cells, but the membrane's permeability to certain chemicals that do *not* dissolve in lipids must depend on the proteins in the bilayer. The ability of cells to transport specific substances actively against their osmotic-concentration gradients must also be accounted for by the properties of the membrane proteins.

It is easy to demonstrate that the membrane is highly selective. If a molecule that can readily enter a cell is slightly altered, but not in such a way as to change its size or electrical "personality," it often loses its capacity to

4.21 Models of membrane transport Many strategies for moving substances across the cell membrane are known. (A) In facilitated diffusion of the simplest kind, a protein channel, or pore, embedded in the membrane provides a direct path for the chemical it passes down its osmotic-concentration gradient. The channel's diameter and the chemical environment it creates (hydrophilic or hydrophobic, for example) serve to prevent all but the correct substance from passing. (B) Other channels pass two substances cooperatively, or exchange two substances. Illustrated here is a hypothetical mechanism for the channel that uses the highly favorable osmotic-concentration gradient for bringing sodium ions into the cell to overcome the osmotic-concentration gradient working against glucose (labeled G). When Na^+ binds to the channel, it may induce an allosteric change that enables glucose also to bind to the channel. This binding of glucose may then result in another change, which causes the channel to close to the outside and open to the inside. This change, which in turn may cause the channel to lose its affinity for glucose, releases the glucose as well as the Na^+ into the interior. Having lost the Na^+ and glucose, the channel can reopen to the outside. (C) An allosteric interaction between a signal molecule (red) and the gated channel causes the gate to open, so that diffusion can take place down a favorable concentration gradient. Other molecular systems, not shown, then inactivate the signal molecule so that the channel can close again. (D) A mobile carrier would not provide transmembrane channels, but would itself migrate back and forth from one surface to the other. The existence of mobile carriers is controversial.

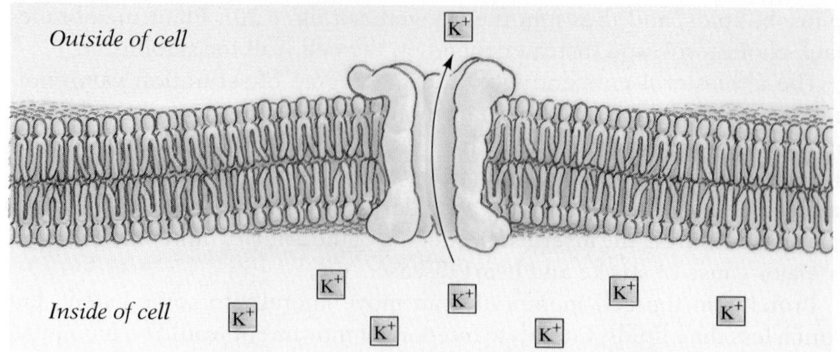

A FACILITATED-DIFFUSION CHANNEL

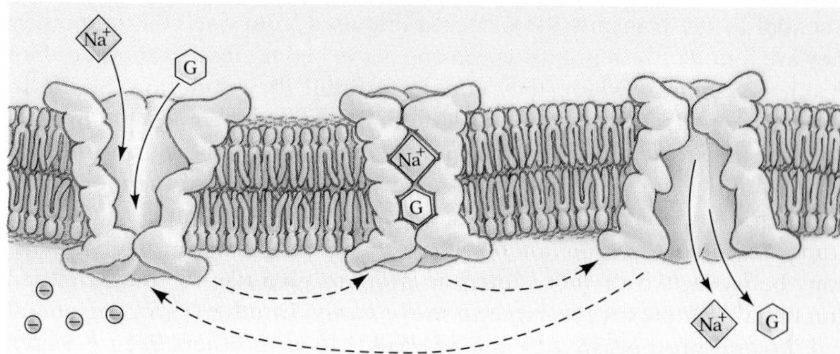

B COOPERATIVE ION CHANNEL (SYMPORT)

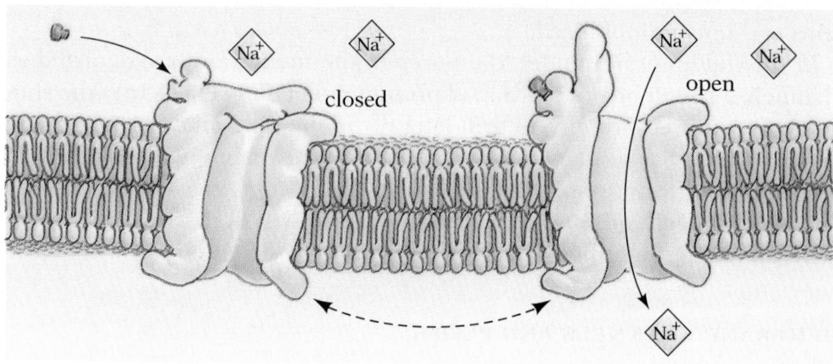

C GATED CHANNEL

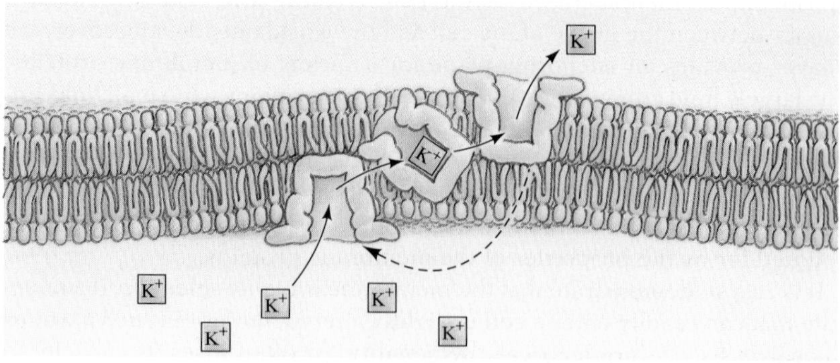

D MOBILE CARRIER

move through the membrane. This selectivity on the part of the membrane suggests that the transport agents, or "carriers," are enzymelike proteins—a hypothesis supported by a variety of experiments. For example, the movement of some substances through the membrane can be competitively inhibited by other substances. Inhibition would not occur if both substances were moving by simple diffusion; apparently the two substances compete for access to specific binding sites on enzymelike carrier molecules in the membrane.

We now know that the transport agents that control molecular traffic in and out of the cell are highly specialized channels and pumps. Each depends on membrane proteins for its operation; in fact, each is usually fabricated from several "cooperating" membrane proteins. Since they enable specific chemicals to permeate the membrane, channels and pumps are known collectively as ***permeases***.

Membrane channels The simplest of the permeases, the ***membrane channels***, provide openings through which specific substances can diffuse across the membrane. These channels, though selective, are passive: they simply permit particular chemicals to move down their concentration gradients. This strategy, known as ***facilitated diffusion***, is the basis of the membrane's highly specific permeability. The protein channel for potassium ions (Fig. 4.21) provides a good example of this strategy. Cellular processes result in the accumulation of K^+ inside most cells. As a charged particle, K^+ is insoluble in the lipid membrane, but the channel specific for K^+ allows it to leak out slowly at a controlled rate. Without such leakage the internal K^+ concentration would become too high for the cell to function properly. The specificity of the K^+ channel is a result of both its internal shape and its charge, but no one really understands in detail just what goes on inside these most simple of all membrane pores.

More complex channels, though passive, frequently move two specific substances in concert. For example, ion-exchange channels, often called ***antiports***, work by trading two similarly charged ions. As one moves into the cell, the other exits, thereby maintaining an electrical balance. The many integral membrane proteins in Fig. 4.18 are channels that exchange Cl^- for HCO_3^- (dissolved carbon dioxide), as part of the red blood corpuscle's job of carrying waste CO_2 from cells to the lungs.

Other cooperative channels move two substances in the same direction; they are often called ***symports***. Coordinated movement of this sort ("cotransport") is important in the transport through the membrane of glucose, the most important energy source for most cells. Sodium ions are 11 times more concentrated outside the cell, and are therefore subject to a highly favorable osmotic gradient to the inside. Yet they must be accompanied by glucose to pass through the appropriate channel (Fig. 4.21B). The channel will not transmit either substance alone. It is as though both must bind to the outside of the channel before this special membrane pore will open. Thus the free energy of the osmotic-concentration gradient of Na^+ can be exploited to overcome the smaller unfavorable concentration gradient of glucose. In thermodynamic terms, the two diffusion "reactions" are linked: moving Na^+ "downhill" releases more free energy than is utilized in moving glucose "uphill," so the cooperative diffusion proceeds.

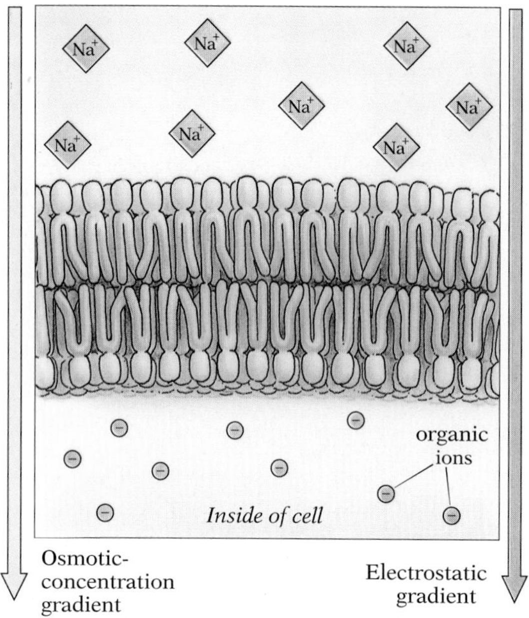

4.22 An electrochemical gradient
Cells have an electrical potential of about 70 mv negative with respect to the fluids that surround them. This gradient arises mainly because large numbers of negatively charged organic ions are trapped inside the cells, while a relatively high concentration of positively charged Na^+ exists outside. Sodium ions are therefore subject to both a strong osmotic-concentration gradient and a sizable electrostatic gradient. The effects of these two gradients combine to create an electrochemical gradient.

This description in terms of osmotic-concentration gradients accounts for the coordinated movement of Na^+ and glucose through a common channel, but the rate of movement is too great to be explained solely by the concentration difference across the membrane. In fact, there is a second important gradient that contributes to the diffusion of ions. As you know, oppositely charged ions attract one another electrostatically, while ions with the same charge repel one another. As a result, if a cell has, say, more negatively charged than positively charged ions, positive ions will be attracted to it from the surrounding fluid. (Most cells actually have a net negative charge of about 70 millivolts relative to the fluids surrounding them.) The difference in charge across the membrane generates an ***electrostatic gradient***, and when appropriate channels are open, positive ions tend to flow into the cell, while negative ions tend to flow out. For ions like Na^+, which are far more concentrated outside, the osmotic and electrostatic potentials combine to create a strong ***electrochemical gradient*** (Fig. 4.22). It is the free energy of the combined potentials that accounts for the particular effectiveness of Na^+ in moving glucose into cells. The mechanism by which the electrochemical gradient is maintained is the sodium-potassium pump, discussed below.

Another strategy for controlling movement across the membrane utilizes a gate across a membrane channel. This strategy is widely used to convert a molecular signal specialized for carrying information between cells into a second signal, more suitable for communicating inside the cell. When a molecular signal—a hormone or one of the transmitter substances that carry messages from one nerve to another—binds to an exposed part of a transmembrane protein, the ***receptor***, an allosteric change in conformation takes place. The change allows the gate to open, and the second signal, usually an ion like Na^+ or Ca^{++}, can then move across, carrying the message into the cell (Fig. 4.21C). This ***gated channel*** strategy underlies the transmission of many chemical messages in both plants and animals, and of the nerve impulses by which animals sense the outside world, move their muscles, and perhaps even think.

Another way for molecules to get across the membrane would be for the permease to act as a ***mobile carrier***, taking them through one by one (Fig. 4.21D). No definite example of such a permease is yet known, but valinomycin acts in just the way we would expect a mobile carrier to act. Valinomycin is a ring-shaped polymer with a hydrophobic exterior and a polar interior (Fig. 4.23). The polar pocket, lined with six oxygen atoms, can hold a single potassium ion. Apparently the complex bobs back and forth randomly, carrying K^+ ions in both directions. The net transfer of K^+ is a statistical consequence of the electrochemical gradient: valinomycin more often picks up K^+ inside the cell and releases it outside simply because there is more K^+ inside than out. Valinomycin is not a normal membrane protein—in fact, technically it is not a protein at all, because one of its peptide subunits is not a normal amino acid. It is an antibiotic produced by certain bacteria to poison competing microorganisms by altering the selective permeability of their membranes. There is some evidence for mobile carriers in normal membranes, but their existence is still an open question.

Membrane pumps Other permeases, known as ***pumps***, do not depend on free-energy gradients. Instead, pumps use the cell's store of energy to move

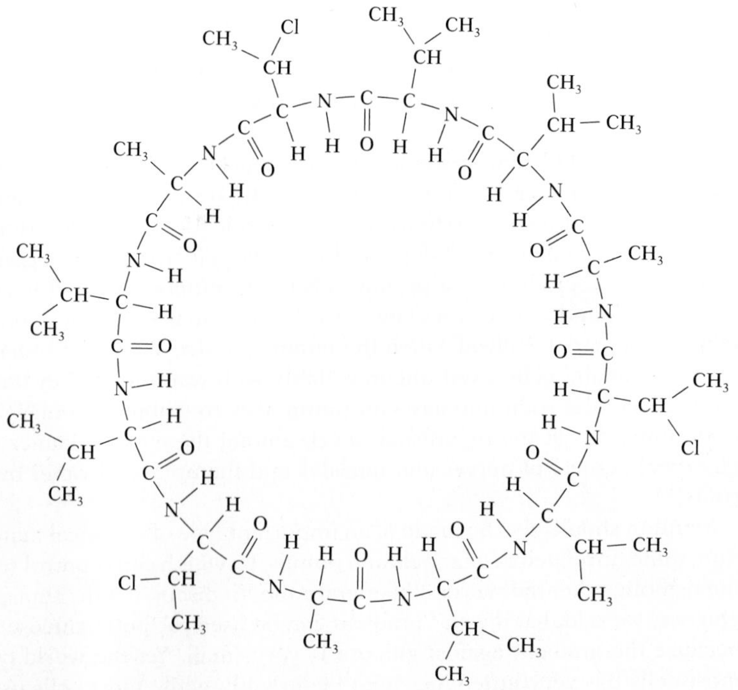

4.23 Valinomycin
This permease is synthesized by some bacteria, and kills cells that take it up. It consists of three kinds of peptides: alanine, valine, and a chlorine-substituted valine. The potassium ion is held to the center by the oxygens while it is carried across the host membrane.

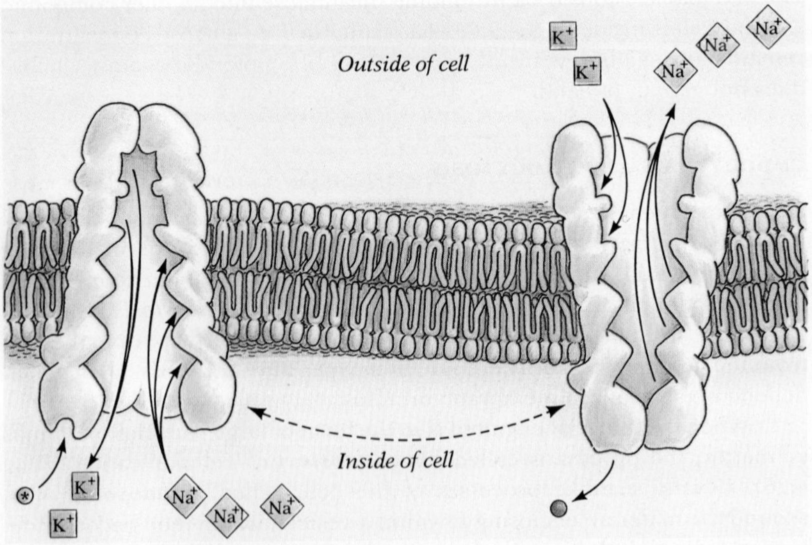

Outside of cell

Inside of cell

4.24 The sodium-potassium pump
In contrast to other methods of membrane transport, the pump strategy uses energy from the cell (rather than the free energy of a concentration gradient) and provides active transport of a substance against its gradient. In this case, three sodium ions are exchanged for two potassium ions; both kinds of ions are already more concentrated on the side to which they are being moved. In the model shown here, the release of K^+ ions brought in during the previous cycle is followed by the binding of three Na^+ ions and an energy source, ATP (circle), on the inside. The resulting conformational changes in the protein open it to the outside, reduce its affinity for Na^+, which is then released, and increase its affinity for K^+ ions. The binding of K^+ then causes the channel to open to the inside, increase its affinity for Na^+, and decrease its affinity for K^+, and the cycle begins again. The net ionic effect is to pump positive charges out of the cell, and the inside of the cell becomes negatively charged with respect to the outside. The electrical and osmotic potential created by the sodium-potassium pump ultimately makes possible the cooperative transport of glucose illustrated in Figure 4.21B.

substances against their gradients. The process, known as ***active transport***, is important in ridding the cell of accumulated substances that are insoluble in the membrane, and of molecules that are simply too big to escape. Pumps also transport many essential building blocks into the cell. The best-understood example of a membrane pump, however, is the complex responsible for maintaining the electrochemical gradient across the membrane: the ***sodium-potassium pump*** (Fig. 4.24). The free-energy source for this pump, and for many cellular processes, is the cellular energy carrier ATP (adenosine triphosphate), which we will discuss in detail in later chapters. The pump utilizes this energy to exchange potassium and sodium ions across the membrane, thereby maintaining the electrochemical gradient.

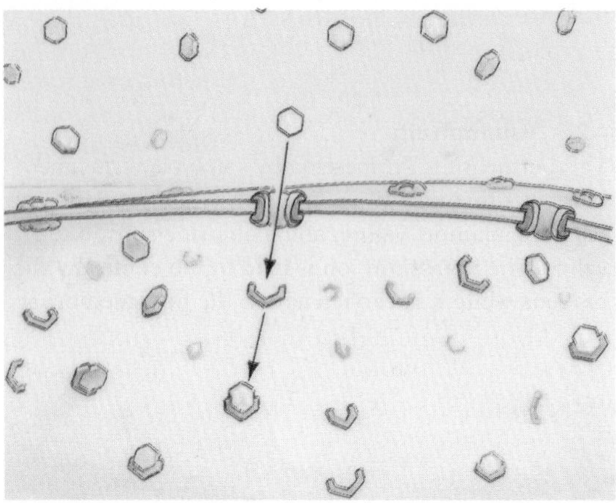

4.25 Formation of a complex inside the cell As fast as glucose molecules (yellow hexagons) enter the cell, they combine with molecules of X (brown) already in the cell to form a new substance. The concentration of free glucose inside the cell therefore remains low, and glucose continues to diffuse inward.

The effects of this pump are far-reaching: it is responsible for the electrical activity of nerves and muscles; it indirectly supplies the free energy for many osmotic transport systems, such as the one for glucose described earlier (which, you may recall, depends on the Na^+ electrochemical gradient); it supplies the gradient that permits a Na^+–H^+ antiport channel to control cellular pH; and it helps regulate the volume of many cells by controlling osmotic potential. Indeed, when the pump is destroyed by the Indian dart poison ouabain, cells swell uncontrollably with water until they burst. As we will see, the sodium-potassium pump also contributes to many metabolic functions at the organismal level, among them maintenance of the electrical activity of nerves and muscles and the uptake of water by plant roots.

Mention should also be made of an important type of chemical manipulation, quite unrelated to channels and pumps, by which cells control some of the osmotic potentials across the membrane. In discussing the transport of glucose, we said that the Na^+ gradient can be used to "pull" glucose inside because the gradient against glucose is very small. Yet the world outside most cells has very little glucose to begin with, while most cells use it in large amounts. How can the concentration inside the cell be kept so low? The cell manages this trick by binding the glucose into another compound as soon as it is inside (Fig. 4.25). As a result, the free-glucose concentration remains artificially low inside, and the osmotic potential against glucose does not get out of hand.

ENDOCYTOSIS AND EXOCYTOSIS

As we have seen, permeases are the means by which substances enter and leave cells through the cell membrane. But cells have ways of admitting substances, usually in larger quantities, without having them actually pass through the membrane. By an active process called ***endocytosis***, a cell encloses the substance in a membrane-bounded vesicle that is pinched off from the cell membrane. There are three types of endocytosis, all of which depend on specialized membrane proteins:

1. When the material engulfed is in the form of large particles or chunks of matter, the process is called ***phagocytosis***, or "cellular eating" (Fig. 4.26). Usually, armlike processes of the cell, called ***pseudopodia***, flow around the material, enclosing it within a vesicle, which then becomes detached from the plasma membrane and migrates into the interior of the cell. Phagocytosis is triggered only when special membrane-mounted receptor proteins bind to a suitable target; this binding is analogous to enzyme-substrate binding. In vertebrates, phagocytosis is generally restricted to scavenger cells in the blood, which ingest large debris and invading microorganisms (Fig. 4.26B).

2. When the engulfed material is liquid the process is termed ***pinocytosis***, or "cellular drinking." Pinocytosis is used to ingest extracellular fluid, or to transport it across a cellular barrier (Fig. 4.27).

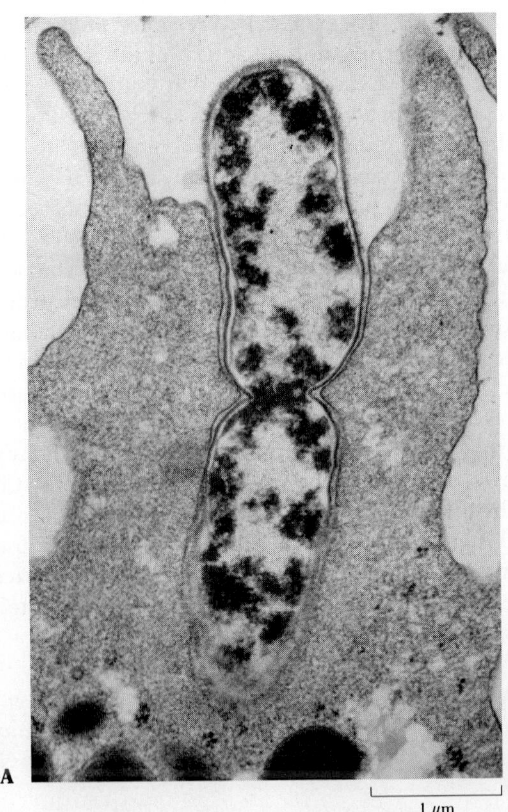

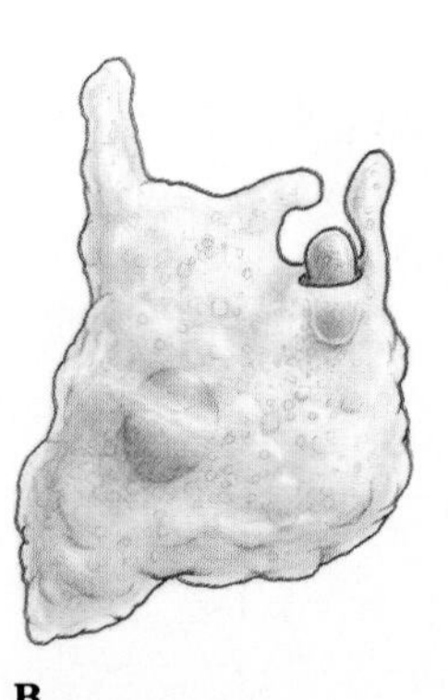

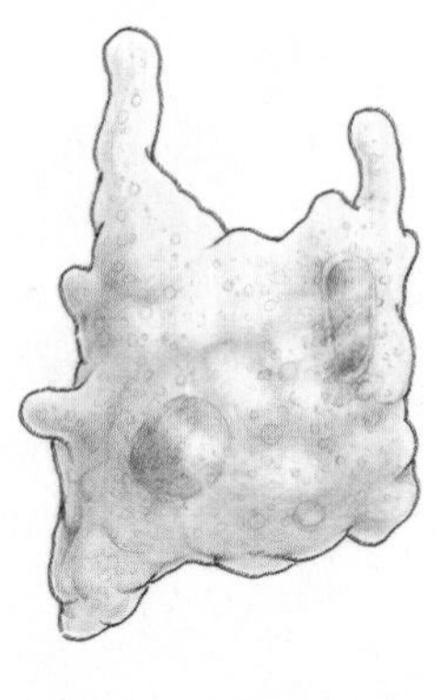

4.26 Phagocytosis
(A) White blood cells, or leukocytes, in the bloodstream use phagocytosis to capture foreign organisms; here the leukocyte is engulfing a dividing bacterium. (B) In *Amoeba* pseudopodia flow around the prey until it is entirely enclosed within a vacuole.

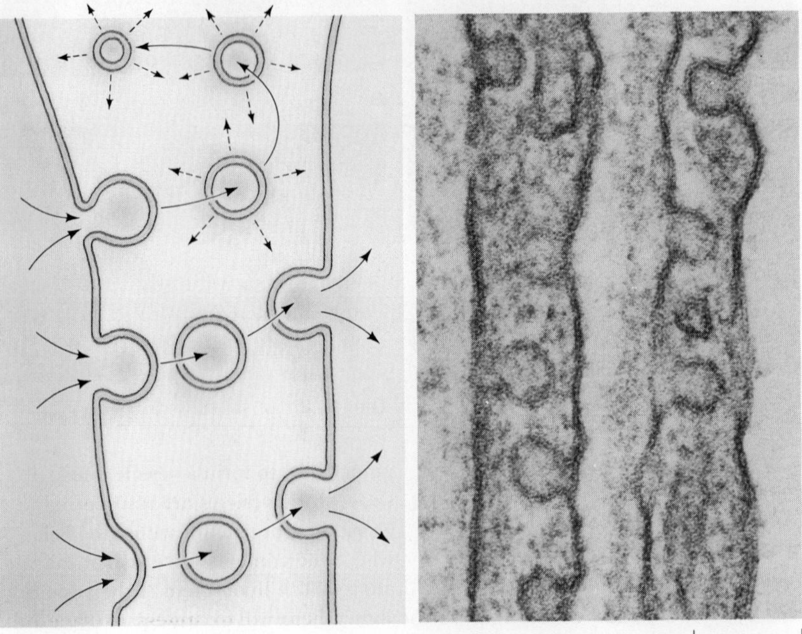

4.27 Pinocytosis
(A) Extracellular fluids are enclosed in vesicles on the cell surface and endocytosed. The vesicles may be used to transport the fluid across the cell, leading to exocytosis on the other side of the cell, or the fluid may be allowed to escape into the cell itself. (B) Complete transcellular movement can be seen in this electron micrograph of the cells lining a blood capillary (left) and a portion of the lung (right). The membrane facing the lung must be kept constantly moist for gases (O_2, CO_2) to dissolve and cross; since the water used constantly evaporates and is exhaled, pinocytotic vesicles are always in use moving water from the blood to the inner surface of the lung, across a basement membrane.

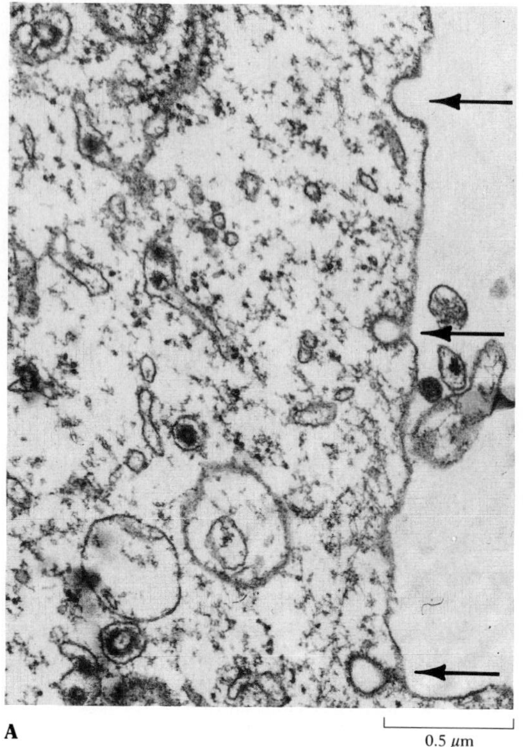

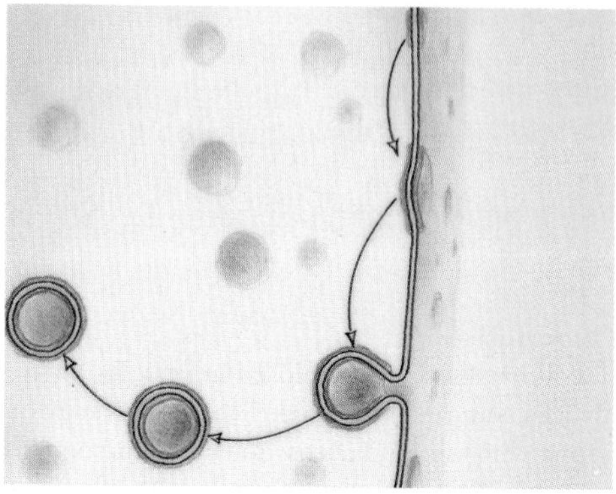

4.28 Receptor-mediated endocytosis
(A) Three stages of receptor-mediated endocytosis of a short polypeptide used in cell–cell communication are seen taking place in a cultured nerve cell. (B) Molecules bind to receptors on the cell surface, triggering vesicle formation and endocytosis.

3. When the material to be ingested is adsorbed on the cell membrane at selective binding sites, the process is called ***receptor-mediated endocytosis***. The loaded vesicles are formed and detach from the membrane at the cell surface (Fig. 4.28). In most cases, the site that collected a particular substance before trapping it in a vesicle appears as a "coated pit," with receptor molecules clustering in one spot in the membrane (Fig. 4.29).

These outward-facing receptors (and the substances to which they bind) produce what looks like a haze on the extracellular surface of the membrane (Fig. 4.29). A similar darkening appears on the intracellular surface of the membrane at the spot destined to form a vesicle. The best-characterized component of this internal patch is the structural protein ***clathrin***, which begins assembling itself into the foundation for a membrane indentation, or pit, and then into the vesicle's coating. The clathrin scaffolding is clearly visible surrounding newly formed vesicles (Fig. 4.30), but is quickly disassembled and the molecules returned to the membrane.

One example that illustrates the importance and specificity of receptor-mediated endocytosis involves cholesterol uptake by cells. Cholesterol is transported in the blood to cells by a carrier complex of low-density lipoprotein (LDL). When the cell needs cholesterol—usually for use in the manufacture of new membrane—LDL receptors are synthesized and incorporated into the cell membrane (Fig. 4.31). The LDL receptors aggregate

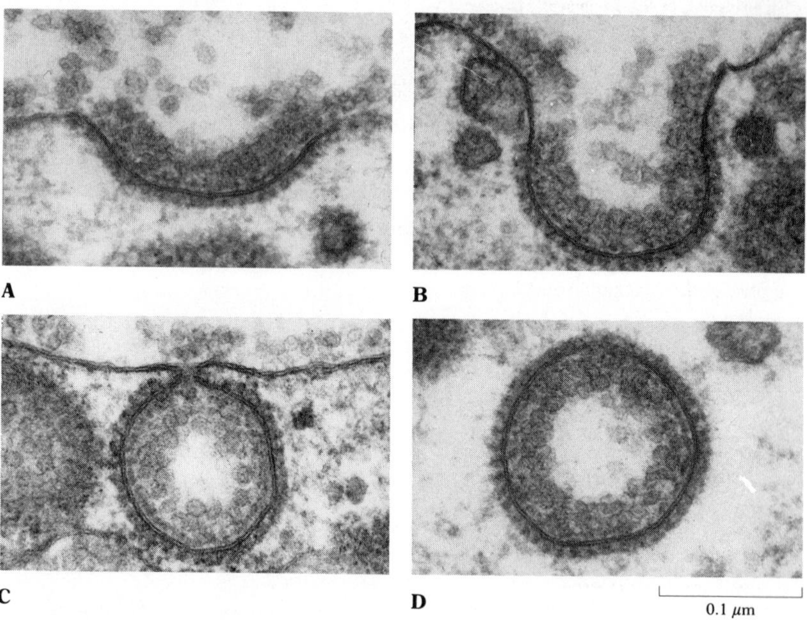

4.29 Endocytosis by means of a coated pit
(A) Specialized receptors for lipoproteins have aggregated to form a coated pit in the membrane of an egg cell. (B–D) The pit is subsequently pinched off to form a vesicle. Most endocytotic vesicles are transported to lysosomes, organelles within the cell where the contents are enzymatically altered. The lipoprotein in the vesicle shown here will become part of the yolk.

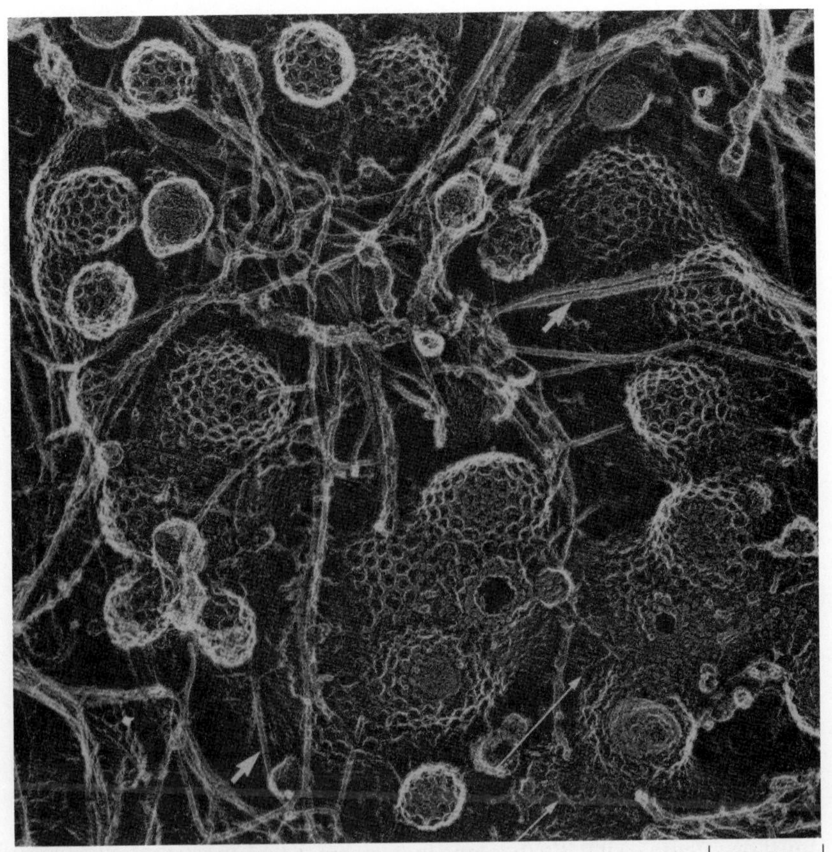

4.30 Clathrin-coated pits
The pits, seen from the inside of a liver cell, are in the process of budding off from the membrane to form intracellular vesicles. The cablelike structures are part of the cytoskeleton, to be discussed in the next chapter.

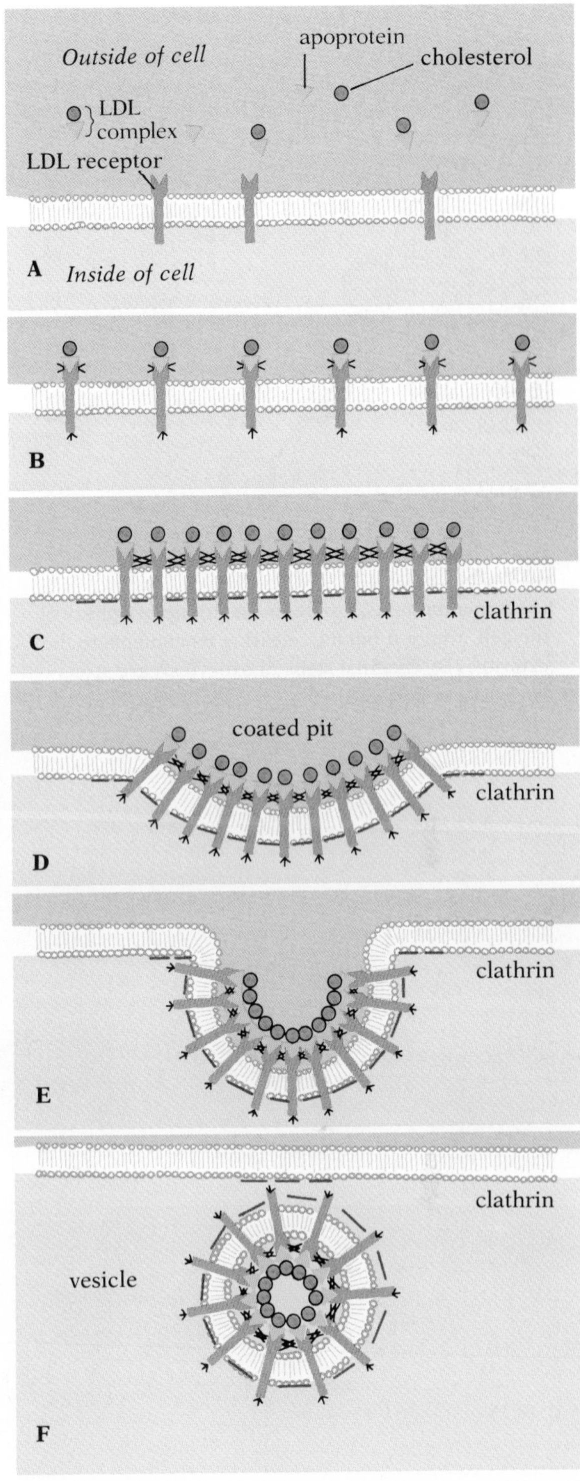

4.31 Endocytosis of cholesterol
When a cell needs cholesterol, it synthesizes receptors for low-density lipoprotein (LDL) and incorporates them into the cell membrane, where they are free to migrate (A). The receptors soon bind LDL, a carrier complex that transports cholesterol in the blood (B). The LDL complex includes some 2000 cholesterol molecules and an associated protein (called apoprotein) that binds to the LDL receptor. Having bound the LDL, the receptors stop drifting in the membrane and stick to one another over a clathrin-rich patch of membrane (C). Even in the absence of bound LDL, many of the receptors eventually aggregate spontaneously. Aggregations of LDL receptors trigger endocytosis, the first step of which is the formation of a coated pit (D–E). The resulting vesicle (F) is transported to the site of membrane synthesis. This method of cholesterol uptake may demonstrate the usual strategy by which cells obtain nutrients that cannot pass through the membrane directly.

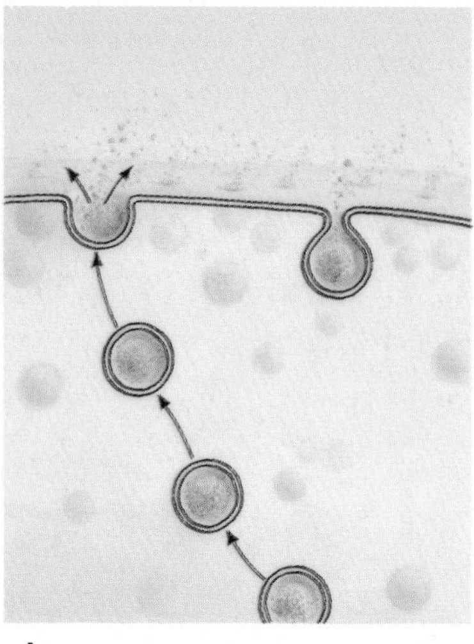

A

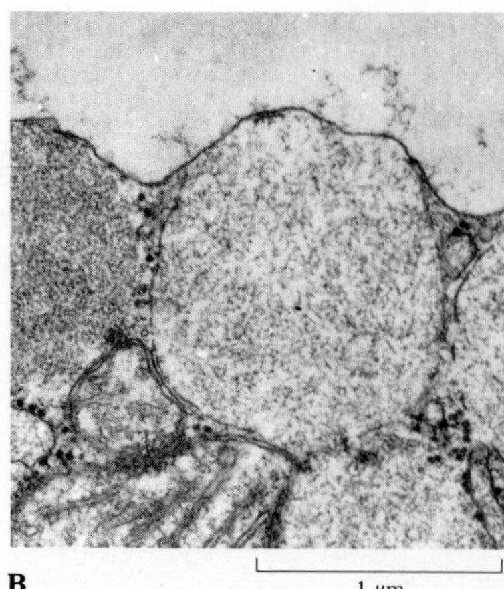

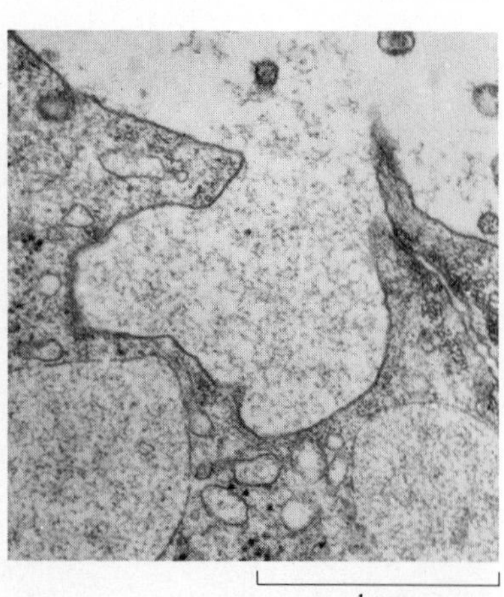

B

4.32 Exocytosis
(A) A membranous vesicle moves to the periphery of the cell, where it bursts, releasing its contents to the exterior. (B) The final steps of exocytosis are seen here, as a vesicle containing tear fluid fuses with the plasma membrane and bursts.

spontaneously over membrane patches rich in clathrin, particularly once they have bound LDL. The cholesterol is then transported in endocytotic vesicles, known as ***endosomes***, for use in the cell's vast complex of membranes. One cause of atherosclerosis involves failure of the receptors to bind LDL; another involves failure of the receptors, having bound LDL, to aggregate and to initiate endocytosis. Either defect can allow cholesterol-rich plaques to form on artery walls.

Material enclosed within endocytotic vesicles has not yet entered the cell in the fullest sense. It is still separated from the cellular substance by a membrane, and it must eventually cross that membrane (or the membrane must disintegrate) if it is to become incorporated into the cell. Normally the vesicle membrane forms from cell membrane, and the vesicle is transported to an endosome. Meanwhile, vesicles full of digestive enzymes form within the cell. These ***lysosomes*** fuse with the endosomes to create miniature cellular stomachs, which break down the endosomal contents. After digestion, many of the products can cross the lysosome membrane into the cytoplasm, while others remain trapped inside. As a result, the cell is able to ingest the substances it requires, while the unwanted parts of the miniature meal, and the lysosome's destructive enzymes, remain segregated from its delicate chemical interior. Endosomes and lysosomes will be described more fully in the next chapter.

In a process essentially the reverse of endocytosis, called ***exocytosis***, materials contained in membranous vesicles are conveyed to the periphery of the cell, where the vesicular membrane fuses with the cell membrane and then bursts, releasing the materials to the surrounding medium (Fig. 4.32). Many glandular secretions are released from cells in this way; the hormone insulin, for example, is released by exocytosis from the pancreatic cells that synthesize it. Exocytosis also functions in the release of waste products from the cell. The undigested remains of materials brought in by endocytotic vesicles, for example, are normally disposed of through exocytosis. And in some cases a coupling of endocytosis and exocytosis moves a substance entirely across a cellular barrier, such as the wall of a blood vessel; the substance is picked up by endocytosis on one side of the cell, and the vesicle then moves through the cell to the other side, where the substance is released by exocytosis (see Fig. 30.14, p. 857).

CELL WALLS AND COATS

For as long as biologists have been examining cells under the microscope, they have been aware that plant cells are encased in conspicuous cell walls. These walls, which are located outside the plasma membrane, are composed primarily of carbohydrates. Biologists have also long known that the cells of fungi and most bacteria have strong, thick walls rich in carbohydrates. But only in recent years have they come to realize that most animal cells, too, have carbohydrates on the outer surface of their membranes. The carbohydrates do not form a wall, but are attached as independent side groups to some lipids and membrane proteins. Though not attached to one another, these carbohydrate groups are usually described as a cell "coat," and the coat plays an important role in determining certain properties of the cells. The presence of carbohydrate materials on their outer surfaces, then, appears to be a general property of cells. Nonetheless, the conspicuous, thick, relatively rigid walls of plant, fungal, and bacterial cells, on the one hand, and the inconspicuous, thin, nonrigid coats of animal cells, on the other, remain among the most striking differences between these groups.

Cell walls of plants, fungi, and bacteria Located outside the cell membrane, the plant cell wall is generally not considered part of the cytoplasm, though it is a product of the cell. The principal structural component of the cell wall of plants is the complex polysaccharide ***cellulose***, which is generally present in the form of long threadlike structures called fibrils. The cellulose fibrils are cemented together by a matrix of other carbohydrate derivatives, including pectin and hemicellulose (a material not structurally related to cellulose). The spaces between the fibrils are not entirely filled with matrix, however, and they generally allow water, air, and dissolved materials to pass freely through the cell wall. The wall does not usually determine which materials can enter the cell and which cannot. This function is mostly reserved for the membrane located below the cell wall.

The first portion of the cell wall laid down by a young growing cell is the ***primary wall***. As long as the cell continues to grow, this somewhat elastic wall is the only one formed. Where the walls of two cells abut, a layer between them, known as the ***middle lamella***, binds them together. ***Pectin***, a complex polysaccharide generally present in the form of calcium pectate, is one of the principal constituents of the middle lamella. If the pectin is dissolved away, the cells become less tightly bound to one another. That is what happens, for example, when fruits ripen. The calcium pectate is partly converted into other more soluble forms, the cells become looser, and the fruit becomes softer. Many of the bacteria and fungi that produce soft rots of the tissues of higher plants do so by first dissolving the pectin, reducing the tissue to a soft pulp, which they can absorb.

Cells of the soft tissues of the plant have only primary walls and intercellular middle lamellae. After ceasing to grow, the cells that eventually form the harder, more woody portions of the plant add further layers to the cell

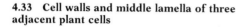

4.33 Cell walls and middle lamella of three adjacent plant cells

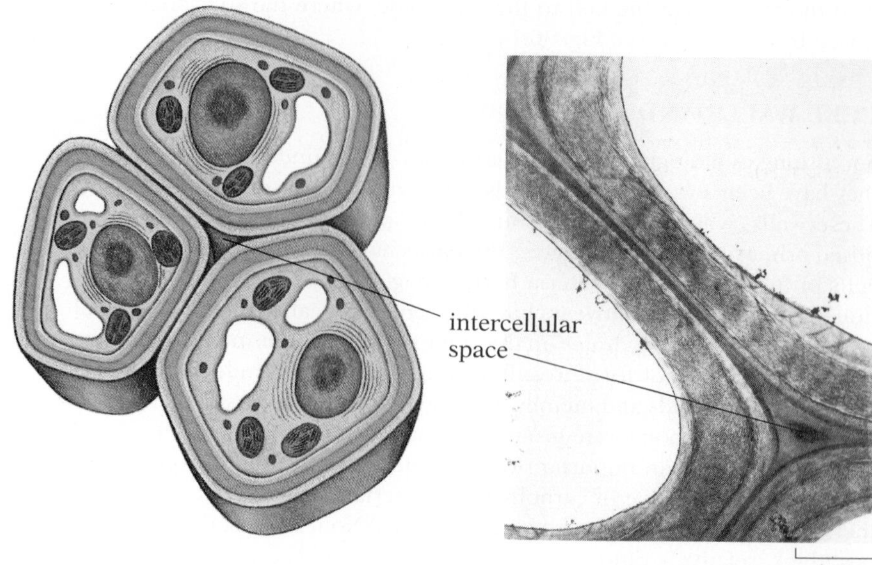

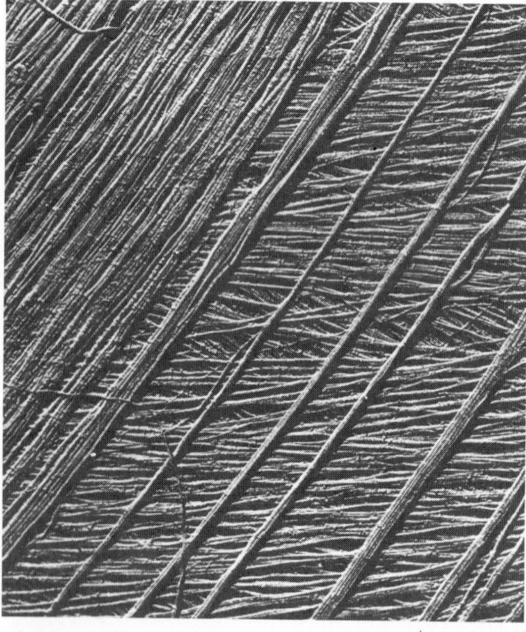

4.34 Electron micrograph of cellulose microfibrils from the cell wall of a green alga
The microfibrils are laid out in parallel lines in two directions; each is about 20 nm wide. Water and ions move freely through this meshwork.

wall, forming the ***secondary wall***. Since this wall, like the primary wall, is deposited by the cytoplasm of the cell, it is located inside the earlier-formed primary wall, lying between it and the membrane (Fig. 4.33). The secondary wall is often much thicker than the primary wall and is composed of a succession of compact layers, or lamellae. The cellulose fibrils in each lamella lie parallel to each other and are generally oriented at angles of 60–90 degrees to the fibrils of the adjacent lamellae (Fig. 4.34). This arrangement gives added strength to the cell wall. In addition to cellulose, secondary walls usually contain other materials, such as ***lignin***, which make them stiffer. Once deposition of the secondary wall is completed, many cells die, leaving the hard tube formed by their walls to function in mechanical support and internal transport for the body of the plant.

The cellulose of plant cell walls is commercially important as the main component of paper, cotton, flax, hemp, rayon, celluloid, and, obviously, wood itself. Lignin extracted from wood is sometimes used in the manufacture of synthetic rubber, adhesives, pigments, synthetic resins, and vanillin.

Plant cell walls generally do not form completely uninterrupted boundaries around the cells. There are often tiny holes in the walls through which delicate connections between adjacent cells may run. These connections are called ***plasmodesmata***.

The cell walls of both fungi and bacteria differ from those of plant cells. In most fungi the main structural component of the wall is not cellulose but ***chitin***, a polymer that is a derivative of the amino sugar glucosamine (see Fig. 3.7, p. 53); chitin is also the major component of insect exoskeletons. In bacteria the cell walls contain several kinds of organic substances, which vary from subgroup to subgroup. (One small group of bacteria lack cell walls altogether). The distinctive responses of these organic substances to diagnostic stains are a regular means of identifying bacteria in the laboratory. Structurally, however, the cell walls of all bacterial groups are alike in one respect. Part of each bacterial wall has a rigid framework of polysaccharide chains covalently cross-linked by short chains of amino acids; the resulting structure can be regarded as a single enormous molecule.

The presence of cell walls means that the cells of plants, fungi, and bacteria can withstand exposure to fluids with low osmotic concentrations without bursting. In such media the cells are, of course, in a condition of ***turgor*** (distention). Water tends to move into them by osmosis, as a result of the high osmotic concentration of the cell contents. The cell swells, building up ***turgor pressure*** against the cell walls. The cell wall of a mature cell can usually be stretched only a minute amount. Equilibrium is reached when the resistance of the wall is so great that no further increase in the size of the cell is possible and, consequently, no more water can enter the cell. Thus the cells of plants, fungi, and bacteria are not so sensitive as animal cells to the difference in osmotic concentration between the cellular material and the surrounding medium. Because of their walls, these cells can withstand much wider fluctuations in the osmotic makeup of the surrounding medium than animal cells. Moreover, turgor pressure actually strengthens the mechanical structure of plants, just as inflating an initially limp balloon or tire produces a much stiffer and stronger structure.

The glycocalyx In plants, fungi, and bacteria, the cell wall is entirely separate from the membrane; if the cell shrinks in a hypertonic medium, the membrane separates from the much more rigid wall (see Fig. 31.2, p. 883). By contrast, the "coat" of an animal cell is not an independent entity. The carbohydrates (short chains of sugars called oligosaccharides) of which it is composed are covalently bonded to protein or lipid molecules in the plasma membrane (Fig. 4.35). The resulting complex molecules are termed glycoproteins and glycolipids, and the cell coat itself is often called the ***glycocalyx***. It is important to realize that the membrane is strictly polarized: the glycolipids (which constitute about 50 percent of the lipids in the outer layer) and the carbohydrate-equipped ends of the glycoproteins are found *only* on the outside of the lipid bilayer.

The glycocalyx provides the recognition sites on the surface of the cell that enable it to interact with other cells. For example, if individual liver and kidney cells are mixed in a culture medium, the liver cells will recognize one another and reassociate; similarly, the kidney cells will seek out their own kind and reassociate. The distinctive composition of the glycocalyx enables the cells to distinguish one from the other in such a situation: the nature of the carbohydrate markers varies consistently from tissue to tissue and from species to species. Cell recognition in the process of embryonic development must also depend, at least in part, on the glycocalyx, and the same is probably true for the control of cell growth. When normal cells grown in tissue culture touch each other, they cease moving and their growth slows down or stops altogether. This phenomenon of ***contact inhibition*** is absent in most cancer cells, which continue growing without restraint because defective glycocalyces prevent them from interacting normally. As a source of identity, the glycocalyx is also important in a variety of infectious diseases: malaria parasites, for instance, recognize their host (the red blood corpuscle) by a distinctive carbohydrate marker that the corpuscle manufactures for a completely different purpose. The recognition of host cells by invading viruses also often depends on the carbohydrate

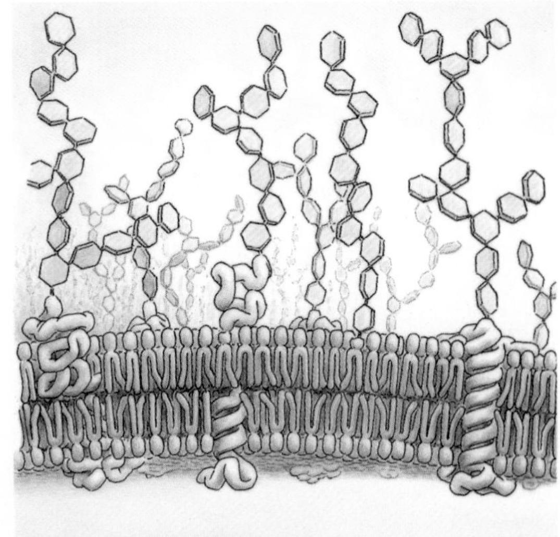

A

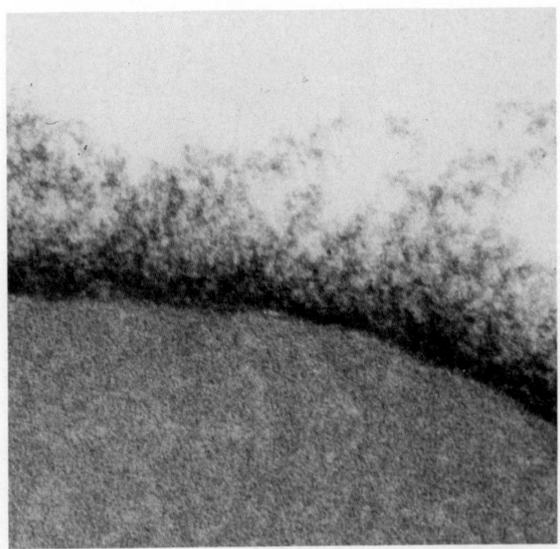

B

4.35 Plasma membrane with glycocalyx
(A) The glycocalyx of an animal cell is composed of oligosaccharides (branching carbohydrates) attached to some of the protein and lipid molecules of the outer surface of the membrane. (B) In this electron micrograph, the glycocalyx of a red blood corpuscle gives the outer surface of the membrane a fuzzy appearance.

markers of the glycocalyx. The markers in the glycocalyces of foreign cells probably provide the cues that the immune system's antibody molecules use to recognize invaders.

MULTICELLULARITY

Having completed our examination of the structure and operation of individual cell membranes, we need to look briefly at how cells can bind to one another to create multicellular organisms like ourselves.

CELL SIZE

Though single-celled organisms contribute roughly half of the earth's biomass (the total weight of all living things on earth), there are enormous benefits to being multicellular. Larger size gives an organism great advantages, such as the ability to capture or harvest smaller organisms efficiently, to move farther and faster, and so on. But large size cannot be achieved by simply increasing the size of a single-celled organism indefinitely. A cell must take in its nutrients and oxygen across the membrane. As a cell triples in volume, so does its need for nutrients and oxygen, and yet its membrane surface area does not even double. Since metabolic needs increase faster than the surface area through which they are satisfied, a point arrives at which the membrane can no longer support its contents. The need for efficient diffusion, therefore, puts a strict limit on the surface-to-volume ratio of a cell, and consequently limits cell size.

Many single-celled organisms are extraordinarily complex, for the one cell must do everything needed for survival. Although simple aggregations of identical cells that assume specialized functions do exist in nature, such as the 32-cell discs of some green algae and the amoeboid clusters of slime molds, in most assemblies of cells more sophisticated specialization becomes possible. The course of evolution has demonstrated that arrangements in which certain cells concentrate on particular functions (propulsion, feeding, reproduction, and so on) can be far more effective than those in which each cell pursues a "jack-of-all-trades" strategy.

CELL ADHESION

There are two dramatically different strategies for giving form to multicellular aggregations, which would otherwise be amorphous lumps of cells. In certain animal tissues specialized cells called ***fibroblasts*** secrete fibrous proteins, among them collagen (see Fig. 3.26, p. 68) and elastin, which are components of an ***intercellular matrix***. Cells are held in place by this structural network, and grow and function there. The resulting tissue, known as connective tissue, is described in detail in a later chapter.

The other strategy for providing shape and strength is for cells to adhere to one another. For this to take place the cells must specifically recognize which other cells are appropriate partners, and then secure their membranes to them. The mechanism of this recognition, a process especially critical during embryological development, when vast numbers of cells must be properly arranged, remains largely a mystery. There is evidence

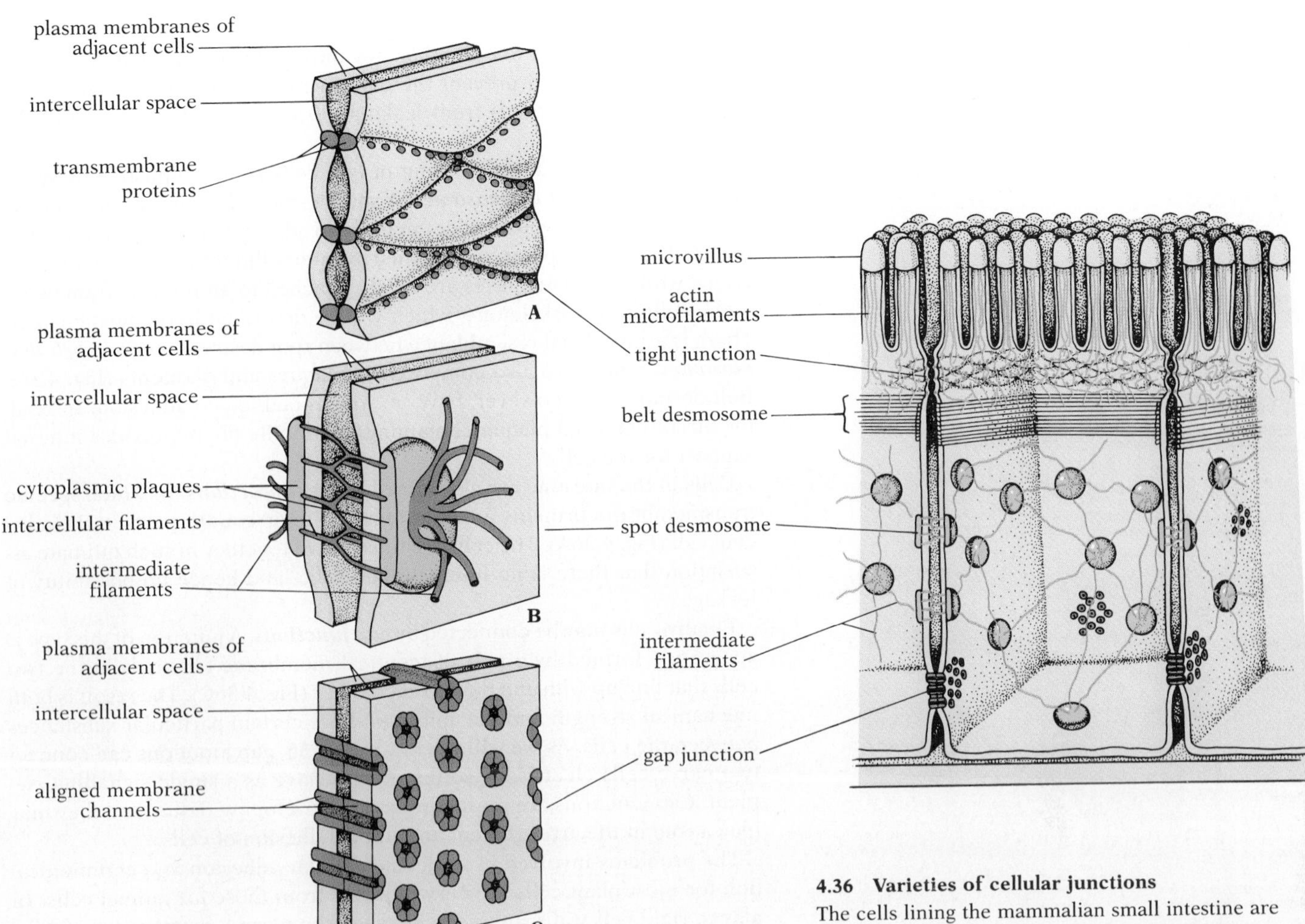

4.36 Varieties of cellular junctions
The cells lining the mammalian small intestine are attached by several types of specialized junctions. An example of each is shown in enlarged detail. (A) A tight junction is composed of rows of transmembrane proteins in adjacent cells that bind to each other. (B) A spot desmosome consists of a pair of cytoplasmic plaques, each just inside the cell membrane and connected to the other across the intermembrane space by specialized intercellular filaments. Each plaque is also attached to fibers of the cytoskeleton within its cell. (C) A gap junction is formed by a pair of membrane channels aligned and bound together to create a specialized pathway between cells. The chemical structure of intermediate filaments and actin microfilaments will be described in Chapter 5.

that at least certain types of cells have characteristic carbohydrate markers on the lipids and proteins of their outer membranes that are recognized by specialized receptors on other cells. Other cells have special glycoproteins, called cell-adhesion molecules, which appear to bind directly to one another. Perhaps this mutual binding involves a match between the carbohydrate part of one molecule and the protein of another, and vice versa. In addition, a curious class of plant proteins, called lectins, recognize cells of specific tissue types on the basis of their carbohydrate identification tags. The role of lectins is unclear; apparently they help many plant cells to adhere to one another, but there is also evidence that they act to glue together and thereby immobilize the bacterial and fungal cells that can otherwise cause plant diseases.

However cells locate each other, they frequently form strong junctions of several sorts. Most of the cellular attachment junctions utilized in multicellular organisms other than plants can be seen in the lining of the small intestine (Fig. 4.36). Intestinal cells are arranged with microvilli projecting into the intestine, where they absorb the nutrients released by digestion. They must adhere to one another not only to form the tubular channel of the gut but also to prevent the digestive enzymes (against which they are specifically armored) from leaking out and digesting the rest of the organism.

The general mechanical joining of two cells is accomplished by structures known as ***spot desmosomes***. A spot desmosome consists of two cytoplasmic plaques, one just inside each of two adjacent cells (Fig. 4.36B). The outside faces of the plaques are joined by intercellular filaments, acting like rivets, while the inside faces are firmly attached to an internal framework —the cellular cytoskeleton—which will be described in the next chapter. There is a superficial resemblance between spot desmosomes and ***belt desmosomes***, which are also composed of plaques and filaments (Fig. 4.36). Belt desmosomes, however, have no role in cell-to-cell adhesion. Instead, the circumferential plaque containing contractile fibers provides internal support for the cell.

Cells in the intestine are also joined by ***tight junctions***, in which specific transmembrane proteins attach directly to their counterparts in the adjacent cell (Fig. 4.36A). The cells are thus drawn together in such intimate association that there is no intercellular space, and hence no possibility of leakage.

Finally, cells may be connected by ***gap junctions***. A junction of this type is apparently formed by a pair of identical membrane channels in the two cells that line up with and bind to each other (Fig. 4.36C). The result is both mechanical strength and the ability to share certain particular substances between the cells. As we will see in Chapter 35, gap junctions can connect two nerve cells electrically so that they behave as a single signalling element. Gap junctions are most common in developing tissue, and they may play a role in the arrangement and initial adhesion of cells.

The problems involved in achieving cellular adhesion and communication for most plant cells are very different from those for animal cells. In plants, rigid cell walls intervene between the plasma membranes of adjacent cells. Hence adhesion must be accomplished for the most part by relatively simple cross-linking between the polysaccharides of the cell walls. Effective adhesion and communication between plant cells is crucial, however, if water and inorganic nutrients are to be passed up from the roots, and the energy-rich products of photosynthesis are to be transmitted from the leaves to other parts of the plant. To accommodate this need, plant cell walls contain specialized openings—***plasmodesmata***—where the membranes of adjacent cells come into contact with each other. Some of these openings take the form of membrane-lined holes through which the cytoplasm of adjoining cells can mix directly. Others retain a double-membrane barrier; these play an important role in controlling the movement of both solutes and solvents between cells.

STUDY QUESTIONS

1. How do water, saturated and unsaturated phospholipid tails, and cholesterol contribute to membrane stability and flexibility? (pp. 102–7)

2. Why does the sodium-potassium pump require chemical energy from the cell? Why does the sodium-glucose symport not require any? (p. 109)
3. List five different cross-membrane transport strategies, and explain how each one works. (pp. 109–16)
4. What are the costs and benefits of the plant strategy of using cell walls to stabilize membranes? (pp. 117–18)
5. What changes in conditions would serve to increase the rate of diffusion of an ionic substance? (The list is not short.) (pp. 93–95)
6. Explain osmosis in terms of probability at the molecular level, and then in terms of thermodynamics. (pp. 96–99)

CONCEPTS FOR REVIEW

- Active transport versus passive diffusion
- Diffusion from the perspective of free energy
- Semipermeable membranes
    - Free-energy basis of osmosis
    - Osmotic concentration versus osmotic potential, osmotic pressure, and water potential
- Cellular membranes
    - Phospholipid-bilayer basis
    - Basis of inherent stability and permeability
    - Cholesterol and membrane fluidity
- Membrane proteins
    - Anchoring
    - Glycocalyx
    - Channels and gated channels
        - selectivity
        - triggering
        - energy sources
    - Membrane pumps
        - polarity of pumping
        - energy sources
- Endocytosis and exocytosis
    - Phagocytosis versus pinocytosis versus receptor-mediated endocytosis
    - Basis of specificity
    - Pattern of vesicle formation
    - Functions
- Role of cell wall

SUGGESTED READING

BRETSCHER, M. S., 1985. The molecules of the cell membrane, *Scientific American* 253 (4). *Reviews the bilayer plasma membrane and membrane proteins, and the process of endocytosis.*

BRETSCHER, M. S., 1987. How animal cells move, *Scientific American* 257 (6). *The role of pinocytosis in the amoeboid movement of cells.*

BROWN, M. S., and J. L. GOLDSTEIN, 1984. How LDL receptors influence cholesterol and atherosclerosis, *Scientific American* 251 (5). (Offprint 1555)

CAPALDI, R. A., 1974. A dynamic model of cell membranes, *Scientific American* 230 (3). (Offprint 1292) *A good discussion of the fluid-mosaic model of membrane structure.*

DAUTRY-VARSAT, A., and H. F. LODISH, 1984. How receptors bring proteins and particles into cells, *Scientific American* 250 (5). (Offprint 1550) *The life cycle of coated pits.*

LODISH, H. F., and J. E. ROTHMAN, 1979. The assembly of cell membranes, *Scientific American* 240 (1). (Offprint 1415) *A good discussion of how the membrane grows and of how and why its two sides differ.*

ROTHMAN, J. E., and J. LENARD, 1977. Membrane asymmetry, *Science* 195, 743–53. *Why some proteins are found on only one side of the membrane.*

SATIR, P., 1975. The final steps in secretion, *Scientific American* 233 (4). (Offprint 1328) *How the membrane of a secretory vesicle interacts with the plasma membrane during exocytosis.*

SHARON, N., 1980. Carbohydrates, *Scientific American* 243 (5). (Offprint 1483) *On the role of carbohydrates in the life of the cell, with particular emphasis on the membrane carbohydrates that are involved in cell recognition.*

SINGER, S. J., and G. NICOLSON, 1972. The fluid-mosaic model of the structure of cell membranes, *Science* 175, 720–31. *The original presentation of the fluid-mosaic hypothesis.*

STAEHELIN, L. A., and B. E. HULL, 1978. Junctions between living cells, *Scientific American* 238 (5). (Offprint 1388) *A freeze-etch exploration of cellular junctions.*

UNWIN, N., and R. HENDERSON, 1984. The structure of proteins in biological membranes, *Scientific American* 250 (2). (Offprint 1547)

Chapter **5**

INSIDE THE CELL

n the last chapter we saw how the cell membrane protects the cell from the world outside, and how it manages to trap some substances and to dispose of others. Some specialized membrane channels make use of the free energy of osmotic-concentration gradients, but most of the cell's selective permeability is the direct or indirect result of energy expenditure. The result is that a homeostatic chemical environment favorable to life is maintained inside the membrane, an environment with optimal ionic concentrations and pH, enough of the right sort of organic building blocks, and a collection of life-supporting enzymes. In this chapter we will examine ***organelles***—the structures inside the cell that use the cell's enzymes, molecular building blocks, and favorable chemical milieu to carry out the instructions encoded in the genes.

Because the chemistry of many of the important processes of life is different from that of the cytoplasm as a whole, some organelles serve as miniature containment vessels for reactions requiring unusual conditions, like high or low pH. For example, digestive enzymes are packaged in special structures called lysosomes to protect the rest of the cell from these destructive but essential biological catalysts.

Organelles that exist to maintain specialized chemical conditions are organized like small cells, complete with their own lipid-bilayer membranes. The membrane of each organelle contains protein channels that maintain the organelle's own unique chemistry, transporting in reactants and exporting products. And, as we will see, current theory suggests a fascinating evolutionary history for some organelles.

SUBCELLULAR ORGANELLES

THE NUCLEUS

The largest and one of the most conspicuous structural areas within the cells of most organisms (though not of bacteria) is the membrane-bounded ***nucleus*** (Fig. 5.1). The nucleus plays the central role in cellular reproduction, the process by which a single cell divides and forms two new cells. It also plays a crucial part, in conjunction with the cell's environment, in determining what sort of differentiation a cell will undergo and what form it will exhibit at maturity. And the nucleus directs the metabolic activities of the living cell. In short, it is from the nucleus that the "instructions" emanate that guide the life processes of the cell as long as it lives.

We have said that bacteria differ from all other kinds of organisms in lacking a membrane-bounded nucleus (though they do possess genetic material that controls the cell's activities). Similarly, this group lacks most of the other cellular structures found in other organisms. These differences are so fundamental that bacteria are classified into two kingdoms of their own (see Chapter 20), and their cells are designated as ***procaryotic*** (before, i.e., without, a nucleus), whereas the cells of all other organisms are designated as ***eucaryotic*** (having a true nucleus). The characteristics of procaryotic cells will be discussed in a later section.

The eucaryotic nucleus contains two kinds of structures, the chromosomes and the nucleolus. With an electron microscope we can see that both are embedded in a mass of amorphous, granular-appearing nucleoplasm. The entire nucleus is bounded by a closely associated pair of membranes called the nuclear envelope.

The ***chromosomes*** (Fig. 5.2) are elongate, threadlike bodies clearly visible only when they "condense" in preparation for cell division; at all other times the chromosomes are in an uncondensed conformation, which, when

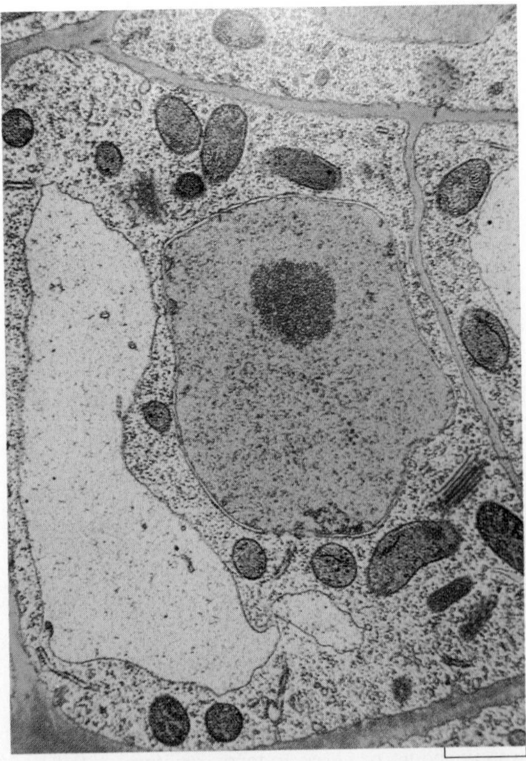

5.1 Electron micrograph of plant cell nucleus The nucleus of most cells is about one-third the diameter of the whole cell, and so occupies 3–4 percent of cellular volume.

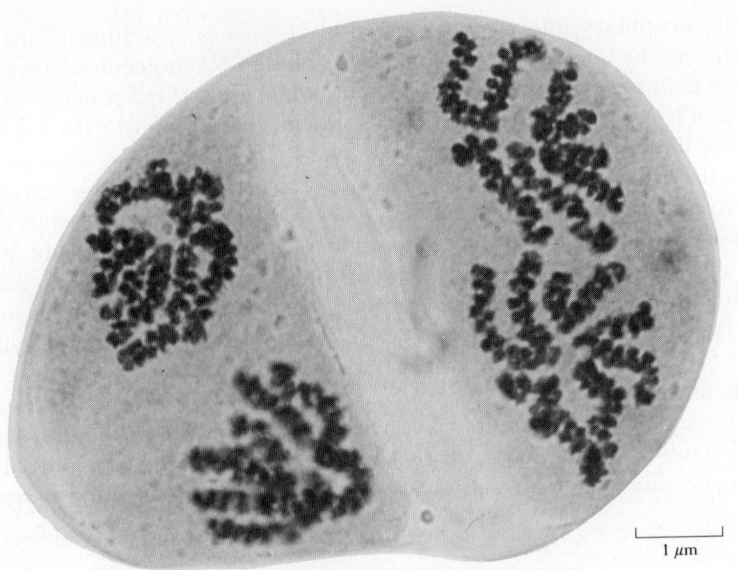

5.2 Chromosomes in a dividing cell of *Trillium* The separate chromosomes in the two nuclei of this spring-flowering plant can easily be distinguished.

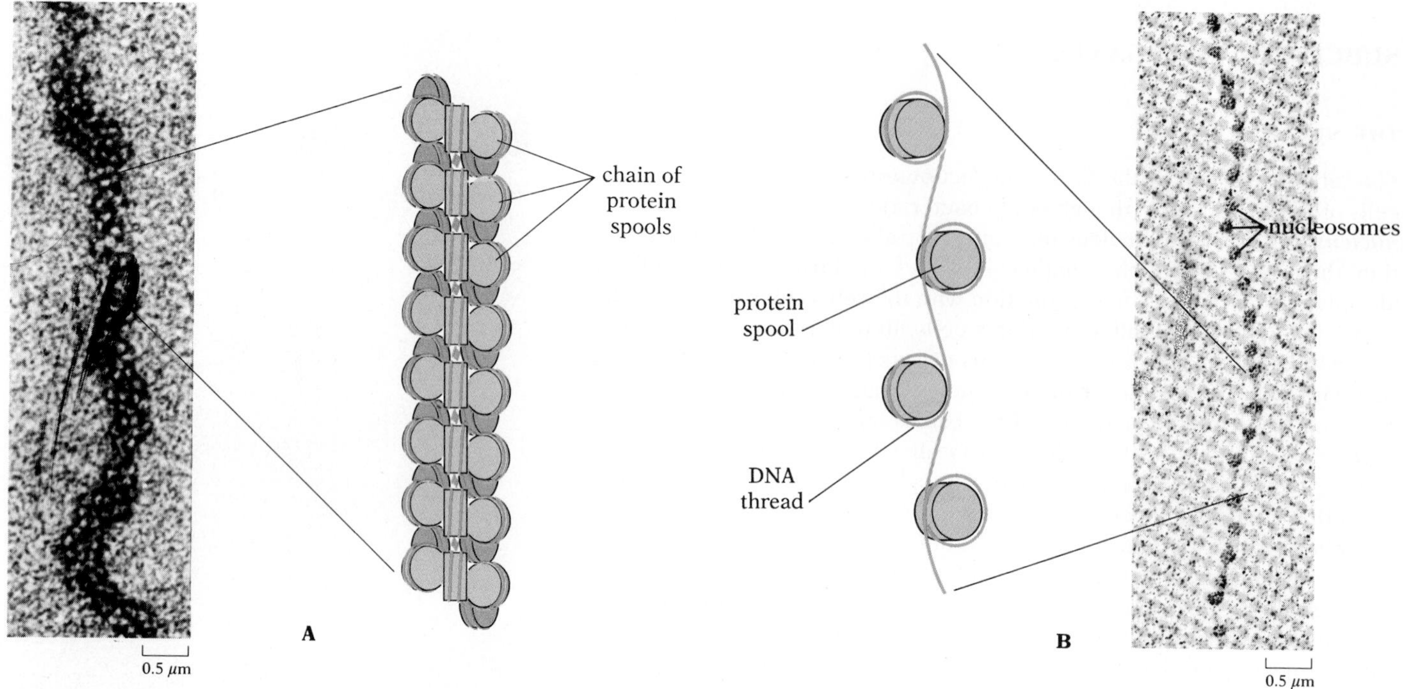

5.3 DNA on nucleosomes

The chromosomal DNA of eucaryotes is wound on protein spools, or cores, to form structures called nucleosomes. Normally the spools adhere to one another in a regular way, giving the DNA the appearance of a piece of yarn (A). When the DNA is treated to break the connections between nucleosomes, the individual protein spools and the thread of DNA can be seen (B).

stained, appears as a dark, hazy material called chromatin. The chromosomes are composed of DNA and protein; the DNA is the substance of the basic units of heredity, called ***genes***, while the protein provides spool-like supports, or cores, on which the DNA is wound to form structures called ***nucleosomes*** (Fig. 5.3). The genes are duplicated whenever a cell divides, and a copy is passed to each new cell. The genes determine the characteristics of cells and act as the units of control in the day-to-day activities of living cells.

The hereditary information carried by the genes is written in the sequence of the nucleotide building blocks of the DNA molecules. Since the genes themselves remain in the nucleus, while most of the processes they control take place in the cytoplasm, some mechanism must exist for conveying the information outside the nucleus. The mechanism is transcription, the process by which the nucleotide sequence in the DNA gives rise to a corresponding nucleotide sequence in RNA. This RNA sequence is then somewhat modified, and the resulting messenger RNA (mRNA) can leave the nucleus and move to the sites of protein synthesis in the cytoplasm. There amino acids are linked by peptide bonds in a sequence corresponding to that of the mRNA nucleotides to form proteins, including enzymes (Fig. 5.4). This process is known as translation. As we saw in Chapter 3, the sequence of amino acids—the primary structure of a protein—determines the three-dimensional conformation of the protein and the biological activity that this conformation bestows. Thus the genes are at the very hub of life; they encode all the information necessary for the synthesis of the enzymes

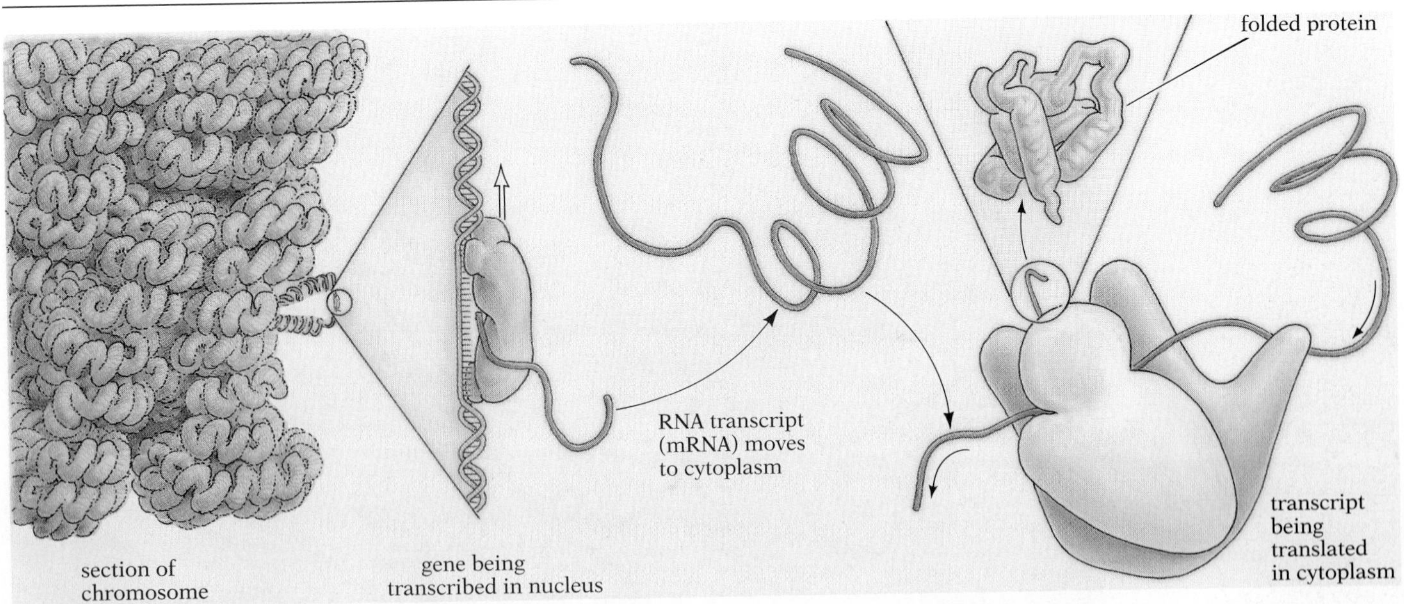

5.4 Flow of information from nucleus to cytoplasm
Information encoded in a region of the chromosome (a gene) is transcribed by an enzyme complex to create an RNA copy. This messenger RNA is then exported to the cytoplasm, where it is decoded by a ribosome and the protein specified by the gene is synthesized.

regulating the myriad interdependent chemical reactions that determine the characteristics of cells and organisms. Later chapters will deal in more detail with the question of how genes encode information and direct protein synthesis, how genes are turned on or off by the cell as required, and how genes reproduce themselves during the process of cell division.

The other structure in the nucleus, besides the chromosomes, is the ***nucleolus***, the dark-staining, generally oval area usually visible within the nuclei of nondividing cells. There may be one or more nucleoli per nucleus, depending on the species of organism. Nucleoli form in association with particular regions of specific chromosomes; they are, in fact, simply specialized parts of the chromosome and, like the rest of the chromosome, are composed of DNA and protein. The DNA of the nucleoli includes multiple copies of the genes from which a type of RNA called ribosomal RNA (rRNA) is transcribed. After the rRNA is synthesized, it combines with proteins, and the resulting complex detaches from the nucleolus, leaves the nucleus, and enters the cytoplasm, where it becomes a part of the protein-synthesizing organelles called ***ribosomes***. Thus the nucleoli are responsible for manufacturing and exporting to the cytoplasm the precursors of the ribosomes on which proteins will be synthesized. Multiple copies of the genes for rRNA make possible rapid manufacture of the ribosomes necessary for active protein synthesis. Nucleoli tend to be small or absent in cells that carry out little protein synthesis.

The ***nuclear envelope*** helps maintain a chemical environment in the nucleus different from that provided by the surrounding cytoplasm. It also appears to provide anchorage for the two ends of each chromosome. Unlike the cell membrane, the complete nuclear envelope consists of distinct

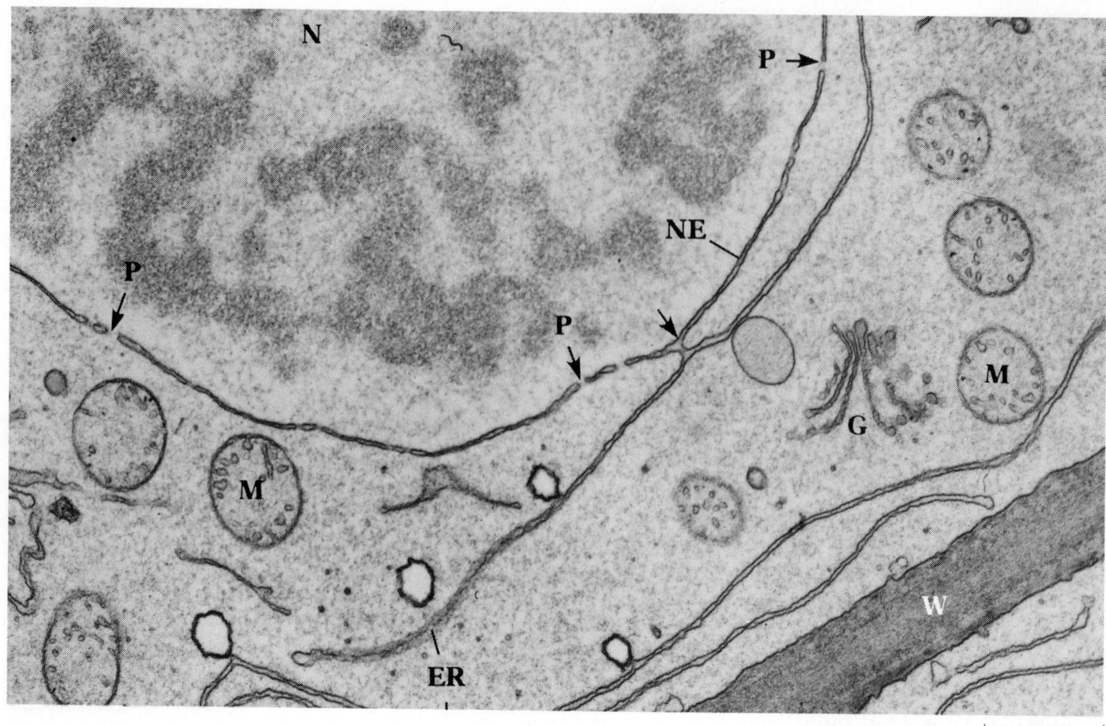

5.5 Electron micrograph showing the nuclear envelope of a cell from corn root
The large structure filling the upper left quarter of the picture is the nucleus. The unlabeled arrow indicates a point where the endoplasmic reticulum and the double nuclear membrane interconnect. ER, endoplasmic reticulum; G, Golgi apparatus; M, mitochondrion; N, nucleus; NE, nuclear envelope; P, pore in nuclear envelope; W, cell wall.

inner and outer membranes, with space enclosed between them (Figs. 5.1 and 5.5).

Electron-microscope studies indicate that the double-membrane envelope is interrupted by fairly large and elaborate pores at points where the outer and inner membrane are continuous (Figs. 5.5, 5.6, and 5.7). Nevertheless, the membrane is highly selective. According to permeability experiments, many substances that can cross the cell membrane into the cytoplasm apparently cannot readily pass through the nuclear envelope into the nucleus and are consequently restricted to the cytoplasm. Even molecules much smaller than the pores fail to move through the pores, yet some large molecules pass readily. These "macromolecules" are primarily substances produced on the genes (such as mRNA) that are moving out of the nucleus, proteins moving into the nucleus to be incorporated into nuclear structures or to catalyze chemical reactions in the nucleus, and various substances from the cytoplasm that move into the nucleus and help regulate gene activity. There is thus a carefully controlled and highly selective two-way exchange through the pores between the nucleus and the cytoplasm. The pores recognize which substances are to be allowed to pass, and in which direction, on the basis of specific "passwords," chemical signal sequences attached to one end of the various molecules. Certain viruses (including the AIDS virus) have broken the molecular code used by nuclear pores; they are able to smuggle in a copy of their chromosome, which can then be incorporated into the cell's DNA library. All the password sequences decoded so far involve a short polypeptide group rich in positively

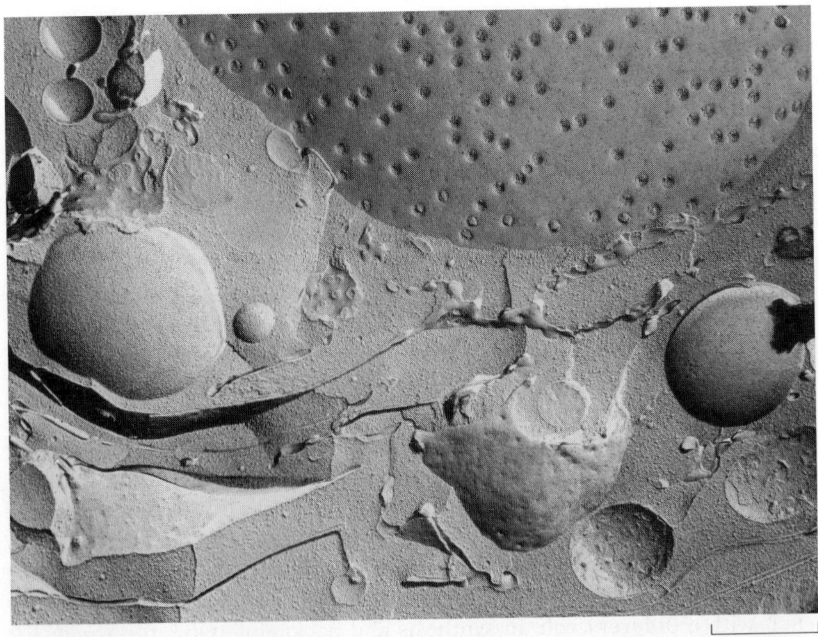

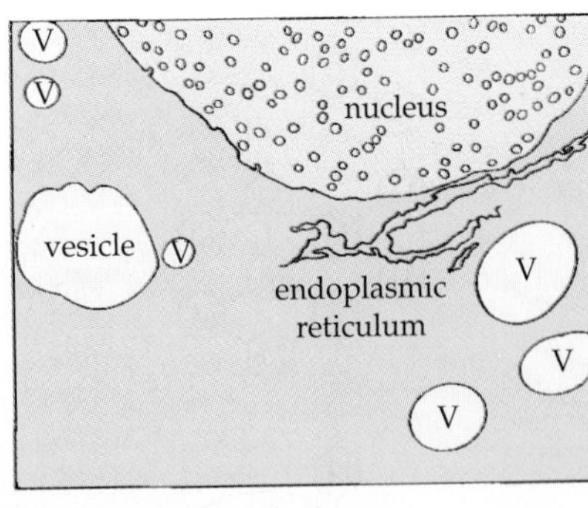

5.6 Electron micrograph of a freeze-fractured cell from the tip of an onion root
At upper right is the surface of the nuclear envelope, with numerous pores; a typical nucleus has a few thousand pores. A variety of vesicles can be seen in the cytoplasm.

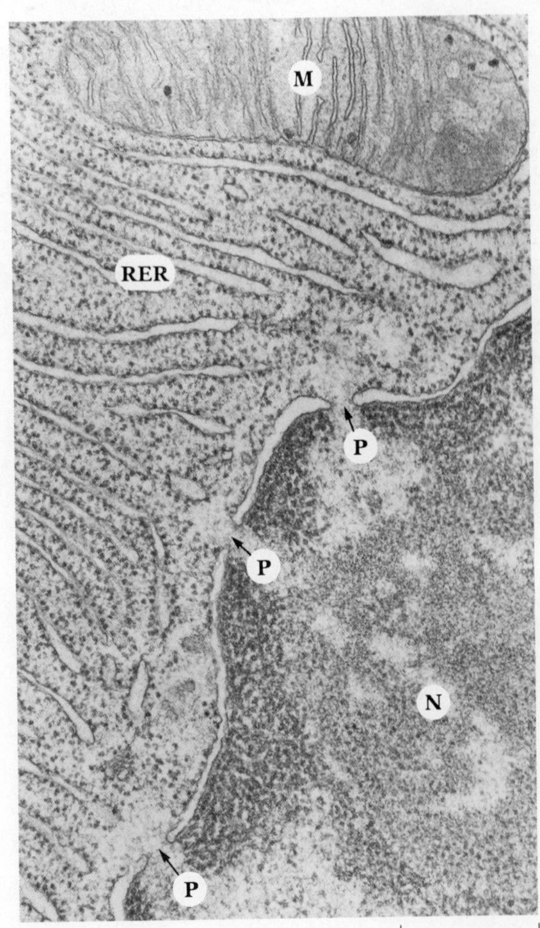

5.7 Electron micrograph of endoplasmic reticulum
Thin section of a pancreatic cell from a bat, showing many flattened cisternae of rough ER; the ribosomes lining the RER membranes can be clearly distinguished. In the lower right portion of the micrograph is part of the nucleus (N); note the very prominent pores (P) in the double membrane of the nuclear envelope. A mitochondrion (M) is at the top.

charged amino acids (lysine and arginine) plus one or more prolines (the unusual amino acid that induces bends in the tertiary structure of proteins).

The electron microscope has revealed another particularly interesting fact about the nuclear envelope—the continuity, at some points, of this double membrane with an extensive cytoplasmic membrane system called the endoplasmic reticulum (Fig. 5.5).

THE ENDOPLASMIC RETICULUM AND RIBOSOMES

Attempts to separate subcellular organelles from the cytoplasm for various research purposes often make use of centrifugation. In this technique, cells are first lysed (broken) by detergents that disrupt the lipids in the membrane, and then they are layered on top of a test tube containing a viscous solution of, say, glucose; finally, the samples are spun in a centrifuge to sort out the cell parts by weight. Each component travels through the solution at a characteristic rate, the densest moving most rapidly.

In 1938 Albert Claude of the Rockefeller Institute isolated some cytoplas-

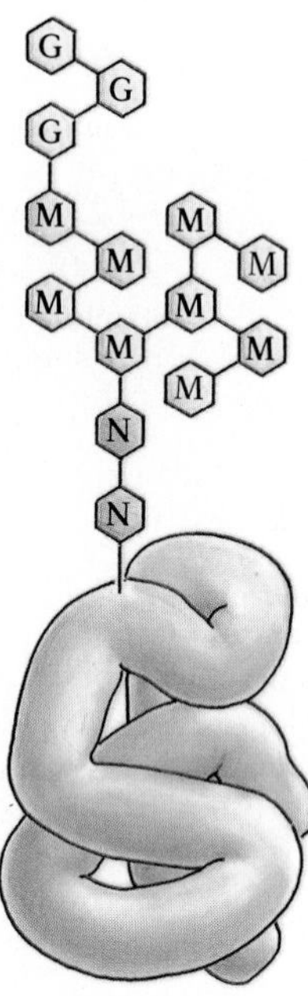

5.8 Oligosaccharide mailing label attached to proteins in the ER

A 14-sugar side chain is attached to nearly all proteins synthesized on the RER, and serves as a "mailing label." Proteins lacking this label remain in the RER. When the four terminal sugars are removed, the protein is exported to the Golgi apparatus in vesicles. The size of the tag in this drawing is exaggerated compared to the protein. (G = glucose, M = mannose, N = *N*-acetyl-glucosamine)

mic components that he called microsomes ("small bodies"). Claude's microsomes made up 15–20 percent of the total cell mass and could be isolated from almost any kind of cell—plant or animal. Chemical analysis showed that the microsomes had a very high nucleic acid content; in fact, they contained almost all the cytoplasmic nucleic acid. They also contained a high percentage of the cytoplasmic phospholipids. However, since the microsomes were not visible under the light microscope, there was much argument about whether they were actually discrete portions of the living cells or simply debris produced by the breaking up of the cells.

Only the advent of the electron microscope allowed Claude's microsomes to become established as part of the cell machinery. In 1945 Keith R. Porter, then of the Rockefeller Institute, described a complex system of membranes that formed a network in the cytoplasm. This system, which Porter named the ***endoplasmic reticulum*** (reticulum being Latin for "network"), has since been shown to be present to some extent in all nucleated cells. Claude's microsomes were actually a mixture of ribosomes and the fragmented endoplasmic reticulum. The endoplasmic reticulum exists in large part to synthesize new membrane phospholipids, to package proteins into vesicles for transport to other parts of the cell, and to store calcium ions. There are two anatomically distinct forms of endoplasmic reticulum, each with a different role in synthesis and packaging. One, the ***rough ER*** (RER) exists as flattened, fluid-filled, membrane-enclosed sacs called ***cisternae***. It seems very likely that in most (perhaps all) cells the RER forms a series of interconnected sheets, so that all the cisternae interconnect. The rough appearance of the RER membrane arises from its association with the numerous protein-synthesizing complexes called ribosomes (Fig. 5.7). Ribosomes also bind to the outer membrane of the nuclear envelope, with which the RER is connected.

The association of the ribosomes with the RER membrane appears to be necessary for the protein they produce to penetrate the membrane to the cisternae or become embedded in the membrane. Most or all of the proteins to be packaged and transported by the ER are synthesized by the ribosomes of the RER. Proteins synthesized on free ribosomes in the cytoplasm are apparently not destined for export from the cell or for incorporation into membranes, but rather are released to function as enzymes in the ***cytosol*** (the more fluid part of the cytoplasm). It seems clear that mRNA from the nucleus that is to produce enzymes for the ER binds first to free ribosomes. The mRNA, however, signals the ribosome (with the help of signal-recognition particle) to bind to special channels in the RER. Once the ribosome binds, the translation begins and the enzyme being synthesized on the ribosome is threaded through the channels into the lumen of the RER. As soon as a ***signal sequence*** in the mRNA is synthesized as part of the growing polypeptide chain, and this sequence (Asparagine–Nonproline–Serine/Threonine) is threaded into the lumen of the ER, enzymes in the RER membrane attach a small sugar complex (an oligosaccharide) that will aid in subsequent sorting (Fig. 5.8). Modifications of this molecular mailing label specify where the protein is to go. Errors in this tagging system can have serious consequences. Individuals suffering from cystic fibrosis, for example, lack a type of chloride channel in their cell membranes; this may account for the abnormally thick secretions produced by many glands (in-

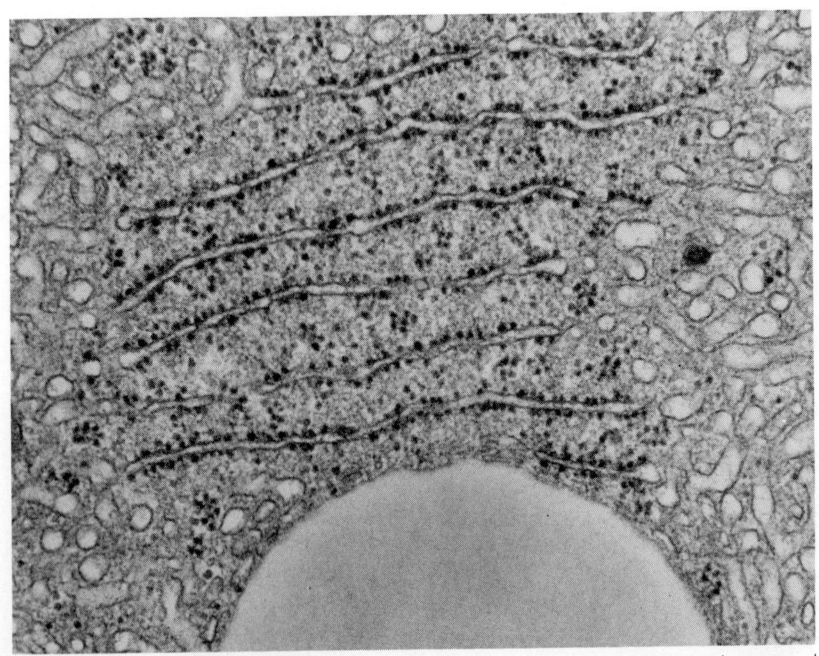

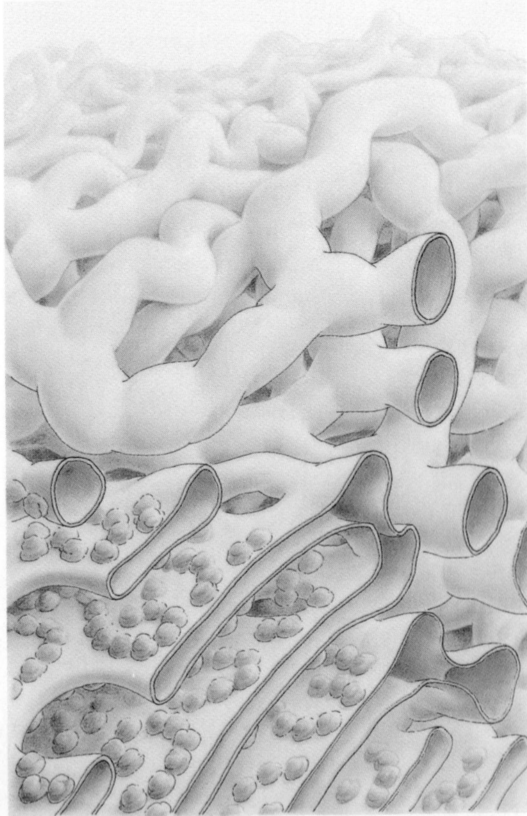

5.9 The endoplasmic reticulum
The endoplasmic reticulum from a steroid-producing cell of a guinea pig testis, shown in this electron micrograph, consists of a complex system of membranes including the RER, with associated ribosomes, and SER. The relationship of rough and smooth ER varies from cell to cell; the sketch shows an association more typical of a cell in which macromolecules are synthesized in the RER and transported to the SER. This representation assumes that the rough and smooth ER are physically continuous; they may actually be separate organelles that communicate via vesicles.

cluding the sweat glands, whose ducts become clogged, creating cysts): absence of this ion channel might prevent the secretion of chloride ions, and thus the failure of water to move osmotically into the secretion. The genetic defect appears to be not in the channel structure itself, but in a region of the protein that enzymes in the RER recognize, and to which they normally attach a label that causes the protein to be transported to the cell membrane.

Despite its importance in intracellular transport, it would be surprising if the RER functioned merely as a passageway. The RER membrane has a large protein content, and proteins, you will remember, may act both as structural elements in cells and as enzymes catalyzing chemical reactions. There is now abundant evidence that at least some of the many protein molecules of which the RER membranes are composed act as enzymes and that the RER functions as a cytoplasmic framework providing catalytic surfaces for some of the biochemical activity of the cell. Its complex folding provides an enormous surface for such activity.

The other form of endoplasmic reticulum, the ***smooth endoplasmic reticulum*** (SER), not only lacks ribosomes, but also has a much more tubular appearance (Fig. 5.9) and a very different set of characteristic membrane proteins. The most obvious functions of the SER are to synthesize membrane phospholipids and to package proteins in the cisternae into membrane-bounded vesicles for transport to other locations in the cell. The vesicles also carry membrane-mounted proteins with them. Though most of the proteins in the cisternae and the SER membrane are synthesized first in the RER, it is not yet clear how they get to the SER. It may be that the RER and SER are continuous, as was once universally assumed; alternatively, the RER may ship membrane-embedded proteins and cisternal proteins to the SER via vesicles.

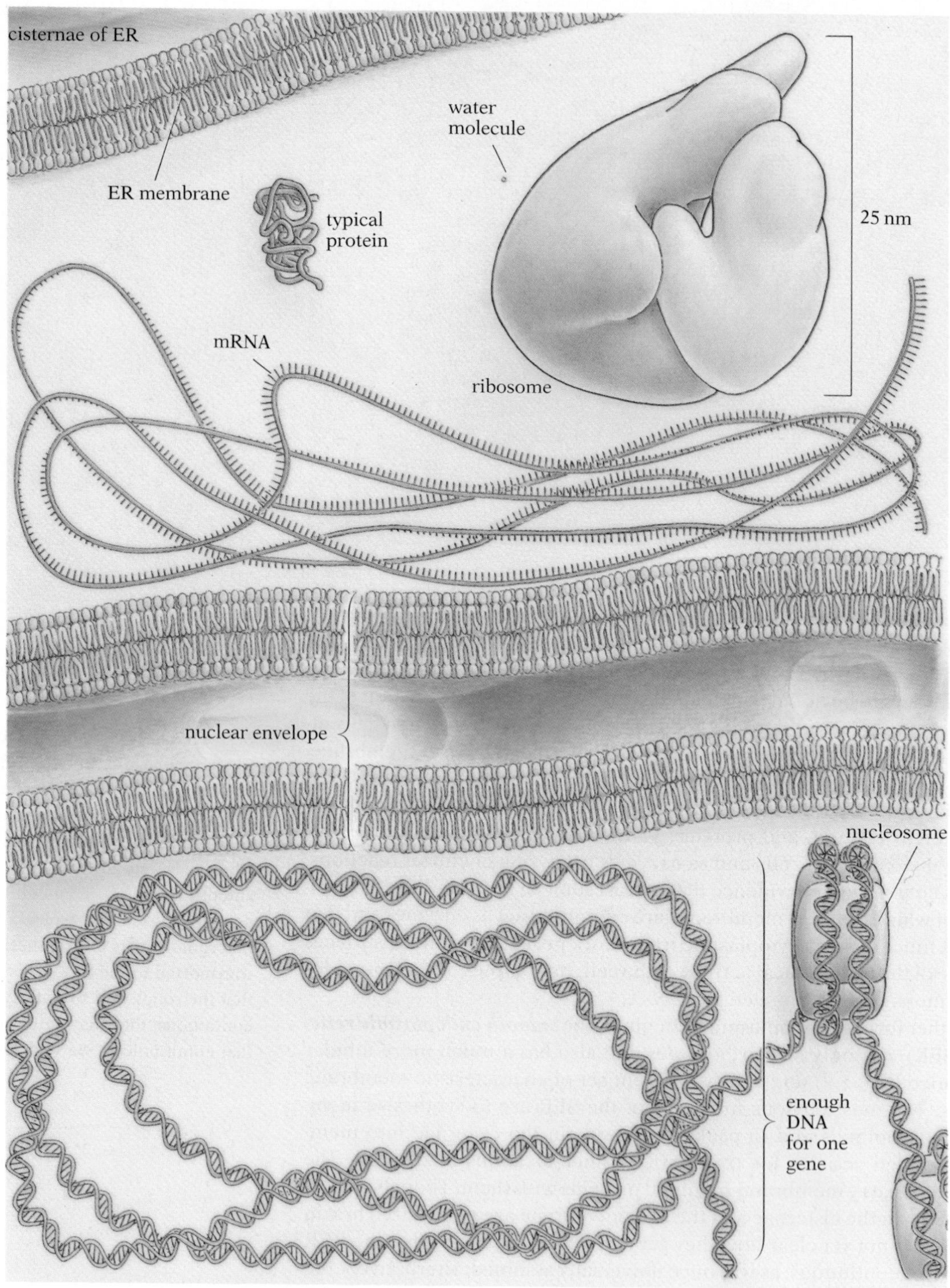

5.10 Relative sizes of subcellular components

Though many of the membrane proteins of the SER are involved in membrane synthesis, as well as preparing and dispatching vesicles, others play a critical role in other aspects of cellular biochemistry.

The SER of liver cells, for example, holds enzymes that detoxify many poisons, including barbiturates, amphetamines, and morphine. Poisons that appear in the bloodstream are transported to the liver. There the rapid synthesis of more phospholipids causes the surface area of the liver cells' SER to double. The positioning of detoxification enzymes in the ER takes maximum advantage of the cell's potential for compartmentalization to protect the cytosol from the poisons. The products of the detoxification process are probably packaged in vesicles that have been formed from the ER itself, and are then transported to other organelles, where they are further broken down. Finally the SER releases calcium ions, whose concentration, as we will see, is important in controlling the building or disassembly of elements of the cytoskeleton.

The relative sizes of some of the subcellular components discussed so far are shown in Figure 5.10.

THE GOLGI APPARATUS

In 1898 the Italian scientist Camillo Golgi first described a new "reticular apparatus" in certain cells of the vertebrate brain. This "apparatus," now recognized as yet another subcellular organelle, became visible under the light microscope only when treated with certain chemicals. Similar cytoplasmic regions have subsequently been found by numerous workers in a great variety of animal and plant cells. Though they vary in form and several different names were at first applied to them, all eventually came to be called ***Golgi apparatus***. This organelle consists of a system of membrane-delimited compartments arranged approximately parallel to each other (Fig. 5.11).

The Golgi apparatus is particularly prominent in cells involved in the secretion of various chemical products; as the level of secretory activity of these cells changes, corresponding changes occur in the morphology of the organelle. In certain cells of the pancreas of guinea pigs, a zymogen (the inactive precursor of an enzyme) synthesized on the ribosomes moves into the channels of the ER; it reaches the Golgi apparatus in vesicles budded off the ER.[1] With the help of the electron microscope, we now know that the zymogen arrives at the compartment nearest the nucleus (the most highly curved cisterna in Figure 5.11, known as the *cis* compartment), and then moves from one layer to the next until it reaches the farthest layer (the *trans* compartment). Movement between layers is by means of vesicles that bud off from one cisterna and then fuse with another, farther from the nucleus. During this *cis–trans* movement the zymogen is modified, and in the final

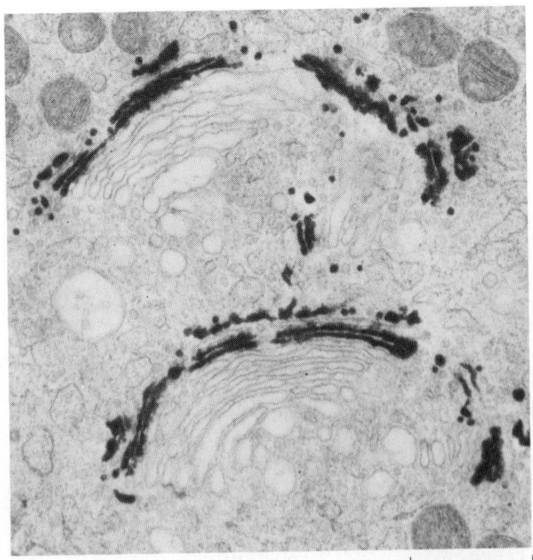

5.11 The Golgi apparatus
Electron micrograph of Golgi apparatus from an amoeba. Vesicles forming at the ends of some of the cisternae can be seen in the EM and in the interpretive drawing. The movement is from the *cis* layer, through the various medial compartments, to the *trans* cistern. *Cis* (meaning "this side of") and *trans* ("other side of") are defined with respect to the nucleus. The Golgi sorts, modifies, relabels, and packages molecules into vesicles for further transport. In this EM, the *trans* compartment and the vesicles they have released are labeled with an electron-dense stain.

[1] There may be some kind of "intermediate compartment" between the ER and the Golgi. Certain proteins that are in the membranes of vesicles leaving the ER are missing when vesicles with identical contents arrive at the Golgi. These proteins are returned to the ER in separate vesicles, implying that some intermediate organelle must exist to sort and repackage ER vesicles.

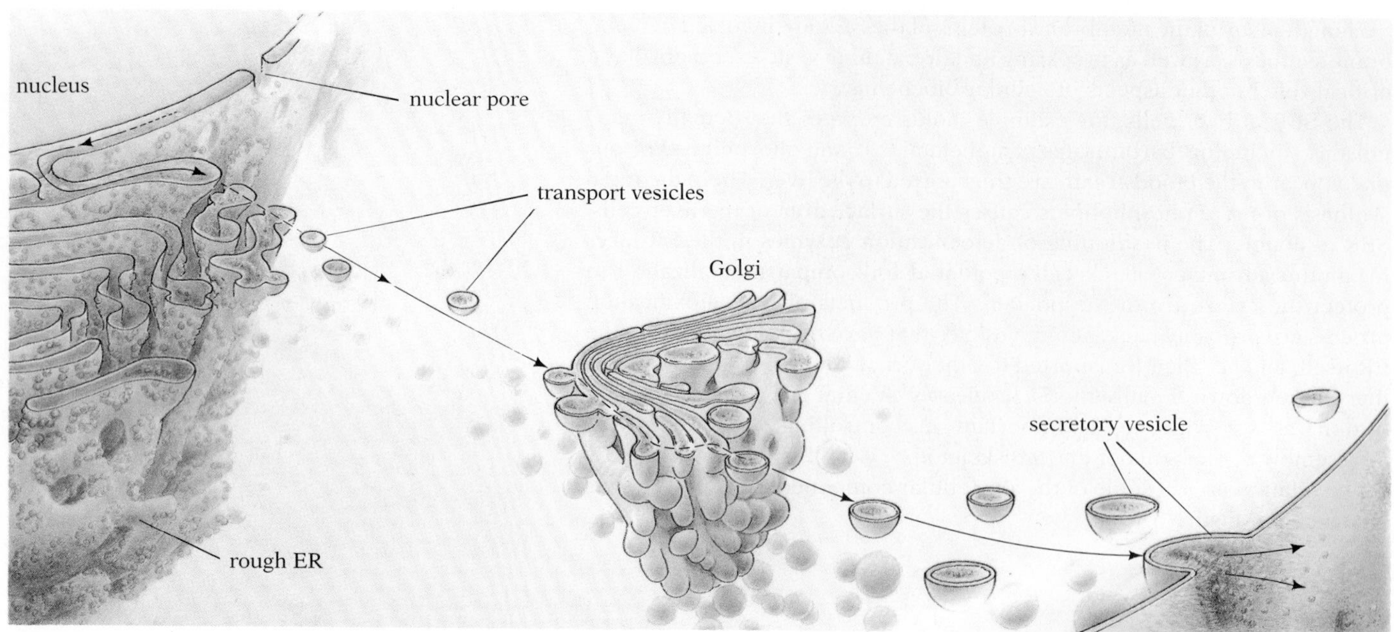

5.12 A probable path of phospholipid movement in the cell
Structural molecules of cellular membranes are constantly on the move. The path traced here shows the movement of a phospholipid from the point of synthesis in the nuclear envelope until it is incorporated into the cell membrane. The lipid first moves through the lumen of the RER and (either directly or via vesicles) into the SER. There it receives a specific carbohydrate marker, which makes it a glycolipid, and becomes part of a vesicle transporting proteins to the Golgi apparatus; it is then incorporated into the membrane of that organelle. Subsequently the same molecule, perhaps with a new or modified carbohydrate marker, moves to the outer layer of the Golgi, where it becomes part of a secretory vesicle that is carried to the plasma membrane. There, in the process of exocytosis, the vesicle membrane (including the glycolipid molecule) fuses with the plasma membrane of the cell. Other membrane molecules follow different paths. The elusive "intermediate compartment" discussed in footnote 1 is not shown.

cisterna it is concentrated and stored; it is eventually released from the cell via secretory vesicles that are produced by this outer compartment of the Golgi and move to the cell surface. Thus the role of the Golgi apparatus in secretion is clear: its functions include storage, modification (for example, removal of water or emulsification of lipids), and packaging of secretory products.

The Golgi apparatus is also the major director of macromolecular transport in cells. Though no protein synthesis takes place in the Golgi, polysaccharides are synthesized there from simple sugars and attached to proteins and lipids to create glycolipids and glycoproteins. Some of these are transported to the glycocalyx as part of vesicle membranes. In addition, proteins already marked with carbohydrate groups (and nearly all proteins transported to the Golgi apparatus from the ER are so marked) usually have their sugar-based carbohydrate tags modified there: these alterations include both further trimming of mannose groups plus the addition of new sugars. How the modified tags aid in subsequent sorting and packaging is still a mystery.

The secretory vesicles produced by the Golgi apparatus probably play an important role in adding surface area to the cell membrane. When one of these vesicles moves to the cell surface, it becomes attached to the plasma membrane and then ruptures, releasing its contents to the exterior in the process of exocytosis (Fig. 5.12). The membrane of the ruptured vesicle may remain as a permanent addition to the plasma membrane, or it may eventually migrate back to the Golgi apparatus or some other organelle as

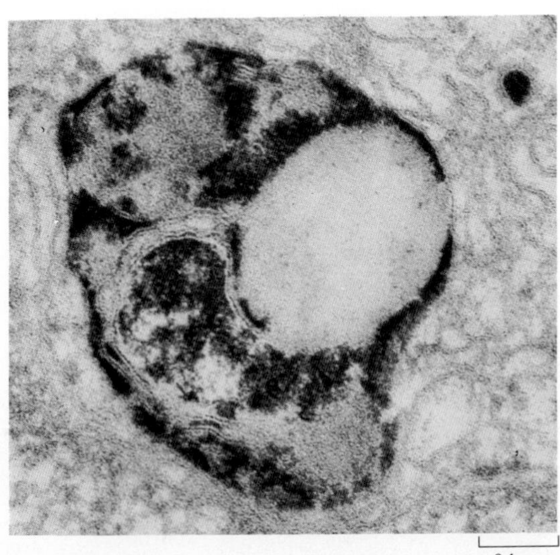

5.13 Lysosomes in a connective-tissue cell from the vas deferens of a rat
The small dark body at upper right is a primary lysosome. The much larger body at left is a secondary lysosome (digestive vacuole) formed by fusion of a primary lysosome with a phagocytic or pinocytic vesicle. (The dark appearance of the lysosomes results from staining for acid phosphatase, a digestive enzyme whose presence is used as the definitive test for these organelles.)

part of an empty vesicle. Indeed, recycling of membrane phospholipids back to the Golgi and ER is essential if the outer membrane is not to grow indefinitely.

LYSOSOMES

First described in the 1950s by the Belgian scientist Christian deDuve of the Catholic University of Louvain, ***lysosomes*** are membrane-enclosed bodies that function as storage vesicles for many powerful digestive (hydrolytic) enzymes (Fig. 5.13). The lysosome membrane contains an ionic pump that maintains a highly acidic internal environment. The membrane permits desirable reaction products to pass through to the cytosol, but it is impermeable to the hydrolytic enzymes and capable of withstanding their digestive action. If the lysosome membrane is ruptured, the hydrolytic enzymes—no longer safely confined—are released into the surrounding cytoplasm and begin immediately to break down the interior of the cell.

As you may have guessed, lysosomes act as the digestive system of the cell, enabling it to process some of the bulk material taken in by endocytosis. Their hydrolytic enzymes are synthesized in the form of zymogens in the rough ER, packaged into transport vesicles in the smooth ER, and carried to the Golgi apparatus. There, receptor proteins on the inside of the Golgi membrane form coated pits to attract and trap these zymogens, recognized by their characteristic carbohydrate markers. When a region has received an appropriate supply of the zymogens, it buds off from the Golgi membrane as a lysosome and the enzymes are activated. Proteins mounted on the exterior of the lysosome's membrane serve as recognition sites to assure that the enzymes it carries are delivered to the appropriate target. The targets of these primary lysosomes include endocytotic vesicles (endosomes) newly arrived from the cell surface, and secondary lysosomes, also known as digestive vacuoles, which are the digestive vesicles already produced by the fusion of primary lysosomes and endocytotic vesicles. When digestion is

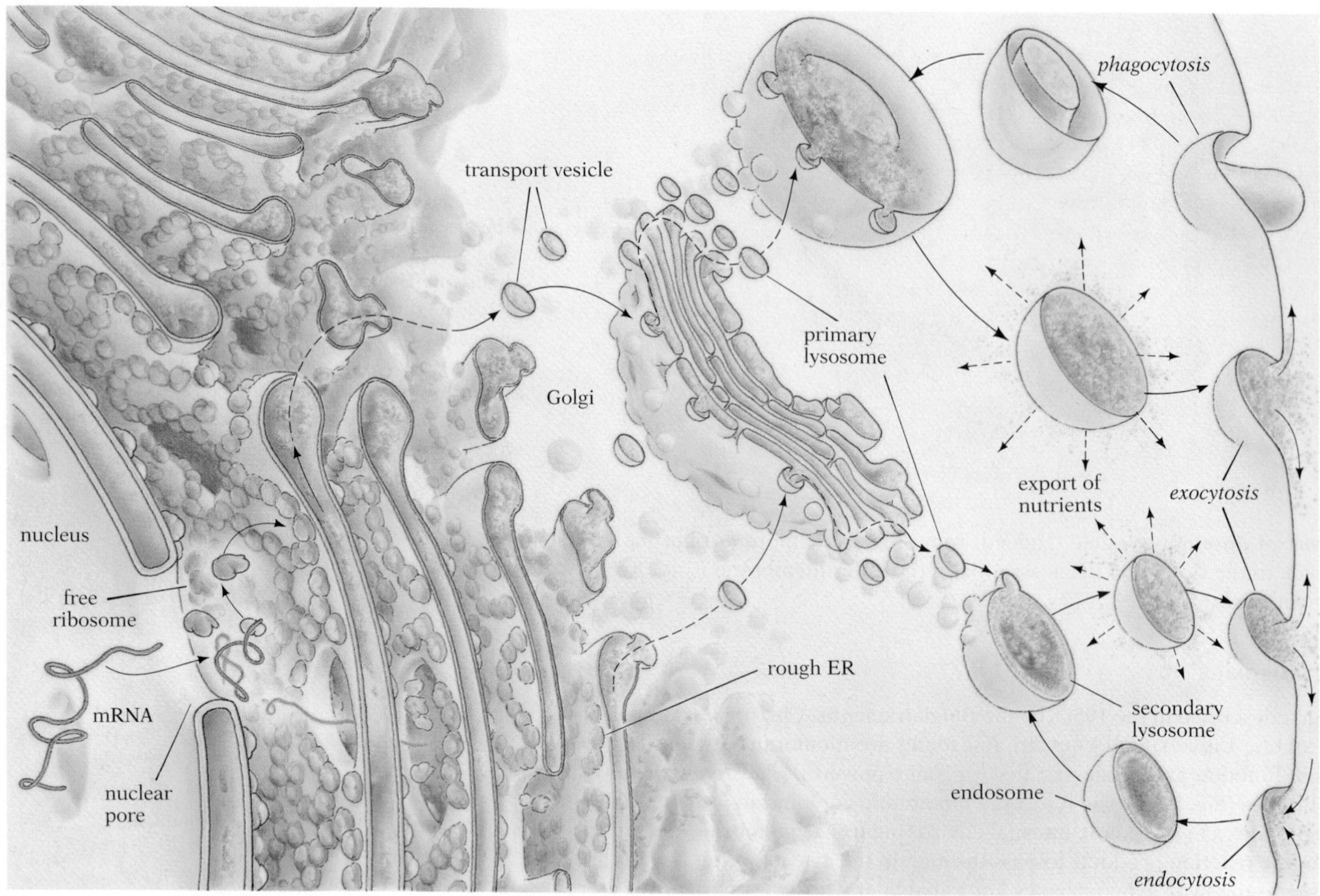

5.14 The hydrolytic enzyme cycle
The role of the ER and the Golgi apparatus in cellular digestion is representative of their functions in the cell. DNA in the cell nucleus produces the mRNA that encodes the hydrolytic enzymes. The mRNA is transported to ribosomes, which then bind to the RER. The enzymes are synthesized there, passed into the ER, and marked with signal sequences for later transport. The enzymes are next collected by receptors in the SER and packaged into transport vesicles. Markers on the outside of the transport vesicles cause them to fuse with the membrane of the Golgi apparatus, where new vesicles (primary lysosomes) are created. These packets of specific mixtures of hydrolytic enzymes fuse with appropriately marked endocytotic vesicles (endosomes) or with secondary lysosomes (produced by previous fusions of this kind) and help digest the contents. Useful products of this digestion pass through the membranes of the secondary lysosomes into the cytosol, while the residue is disposed of through exocytosis. The same pattern of synthesizing and marking enzymes in the ER, transporting them to the Golgi apparatus, sorting, chemically modifying, and repackaging them there, and then dispatching them to various intracellular targets seems to be the general strategy for managing macromolecules in cells.

complete, the useful products pass into the cytosol, while the residue is discharged by exocytosis and the vesicle membrane and any receptors it contains are reincorporated into the cell membrane (Fig. 5.14).

Several diseases are caused by lysosome disorders. In human inclusion cell disease, for instance, the enzyme that adds the distinctive sugar (mannose-6-phosphate) to the mailing label is defective; as a result most of the 40 or so hydrolytic enzymes are mismarked for secretion. Undigested "food" builds up in cells, resulting in general debilitation, followed by death when one or more of the body's organs fail. In the devastating nervous disorder known as Tay-Sachs disease, lipid-digesting lysosomes lack a particular enzyme. When these deficient lysosomes fuse with lipid-containing vesicles, they do not fully digest their contents. The resulting defective secondary lysosomes can accumulate and block the long thin parts of nerve cells responsible for transmitting nerve impulses.

The elaborate organization of membrane movement in cells is vulnerable to many infectious diseases. Semliki Forest virus, for instance, infects a wide range of hosts from invertebrates to humans. The virus carries a marker on its surface that mimics a substance for which one kind of coated pit has specific receptors. SFV is carried into the cell by endocytosis, after which its chromosome escapes into the cytosol. The virus uses the host cells' ribosomes, ER, Golgi apparatus, and the whole system of carbohydrate tags to reproduce itself and direct the subsequent exocytosis of its viral progeny. The virus thus ensures its own propagation by exploiting much of the specialized membrane machinery of the cell.

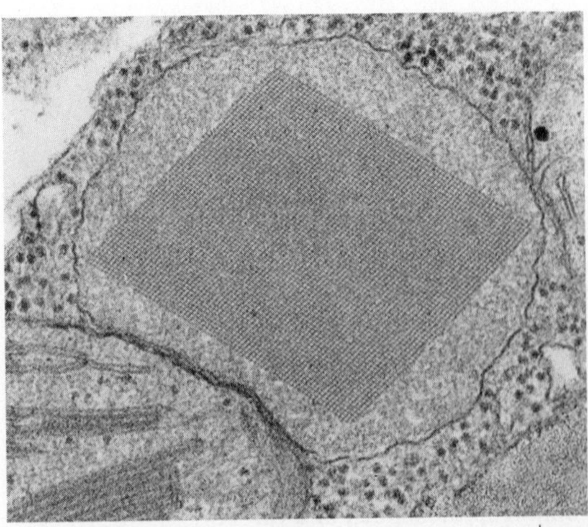

5.15 Electron micrograph of a tobacco-leaf peroxisome
The tobacco-leaf cell has been treated with stain, which reveals the crystalline core of this peroxisome and the presence in it of oxidizing enzyme. In mammals, peroxisome enzymes neutralize hydrogen peroxide, alcohol, and other potentially harmful substances.

PEROXISOMES

Improved methods for separating cell contents have revealed that certain membrane-bounded organelles previously confused with lysosomes actually have their own distinct identity: they are now known collectively as microbodies. The most thoroughly understood are the ***peroxisomes*** (Fig. 5.15). Like lysosomes, peroxisomes contain an assortment of powerful enzymes. But where the enzymes of the lysosomes are hydrolytic (water-splitting), these enzymes catalyze condensation reactions, such as the oxidative removal of amino groups from amino acids, the detoxification of alcohol, the oxidation of the dangerous compound hydrogen peroxide into water and oxygen, and the reactions involved in the production of macromolecules used in respiration and other synthetic pathways. In the leaves of green plants, peroxisomes are involved in photorespiration (to be discussed in Chapter 7).

Though similar in appearance to lysosomes, peroxisomes are not produced by budding from the Golgi apparatus. In fact, peroxisomes grow through accretion of lipids and enzymes, and multiply by fissioning. A cell that fails to inherit at least one peroxisome from the cytoplasm of its parent cell cannot make its own from scratch, and will die. Peroxisome growth requires elaborate signaling and control. The necessary new membrane phospholipids are synthesized in the SER, then recognized and picked up by transfer proteins, carried through the cytoplasm, and deposited in the peroxisomal membrane. The precursors of the enzymes peroxisomes contain are produced in the cytoplasm rather than the ER, and are recognized and transported to these tiny organelles. Just as proteins synthesized on the ER must be supplied with labels to assure accurate delivery, so too proteins built in the cytoplasm must have signals to assure proper routing, whether to peroxisomes or other destinations. In the cytoplasm, however, the signals are embedded in the amino acid sequence at one end of the protein

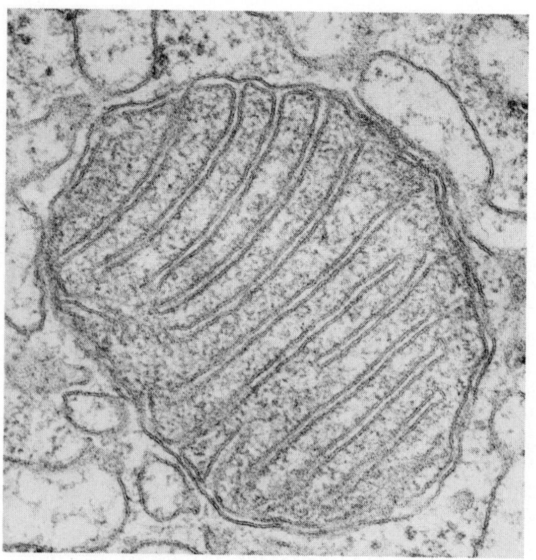

5.16 Electron micrograph of a mitochondrion from an epithelial cell of a rat
Note the double outer membrane and the numerous cristae, which can be seen to arise as folds of the inner membrane.

rather than in a carbohydrate side chain. Similarly, an amino acid signal sequence embedded at one end of proteins synthesized in the cytoplasm and destined for mitochondria assures transport to that organelle.

We have now discussed the functioning of a variety of subcellular organelles, ranging from the nucleus to the lysosomes. But we have yet to discuss the organelles that provide the energy necessary to build and fuel all the others. These are the mitochondria and the chloroplasts.

MITOCHONDRIA

Mitochondria, often thought of as the powerhouses of the cell, are the sites of the chemical reactions known as respiration, which extract energy from food and make it available for innumerable energy-demanding activities. Each mitochondrion is bounded by a double membrane; the outer membrane is smooth, while the inner membrane has many inwardly directed folds (Fig. 5.16; see also Fig. 6.10, p. 179). These folds, called ***cristae***, extend into an amorphous semifluid matrix. Reactants—fatty acids and pyruvic acid (an energy-rich product of glucose specially suitable for "burning" in mitochondria)—are concentrated in this organelle, and here, with the help of the appropriate enzymes, either of these fuels can combine with oxygen to generate water, carbon dioxide, and the energy that runs the cell. We will discuss the operation of this important organelle more fully in Chapter 6.

PLASTIDS

Plastids are large cytoplasmic organelles found in the cells of plants, but not in those of fungi or animals. Plastids are clearly visible through an ordinary light microscope. There are two principal categories: ***chromoplasts*** (colored plastids) and ***leucoplasts*** (white or colorless plastids).

Chloroplasts are chromoplasts that contain the green pigment ***chlorophyll***, along with various yellow or orange pigments called ***carotenoids***. We shall explore in detail in Chapter 7 how the radiant energy of sunlight is trapped in the chloroplasts by molecules of chlorophyll and is then used in the manufacture of complex organic molecules (particularly glucose) from simple inorganic raw materials such as water and carbon dioxide. Oxygen is a by-product of this photosynthetic reaction upon which nearly all organisms depend.

The electron microscope reveals that the typical chloroplast is bounded by two concentric membranes and has, in addition, a complex internal membranous organization (Fig. 5.17). The fairly homogeneous internal proteinaceous matrix is called the ***stroma***. Numerous flat compartments called ***thylakoids***, or lamellae, are embedded in it. In most higher plants these thylakoids come in two varieties: separate thylakoids that run through the stroma, and stacks of platelike thylakoids known as ***grana*** (Fig. 5.17). Some photosynthetic organisms—the brown algae, for example—have no grana: all the thylakoids are stromal. The chlorophyll and carotenoids are bound to the proteins and lipids in the membranes of the thylakoids. The precise arrangement of the protein, lipid, and pigment components in the thylakoids is essential for photosynthesis. This structure will be considered in more detail in Chapter 7, when we examine photosynthesis.

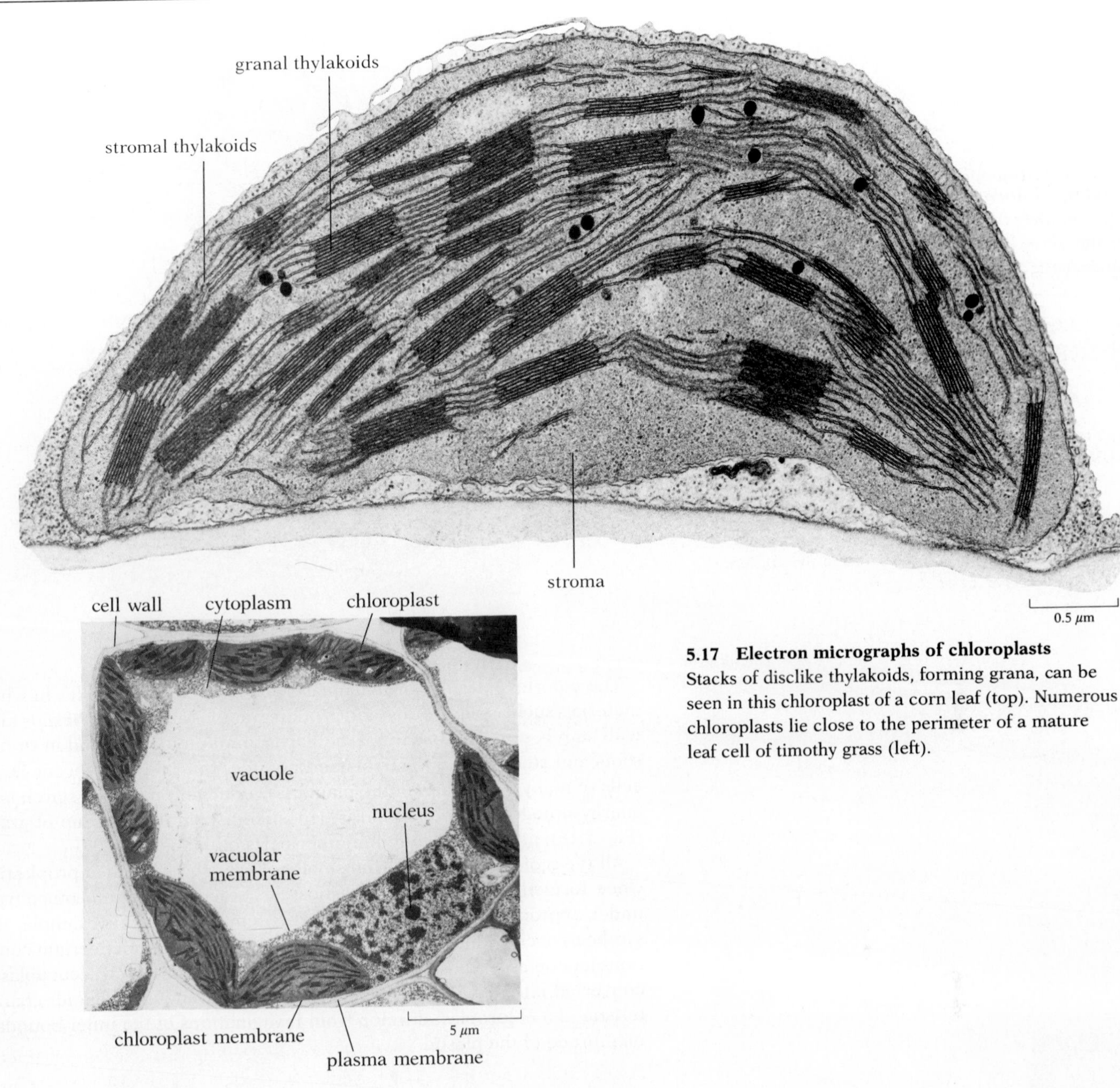

5.17 Electron micrographs of chloroplasts
Stacks of disclike thylakoids, forming grana, can be seen in this chloroplast of a corn leaf (top). Numerous chloroplasts lie close to the perimeter of a mature leaf cell of timothy grass (left).

Chromoplasts with little or no chlorophyll are usually yellow or orange (occasionally red) because of the carotenoids they contain. It is these kinds of plastids that give many flowers, ripe fruits, and autumn leaves their characteristic yellow or orange color. Some of these chromoplasts have never contained chlorophyll, while others are formed from chloroplasts whose chlorophyll has been lost. The latter are particularly common in ripe fruits and autumn leaves, structures that were once green.

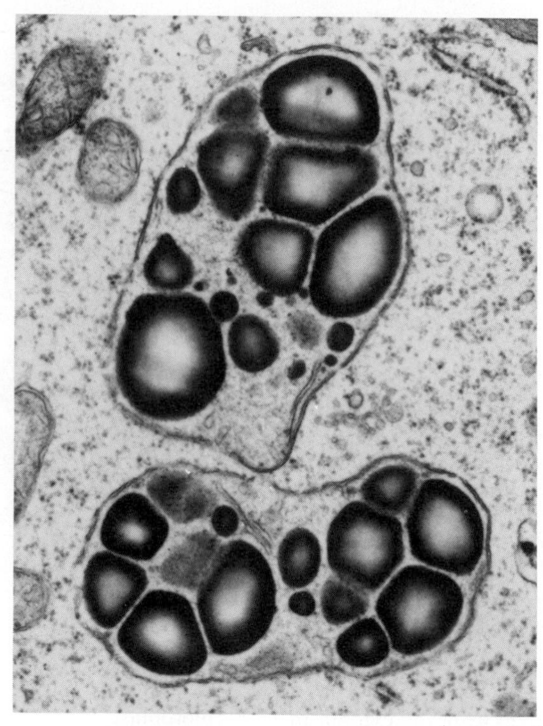

5.18 Electron micrograph of leucoplasts from the root tip of *Arabidopsis*
Because they contain numerous prominent starch grains, these leucoplasts from the small desert plant *Arabidopsis* are called amyloplasts.

The colorless plastids, or leucoplasts, are primarily organelles in which materials such as starch, oils, and protein granules are stored. Plastids filled with starch, ***amyloplasts***, are particularly common in seeds and in storage roots and stems, such as carrots and potatoes, but they also occur in the cells of many other parts of the plant. As we saw in Chapter 3, starch is an energy-storage compound, and is deposited as a grain or group of grains (Fig. 5.18); no starch is found in other parts of the cell.

All types of plastids form from small colorless bodies called proplastids. Once formed, many kinds of plastids can be converted into other types under appropriate conditions. It can be demonstrated, for example, that synthesis of chlorophyll is dependent on light and that under certain conditions leucoplasts exposed to light develop chlorophyll. When a leucoplast is converted into a chloroplast, the internal membranous thylakoids characteristic of a chloroplast develop from invaginations of the inner boundary membrane of the plastid.

VACUOLES

Membrane-enclosed, fluid-filled spaces called ***vacuoles*** are found in both animal and plant cells, though they have their greatest development in plant cells. There are various kinds of vacuoles, with a corresponding variety of functions. In some protozoans, specialized vacuoles, called contractile vacuoles, play an important role in expelling excess water and some wastes from the cell; we will discuss them in greater detail in a later

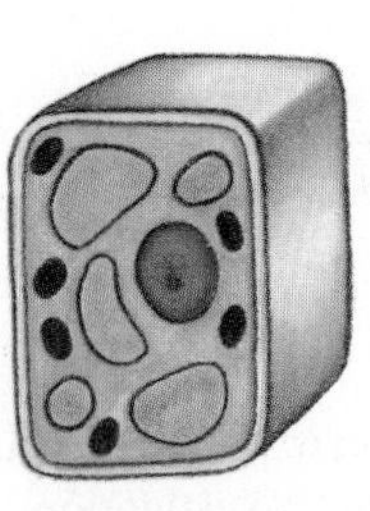
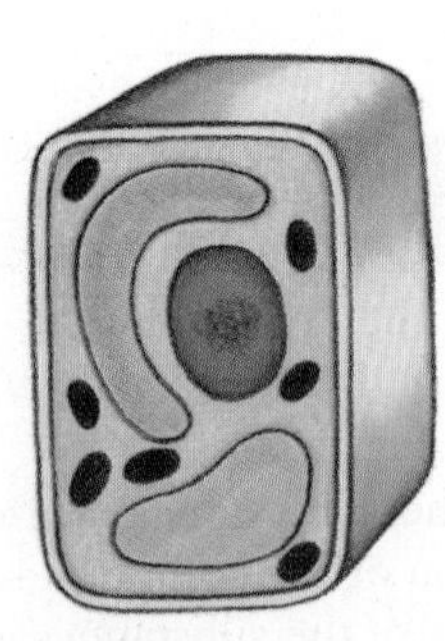
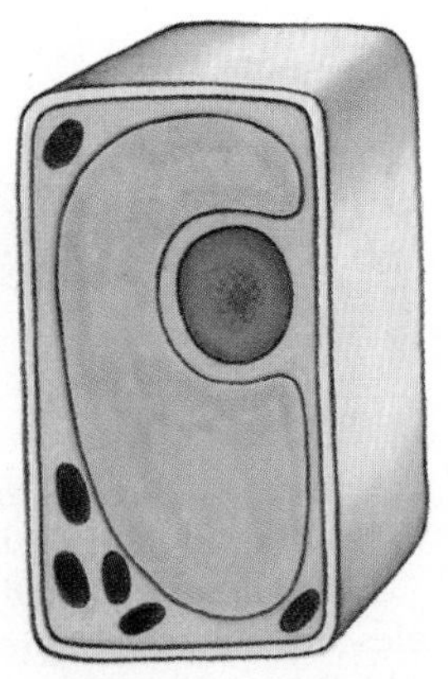

5.19 Development of the plant-cell vacuole The immature cell (left) has many small vacuoles. As the cell grows, these vacuoles fuse and eventually form a single large vacuole, which occupies most of the volume of the mature cell (right), the cytoplasm having been pushed to the periphery.

chapter. Many protozoans also have food vacuoles, chambers that contain food particles. They are similar to the vesicles formed by many cells when they take in material by endocytosis.

The distinction between vesicles and vacuoles, both of which are membrane-bounded, is hard to draw with any precision, particularly since vesicles may fuse with or bud off from vacuoles. The most obvious differences are in permanence, activity, and size: vesicles are relatively short-lived transport vehicles, while vacuoles tend to be long-lived; vesicles usually move quickly, while vacuoles are relatively static; finally, vesicles are usually small, while vacuoles are most often quite large.

In most mature plant cells, a large vacuole occupies much of the volume of the cell. The immature cell usually contains many small vacuoles. As the cell matures, the vacuoles take in more water and become larger, eventually fusing to form the very large definitive vacuole of the mature cell (Fig. 5.19). This process pushes the cytoplasm to the periphery of the cell, where it forms a relatively thin layer.

The plant vacuole contains a liquid called cell sap—primarily water, with a variety of substances dissolved in it. Since the cell sap is generally hypertonic relative to the external medium, the vacuole tends to take in water by osmosis. As the vacuole swells, the vacuolar membrane (or tonoplast, as it is often called) pushes outward against the cytoplasm, which, being essentially fluid, resists compression and transmits the pressure to the cell wall. The wall is strong enough to limit the swelling and prevent the cell from bursting, but the outward push of the vacuolar membrane is sufficient to maintain cell turgidity and stiffness.

Many substances of importance in the life of the plant cell are stored in the vacuoles, among them high concentrations of soluble organic nitrogen compounds, including amino acids; vacuoles also store sugars, various organic acids, and some proteins.

The vacuoles also function as dumping sites for noxious wastes. Enzymes secreted into them degrade some of these wastes into simpler substances that can be reabsorbed into the cytosol and reused. The presence of poisonous wastes and powerful precipitating or denaturing agents in the vacuoles has greatly complicated studies of plant biochemistry, for when entire cells

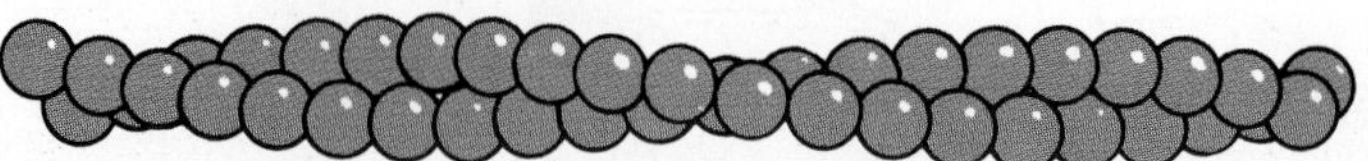

5.20 An actin microfilament
This portion of an actin microfilament shows the helically intertwined chains of protein subunits.

are disrupted, the mixing of these substances with the cytosol often results in the alteration or destruction of the compounds under investigation.

As might be expected, many of the substances accumulated in the vacuoles are selectively prevented from leaving by the vacuolar membrane, which must have its own distinctive permeability characteristics and must be capable of regulating the direction of movement of substances across it. If living beet cells are placed in distilled water, the pigment in the cell sap, one of a group of red pigments called betacyanins, does not diffuse out, even though it is in much higher concentration inside the vacuoles than outside. As soon as the beet cells die, the vacuolar membranes lose their selectivity and the betacyanin diffuses out.

Anthocyanins, another group of red pigments in the cell sap, are responsible for many of the purples, blues, and dark reds commonly seen in flowers, fruits, and autumn leaves (we have already seen that the carotenoids in the plastids are responsible for orange, yellow, and sometimes light red in these same structures). The relative amounts of anthocyanins and carotenoids in autumn leaves differ for different species of plants and also for the same species under different conditions. A high accumulation of sugars, low temperatures, and adequate light favor anthocyanin formation.

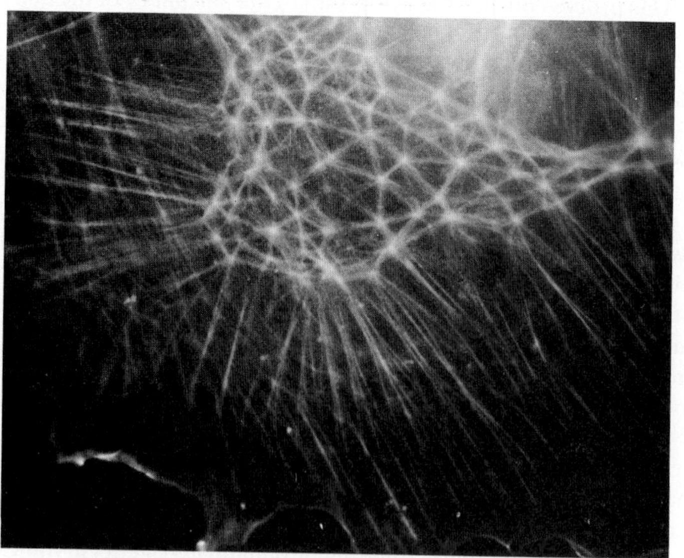

5.21 Actin in a rat fibroblast cell
Actin microfilaments appear to run between distinct anchor points on the membrane, providing structure and a scaffolding for movement.

THE CYTOSKELETON

As we have seen, the cell is full of specialized, membrane-lined organelles that mediate much of cellular chemistry. But the internal architecture of the cell also includes a variety of important protein-based components that help organize movement, not only within the cell but by the cell itself, and that aid in defining and controlling cell shape.

MICROFILAMENTS

Molecules of a protein called ***actin*** polymerize spontaneously under conditions of elevated Ca^{++} and Mg^{++} concentration, or normal levels of K^+ and Na^+. The resulting long, extremely thin polymers, helically intertwined, form actin microfilaments (Fig. 5.20), and account for 2–20 percent of the protein in most cells. A filament spanning the entire length of a cell can form in a matter of minutes.

Actin microfilaments can play a purely structural role; when cross-linked for strength they are a component of the cytoskeleton, the complex weblike array of molecules that helps maintain cell shape. This network is densest just under the cell membrane (Fig. 5.21), to which it is anchored by special proteins. Parallel, cross-linked arrays of actin microfilaments provide reinforcement for various stiff cellular protuberances, including the dense forests of rodlike projections known as microvilli that line the intestine (Fig. 5.22).

In addition to its structural role, actin is involved in cell movement through interactions with a second cytoskeleton protein, myosin. ***Myosin*** consists of a two-part head plus an anchor. The tip of the head can bind to actin, while the base can accept energy from ATP, which causes the head to twist relative to the anchor. This twisting is harnessed to effect movement (Fig. 5.23).

In muscle cells, which we will discuss in detail in Chapter 37, the anchor is actually a long tail that binds to other myosin tails to create a highly specialized bundle that makes coordinated contractions possible. In nonmuscle cells, the myosin anchor is relatively short, and is often (perhaps always) bound to membranes.

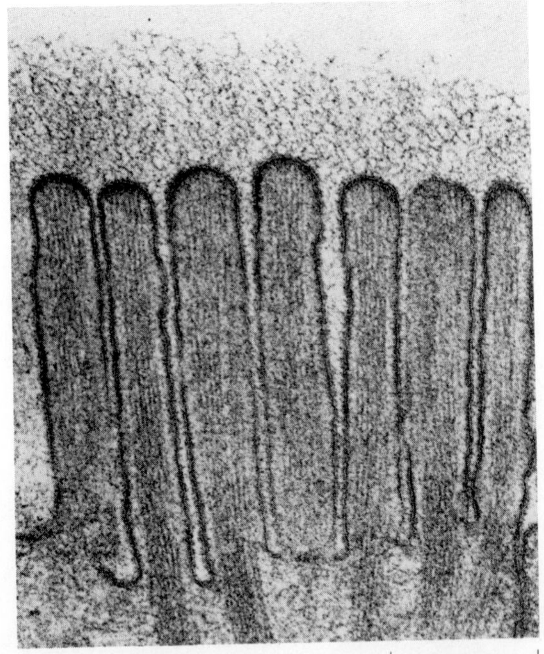

5.22 Microfilaments in the microvilli
Cross-linked into tight bundles, actin microfilaments in the microvilli provide rigidity.

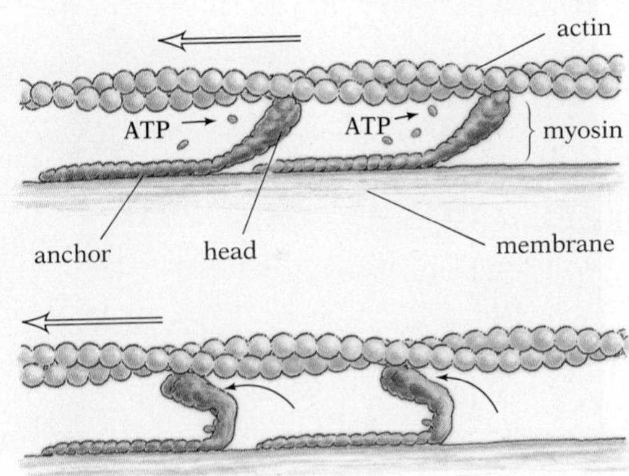

5.23 Interaction of actin and myosin
Some (perhaps all) nonmuscle myosin is anchored in membranes and, when fueled by ATP, can bind to actin microfilaments and undergo a conformational change. The resulting movement of the myosin head (described in detail in Chapter 37) moves the membrane (to which the anchor is permanently connected) and the microfilament (to which the head is temporarily attached) past one another in opposite directions. Myosin may also be able to anchor itself to vesicles or actin filaments, and so move them along the microfilament network. Myosin can create movement of 1–9 microns per second.

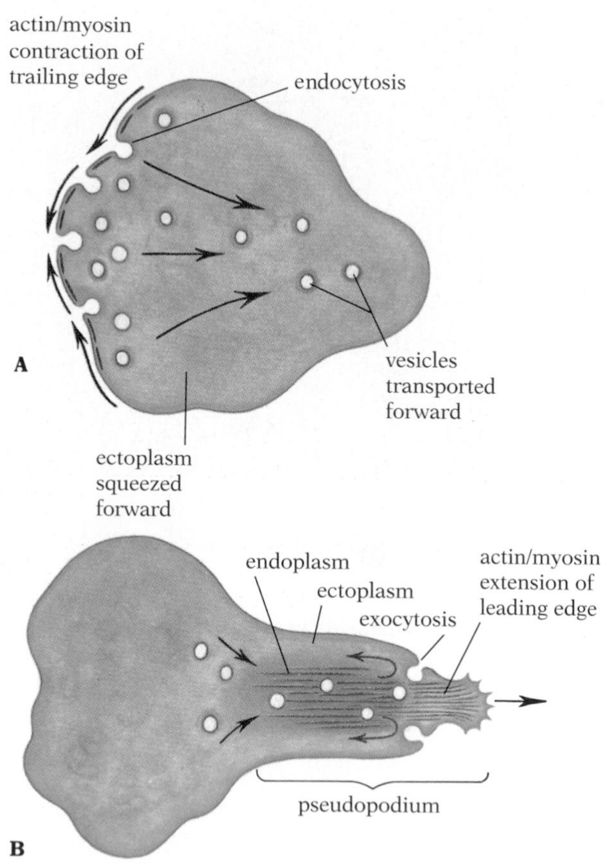

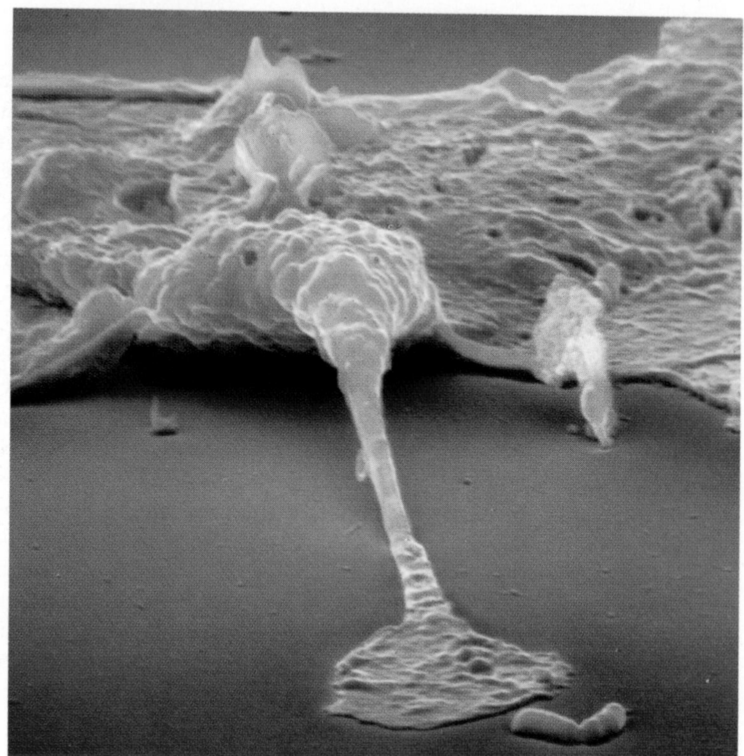

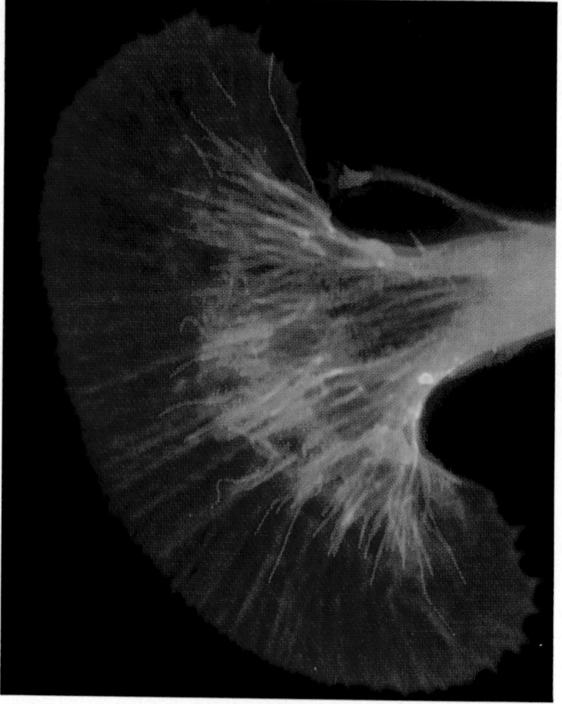

5.24 Amoeboid movement
(A) As the trailing edge of the cell contracts, squeezing the cytoplasm forward, the cytoplasm moving into the pseudopod loses its actin cross-linking, and becomes more fluid (endoplasmic). This endoplasmic core slides into the pseudopodium between the peripheral layers of stationary ectoplasm. (B) The pseudopod extends as actin filaments push it forward and vesicles transported from the rear provide more membrane. As the core moves forward, ectoplasm at the rear of the pseudopod and from the rest of the cell is converted into endoplasm, while endoplasm at the front of the pseudopod is converted into ectoplasm. (C) In this dramatic example of actin-based cell movement, a phagocyte (an amoeboid cell in the blood that captures foreign material by endocytosis) is extending a pseudopodium toward a bacterium, which it has detected on the basis of a waste chemical released by the bacterium. (D) Actin bundles are visible in the tip of the pseudopodium of this migrating nerve cell.

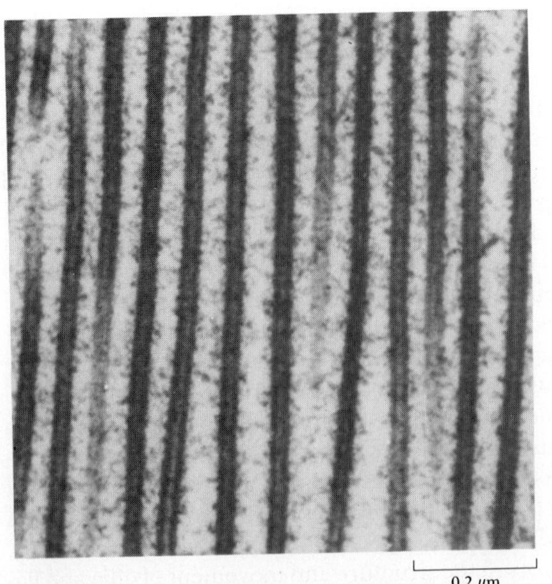

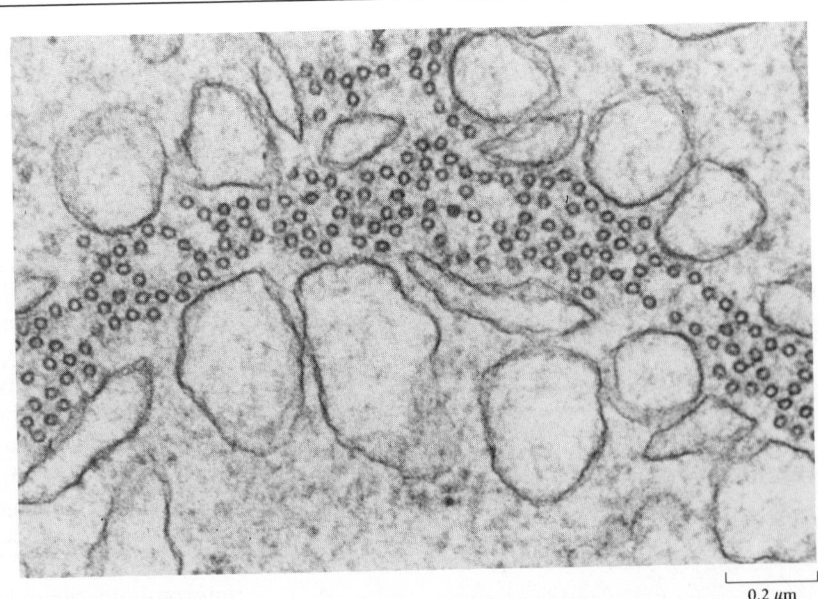

5.25 Electron micrographs of microtubules
Left: Longitudinal section, from bovine brain. Right: Cross section, from hamster spermatid.

The best understood case of myosin-mediated movement in nonmuscle cells occurs during cell division, when the daughter cells are created (a process described more in Chapter 12). Actin filaments are formed along the midline of the parent cell, and myosins (perhaps mounted on the membrane) "pinch off" the two cells from one another.

Similar actin/myosin interactions are responsible for cell movement (Fig. 5.24). At the trailing edge of the cell the membrane is being constantly tightened; this myosin-based activity squeezes the cytoplasm toward the front. Vesicles produced from the rear membrane are sent forward to enable the leading edge to grow. Myosin-mediated movement of actin filaments is believed to cause extension of the pseudopod at the front, though the details of this process are still a mystery. What is clear is that while most of the cytoplasm in the cell, a region often called ectoplasm, contains highly cross-linked actin filaments and is relatively rigid, the cytoplasm flowing into the leading edge, the endoplasm, is unlinked, and so is much more fluid. Once this endoplasm has been squeezed into the extension provided by the myosin-moved actin filaments, the actin becomes cross-linked again; other regions of ectoplasm are then converted into endoplasm to feed further movement.

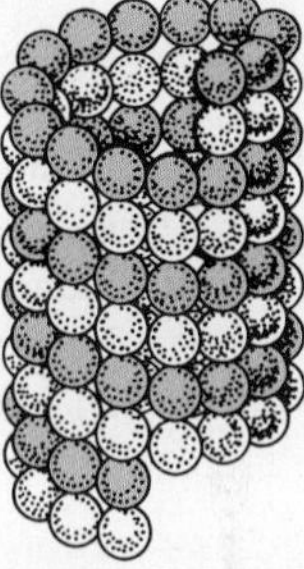

5.26 Structure of portion of a microtubule
The subunits of tubulin, each consisting of two proteins (one kind shown colored, the other white), are helically stacked to form the wall of the tubule, which is usually 13 subunits in circumference.

MICROTUBULES

Microtubules can be thought of as heavy-duty versions of microfilaments. They are long, hollow, cylindrical structures (Fig. 5.25) that, like microfilaments, form spontaneously in response to the proper ionic signals. Molecules of a globular protein called ***tubulin***, each molecule consisting of two proteins (α and β), polymerize to form a helical stack (Fig. 5.26). Microtubules radiate from ***microtubule organizing centers*** (usually near the Golgi) and play a critical role in general cell structure, in vesicle and organelle movement, and in cell division. During cell division the microtubules radi-

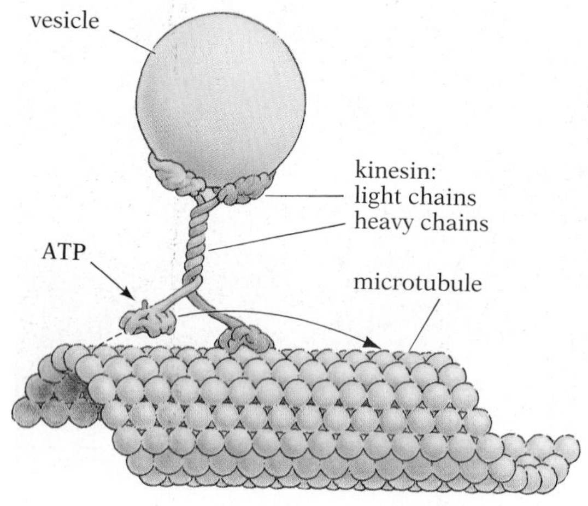

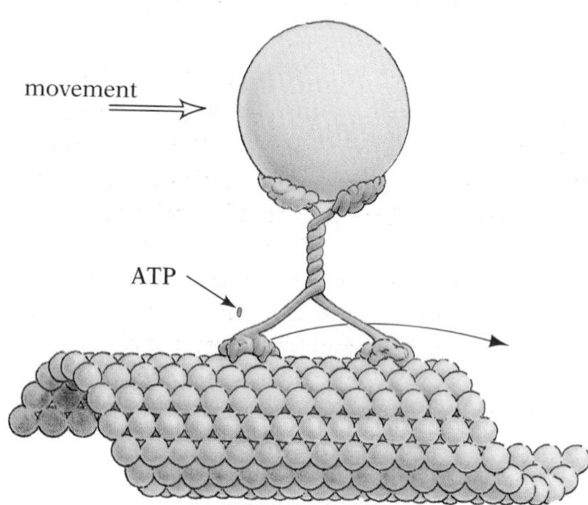

5.27 Kinesin-mediated movement
Kinesin has two identical heavy chains (which bind to the microtubule and "walk" along it in a particular direction) and two identical light chains (which bind the vesicle to the heavy chains).

ate from ***centrosomes***, areas near structures called centrioles at each end of the cell, to form a basketlike arrangement (the spindle), which is instrumental in moving the chromosomes to the locations of the new nuclei (see Fig. 12.9, p. 316). Most of the tubules in this spindle interact to push on one another, thus pushing the poles of the cell apart. A special ATP-powered protein connects adjacent microtubules radiating from different poles and causes them to "walk" past each other.

Similar proteins (called kinesins) move vesicles away from the organizing center; one end binds to the vesicle while the other, which has two "feet," steps along the microtubule at speeds of 1–3 microns per second (Fig. 5.27). Other kinesin-like proteins move mitochondria and other cargo. Yet other molecules (especially dynamin) move various loads *toward* the organizing center (Fig. 5.28). The positioning of the organizing center of this elaborate transportation network near the cellular mailroom—the Golgi—is no accident. Like microfilaments, microtubules also help provide shape and support for the cell and its organelles as part of the cytoskeleton. When colchicine, a compound that prevents the stacking of tubulin, is added to a cell, it quickly loses its distinctive shape. Finally, as we will see presently, microtubules play a key role in the structure and movement of cilia and flagella.

INTERMEDIATE FILAMENTS

EM studies of the cytoplasm show that there are numerous tubular fibers larger than actin but smaller in diameter than microtubules, with (in general) a much lower degree of organization. Although there are several chemically distinct varieties of these fibers, they are lumped together and included in the category known as ***intermediate filaments*** (Fig. 5.29). (Had their hollow structure been evident when they were named, they would doubtless have been called "intermediate tubules" instead.) Other intermediate filaments are associated with the cell membrane (where they provide structure and aid in the formation of cell–cell junctions, as we saw in Chapter 4) and the nuclear envelope (Fig. 5.30).

The collection of intermediate filaments in the cytoplasm is often referred to as a lattice, implying a structural role that is yet to be firmly proved. The information we have about the cytoplasmic lattice can be interpreted in two quite different ways. For example, the proteins in the cytosol are concentrated on these fibers, while the surroundings are basically aqueous. Perhaps in living cells the proteins are mounted on the fibers and organized into arrays for the sequential processing of reactants along an enzymatic pathway, or perhaps they merely stick to the fibers when the specimen is undergoing fixation for viewing. Similarly, the free ribosomes of the cytoplasm are concentrated at lattice intersections, perhaps in living cells, perhaps as a result of fixation (Fig. 5.31). The role of the various sorts of intermediate filaments is a focal point of contemporary research.

5.28 Dynamin-mediated movement
A mitochondrion from the giant amoeba *Reticulomyxa* is being moved along a microtubule. The dynamin molecule attaching the organelle to the microtubule is clearly visible.

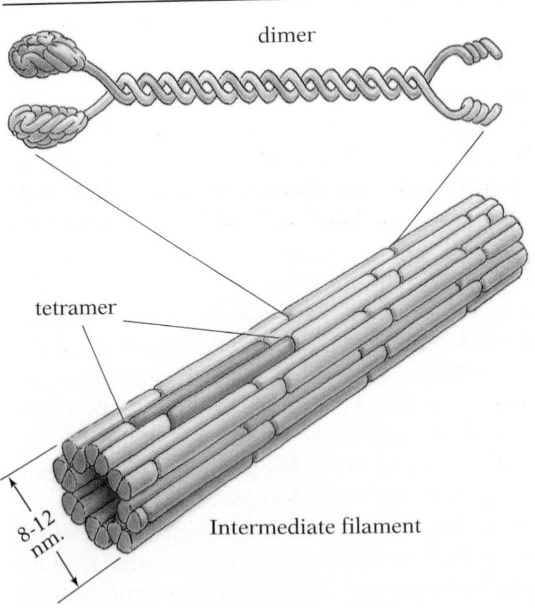

5.29 Structure of intermediate filaments
The basic subunit is a protein dimer formed from two identical molecules coiled together (top). Each dimer is paired with another, slightly offset dimer to form a tetramer. The tetramers bind head to tail to create strands. Eight strands bound together in the form of a hollow tube (bottom) generate the structure typical of intermediate filaments.

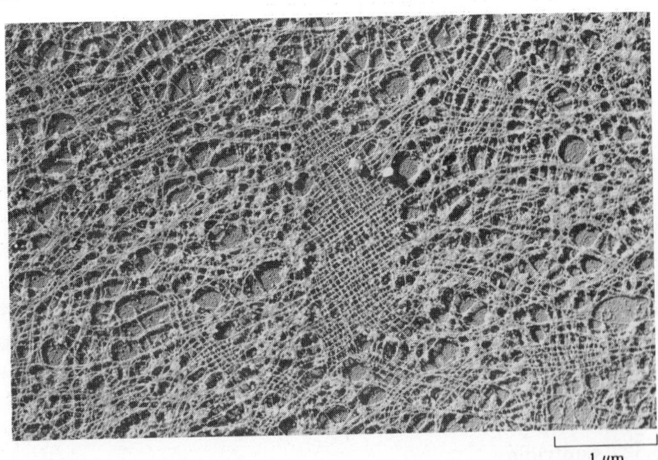

5.30 The nuclear lamina
This meshlike array of intermediate filaments lies just inside the inner membrane of the nuclear envelope in frog eggs.

5.31 The cytoplasmic lattice
The cytoplasmic lattice is thought by some researchers to be composed of a network of fibers that help give shape to the cell and hold various cellular organelles in position. The hypothesis suggests that fibers are anchored to the cell membrane, as shown in the interpretive drawing, and are also linked to cellular microtubules and microfilaments (not shown). But other researchers argue that the lattice is merely an artifact of the fixation technique used to prepare the specimen for the EM, and that cell movement, cell division, transport of vesicles along internal cell pathways, and similar phenomena are determined solely by the interactions of microtubules, intermediate filaments, and microfilaments.

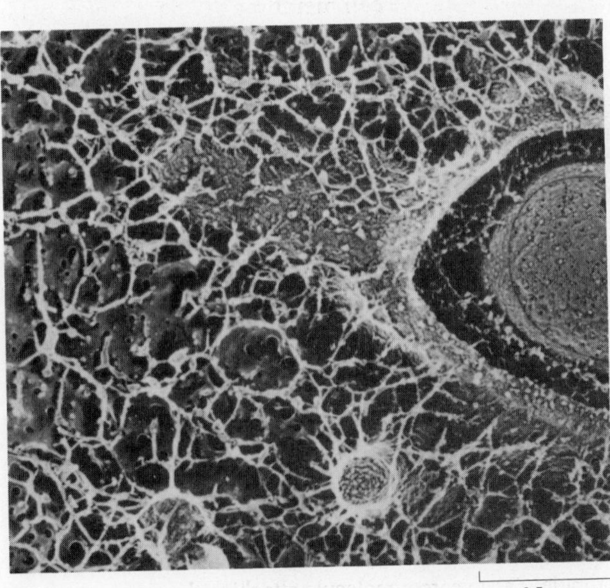

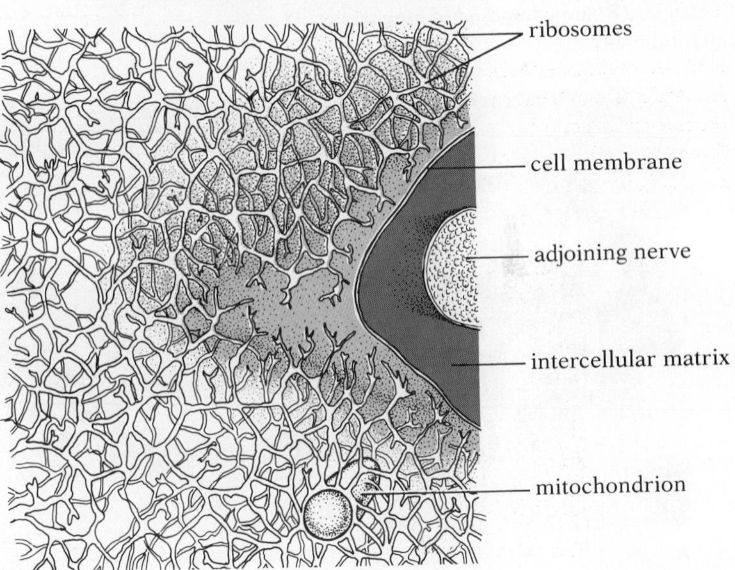

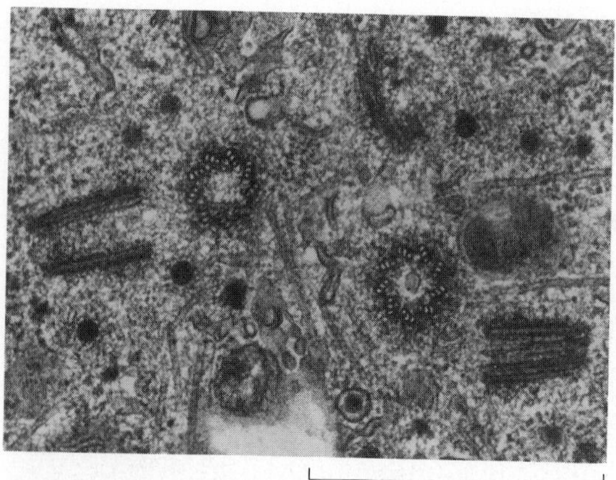

5.32 Centrioles
This electron micrograph shows newly replicated centrioles. Since the centrioles of each pair lie at right angles to each other, the sectioning of the specimen results in one from each pair being cut longitudinally and one being cut in cross section.

CENTRIOLES AND BASAL BODIES

Centrioles are found in pairs, oriented at right angles to each other, just outside the nuclei of many kinds of cells (Fig. 5.32). From the neighborhood of these organelles projects an array of microtubules. When centrioles are present, they are the focus of the microtubule spindle during cell division. In cross section, centrioles display a uniform structure, with nine triplets arranged in a circle and each triplet composed of three fused microtubules (Fig. 5.33B).

Basal bodies anchor the many hairlike cilia and flagella (discussed next) to the cell membrane. As Figure 5.33A indicates, they have exactly the same structure as centrioles and are probably in truth the same organelle: basal bodies can become centrioles, and vice versa. During cell division in the mobile alga *Chlamydomonas,* for example, the basal bodies of the two flagella abandon their posts and migrate to the poles of the cell, where they take up their positions in the spindle. Similarly, the basal body of the flagellum in many kinds of sperm becomes one centriole of the egg after fertilization. And the centrioles of cells that differentiate to line the oviduct multiply to become the basal bodies of the cilia whose rhythmic beating moves eggs from the ovary to the site of fertilization.

In the past, centrioles have often been thought of as the organizing centers of the spindle, but now a great deal of evidence suggests a more passive role. In at least some cells the absence or destruction of the centrioles does not prevent subsequent spindle formation and cell division. In fact, centrioles are typically surrounded by a dense material known as satellite bodies, and this same material is found at the organizing centers in the centrosomes of cells lacking centrioles. It may be that centrioles are present merely to facilitate the production of basal bodies, and that they follow the satellite bodies into the separating cells during cell division.

5.33 Basal bodies and centrioles
Basal bodies (A), such as the three shown in cross section in this electron micrograph of a protozoan, are essentially identical to centrioles in structure. Centrioles (B) are composed of nine triplet microtubules.

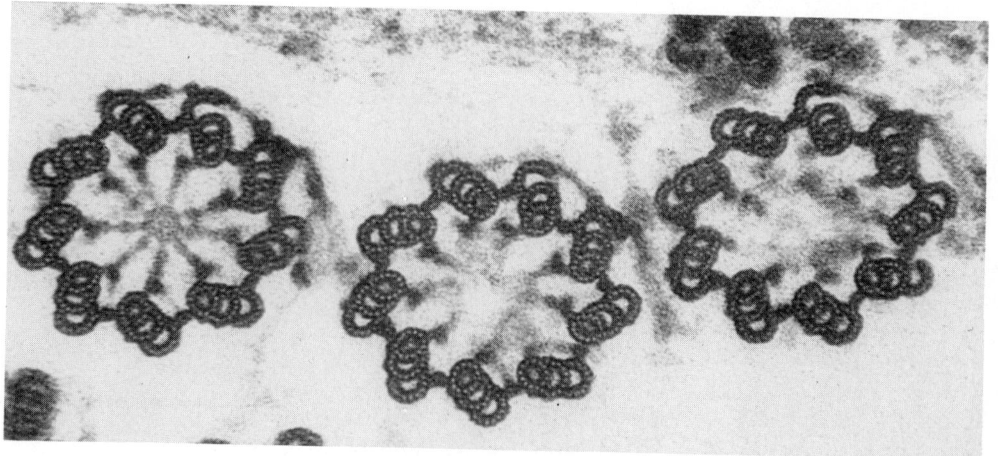

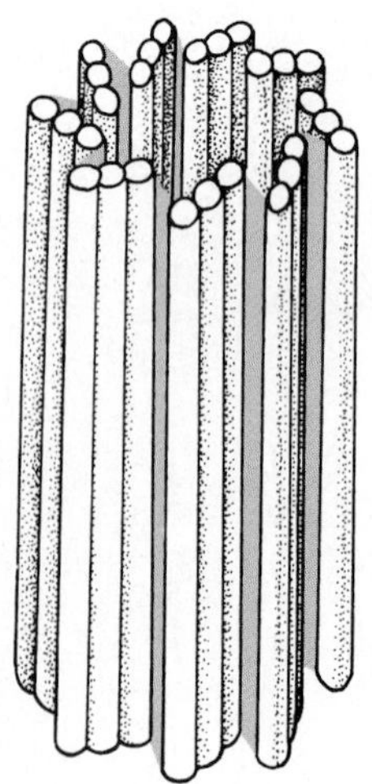

CILIA AND FLAGELLA

Some cells of both plants and animals have one or more movable hairlike structures projecting from their free surfaces. If there are only a few of these appendages and they are relatively long in proportion to the size of the cell, they are called ***flagella***. If there are many and they are short, they are called ***cilia*** (Fig. 5.34). Actually, the basic structure of flagella and cilia in eucaryotes is the same, and the terms are often used interchangeably. Both usually function either in moving the cell, or in moving liquids (or small particles) across the surface of the cell. They occur commonly on unicellular and small multicellular organisms and on the male reproductive cells of most animals and many plants, in both of which they may be the principal means of locomotion. They are also common on the cells lining many internal passageways and ducts in animals, where their beating aids in moving materials through the passageways. In the trachea they reach densities of a billion per square centimeter.

The flagella and cilia of eucaryotic cells are an extension of the cell membrane, containing a cytoplasmic matrix, with eleven groups of microtubules embedded in the matrix. Almost invariably, nine of these groups are fused pairs arranged around the periphery of the cylinder, while the other two are isolated microtubules lying in the center (Fig. 5.35). Each cilium

5.34 Scanning electron micrograph of ciliated surface of the trachea of a hamster
Coordinated movements of the many cilia function to sweep dust particles and other foreign material from the respiratory surfaces to the mouth.

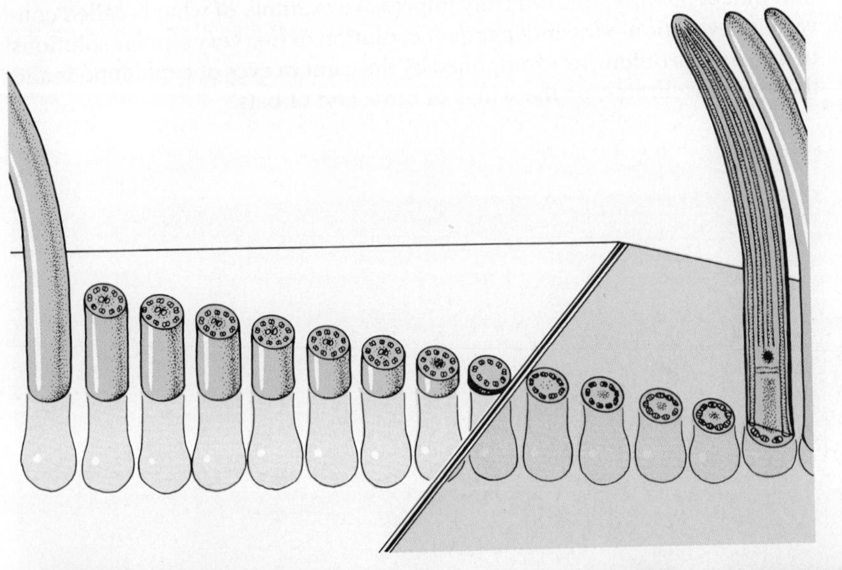

5.35 Cross section of cilia from the protozoan *Tetrahymena*
The electron micrograph of an oblique section of surface tissue reveals both the "nine plus two" arrangement of microtubules in cilia and the nine triplet microtubules characteristic of basal bodies. The interpretive drawing shows the plane of the section and the position of microtubules in a lateral section of an adjacent cilium.

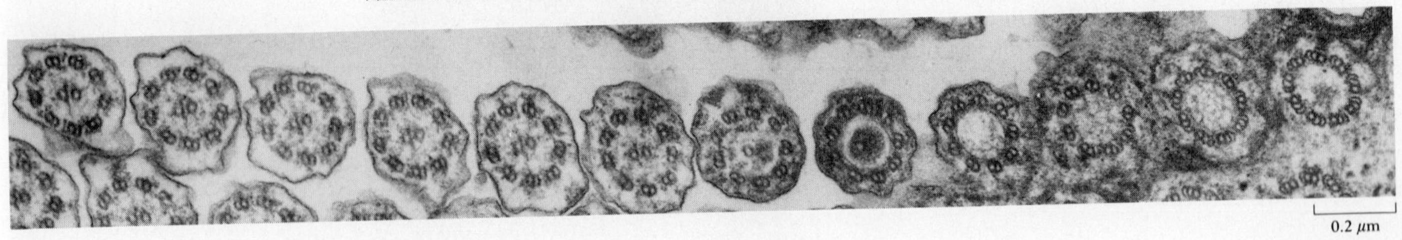

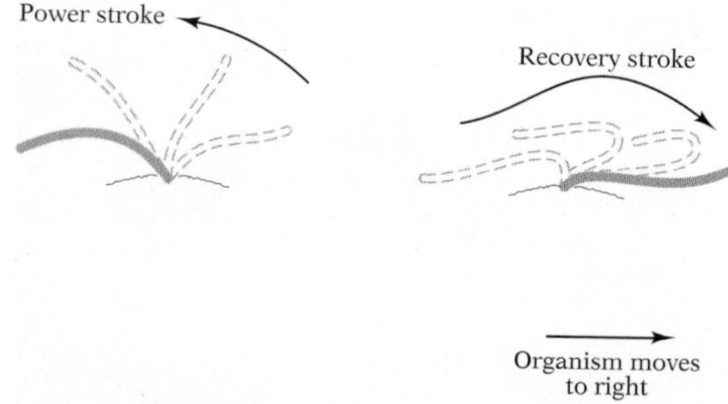

5.36 The stroke of a cilium
In the power stroke the stalk is extended fairly rigidly and swept back by bending at its base. The recovery stroke brings the cilium forward again, as a wave of bending moves along the stalk from its base. At no time during the recovery stroke is much surface opposed to the water in the direction of movement.

and flagellum is anchored to the cell by a basal body; the similarities between these organelles and the basal body in the arrangement of microtubules is obvious. The beating of cilia and flagella (Figs. 5.36 and 5.37) depends on two clawlike arms that allow each of the nine outer doublets to interact with an adjacent doublet (Fig. 5.38).

In a series of definitive experiments, Ian Gibbons of Harvard University showed that a protein called ***dynein***, found in cilia, can extract energy from ATP and, further, that when all the dynein is extracted from cilia the doublets are left armless, and when the dynein is restored the arms reappear—a clear demonstration that the arms are the site of ATP hydrolysis for the cilia. Current models of ciliary movement postulate that the arms provide the basis for a rachetlike mechanism that enables ciliary microtubules to "walk along" or slide over one another. The models postulate, further, that because of the shear resistance within the cilium, due in large part to the radial spokes that bind all the doublets to the central sheath, the sliding of some doublets past others brings about a bending of the ciliary stalk (Fig. 5.39).

The remarkable resemblance in mode of action between the tubulin-dynein system and the actin-myosin system might suggest that the two are evolutionarily related, and that one is derived from the other. But there is no evidence to support this view. The amino acid sequences in tubulin and dynein show no similarities to those of actin and myosin. These two systems evolved their similar mechanisms for producing ratchet-driven sliding motion independently; this is a truly impressive example of what is called convergent evolution—the independent evolution of two very similar solutions to the same problem, as exemplified by the camera eyes of cephalopods and those of vertebrates, or the wings of birds and of bats.

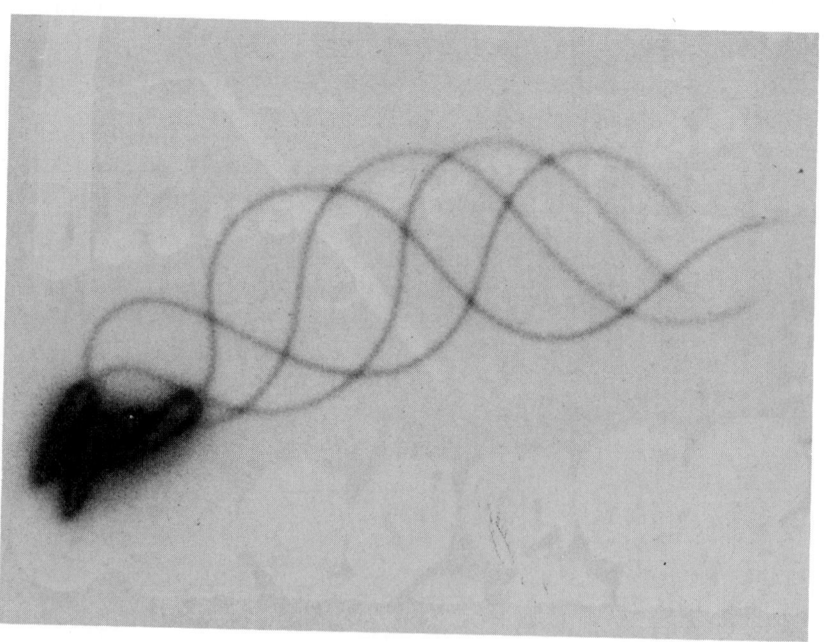

5.37 Multiple-exposure photograph of flagellum movement in a sea-urchin spermatozoon
Successive waves of bending that move along the flagellum, from its base toward its tip, push against the water and propel the sperm forward. This photograph was taken using four light flashes 10 milliseconds apart. In some protozoans, the flagella have lateral "hairs" that reverse the thrust; as a result, the waves move in the opposite direction, and so "pull" the cell along.

EUCARYOTIC VERSUS PROCARYOTIC CELLS

In spite of the extreme variety among cells, they break down into two fundamental categories: eucaryotic cells, which compose the great majority of organisms, and procaryotic cells—the bacteria. The preceding discussion has been almost wholly concerned with eucaryotic cells. Before turning to the contrasting characteristics of procaryotic cells, let us summarize the main features of eucaryotic cells in an effort to achieve an integrated picture of cell structure and function.

"TYPICAL" EUCARYOTIC CELLS

How complex cells are was not truly appreciated until the advent of the electron microscope and modern biochemical techniques, which combined to change our whole picture of the cell. Today we know there is no such thing as a "typical" cell, not even a typical eucaryotic cell. Plant and animal cells differ from one another; cells of particular plants or animals differ from those of other plants or animals; and within the body of any one plant or animal the various cells often differ strikingly in shape, size, and function. This much, of course, has been known for a long time. But now that the number of known cellular components has grown so large and their great variability has been so well demonstrated, it becomes even more obvious that no single diagram, or even series of diagrams, can really portray a typical cell. Nevertheless, to help summarize and visualize the arrangement of the organelles discussed in the preceding pages, two diagrams of relatively unspecialized cells (Fig. 5.40 and 5.41) are given here. As you examine them, keep in mind that not all the components shown always occur together in any one real cell.

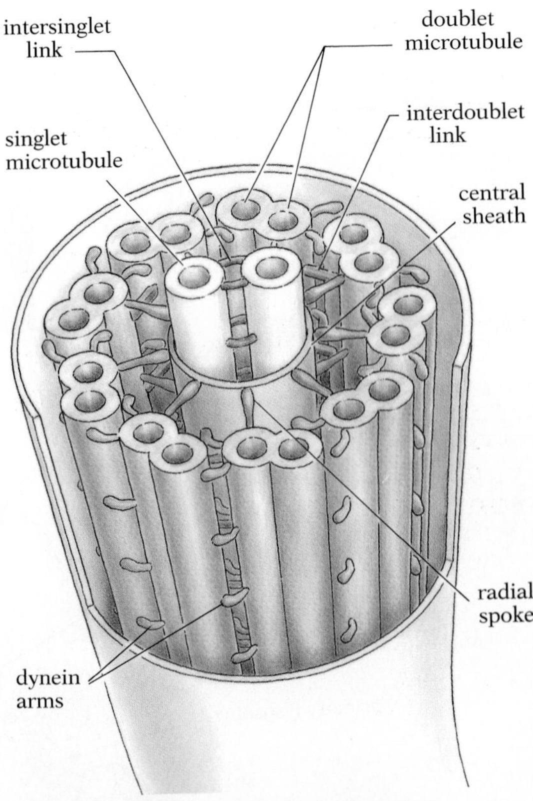

5.38 Internal structure of a cilium
The nine microtubule doublets are arranged around two singlet microtubules in the center of the cilium. Movement is generated when the dynein arms of one subset of doublets begin to "walk" along the length of other doublets. The doublets moving the most force the other doublets to bend. The intersinglet links may prevent lateral bending, and so explain why ciliary beats are constrained to a fixed plane.

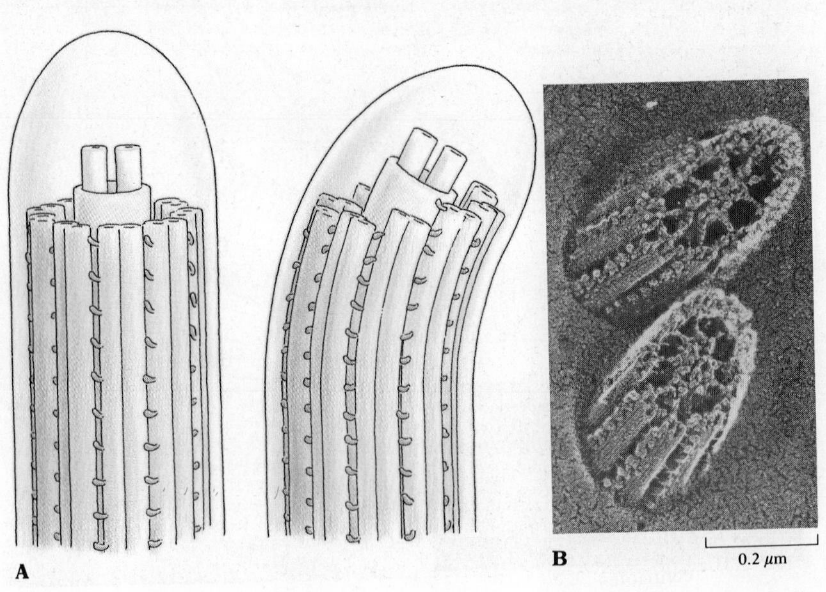

5.39 Bending of a cilium produced by sliding of microtubules past each other
(A) The microtubules on the concave side of the bend (green, right) have slid tipward. Because the tubules are all interconnected, their changes in relative position can be accommodated only if the stalk of the cilium bends. (B) Rows of outer dynein arms are visible in this electron micrograph of two *Tetrahymena* cilia that have been stripped of their outer cell membranes. The cilia were first frozen and then fractured obliquely to their axes. The surrounding matrix was subsequently etched back to expose the internal structures.

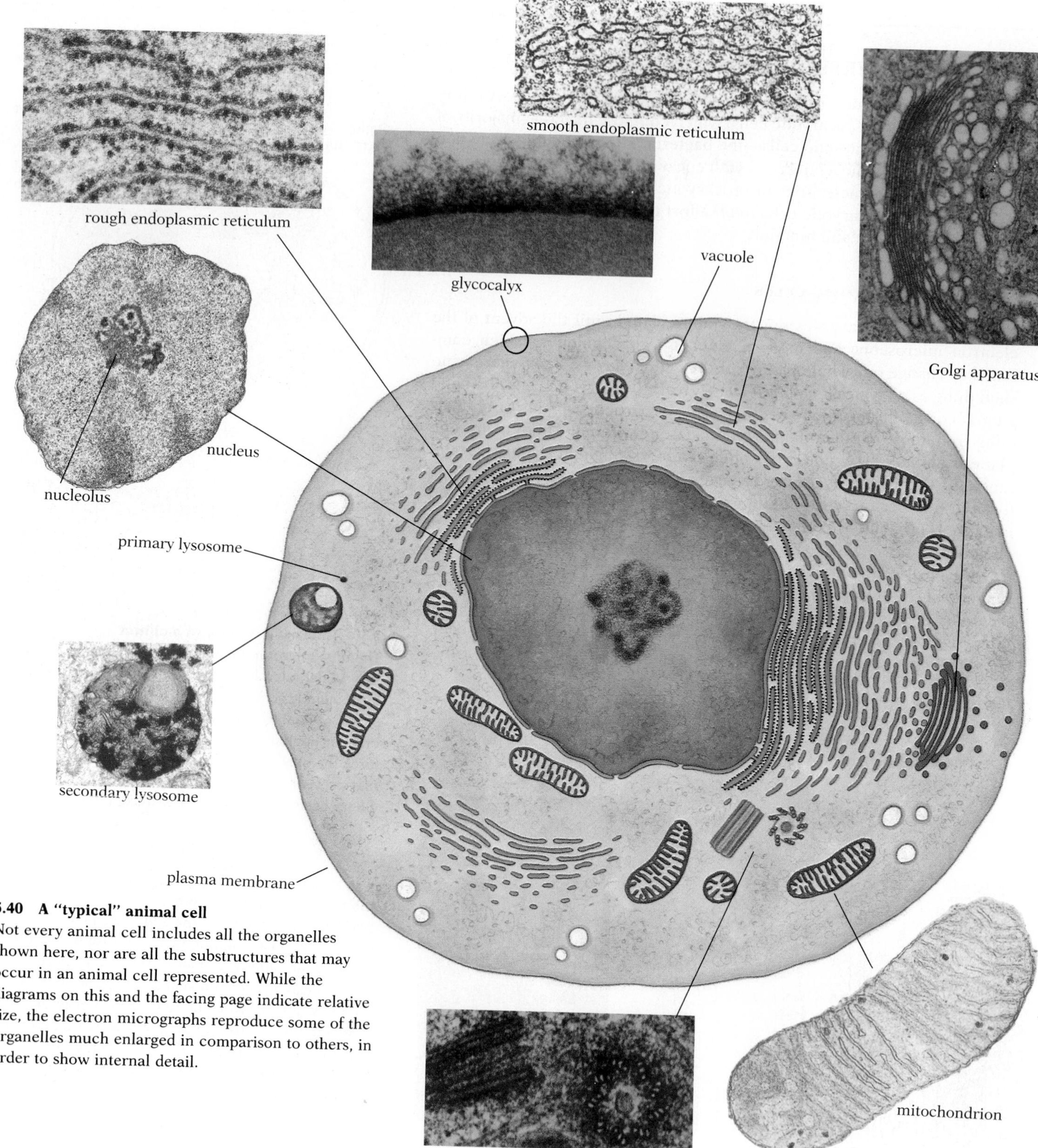

5.40 A "typical" animal cell
Not every animal cell includes all the organelles shown here, nor are all the substructures that may occur in an animal cell represented. While the diagrams on this and the facing page indicate relative size, the electron micrographs reproduce some of the organelles much enlarged in comparison to others, in order to show internal detail.

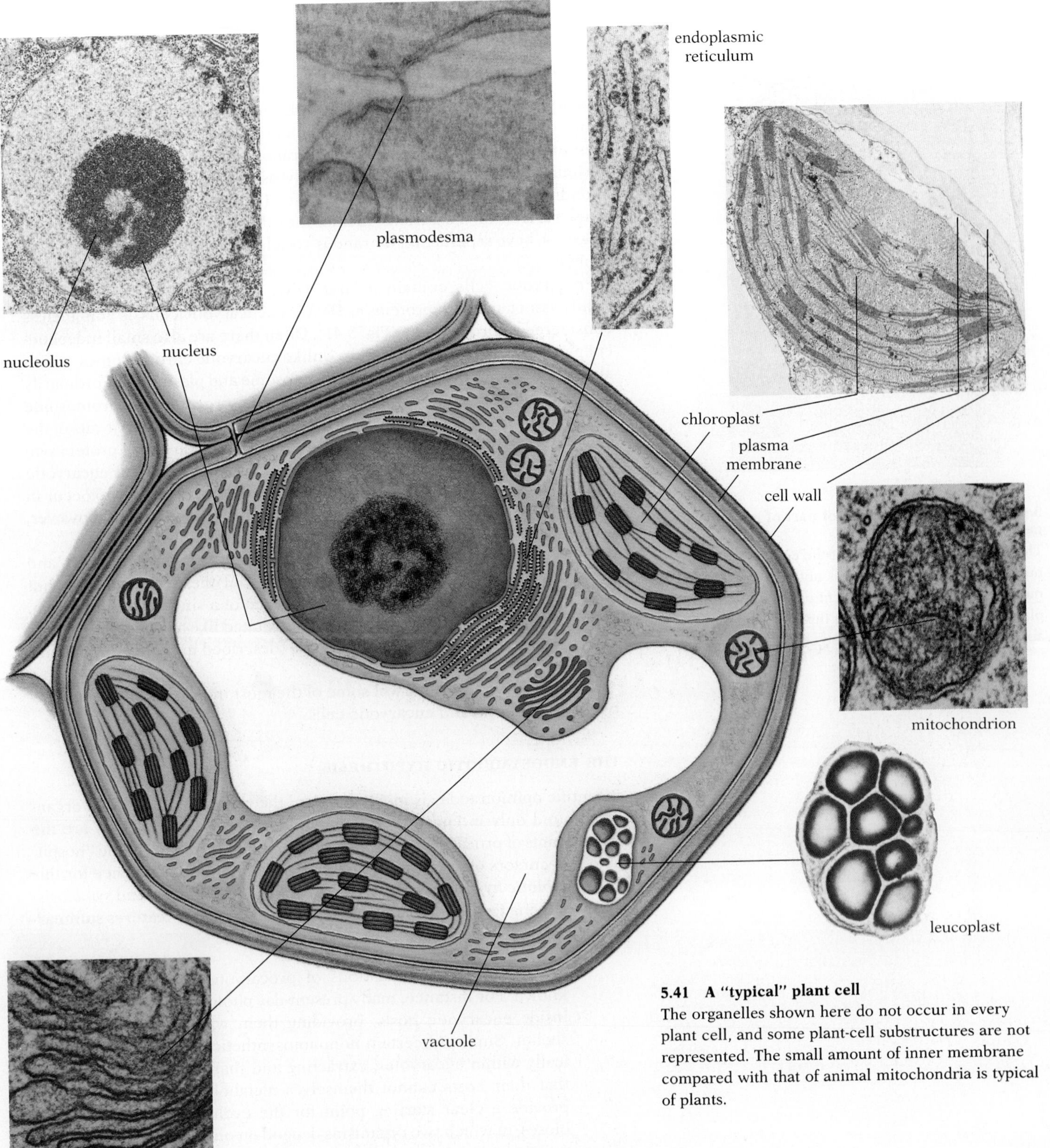

5.41 A "typical" plant cell
The organelles shown here do not occur in every plant cell, and some plant-cell substructures are not represented. The small amount of inner membrane compared with that of animal mitochondria is typical of plants.

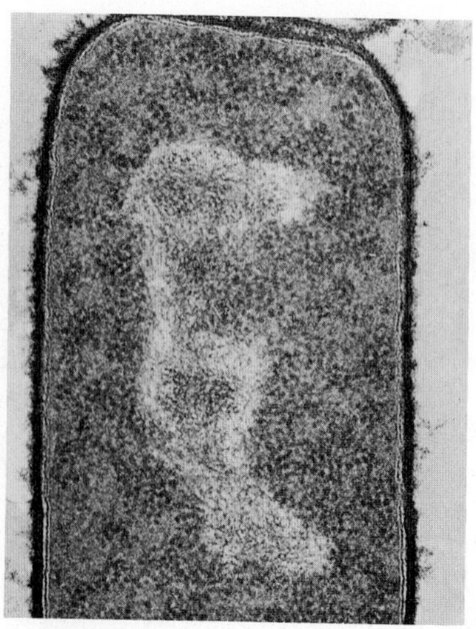

5.42 Electron micrograph of part of a bacterial cell
The light area in the center of the cell, called the nucleoid, contains DNA, but is not bounded by a membrane. Note the prominent cell wall, with the plasma membrane visible just inside it, and the absence of membrane-bounded cellular organelles.

PROCARYOTIC CELLS (BACTERIA)

Procaryotic cells lack most of the cytoplasmic organelles present in eucaryotic cells (Fig. 5.42). We have already mentioned that they have no nuclear membrane; they also lack other membranous structures such as an endoplasmic reticulum, a Golgi apparatus, lysosomes, peroxisomes, and mitochondria (many of the functions of mitochondria are carried out by the inner surface of the plasma membrane). Many photosynthetic bacteria, however, have separate membranous vesicles or lamellae containing chlorophyll.

Procaryotic cells contain a large DNA molecule, which, though not tightly associated with proteins as DNA is in eucaryotic cells, is nonetheless considered a chromosome (Fig. 5.43). Often there are also small independent pieces of DNA, called plasmids. Unlike eucaryotic chromosomes, which are usually linear, the procaryotic chromosome and plasmids are ordinarily circular. Like eucaryotic chromosomes, the procaryotic chromosome bears, in linear array, the genes that control both the hereditary traits of the cell and its ordinary activities. The DNA functions by directing protein synthesis on ribosomes via mRNA, in the way already described for eucaryotic cells. Ribosomes are the most prominent cytoplasmic structure to occur in both eucaryotic and procaryotic cells. Those of procaryotic cells, however, are somewhat smaller than those of eucaryotic cells.

Some bacterial cells possess hairlike organelles used in swimming, and these have traditionally been called flagella. But these structures do not have microtubules; instead they are composed of a single kind of protein, flagellin. Flagellin forms a stiff helix that is rotated like a propeller by a special structure in the membrane at its base (described more fully in Chapter 20).

Table 5.1 gives a summary of some of the most important differences between procaryotic and eucaryotic cells.

THE ENDOSYMBIOTIC HYPOTHESIS

Scientific opinion today is moving toward the view that at least two organelles found only in eucaryotes—mitochondria and chloroplasts—are the descendants of procaryotic organisms that took up residence in the "hospitable" precursors of eucaryotes. There are several lines of evidence for this ***endosymbiotic hypothesis*** (the term is from *endo-*, "within," and *symbiosis*, "state of living together"); some of these are based on the features summarized in Table 5.1.

1. Many symbiotic associations of procaryotes and eucaryotes are known. For instance, many present-day photosynthetic bacteria live inside eucaryotic hosts, providing them with food in return for shelter. Similarly, certain nonphotosynthetic bacteria live symbiotically within eucaryotes, extracting and sharing energy from foods that their hosts cannot themselves metabolize. Such associations provide a clear starting point for the evolution of ***obligate*** symbioses, in which two organisms depend on one another for their mutual survival.

TABLE 5.1 *A comparison of typical procaryotic and eucaryotic cells, and certain eucaryotic organelles*

Characteristic	*Procaryotic cells*	*Eucaryotic cells*	*Mitochondria and chloroplasts*
Size	1–10 μm	10–100 μm	1–10 μm
Nuclear envelope	Absent	Present	Absent
Chromosomes	Single, circular, with no nucleosomes	Multiple, linear, wound on nucleosomes	Single, circular, with no nucleosomes
Golgi apparatus	Absent	Present[a]	Absent
Endoplasmic reticulum, lysosomes, peroxisomes	Absent	Present[a]	Absent
Mitochondria	Absent	Present[a]	
Chlorophyll	Not in chloroplasts	In chloroplasts	
Ribosomes	Relatively small	Relatively large	Relatively small
Microtubules, intermediate filaments, microfilaments	Absent[b]	Present	Absent
Flagella	Lack microtubules	Contain microtubules	

[a] Certain parasites and anaerobic organisms lack mitochondria, having lost them either because high-energy compounds are available directly from the host cell, or because they cannot use the oxygen-dependent enzymatic pathways unique to mitochondria. One kingdom of primitive eucaryotes (discussed in Chapter 21) lacks mitochondria, ER, and Golgi.

[b] Microtubules have been reported in certain spirochete bacteria inhabiting the digestive tract of termites; if the report proves correct, this may be the evolutionary source of eucaryotic flagella and cilia.

2. Both mitochondria and chloroplasts contain their own ribosomes and their own chromosomes; the chromosomes code for their ribosomal RNA and ribosomal proteins, and for some (though not all) of their enzymes. Mitochondria and chloroplasts also build their own membranes.
3. The organelle chromosomes resemble those of procaryotes in that they are circular, not wound on special protein spools, and not enclosed by a nuclear envelope.
4. The internal organization of the organelle genes is similar to that of procaryotes, but very different from that of eucaryotes. We will examine the details of this organization in later chapters.
5. Cell division in eucaryotes involves a spindle apparatus, whereas bacteria, mitochondria, and chloroplasts each divide by fissioning.
6. The ribosomes of mitochondria and chloroplasts are more similar to those of procaryotes than to the ribosomes in the cytosol of the cells in which they live. Indeed, the large ribosomal subunit of the bacterium *Escherichia coli* and the large subunit of chloroplasts are so similar that they may be substituted for one another without affecting the hybrid ribosome's capacity to carry out protein synthesis.

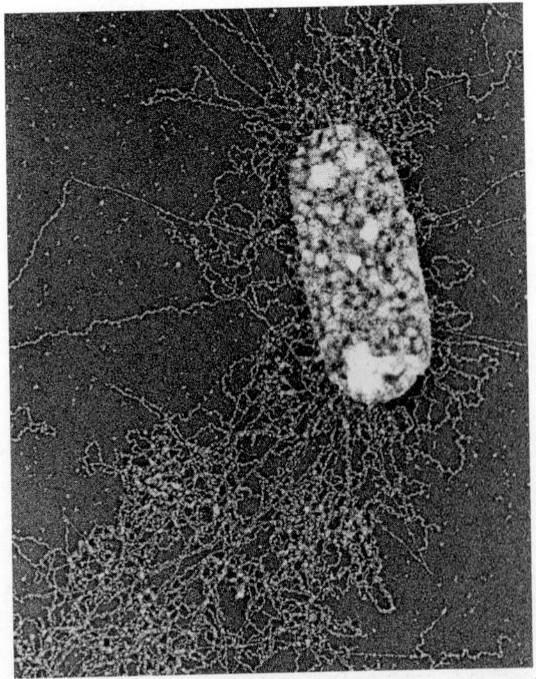

5.43 A disrupted cell of *Escherichia coli* Exposure to detergent releases the DNA of the common intestinal bacterium *E. coli*, most of which can be seen outside the cell in this electron micrograph. Though not readily recognizable here, the main chromosome and the small accessory chromosomes (plasmids) of a bacterium each form a circle rather than individual strands in the cell. The mottled appearance of the surface of the bacterium results from the effects of alcohol, drying, and shadowing with platinum.

In summary, the endosymbiotic hypothesis assumes that a bacterium with the unique metabolic capabilities exhibited by mitochondria was captured through endocytosis about 1.5 billion years ago, resisted digestion, lived symbiotically inside its host, and divided independently of its host. Later, some of the symbiont's genes were moved to the host nucleus, which took control from its guest. From this partnership all present-day plants, animals, fungi, and protozoans must have evolved. Subsequently, according to this hypothesis, various photosynthetic bacteria met the same fate, the result being the chloroplasts found in algae and plants.

Some current speculation, less well supported but thought-provoking, holds that peroxisomes, basal bodies/centrioles, and nematocysts (the devices by which jellyfish and similar creatures harpoon potential prey and careless swimmers) may also have had an endosymbiotic origin. The recent report that basal bodies have their own chromosomes adds strength to such suggestions.

STUDY QUESTIONS

1. Trace the pathway of information flow from the chromosomes to the outer membrane during the secretion of a digestive enzyme. (pp. 125–37)

2. Are the carbohydrate markers that are placed on proteins destined to be mounted in the cell membrane ever exposed to the cytosol? (pp. 113–16, 134–36)

3. What would happen to a cell if it lost its Golgi apparatus? Its nucleolus? Its peroxisomes? (pp. 127, 133–34, 137–38)

4. List five important differences between procaryotes and eucaryotes. How do these contrasts bear on the endosymbiotic hypothesis? (pp. 151–56)

CONCEPTS FOR REVIEW

- Anatomy and function of:
    - Organelle membranes
    - Nucleus
    - Nucleoli
    - Rough ER
    - Smooth ER
    - Golgi apparatus
    - Lysosomes
    - Mitochondria
    - Chloroplasts
    - Microfilaments
    - Microtubules
    - Centrioles/basal bodies
    - Cilia/flagella
- Digestive cycle in cells
- Flow of membrane in cells
- Role of and action of microfilaments and microtubules
    - Movement of vesicles
    - Movement of the cell
- Eucaryotes versus procaryotes
    - Essential distinctions
    - Endosymbiotic hypothesis
        - scenario
        - evidence

SUGGESTED READING

ALBERTS, B., D. BRAY, J. LEWIS, M. RAFF, K. ROBERTS, and J. D. WATSON, 1989. *Molecular Biology of the Cell*, 2nd ed. Garland, New York.

This massive tome is the most complete and up-to-date summary of cell biology available.

ALLEN, R. D., 1987. The microtubule as an intracellular engine, *Scientific American* 256 (2).

BRETSCHER, M. S., 1987. How animal cells move, *Scientific American* 257 (6). *On the role of endocytosis in movement.*

DEDUVE, C., 1963. The lysosome, *Scientific American* 208 (5). (Offprint 156) *The discoverer of lysosomes describes his work.*

DEDUVE, C., 1983. Microbodies in the living cell, *Scientific American* 248 (5). (Offprint 1538) *On the class of specialized enzymatic organelles, such as peroxisomes, that are not produced by the Golgi.*

DEDUVE, C., 1986. *The Living Cell.* Scientific American Library, New York. *This two-volume set provides a well-illustrated, up-to-date tour of the cell.*

DUSTIN, P., 1980. Microtubules, *Scientific American* 243 (2). (Offprint 1477) *An excellent summary of the formation of microtubules and the diverse roles they play in the cell.*

MARGULIS, L., 1981. *Symbiosis in Cell Evolution.* W. H. Freeman, San Francisco. *A leading advocate of the endosymbiotic hypothesis argues the case.*

NEUTRA, M., and C. P. LEBLOND, 1969. The Golgi apparatus, *Scientific American* 220 (2). (Offprint 1134) *How radiography was used in working out the function of the Golgi apparatus.*

PORTER, K. R., and J. B. TUCKER, 1981. The ground substance of the living cell, *Scientific American* 244 (3). (Offprint 1494) *Describes the cytoskeleton and the evidence for a microtrabecular lattice.*

ROTHMAN, J. E., 1985. The compartmental organization of the Golgi apparatus. *Scientific American,* 253 (3). *An incisive analysis of the fine structure of this important organelle.*

SIMONS, K., H. GAROFF, and A. HELENIUS, 1982. How an animal virus gets into and out of its host cell, *Scientific American* 246 (2). (Offprint 1511) *A fascinating account of how Semliki Forest virus subverts the membrane system of its many vertebrate hosts.*

STOSSEL, T. P., 1990. How cells crawl. *American Scientist* 78, 407–423. *A detailed look at the mechanisms and control of amoeboid movement.*

WEBER, K., and M. OSBORN, 1985. The molecules of the cell matrix, *Scientific American* 253 (4). *Reviews the structure and function of microfilaments and microtubules.*

Chapter **6**

ENERGY TRANSFORMATIONS: RESPIRATION

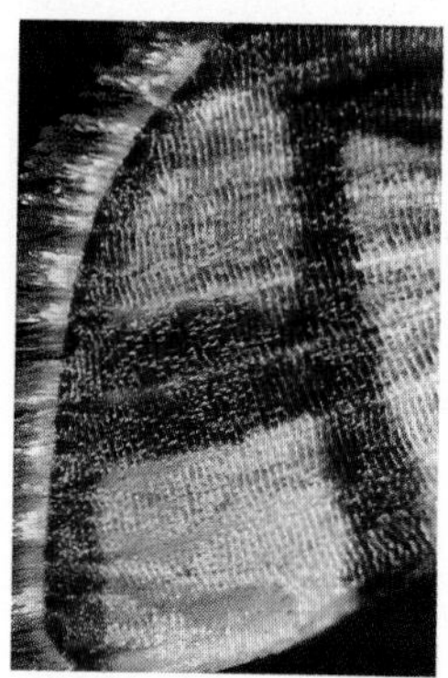

basic principle of physics, as we saw in Chapter 3, is that all systems have a natural tendency toward disorder. The more orderly an arrangement of matter, the less probable it is, and the less likely to endure if energy is not expended to counteract the tendency toward disorder. Within a cell, therefore, energy is needed at every stage to drive the reactions that maintain life: to read, copy, and repair the genetic instructions in the chromosomes; to construct, repair, and move organelles; to bring in nutrients, expel wastes, preserve the proper pH and ionic balance; and so on. Without a constant supply of energy, these reactions cannot take place; the cell spontaneously decays, and life ceases. But where does the energy for maintaining life come from, and how is it used by the cell?

THE FLOW OF ENERGY

Virtually all the energy that fuels life today comes from the sun and is captured in the process known as photosynthesis by those organisms, notably plants, that are able to use it to build energy-rich compounds like glucose. Most creatures that do not capture the energy of sunlight directly obtain their energy by ingesting or absorbing photosynthetic organisms or by eating those that eat photosynthetic organisms.

In today's cells, stored energy is usually released through a process

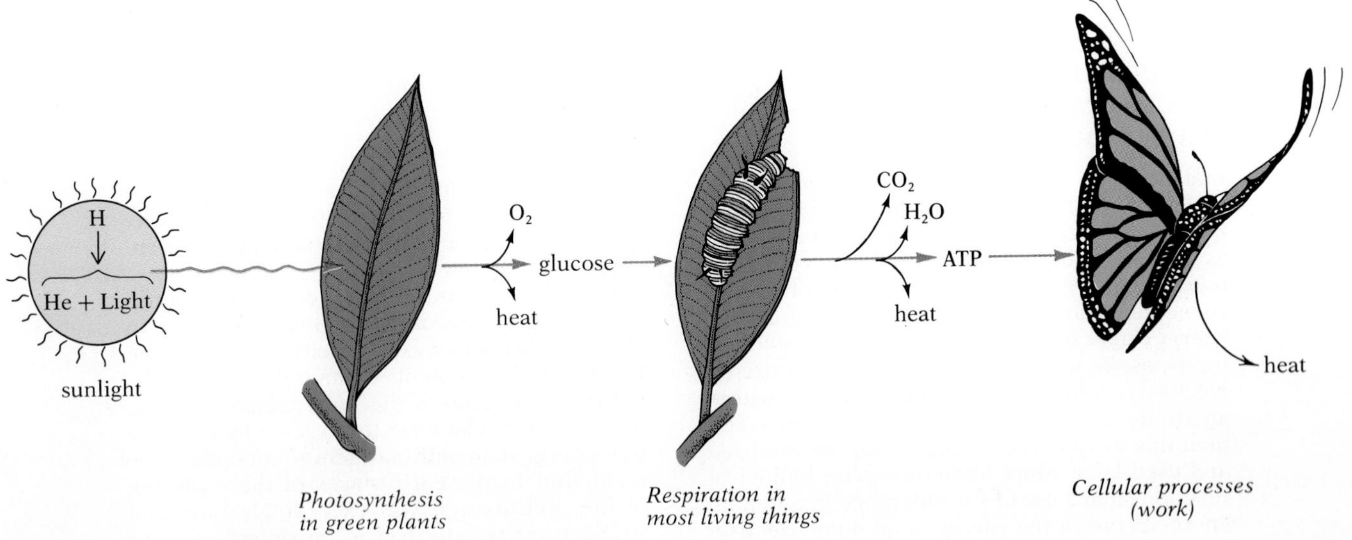

6.1 Summary of biological energy flow
Today, nearly all the energy for life originates in the sun, where hydrogen is converted by fusion into helium and light is produced. In the process of photosynthesis, green plants convert the radiant energy of sunlight into chemical energy, which is most often stored initially in glucose. When cells in most organisms need energy, the glucose is broken down and some of its chemical energy is recovered by the process of aerobic respiration; the resulting product —ATP—supplies the energy in a more manageable form, making it available for muscular contraction, nerve conduction, active transport, and other work. Aerobic respiration utilizes oxygen, a by-product of photosynthesis, and photosynthesis utilizes carbon dioxide and water, by-products of respiration. With each of the transformations shown, much energy is lost as waste heat.

known as ***aerobic respiration***, in which, ultimately, oxygen and glucose are combined to yield CO_2 and water; the energy is then used to run all of the cell's enzyme-mediated reactions (Fig. 6.1). These reactions, collectively known as ***metabolism***, may be viewed as belonging to one of two phases—***anabolism***, encompassing the processes by which complex organic molecules are assembled, and ***catabolism***, the processes by which living things extract energy from food.

EVOLUTION OF ENERGY TRANSFORMATIONS

Metabolism evolved under conditions very different from those that exist over most of the earth today. The major dissimilarities were the lack of oxygen in the ancient atmosphere, which made aerobic respiration impossible, and the absence of photosynthesis, which had yet to evolve. Relict species still living today in bogs, volcanic pools, and on the ocean bottom provide invaluable insights into the nature of early metabolic adaptations and the evolution of modern chemical pathways. As we will see in more detail in Chapter 19, a limited supply of energy-rich organic molecules developed on the early earth in the absence of living things, and the first living organisms that were capable of metabolizing this energy source arose 3.5 billion years ago. But natural selection strongly favored organisms capable of synthesizing their own food. To create organic compounds, this second group of early organisms used energy drawn for the most part from the covalent bonds of molecular hydrogen. Organisms that obtain their energy in this way from inorganic energy sources are said to be chemosynthetic. Some of the early chemosynthetic organisms combined naturally occurring CO_2 with energy-rich H_2 to produce methane and water by the condensation reaction shown in the following equation, which summarizes several intermediate steps:

$$CO_2 + 4H_2 \longrightarrow CH_4 + 2H_2O + \text{energy}$$

Exploring Further

THE HIGH ELECTRONEGATIVITY OF OXYGEN: ITS ROLE IN ENERGY TRANSFORMATIONS

Life depends on the efficient management of electrons. With the aid of highly specialized enzymes, each cell harvests the energy of electrons as they move from a position of relatively great potential energy to one of less potential energy. In biochemical pathways, an electron's potential energy can be unlocked in two ways. In photosynthesis, as we will see, a photon can energize an electron to a higher electron energy level within an atom—from level *L* to *M*, for example—and then this extra potential energy can be captured and used later. More often, however, biological reactions make use of the difference in potential energy between the energy of an outer electron in one atom and a vacancy with a lower potential energy in another atom. When an electron is shifted between atoms, energy is released. This movement need not be—indeed, usually is not—between different electron levels; instead, because the energy of, for instance, the *L* level of carbon is significantly higher than that of the *L* level of oxygen (Fig. A), movement of an electron from one to the other is highly exergonic (producing ionized reactants and products). The same holds for electrons in covalent bonds within a molecule: the shared electrons in a C—H bond, for example, have about 11 kcal per mole more potential energy than those in an O—H bond. Transferring a hydrogen from a carbon to an oxygen, either within a single covalently bonded molecule or between two different molecules, thus releases a substantial amount of energy.

The difference in potential energies of outer-shell electrons accounts for the differences in electronegativities we discussed in Chapter 2 (see Fig. B). Of the elements commonly found in living things, oxygen is the most electronegative; hence, moving electrons to oxygen releases more free energy than shifting them to any other element. But in the early stages of the evolution of life, free oxygen was exceedingly rare, and so electrons usually had to be shifted to other less electronegative electron "acceptors." Even when oxygen that was part of a larger molecule could be used, the result was not optimal: though a substitution of electrons can result in a net liberation of energy (replacing a C—O bond with an O—H bond, for instance, is exergonic), it releases only about 25 percent of the energy available from the direct formation of an O—H bond from free oxygen (Fig. B). Using sulfur as an alternative electron acceptor, producing an S—H bond rather than an O—H bond, is also exergonic, but yields less than 30 percent of the electron's potential energy. Some bacteria still use sulfur, nitrogen, or even carbon as their electron acceptors.

As a result of this reaction, in which carbon and (especially) oxygen act as electronegative acceptors of energy-rich electrons, some of the energy stored in the bonds of H_2 is released, and can be used to do work in the cell.

Another energy-liberating reaction uses sulfur as an acceptor to liberate the energy in H_2; indeed, many present-day bacteria thrive in the sulfur-rich waste found in sewers and bogs. They produce the foul-smelling gas hydrogen sulfide (H_2S):

$$H_2 + S \longrightarrow H_2S + \text{energy}$$

Early organisms must have lived from hand to mouth, so to speak, eating whatever foods were available, but having little ability to store energy for later use. The evolution in chemosynthetic organisms of enzyme pathways that direct the synthesis of more complex organic molecules made possible long-term energy storage. The primary storage substance was probably glucose, or polysaccharides that could be converted into glucose. When needed, some of the energy of the glucose might then have been extracted,

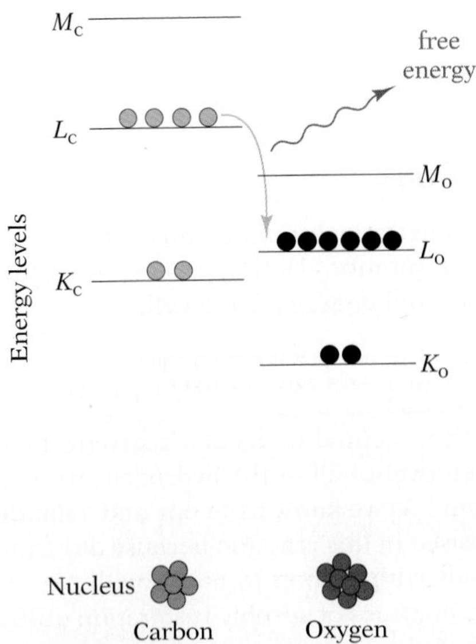

A Relative energy levels
The corresponding electron levels in two elements never have the same energy. As a result, an electron moving from the *L* level of carbon to the *L* level of oxygen will lose (liberate) energy in the process.

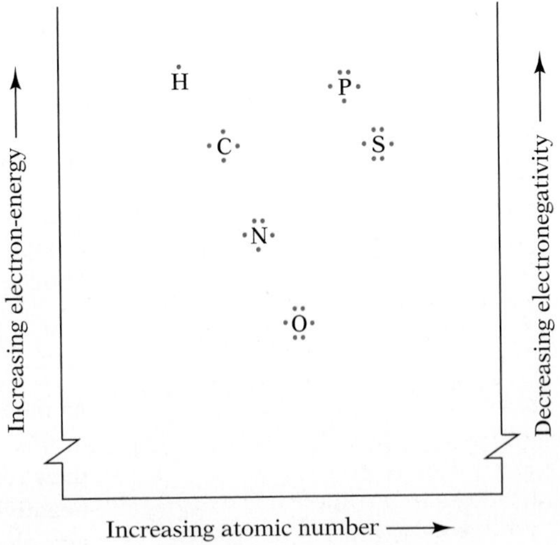

B Electronegativity and potential energy
The electronegativity chart has been inverted to show differences in electron potential energy—the essential factor controlling whether a reaction is endergonic (that is, requires electrons to move up on the graph) or exergonic (in which electrons move down to lower energy values). Only the six most important elements in living tissue are shown. Note that hydrogen is the least electronegative of these elements; electron transfers from hydrogen are therefore inevitably exergonic. Oxygen, on the other hand, is the most electronegative; transfers of electrons *to* it always release energy.

as energy still is today, by the series of enzyme-mediated reactions called glycolysis, or the glycolytic pathway.

About 3 billion years ago an enormously important biochemical event occurred: certain organisms acquired a primitive ability to capture the sun's energy directly and use it to synthesize glucose and other important organic compounds. Then, about 2.5 billion years ago, the forerunner of aerobic respiration evolved. This pathway, which includes glycolysis as its first stage, extracts large amounts of energy from the end products of glycolysis, but requires molecular oxygen (O_2) to be really efficient. Since molecular oxygen was in short supply at first, less electronegative acceptors (sulfur, nitrogen, and carbon) had to suffice, and so cells had to waste most of the potential energy in the chemical bonds of storage compounds. However, oxygen became increasingly available when the descendants of the early photosynthetic organisms began producing it as a by-product of a more advanced form of photosynthesis; thus, about 2.3 billion years ago, oxygen began to accumulate in the atmosphere. Since oxygen (in the form of atmospheric ozone, or O_3) filters out biologically destructive X rays and ultraviolet light from the sun, organisms could now emerge from beneath the radiation shielding provided by earth and water to colonize the surface of the land. In addition, the abundance of oxygen made aerobic respiration the dominant catabolic pathway: aerobic respiration is almost twenty times as efficient in extracting the stored energy of glucose as glycolysis alone.

OXIDATION AND REDUCTION

Sugars like glucose are energy-rich compounds, while carbon dioxide is energy poor. Aerobic respiration is the process by which glucose is "burned" to produce CO_2 and energy for the cell:

$$C_6H_{12}O_6 + 6O_2 \longrightarrow 6CO_2 + 6H_2O + \text{energy}$$

As the arrows indicate, the essential molecular rearrangement that occurs in this reaction is the removal of all of the hydrogens from the sugar, and their rebonding to oxygen[1]. As we know from our understanding of electronegativity, energy is released in this reaction because *any* transfer of hydrogen electrons from bonds with carbon to bonds with oxygen results in a liberation of energy. The enzymes of aerobic respiration utilize some of this energy to do work in the cell, while the rest of the energy is dissipated as heat. This transfer of hydrogen atoms from carbon to oxygen is an example of ***oxidation***. Stated more generally, a substance is oxidized when it loses electrons (almost always hydrogen electrons), with or without the rest of the atom, to a more electronegative substance (usually oxygen); as a result, oxidation inevitably liberates energy.

The converse process—storing energy in a compound—is called, oddly enough, ***reduction***. Consider the central energy-storing reaction in nature, photosynthesis. As we will see in Chapter 7, photosynthesis basically reverses aerobic respiration: solar energy is used to power the synthesis of high-energy sugar from energy-poor CO_2. The solar energy is stored by the reduction of CO_2—that is, by removing hydrogens from low-energy bonds with electronegative oxygen, and moving them into high-energy bonds with carbon:

$$6CO_2 + 6H_2O + \text{light} \longrightarrow C_6H_{12}O_6 + 6O_2$$

Though reduction in cellular reactions usually involves the transfer of hydrogen atoms from oxygen to carbon, reduction occurs whenever an electron (with or without the rest of its atom) is moved to a less electronegative atom. This broader definition is important because we will see cases in both photosynthesis and aerobic respiration in which oxidation and reduction entail the movement of isolated electrons, and involve elements other than oxygen and hydrogen.

We can see that oxidation and reduction are mirror-image processes, one

[1] Five of the 12 hydrogens in glucose are bonded to oxygens to begin with; these attachments are broken and new ones made with different oxygen atoms. This rearrangement involves no net change in potential energy; the seven hydrogens initially bound to carbons are the ones crucial to harvesting energy in respiration.

[2] The converse is not necessarily true: energy can be lost from the reactants as heat with no corresponding gain in the energy of the products.

lowering, the other raising an electron's potential energy. Since all reactions involve making or rearranging bonds, it follows that if one atom or molecule gains energy, another must have lost energy—that is, whenever one substance is reduced, another is oxidized[2]. We can follow this so-called ***redox*** reaction from the perspective of an electron, as it moves from one atom to another:

$$A^{e-} + B \longrightarrow \underset{\substack{\text{has been}\\ \text{oxidized}\\ \text{(lost energy)}}}{A} + \underset{\substack{\text{has been}\\ \text{reduced}\\ \text{(gained energy)}}}{B^{e-}}$$

More often though, the electron undergoing a change in potential is part of an atom that is rebonded; no ions are created. As we have seen, the most common reduction reactions involve removal of oxygen or addition of hydrogen; conversely, oxidation usually involves addition of oxygen or removal of hydrogen:

$$A + BO \longrightarrow \underset{\substack{\text{has been}\\ \text{oxidized}}}{AO} + \underset{\substack{\text{has been}\\ \text{reduced}}}{B}$$

or:

$$AH + B \longrightarrow \underset{\substack{\text{has been}\\ \text{oxidized}}}{A} + \underset{\substack{\text{has been}\\ \text{reduced}}}{BH}$$

In biological systems, removal or addition of an electron derived from hydrogen is the most frequent mechanism of redox reactions. A summary of redox reactions appears in Table 6.1.

It now becomes clear why the synthesis of sugar from carbon dioxide constitutes reduction of the CO_2: hydrogen obtained by splitting water molecules is added to the CO_2, thereby storing energy that is later recovered through respiration. The major product of respiration is ATP, the major energy currency of the cell. It is the energy of ATP that is used to fuel endergonic reactions.

TABLE 6.1 *Redox reactions*

Oxidation	*Reduction*
Adds oxygen	Removes oxygen
Removes hydrogen	Adds hydrogen
Removes electron(s)	Adds electron(s)
Liberates energy	Stores energy

ADENOSINE TRIPHOSPHATE (ATP)

One of the essential substances of life is the compound ***adenosine triphosphate***, or ***ATP***, which plays a key role in nearly every energy transformation in every living thing (Fig. 6.2).

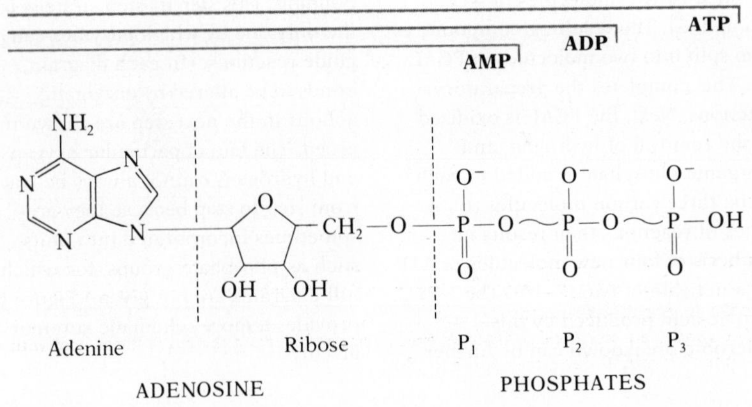

ATP
ADP
AMP
NH_2
N
N
O
N
N
CH_2—O—P—O~P—O~P—OH
O^- O^- O^-
O O O
OH OH
Adenine
Ribose
P_1 P_2 P_3
ADENOSINE
PHOSPHATES

6.2 The ATP molecule
Adenosine triphosphate (ATP) is composed of an adenosine unit (a complex of adenine and ribose sugar) and three phosphate groups arranged in sequence. The last two phosphates are attached by so-called high-energy bonds (wavy lines). The cell stores energy by adding a phosphate group to ADP (adenosine diphosphate) to make ATP, and later recovers some of this energy by hydrolyzing ATP into ADP and inorganic phosphate. Occasionally ADP is hydrolyzed to make AMP (adenosine monophosphate), one of the nucleotides of which RNA is composed. The widely used term "high-energy bonds" is somewhat misleading: the energy is actually stored in internal bonds of the phosphate group itself.

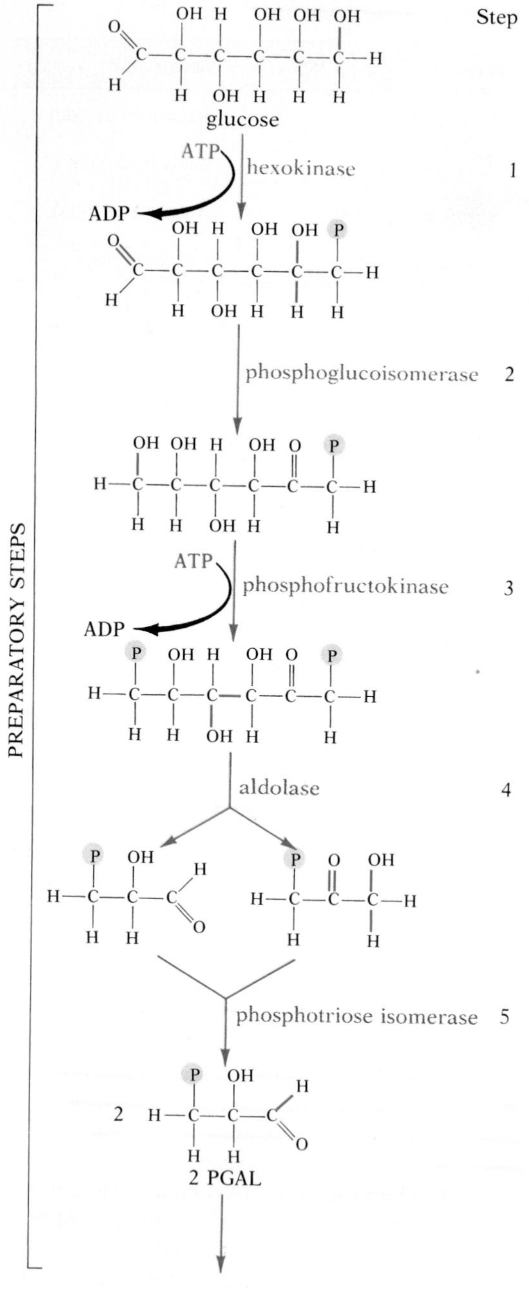

PREPARATORY STEPS

Step

glucose

ATP ADP hexokinase 1

phosphoglucoisomerase 2

ATP ADP phosphofructokinase 3

aldolase 4

phosphotriose isomerase 5

2 PGAL

The ATP molecule is composed of a nitrogen-containing compound (adenosine) plus three phosphate groups bonded in sequence:

$$\text{adenosine}—Ⓟ \sim Ⓟ \sim Ⓟ$$

According to convention, Ⓟ stands for the entire phosphate group, and the wavy lines between the first and second and the second and third phosphate groups represent so-called high-energy bonds.[3]

Actually, it is often only the terminal phosphate bond of ATP that is involved in energy conversions. The exergonic reaction by which this bond is hydrolyzed and the terminal phosphate group removed leaves a compound called adenosine diphosphate, or ***ADP*** (adenosine plus two phosphate groups), and inorganic phosphate (symbolized by P_i):

$$ATP + H_2O \xrightarrow{\text{enzyme}} ADP + P_i + \text{energy}$$

If both the second and third phosphate groups are removed from ATP, the resulting compound is adenosine monophosphate, or ***AMP***.

New ATP can be synthesized from ADP and inorganic phosphate if adequate energy is available to force a third phosphate group onto the ADP. Addition of phosphate is termed ***phosphorylation***:

$$ADP + P_i + \text{energy} \xrightarrow{\text{enzyme}} ATP + H_2O$$

[3] The widely used term "high-energy bond" does not mean that energy is actually stored in phosphate bonds. Instead, it means that in comparison to the other bonds in an ATP molecule, the bonds between the negatively charged phosphate groups are very unstable, and so are thermodynamically less expensive to break. It also means that in a series of chemical reactions involving ATP, the energy needed to break phosphate bonds is less than the energy released when new bonds are formed.

6.3 Glycolysis and fermentation

The entire reaction series for glycolysis is shown to illustrate how biochemical pathways work; there is no reason to try to memorize each step. Glucose in solution normally exists as a ring structure, but the straight-chain form is adopted here for clarity. Energy to initiate the breakdown of glucose is supplied by two molecules of ATP (Steps 1–3). The resulting compound is then split into two molecules of PGAL (5). This completes the preparatory reactions. Next, the PGAL is oxidized by the removal of hydrogen, and inorganic phosphate is added to each of the three-carbon molecules (6). A series of reactions then results in synthesis of four new molecules of ATP, for a net gain of two (7–10). The pyruvic acid produced by this anaerobic breakdown can be further oxidized in most cells if O_2 is present (by reactions not shown here). But in the absence of sufficient O_2, the pyruvic acid may be converted to lactic acid in some kinds of organisms, or CO_2 and ethanol in others (11–12). At each step a particular enzyme catalyzes a specific redistribution of electrons, and thereby brings about changes in bonding. This step-by-step strategy is the only one by which enzymes can guide reactions. (In each diagram, bonds to be altered by enzymatic actions in the next step are shown in green. The fate of particular oxygens and hydrogens cannot always be traced from step to step because they are sometimes incorporated into units, such as phosphate groups, for which full diagrams are not given.) Figure 6.4 provides a more schematic summary of glycolysis.

ATP is often called the universal energy currency of living things. Cells initially store energy in the form of carbohydrates such as glucose, and lipids, such as the many kinds of fats. But the amount of energy in even a single glucose molecule is inconveniently large for driving most reactions: 670 kcal/mole. The hydrolysis of ATP to ADP releases a more useful amount of energy: 7.3 kcal/mole. Glucose molecules are the hundred-dollar bills of the cellular economy, while ATP molecules are the everyday denominations, sufficient for making and breaking covalent bonds. Though some other compounds can supply energy, ATP is the one most often used by cells in the various kinds of work they perform: it powers synthesis of more complex compounds, muscular contraction, nerve conduction, active transport across cell membranes, light production, and so on. The energy price of the work is paid through the energy-releasing hydrolysis of ATP to ADP.

CELLULAR RESPIRATION

The energy stored in lipids and carbohydrates is not liberated through a single large reaction; rather, the universal catabolic process by which the molecules are broken down occurs as a series of smaller reactions, each catalyzed by its own specific enzyme. These result in the release of small amounts of energy, some of which is transferred to ATP by the phosphorylation just described. We will examine the most important steps in the catabolism of glucose, part or all of which constitutes the central energy-liberating pathway in all organisms. In the next chapter we will look at the major pathway for the synthesis of glucose: photosynthesis.

ANAEROBIC RESPIRATION

Glycolysis (Stage I of aerobic respiration) The complete catabolism of glucose involves five stages, divided between anaerobic and aerobic series of reactions. The anaerobic portion of the process, the breakdown of glucose to pyruvic acid, known as ***glycolysis***, is the most ancient series of reactions in the pathway and evolved long before free oxygen became available. We will trace the steps of glycolysis in detail because they illustrate clearly how reaction pathways are organized into small, enzyme-mediated steps, and made to work through the careful management of free energy changes. You will do well to concentrate on these principles and the overall sequence of steps (summarized in Fig. 6.3), rather than on the exact thermodynamics and the structural details of the molecular intermediates.

Glucose is a stable compound—one with little tendency to break down spontaneously into simpler products. If its energy is to be harvested, glucose must first be activated by the investment of a small amount of energy. The first steps of glycolysis, therefore, are preparatory, enabling the later steps to extract the stored energy.

The energy for initiating glycolysis (Fig. 6.3) comes from ATP. The initial reaction, like the succeeding ones, is made possible by its own specific enzyme, which binds (via weak bonds) to the reactants, activates them (by causing the electrons of the bound molecules to redistribute themselves), and joins or rearranges the reactants before releasing the products. In the

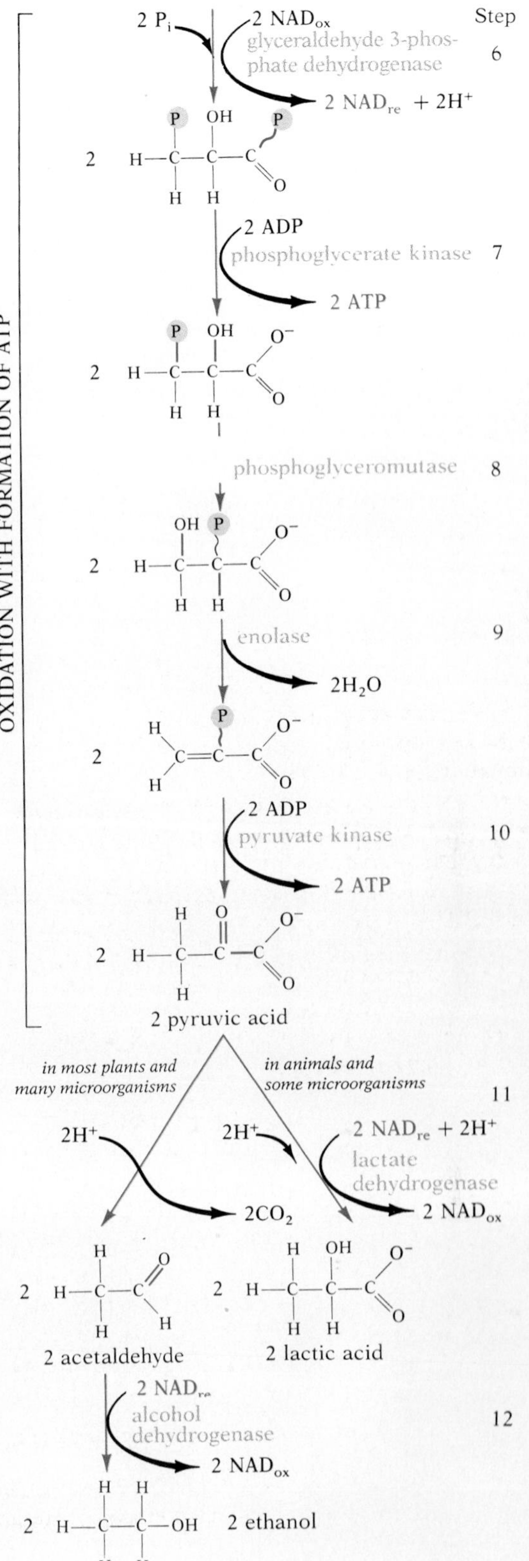

first reaction, a molecule of ATP donates its terminal phosphate group to the glucose.

$$\text{(1)}\quad \underset{\text{glucose}}{\text{C—C—C—C—C—C}} + \text{ATP} \xrightarrow{\text{enzyme}} \underset{\text{glucose-6-phosphate}}{\text{C—C—C—C—C—C—Ⓟ}} + \text{ADP}$$

$$(\Delta G = -4.0 \text{ kcal/mole})$$

(The simplified equations given here show only the carbon skeleton; the more complete molecular structure is shown in Figure 6.3.) The name of the product tells us that it is a glucose with a phosphate group attached to its sixth carbon atom.

Let's look carefully at what has happened in this reaction. An enzyme, hexokinase, has bound glucose and ATP, catalyzed the transfer of a phosphate group to the glucose, and released the products. The overall change in free energy of the electrons rearranged in this reaction is −4.0 kcal/mole; the free energy is liberated primarily as heat. As always, the negative ΔG means that this is a strongly exergonic (downhill) reaction.[4] In fact, the reaction has an equilibrium constant (K_{eq}) of about 1000—that is, the products outnumber the reactants by 1000 to 1. The free energy for this reaction comes from ATP: the energy available from the terminal phosphate of the ATP is 7.3 kcal/mole. Only 4.0 kcal/mole is liberated to drive this first step of glycolysis, and the other 3.3 kcal/mole from the ATP is stored in the electrons of the product—the activated glucose.

The next step in glycolysis converts glucose-6-phosphate into the nearly identical compound fructose-6-phosphate:

$$\text{(2)}\quad \underset{\text{glucose-6-phosphate}}{\text{C—C—C—C—C—C—Ⓟ}} \xrightarrow{\text{enzyme}} \underset{\text{fructose-6-phosphate}}{\text{C—C—C—C—C—C—Ⓟ}}$$

$$(\Delta G = +0.4 \text{ kcal/mole})$$

The positive ΔG indicates that this is an uphill, endergonic (energy-requiring) reaction, one that cannot proceed spontaneously. How is it, then, that glycolysis does not grind to a halt? The answer lies in the operation of ***coupled reactions:*** two reactions that share a common intermediate molecule —in this instance, glucose-6-phosphate, the product of Step 1 *and* the reactant of Step 2—can proceed as a single reaction. The 1000-to-1 ratio of products to reactants in Step 1 supplies an enormous number of reactant molecules to Step 2; this abundance insures that, though the equilibrium constant of the second reaction is only about 0.2 (that is, a 1-to-5 ratio of products to reactants), many product molecules are nevertheless created to feed Step 3.

From the perspective of thermodynamics, these two steps can be treated as one reaction. The −4.0 kcal/mole liberated by Step 1 is combined with

[4] As noted in Chapter 3, we use ΔG (rather than the formal symbol $\Delta G'^{\circ}$) for the change in the free energy of a reaction under standard conditions. In addition, we divide ΔGs simply between products and liberated free energy, even though, in fact, a more complicated but poorly understood division is probably occurring. The resulting approximation is reasonably accurate. The equilibrium constant (K_{eq}) is being treated for simplicity as though there is only a single product. This is possible because glycolysis has relatively little effect on the concentration of water, ATP, ADP, and so on; the concentration of these compounds is maintained through various homeostatic mechanisms in the cell.

the +0.4 kcal/mole consumed by Step 2 to yield a net ΔG of −3.6 kcal/mole. Taken together, the two steps are strongly exergonic, and so the reaction proceeds. The glycolytic pathway is a series of such coupled reactions, in which exergonic steps push or pull endergonic steps, with the favorable *net* free-energy change of the steps taken together enabling the sequence of reactions to proceed.

The conversion of glucose-6-phosphate into fructose-6-phosphate also provides a good illustration of how enzymatic pathways work. You may recall that when a substrate binds to an enzyme by means of weak bonds, a slight shift in the electron distribution of the substrate is induced, which lowers the activation energy required for a particular change in bonding and thereby catalyzes the reaction. The result of the redistribution of electrons can be seen in Step 2, in which two hydrogens, bonded to the fifth carbon and its oxygen, are transferred to the first carbon and its oxygen. This trivial change is essential to prepare for the next step in glycolysis. Once the change has taken place, the substrate no longer "fits" the enzyme, and so drifts away to be captured by the next enzyme in the series. Each step in the glycolytic pathway is therefore very small and is mediated by a highly specific enzyme.

After the formation of fructose-6-phosphate in Step 2, another molecule of ATP is consumed, to add a phosphate to the other end of the molecule. Of the ATP's energy, 3.9 kcal/mole is stored in the product, while the remaining 3.4 kcal/mole is liberated as heat; hence the reaction is exergonic:

$$(3)\quad \underset{\text{fructose-6-phosphate}}{\text{C—C—C—C—C—C—Ⓟ}} + \text{ATP} \xrightarrow{\text{enzyme}} \underset{\text{fructose-1,6-bisphosphate}}{\text{Ⓟ—C—C—C—C—C—C—Ⓟ}} + \text{ADP}$$

$$(\Delta G = -3.4 \text{ kcal/mole})$$

Since in its turn this reaction is coupled to the previous one (with which it shares the intermediate compound fructose-6-phosphate), we can add the free energy of the separate steps along the way. The overall reaction chain so far has liberated 7.0 kcal/mole, and so has a highly favorable K_{eq} of more than 10^5.

Next, the fructose-1,6-bisphosphate is split between the third and fourth carbons, forming two essentially similar three-carbon molecules (Step 4). One is phosphoglyceraldehyde—***PGAL***—and the other, an intermediate compound, is usually converted immediately to PGAL, in Step 5. (The cell can also use PGAL to synthesize fat if conditions warrant.) PGAL, a phosphorylated three-carbon sugar, is a key intermediate in both glycolysis and photosynthesis.

We can summarize these reactions simply:

$$(4\text{–}5)\quad \underset{\text{fructose-1,6-bisphosphate}}{\text{Ⓟ—C—C—C—C—C—C—Ⓟ}} \xrightarrow{\text{enzyme}} 2\ \underset{\text{PGAL}}{\text{C—C—C—Ⓟ}}$$

$$(\Delta G = +7.5 \text{ kcal/mole})$$

To this point, instead of releasing energy from glucose to form new ATP molecules, glycolysis has actually cost the cells two ATPs. Indeed, Steps 4 and 5 are so unfavorable energetically that the net change in free energy is now +0.5 kcal/mole. For subsequent reactions to proceed, significant

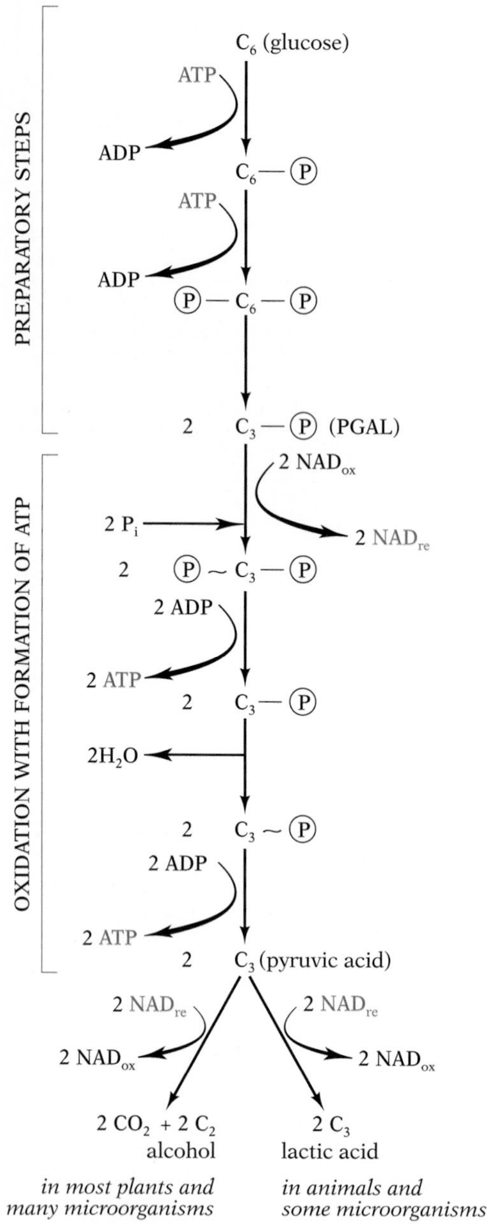

6.4 Summary of glycolysis and fermentation

amounts of free energy must be liberated to pull the reactants past the five preparatory steps.

The next reaction actually involves two separate molecular changes, which we summarize for simplicity in one step. The first change is oxidation of the PGAL by reduction of nicotinamide adenine dinucleotide, or ***NAD***. (The characteristic function of NAD is temporary storage of high-energy electrons; it transports energy from one pathway to another, or from one step in a pathway to another step, elsewhere in the pathway.) Each NAD_{ox} accepts two hydrogens, keeping one and the electron of the other to produce NAD_{re} and an H^+ ion. The second change is phosphorylation of the PGAL:

$$(6)\quad 2\ \text{Ⓟ—C—C—C} + 2\ NAD_{ox} + 2\ P_i \xrightarrow{\text{enzyme}} 2\ \text{Ⓟ—C—C—C} \sim \text{Ⓟ} + 2\ NAD_{re} + 2H^+$$
$$(\Delta G = +3.0 \text{ kcal/mole})$$

The oxidation phase of this reaction, taken alone, is strongly exergonic, while the phosphorylation phase is strongly endergonic. Since the two processes occur together, the energy that would have been released by the oxidation (more than 100 kcal/mole) is conserved in the reduced NAD (NAD_{re}) and the phosphorylated PGAL. The consequence of Step 6 is to make the net change in free energy for the overall reaction chain even more unfavorable (+3.5 kcal/mole), but the next downhill reaction, to which Step 6 is coupled, begins once again to turn the balance, as the high-energy phosphate bond is broken; some of the free energy is harvested by transferring the phosphate group to the ADP, to form ATP:

$$(7)\quad 2\ \text{Ⓟ—C—C—C} \sim \text{Ⓟ} + 2\ \text{ADP} \xrightarrow{\text{enzyme}} 2\ \text{Ⓟ—C—C—C} + 2\ \text{ATP}$$
$$(\Delta G = -9.0 \text{ kcal/mole})$$

At this point, then, the cell regains the two ATP molecules invested to activate the glucose in Steps 1 and 3, and the overall net change in free energy is again favorable: −5.5 kcal/mole. Moreover, a great deal of energy has been stored in NAD_{re}.

Next comes a reaction that energizes the remaining phosphate groups:

$$(8)\quad 2\ \text{Ⓟ—C—C—C} \xrightarrow{\text{enzyme}} 2\ \text{Ⓟ} \sim \text{C—C—C} + H_2O$$
$$(\Delta G = +1.5 \text{ kcal/mole})$$

After a reaction that rearranges the substrate and has a ΔG of −0.4 kcal/mole (Step 9), these energized phosphate groups are transferred to ADP; the products are ***pyruvic acid*** (also referred to as pyruvate) and ATP:

$$(10)\quad 2\ \text{Ⓟ} \sim \text{C—C—C} + 2\ \text{ADP} \xrightarrow{\text{enzyme}} 2\ \underset{\text{pyruvic acid}}{\text{C—C—C}} + 2\ \text{ATP}$$
$$(\Delta G = -15.0 \text{ kcal/mole})$$

Obviously this last reaction is overwhelmingly favorable; indeed, the K_{eq} is greater than 10^9. Moreover, a profit of two ATPs is generated by this step of glycolysis. Because the two ATP molecules used in Steps 1 and 3 have already been regained (in Step 7), the two additional molecules formed here represent a net gain in ATP for the cell. The highly exergonic last step results in an overall ΔG of -19.4 kcal/mole. It is the liberation of this energy, primarily as heat, that causes this series of coupled reactions to proceed.

Figure 6.4 summarizes the essential steps of glycolysis; Figure 6.5 shows graphically both the thermodynamics of the reaction steps and the changes in free-energy content at each successive step in glycolysis, from glucose to pyruvic acid (as well as in the process of fermentation, which will be discussed shortly).

We can now summarize the most important features of glycolysis:

1. Each molecule of glucose (a six-carbon compound) is broken down into two molecules of pyruvic acid (a three-carbon compound).
2. Two molecules of ATP are used to initiate the process. Later, four new molecules of ATP are synthesized, for a *net* gain of two molecules of ATP from each glucose molecule broken down. The energy stored in the new ATP molecules represents only about 2 percent of the energy initially present in the glucose molecule.
3. Two molecules of reduced NAD (NAD_{re}) are formed.
4. Because no molecular oxygen is used, glycolysis can occur whether or not O_2 is present. It is a process encountered in the cytoplasm of all living cells, whatever their mode of life.

Fermentation We have seen that in glycolysis two molecules of NAD_{ox} are reduced to NAD_{re}, and that NAD functions in the cell as an energy-transport compound, shuttling high-energy electrons between one substance and another. Thus NAD is only a temporary acceptor of electrons, promptly passing its extra electrons to some other compound and then going back for another load. The cell has only a limited supply of NAD molecules, and

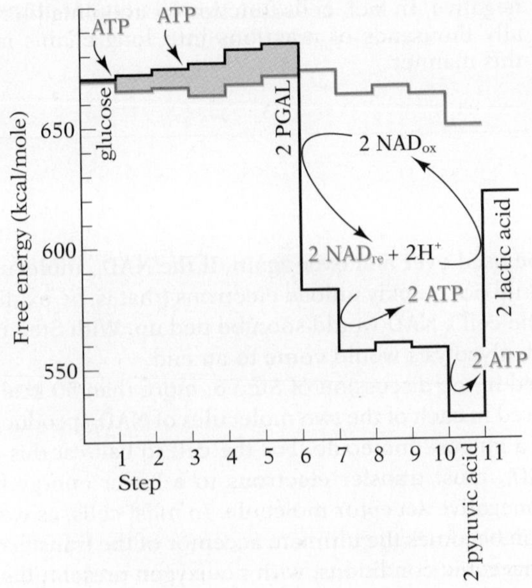

6.5 Changes in free energy at successive steps in glycolysis and fermentation
The graph shows two very different free-energy summaries. The first (red) traces only the free-energy changes resulting from each reaction; these values correspond to the ΔGs cited in the text, and show the net change of -19.5 kcal/mole that accompanies glycolysis. The black line represents the energy in the chemical intermediates; it includes the cost of the ATPs invested in Steps 1 and 3 to activate the glucose, and the large drops in free energy correspond to steps in which energy is removed from those intermediates and stored in ATP or NAD_{re}. In the five preparatory steps, which convert glucose into PGAL, the free-energy content of the intermediate substances is slightly increased owing to the investment of two molecules of ATP. In Step 6 there is a sharp drop in free energy associated with the formation of two molecules of NAD_{re}. There are also major drops in Steps 7 and 10, each associated with the formation of two molecules of ATP. The space between the two free-energy summaries indicates the cellular energy profit; the process runs at a loss until Step 6.

Exploring Further

COUPLED REACTIONS

Many important reactions are endergonic; outside energy is required to build molecules like DNA, RNA, and protein, to form structures like cell membrane, and to activate energy sources like glucose and fat so their stored energy can be utilized. This outside energy is supplied by a coupling of the unfavorable endergonic reaction to a strongly exergonic reaction. The first two reactions of glycolysis provide an example of how coupling works:

$$(1)\quad \text{glucose} + \text{ATP} \xrightleftharpoons{K_{eq}=610} \text{glucose-6-phosphate} + \text{ADP} \quad (\Delta G = -4.0 \text{ kcal/mole})$$

$$(2)\quad \text{glucose-6-phosphate} \xrightleftharpoons{K_{eq}=0.54} \text{fructose-6-phosphate} \quad (\Delta G = +0.4 \text{ kcal/mole})$$

(Changes in ADP, being affected also by reactions unrelated to glycolysis, need not be considered and are omitted in the equation for Step 2.)

Reaction 1 is downhill, with the products eventually outnumbering the reactants by 610 to 1, while reaction 2 is uphill, with the reactant eventually outnumbering the product by about 2 to 1. Notice that one of the products of reaction 1 is the reactant of reaction 2: the reactions are coupled—they share an ***intermediate compound***, glucose-6-phosphate. The table, a molecular scorecard, shows these coupled reactions set in motion by the combination of 10^5 molecules of glucose and an equal number of molecules of ATP (*a*). The equilibrium constant of reaction 1 (610) means that it will proceed until a ratio of roughly 99,836 molecules of glucose-6-phosphate to 164 of glucose has been reached (*b*). Now, since glucose-6-phosphate is a reactant of reaction 2, some of it will be converted into fructose-6-phosphate. The equilibrium constant of this ratio is 0.54, so about 35,007 molecules of product will then be produced (*c*).

With the consequent reduction in the total amount of glucose-6-phosphate, reaction 1 is no longer at equilibrium, so additional glucose and ATP react to produce glucose-6-phosphate (*d*), only to have some of this intermediate promptly converted into fructose-6-phosphate (*e*). At this point equilibrium is reached, with the ratios of product to reactant for both reaction 1 and reaction 2 approximating the corresponding equilibrium constants. The addition of more glucose and ATP (which happens almost continuously in cells) or the removal of fructose-6-phosphate (which also happens continuously, as it is converted into another compound in the third step of glycolysis) will cause more molecules to move along this reaction pathway (*f–j* and *k–n*).

Thus favorable and unfavorable reactions are coupled in the cell: reactants are pushed through the uphill steps as long as appropriate intermediate compounds are present and the *net* ΔG is negative. In fact, cells function by coupling literally thousands of reactions into long chains in this manner.

these must be used over and over again. If the NAD_{re} molecules formed in glycolysis could not quickly unload electrons (that is, be oxidized back into NAD_{ox}), all the cell's NAD would soon be tied up. With Step 6 of glycolysis thus blocked, glycolysis would come to an end.

As we noted in the discussion of Step 6, more than 50 kcal/mole of free energy is stored in each of the two molecules of NAD_{re} produced during the glycolysis of a glucose molecule. For the cell to harvest this abundant energy, the NAD_{re} must transfer electrons to a lower energy level in some more electronegative acceptor molecule. In most cells, as we will see, molecular oxygen becomes the ultimate acceptor of the transferred electrons. But under anaerobic conditions, with no oxygen present, the pyruvic acid

Coupled Reactions

	Reactants	Reaction 1 $K_{eq} = 610$	Intermediate compound	Reaction 2 $K_{eq} = 0.54$	Product
	glucose + ATP	$\rightleftharpoons$	glucose-6-phosphate	$\rightleftharpoons$	fructose-6-phosphate
a	100,000		0		0
b	164		99,836		0
c	164		64,829		35,007
d	106		64,887		35,007
e	106		64,866		35,028
Addition of 10^4 molecules of glucose and 10^4 molecules of ATP					
f	10,106		64,866		35,028
g	123		74,849		35,028
h	123		71,349		38,528
i	117		71,355		38,528
j	117		71,353		38,530
Removal of 10^4 molecules of fructose-6-phosphate					
k	117		71,353		28,530
l	117		64,857		35,026
m	106		64,868		35,026
n	106		64,864		35,030

formed by glycolysis accepts electrons from NAD_{re}. This reduction of pyruvic acid results in the formation of ***lactic acid*** in animal cells and some unicellular organisms, or of ***ethanol*** (ethyl alcohol) and carbon dioxide in most plants and many unicellular organisms:

$$\text{(11)} \quad \text{pyruvic acid} + NAD_{re} + H^+ \xrightarrow{\text{enzyme}} \text{lactic acid} + NAD_{ox}$$

or

$$\text{(11)} \quad \text{pyruvic acid} \xrightarrow{\text{enzyme}} \text{acetaldehyde} + CO_2$$

$$\text{(12)} \quad \text{acetaldehyde} + NAD_{re} + H^+ \xrightarrow{\text{enzyme}} \text{ethanol} + NAD_{ox}$$

Thus, under anaerobic conditions, NAD shuttles back and forth, picking up electrons—becoming NAD_{re}—in Step 6 and giving up the electrons—becoming NAD_{ox}—in Step 11 or 12.

The process that begins with glycolysis and ends with the transformation of pyruvic acid into ethanol or lactic acid is called ***fermentation***.[5] We can thus speak of alcoholic fermentation or lactic acid fermentation, depending on the end product of the process. (In a few organisms, fermentation leads to products other than ethanol or lactic acid, but these are of less general importance and will not be discussed here.)

Whatever the end product, fermentation enables a cell to continue synthesizing ATP by breakdown of nutrients under anaerobic conditions. But, because the electrons transferred from NAD_{re} remain at a relatively high energy level in the reduction of pyruvic acid, fermentation extracts only a very small portion (about 2 percent) of the energy present in the original glucose.

Fermentation by yeast cells and other microorganisms is, of course, the basis for the extensive and economically vital industries that produce breads and both commercial alcohol and alcoholic beverages. Bacterial fermentations are also essential to the production of most cheeses, yogurt, and a variety of other dairy products.

AEROBIC RESPIRATION

The more efficient process of energy extraction that occurs in the presence of abundant molecular oxygen was probably perfected in the tiny, single-celled organisms whose descendants, as noted in Chapter 5, may have given rise to the mitochondria of eucaryotic cells. In eucaryotic cells, aerobic respiration now takes place exclusively in the mitochondria. When O_2 is present, pyruvic acid need not then act as the electron acceptor and become converted into lactic acid or ethanol. Instead, oxygen acts as the electron acceptor; mitochondrial enzymes move the transient electrons to the oxygen, thereby releasing the free energy of NAD_{re}:

$$O_2 + 2\ NAD_{re} + 2H^+ \rightleftharpoons 2H_2O + 2\ NAD_{ox} \qquad (\Delta G = -52.4\ \text{kcal/mole})$$

[5] The term "fermentation" has been used in countless ways in the scientific literature. It is often restricted to the breakdown of glucose to ethanol. It is also applied to the production of either ethanol or lactic acid by microorganisms—lactic acid production in animal cells being called glycolysis. Both these uses lead to confusion between the terms "fermentation" and "glycolysis," and both tend to obscure the general occurrence of the same basic fermentation process in all living cells. Accordingly, "fermentation" is here applied to any process in which glucose is catabolized and organic molecules are used as the electron acceptors of glycolysis. The glycolytic pathway to pyruvic acid is taken both as a preparatory reaction sequence leading to the Krebs citric acid cycle when sufficient oxygen is present, and as the initial portion of fermentation in the absence of sufficient oxygen. (A few microorganisms carry out fermentation in the presence of abundant oxygen.)

The free energy is then utilized in the formation of ATP. Moreover, the pyruvic acid (which still has 590 kcal/mole of free energy at Step 10) can be broken down to yield additional energy for the synthesis of still more ATP. If lactic acid has already been formed, it can be reconverted into pyruvic acid (with consequent regeneration of the lost NAD_{re}) when sufficient oxygen becomes available. This pyruvic acid too may then be oxidized.

The process of aerobic breakdown of nutrients with accompanying synthesis of ATP is called ***aerobic respiration***. Whereas anaerobic respiration consists of fermentation (glycolysis, followed by the transfer of electrons to pyruvic acid, producing lactic acid or ethanol), aerobic respiration consists of a longer sequence of events: glycolysis, oxidation of pyruvic acid to acetyl-CoA, the reactions of what is known as the Krebs citric acid cycle, a series of reactions involving an electron-transport chain, and finally the processes that culminate in the synthesis of ATP. It is the transfer of electrons from NAD_{re} to highly electronegative oxygen atoms—a transfer mediated by the electron-transport chain—that ultimately produces most of the ATP from glucose.

Oxidation of pyruvic acid to acetyl-CoA (Stage II of aerobic respiration) The net effect of the aerobic oxidation of pyruvic acid is to break down the three-carbon pyruvic acid to CO_2 and the two-carbon compound acetic acid, which is connected by a high-energy bond to a coenzyme called CoA, for short; the complete compound is called ***acetyl-CoA***. When a molecule of pyruvic acid is oxidized to acetyl-CoA and CO_2, hydrogen is removed and a molecule of NAD_{re} is formed. Since two molecules of pyruvic acid were formed from each glucose molecule, two molecules of NAD_{re} are formed here. This complicated series of reactions can be summarized by the following equation:

$$\text{2 pyruvic acid} + \text{2 CoA} + 2\ NAD_{ox} \rightleftharpoons \text{2 acetyl}{\sim}\text{CoA} + 2CO_2 + 2\ NAD_{re} + 2H^+$$

Note that, at this stage, two of the six carbons present in the original glucose have been released as CO_2. Note also that the newly formed NAD_{re} must be oxidized if the breakdown process is to continue; we will return to this problem shortly.

The Krebs citric acid cycle (Stage III of aerobic respiration) The acetyl-CoA is next fed into a complex series of reactions called the Krebs citric acid cycle, after the British scientist Sir Hans Krebs, who was awarded a Nobel Prize for his elucidation of this system. Briefly, each of the two two-carbon acetyl-CoA molecules formed from one molecule of glucose is combined with a four-carbon compound (oxaloacetic acid) already present in the cell, to form a new six-carbon compound called ***citric acid***. Each of the citric acid molecules is then oxidized to a five-carbon compound plus CO_2. The five-carbon unit, in turn, is oxidized to a four-carbon compound plus CO_2. This four-carbon compound is then converted into the four-carbon compound—oxaloacetic acid—to which acetyl-CoA was originally at-

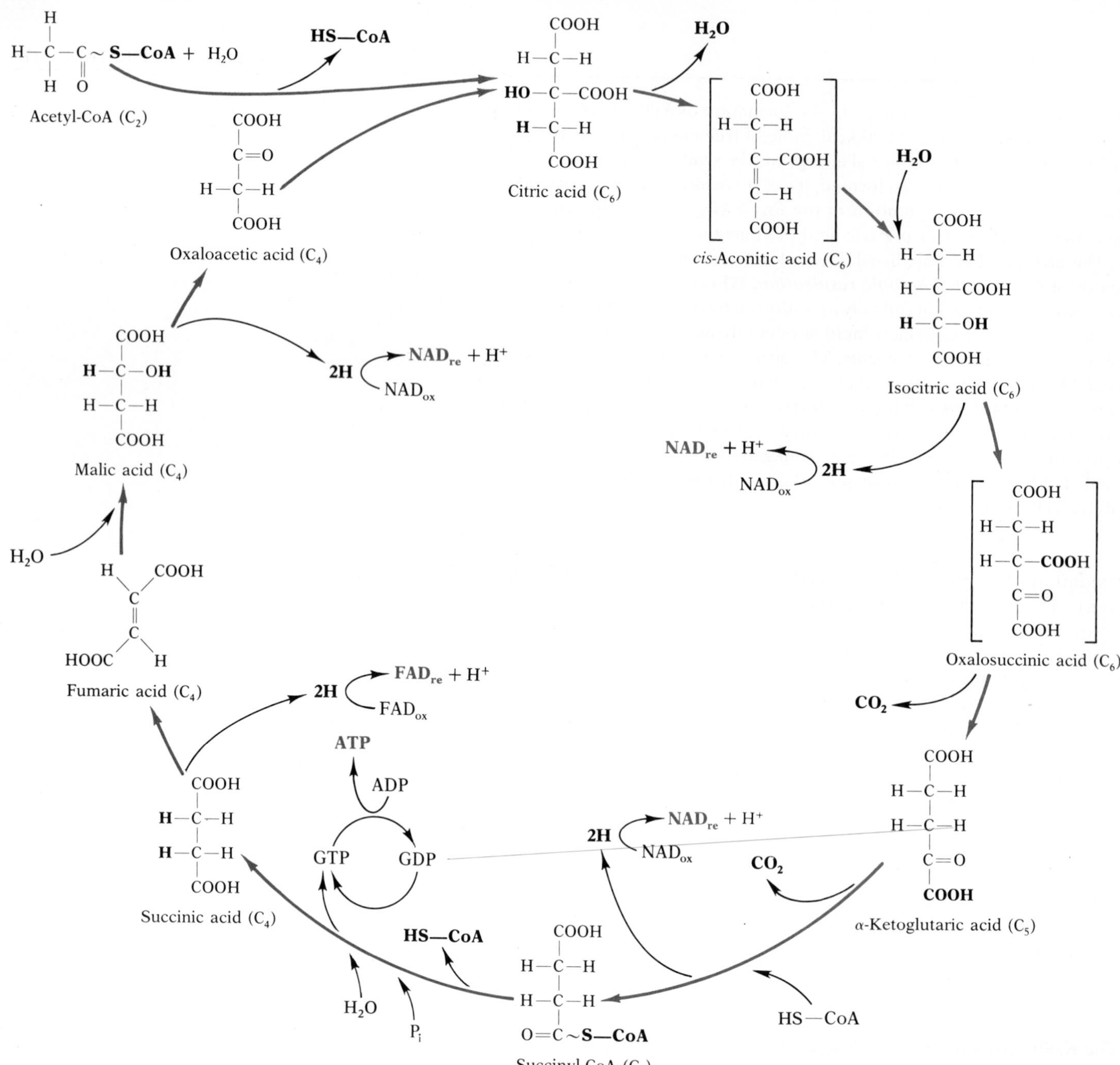

6.6 The Krebs citric acid cycle
The complete cycle is shown here only to illustrate the characteristic complexity of metabolic pathways; there is no point in trying to learn all the reactions involved. As shown at upper left, the acetyl group (two carbons) from acetyl-CoA enters the cycle by combining with oxaloacetic acid (four carbons) to form citric acid (six carbons). During subsequent reactions two of the carbons are removed as CO_2 (between oxalosuccinic acid and α-ketoglutaric acid, and between α-ketoglutaric acid and succinyl-CoA), and a total of eight hydrogens are removed. These hydrogens are picked up by NAD (or by the related acceptor molecule FAD). One molecule of ATP is synthesized (bottom). Finally, oxaloacetic acid is regenerated and can combine with a new acetyl group to start the cycle over again. The cycle is completed twice for each molecule of glucose oxidized. (The atoms removed at each step are shown in boldface in the structural formulas. The two substances in brackets—*cis*-aconitic acid and oxalosuccinic acid—are enzyme-bound intermediates that seldom exist as free compounds.)

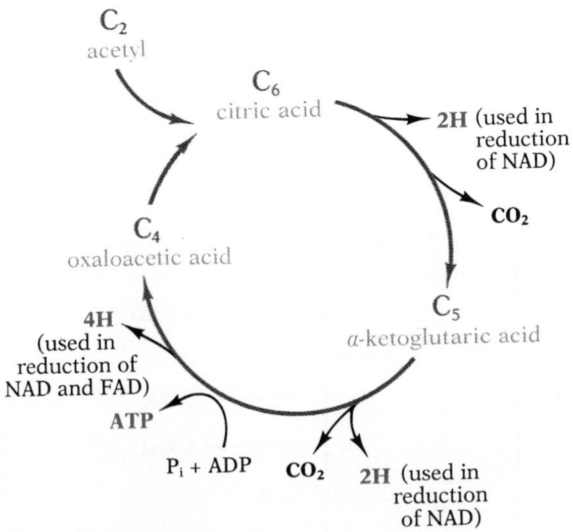

C_2
acetyl
C_6
citric acid
2H (used in reduction of NAD)
CO_2
C_5
α-ketoglutaric acid
2H (used in reduction of NAD)
CO_2
P_i + ADP
ATP
4H (used in reduction of NAD and FAD)
C_4
oxaloacetic acid

6.7 Simplified version of the Krebs citric acid cycle
The two carbons of the acetyl group combine with a four-carbon compound to form citric acid, a six-carbon compound. Removal of one carbon as CO_2 leaves a five-carbon compound. And removal of a second carbon as CO_2 leaves a four-carbon compound, which can combine with another acetyl group and start the cycle over again. In the course of the cycle, one molecule of ATP is synthesized, and eight hydrogens are released, which are used in reduction of NAD and FAD. Since one molecule of glucose gives rise to two acetyl units, two turns of the cycle occur for each molecule of glucose oxidized, with production of four molecules of CO_2, two molecules of ATP, and 16 hydrogens.

tached; it can now pick up more acetyl-CoA, forming new citric acid and beginning the cycle again (Fig. 6.6). The cycle is simplified in Figure 6.7.

We see, then, that two carbons are fed into the Krebs cycle as the acetyl group and two are released as CO_2. Since each glucose molecule being oxidized yields two molecules of acetyl-CoA, two turns of the cycle are required, and a total of four carbons are released as CO_2 during this stage of glucose breakdown. With the two carbons already released as CO_2 during the oxidation of pyruvic acid to acetyl-CoA, all six carbons of the original glucose are accounted for.

The oxidative breakdown of each molecule of acetyl-CoA via the Krebs citric acid cycle also involves the removal of eight hydrogens, which are picked up by NAD_{ox} (or by FAD_{ox}, the oxidized form of a related electron-carrier protein called flavin adenine dinucleotide); four units of reduced carrier are thus formed (Fig. 6.6). Since the breakdown of one molecule of glucose leads to two turns of the Krebs cycle, a total of eight molecules of reduced carrier (six NAD_{re} and two FAD_{re}) are formed during this stage of the breakdown of glucose. Two molecules of ATP are also synthesized in two turns of the Krebs cycle.

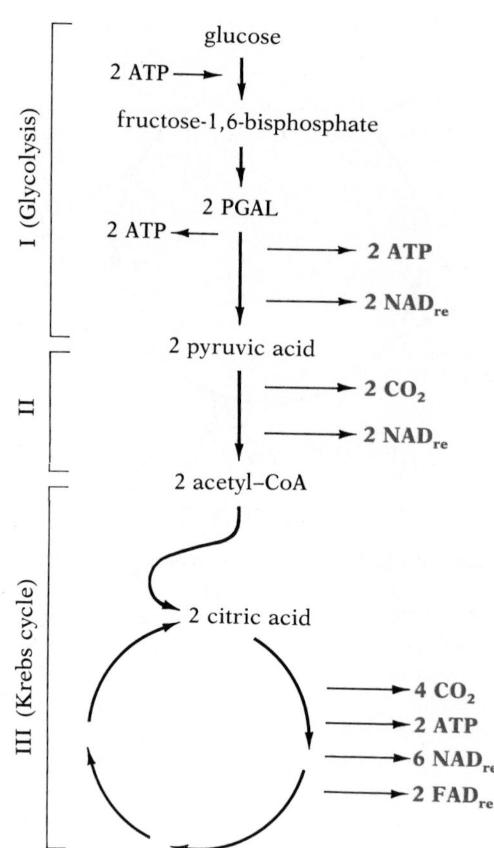

6.8 Summary of the most important products of Stages I, II, and III in the complete breakdown of one molecule of glucose
Stage I (glycolysis) begins with the expenditure of two molecules of ATP to activate glucose and produce two molecules of PGAL. The two PGAL molecules are then broken down to two molecules of pyruvic acid, in a process that first pays back the two ATP molecules originally invested and then yields two molecules each of ATP and NAD_{re} (red). Stage II yields two molecules each of CO_2 and NAD_{re}. Stage III, in which the two molecules of acetyl-CoA are fed into the Krebs cycle and further broken down, yields four CO_2 molecules, two ATP molecules, six NAD_{re} molecules, and two FAD_{re} molecules. (The H^+ ions liberated in the production of NAD_{re} and FAD_{re} are not shown.)

Figure 6.8 summarizes the yield of ATP, NAD_{re}, FAD_{re}, and CO_2 from the three stages of the breakdown of glucose.

The respiratory electron-transport chain (Stage IV of aerobic respiration) We pointed out earlier that glucose is energy-rich, and that its breakdown enables the cell to synthesize new ATP, the cell's energy currency. But in our examination of the first three stages of catabolism we have seen a net gain of only four new ATP molecules (two in glycolysis and two in the Krebs cycle). These represent only a small fraction of the energy originally available in the glucose. Of the remainder, some is liberated during the first three stages (mostly as heat, which is essential for the reactions to proceed); the rest is stored in the high-energy intermediates NAD_{re} and FAD_{re}. Twelve of these molecules are synthesized in the breakdown of each molecule of glucose (Fig. 6.8).

How is this energy used to synthesize ATP? We said earlier that under aerobic conditions the regeneration of NAD_{ox} from NAD_{re} is achieved by the

Exploring Further

REGULATION OF GLUCOSE BREAKDOWN

There are several major points at which the interconnected enzyme pathways shown in the accompanying figure can be regulated. One of the most important is at the step in glycolysis where fructose-1,6-bisphosphate is produced. The allosteric enzyme that catalyzes this reaction is positively modulated (made more reactive) by ADP and AMP; it is negatively modulated (made less reactive) by ATP and by citric acid. Hence the enzyme is most active, and glucose breakdown is greatest, when there is a shortage of ATP and a buildup of ADP and AMP. The enzyme is least active, and glucose breakdown is slowed, when there is an accumulation of ATP and citric acid. Regulation of the enzyme for fructose-1,6-bisphosphate by citric acid is a good example of one type of ***feedback inhibition***—the inhibition of an allosteric enzyme by one of the products of a later reaction in the biochemical pathway. The positive modulation by ADP and AMP, on the other hand, constitutes ***activation***. Both types of control are illustrated in the figure.

Other steps of this pathway for glucose breakdown are also regulated. The product of the first step of glycolysis, glucose-6-phosphate, inhibits the enzyme that produces it, thereby assuring that additional glucose is not activated until the glucose-6-phosphate has been consumed in subsequent reactions. In the citric acid cycle, an excess of ATP or NAD_{re} inhibits the enzyme that converts isocitric acid into α-ketoglutaric acid, while an excess of ADP or NAD_{ox} activates it. Hence, the cycle continues only if there are electron acceptors ready to be charged. Each of these control mechanisms assures that the rate at which this glucose pathway operates is precisely matched to the current needs of the cell.

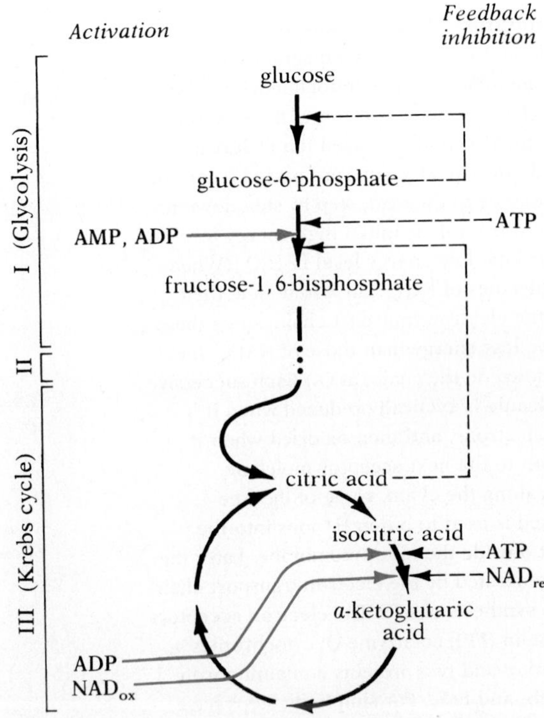

passage of electrons from NAD_{re} to O_2, with oxygen thus acting as the ultimate acceptor of electrons:

$$O_2 + 2\ NAD_{re} + 2H \longrightarrow 2H_2O + 2\ NAD_{ox}$$

The NAD_{re} does not, however, pass its electron directly to the oxygen, as this summary equation might seem to indicate. The electrons and their associated proton reach their ultimate targets indirectly. In particular, the hydrogen electrons are passed down a "respiratory chain" of electron-transport compounds, many of which are iron-containing enzymes called

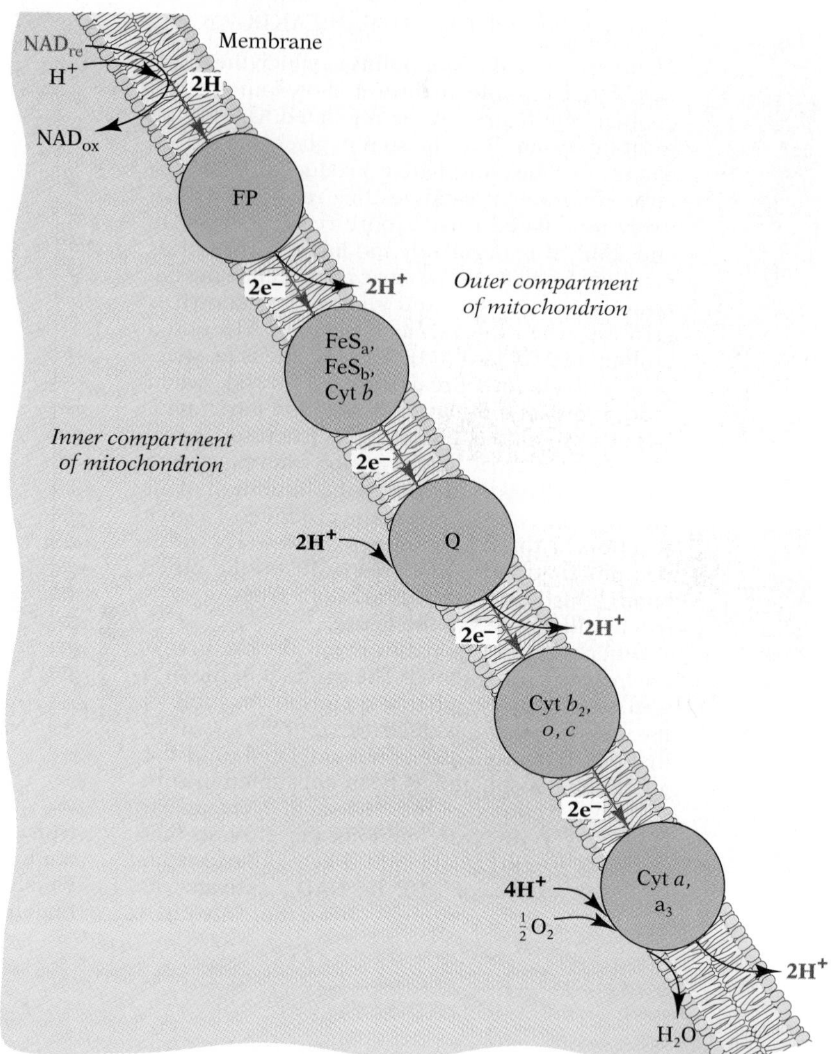

6.9 The respiratory electron-transport chain The reactions summarized in this diagram take place on the inner membrane of the mitochondrion. NAD_{re} donates two electrons and a proton to the electron-transport chain; a second hydrogen ion is drawn from the medium. The electrons are passed from one acceptor substance to the next, step by step down an energy gradient from their initial high energy level in NAD_{re} to their final low energy level in H_2O. (When available, molecules of FAD_{re} can also donate their electrons to the electron-transport chain; since these electrons have less energy than those of NAD_{re}, they enter lower down on the chain, at Q.) Each successive acceptor molecule is cyclically reduced when it receives the electrons, and then oxidized when it passes them on to the next acceptor molecule. At three sites along the chain, some of the free energy released is used to pump H^+ ions into the compartment outside the inner membrane. Later the H^+ gradient generated by the electron-transport chain is used in the synthesis of ATP. The electron acceptors are a flavoprotein (FP); coenzyme Q; cytochromes *a*, a_3, *b*, b_2, *c*, and *o*; and two proteins containing iron and sulfur, FeS_a and FeS_b. For simplicity, steps have been combined wherever possible. A more complete sequence is shown as part of Figure 6.11.

cytochromes (Fig. 6.9). The electrons move to ever-lower energy levels with each transfer. As we will see, the energy extracted by way of the electron-transport chain is then used for the production of ATP.

THE ANATOMY OF AEROBIC RESPIRATION

The elaborate internal structure of the mitochondrion, mentioned in the preceding chapter, plays a crucial role in respiration. You may recall that an inner membrane divides each mitochondrion into two compartments, and that this membrane, being extensively folded, has a very large surface area (Fig. 6.10). Since the usual role of biological membranes is to create and keep separate two different chemical environments and to serve as scaffolding for gates, pumps, and organized enzyme arrays, you can probably guess

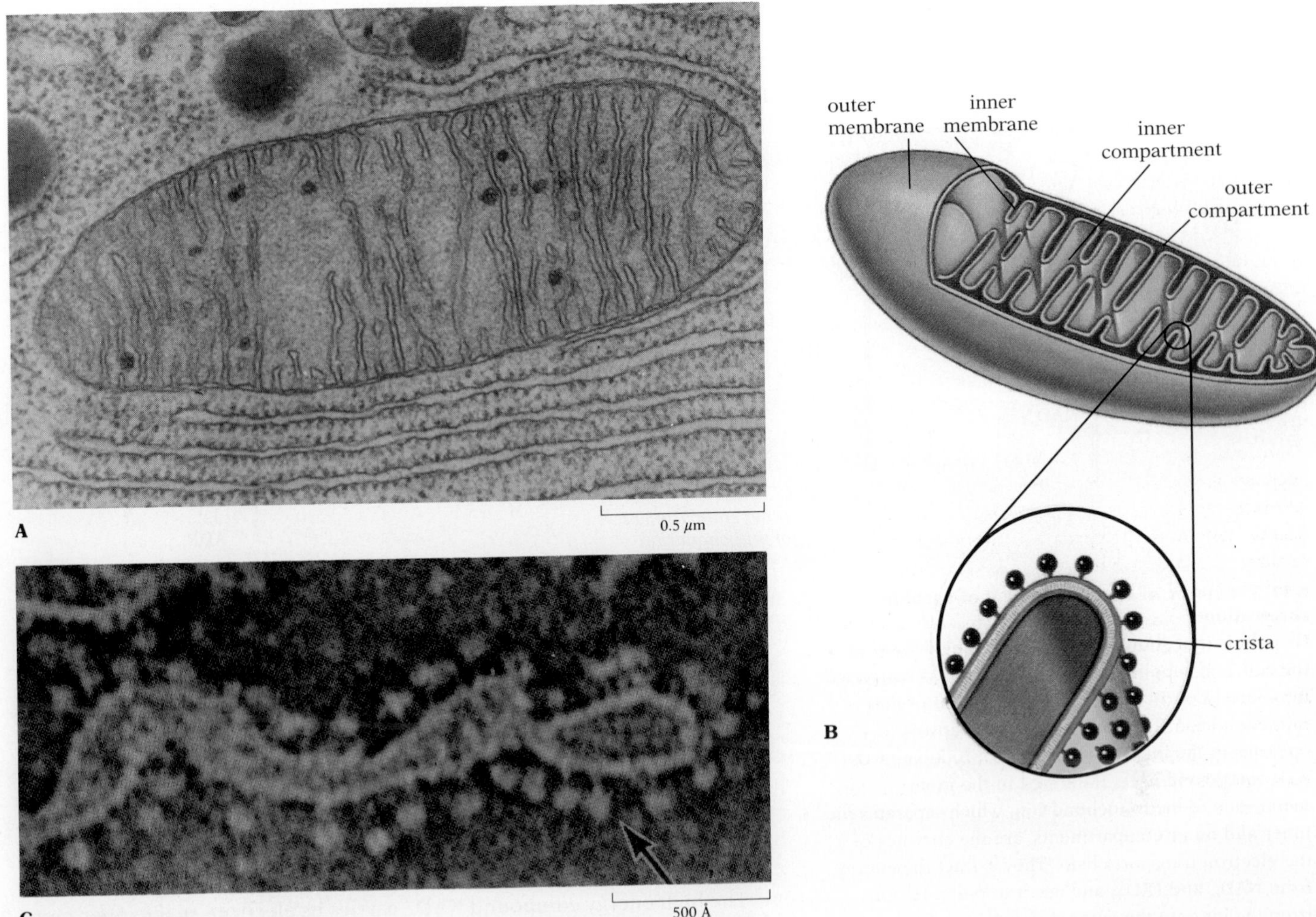

6.10 Structure of a mitochondrion (A) A typical animal mitochondrion. (B) In this interpretive drawing, much of the outer membrane has been cut away, and the interior has been sectioned to show how the inner membrane folds into cristae. The mitochondria of metabolically very active cells have more cristae than those of less active cells. The inner compartment is within the inner membrane, while the outer compartment is the space between the two membranes. (C) A close-up of the inner membrane. As shown in Figure 6.11, much of the activity of cellular respiration, such as electron transport, occurs across the inner membrane between the inner and outer compartments. The knoblike structures are enzyme complexes responsible for synthesizing ATP.

that the capacious inner membrane of the mitochondrion is the key to understanding how this organelle works.

To begin with, we now know that the inner membrane is clearly polarized; its inner face is studded with 9-nm spheres (thought to be the exposed parts of ATP-synthesizing enzyme complexes), and its outer face is smooth. And while the enzymes of the Krebs citric acid cycle are contained in the fluid of the inner compartment, the enzymes of the electron-transport chain are found in the inner membrane.

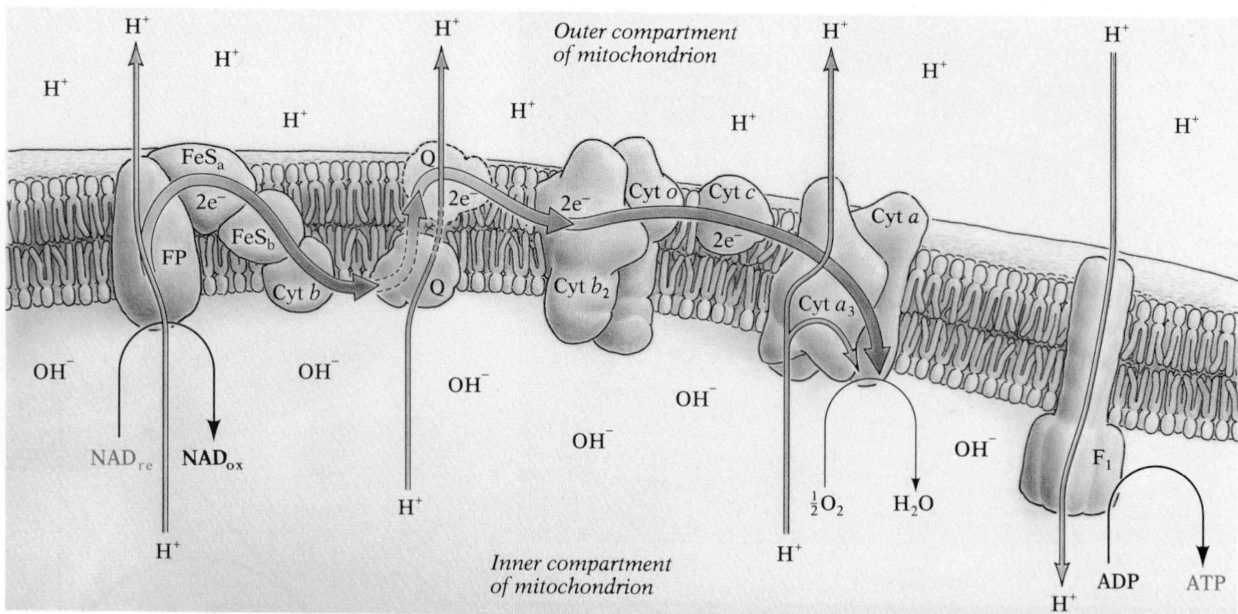

6.11 Anatomy of stages IV and V of aerobic respiration
Glycolysis (not shown) takes place in the cytosol of the cell, and supplies pyruvic acid, which is converted into acetyl-CoA in the inner compartment of the mitochondrion. The Krebs cycle (not shown) operates in the inner compartment, producing CO_2, ATP, NAD_{re}, and FAD_{re}. Mounted in the inner membrane of the mitochondrion, which separates the inner and outer compartments, are the enzymes of the electron-transport chain. They extract the energy from NAD_{re} and FAD_{re}, and use it to pump H^+ ions from the inner to the outer compartment. Some of the ions are carried across by Q, which shuttles between the inner and outer surface of the membrane. The energy of the resulting electrochemical gradient is then used by the enzyme complex F_1 (right) to make ATP from ADP. The ATP is then exported from the cytoplasm (not shown).

Approximately two H^+ ions must pass through the F_1 complex to create one ATP from ADP. Since the energy of the two electrons donated by each NAD_{re} molecule is used to transport six H^+ into the outer compartment, the oxidation of an NAD_{re} produces about three ATPs. Two electrons from a molecule of FAD_{re}, which have less energy than those of NAD_{re}, can enter the electron-transport chain at Q; the oxidation of the lower-energy compound FAD_{re} produces about two ATPs.

The anatomy of the electron-transport chain is shown in Figure 6.11. The major raw material, NAD_{re}, was generated by the sequence of events already described: glycolysis (which takes place outside the mitochondrion in the cytosol) ⟶ pyruvic acid (which is transported through both the outer and inner membranes to the inner compartment of the mitochondrion) ⟶ acetyl-CoA ⟶ Krebs cycle ⟶ NAD_{re} (as well as CO_2, ATP, and FAD_{re}). The electron transport chain must now harvest this energy.

The high-energy compound NAD_{re} carries its electrons to the inner membrane. Here NAD_{re} is oxidized by losing a hydrogen and an electron to FP (part of a four-enzyme complex), which is reduced; simultaneously, FP accepts an H^+ from the medium of the inner compartment. The hydrogen from NAD_{re} is split into a hydrogen ion (H^+) and an electron (e^-). The H^+ ion, along with the H^+ accepted from the medium, is deposited in the outer compartment; the two electrons are passed immediately to another enzyme in the complex, FeS_a. The result is that the H^+ concentration of the outer compartment is raised, the H^+ concentration of the inner compartment is lowered, high-energy-level electrons (two from each NAD_{re}) are inserted into the transport chain (center) where they will be used to do work, and NAD_{ox} is returned to the citric acid cycle to be "recharged."

The transfer of electrons from one acceptor to another proceeds because free energy is liberated (that is, lost by electrons) at each step. For each electron reaching enzyme Q (which shuttles back and forth across the membrane), another hydrogen ion enters the chain from the inner compartment and is deposited in the outer compartment. (At this stage, the lower-energy electrons of FAD_{re} can also enter the electron-transport chain.) The electrons, whatever their origin, now move through the cytochrome series (left) to cytochrome a_3. This last enzyme group uses the energy of the electrons to split molecular oxygen (O_2) and catalyze the reaction

$$\tfrac{1}{2}O_2 + 4H^+ + 2e^- \xrightarrow{\text{cytochrome } a_3} H_2O + 2H^+$$

This reaction reduces by four the number of hydrogen ions in the inner compartment. Two of these hydrogen ions are exported to the outer compartment, while the other two are incorporated into the water molecule. About 100 electrons can pass through each transport chain each second.

Chemiosmotic synthesis of ATP (Stage V of aerobic respiration) All the energy stored in the NAD_{re} has now been used up without generating any new ATP. However, an electrostatic and osmotic-concentration gradient has been built up across the inner membrane, which is relatively impermeable to H^+ ions: the inner compartment, having lost H^+, has become negatively charged, while the outer compartment—the space between the inner and outer membrane—is filled with H^+ ions and therefore positively charged. The result is like a battery, in which the difference in charge across the inner mitochondrial membrane is used to make ATP from ADP. Most of the H^+ ions return from the outer to the inner compartment by way of the F_1 complex (Fig. 6.11), passing down what has come to be called the ***chemiosmotic gradient***. As they do so, the energy of the the electrochemical gradient—the combined osmotic and electrostatic gradients—is harvested. The abundant energy generated by this elegant system is converted and stored in the phosphate groups of ATP. The same H^+ gradient probably also powers the pump that exports ATP to the cytosol. Peter Mitchell of the Glynn Research Laboratories in England won a Nobel Prize in 1978 for working out this indirect route, by which mitochondria convert the energy of NAD_{re} into ATP.

It is a continuing challenge to explain how such efficient, highly refined examples of biological engineering as the electron-transport chain might have evolved. How do the first elements in such a complicated array, which seem useless without the rest, ever come into existence? To answer the question, biologists look at a wide variety of species for "preadaptations," fragments of the system as we know it now that are employed in other ways, and which might have been present in ancestral organisms and available to natural selection for "remodeling."

Consider the most intricate element of the mitochondrial system, the F_1 complex, which uses the H^+ gradient to make ATP. Relict species of anaerobic bacteria living today still lack an electron-transport chain, but nevertheless have an F_1 assembly that works in the reverse direction of the mitochondrial F_1. These anaerobes burn ATP to pump H^+ out of their cells, eliminating the acidic by-products of fermentation. There is similar evidence to suggest that the cytochrome complexes evolved later in ancient bacteria, where they served similar detoxification functions. Natural selection must then have favored combining these various systems into a rudimentary electron-transport chain using sulfur, carbon, or nitrogen as the

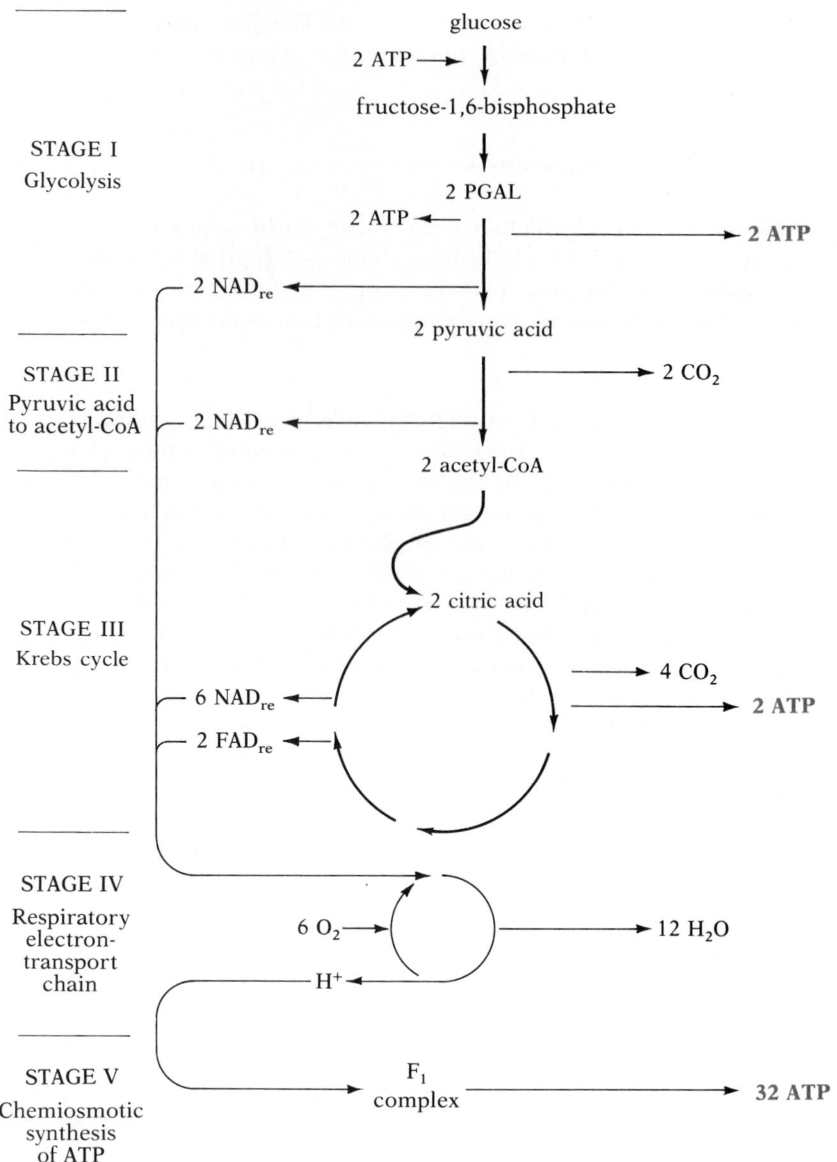

6.12 The ATP yield from complete breakdown of glucose to carbon dioxide and water
In the first three stages, four molecules of ATP are directly synthesized, along with ten of NAD$_{re}$ and two of FAD$_{re}$. In Stage IV, the NAD$_{re}$ and FAD$_{re}$ are fed to the electron-transport chain, and their stored energy is utilized to create a difference in charge across the inner membrane of the mitochondrion. Finally, in Stage V, the F_1 enzyme complex uses the energy of this chemiosmotic gradient across the membrane to make approximately 32 more molecules of ATP.

ultimate electron acceptor. And, as we will see in Chapter 7, the evolution of the respiratory electron-transport chain made possible the evolution of photosynthesis.

SUMMARY OF AEROBIC RESPIRATION ENERGETICS

Now that we have looked at all five stages in the aerobic breakdown of glucose, we can summarize their combined energy yield. The overall flow of energy in aerobic respiration is summarized in Figure 6.12. As we have

Exploring Further

TESTING THE MITCHELL HYPOTHESIS

Convincing evidence for Peter Mitchell's brilliant chemiosmotic hypothesis comes from several researchers, including (besides Mitchell) Efraim Racker, Andre T. Jagendorf, and Peter C. Hinkle, all of Cornell University. Much of their research involves artificially reconstructed vesicles. The technique, widely used in the study of membrane function in general, takes advantage of the ability of phospholipids in an aqueous solution to form spontaneously into spherical liposomes. When mitochondria are exposed to ultrasound (sound of very high frequency), both the outer and inner membranes break into sheetlike fragments; the inner membranes subsequently "heal" by forming spheres about 100 nm in diameter, as shown in the accompanying figure. Curiously enough, in these liposomes—called submitochondrial vesicles—the 9-nm spheres that are thought to be parts of the ATP-synthesizing F_1 complexes are found on the outer face, rather than on the inside, as in mitochondria. This characteristic inversion is the basis of a technique whereby researchers disrupt the mitochondria, let the submitochondrial vesicles form in one kind of solution—trapping that chemical milieu inside the vesicles—and then transfer the vesicles to a different chemical environment. In this way, the chemistry of the two sides of the membrane can be altered at will.

According to Mitchell's chemiosmotic battery model, in a normal mitochondrion the necessary gradient exists because the H^+ concentration is higher on the outside of the inner membrane than on the inside. If the model is correct, the inside-out patches of membrane should function when the H^+ concentration is higher on the inside. And in fact, when the vesicles are created with more H^+ inside than out, they begin producing ATP, but when the H^+ concentration is higher on the outside, they do nothing. The model also predicts that if these vesicles are supplied with NAD_{re} on the outside, they will increase the gradient by raising the H^+ concentration on the inside. This too does indeed happen. In addition, the model predicts that the insertion into these vesicles of channels permeable to H^+ should short-circuit the battery by allowing H^+ to cross the membrane freely, and should thereby prevent ATP synthesis. Again, this is exactly what is observed experimentally.

Mitchell's chemiosmotic hypothesis has helped biologists understand a great deal. It clarifies the structure and functioning of the mitochondrial electron-transport chain. It suggests why the mitochondria are separate, highly organized organelles. The Mitchell hypothesis, as discussed in the next chapter, also accounts for the workings of chloroplasts, and adds credibility to the hypothesis that life evolved from anaerobic procaryotes to aerobic (mitochondrialike) procaryotes, and that photosynthetic chloroplastlike procaryotes evolved from aerobic procaryotes.

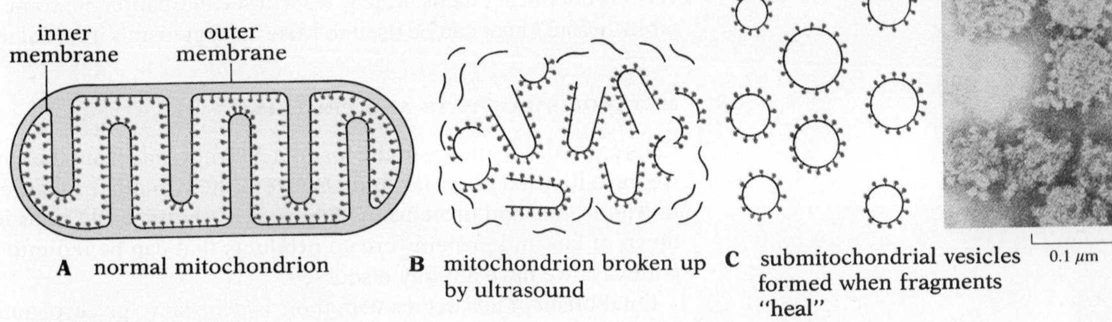

A normal mitochondrion **B** mitochondrion broken up by ultrasound **C** submitochondrial vesicles formed when fragments "heal"

Formation of submitochondrial vesicles

Submitochondrial vesicles are made by breaking the mitochondrion into pieces with ultrasound (B), and then allowing the bits to "heal" (C). By controlling the medium in which vesicle formation takes place, and the solution into which the resulting inside-out membrane spheres are then put, the researcher can create virtually any desired combination of internal and external chemical environments.

seen, glycolysis generates two molecules of ATP per molecule of glucose, along with two NAD_{re}; the conversion of pyruvic acid into acetyl-CoA yields another two NAD_{re}; the Krebs citric acid cycle produces an additional two ATP, six NAD_{re}, and two FAD_{re}; finally, and most critically, the charging of the mitochondrial battery—the expulsion of H^+ across the inner membrane by means of the electron-transport chain (driven by oxidation of the NAD_{re} and FAD_{re})—stores enough energy to synthesize roughly another 32 ATP. Each NAD_{re} provides the energy to pump enough H^+ ions for the synthesis of almost 3 ATPs (10 $NAD_{re} \longrightarrow$ 28 ATP), while each FAD_{re} pumps enough H^+ to power the synthesis of 2 ATPs (2 $FAD_{re} \longrightarrow$ 4 ATP).

It should now be clear why atmospheric oxygen, and the unimpeded operation of the electron-transport chain, are so important to life as we know it: aerobic respiration extracts 18 times as much energy from glucose as does anaerobic metabolism (glycolysis followed by fermentation). We can see, then, why a metabolic poison like cyanide (which binds irreversibly with a cytochrome and thereby blocks the electron-transport chain) is fatal: the cell is suddenly denied 94 percent of its normal energy. This loss is normally disastrous. However, parts of our bodies *can* operate anaerobically for short periods. During intensive exercise, for instance, muscles often need so much energy that the oxygen supplied from breathing is insufficient. In such cases glycolysis and fermentation provide the needed energy for a time, but they are inefficient and fatigue soon results. Later, the oxygen debt is paid back by deep breathing or panting, and the lactic acid that accumulated in the muscles as a result of fermentation is removed to the liver and reconverted into glucose.

It is now simple to calculate the overall efficiency of aerobic respiration. We know that a molecule of glucose has a free-energy content of about 670 kcal/mole, while a molecule of ATP stores about 7.3 kcal/mole. Since the 36 ATPs that are generated therefore represent just under 270 kcal/mole, the cell has retained only 39 percent of the energy originally stored in the glucose; the other 61 percent is released, primarily as heat. This liberated energy is essential to the efficient shuttling of the reactants through the various chemical chains. And as we will see in Chapter 30, some of the inevitable "waste" heat can be used to raise an organism's internal temperature.

METABOLISM OF FATS AND PROTEINS

Cells can extract energy in the form of ATP not only from the carbohydrates we have focused on so far, but also from the two other major categories of nutrients: fats and proteins. As Figure 6.13 shows, early steps in the breakdown of fats and proteins create products that can be fed into the enzyme pathways we have already discussed.

Catabolism of fats begins with their hydrolysis to glycerol and fatty acids. The glycerol (a three-carbon compound) is then converted into PGAL and fed into the glycolytic pathway at the point where PGAL normally appears. The fatty acids, on the other hand, are broken down to a number of two-carbon fragments, which are converted into acetyl-CoA and fed into the respiratory pathway at the appropriate point. Since fats are more completely reduced compounds than carbohydrates (that is, they have a higher proportion of hydrogens), their full oxidation yields more energy per unit weight; one gram of fat yields more than twice as much energy as one gram of carbohydrate.

The amino acids produced by hydrolysis of proteins are catabolized in a variety of ways. After removal of the amino group (deamination) in the form of ammonia (NH_3), some amino acids are converted into pyruvic acid, some into acetyl-CoA, and some into one or another compound of the citric acid cycle. Complete oxidation of a gram of protein yields roughly the same amount of energy as that of a gram of carbohydrate.

Such compounds as pyruvic acid, acetyl-CoA, and the compounds of the citric acid cycle, which are common to the catabolism of several different types of substances, not only play a crucial role in the oxidation of energy-rich compounds to carbon dioxide and water but also function in the anabolism of amino acids, sugars, and fats. They serve as biochemical crossroads, at which several enzyme pathways intersect. By investing energy, the cell can reverse the direction in which substances move along some of these pathways; for example, the PGAL and acetyl-CoA produced at different points in the breakdown of carbohydrate can be moved up the pathways to glycerol and fatty acids, for use in building fats. Similarly, many amino acids can be converted into carbohydrate via the common intermediates in their metabolic pathways. Not all pathways are two-way, however; most higher animals lack a reversible pyruvic-acid-to-acetyl-CoA enzyme, as well as any alternative enzyme, and so cannot convert fatty acids into carbohydrate.

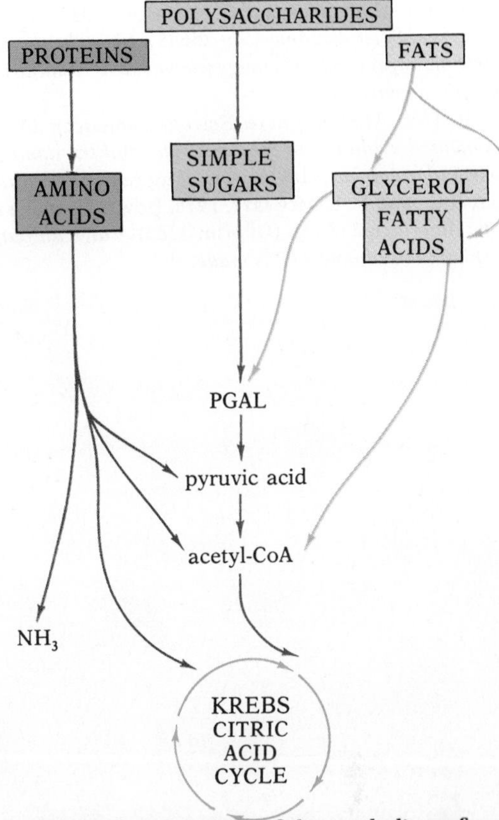

6.13 The relationships of the catabolism of proteins and fats to the catabolism of carbohydrates

STUDY QUESTIONS

1. What is the source of energy in the glucose molecule that is extracted in fermentation? How exactly can you localize it? (pp. 165–72)
2. In the absence of oxygen, which element is the best to use as an electron acceptor in order to extract the most energy: carbon, nitrogen, or sulfur? Why? (p. 160)
3. Why are there two separate entry points into the electron-transport chain, one for electrons from NAD and one for electrons from FAD? (p. 178)
4. What is the purpose of the space between the inner and outer membranes of the mitochondrion? Why not have just one membrane, as the procaryotic precursor of this organelle must have had? (pp. 178–81)
5. Why does reaction coupling work? How does it help explain the operation of fermentation? (pp. 169–71)

CONCEPTS FOR REVIEW

- Role of ATP
- Role of NAD and FAD
- Roles of acetyl-coenzyme A
- Logic of glycolysis
  - Initial investment
  - Series of small, enzyme-mediated steps
  - Net input and output
  - Thermodynamic basis
  - Role of coupled reactions
- Fermentation
  - Need
  - Efficiency
- Krebs cycle
  - Function
  - Efficiency
  - Input and output
- Electron-transport chain
  - General anatomy
  - Electrochemical basis

SUGGESTED READING

CHILDRESS, J. J., H. FELBECK, and G. N. SOMERO, 1987. Symbiosis in the deep sea, *Scientific American* 256 (5). *How chemosynthetic bacteria manage to extract energy from the 250°C sulfurous water at deep sea vents.*

CLOUD, P., 1983. The biosphere, *Scientific American* 249 (3). *On the combined evolution of the earth, life, and the atmosphere, with particular emphasis on the role of oxygen concentration.*

HINKLE, P. C., and R. E. MCCARTY, 1978. How cells make ATP, *Scientific American* 238 (3). (Offprint 1383) *A difficult but rewarding explanation of how ATP is made.*

LEHNINGER, A. L., 1965. *Bioenergetics.* W. A. Benjamin, New York. *A brief, relatively elementary treatment of energy transformations in organisms.**

RACKER, E., 1968. The membrane of the mitochondrion, *Scientific American* 218 (2). (Offprint 1101)

STRYER, L., 1988. *Biochemistry,* 3rd ed. W. H. Freeman, New York. *Traces in great detail the biochemical pathways of respiration, and their regulation.*

* Available in paperback.

Chapter 7

ENERGY TRANSFORMATIONS: PHOTOSYNTHESIS

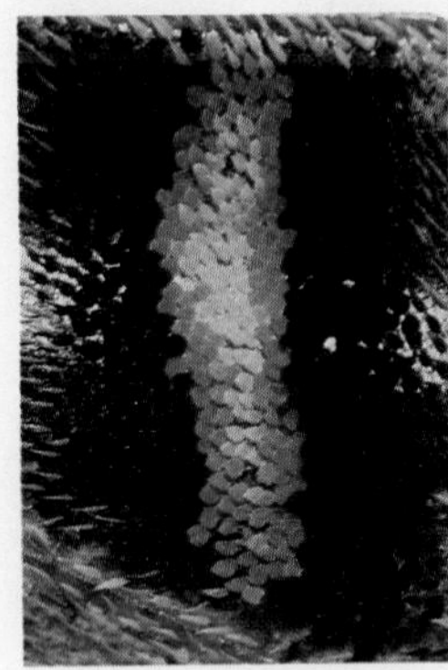

In the last chapter we saw how cells metabolize glucose, first through glycolysis in the cytosol, and then through the remaining steps of aerobic respiration in the mitochondrion. We saw also that from the point of view of bioenergetics, the anaerobic pathway culminating in fermentation is extremely wasteful, because no strongly electronegative substance is available to act as the final acceptor of electrons in the reactions that yield up the energy stored in glucose. Though life existed on the earth for almost 2 billion years in the absence of significant quantities of molecular oxygen, and therefore in the absence of large-scale aerobic respiration, it can hardly be said to have flourished.

According to recent evolutionary evidence, an enormously important biochemical innovation occurred about 3 billion years ago: certain primitive organisms developed for the first time the ability to capture the sun's energy directly and to use it to synthesize foods such as glucose. In these organisms, photons interacting with electrons in special pigment molecules pushed the electrons up to higher levels, and the molecules thus activated passed the energy to still other molecules (Fig. 7.1). According to present evidence, pre-existing enzymes were modified to form an elaborate chemical pathway that made some of this energy from the photons available to drive cellular metabolism.

This process was probably the earliest form of photosynthesis, the trans-

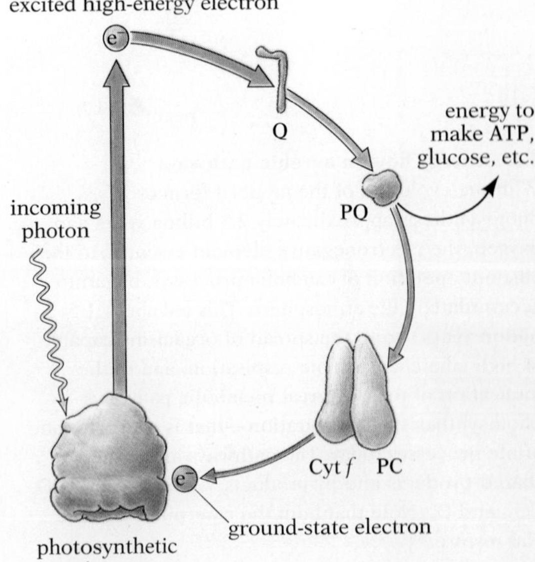

7.1 Energetic basis of life
The basic energy equation of life begins when a photon from the sun excites an electron in a pigment molecule by moving the electron up to a higher level. The energy thus acquired ultimately excites an electron in an acceptor molecule. Subsequently, as the excited electron is returned to a low-energy level in a pigment molecule by way of a series of acceptors, its free energy is used for generating an energy-storage compound (usually glucose or ATP). The pathway shown here is the oldest and most basic form of photosynthesis.

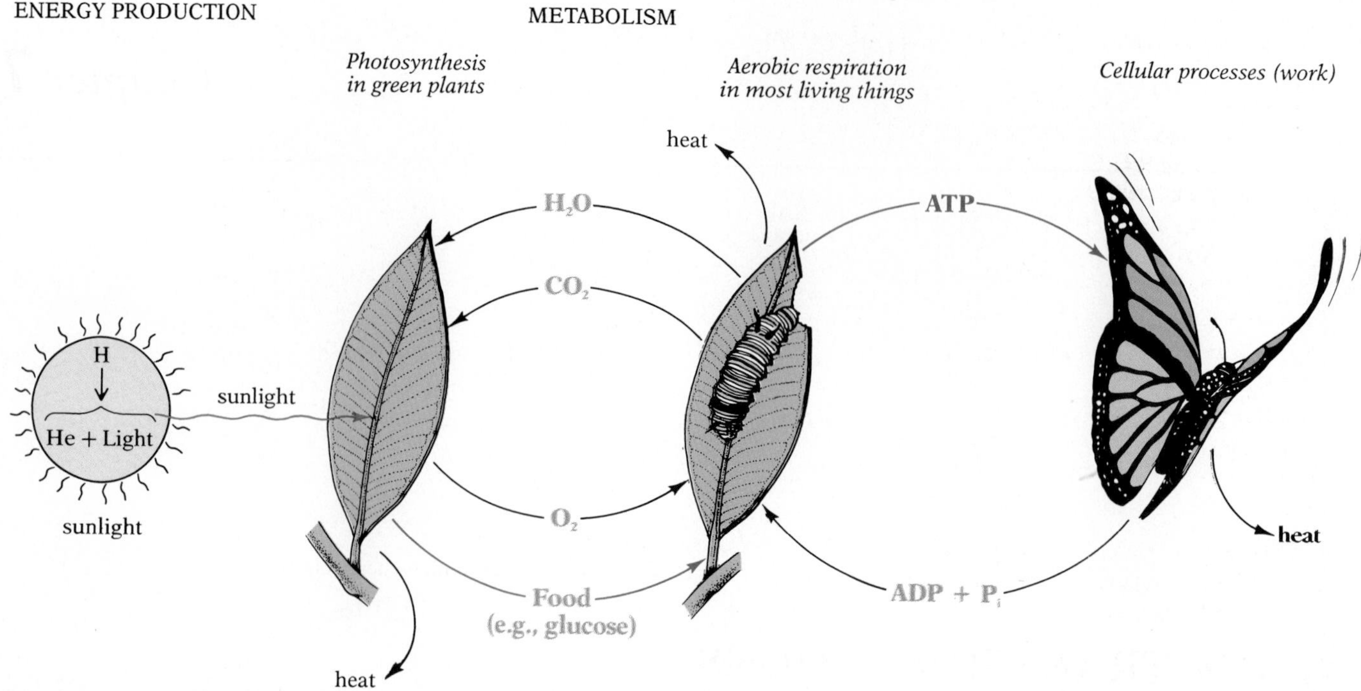

7.2 Energy flow in aerobic pathways
With the evolution of the modern form of photosynthesis approximately 2.3 billion years ago, oxygen, the electronegative element essential to the efficient operation of catabolic processes, began to accumulate in the atmosphere. This led about 1.5 billion years ago to the spread of organisms capable of high-efficiency aerobic respiration, and to the unification of the two great metabolic pathways—photosynthesis and respiration—that is characteristic of life processes today. The pathways are unified by shared products and by-products: organic food, H_2O, CO_2, and O_2. Note that both the caterpillar and the leaf respire.

formation of light energy into chemical energy. Though it did not liberate oxygen into the early atmosphere, early photosynthesis did free organisms from dependence on inorganically generated food. Further evolution resulted in a form of photosynthesis that produced molecular oxygen as a by-product. As this highly electronegative substance began to accumulate in the atmosphere 1.5–2 billion years ago, high-efficiency aerobic respiration became the main mechanism for extracting energy from food, and the two major chemical pathways—photosynthesis and respiration—were joined through shared by-products (Fig. 7.2). Life became the dominant feature of the earth.

Today, virtually all organisms depend directly or indirectly on photosynthesis to fill their energy needs. ***Autotrophs***, organisms such as plants that are capable of making organic nutrients from inorganic materials, depend almost exclusively on photosynthesis, while ***heterotrophs***, organisms such as animals that must obtain organic nutrients from the environment, depend indirectly on photosynthesis, since they consume autotrophs, or heterotrophs that have eaten autotrophs, or both. Photosynthesis, then, is life's single most important biochemical process. Much about photosynthesis is still not known, but in the last three decades the chemical pathways have become much clearer. This chapter will acquaint you with the broad outlines and thermodynamic logic of photosynthesis.

EARLY RESEARCH IN PHOTOSYNTHESIS

Considering how much is now known of anabolic processes, it is easy to forget that for most of recorded history scientists had no idea that the sun supplies the surface of the earth with virtually all its energy, or that green plants trap that energy and produce the invisible gas we breathe. In fact, only in 1772 did the English clergyman Joseph Priestley demonstrate that green plants affect air in such a way as to reverse the effects of burning or of breathing. As he reported it:

> I flatter myself that I have accidentally hit upon a method of restoring air which has been injured by the burning of candles, and that I have discovered at least one of the restoratives which nature employs for this purpose. It is vegetation. In what manner this process in nature operates, to produce so remarkable an effect, I do not pretend to have discovered; but a number of facts declare in favour of this hypothesis. I shall introduce my account of them, by reciting some of the observations which I made on the growing of plants in confined air, which led to this discovery.
>
> One might have imagined that, since common air is necessary to vegetable, as well as to animal life, both plants and animals had affected it in the same manner, and I own I had that expectation, when I first put a sprig of mint into a glass-jar, standing inverted in a vessel of water; but when it had continued growing there for some months, I found that the air would neither extinguish a candle, nor was it at all inconvenient to a mouse, which I put into it.
>
> Finding that candles burn very well in air in which plants had grown a long time, and having had some reason to think, that there was something attending vegetation, which restored air that had been injured by respiration, I thought it was possible that the same process might also restore the air that had been injured by the burning of candles.
>
> Accordingly, on the 17th of August, 1771, I put a sprig of mint into a quantity of air, in which a wax candle had burned out, and found that, on the 27th of the same month, another candle burned perfectly well in it. This experiment I repeated, without the least variation in the event, not less than eight or ten times in the remainder of the summer. Several times I divided the quantity of air in which the candle had burned out, into two parts, and putting the plant into one of them, left the other in the same exposure, contained, also, in a glass vessel immersed in water, but without any plant; and never failed to find, that a candle would burn in the former, but not in the latter. I generally found that five or six days were sufficient to restore this air, when the plant was in its vigour; whereas I have kept this kind of air in glass vessels, immersed in water many months without being able to perceive that the least alteration had been made in it.

Priestley's important experiments were the first demonstration that plants produce oxygen, though he himself did not realize that this was what was happening; nor did he realize that light was essential for the process he observed. But his findings stimulated interest in photosynthesis, as we now call it, and led to further investigation. Only seven years later, the Dutch physician Jan Ingenhousz demonstrated the necessity of sunlight for oxygen production (though, like Priestley, he knew nothing about oxygen at that time and explained his results in other terms), and he also showed that only the green parts of the plant could photosynthesize. He reported his results in a book with the richly descriptive title *Experiments upon Vegetables, Discovering Their Great Power of Purifying the Common Air in the Sun-Shine, and of Injuring It in the Shade and at Night.* In 1782 a Swiss pastor and part-time scientist, Jean Senebier, showed that the process depended on a particular kind of gas, which he called "fixed air" (and we call carbon dioxide). Finally, in 1804, another Swiss researcher, Nicolas Théodore de Saussure, found that water is necessary for the photosynthetic production of organic materials.

Thus, early in the nineteenth century, all the important components of the photosynthetic process were at least vaguely known, and could be summarized by the following equation:

$$\text{carbon dioxide} + \text{water} + \text{light} \xrightarrow{\text{green plants}} \text{organic material} + \text{oxygen}$$

Later, scientists came to believe that light energy splits carbon dioxide, CO_2, and that the carbon is then combined with water, H_2O, to form the group $—CH_2O$, on which carbohydrates are based. According to this view, the oxygen released by the plant during photosynthesis comes from CO_2. This idea received a severe blow when, about 1930, C. B. van Niel of Stanford University showed that some photosynthetic bacteria, which use hydrogen sulfide, H_2S, instead of water as a raw material for photosynthesis, give off sulfur instead of oxygen as a by-product. Now, H_2S and H_2O have obvious chemical similarities, and if the sulfur produced by the bacteria during photosynthesis came from H_2S, it seemed reasonable to suppose that the oxygen produced by plants during photosynthesis might come from H_2O rather than CO_2 (Table 7.1). This was eventually shown by use of a heavy isotope of oxygen (^{18}O instead of the usual ^{16}O). If photosynthesizing plants are given normal carbon dioxide plus water containing heavy oxygen, the heavy isotope does indeed appear as molecular oxygen:

TABLE 7.1 *Comparison of photosynthesis based on water and on hydrogen sulfide*

$CO_2 + 2H_2S \longrightarrow S_2 + (CH_2O) + H_2O$
$CO_2 + 2H_2O \longrightarrow O_2 + (CH_2O) + H_2O$
(The parentheses indicate that CH_2O is not itself a molecule, but rather a subunit of a sugar.)

$$CO_2 + 2H_2{}^{18}O \longrightarrow {}^{18}O_2 + (CH_2O) + H_2O$$

We now know that ***chlorophyll***, the green pigment of plants, traps the energy required to split the water. The currently accepted equation for green-plant photosynthesis is given in more detail below; the dashed lines indicate the fates of all the atoms involved:

$$CO_2 + 2H_2O + \text{light} \xrightarrow{\text{chlorophyll}} O_2 + (CH_2O) + H_2O$$

Multiplying this summary equation by 6 has the advantage of showing that glucose, a six-carbon simple sugar, is often the end product:

$$6CO_2 + 12H_2O + \text{light} \xrightarrow{\text{chlorophyll}} 6O_2 + C_6H_{12}O_6 + 6H_2O \quad (\Delta G = -1300 \text{ kcal/mole})$$

It may seem curious that water should appear on both sides of the equation. The reason is that the water produced by the photosynthetic process is new; it is not the water used as a raw material. You may recall from the last chapter that the free energy stored in glucose is about 670 kcal/mole. The storage of this energy requires capturing roughly 1970 kcal/mole of light energy, the remaining 1300 kcal/mole being liberated during the process, which is thus strongly downhill, or exergonic.

Though the above equation is a convenient summary of photosynthetic carbohydrate synthesis, it tells us nothing about how the synthesis is actually achieved. The process is certainly not one gross chemical reaction, as

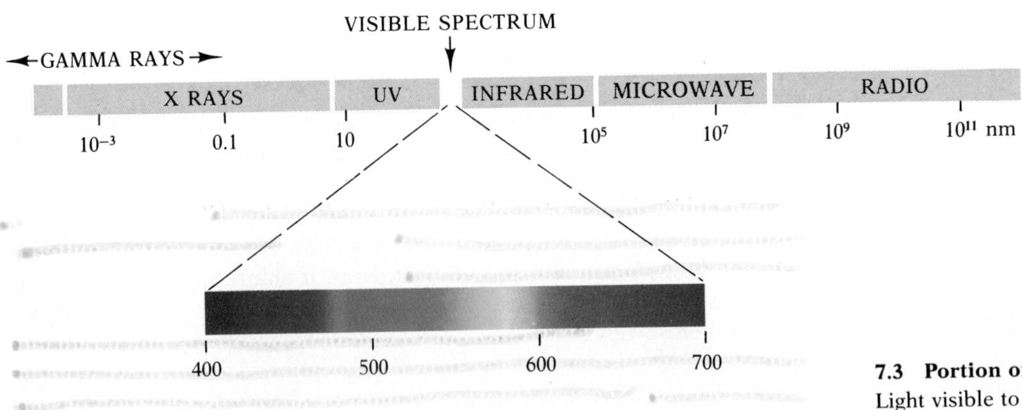

7.3 Portion of the electromagnetic spectrum Light visible to humans constitutes only a very small portion of the total spectrum. Within the visible spectrum, light of different wavelengths stimulates different color sensations in us. Not only vision and photosynthesis, but also other radiation-dependent biological processes, rely on this same small portion of the electromagnetic spectrum (sometimes extended into the ultraviolet or infrared). Light of wavelengths shorter than about 300 nm (nanometers) is absorbed by the atmosphere, while wavelengths longer than 800 nm have too little energy to drive biological reactions.

the summary equation might imply. Many reactions are involved, some that require light, others that require it only indirectly—so-called "dark" reactions. The "light" reactions convert and store the energy from light in specialized energy-transfer molecules like ATP, while the dark reactions use that stored energy to convert (or "fix") carbon dioxide into carbohydrates such as glucose.

THE LIGHT REACTIONS: PHOTOPHOSPHORYLATION

The term ***photophosphorylation*** is often used to describe the light-dependent reactions of photosynthesis. Photophosphorylation means the use of light energy to phosphorylate (add inorganic phosphates to) a molecule, usually ADP:

$$ADP + P_i + \text{energy} \xrightarrow{\text{enzyme}} ATP + H_2O$$

Like so many terms, "photophosphorylation" became part of the scientific vocabulary before the process it describes was well understood. We now know that light-energy absorption and photophosphorylation are separate reactions in just the way the operation of the mitochondrial electron-transport chain and the subsequent synthesis of ATP by the F_1 complex are separate. In fact, as we will see, the two parts of the light reactions, which occur in chloroplasts, directly parallel these two stages of respiration in the mitochondria.

LIGHT AND CHLOROPHYLL

Light waves constitute one small region of the spectrum of electromagnetic radiations (Fig. 7.3). Each radiation in this spectrum has a characteristic wavelength and energy content. These two characteristics are inversely related: the longer the wavelength, the smaller the energy content. Within the narrow band visible to human beings, the shortest light waves produce the sensation of violet and the longest produce the sensation of red. Radiations such as ultraviolet, X rays, and gamma rays, which are of shorter wavelengths than violet, are invisible to us, as are infrared, microwave, and radio-TV radiations, which are of longer wavelengths than red.

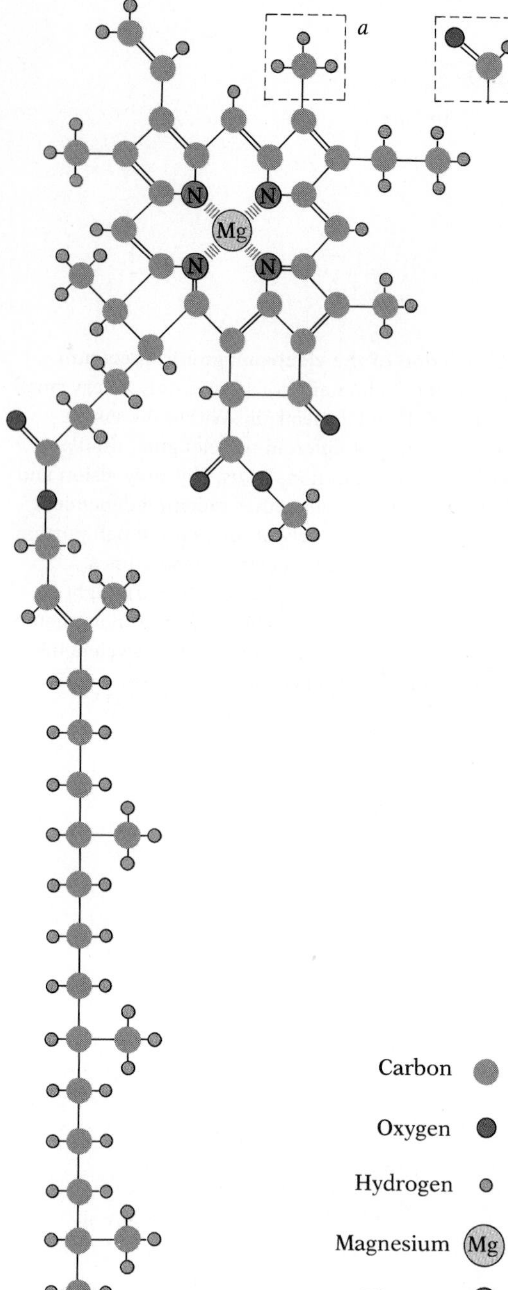

Is all light, regardless of wavelength, equally effective for photosynthesis? To answer this question, we must turn to the all-important green pigment chlorophyll, which, we have said, traps light energy and helps convert it into chemical energy. Of the several slightly different kinds of chlorophyll, the most widespread is chlorophyll *a*, and our discussion will primarily refer to this compound (Fig. 7.4).

Light falling on an object may pass through the object (be transmitted), be absorbed by it, or be reflected from it (Fig. 7.5). We can see transmitted or reflected light, but not absorbed light. Now, if chlorophyll is the material that traps the incident light, and if it appears green to our eyes, several facts suggest themselves. First, chlorophyll cannot be absorbing much radiation of the wavelengths that produce in us the sensation of green, or we wouldn't see green color. Second, chlorophyll must be absorbing radiation of some wavelengths within the visible part of the spectrum, or the light transmitted or reflected to us would appear white (when all visible wavelengths are combined, they produce the sensation of white). At this point we have already partly answered the question of whether all light is equally effective for photosynthesis: green light is not as effective as light of some other colors, since it is not absorbed as readily by chlorophyll.

More precise information can be obtained if chlorophyll is extracted from the leaf and exposed to light of varying wavelengths to determine the amount of absorption at each wavelength. The absorption spectrum of chlorophyll *a* thus obtained (Fig. 7.6A) shows that primarily light in the violet, blue-violet, and red regions is absorbed, while green, yellow, and orange light is absorbed only very slightly. It must be noted that the action spectrum of photosynthesis—a measure of the effectiveness of light of various wavelengths in driving photosynthesis—is somewhat different from the absorption spectrum of chlorophyll *a*. That there is relatively high activity in parts of the spectrum where the chlorophyll absorbs very little light, as Figure 7.6B shows, is an indication that light of some wavelengths not readily absorbed by chlorophyll *a* is still effective in driving photosynthesis. Apparently other pigments that are present in green plants, principally the yellow and orange carotenoids and other forms of chlorophyll, absorb light in these regions of the spectrum and then pass the energy to the chlorophyll *a*. ***Accessory pigments*** like the carotenoids thus enable the plants to use light

7.4 Molecular structure of chlorophyll
The structure of chlorophyll *a* is shown. The structure of the other major type of chlorophyll in green plants, chlorophyll *b*, differs only in the side group shown in the inset: a formyl group (—CHO) is substituted for the methyl group (—CH_3) of the *a* form. As you can probably guess, the long hydrophobic tail of the molecule serves to anchor the chlorophyll in the appropriate membrane. The electron that is excited by light energy is in the "head" region, near the magnesium atom. Note the similarity of chlorophyll's prosthetic group to the heme of hemoglobin.

of more different wavelengths than could be trapped by chlorophyll *a* alone.

What happens when light of a proper wavelength strikes a chlorophyll molecule? The exact answer is not known, but many aspects of the process have been elucidated.

In the chloroplasts of photosynthesizing cells in green plants, the chlorophyll and accessory pigments are organized into functional groups called ***photosynthetic units***. Each unit contains some 300 pigment molecules, including chlorophyll *a*, chlorophyll *b*, and carotenoids. One group of four pigment molecules in each unit is distinct from all the rest; it is part of a complex that acts as a ***reaction center***. The other pigment molecules function something like antennas responsive to light energy.

As we have seen, light energy comes in discrete units called photons. When a photon strikes a chlorophyll (or a carotenoid) molecule and is absorbed, its energy is transferred to an electron of the pigment molecule; the

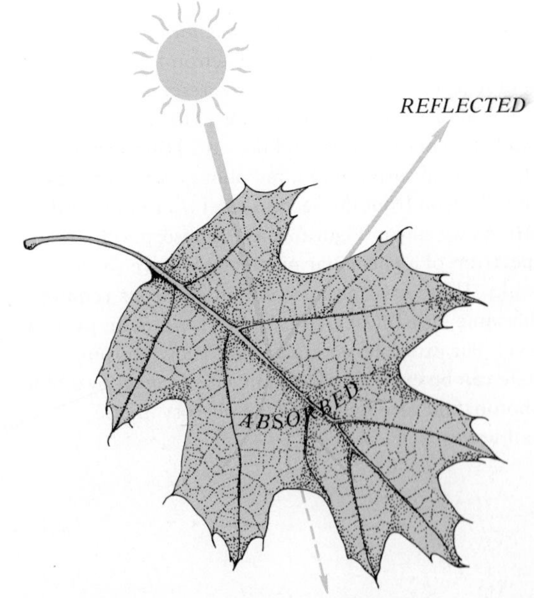

7.5 Light striking a leaf
Light striking an object, such as a leaf, may be reflected, absorbed, or transmitted.

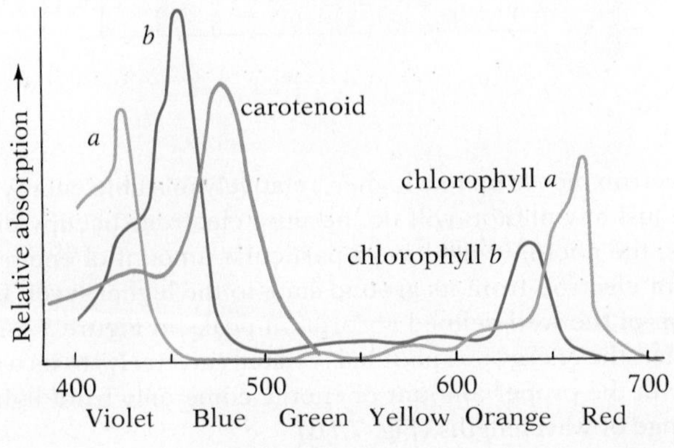

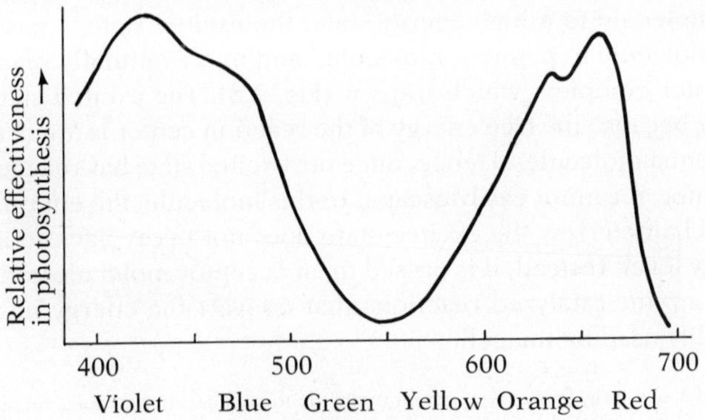

7.6 Absorption and action spectra of photosynthetic pigments
(A) Absorption spectra of chlorophyll *a*, chlorophyll *b*, and a carotenoid. Taken together, the absorption spectra of the two chlorophylls and the carotenoid cover more of the range of wavelengths available to the plant than does the spectrum of chlorophyll *a* alone. (B) Action spectrum of photosynthesis, indicating the relative effectiveness of different colors of light in driving photosynthesis. Light of intermediate wavelengths is more effective in driving photosynthesis than would be predicted on the basis of the absorption spectrum of chlorophyll *a* alone. Other pigments, such as carotenoids, absorb light of these intermediate wavelengths and pass the energy to the photosynthetic pathway.

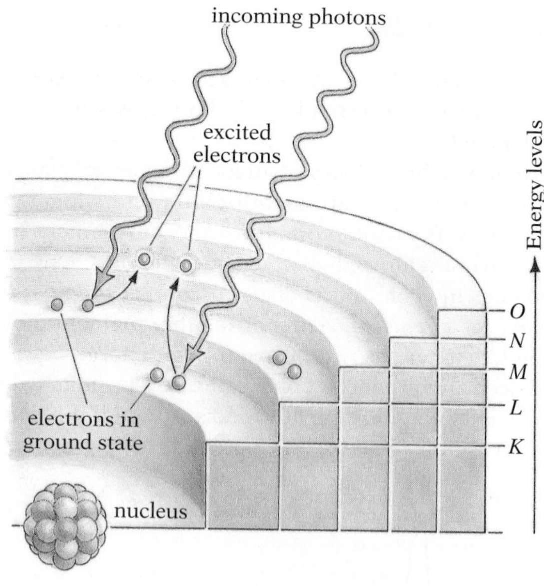

7.7 Effect of light on chlorophyll
When a photon is absorbed by a chlorophyll molecule, the photon's energy raises an electron to a higher energy level. In this simplified representation we see a typical distribution of electrons at their lowest available energy levels, and the distribution as it is altered by absorption of a red photon, which raises one electron from the second level (*L*) to the third (*M*). As we saw in Figure 7.6A, the absorption spectrum of a particular pigment can have several peaks. These exist because not all electrons require the same amount of energy to be raised to an excited level. For example, an electron at a lower ground state can be excited to the same energy level by a blue photon, and can then enter the photosynthetic pathway.

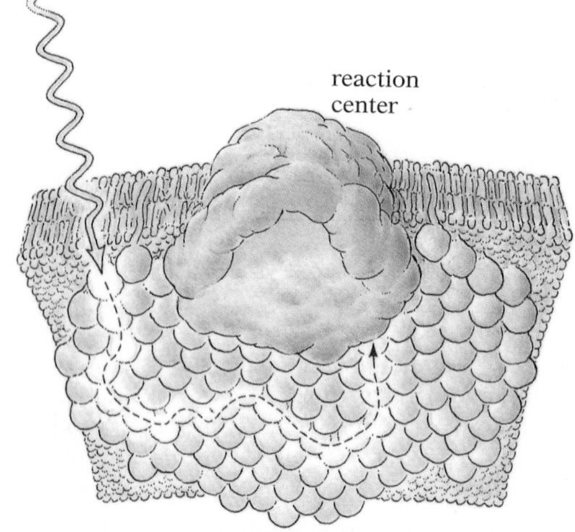

7.8 Flow of excited state within a photosynthetic unit
A photon strikes one of the antenna pigments (green circles) and raises an electron in the pigment to a higher energy level. This excited state is then passed from pigment molecule to pigment molecule in a random sequence (dotted pathway) until it eventually reaches the reaction center (large green unit), where it is trapped.

excited electron moves up to a higher, relatively unstable energy level (Fig. 7.7A). Not just any photon will do: because electrons occupy discrete energy levels, the photon must have a particular amount of energy to excite the pigment electron from its ground state to the higher level. Here is the explanation of the well-defined absorption peaks in Figure 7.6A: since, as noted earlier, the energy of a photon is related (inversely) to its wavelength, photons with the proper amount of energy come only from light within a certain range of wavelengths (Fig. 7.7B).

An excited and therefore unstable electron will return spontaneously to its inactive state almost immediately, giving up its absorbed energy. Isolated chlorophyll in a test tube, for instance, promptly loses the energy it captures by reemitting it as visible light, in a process known as fluorescence: the chlorophyll pigment molecules alone, separated from their photosynthetic units, are incapable of converting light energy into chemical energy. But in the functioning chloroplast, once light energy has raised an electron in an antenna molecule to a high-energy state, the excited state is passed from pigment molecule to pigment molecule,[1] and may eventually reach the reaction-center complex, which traps it (Fig. 7.8). The excited state is captured here because the free energy of the reaction center is *lower* than that of the antenna molecules. Hence, once the excited state has reached the reaction center, it cannot easily escape. In this molecule, the energized electron that characterizes the excited state does not decay back to a normal low-energy level. Instead, it is passed to an acceptor molecule and enters a series of enzyme-catalyzed reactions that convert the energy into a form more readily used by the cell.

[1] The transfer of energy from one pigment molecule to an adjacent one does not appear to involve the physical transfer of an excited electron. Instead, when the excited electron in one molecule falls back to a lower energy level, an electron in an adjacent pigment molecule is boosted to a higher level, thus taking on the excited state. Some researchers refer to this process as a transfer of excitation energy, or inductive resonance.

CYCLIC PHOTOPHOSPHORYLATION

There are two general pathways by which the energy from excited electrons is harvested—the cyclic pathway and the noncyclic pathway. The cyclic pathway involves only one of the two types of photosynthetic units found in most plants; the noncyclic pathway involves both. We will first examine the cyclic pathway, which is the simpler and more ancient of the two.

Electron transport The specialized chlorophyll complex that serves as the reaction center of a photosynthetic unit is capable of passing an energized electron to an acceptor molecule. Once cleared of the electron it received from an antenna molecule, the reaction-center complex is ready to be activated again, so photosynthesis can continue. In cyclic photophosphorylation the reaction center is designated P700 (because it cannot absorb significant quantities of light of wavelengths above 700 nm). The acceptor molecule to which it transfers the energized electron is an enzyme (FeS) containing an iron and sulfur prosthetic group. The acceptor is reduced and P700 is oxidized by this electron transfer. The electron is then passed along by a series of membrane-mounted enzymes (Fig. 7.9) very like the electron-transport chain of the mitochondrion. Eventually, the electron reaches a molecule of the enzyme plastocyanin (PC); there it waits to fill an opening in the P700 complex. When another electron is energized and transferred to FeS, the slot opens and our first electron fills it, thereby completing the cycle.

When a light-energized electron leaves a chlorophyll molecule, it is energy-rich; when it finally returns, it is energy-poor. The transition is gradual. As the electron is passed from transport molecule to transport molecule in the chain, it releases some of its extra energy with each transfer; in other words, it is eased down the energy gradient step by step from the excited state to the normal state. Hence when the electron finally falls back into the chlorophyll, it has discharged all its extra energy, but it has not discharged it all at once, as would happen with isolated chlorophyll in a test tube. Instead, the energy has been released in a series of small portions of manageable size. The whole process is called ***cyclic photophosphorylation*** because the electron is returned to the chlorophyll and some of its energy is used (indirectly, as we will see) to phosphorylate ADP into ATP.

Cyclic photophosphorylation was probably the first form of photosynthesis to evolve, and it is still the only form available to most photosynthetic bacteria. As Figure 7.9 indicates, however, the cyclic system is not very efficient: of the 25 kcal/mole of energy gained as a result of the excitation of P700, only the energy liberated in the passage from PQ to cytochrome *f*—3.4 kcal/mole—is actually made available to the cell. The energy released in the other steps is not utilized; it goes to waste. Though even 3.4 kcal/mole is far better than nothing (particularly since photons do not cost the cell anything), most photosynthesis now follows a highly modified, noncyclic pathway, which is more efficient under many conditions. Indeed, just as aerobic respiration represents an advance over the anaerobic processes of the glycolytic pathway to fermentation, so noncyclic photophosphorylation is a significant advance over the more primitive cyclic form. As we will see, in most organisms today cyclic photophosphorylation serves only as a supplement to the noncyclic pathway.

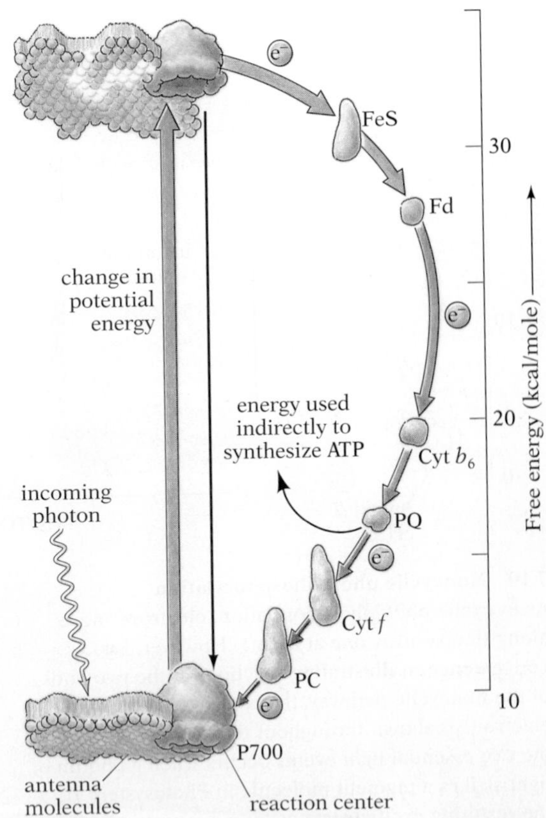

7.9 Cyclic photophosphorylation
A photon of light strikes a pigment molecule in the P700 antenna system. The excited state eventually reaches the P700 complex, which is thereby energized. An energized electron from P700 then begins passage from one acceptor enzyme to the next, each more electronegative than the previous one. Free energy is released at each step, and the electron ultimately returns to the ground state at which it began in P700. Only the energy released in the passage from plastoquinone (PQ) to cytochrome *f* is used by the cell. The other electron acceptors are FeS, an enzyme containing iron and sulfur; ferredoxin (Fd); cytochrome b_6; and plastocyanin (PC).

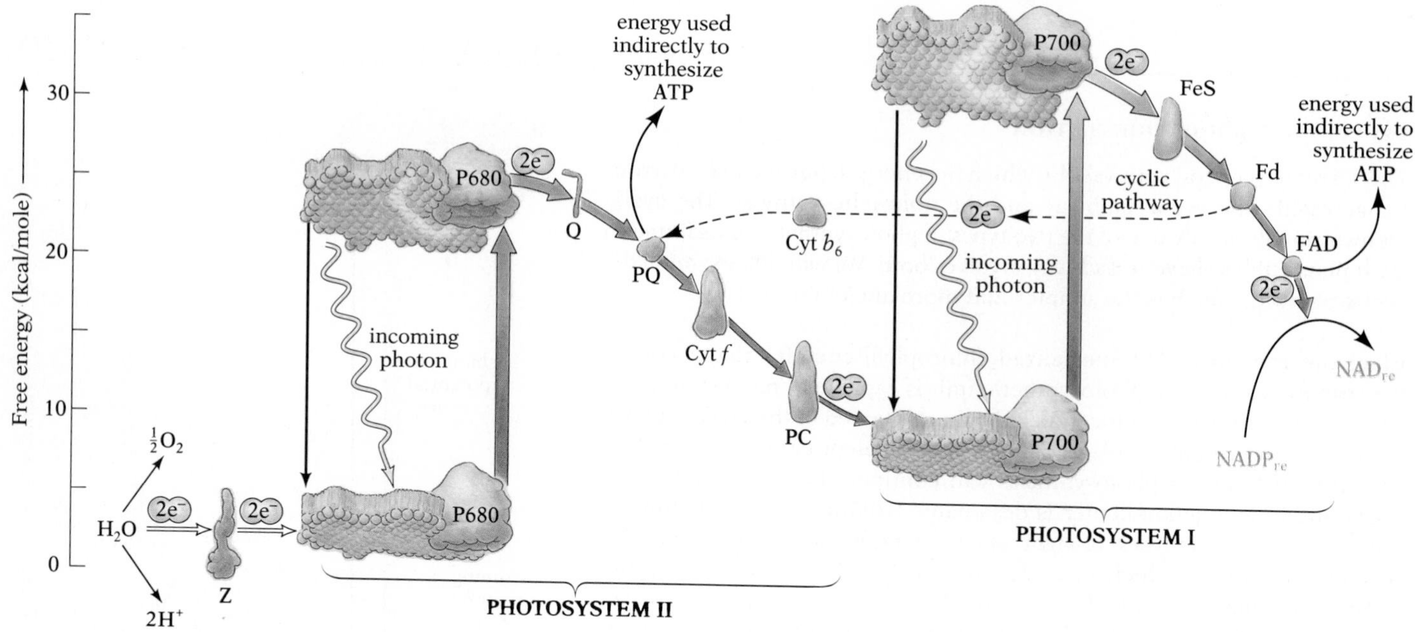

7.10 Noncyclic photophosphorylation
As in cyclic photophosphorylation, electrons move along the pathway one at a time. However, for convenience in illustrating reactions at the two ends of the noncyclic pathway, the passage of a pair of electrons is shown throughout the diagram. One of the two essential light events occurs when a photon of light strikes a pigment molecule in Photosystem II; the resulting excited state eventually reaches P680, which donates an energized electron to substance Q. The electron moves down the Photosystem II electron-transport chain from Q to PQ to cytochrome *f* to PC, and as in cyclic photophosphorylation, the energy released in the passage from PQ to cytochrome *f* is used by the cell. At PC the electron waits to fill an opening in the P700 complex, which occurs when a photon strikes Photosystem I and P700 transfers an energized electron to the Photosystem I electron-transport chain. This electron passes from FeS to Fd to FAD, and finally to NADP. Energy released in the step down from FAD is used by the cell, and for each two electrons transported, one NADP$_{re}$ is generated to supply energy for carbon fixation. The vacancy in P680 that was created by the passage of an electron to Q is filled by an electron released in the splitting of water (lower left), with one molecule of water yielding two electrons, along with two H^+ ions and one atom of oxygen. As indicated by the changes in relative free energy, the electron that has moved through Photosystem II still has much of its original energy when it enters Photosystem I. The primitive cyclic pathway (which connects Fd to PQ via cytochrome b_6) is also shown.

NONCYCLIC PHOTOPHOSPHORYLATION

Like the cyclic process already described, noncyclic photophosphorylation (Fig. 7.10) begins when a photon of light strikes an antenna molecule of chlorophyll and raises an electron to an excited state, which may be trapped by the reaction center P700. As before, an energized electron is then led away from the P700 complex by electron acceptors, the first two being FeS and ferredoxin (Fd). But here the similarity to cyclic photophosphorylation ends. Instead of continuing down the cyclic transport chain, the electron is passed from Fd to a different acceptor molecule: the flavoprotein FAD. FAD passes the electron to an extremely important substance called nicotinamide adenine dinucleotide phosphate, or ***NADP***, which is closely related to the NAD in mitochondria. The antenna molecules and the P700 reaction center, together with the electron-transport chain from FeS to Fd to FAD, constitute ***Photosystem I***.

Unlike the reduced electron-acceptor molecules in cyclic photophosphorylation, NADP does not promptly pass along the electrons it receives from the electron-transport chain to another acceptor molecule. Instead, it retains a pair of energized electrons and their associated protons. Eventually NADP$_{re}$ acts as an electron donor in the reduction of CO_2 to carbohydrate such as glucose, a process known as ***carbon fixation***. Thus electrons move from the chlorophyll through an electron-transport chain to NADP to carbohydrate (via other intermediate compounds). In addition, some of the energy liberated in the passage from FAD to NADP is used, indirectly, in the synthesis of ATP.

Now, if the energized electrons from P700 are retained by NADP$_{re}$ and eventually incorporated into carbohydrate, it follows that Photosystem I is left short of electrons: it is left with "electron holes." These electron holes are filled, indirectly, by electrons derived from water through a process we will now examine.

At one time, it was thought that the electrons from the water passed, via a few transport molecules, directly to Photosystem I, and hence that there was only one light-driven event in photosynthesis—the one that initially excited the chlorophyll electrons of Photosystem I. Later, however, it became apparent that there are two light events, the one we have already discussed and a second one more intimately related to the splitting of water. This second light event involves a different type of photosynthetic unit, which contains about 200 molecules of chlorophyll *a*, about 200 molecules of chlorophyll *b*, *c*, or *d*, depending on the species of plant,[2] and a reaction-center complex called P680. (The designation P680 indicates that this system cannot absorb any significant amount of light of wavelengths above 680 nm.) The antenna molecules and the P680 reaction center, plus its special set of electron-transport molecules, constitute ***Photosystem II***.

When light of the proper wavelength strikes a pigment molecule of Photosystem II, the energy is passed around within the photosynthetic unit until it finally reaches the P680 complex. This unit, in turn, donates a high-energy electron to an acceptor, designated Q (Fig. 7.10). Substance Q then passes the electron to a chain of acceptor molecules, which transports the electron, by some of the same steps we saw in cyclic photophosphorylation, down an energy gradient to the electron hole in the P700 molecule of Photosystem I. As the electron moves down the transport chain, some of the energy released along the way is used by the cell indirectly to synthesize ATP.

Thus the electron holes created in Photosystem I by the first light event are refilled by electrons moved from Photosystem II by the second light event. But this process alone would leave electron holes in Photosystem II; the electron deficit would simply have been shifted from Photosystem I to Photosystem II. It is at this point that the electrons from water, mentioned earlier, play their role. As Figure 7.10 indicates, it is thought that P680 (with the aid of an enzyme complex referred to as Z) pulls replacement electrons away from water, leaving behind free protons and molecular oxygen:

$$2H_2O \longrightarrow \underset{\substack{\downarrow \\ \text{to P680}}}{4e^-} + 4H^+ + O_2$$

The oxygen from H_2O is released as a gaseous by-product (note that two molecules of H_2O must be split to yield one molecule of O_2). The protons, as we will see, also play an important role.

The electrons involved in the second light event move from water to the P680 complex of Photosystem II to Q to the transport chain of Photosystem II and to Photosystem I (Fig. 7.10). If we combine these steps with the electron movement associated with the first light event, as traced above, we obtain the following abbreviated sequence showing the overall electron movement:

$$H_2O \longrightarrow \text{P680} \longrightarrow \text{Photosystem II transport chain} \longrightarrow \text{P700} \longrightarrow$$
$$\text{Photosystem I transport chain} \longrightarrow \text{NADP}_{re} \longrightarrow \text{carbohydrate}$$

This sequence shows that the electrons necessary to reduce carbon dioxide

[2] Chlorophyll *b* is found in green algae, bryophytes, and vascular plants. Chlorophyll *c* occurs in brown algae, and chlorophyll *d* in red algae.

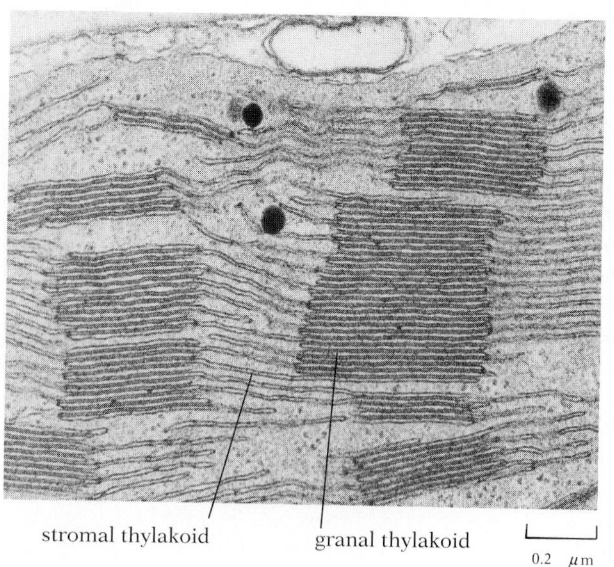

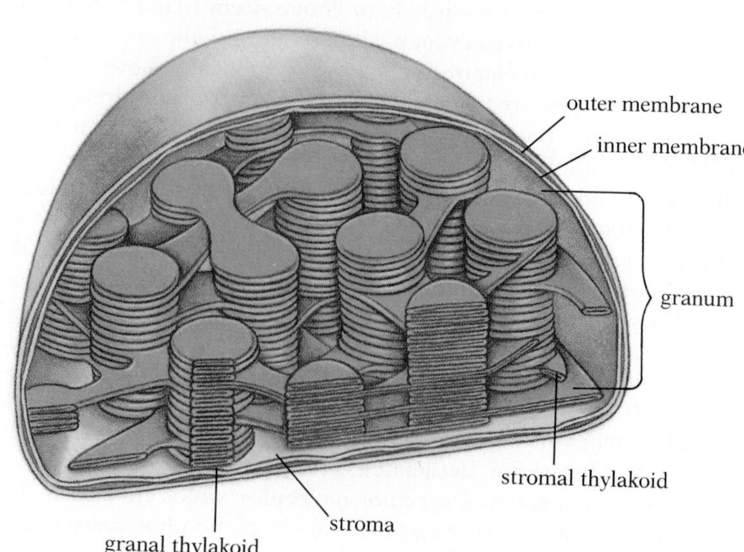

7.11 Structure of a chloroplast
Left: Electron micrograph of a section of a chloroplast of timothy grass showing several grana; note the continuity between granal and stromal thylakoids. Right: This cutaway view of a typical chloroplast shows the inner and outer membranes lying close together, enclosing the large compartment known as the stroma. Inside the stroma can be seen a third distinct membrane, which forms the interconnected compartments called thylakoids. The stacks of flat, disclike thylakoids are grana. Chlorophyll molecules and most of the electron-transport-chain molecules are located in the thylakoid membrane.

to carbohydrate come from water, and that the movement of electrons from water to carbohydrate is an indirect and complex process.

Since electrons are not passed in a circular chain in this process, some leaving the system via $NADP_{re}$ and others entering the system from water as replacements, this series of reactions is termed ***noncyclic photophosphorylation***. The whole process results in the formation of both ATP and $NADP_{re}$ and in the release of molecular oxygen. The reactions of photophosphorylation have traditionally been known as the "light" reactions of photosynthesis, but, as we have seen, only two steps are directly light-dependent.

We should point out that the details given above apply only to photosynthesis by plants and a few bacteria. Some other bacteria possess a form of chlorophyll and can use light energy in synthesis of ATP and $NADP_{re}$, but do not use water as the source of electrons. As we have noted, some of these use hydrogen sulfide (H_2S) and give off sulfur instead of oxygen. Others use nitrogen-based compounds instead. Plants themselves can be experimentally induced to use a source of electrons other than water. If oxidation of water is blocked by chemical inhibitors, and a strong electron donor is provided as a substitute, noncyclic photophosphorylation can continue without production of oxygen; in other words, plant photosynthesis has been experimentally converted into an essentially bacterial type of photosynthesis. Water is therefore only one of many possible electron sources for photosynthesis. It is not surprising, however, that the most successful photosynthetic organisms, the plants, evolved utilizing photosynthesis based on water, which is ubiquitous in the chemistry of life.

THE ANATOMY OF PHOTOPHOSPHORYLATION

By now you have probably noticed many similarities between the electron-transport strategies of photosynthesis and those of respiration. We saw in the last chapter that the functioning of the respiratory electron-transport chain depends on ordered arrays of enzyme molecules embedded in the highly folded inner membrane of the mitochondrion. And we saw too that the energy made available by this stage is used in large part to transport H^+ out of the inner compartment of the mitochondrion, with the resulting electrochemical gradient across the inner membrane acting like a battery to supply energy for the synthesis of ATP. The chloroplast is organized in a similar way, and also establishes and maintains a chemiosmotic gradient.

Like the mitochondrion, the chloroplast is a membrane-bounded organelle with its own genes, which probably derives from an ancient endosymbiotic bacterium. However, unlike the mitochondrion, which has a heavily folded inner membrane containing electron-transport chains, the chloroplast has an inner membrane that is relatively smooth and flat, and follows the contours of the outer membrane (Fig. 7.11); it has many selective gates and pumps, but no electron-transport complexes. It encloses the main volume, or stroma, of the chloroplast. The antenna pigments, reaction centers, and electron-transport-chain molecules of the chloroplast are in fact embedded in a third membrane inside the stroma, which forms a series of flattened, interconnected compartments known as ***thylakoids***. Thylakoids occur in two distinct but interconnected arrangements, each with its own functional specializations (Fig. 7.11). When loose in the stroma, the proportion of the surface area of the ribbonlike ***stromal thylakoids*** that is in contact with stromal fluids is essentially 100 percent, thus maximizing their ability to take up CO_2. When stacked neatly in piles called ***grana***, the densely packed, chlorophyll-laden membranes of granal thylakoids are highly efficient at capturing photons, but only a small proportion of the granal membrane is exposed to stromal fluids; as a result, the ability to take up CO_2 is reduced. Stromal thylakoids, then, are ideal for conditions of bright light but limited CO_2; granal thylakoids are well adapted to take advantage of moderate light and abundant CO_2. Like the inner mitochondrial membrane, the thylakoid membrane makes possible an electrochemical gradient, which, functioning like a battery, supplies energy for the synthesis of ATP. However, in the chloroplast the H^+ ions accumulate in the innermost compartment of the organelle—that is, the interior of the thylakoids—while the outer compartment, the stroma, becomes negatively charged.

If we look closely at the anatomy of the chloroplast, we can see how photosynthesis works to provide the growing plant with the carbohydrates it needs. A photon is absorbed in the antenna complex of Photosystem II, exciting an electron to a higher energy level. This energy is transferred from one antenna chlorophyll to another in the membrane until it reaches the reaction-center complex. There an energized electron in a normal chlorophyll is moved through a short series of slightly more electronegative mole-

Exploring Further

THE P680 REACTION CENTER

Painstaking research has revealed the structure of a photosynthetic reaction center, and in the details lies the explanation of how the center complex manages to capture and keep electron energy that is transferred freely between the conventional chlorophyll molecules in the antenna. The reconstruction of a P680 chloroplast reaction center shown here is based on the structure of the evolutionary precursor in photosynthetic bacteria; the two are thought to be essentially identical, but may differ in minor details. A photon is shown activating an antenna chlorophyll in the chloroplast membrane immediately adjacent to the reaction center (1); this molecule transfers its energy to one of the two central reaction-center chlorophylls (2). The energized electron is moved to a peripheral reaction-center chlorophyll (3), and then to a pheophytin molecule (4); it moves immediately to a bound quinone (5), which is closely related to enzyme Q of the electron-transport chain. The high-energy electron then moves more slowly to another quinone (6).

The entire process is then repeated (not shown here) so that there are two electrons waiting to leave, and both central chlorophylls have electron vacancies. Note that each of these molecules has a prosthetic group (which holds a metal atom and so creates the precise electronegativity necessary for smooth, efficient, unidirectional transfer) and a hydrophobic tail for anchoring.

The electrons leave the reaction-center complex by way of the membrane quinone, Q (8). The electrons missing from the central chlorophylls are replaced (10) by enzyme Z when it splits a water molecule (9). Because the movement of only one electron is pictured here, the water-splitting reaction is shown as involving only "half" a water molecule. The electron-capture ring of enzyme Z is represented in this illustration as a cytochrome with a manganese-based prosthetic group; its exact identity is not yet certain. The eight molecules of the reaction-center complex are bound into a unit by a large protein, shown here in outline.

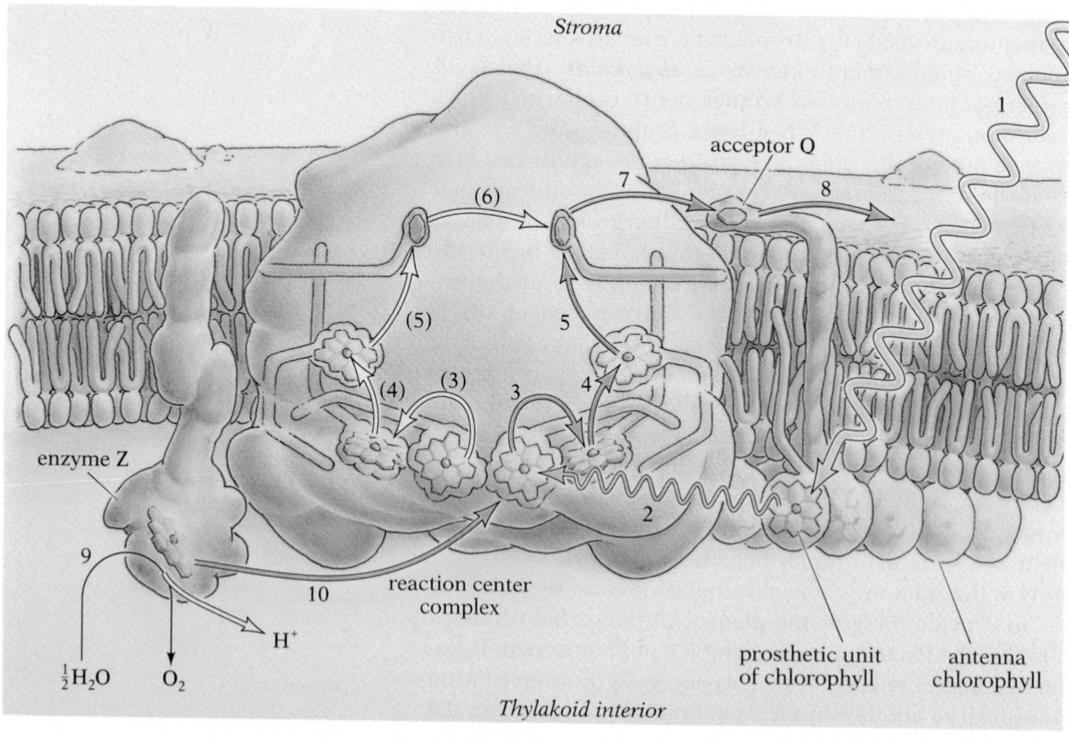

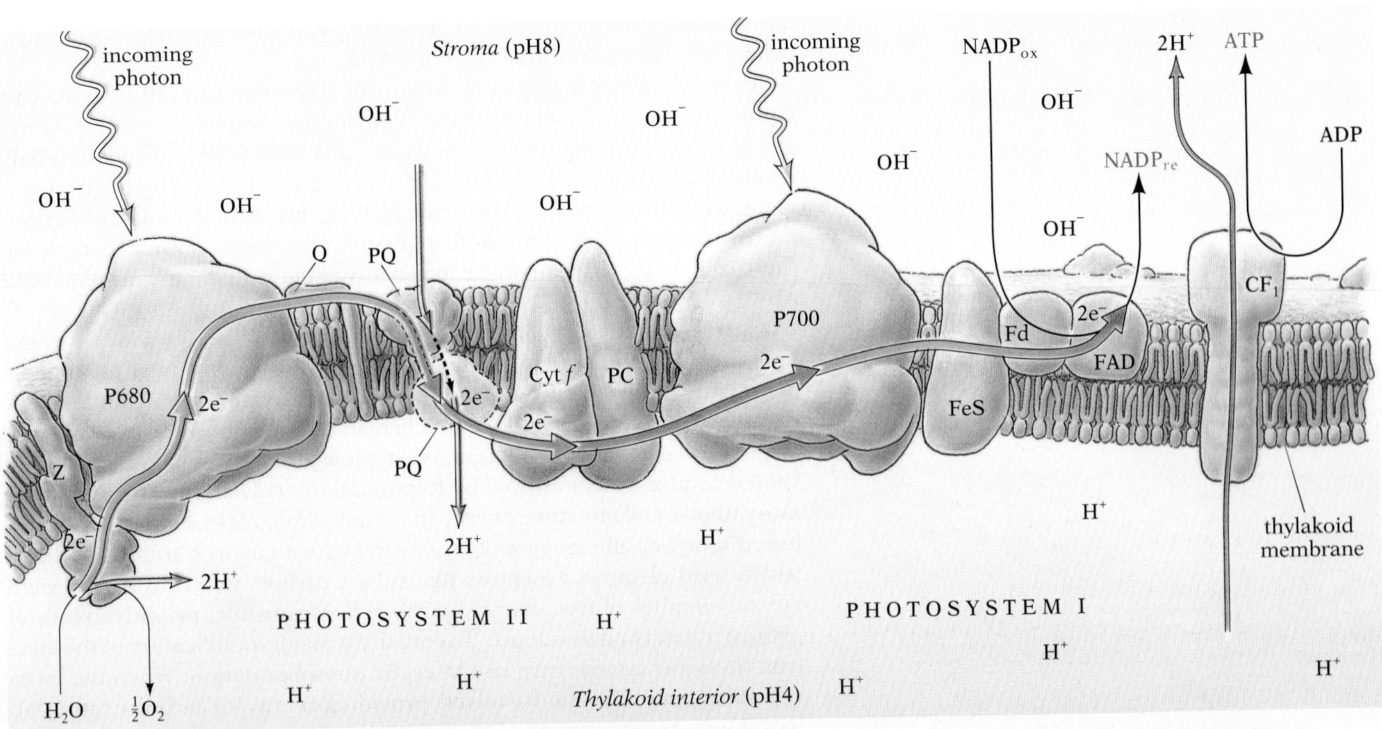

7.12 Anatomy of photophosphorylation
The enzyme molecules of photophosphorylation are shown here for simplicity as a linear chain; in the membrane the chain is thought to be folded back on itself, so that cytochrome b_6 (not shown) connects Fd and PQ, thereby making possible cyclic photophosphorylation. The antenna molecules of the two photosystems are omitted. This representation makes it clear how electrons from water are moved along the two electron-transport chains, and how the thylakoid battery is charged. Since the pH scale is logarithmic (see Fig. 2.15, p. 35), the values shown indicate that the concentration of H^+ ions is 10,000 times greater within the thylakoid than on the other side of the membrane. Note that PQ migrates from one side of the membrane to the other to transport H^+ ions. The CF_1 complex, which uses the energy of the electrochemical gradient to make ATP, is also shown.

cules. After another electron follows the same path, they are fed to an acceptor in the membrane (Fig. 7.12). The resulting electron vacancies in the normal chlorophylls in the reaction center are filled by enzyme Z, which splits water and deposits two H^+ ions in the thylakoid interior for each pair of electrons energized in P680. The entire process takes about a thousandth of a microsecond.

The two energized electrons that left the reaction center via the acceptor (Q) are delivered, one at a time (and we will follow only one from now on), to PQ, a molecule that can shuttle between the stromal side of the membrane and the thylakoid side. Some of the energy of each electron is used here to move an H^+ ion from the stroma to the thylakoid interior. The electron now moves to cytochrome f and then to PC, where it waits for a vacancy in P700.

When a photon is absorbed by an antenna molecule in Photosystem I and excites an electron, the excitation energy is passed to the P700 complex. The energized electron of P700 is immediately passed to an acceptor (FeS). The resulting vacancy in P700 can now be filled by the electron from Photosystem II waiting in PC.

The energized electron of Photosystem I, meanwhile, moves on to another acceptor (Fd), from which it can follow one of two pathways. Under normal conditions, the most likely fate for this electron is to reduce NADP$_{ox}$. This process consumes an H^+ ion from the stroma, so that in all, two H^+ ions are now gone from the stroma while two H^+ ions have been added to the thylakoid. The alternative (not shown in Fig. 7.12) is the cyclic pathway (dashed line in Fig. 7.10). The cyclic alternative represents a less efficient use of energy, but is favored when $NADP_{ox}$ is in short supply, or when the cell is more in need of ATP than of $NADP_{re}$.

The result of this highly ordered flow of electrons is twofold: on the one hand, it generates a supply of the high-energy carrier molecule $NADP_{re}$, whose role in carbohydrate synthesis we will examine shortly; on the other, it helps create, through the flow of H^+ ions, a powerful electrochemical gradient, which is then used to generate ATP. For just as in the mitochondrion, the membrane in the thylakoid contains numerous enzyme complexes (here called CF_1) that can utilize the energy of the gradient to phosphorylate ADP.

A brief comparison of Figure 7.12 and Figure 6.11 (p. 180) shows that the organization and anatomy of the electron-transport chains of photophosphorylation and of respiration are very similar. In fact, there is little doubt that the anabolic and catabolic mechanisms have a common evolutionary origin. Current evidence indicates that a low-efficiency anaerobic pathway similar to glycolysis followed by fermentation evolved first, enabling chemosynthetic cells to store energy in carbohydrates for later use. Then followed a higher-efficiency membrane-embedded electron-transport chain, which used electron acceptors like sulfur, carbon, and nitrogen to yield greater supplies of free energy for the cell. Meanwhile, primitive kinds of photosynthesis had developed. Eventually, a major modification of the electron-transport system gave rise to cyclic phosphorylation. When the more versatile noncyclic system evolved, providing energy for carbon fixation, atmospheric oxygen was produced. For all its potential advantages, oxygen was toxic to much of the anaerobically evolved chemistry of the cell; the evolution of oxygen-tolerant enzymes must have been the next step. With this, the evolution and rapid spread of the respiratory pathway using oxygen as the electron acceptor became possible. The plausibility of this scenario is bolstered by the recent discovery of procaryotes that may represent the "missing links." The gradual emergence of varied but related systems for obtaining the energy essential to life illustrates how evolution builds on modifications of what is already at hand, and how one change, such as the evolution of noncyclic photophosphorylation in the forerunners of chloroplasts, can suddenly make possible (or necessary) other rapid advances. Once the oxygen produced by the new photosynthetic strategy began to accumulate in the atmosphere, and the problems of oxygen toxicity had been overcome, aerobic respiration allowed the evolution of vast numbers of highly efficient nonphotosynthetic organisms.

THE DARK REACTIONS: CARBON FIXATION

So far, we have seen how the energy of photons is captured and used to make ATP and $NADP_{re}$, but not how this energy is used in turn to transform the low-energy substance CO_2 into high-energy compounds like glucose. The entire process of photosynthesis, as we have noted, is frequently divided into the light reactions of photophosphorylation and the dark reactions of carbon fixation. This dichotomy is not quite accurate, since the energy stored by the thylakoid battery can be used to make ATP regardless of illumination, but it is a useful way of distinguishing between the energy-accumulating reactions of the thylakoids, which culminate in the production of ATP and $NADP_{re}$, and the energy-consuming carbon fixation that takes place in the stroma.

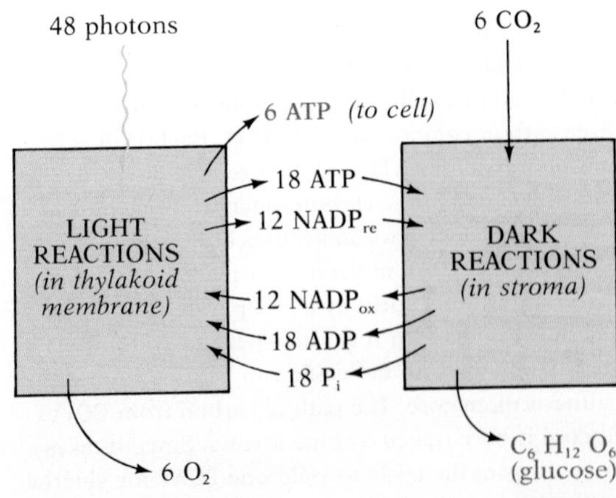

7.13 The light and dark reactions
Photosynthesis consists of two physically separate but interlocking sets of reactions. The light reactions—photophosphorylation—use light to generate the energy intermediates ATP and $NADP_{re}$. These reactions, as we have seen, take place in the thylakoid membrane. The dark reactions—carbon fixation—use these energy intermediates to turn carbon dioxide into carbohydrates like glucose. Noncyclic photophosphorylation, summarized in the left half of this figure, produces ATPs for the cell above and beyond those utilized in the synthesis of glucose. Cyclic photophosphorylation goes on simultaneously to provide additional ATP for the cell, but since it does not produce $NADP_{re}$ for carbon fixation, we have ignored it in this summary.

CARBOHYDRATE SYNTHESIS BY THE CALVIN CYCLE

Carbohydrates contain much chemical energy, while CO_2 contains very little. The reduction of CO_2 to form glucose proceeds by many steps, each catalyzed by a specific enzyme. Like a person pushing a heavy chest up a flight of stairs, CO_2 is pushed up an energy gradient through a series of intermediate compounds, some of them unstable, until the carbohydrate end product is formed. Just as a person might be able to lift the chest just high enough to get it up one step at a time, balancing it on each step long enough to marshal strength before the next heave, so do the enzyme-mediated biochemical steps make it possible to move CO_2 up an energy gradient. The energy for the stepwise synthesis of carbohydrates from CO_2 comes from light via ATP and $NADP_{re}$ (Fig. 7.13).

Since there are many sequential steps in the reduction of CO_2 to carbohydrates, and since many of the intermediate compounds occur also in other processes, leading to different end products, you may well wonder how the exact sequence of steps was ever discovered. The tool that made such discoveries possible was a radioactive isotope of carbon, designated ^{14}C. Samuel Ruben and Martin D. Kamen at the University of California, Berkeley, who discovered this isotope about 1940, immediately recognized its potential as a research tool in photosynthesis. They showed that plants exposed to carbon dioxide containing the radioactive isotope ($^{14}CO_2$ instead of the normal $^{12}CO_2$) incorporated the isotope into a variety of compounds. Later, in 1946, Melvin Calvin and his associates, also at Berkeley, began an intensive long-term investigation of carbon dioxide fixation in photosynthesis, using ^{14}C as their principal tool. They exposed algal cells to light in an atmosphere of $^{14}CO_2$ for a few seconds and then killed the cells by immersing them in alcohol. The alcohol not only killed the cells but also inactivated the enzymes that catalyze the reactions of photosynthesis. With the enzymes inactivated, whatever amount of each intermediate compound existed in the cell at the moment of inactivation was, in effect, locked in. Calvin and his co-workers could then determine which of these locked-in intermediate compounds contained ^{14}C. How long the algal cells were exposed to the $^{14}CO_2$ before being killed determined the number of compounds in which ^{14}C was detected: when the time was very short, the ^{14}C reached only the first few compounds in the synthetic sequence; when the time was longer, the isotope moved through more steps in the sequence and appeared in a great variety of compounds. After years of painstaking research, Calvin, who in 1961 was awarded the Nobel Prize for his critically important investigations, worked out the sequence of reactions now called the ***Calvin cycle***.

According to Calvin, the CO_2 first combines with a five-carbon sugar called ribulose bisphosphate, or ***RuBP***, to form a highly unstable six-carbon compound, which is promptly broken into two three-carbon molecules called phosphoglyceric acid, or PGA. Each molecule of PGA is then phosphorylated by ATP and reduced by hydrogen from $NADP_{re}$. The resulting energy-rich three-carbon compound is phosphoglyceraldehyde, or ***PGAL***.

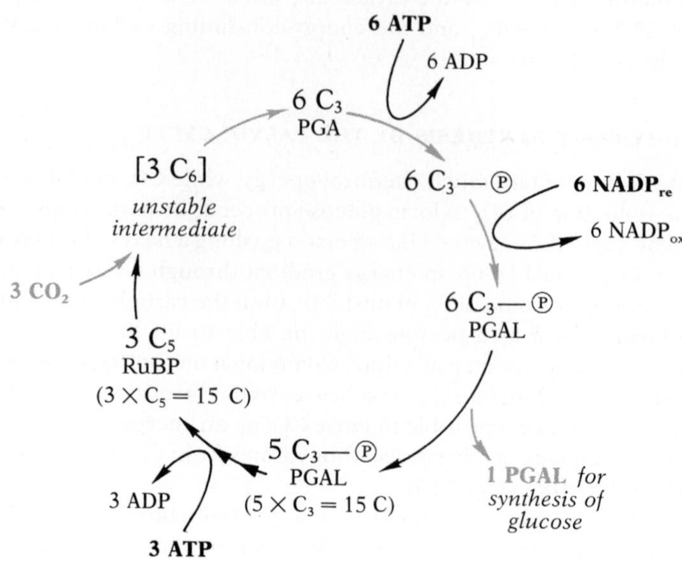

7.14 Synthesis of carbohydrate by the Calvin cycle
Each CO_2 molecule combines with a molecule of ribulose bisphosphate (RuBP), a five-carbon sugar, to form a highly unstable six-carbon intermediate, which promptly splits into two molecules of a three-carbon compound called PGA. Each PGA is phosphorylated by ATP and then reduced by $NADP_{re}$ to form PGAL, a three-carbon sugar. Thus each turn of the cycle produces two molecules of PGAL. Five of every six new PGAL molecules formed are used in synthesis of more RuBP by a complicated series of reactions (not shown separately here) driven by ATP. The sixth new PGAL molecule can be used in the synthesis of glucose. The path of carbon from CO_2 to glucose is here traced by blue arrows. Since it takes three turns of the cycle to yield one PGAL for glucose synthesis, the diagram begins with three molecules of CO_2; it would require a total of six turns to produce one molecule of glucose, a six-carbon sugar. Note that the cycle is driven by energy from ATP and $NADP_{re}$, formed by the light reactions of photophosphorylation.

This compound, which we have already encountered in our study of glycolysis, is a true sugar and, in a sense, is the stable end product of photosynthesis. Because PGAL is a three-carbon compound, as are the intermediate compounds leading to its formation, the Calvin cycle is often called ***C_3 photosynthesis***. We outline this sequence in very abbreviated form in Figure 7.14.

Five of every six molecules of PGAL are used in the formation of new RuBP (by a complicated series of reactions powered by ATP), with which more CO_2 can be processed. But one molecule out of six can be combined by a series of steps with another molecule of PGAL (produced in another turn of the cycle) to form the six-carbon sugar glucose. Thus, it takes six carbon dioxide molecules and six turns of the Calvin cycle to produce one molecule of glucose.

Though glucose has traditionally been considered the end product of photosynthesis, free glucose is not present in significant amounts in most higher plants. Some of the PGAL produced by the Calvin cycle is at once utilized in the formation of lipids, amino acids, and nucleotides. Even when glucose is synthesized, it is normally used almost immediately as a building-block unit for double sugars, starch, cellulose, or other polysaccharides. As we mentioned in Chapter 3, carbohydrates are generally stored in higher plants in the form of starch. One of the most important advantages of storing carbohydrates in this form is that starch, which is insoluble in water, has much less osmotic activity than sugar. An excessive accumulation of sugar, which is hydrophilic and so dissolves in the cellular cytoplasm, would raise the osmotic concentration of the cytoplasm relative to the environment and severely upset the osmotic balance between the cell and the surrounding fluid. The result would be the intake of too much water by the cell, leading to extreme swelling.

Photorespiration One property of the Calvin cycle is perplexing in that it has no obvious biological function: apparently RuBP carboxylase, the enzyme that catalyzes the carboxylation of ribulose bisphosphate (that is, the addition of CO_2 to RuBP) at the start of the Calvin cycle, can also catalyze the oxidation of RuBP by molecular oxygen (the addition of O_2 to RuBP). In other words, CO_2 and O_2 are alternative substrates that compete with each other for the same binding sites on this enzyme. When the concentration of CO_2 is high and that of O_2 is low, carboxylation is favored and carbohydrate synthesis by the Calvin cycle proceeds. But when the reverse conditions prevail—when the concentration of CO_2 is low and that of O_2 is high—oxidation is favored. Higher than normal temperatures also favor the alternative oxidation pathway.

Oxidation of RuBP results in formation of a two-carbon compound called phosphoglycolate, which can then be further broken down to CO_2. Under certain conditions, then, high-energy molecules such as RuBP, themselves produced by photosynthesis, are destroyed by a series of reactions initiated by the very same enzyme that under more favorable conditions would facilitate photosynthesis. This breakdown of photosynthetic intermediates to CO_2 is called ***photorespiration***. Since it does not result in synthesis of ATP, as other types of respiration do, it would appear to be a wasteful process, short-circuiting the Calvin cycle to no purpose.

Things could be worse, however. By means of a complex series of reactions involving chloroplasts, mitochondria, and peroxisomes, plant cells salvage much of the energy they stand to lose from the breakdown of phosphoglycolate. Only one of every three carbons entering photorespiration is actually lost as CO_2.

Because photorespiration predominates over photosynthesis at low concentrations of CO_2, plants that depend exclusively on the Calvin cycle for CO_2 fixation cannot synthesize carbohydrates unless the CO_2 concentration in the air is above a critical level (commonly about 50 parts per million); even at normal levels much of the production of photosynthesis is undercut by concurrent photorespiration. At atmospheric CO_2 concentrations, net photosynthesis by such plants could be increased by as much as 50 percent if oxygen inhibition of photosynthesis and the associated photorespiration could be stopped. We will return to this idea later in this chapter.

THE LEAF AS AN ORGAN OF PHOTOSYNTHESIS

As we have seen repeatedly, life depends both on the precisely catalyzed reactions of various biochemical pathways, with the accompanying interplay of their products, and on the particular anatomy of cells and their organelles. Nowhere is this intimate relationship between form and function more obvious than in the specialized tissues responsible for photosynthesis.

THE ANATOMY OF LEAVES

Photosynthesis can occur in all green parts of the plant, but in most plants the leaves expose the greatest area of green tissue to the light and are therefore the principal organs of photosynthesis.

7.15 Leaf types
A leaf usually consists of a blade and a petiole, which sometimes has stipules at its base. Veins run from the petiole into the blade. The main veins may branch in succession off the midvein (pinnate venation); or they may all branch from the base of the blade (palmate venation); or they may be parallel. The blade may be simple, or it may be compound—that is, divided into leaflets that may be pinnately or palmately arranged. Leaves with parallel venation (lower left) are characteristic of the monocot group, plants whose seedlings have only one "seed leaf" or cotyledon; they include grasses, grains, and spring bulbs, for example. The other leaves are from the dicot group.

VENATION

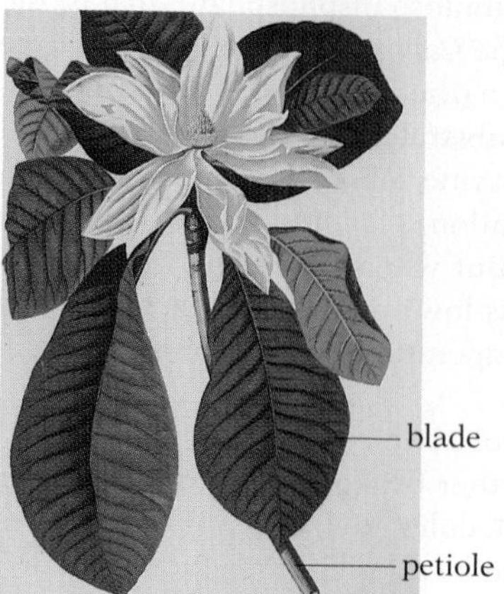

Pinnate venation

Palmate venation

Parallel venation

LEAF TYPE

Simple

Pinnately compound

Palmately compound

Figure 7.15 shows leaves of a variety of familiar land plants. Most dicot leaves consist of a stalk, or ***petiole***, and a flattened ***blade***. In addition, the petioles of some leaves bear small appendages, called stipules, at their bases. However, some leaves, particularly those of monocots (the grasslike plants), lack even petioles, the base of the blade being attached directly to the stem. The blades of most leaves are broad and flat and contain a complex system of veins. Because of the flatness of the blade, the leaf exposes to the light an area that is very large in relation to its volume.

When we examine a transverse section of a leaf under the microscope (see Fig. 7.16), it becomes clear that the outer surfaces are formed by layers of epidermis, usually only one but sometimes two or more cells thick. A waxy layer, the ***cuticle***, usually covers the outer surfaces of both the upper and the lower epidermis, but is generally thicker on the upper side. The chief function of the epidermis is protection of the internal tissues of the leaf from excessive water loss, from invasion by fungi, and from mechanical injury. Most epidermal cells do not contain chloroplasts.

The entire region between the upper and lower epidermis constitutes the ***mesophyll*** portion of the leaf. The mesophyll is commonly (but by no means always) divided into two fairly distinct parts: an upper palisade mesophyll, consisting of cylindrical cells arranged vertically, and a lower spongy mesophyll, composed of irregularly shaped cells. The cells of both parts of the mesophyll are very loosely packed and have many intercellular air spaces between them. These spaces are interconnected and communicate with the atmosphere outside the leaf by way of holes in the epidermis called ***stomata*** ("stoma" is the singular). The size of the stomatal openings is regulated by a pair of modified epidermal cells called ***guard cells***.

A conspicuous system of veins (also called vascular bundles) branches into the leaf blade from the petiole (Fig. 7.15). The veins form a structural framework for the blade and also act as transport pathways, being connected with the transport system of the rest of the plant. Each vein contains cells of the two principal vascular tissues, xylem and phloem; and each is usually surrounded by a ***bundle sheath***, composed of cells packed so tightly together that there are few spaces between them. In most cases the branching of the veins is such that no mesophyll cell is far removed from a veinlet; in one study the veins were found to attain a combined length of 102 cm per square centimeter of leaf blade.

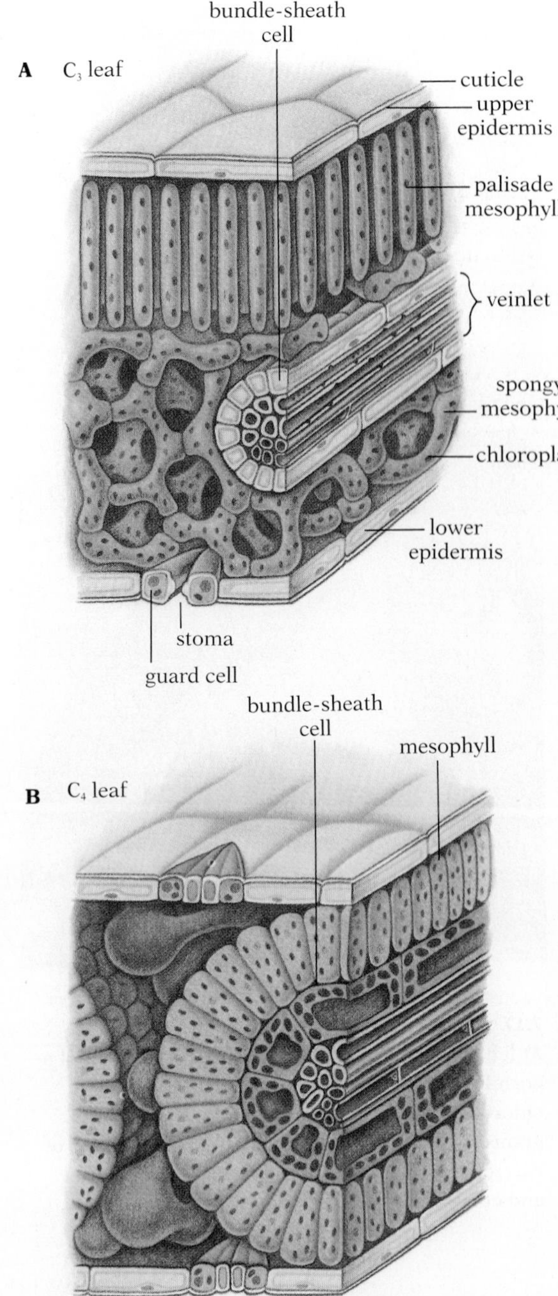

7.16 The anatomy of C_3 and C_4 (Kranz) leaves
In a C_3 leaf the palisade mesophyll cells typically form a layer in the upper part of the leaf; the corresponding mesophyll cells in a C_4 leaf are usually arranged in a ring around the bundle sheath. While the bundle-sheath cells of C_4 leaves have chloroplasts (dark green), those of C_3 leaves usually lack them.

LEAVES WITH KRANZ ANATOMY

As early as 1904 German plant anatomists observed that the leaf anatomy of some plants of tropical origin—plants associated with bright, hot, but especially xeric (dry) habitats—showed a combination of features not generally found in plants native to the temperate zones. This unusual complex of features came to be called Kranz anatomy. *Kranz*, which means "wreath" in German, refers to the ringlike arrangement of photosynthetic cells around the leaf veins of these plants.

The bundle-sheath cells of plants with Kranz anatomy (also called C_4 plants, for reasons we will see shortly) contain numerous chloroplasts, whereas those of other plants (C_3 plants) often do not. In plants with Kranz anatomy the mesophyll cells that correspond to the palisade layer tend to be

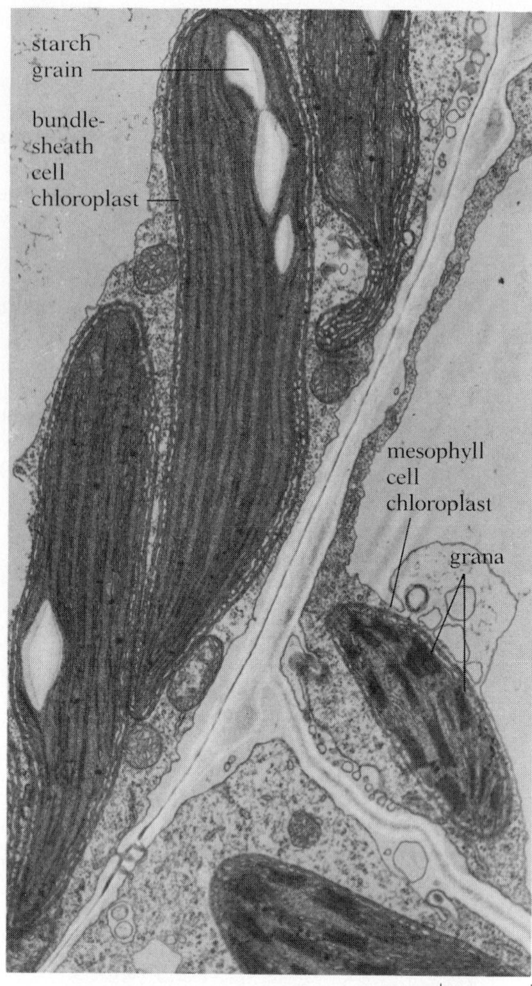

7.17 Two kinds of chloroplasts in a C_4 leaf At left in this electron micrograph of a segment of a corn leaf is part of a bundle-sheath cell; its chloroplasts have small grana and contain starch grains (light areas). At right and bottom are parts of two mesophyll cells; their chloroplasts are smaller and contain numerous grana but no starch grains.

clustered in a ringlike arrangement around the veins, just outside the bundle sheaths (Fig. 7.16B). These mesophyll cells contain numerous chloroplasts, but the spongy mesophyll cells outside the rings often have reduced numbers of chloroplasts, or even none at all. In Kranz plants the chloroplasts of the bundle-sheath cells and mesophyll cells usually differ in a number of ways. In the bundle-sheath cells the chloroplasts are bigger, they accumulate large amounts of starch in the light, and the grana are few and poorly developed; in the mesophyll cells the chloroplasts are smaller, they usually do not accumulate much starch in the light, and they have numerous large grana (Fig. 7.17).[3]

Most suggestions concerning the functional significance of Kranz anatomy put forward between 1904 and 1965 turned either on enhanced conservation of water under dry conditions or on unusually rapid transport of photosynthetic products away from the sites of synthesis. Though Kranz anatomy may well serve both these functions, investigations of the late 1960s and early 1970s suggest that it is also associated with special biochemical adaptations that enhance the ability of the plants to carry out photosynthesis under conditions of high temperature, intense light, low moisture, low CO_2 and high O_2 concentrations—all conditions far from optimal for plants that depend entirely on the Calvin cycle for CO_2 fixation. We will next examine the special photosynthetic pathways found in plants with Kranz anatomy.

C_4 PHOTOSYNTHESIS

As Figure 7.18A shows, corn, which originated in the tropics and has Kranz anatomy, can carry out photosynthesis at very low concentrations of CO_2, whereas bean plants, which are native to the temperate zone, cannot, because of photorespiration. When the CO_2 concentration falls below about 50 parts per million (in 21 percent O_2 at 20°C), bean plants become incapable of CO_2 fixation; and at CO_2 concentrations of 200 or 300 ppm, where corn approaches its maximum photosynthetic capacity, beans perform below their potential capacity. Again because of photorespiration, the concentration of O_2 that inhibits photosynthesis is far lower for beans than for corn (Fig. 7.18B). As was first realized in 1965, these contrasts between corn and beans are characteristic of the differences in photosynthetic capacity between Kranz plants and other plants.

Now, consider a plant exposed to great heat, dryness, and brilliant light, as in a desert or on a dry savanna. Under such conditions, when the moist walls of the mesophyll cells risk losing an excess of water by evaporation through the stomata, the guard cells close the stomata almost completely. Water loss is thus reduced, but now gases can no longer move freely between the atmosphere and the air spaces inside the leaf. As CO_2 is used up in photosynthesis, the nearly closed stomata prevent the supply inside the leaf from being fully replenished, and the CO_2 concentration in the air spaces around the mesophyll cells falls. Under such conditions, as we have seen,

[3] The frequent use of "usually," "often," and other qualifiers, you probably have already gathered, indicates that there is considerable variation among species exhibiting Kranz anatomy, and that the various Kranz characteristics do not always occur together.

non-Kranz plants like the bean will carry out so much photorespiration that their ability to synthesize carbohydrate from CO_2 will be greatly reduced. By contrast, corn and other Kranz plants *can* synthesize carbohydrate under dry conditions; they are thus able to survive in climates that would be fatal to other plants. Even under less extreme conditions they can often carry out photosynthesis at a higher rate than other plants.

Surely it is not the distinctive anatomy of Kranz plants, by and of itself, that enables them to carry out photosynthesis under conditions inhospitable to other plants. It must be, rather, some special biochemical capability correlated with their anatomy—perhaps an ability to avoid the photorespiration that limits the photosynthetic ability of many plants under conditions of low CO_2 and high O_2. A finding in 1965 suggested that in the presence of light, some plants with Kranz anatomy do not undergo photorespiration. But how do they avoid it? Here a third piece of the structure-function puzzle must be added to the other two. This third piece is a special way of fixing CO_2 initially.

In 1954 Hugo Kortschak of Hawaii, using the ^{14}C tracer technique much as Calvin did, found that one of the main early products of photosynthesis in sugarcane is not one of the three-carbon intermediates (C_3) of the Calvin cycle, but a four-carbon compound (C_4) instead. It was not until the late 1960s, however, that M. D. Hatch of Queensland, Australia, and C. R. Slack of Liverpool, England, worked out the biochemical pathway responsible for this C_4 type of photosynthesis. They found that in the mesophyll cells of sugarcane and other Kranz plants (where, you will recall, the mesophyll cells richest in chloroplasts are arranged in rings around the veins) CO_2 is combined, not with ribulose bisphosphate as in the Calvin cycle, but rather with a three-carbon compound called phosphoenolpyruvate or PEP, to form a four-carbon compound. The enzyme that catalyzes this carboxylation of PEP, unlike the one that catalyzes the carboxylation of RuBP in the Calvin cycle, does not have O_2 as an alternative substrate and is not inhibited by high O_2 concentrations. Thus it enables Kranz plants (which are therefore

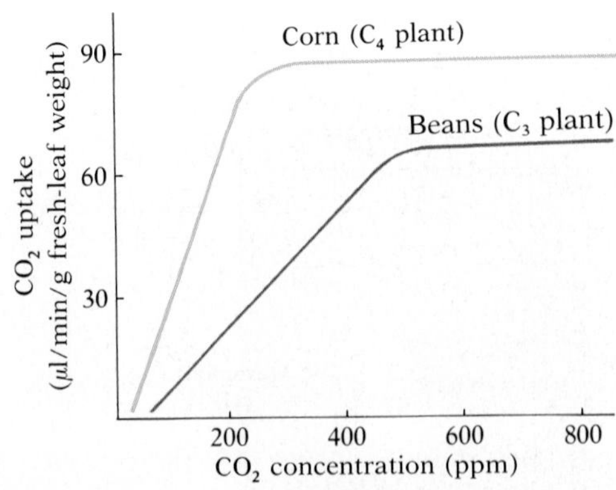

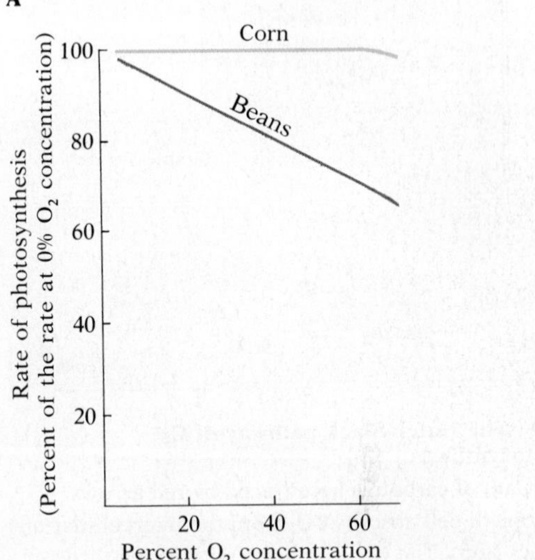

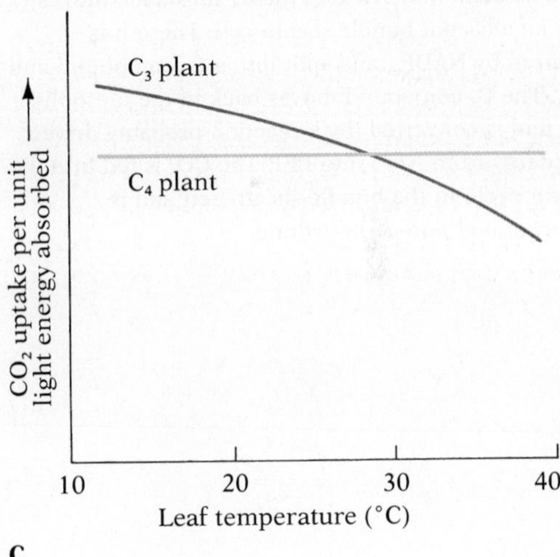

7.18 Comparison of photosynthetic efficiency in C_3 and C_4 plants
(A) Corn can fix carbon at CO_2 concentrations as low as one part per million, and it carries out photosynthesis at a very high rate at concentrations of 200–300 ppm (a normal concentration of CO_2 in the atmosphere is about 330 ppm). By contrast, beans perform no net carbon fixation at CO_2 concentrations below about 50 ppm, and their rate of photosynthesis at concentrations of 200–300 ppm is not very high.
(B) Photosynthesis in corn shows no inhibition at all at O_2 concentrations below 65 percent, whereas the photosynthetic rate of beans falls steadily as the O_2 concentration rises (a normal concentration of O_2 in the atmosphere is about 21 percent). Both A and B are for a temperature of 20°C and a light intensity of 2,000 foot-candles. Obviously, C_4 photosynthesis is superior under these conditions. However, when the temperature or illumination varies, while the O_2 and CO_2 concentrations remain at normal levels, a very different picture emerges. For example, as the temperature drops (C), C_3 plants clearly perform more efficiently, so they have the advantage in cooler (that is, more temperate) climates.

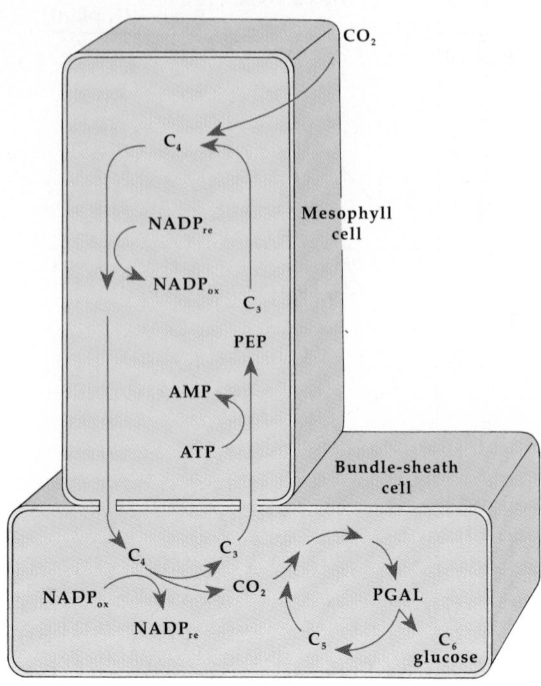

7.19 The Hatch-Slack pathway of C_4 photosynthesis

The path of carbon is here traced by red arrows. A mesophyll cell absorbs CO_2 from the intercellular air spaces (top). The CO_2 combines with PEP, a three-carbon compound, to form a four-carbon compound. After reduction by $NADP_{re}$, the C_4 substance moves into an adjacent bundle-sheath cell. There it is oxidized by $NADP_{ox}$ and split into a C_3 compound and CO_2. The C_3 compound moves back to the mesophyll cell and is converted (by a reaction probably driven by energy from ATP) into PEP. The CO_2 is fed into the Calvin cycle in the bundle-sheath cell and is incorporated into carbohydrate.

also called C_4 plants) to fix CO_2 under conditions when photorespiration would predominate over photosynthesis in C_3 plants, which use only the Calvin cycle.

Curiously enough, the C_4 compound formed in the mesophyll cells is not used for growth or nutrition by the plant. Instead, it is passed in reduced form into the bundle-sheath cells (which in C_4 plants, you will remember, are very well developed and contain chloroplasts), where it is decarboxylated: the C_4 compound is broken down to CO_2 and a C_3 compound (Fig. 7.19). The C_3 residue moves back to the mesophyll cells, where it is reconverted into PEP and starts the C_4 cycle over again. But the CO_2 remains in the bundle-sheath cells, where it is picked up by the RuBP carboxylase in the chloroplasts and incorporated into carbohydrate via the Calvin cycle.

Note, then, that in both C_3 and C_4 plants the ultimate assimilation of CO_2 into carbohydrate is by the Calvin cycle. The difference is that in C_3 plants the Calvin cycle is the only pathway of CO_2 fixation, whereas in C_4 plants there is another, preliminary fixation pathway. It may at first seem strange that a plant would have evolved a special mechanism for fixing CO_2 as C_4 in the mesophyll, only to combine it with a mechanism for promptly breaking off the CO_2 again and refixing it as carbohydrate in the bundle-sheath cells. But remember that the C_4 fixation in the mesophyll cells, because it is insensitive to O_2 concentration, cannot be short-circuited by photorespiration and can therefore operate under conditions in which C_3 plants could not carry out net photosynthesis. The mesophyll cells can then pump enough CO_2 (via the C_4 intermediate) into the bundle-sheath cells to maintain an artificially high CO_2 concentration in which the Calvin cycle is able to function. Moreover, since the bundle sheath is surrounded by mesophyll cells, any CO_2 lost from the sheath cells as a result of photorespiration may be reclaimed by the mesophyll cells.

In summary, C_4 plants have an advantage over C_3 plants under conditions of high temperature and intense light, when stomatal closure results in low CO_2 and high O_2 in the air spaces inside the leaf. Under such conditions, C_3 plants are unable to use CO_2 effectively because O_2 competes for RuBP. But C_4 plants can fix CO_2, because the mesophyll cells, acting as CO_2 pumps, can elevate the CO_2 concentration in the bundle-sheath cells to a level at which carboxylation of ribulose bisphosphate (leading into the Calvin cycle) exceeds its oxidation. The Kranz anatomy, with its concentric rings of mesophyll and bundle-sheath cells, facilitates the compartmentalization on which the process of CO_2 pumping depends.

The combination of Kranz anatomy and C_4 photosynthesis has evolved independently in a variety of unrelated plants, including both monocots like corn, sugarcane, sorghum, and crabgrass, and dicots like saltbush and portulaca. It is therefore an especially impressive illustration of the intimate relationship between structure and function in living systems.

CRASSULACEAN ACID METABOLISM (CAM)

Another interesting dry-climate variation of photosynthesis is found in many succulents—plants that store water in fleshy leaves (Fig. 7.20)—and a few other plants, including pineapple, Spanish moss, and some cacti. Like C_4 plants, these organisms avoid water loss in their hot environment by

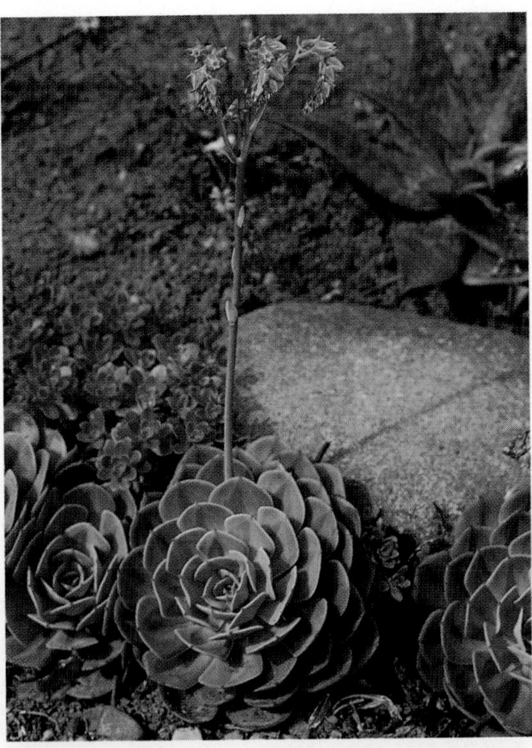

7.20 Stonewort, a member of the Crassulaceae The thick, fleshy leaves of this plant are typical of succulents.

closing their stomata during the day and opening them at night. The CO_2 necessary for photosynthesis is stored at night in the form of malic acid and isocitric acid, and then released in the cells during the day, to be fixed by C_3 photosynthesis. Though CAM seems to have evolved independently of C_4 photosynthesis, they serve the same purpose. The essential difference is that Kranz plants solve the CO_2-buildup problem by segregating the two steps of carbon fixation anatomically, while CAM plants separate them temporally. The existence of these variations illustrates again how natural selection can lead to several solutions of the same problems.

STUDY QUESTIONS

1. Make a list of the analogies between the strategy behind the electron-transport chain of respiration and the chain in photosynthesis. Do the same for the enzymes and the chemicals involved in the two processes. (pp. 176–82, 199–204)

2. Distinguish between C_3 photosynthesis, C_4 photosynthesis, and CAM photosynthesis. Why hasn't C_4 photosynthesis taken over? (pp. 205, 207–11)

3. If the predictions of some climatologists are borne out, human activity (especially burning of trees and fossil fuels) will increase the CO_2 concentration in the atmosphere significantly and, through a greenhouse

effect, raise the average temperature of the earth. What might this do for the various strategies of photosynthesis? (pp. 208–11)

4. Is there some obvious reason that chloroplasts have three sets of membranes, when mitochondria manage with two?

5. Compare the dark reactions of photosynthesis with the Krebs citric acid cycle. Is there any hint of a common evolutionary origin? (pp. 173–76, 203–4)

CONCEPTS FOR REVIEW

- Overall input and output of photosynthesis
  - Whole reaction
  - Light versus dark reactions
- Antenna function and reaction center
- General organization of cyclic photosynthesis
- General organization of noncyclic photosynthesis
- Anatomy of photosynthesis
  - Organization in thylakoids
  - Parallels with mitochondria
- General operation of Calvin cycle

SUGGESTED READING

BASSHAM, J. A., 1962. The path of carbon in photosynthesis, *Scientific American* 206 (6). (Offprint 122) *An old but informative account of how the Calvin cycle was worked out.*

BJÖRKMAN, O., and J. BERRY, 1973. High-efficiency photosynthesis, *Scientific American* 229 (4). (Offprint 1281) *The photosynthetic pathway and leaf anatomy of a group of* C_4 *plants.*

GOVINDJEE and R. GOVINDJEE, 1974. The primary events of photosynthesis, *Scientific American* 231 (6). (Offprint 1310) *Summary account of photosynthesis, including some intermediate steps for which evidence is scant.*

GOVINDJEE and W. J. COLEMAN, 1990. How plants make oxygen, *Scientific American* 262(2). *On the operation of the water-splitting enzyme of noncyclic photosynthesis.*

LEVINE, R. P., 1969. The mechanism of photosynthesis, *Scientific American* 221 (6). (Offprint 1163) *A clear explanation of the light reactions.*

MILLER, K. R., 1979. The photosynthetic membrane, *Scientific American* 241 (4). (Offprint 1448) *An excellent discussion relating the chemiosmotic theory of chloroplast function to the structure of thylakoid membranes as shown by freeze-etch microscopy.*

STOECKENIUS, W., 1976. The purple membrane of salt-loving bacteria, *Scientific American* 234 (6). (Offprint 1340) *On a unique photosynthetic mechanism based on a pigment very similar to the visual pigments of animals rather than on chlorophyll. Provides food for thought about the evolution of our visual sensitivity.*

YOUVAN, D. C., and B. L. MARRS, 1987. Molecular mechanisms of photosynthesis, *Scientific American* 256(6). *An account of the molecular events occurring during the first 200 microseconds following photon absorption.*

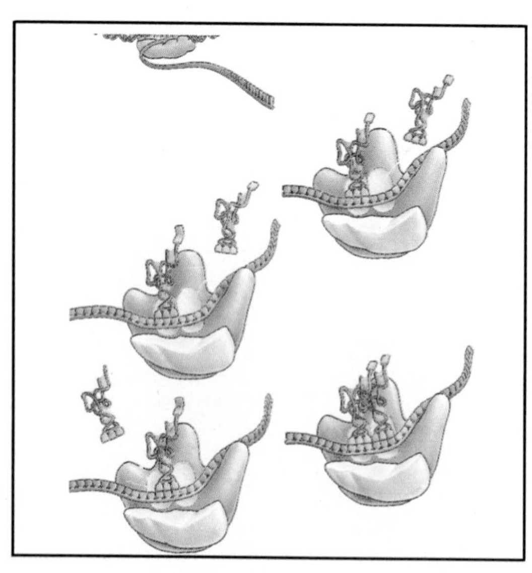

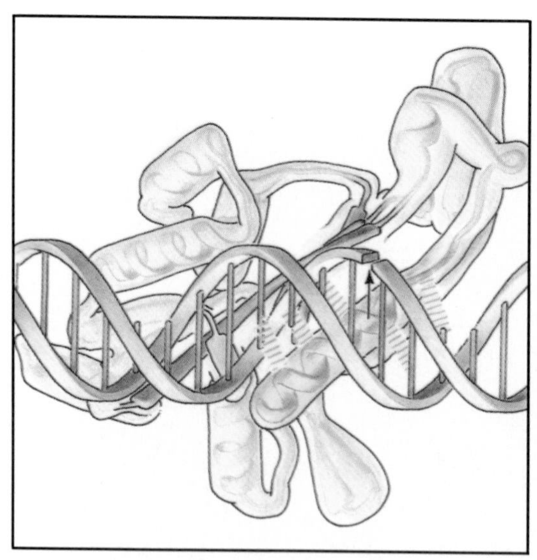

PART II

THE PERPETUATION OF LIFE

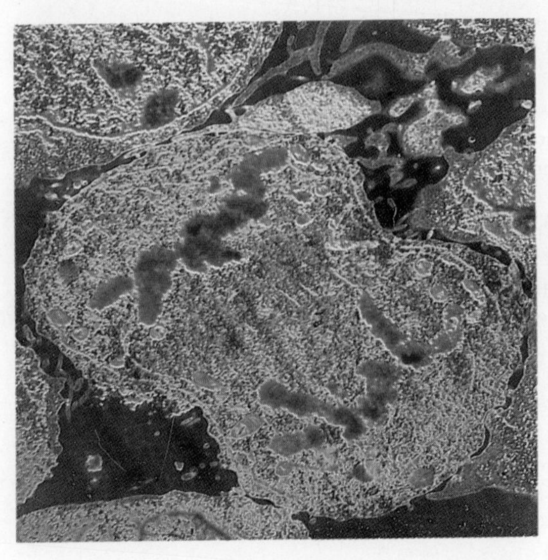

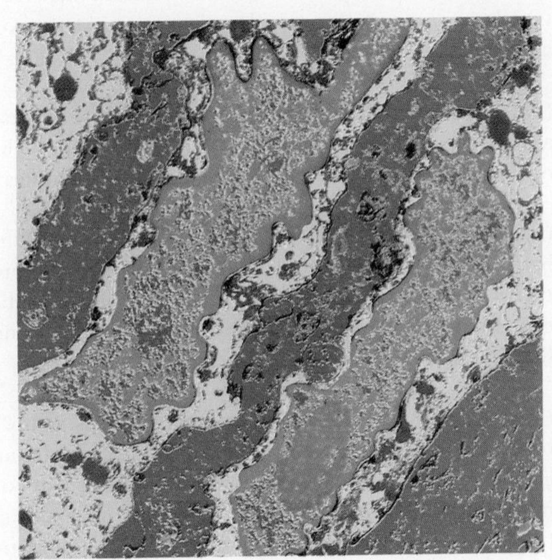

Part II

Part Opening Photographs:

(1) The translation cycle builds proteins from instructions in RNA. Here one cycle of translation on a ribosome adds an amino acid to a growing string that will constitute a protein.

(2) Proteins often bind to DNA and regulate the expression of a gene. In this case, a protein called *Eco*RI digests intrusive viral DNA by breaking it at specific nucleotides. The blue portion of that protein identifies the target sequence.

(3) Color-enhanced TEM of a human lymphocyte in anaphase showing chromosomes being moved into the two daughter cells. Cellular reproduction is an exquisitely controlled series of molecular events that involves subdividing the cytoplasm, expanding, breaking, and reforming the cell membrane, and accurately subdividing like chromosomes to daughter cells.

(4) Color-enhanced TEM of human cancer cells. Many cancer cells proliferate wildly because they suffer from a variety of failures of cellular control. The cellular control centers, the nuclei (green), are unusually elongated compared to normal cells, which may indicate the degradation of the nuclei or the cell's attempt to increase the area of contact between nucleus and cytoplasm for rapid growth.

Chapter 8

THE STRUCTURE AND REPLICATION OF DNA

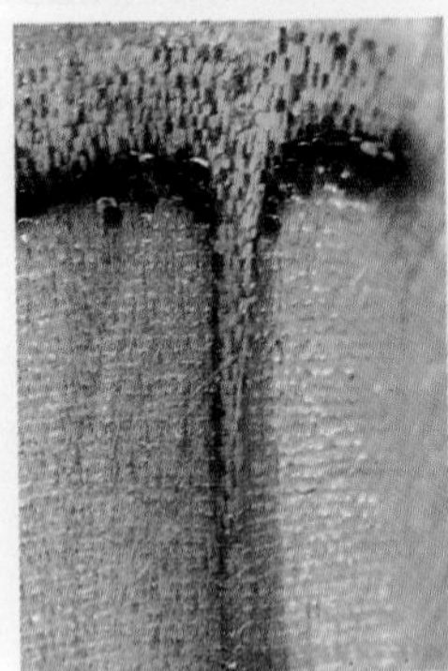

The processes of life are guided by an elaborate and precise series of information transfers. The genes of an organism's DNA contain all the information necessary to build and operate the organism—information that orchestrates its development from a single, unspecialized cell into a complex corporation of tissues and organs in which cells have been differentiated according to function, and that also direct events ranging from glycolysis and the Krebs cycle to the organism's behavior.

The transfer of this genetic information occurs in two directions. First, the instructions in the DNA are used to build cells and direct biochemical events. Second, and equally important, the instruction set is duplicated and transferred to each new cell, whether it be a cell added as the organism grows, or one of the ***gametes*** (sex cells, like sperm and egg) that unite to form an entirely new organism. It is this information that makes possible the perpetuation of life, both during an organism's own lifetime, and from one generation to another (Fig. 8.1).

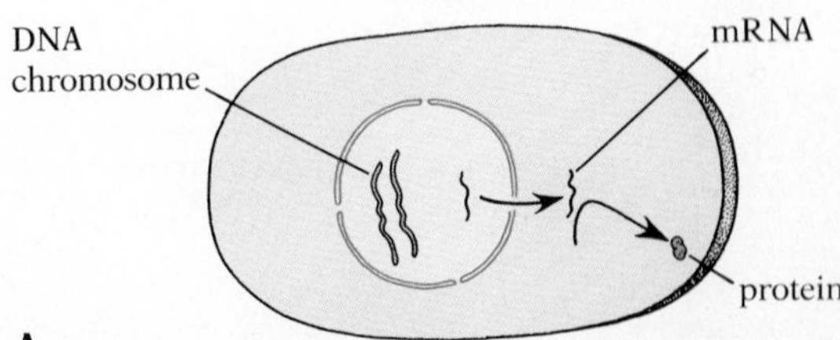

A
Instructions in the DNA are used to build cellular components and direct biochemical events. In most eucaryotes, there are two copies of each chromosome.

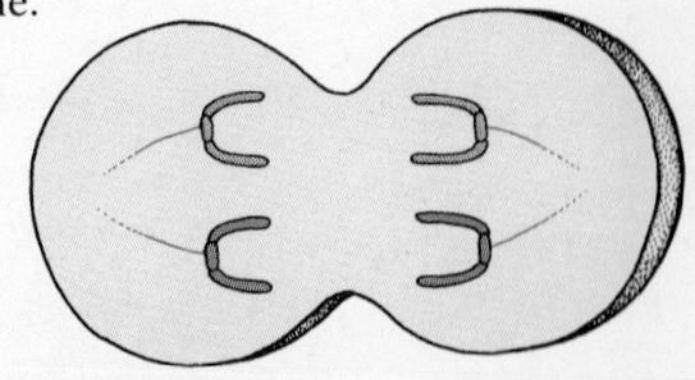

B
Chromosomes are duplicated prior to cell division and a complete copy of each pair is passed to each new cell.

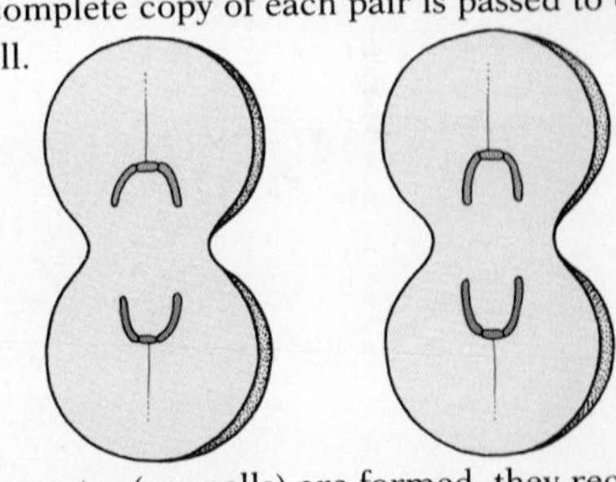

C
When gametes (sex cells) are formed, they receive only one chromosome from each pair. The normal number of chromosomes is restored by fusion of a male and female gamete.

8.1 Transfer of genetic information
The set of instructions in the DNA is transcribed and used to direct events within the cell (A), is duplicated and passed on to daughter cells during normal cell division (B), and is halved and passed to gametes in preparation for sexual reproduction (C).

It is difficult to appreciate how far the study of genetics has come in the last four decades. By 1950, micrographs of stained chromosomes and breeding experiments had shown little more than that the physical units of heredity—the genes—are arranged in linear sequence on the chromosomes; it was still virtually impossible to say anything useful about the structure of genes or how their properties might control life processes. Since then, however, unprecedented progress in uncovering the molecular basis of genetics has produced a revolution whose effects have yet to be fully measured. We know now that ***genes*** are sequences of DNA that encode proteins or, occasionally, the nucleic acid RNA, which can be used directly in the construction of ribosomes. The genes are organized in a linear sequence on one or more chromosomes, collectively called the ***genome***.[1] We can describe in remarkable detail how genes duplicate themselves, direct the synthesis of specific proteins, regulate their own activity, orchestrate the development of complex multicellular organisms, and both cause and prevent disease; we can even describe how evolution at the molecular level may take place.

Modern molecular genetics begins with the study of DNA structure; in this chapter we will look first at the structure of the genetic material and then go on to examine the process of chromosome duplication, the essential step of which is more formally known as ***replication***. This process, which includes self-repair of replication errors, enables each daughter cell produced during cell division to receive its own copy of each chromosome (and thus a complete set of genes) from the parental cell. In the next chapter we will look at how the information in the genes is transcribed into RNA and translated on the ribosomes to produce enzymes and structural proteins.

THE DISCOVERY OF DNA STRUCTURE AND FUNCTION

THE COMPOSITION OF CHROMOSOMES

The revolution brought about by molecular biology has its roots in early attempts to ascertain the types of compounds present in cell nuclei. In 1868 the young Swiss biochemist Friedrich Miescher showed that when a cell is treated with pepsin to digest its proteins, the nucleus shrinks but remains essentially intact. He showed that the same nuclear material that can withstand peptic digestion also behaves totally unlike protein when treated with a variety of other reagents, and that it contains phosphorus in addition to the carbon, oxygen, hydrogen, and nitrogen that would be expected if it were a protein. Taken together, these results meant that the nucleus contains large quantities of both protein and some hitherto unrecognized nonproteinaceous compound. The latter, which Miescher called nuclein, has since been named nucleic acid. Further research has shown that cells contain several sorts of nucleic acids, some not restricted to the nucleus; the type studied by Miescher was DNA (deoxyribonucleic acid).

In 1914 Robert Feulgen, a German chemist, devised a method of selectively staining nuclein a brilliant crimson. When, ten years later, Feulgen

[1] The genome also includes the DNA in organelles, as well as tiny chromosomelike elements (plasmids) to be discussed in Chapter 10.

applied his technique to whole cells, he found that the nuclear DNA is restricted to the chromosomes. Feulgen's method is still employed, and has been used to measure the DNA content of nuclei from many types of cells (Fig. 8.2). Researchers have shown conclusively that all the ***somatic*** cells of a given organism (all the cells of an organism except the gamete-producing ***germ*** cells) ordinarily contain the same amount of DNA—even though cells from such different tissues as liver, kidney, heart, nerve, and muscle differ drastically in the amounts of other substances they contain—and, further, that egg and sperm cells contain only half as much DNA as the somatic cells. Since biologists had already concluded that cell division distributes a complete set of genes to every somatic cell, regardless of its eventual role, and that the process of gamete production distributes to every sperm and egg cell exactly half the amount of genetic material found in the somatic cells, the discovery that the amount of DNA is usually constant in all somatic cells within a species, but is halved in the gametes, suggested that DNA rather than protein might be the essential material of the genes. But many workers refused to take this possibility seriously. The chromosomes of most organisms contain both protein and DNA, and most biologists assumed that the protein must be the genetic material, because only protein seemed to have the chemical complexity necessary to encode so much information.

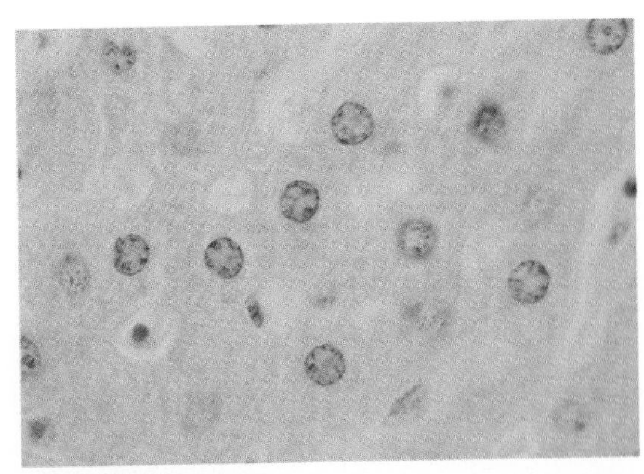

8.2 Feulgen staining
Feulgen staining was used to distinguish the DNA in this sample of liver cells.

DNA VERSUS PROTEIN

In 1928 Fred Griffith, an English medical bacteriologist, published a paper describing some experiments, now considered classic, on pneumococci, the bacteria that cause pneumonia. Griffith studied the effects on mice of a virulent (disease-causing) strain of bacteria (S, for "smooth colonies") and a nonvirulent strain (R, for "rough"). He showed that mice injected with live strain-R bacteria survived, mice injected with live strain-S bacteria soon died, but mice injected with heat-killed strain-S bacteria survived. These results (Table 8.1) were readily understandable. But the results of another of his experiments were thoroughly perplexing: mice injected with a mixture of live strain-R and heat-killed strain-S bacteria also died. How could a mixture of nonvirulent and dead bacteria have killed the mice? Griffith examined the bodies of the dead mice and found that they were full of live strain-S bacteria! Where had they come from? After many careful experiments, he became convinced that somehow the live strain-R bacteria had been transformed into live strain-S bacteria by material from the dead strain-S cells. The transformed bacteria, when cultured, reproduced new strain-S bacteria. Presumably, hereditary material from the dead bacteria had entered the live strain-R cells and changed them into strain-S cells.

By 1931 other workers had shown that the rodent host was not essential for bacterial ***transformation;*** it could occur just as well in test-tube cultures. Two years later James L. Alloway of the Rockefeller Institute (now Rockefeller University) showed that not even whole strain-S cells were necessary; live strain-R cells in a test tube could be transformed into strain-S bacteria by a cell-free extract of S bacteria. Apparently the hereditary characteristics of the virulent strain were transmitted to the cells of the nonvirulent strain by some substance that could withstand both the killing and extracting of the cells in which it had originally been contained.

TABLE 8.1 *Griffith's results*

Bacteria injected	*Reaction of mice*
Live strain R	Survived
Live strain S	Died
Dead strain S	Survived
Live strain R plus dead strain S	Died

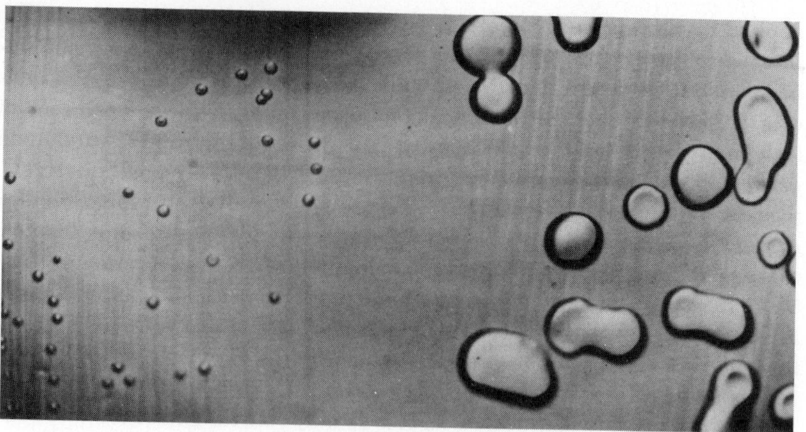

8.3 Rough versus smooth pneumococci
Rough (untransformed) colonies are on the left; smooth (transformed) colonies are on the right. This photograph is from the 1944 paper by Avery, MacLeod, and McCarty.

The work of Griffith, Alloway, and others in the late 1920s and early 1930s was interesting to other biologists, but its full significance was not appreciated at that time. It was not until 1944 that O. T. Avery, Colin MacLeod, and Maclyn McCarty of the Rockefeller Institute demonstrated through purification techniques that the transforming agent was DNA; nothing else was necessary (Fig. 8.3). From our present perspective this seems like strong evidence that DNA rather than protein or a nucleoprotein complex is the essential genetic material, but at that time many scientists remained unconvinced. For all anyone knew, the DNA from the virulent strain might merely activate the protein-based genes in the nonvirulent strain.

During the next ten years, however, the evidence for DNA steadily became stronger. At least 30 different examples of bacterial transformation by purified DNA were described. And strong evidence came from another source: the studies by Alfred D. Hershey and Martha Chase, of the Carnegie Laboratory of Genetics, of a special type of virus that attacks the bacterium *Escherichia coli,* which is abundant in the human digestive tract. This type of bacteria-destroying virus is called ***bacteriophage***—"phage" for short (from the Greek *phagein,* "to eat").

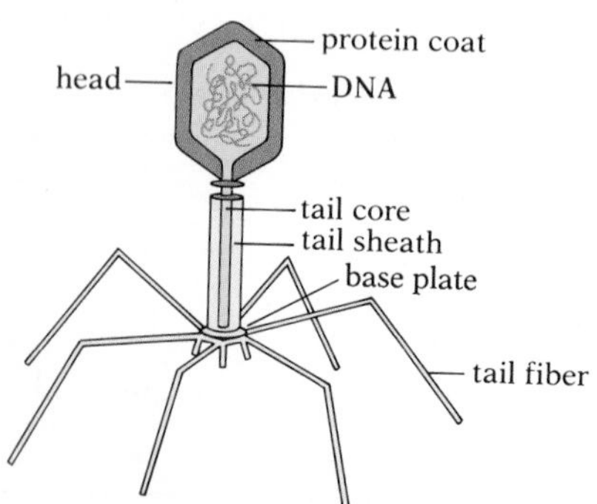

8.4 A complex bacteriophage

Viruses, tiny parasites that subvert the cellular machinery of host organisms to reproduce their own genes, are composed primarily of a protein coat and a nucleic acid core. The electron microscope reveals that certain phage viruses are structurally more complex than many other types. Their protein coat is divided into a head region (containing the genetic material) and an elongate tail region made up of a hollow core, a surrounding sheath, and six distal fibers (Fig. 8.4). Electron micrographs show that when such a phage attacks a bacterial cell it becomes attached by the tip of its tail fibers and its base plate to the wall of the bacterial cell (Fig. 8.5); the plate and fiber tips contain a protein that binds with a specific component of the bacterial wall. The protein coat of the phage never enters the bacterial cell, but within an hour after the phage becomes attached to the cell wall, new phage appear within the bacterium. At the appropriate time, the phage activates its genes encoding a digestive enzyme called lysozyme, and other enzymes; as a consequence of their action, the bacterial cell lyses, or ruptures, releasing dozens or even hundreds of new phage into the surrounding medium. The

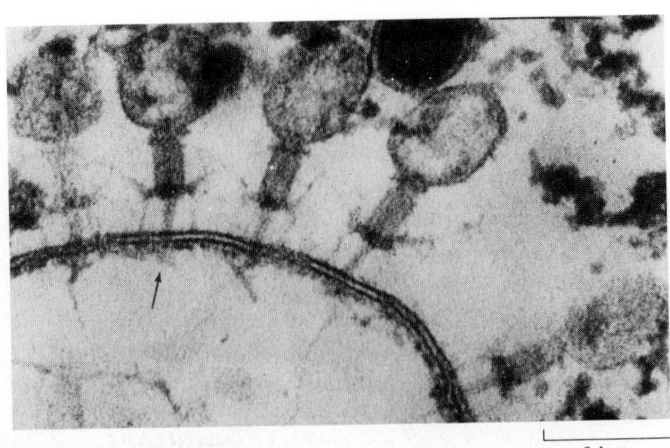
0.1 μm

8.5 Bacteriophage replication
Each bacteriophage attaches to the bacterial cell wall by its tail fibers and base plate, and injects its genetic material into the cell. Once inside, the genetic material takes over the metabolic machinery of the cell and puts it to work making new phage.

new phage released from a lysed bacterium are genetically identical with the one that initiated the infection. Hereditary material must have been injected into the bacterial cell by the phage attached to its wall, and this hereditary material must have usurped the metabolic machinery of the bacterium and put it to work manufacturing new phage genes as well as the many proteins necessary to construct the complex protein coat and the several enzymes that shut off much of the host's own protein synthesis and in due time lyse the cell.

Hershey and Chase designed an experiment to determine whether the infecting phage injects into the bacterium only DNA or only protein or some of both. Since DNA contains phosphorus but no sulfur, whereas protein contains sulfur (in the disulfide bonds described in Chapter 3) but no phosphorus, they were able to distinguish DNA from protein by using radioactive isotopes of phosphorus and sulfur. They cultured phage on bacteria grown on a medium containing radioactive phosphorus (^{32}P) and radioactive sulfur (^{35}S). The phage incorporated the ^{35}S into their protein and the ^{32}P into their DNA (Fig. 8.6B). Hershey and Chase then infected nonradioactive bacteria with the radioactive phage. They allowed sufficient time for the phage to become attached to the walls of the bacteria and inject hereditary material (Fig. 8.6D). Next they agitated the bacteria in a blender in order to detach what remained of the phage from their surfaces. They then centrifuged the mixture to separate the infected bacteria from the phage coats (Fig. 8.6G). Analysis showed that the fluid left behind after the bacteria had been centrifuged into a pellet at the bottom of the sample tube con-

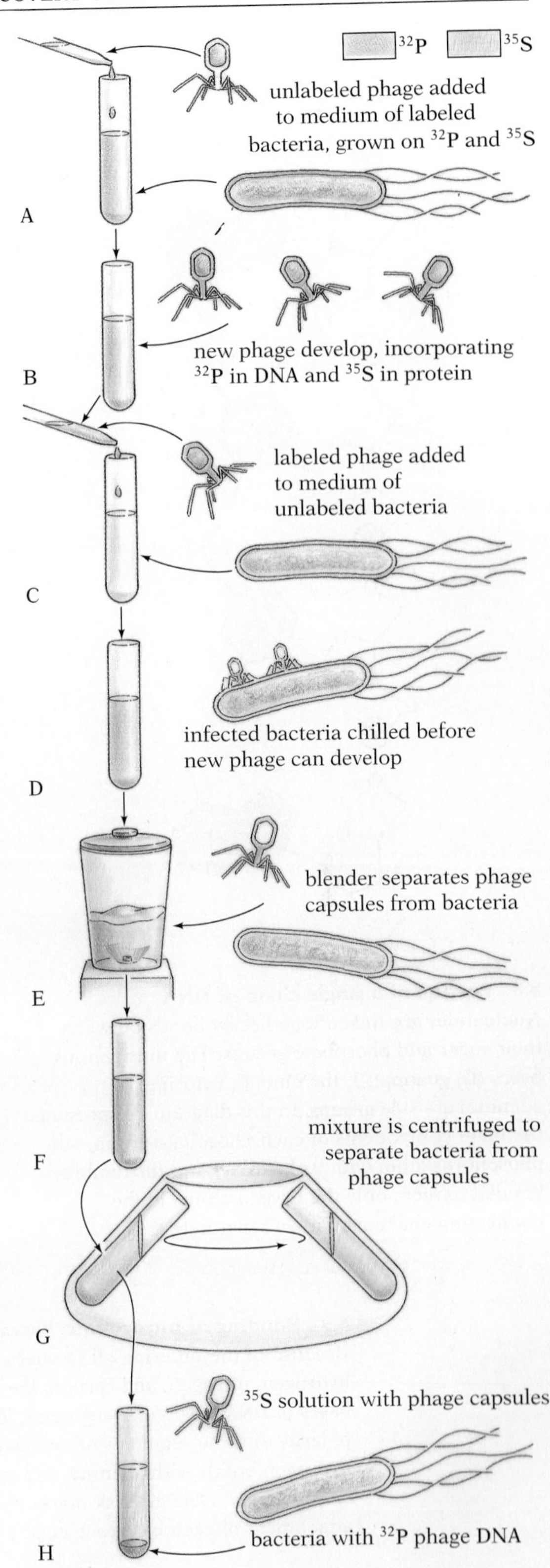
^{32}P
^{35}S
unlabeled phage added to medium of labeled bacteria, grown on ^{32}P and ^{35}S
A
new phage develop, incorporating ^{32}P in DNA and ^{35}S in protein
B
labeled phage added to medium of unlabeled bacteria
C
infected bacteria chilled before new phage can develop
D
blender separates phage capsules from bacteria
E
mixture is centrifuged to separate bacteria from phage capsules
F
G
^{35}S solution with phage capsules
bacteria with ^{32}P phage DNA
H

8.6 The Hershey-Chase experiment
For details of this experiment, which demonstrated that DNA rather than protein carries genetic information, see the text.

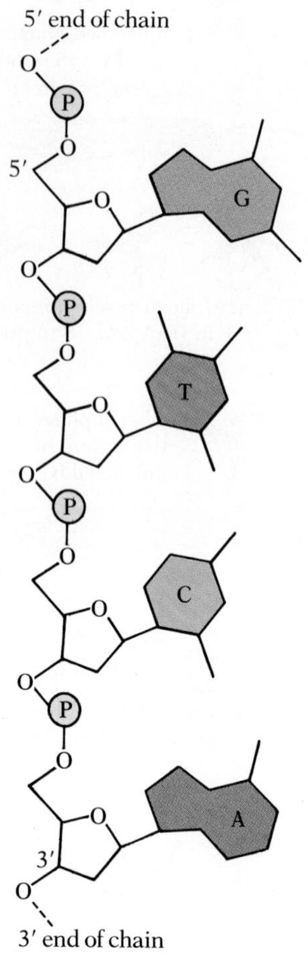

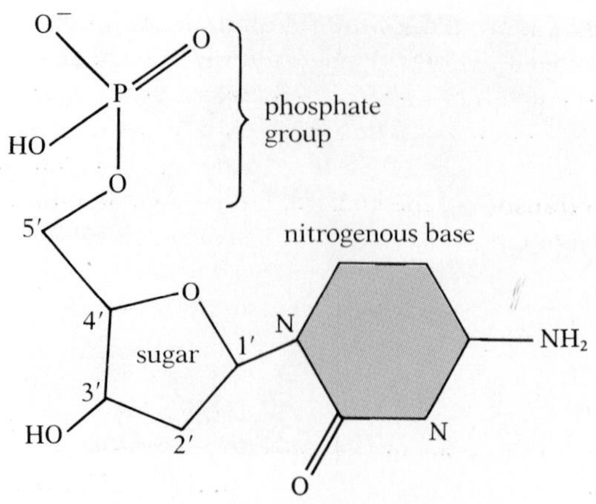

8.7 Diagram of a nucleotide from DNA
A phosphate group and a nitrogenous base are attached to deoxyribose, a five-carbon sugar. The carbons in deoxyribose are designated 1′–5′, as shown here, though the numbers are not normally included in molecular diagrams. The phosphate group is bound to the 5′ carbon of the sugar, while a hydroxyl group is bound to the 3′ carbon. Of the four different nitrogenous bases, cytosine is illustrated here. RNA nucleotides have a slightly different sugar, identical to the one shown except that the 2′ carbon has a hydroxyl group bound to it.

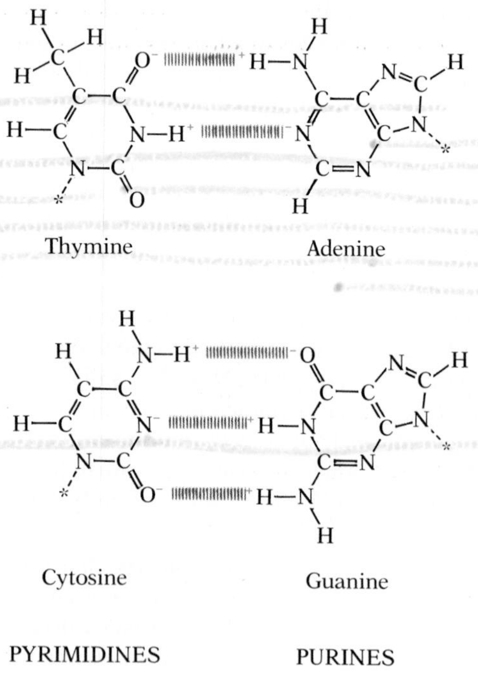

8.8 Portion of a single chain of DNA
Nucleotides are linked together by bonds between their sugar and phosphate groups. The nitrogenous bases (G, guanine; T, thymine; C, cytosine; A, adenine) are side groups. In this diagram P represents the main components of each phosphate group—the phosphorus atom with its hydroxyl and the double-bonded oxygen; only the oxygen atoms in the connecting chain are shown separately.

8.9 Bonding of nitrogenous bases in nucleotides
Because of the differing electronegativities of oxygen, hydrogen, nitrogen, and carbon, the nitrogenous bases of DNA have polar segments. The spacing and polarity of these segments allow thymine to form hydrogen bonds with adenine, and cytosine to bond with guanine. (The asterisk marks the point of attachment of each base to a sugar.)

tained a substantial amount of ^{35}S but little ^{32}P, an indication that only the empty protein coat had been left outside the bacterial cell. Analysis of the bacterial fraction showed that it contained much ^{32}P but little ^{35}S, an indication that only DNA had been injected into the bacteria by the phage (Fig. 8.6H). Since new phage were produced in these bacteria, DNA alone had to have been sufficient to transmit to the bacteria all the genetic information necessary for their production. This experiment, reported in 1952, strongly supported the earlier conclusions based on transformation experiments that nucleic acids, not proteins, constitute the genetic material.

THE MOLECULAR STRUCTURE OF DNA

We have already seen that a molecule of DNA is composed of building blocks called nucleotides, each of which is itself composed of a five-carbon sugar bonded to a phosphate group and a nitrogenous base (Fig. 8.7). By convention, the five carbons are designated by numbers, 1′–5′. There are four kinds of nucleotides in DNA, which differ from one another in their nitrogenous bases. Two of the bases, ***adenine*** and ***guanine***, are purines, which are double-ring structures; the other two, ***cytosine*** and ***thymine***, are pyrimidines, which are single-ring structures (see Fig. 8.8). In a DNA molecule the nucleotides are arranged in a specific sequence; the sugars are held together by the phosphate groups that link the 3′ carbon of one sugar to the 5′ carbon of the next; the nitrogenous bases are arranged as side groups off the chains (Fig. 8.8). DNA molecules ordinarily exist as double-chain structures, with the two chains, or strands, held together by hydrogen bonds between their nitrogenous bases; such bonding can occur only between cytosine and guanine or between thymine and adenine (Fig. 8.9). Thus the sequence of bases in one strand determines the complementary sequence in the other (Fig. 8.10). Notice that the polarities of the two strands are opposite: one runs from 5′ to 3′, while the other goes from 3′ to 5′.[2] Finally, the

[2] This strand "polarity" has nothing to do with the unequal distribution of charge that gives rise to polar molecules.

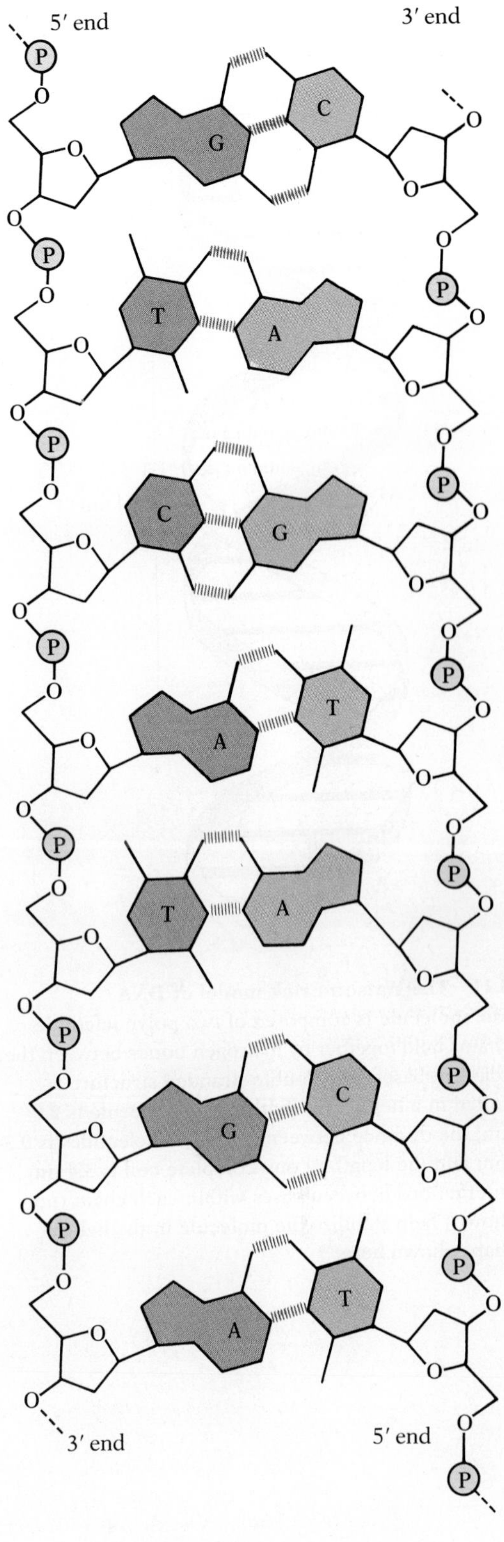

8.10 Portion of a DNA molecule uncoiled
The molecule has a ladderlike structure, with the two uprights composed of alternating sugar and phosphate groups and the cross rungs composed of paired nitrogenous bases. Each cross rung has one purine base (a pentagon attached to a hexagon) and one pyrimidine base (a hexagon). When the purine is guanine (G), the pyrimidine with which it is paired is always cytosine (C); when the purine is adenine (A), the pyrimidine is thymine (T). Adenine and thymine are linked by two hydrogen bonds (striped bands), guanine and cytosine by three. Note that the two chains run in opposite directions: the free phosphate is linked to the 5′ carbon at the upper end of the left chain and to the corresponding carbon at the lower end of the right chain.

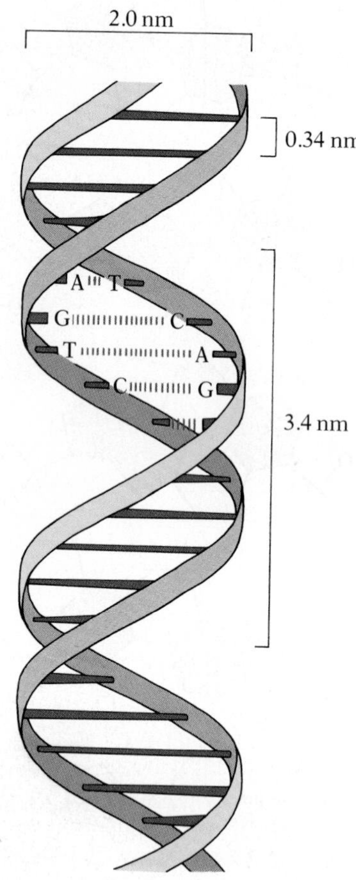

8.11 The Watson-Crick model of DNA
The molecule is composed of two polynucleotide chains held together by hydrogen bonds between their adjacent bases. The double-stranded structure is coiled in a helix. The width of the molecule is 2.0 nm; the distance between adjacent nucleotides is 0.34 nm; and the length of one complete coil is 3.4 nm. Interactions between bases within each chain (not shown) help stabilize the molecule in the helical shape shown here.

ladderlike double-chain molecule is coiled into a double helix (Fig. 8.11), and stabilized further by hydrogen bonds aligned with the chain; these bonds between separate rungs of the nucleotide ladder are analogous to those that stabilize the alpha helix of proteins (see Fig. 3.22, p. 66).

Determining the structure of so complicated—and important—a molecule as DNA had become an irresistible challenge to many scientists. In 1950 almost nothing was known about the spatial arrangement of the atoms within the DNA molecule; nor was it known how this molecule could contain within it the necessary information for replicating itself and for controlling cellular function. About this time, several workers began applying the techniques of X-ray diffraction analysis to DNA. Outstanding among them were Rosalind Franklin and Maurice H. F. Wilkins of King's College, London, who succeeded in producing much sharper X-ray diffraction patterns than had previously been obtainable (Fig. 8.12). Francis H. C. Crick of Cambridge University had just developed mathematical methods for interpreting X-ray patterns of protein helices and, working from the Franklin and Wilkins photographs, was able to show that crystalline DNA had to be a helix with three major periodicities: repeating patterns of 0.34 nm, 2.0 nm, and 3.4 nm.

Now began a collaboration whose outcome would rank as one of the major milestones in the history of biology. James D. Watson and Crick (Fig. 8.13), working at Cambridge University, decided to try to develop a model of the structure of the DNA molecule by combining what was known about the chemical content of DNA with the information gained from Crick's analysis of Franklin and Wilkins's X-ray diffraction studies, as well as with data on the exact distances between bonded atoms in molecules, the angles between bonds, and the sizes of atoms. Watson and Crick built scale models of the component parts of DNA and then attempted to fit them together in a way that would agree with the information from all these separate sources.

They were certain that the 0.34 nm periodicity corresponded to the distance between successive nucleotides in the DNA chain, the 2.0 nm periodicity to the width of the chain, and the 3.4 nm periodicity to the distance between successive turns of the helix. Since 3.4 is exactly ten times the distance between successive nucleotides, each turn of the helix had to be ten nucleotides long.

Having made these assumptions about the meaning of the X-ray diffraction data, Watson and Crick tried to correlate them with the information from other sources and immediately ran into a discrepancy: they calculated that a single chain of nucleotides coiled in a helix 2.0 nm wide with turns 3.4 nm long would have a density only half as great as the known density of DNA. An obvious inference was that the DNA molecule is composed of two nucleotide chains rather than one. Now they had to determine the relationship between the two chains within the double helix. They tried several arrangements of their scale model; the one that best fitted all the data had the two nucleotide chains wound in opposite directions within a hypothetical cylinder of appropriate diameter, with the purine and pyrimidine bases oriented toward the interior of the cylinder (Fig. 8.11). With the bases oriented in this manner, hydrogen bonds between the bases of opposite chains could supply the force to hold the two chains together and to maintain the helical configuration. In other words, the DNA molecule, when unwound, would have a ladderlike structure, with the uprights of the ladder formed by the two long chains of alternating sugar and phosphate groups, and with each of the cross rungs formed by two nitrogenous bases loosely bonded to each other by hydrogen bonds (Fig. 8.10).

Watson and Crick soon realized that each cross rung must be composed of one purine base and one pyrimidine base. Their scale model showed that the available space between the sugar-phosphate uprights was just sufficient to accommodate three ring structures. Hence two purines opposite each other occupied too much space, because each had two rings, for a total of four, and two pyrimidines opposite each other did not come close enough to bond properly, because each had only one ring. This left four possible pairings: A–T, A–C, G–T, and G–C. Further examination revealed that though adenine (A) and cytosine (C) were of the proper size to fit together into the available space, they could not be arranged in a way that would permit hydrogen bonding between them; the same was true of guanine (G) and thymine (T). Therefore neither A–C nor G–T cross rungs could occur in the DNA molecule. This left only A–T and G–C. Both of these base pairs seemed to fulfill all requirements. It did not seem to matter in which order the bases occurred; the essential requirement was that adenine and thymine always be paired with each other and that guanine and cytosine always be paired. This pairing quite unexpectedly explained an earlier finding by Erwin Chargaff and his colleagues at Columbia University. They had noted that while the proportion of, say, cytosine in DNA varies from species to species, it is constant between individuals of a given species, and always matches the proportion of guanine in that species. Similarly, the DNA of any given species always contains exactly equal amounts of adenine and thymine nucleotides. The amounts of adenine and thymine are always equal because these two bases are always paired, and, similarly, the amounts of guanine and cytosine are always equal because these are always paired.

In summary, then, the Watson-Crick model of the DNA molecule shows a double helix in which the two chains, composed of alternating sugar and phosphate groups, are bonded together by hydrogen bonds between adenine and thymine from opposite chains and between guanine and cytosine from opposite chains. This model, in essentially the same form in which it was first proposed in 1953, has been consistently supported by later research, and it has received general acceptance. Watson, Crick, and Wilkins were awarded the Nobel Prize in 1962 for their critically important work. Franklin, who would probably have shared this honor, had died in 1958.

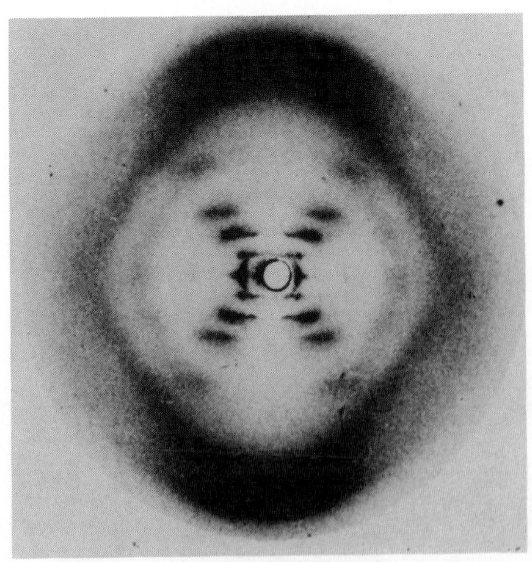

8.12 X-ray diffraction image of DNA
As Crick had discovered, the X-shaped cross radiating from the center of this pattern is diagnostic of a helix. The position of the strong bands at the top and bottom indicate a periodicity of 3.4 Å. More subtle patterns suggest a 34-Å period as well.

THE REPLICATION OF DNA

THE TEMPLATE THEORY OF WATSON AND CRICK

DNA, if it is the genetic substance, must have built into it the information necessary to replicate itself and to control the cell's attributes and functions. One of the most satisfying things about the Watson-Crick model of

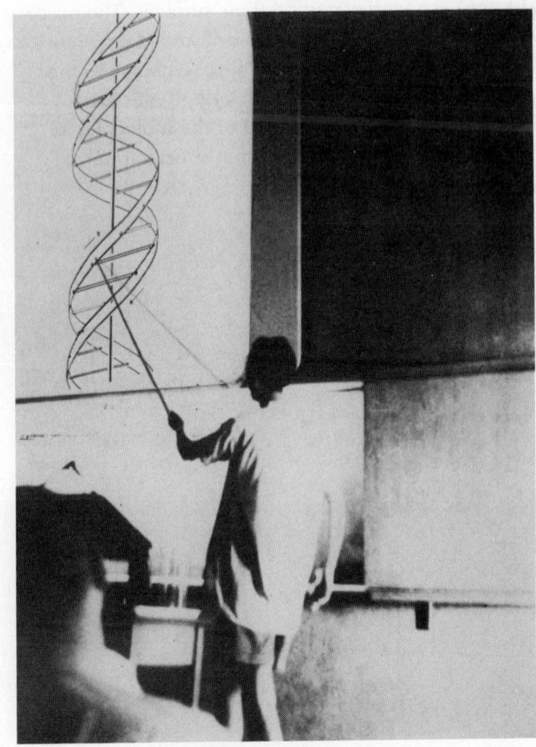

8.13 Watson describing the Watson-Crick model of DNA
This photograph was taken at a 1953 seminar at the Cold Spring Harbor Laboratory.

DNA is that it immediately suggests a way in which the first of these two requirements may be met.

Since the DNA of all organisms is alike in being a polymer composed of only four different nucleotides, the essential distinction between the DNA of one gene and the DNA of another gene must be—aside from the total number of nucleotides—the sequence in which the four possible types of base-pair cross rungs (A–T, T–A, G–C, and C–G) occur. The basic question of genetic replication is, then, assuming that an adequate supply of the four nucleotides has already been synthesized, what tells the cell's biochemical machinery how to put these nucleotide building blocks together in exactly the sequences and quantities characteristic of the DNA already present in the cell?

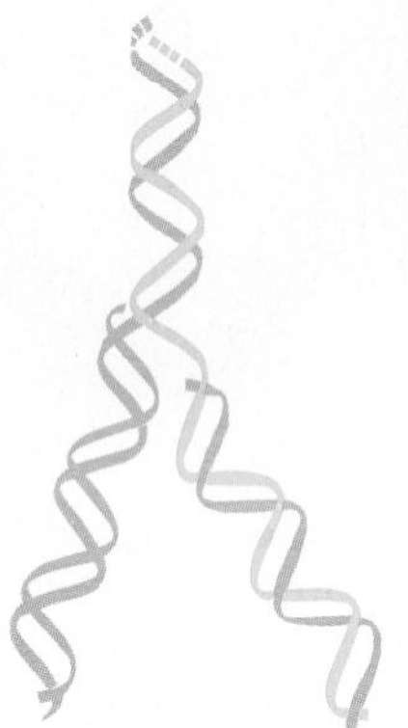

8.14 Replication of DNA
As the two polynucleotide chains of the old DNA (yellow and blue) uncoil, new polynucleotide chains (green) are synthesized on their surfaces. The process produces two complete double-stranded molecules, each of which is identical in base sequence to the original double-stranded molecule. Replication actually begins not at the end of the molecule, as shown here for simplicity, but at specific internal points.

Watson and Crick pointed out that if the two chains of a DNA molecule are separated by rupturing the hydrogen bonds between the base pairs, each chain provides all the information necessary for synthesizing a new partner identical to its previous partner. Since an adenine nucleotide must always pair with a thymine nucleotide, and since a guanine nucleotide must always pair with a cytosine nucleotide, the sequence of nucleotides in one chain, or strand, precisely specifies what the sequence of nucleotides in its complementary strand must be. In other words, if the cell could separate the two chains in its DNA molecules—much as one might unzip a zipper—it could line up nucleotides for a new chain next to each of the old chains, putting each type of nucleotide opposite its proper partner. As the nucleotides were arranged in the proper sequence, they could be bonded together to form the new chain. Thus, separating the two chains of a DNA molecule and using each chain as a template or mold against which to synthesize a new partner for it would result in two complete double-chained molecules identical to the original molecule (Fig. 8.14).

Experimental support for the theory Satisfying as it was, this template theory of DNA replication was pure speculation, unsupported by any experimental evidence, when it was first put forward by Watson and Crick in 1953. Since then, convincing evidence has come from the work of a number of investigators.

In 1957 Arthur Kornberg and his associates at Washington University in St. Louis developed a method for achieving DNA synthesis in a test tube. They extracted from cells of *Escherichia coli* an enzyme complex that can catalyze the synthesis of DNA—a DNA polymerase (literally, an enzyme that creates a DNA polymer) —and combined it with a plentiful supply of the four nucleotides as raw material. These nucleotides had been activated by ATP—that is, a high-energy phosphate group had been attached, to provide the energy for the later reaction steps. The nucleotides also contained a radioactive isotope of carbon (^{14}C). After the experimenters had added some DNA to serve both as a primer—a starting point for the expected reactions—and as a potential template, they incubated the preparation, and found that new DNA containing ^{14}C appeared; the labeled nucleotides had been built into new DNA chains. This synthesis would not occur unless DNA was present at the beginning of the reaction. Kornberg was able to show that the ratio of adenine and thymine to guanine and cytosine in the new DNA was precisely the same as that in the primer DNA. These experiments and others indicated that the newly synthesized DNA was identical to the DNA added to the original mixture—that the primer DNA had indeed functioned as a template. Kornberg received a Nobel Prize for this work.[3]

Kornberg's experiment seemed to demonstrate that DNA synthesis cannot take place in the absence of template DNA, which presumably provides the base-sequence information necessary to guide the synthesis. It also strongly suggested that new DNA is always a copy of the template DNA. But it did not actually provide any evidence that the copy mechanism works in the way proposed by Watson and Crick. More direct support for their model came from an experiment reported in 1958 by Matthew S. Meselson and Franklin W. Stahl, then of the California Institute of Technology.

Meselson and Stahl grew *E. coli* for many generations on a medium whose only nitrogen source was the heavy isotope ^{15}N. Eventually all the DNA in these bacteria contained the heavy isotope instead of the normal isotope ^{14}N. Then the nitrogen source was abruptly changed from ^{15}N to ^{14}N. Cell samples were removed at regular intervals thereafter, and the DNA was extracted from them and subjected to a complicated procedure designed to separate DNA of different densities.

The experiment showed that when cells containing only heavy DNA (DNA in which both chains had only ^{15}N in their purine and pyrimidine bases) were allowed to undergo one division in the ^{14}N medium, the DNA of the new cells was intermediate in density between heavy DNA and light DNA. In other words, the nitrogen in the DNA of the new cells was half ^{15}N and half ^{14}N. This is precisely what would be expected if the two chains of the heavy parental DNA separated and acted as templates for the synthesis of new partners from nucleotides containing only ^{14}N (Fig. 8.15): each new DNA molecule should be composed of one heavy chain from the parent and one light chain newly synthesized, the new molecule thus having intermediate density.

To prove that all the ^{15}N really was in one chain of the intermediate-density DNA and all the ^{14}N in the other chain, Meselson and Stahl subjected the DNA to a treatment that breaks the hydrogen bonds between the bases and separates the chains. Sure enough, this procedure produced heavy chains and light chains; the two isotopes had not been distributed randomly throughout the DNA molecule, but had been localized each in one of the chains, just as would be predicted from the Watson-Crick theory.

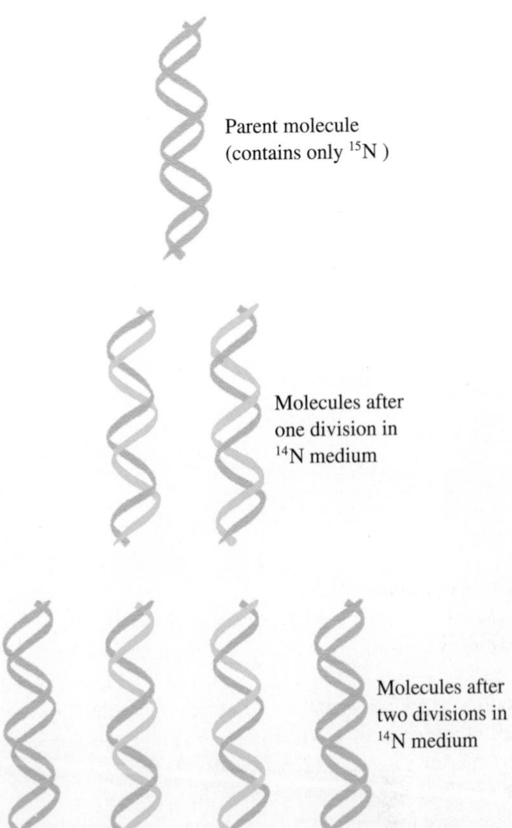

8.15 Results of the Meselson-Stahl experiment The parent DNA molecule (blue chains) contained only heavy nitrogen. After one division, the DNA had an intermediate density, an indication that half the nitrogen in each molecule was heavy and half was light; the two heavy parental chains had separated, and each had acted as the template for synthesis of a complementary light chains (yellow and green). Even after several additional duplications, the two original heavy parental chains remained intact.

MECHANISMS OF DNA REPLICATION

The process of replication is complex and anything but haphazard: hydrogen bonds stabilizing the helical shape and linking the two chains of the DNA molecule must be broken, and the chains separated; complementary nucleotides must be paired with the nucleotides of each existing chain; the

[3] Ironically, we now know that there is more than one kind of DNA polymerase in cells, and that the one Kornberg isolated, now called DNA polymerase I to indicate its priority of discovery, is primarily used for repairing damage to chromosomes and cleaning up loose ends created during replication (see *Exploring Further* pp. 228–229). The second DNA polymerase found is involved in creating the nucleolus (see pp. 126 and 290). The main enzyme complex responsible for replication, DNA polymerase III, was discovered last.

new nucleotides must be covalently linked to form a chain; and so on. Every step is managed by specific enzymes, and takes place quickly and accurately. In *E. coli*, for instance, a complex of several enzymes, collectively known as DNA polymerase III, adds an average of 500 base pairs each second, and there is only one error in every billion pairs copied.

The basic task of DNA replication is the same in procaryotes and eucaryotes: the process in bacteria and the process in humans display many similarities. For example, though the chromosomes differ in shape—bacterial chromosomes are circular while ours are linear—replication begins in both at particular spots in the DNA and proceeds in both directions away from the initiation site; replication in eucaryotes never begins at an end. In both procaryotes and eucaryotes, the individual chains are copied in only one direction: the replication enzymes move along the parental strands from 3′ to 5′, generating a complementary strand running from 5′ to 3′; this means that one chain can be copied continuously but the other must be replicated "backward," in discontinuous segments (as described in the *Exploring Further* box). Both groups of organisms have mechanisms that locate and correct errors; both have special mechanisms that prevent the chains from tangling.

But along with these general similarities come differences in detail. For example, because eucaryotic chromosomes are packaged on protein spools to form nucleosomes (see Fig. 5.3, pp. 126), replication proceeds at a rate of only 50 base pairs per second; it takes time for the DNA to be unwound from the spools and, after a duplicate set of spools has been synthesized, to be rewound. And because eucaryotic chromosomes are much larger than procaryotic chromosomes, replication is initiated at many independent sites on each chromosome simultaneously; otherwise, it might take weeks or even months for a complex eucaryotic cell to divide.

Another major difference between procaryotes and eucaryotes comes in their respective strategies for sorting newly replicated chromosomes into two daughter cells. Procaryotes solve this problem of ***segregation*** by attaching their single chromosome to the cell membrane; after it is replicated, a membrane grows in from between these two anchor points, enclosing one copy in each daughter cell (Fig. 8.16A). For eucaryotes, which typically have 10–20 times as many genes, and two copies of each of their several chromosomes, the problem is more difficult. In these organisms, a special spindle structure lines up each replicated pair along the plane of the cell's coming division. Microtubules run from each pole to one member of each replicated pair, and serve to draw each new cell's complement of chromosomes away from the midline as cell division proceeds (Fig. 8.16B).

Another unique feature of nearly all eucaryotes is sexual reproduction. Most eucaryotes reproduce by means of special sex cells (eggs and sperm, for example), which contain only one copy of each chromosome; fusion of two gametes restores the normal number. Clearly, the pattern of replication or segregation necessary to produce gametes must be different from that used in normal cell division (Fig. 8.16C). Sexual reproduction serves, at least in part, to promote variation between parent and offspring. This variation occurs because the two copies of each chromosome in an individual are not absolutely identical, since one of the chromosomes came from the female gamete—the egg—while the other was contributed by the male gamete. Although each member of a chromosome pair contains the same kinds of genes in the same order, the base sequence of a gene on one chromosome is often slightly different from that of the same gene on the other, and thus may encode a product with slightly different characteristics. Alternative forms of a gene are called ***alleles***. The precise set of alleles in one gamete of an individual usually differs from that in most of its others, and usually differs substantially from the set of alleles in gametes produced by other individuals. The result is that when two gametes from different individuals fuse and begin developing into an organism, the resulting individual is likely to be at least somewhat different from either of its parents, and even from its siblings.

Though there are many differences between procaryotes and eucaryotes, we will in later chapters stress the common features of genetic information flow, focusing first on the relatively well-understood mechanisms used by procaryotes, and then on the modifications seen in eucaryotes. *The Exploring Further* box summarizes the process in bacteria.

REPLICATION IN ORGANELLES

As we saw in Chapter 5, mitochondria and chloroplasts share several characteristics with procaryotes, including (usually) circular chromosomes. The most interesting interpretation of such similarities is that these organelles were once free-living procaryotes that took up symbiotic residence in primitive eucaryotic cells.

The original observation that led to the discovery that organelles have their own genes came in 1909, when Carl Correns reported that in four o'clocks, the bushy, evening-blooming plant *Mirabilis jalapa*, variegation—the appearance of white patches on green leaves—was transmitted through the cytoplasm of the maternal gamete. Correns correctly guessed that chlo-

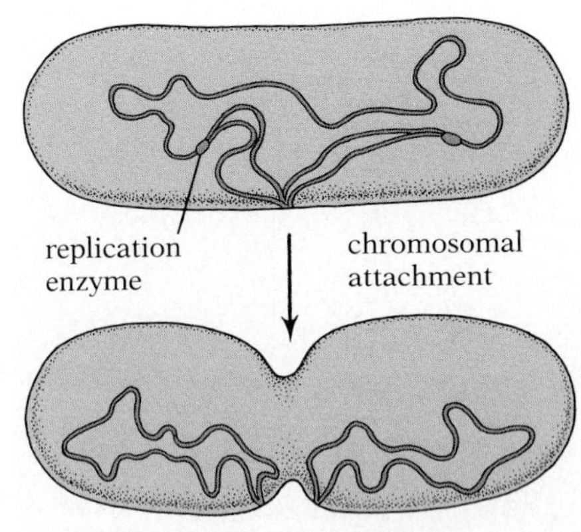

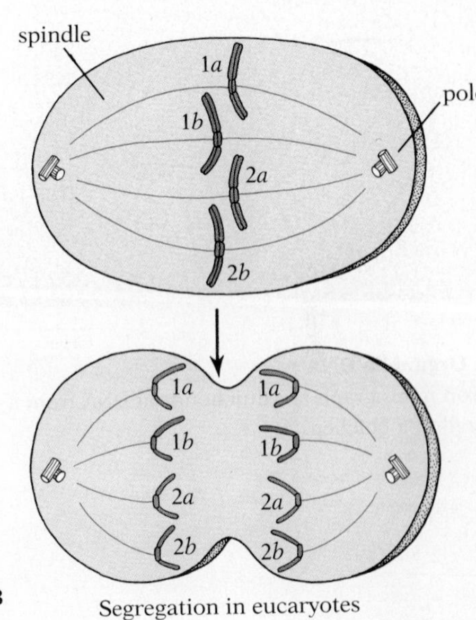

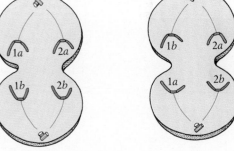

1a 2a 1b 2b 1b 2a 1a 2b

C Gamete production

8.16 Alternative patterns of segregation after DNA replication

(A) In procaryotes, the chromosome remains attached to the cell membrane. After cell replication, the two copies are separated by the growth of the new membrane. (B) In most eucaryotes, there are two copies of each chromosome before replication, and each copy is used to produce a replica of itself; in this example, there was copy *a* and *b* of chromosome 1 and chromosome 2, each of which was then replicated to create the four pairs shown. As cell division proceeds, each pair of replicates becomes aligned along a plane, and one member of each pair is drawn to each end of the cell. (C) In preparation for the production of gametes, each copy of a chromosome is replicated, and then the replicated pairs join into a tetrad. Two rounds of cell division follow, generating four gametes, each with only one of the two chromosomes in each pair.

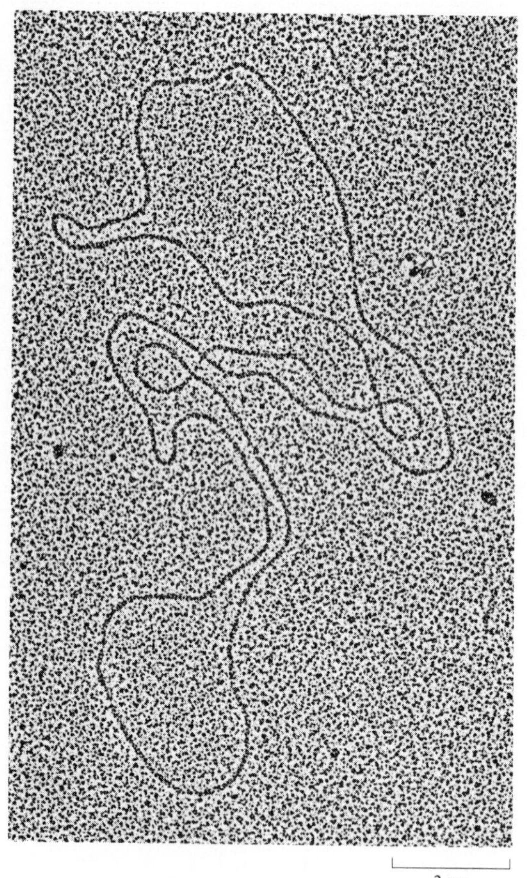

8.17 Organelle DNA
Electron micrograph of mitochondrial DNA from a liver cell of a chicken.

roplasts must contain genes and replicate themselves, and that variegation can occur when one or more of the genes involved in chlorophyll synthesis are defective. Since plants cannot survive without some functioning chloroplasts, the mutations must be carried by some but not all of the chloroplasts supplied by the seed-producing plant. (Pollen grains lack chloroplasts.) During the course of rapid cell division in growing leaves, certain cells, by chance, wind up only with chloroplasts carrying the mutant gene; the cells derived from the first cell that lacks at least one chlorophyll-producing chloroplast appear as an elongated streak on the leaf.

The first evidence of mitochondrial genes did not come until 1938, when T.M. Sonneborn discovered that some strains of *Paramecium aurelia* carry a cytoplasmic gene which produces a poison (originally called the kappa factor) that kills other strains. It subsequently became clear that the poison is produced by a mitochondrial gene, and that mitochondria, like chloroplasts, both replicate themselves and are inherited from the female (egg-producing) parent in nearly all species.

We now know that the DNA replication and the division of mitochondria and chloroplasts occur out of phase with chromosome replication in the nucleus, though the pace of cell division limits organelle division in some way. The DNA is almost always circular (Fig. 8.17), an exception being the linear mitochondrial DNA of some fungi and protozoans. Each organelle contains several copies of its DNA, and this DNA is more similar to procaryotic DNA than to the nuclear DNA of eucaryotic cells. It lacks the histone nucleosome cores characteristic of nuclear DNA, for example, and is attached to the inner membrane of the organelle much as the procaryotic chromosome is attached to the cell membrane. However, it has considerably fewer base pairs than bacterial DNA. For instance, the *E. coli* chromosome encodes about 3000 products, while the genes of a typical animal mitochondrion encode about 40. Since these are insufficient to carry out organelle synthesis and operation (even in bacteria at least 90 gene products are required just for replication, transcription, and translation), it appears that in the course of evolution most of the genes necessary for organelle function have come to reside in the cell's nucleus.

DNA REPAIR

The precise replication of DNA is essential for normal cell function. ***Mutations***—random changes in genes—are far more likely to disrupt a pathway, by destroying the delicate architecture of an essential enzyme, for instance, than to improve it. Since the genetic instructions for the synthesis of the thousands of structural proteins, enzymes, regulatory proteins, and so on are each hundreds or thousands of bases long, an error rate as low as one in a thousand bases, though it may seem insignificant, is far too great. Not surprisingly, then, special enzymes have evolved in both procaryotes and eucaryotes to detect and repair mutations. These enzymes keep uncorrected replication errors at a very low level, and other enzymes locate and repair most of the damage that occurs to the DNA *between* bouts of replication. Like the other enzymes we have discussed, repair enzymes attach themselves to particular substrates (in this instance faulty or damaged areas of the DNA) that display the patterns of spacing and polar charges to which their active sites bind; they loosen specific bonds, and catalyze the formation of new ones.

Repair during replication The initial error rate of the replication enzymes of both procaryotes and eucaryotes is about 3 in 10^5 pairs. (This value, like much of what is known about how DNA polymerases work, is obtained from studies of polymerase complexes with one or more inactive component enzymes. From the change in the operation of the complex, the function of the inactive enzyme can often be deduced. In this case, replication proceeds when the enzymes under study are inactive, but most errors are not corrected.) If the errors produced at this rate were left uncorrected, the result would be a mistake in roughly 3 percent of each cell's proteins—perhaps 1000 changed proteins in every human cell after each replication. Fortunately, the DNA polymerase complex includes one or more enzymes that successively "proofread" each base, clipping out mistakes. Other enzymes in the polymerase complex then substitute a rematched base for the excised unit; the complex then moves on without checking a second time. This second pass reduces the chance of an error to roughly 1 in 10^9. For humans, with our 50,000 functional genes, each of which has an average of about 1500 bases on each of its two strands, this corresponds to an error in a gene somewhere in the chromosomes once in every ten cell divisions. As we will see, this rate is usually dwarfed by other sources of mutation.

Repair of other mutations The integrity of the genetic message is also threatened by alterations in base sequences induced by heat, radiation, and various chemical agents. The rate at which these mutations occur is astoundingly high: thermal energy alone, for instance, breaks the bonds between roughly 5000 purines (adenine and guanine) and their deoxyribose backbones in each human cell every day. Cytosine is chemically converted to uracil (a nucleotide normally found only in RNA, and misread by DNA replication and transcription enzymes) at a rate of about 100 per cell per day. Ultraviolet radiation from sunlight fuses together adjacent thymines at a high rate in exposed epidermal cells (see Fig. 9.18, p. 251). And yet, because of the continuous operation of repair enzymes, the rate at which mutations accumulate in cells, on average, is even lower than that of uncorrected errors in replication.

The strategy for repairing chromosomes is basically the same for the many sorts of mutations as for replication errors: enzymes locate and bind to the faulty sequences and clip out the flaws, and the intact complementary strand guides repair. A remarkable array of specific enzymes is involved; for instance, the chromosomes are scanned for chemically altered bases by fully 20 different enzymes, each specific for a particular class of problems. Another five or so enzymes are specific for faulty covalent bonding between a base and some other chemical, or between adjacent bases on one strand; the fusion of thymines induced by ultraviolet radiation is the most common error of this type. Other enzymes bind at the sites of missing bases, like the purines that are easily lost to thermal energy. In all, some 50 enzymes locate and correct errors.

In spite of this elaborate repair system, some mutations manage to survive to generate phenomena ranging from blue eyes (which result from a

Exploring Further

REPLICATION OF THE *E. COLI* CHROMOSOME

The process of replication in *E. coli* illustrates the complex series of steps necessary even in a relatively simple organism to make an accurate copy of a chromosome. Highly specific enzymes underlie each event. Like all enzymes they recognize specific reactants by their complementary shape and pattern of polar charges. And as enzymes often do, they work together to rearrange the bonds in various reactants and produce a final product—in this case, a complete copy of the circular bacterial chromosome. Enzymes in complex pathways work nearly simultaneously, but for clarity we will discuss the events of this process in sequence.

1. In *E. coli,* the process begins when a protein, DNA B, recognizes an initiation site on the chromosome by its particular sequence of bases and binds to it (top figure).

2. Next, molecules of the enzyme group known as DNA gyrases (or topoisomerases, enzymes that change the shape but not the chemistry of other molecules) begin to relax the supercoiling of the chromosome on each side of the DNA B protein.

3. As the two DNA gyrase molecules move away from the initiation site, two molecules of ***rep*** enzyme unzip the double helix, using energy from ATP to break the hydrogen bonds that hold the bases together.

4. Single-strand binding proteins (SSB) then form a scaffolding, which holds the two strands apart and prevents them from rebinding to each other spontaneously (A, bottom figure).

5. The last of the five steps that must occur before the actual replication of DNA can begin involves a primer enzyme (the primase). Because the replication enzyme can only add bases to the end of an incomplete strand as it reads the template strand, there must be at least a few new bases in place. The primase binds to the initiation site and adds a complementary sequence about ten bases long. Oddly enough, the primer is made out of RNA, which must later be replaced by DNA; we will see how this substitution is accomplished presently.

6. Next, ***DNA polymerase III***, a complex consisting of several proteins bound together, begins to replicate one of the two DNA strands by binding to it and adding complementary bases, creating, as its name implies, a polymer of DNA nucleotides. A complex is necessary because DNA polymerase must catalyze several different reaction steps and must be able to use four different nucleotides as reactants, depending on what it "reads" from the strand it is copying (the template strand). Presumably there are four active sites, one each for adenine, cytosine, guanine, and thymine. Once the complex has read the template strand and brought the complementary nucleotide into place, it catalyzes the binding of the nucleotide to the growing complementary strand. The nucleotides that are added have been activated by the addition of a high-energy phosphate group from ATP, which provides the energy for this step.

Because of its enzymatic specificity, DNA polymerase III can add nucleotides only to the 3′ end of a nucleotide strand, the end without the phosphate group. This creates a serious problem: since the two strands of the double helix have opposite polarities, with one running from 5′ to 3′ and the other from 3′ to 5′ (Fig. 8.10), they must be copied in opposite directions. Only one strand can be copied by DNA polymerase III, following along behind the rep enzyme as it unzips the DNA (A, bottom figure); the other strand must somehow be copied "backward." The DNA formed by the DNA polymerase that follows the rep enzyme is known as the ***leading strand***, while the DNA synthesized backward is known as the ***lagging strand***. The latter is formed bit by bit in a looping "backstitch" pattern.

7. Backstitching begins when the primer enzyme synthesizes short lengths of complementary RNA nucleotides at intervals along the single-stranded DNA (A).

8. The short segments created in Step 5 provide the 3′ free end for the DNA polymerase III; from here the polymerase works backward, copying the strand, until it reaches the preceding RNA primer segment (B). One strand, then, is copied continuously while the other is copied in sections, known as Okazaki fragments, that are 1000–2000 bases long and are flanked by the RNA primers.

9. Now a series of enzymes must patch the fragments together into a continuous strand. The repair complex ***DNA polymerase I*** removes the anomalous 10-base RNA primer segments and replaces then with DNA (C).

10. Finally, the fragments are welded together by ***DNA ligase*** (D), and the new strand is finished.

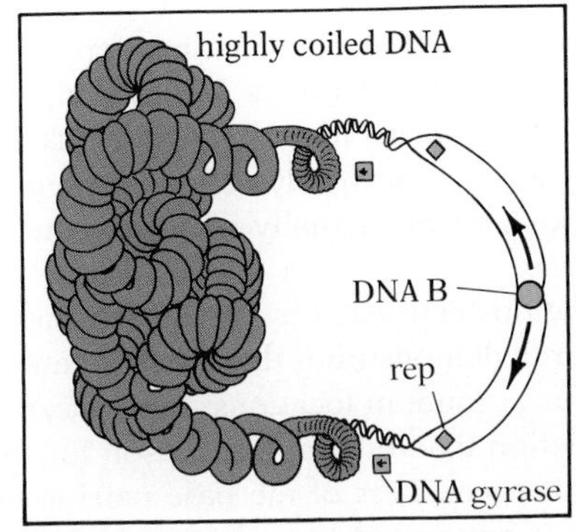

Step	*Symbol*	*Substance*	*Function*
1	○	DNAB	finds and marks initiation site
2	□	DNA gyrase	relaxes coiling
3	◇	rep	separates DNA strands
4	⬭	SSB	hold strands apart
5,7	○	RNA primase	primes lagging strand for replication
6,8	◁	DNA polymerase III	synthesizes complementary DNA
9	◊	DNA polymerase I	erases primer and replaces with DNA
10	▽	DNA ligase	welds gaps

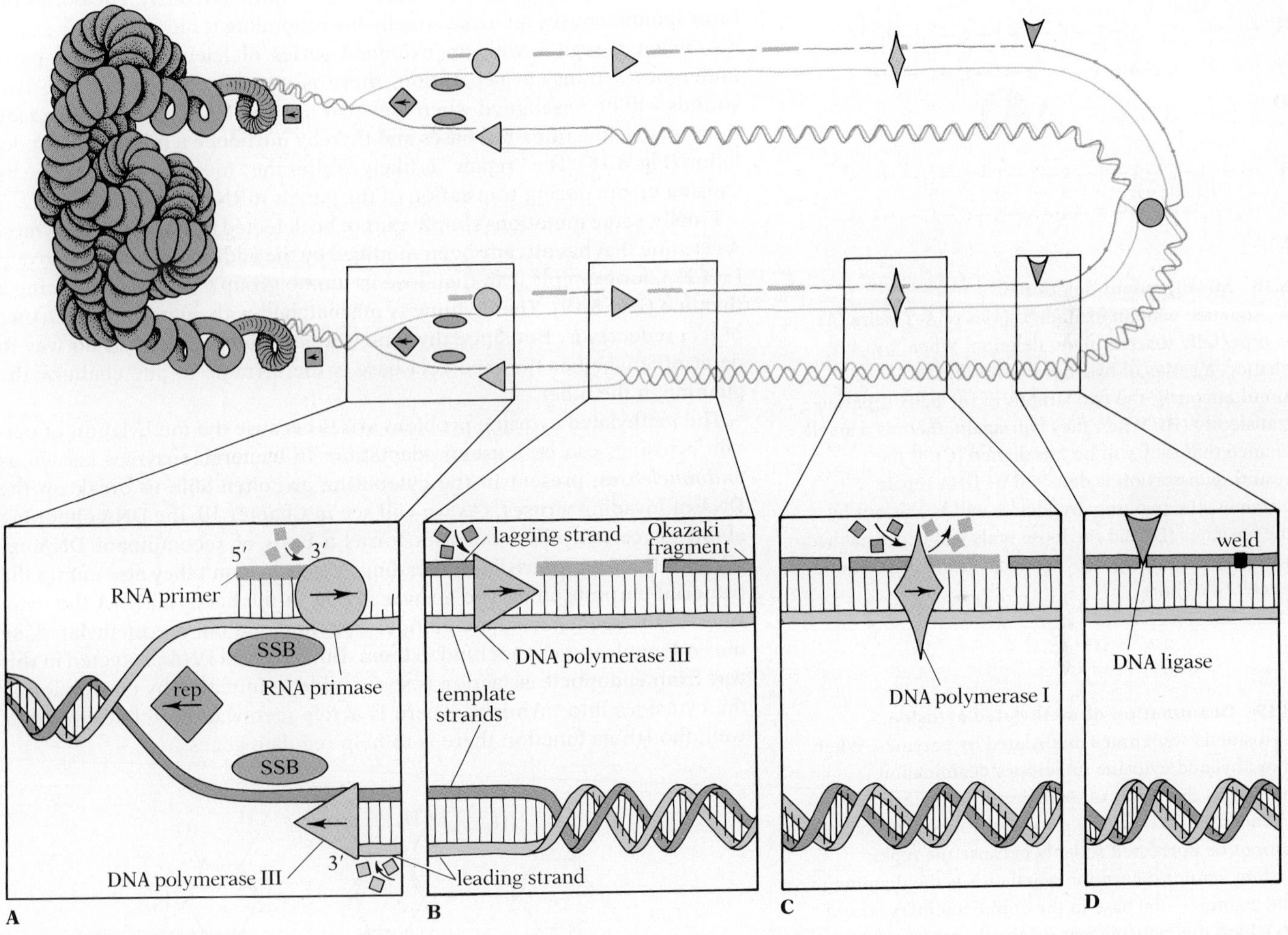

5′
3′
RNA primer
SSB
rep
RNA primase
SSB
3′
DNA polymerase III
A

lagging strand
Okazaki fragment
DNA polymerase III
template strands
leading strand
B

DNA polymerase I
C

weld
DNA ligase
D

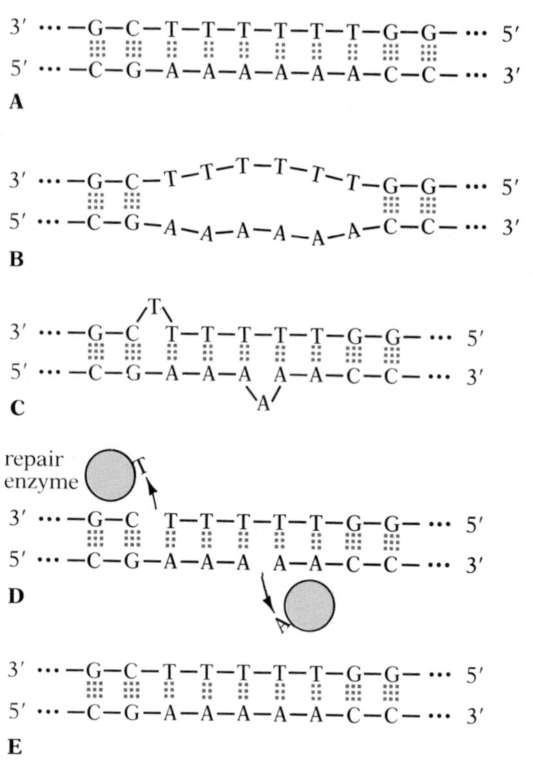

defective pigment gene) to genetic diseases. How does that happen? Most obviously, no enzymatic system is perfect: some errors are missed, and others are repaired incorrectly. Then, too, when a mutation occurs just before or during replication, there may not be time for detection and repair. Still other mutations are not overlooked, but are actually created by the repair system.

One well-understood class of mutations generated by repair enzymes is ***misalignment deletion***. Recent research demonstrates that small deletions (loss of a few base pairs) do not occur at random locations; some base sequences are more susceptible to deletion than others. The reason for this differential susceptibility is the relative weakness of the base pairing between adenine and thymine: they are connected by only two hydrogen bonds, whereas cytosine and guanine are connected by three. The hydrogen bonds are continually breaking and re-forming, and it frequently happens, particularly in regions rich in adenine–thymine pairs, that by chance all the bonds in a small segment of DNA are broken simultaneously. The bonds re-form spontaneously, but occasionally the rebonding is incorrect. For example, when a region with an extended series of adenine–thymine pairs undergoes a transient separation, there is a small chance that the two strands will be misaligned when they pair again. DNA repair enzymes may then remove the unpaired bases and thereby introduce a misalignment deletion (Fig. 8.18). The "repair" is likely to alter the "meaning" of the gene by causing errors during translation of the gene's mRNA.

Finally, some mutations simply cannot be detected by the repair enzymes. A cytosine that has already been modified by the addition of a methyl group ($—CH_3$), for example, can then lose its amino group ($—NH_2$), producing a thymine (Fig. 8.19). The thymine is mismatched with guanine, the partner of its predecessor. But since thymine is a normal base, there is no way to determine whether the incorrect base is the thymine in one chain or the guanine in the other.

The methylated cytosine problem arises because the methylation of certain cytosines can be a useful adaptation. In bacteria, enzymes known as ***endonucleases***, present in the cytoplasm, are often able to break up the DNA of invading viruses. (As we will see in Chapter 10, the DNA-chopping ability of certain endonucleases forms a basis of recombinant DNA research.) If the enzymes digest invading DNA, why don't they also cut up the bacterial chromosome? The answer is that in the bacterial DNA the cytosines in the sequences the endonucleases find and cut are methylated, so the endonucleases cannot bind to them. But bacterial DNA protected in this way from endonuclease action is susceptible to mutation by conversion of the cytosines into thymines. There is active methylation in eucaryotes as well, though its function there is to help regulate genes.

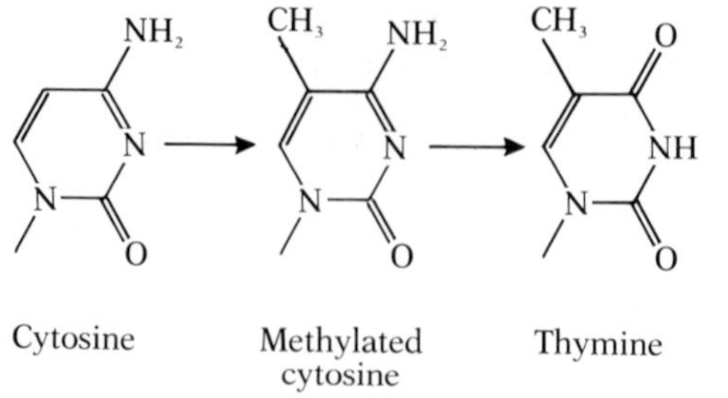

3′ ··· —G—C—T—T—T—T—T—T—G—G— ··· 5′
5′ ··· —C—G—A—A—A—A—A—A—C—C— ··· 3′
A

3′ ··· —G—C—T—T—T—T—T—T—G—G— ··· 5′
5′ ··· —C—G—A—A—A—A—A—A—C—C— ··· 3′
B

3′ ··· —G—C T—T—T—T—T—G—G— ··· 5′ (unpaired T)
5′ ··· —C—G—A—A—A A—A—C—C— ··· 3′ (unpaired A)
C

repair enzyme
3′ ··· —G—C T—T—T—T—T—G—G— ··· 5′
5′ ··· —C—G—A—A—A A—A—C—C— ··· 3′
D

3′ ··· —G—C—T—T—T—T—T—G—G— ··· 5′
5′ ··· —C—G—A—A—A—A—A—C—C— ··· 3′
E

8.18 Misalignment deletion
A sequence with an extended series of A–T pairs (A) is especially susceptible to deletion. When, by chance, a series of hydrogen bonds breaks simultaneously, the two strands of the helix separate transiently (B). When they pair again, there is a small chance that they will be misaligned (C). If the resulting distortion is detected by DNA repair enzymes, the two unpaired bases will be cut out by the enzymes (D) and the loose ends will be reattached with DNA ligase (E), leaving each strand of DNA one nucleotide short.

Cytosine → Methylated cytosine → Thymine

8.19 Deamination of methylated cytosine
Cytosine is sometimes methylated by enzymes. When a methylated cytosine undergoes deamination, usually as the result of oxidation by a mutagenic chemical, the result is a normal thymine. This change cannot be corrected reliably because the repair system cannot determine whether it is the thymine or the guanine—the base in the complementary strand to which the cytosine was originally paired—that is incorrect.

STUDY QUESTIONS

1. Compare and contrast ATP and the nucleotide having adenine for its nitrogenous base (i.e. adenosine monophosphate, or AMP). (pp. 163, 218)

2. Bacteria have only one copy of each gene, whereas most eucaryotes have two. If most serious mutations destroy the activity of the product encoded by a gene, what effect does having two copies of a gene have on the chance of suffering a complete loss of a gene?

3. The bacterial chromosome has about four million base pairs. Assuming all bases occur with equal frequency, calculate the approximate total energy of the hydrogen bonds holding the two strands together. (p. 39)

4. Assuming one ATP is needed to prime each of the eight million bases added during *E. coli* replication, how many glucose molecules must be metabolized to fuel a single such cell division, ignoring the energy needed for the other steps? (p. 182)

5. In the 1958 experiment of Meselson and Stahl, what pattern of DNA densities would be observed halfway through the first replication? halfway through the second replication? (p. 223)

CONCEPTS FOR REVIEW

- Pattern of information flow
  - Replication
  - Translation and transcription
- Structure of DNA
  - Bases
  - Molecular backbone
  - Complementary binding
- Replication
  - Steps, and roles of enzymes
  - Proofreading
- Repair

SUGGESTED READING

BAUER, W. R., F. H. C. CRICK, and J. H. WHITE, 1980. Supercoiled DNA, *Scientific American* 243 (1). (Offprint 1474) *On how the genetic material is compacted, so that DNA a thousand times the length of a bacterium can be packed into the space available, without tangling, and leaving most of the cell volume free for the cytoplasm.*

CRICK, F. H. C., 1954. The structure of the hereditary material, *Scientific American* 292 (4). (Offprint 5) *On the discovery of the structure of DNA.*

HOWARD-FLANDERS, P., 1981. Inducible repair of DNA, *Scientific American* 245 (5). (Offprint 1503) *On how cells recognize when the DNA has been damaged, how they switch on genes for repair enzymes, and how the enzymes work.*

RADMAN, M., and R. WAGNER, 1988. The high fidelity of DNA replication, *Scientific American* 259 (2). *On how the proofreading component of DNA polymerase works.*

UPTON, A. C., 1982. The biological effects of low-level ionizing radiation, *Scientific American* 246 (2). (Offprint 1509) *A superb summary of how DNA is damaged by radiation.*

WANG, J. C., 1982. DNA topoisomerases, *Scientific American* 247 (1). (Offprint 1520) *On the enzymes responsible for untangling DNA during replication.*

WATSON, J. D., 1980. *The Double Helix.* A Norton Critical Edition, ed. Gunther S. Stent, W. W. Norton, New York. *A fascinating account of the elucidation of the structure of DNA by one of the two discoverers. Watson spares neither himself nor others in giving a rare behind-the-scenes look at the dynamics of research in a very competitive field. This edition of the original 1968 book includes articles, relevant to the discovery, by other scientists.*

WATSON, J. D., N. H. HOPKINS, J. W. ROBERTS, J. A. STEITZ, and A. M. WEINER, 1987. *Molecular Biology of the Gene*, 4th ed. Benjamin Cummings, Menlo Park, Calif. *A particularly well-written text on molecular genetics. Makes even difficult topics easy to understand.*

YUAN, R., and D. L. HAMILTON, 1982. Restriction and modification of DNA by a complex protein, *American Scientist* 70, 61–69. *On how some endonucleases can cut DNA at a specific site if it is fully unmethylated, or finish methylating (and thereby protect) the same site if it is partially methylated.*

Chapter 9

TRANSCRIPTION AND TRANSLATION

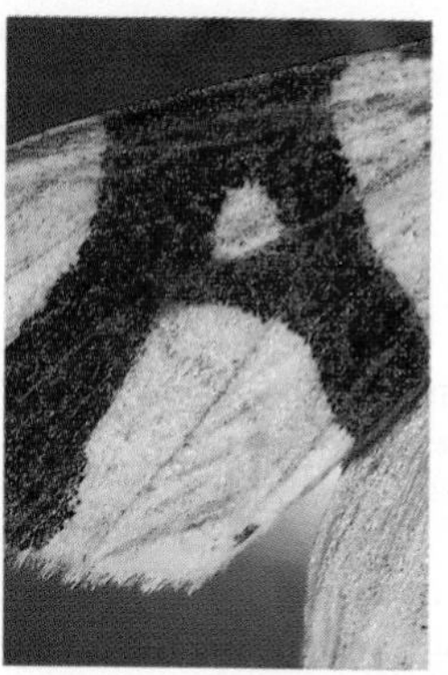

s we saw in the last chapter, a complex set of enzymes makes new copies of the DNA in chromosomes and corrects most errors that arise. But the all-important sequence of bases in the DNA must not only be replicated and repaired, it must be used to produce both the proteins that form cellular structures and the enzymes that direct cellular metabolism. In this chapter we will look at how information is actually encoded in DNA, how the information is transcribed into ribonucleic acid (RNA), and how some of this RNA is used directly while the rest—messenger RNA—is translated into protein. Finally we will look at some of the consequences of transcription and translation errors.

TRANSCRIPTION

With only a few exceptions, every cell has a full set of chromosomes—a set replicated and repaired and passed on at every cell division. But at any given moment, only a small proportion of the thousands of genes in the cell are active. Which genes are needed depends not only on what the cell is doing—dividing, growing, resting, moving—but on the cell's environment and, if it is part of a complex multicellular organism, on its specialty. The small fraction of the information in the DNA that is necessary at any particular time is read out by enzymes very selectively.

In this chapter we will look at how the enzymes recognize the active genes and ignore the rest, and how they know where genes begin and end. (Replication, which is an all-or-none process, does not make any distinction between genes.) In a later chapter we will take up the important question of ***gene expression***: how genes are activated and inactivated according to the needs of the cell.

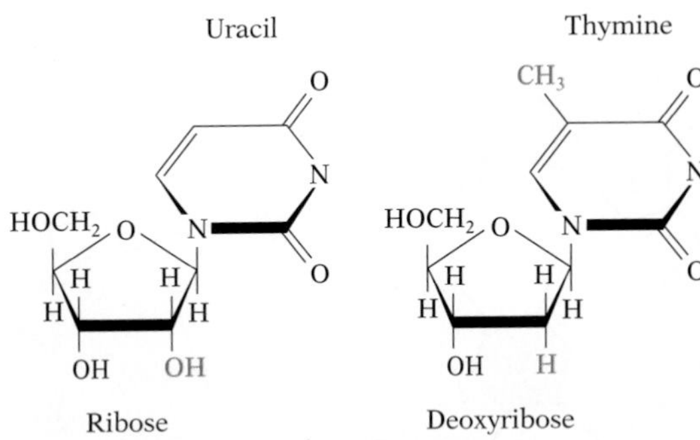

9.1 Uracil in RNA and thymine in DNA compared
The five-carbon sugars of RNA and DNA differ only at the site shown in color, where deoxyribose lacks an oxygen atom that is present in ribose. Uracil differs from thymine only in lacking a methyl group (CH_3).

MESSENGER RNA

The first question to arise when the flow of information from DNA to protein began to be studied in the 1950s was whether proteins are synthesized directly off the DNA through the intervention of appropriate enzymes, or indirectly by means of some intermediary substance. Since in eucaryotes nearly all the DNA is found in the nucleus, while the process of protein synthesis was known to take place in the cytoplasm, most researchers suspected that a molecular middleman must exist, but the identity of this substance was unknown.

In the early 1940s several molecular biologists had shown that cells in tissues such as the vertebrate pancreas, where protein synthesis is particularly active, contain large amounts of RNA. Since this nucleic acid is present in only limited quantities in cells that do not produce protein secretions, such as those in muscle and kidney, there seemed to be a strong correlation between protein synthesis and RNA. Moreover, it had long been known that RNA, unlike DNA, occurs in the cytoplasm as well as in the nucleus. Subsequent radioactive-tracer experiments demonstrated that RNA is synthesized in the nucleus and moves from the nucleus into the cytoplasm. All these lines of evidence suggested that RNA might be the chemical messenger between the DNA of the nucleus and the protein-synthesizing cytoplasmic ribosomes.

Though RNA and DNA are very similar compounds, they differ in three important ways, as we saw in Chapter 3: (1) The sugar in RNA is ribose, whereas that in DNA is deoxyribose (Fig. 9.1). (2) RNA has uracil where DNA has thymine (Fig. 9.1).[1] (3) RNA is ordinarily single-stranded, whereas DNA is usually double-stranded.

These differences aside, it was obvious nevertheless that DNA could easily act as a template for the production of RNA. We now know that the synthesis proceeds in essentially the same way as that of new DNA: the two strands of a DNA molecule uncouple and RNA is synthesized along one of the DNA strands by RNA polymerase (Fig. 9.2). For every adenine in the DNA template, a uracil ribonucleotide, rather than a thymine, is added to the growing RNA strand; for every thymine in the DNA, an adenine ribonucleotide is added; for every guanine, a cytosine; and for every cytosine, a

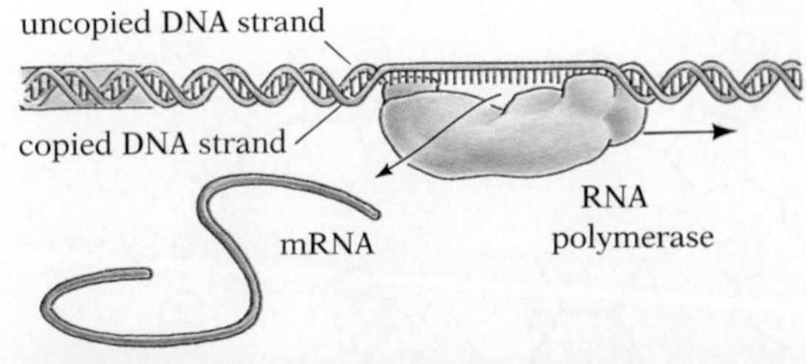

A

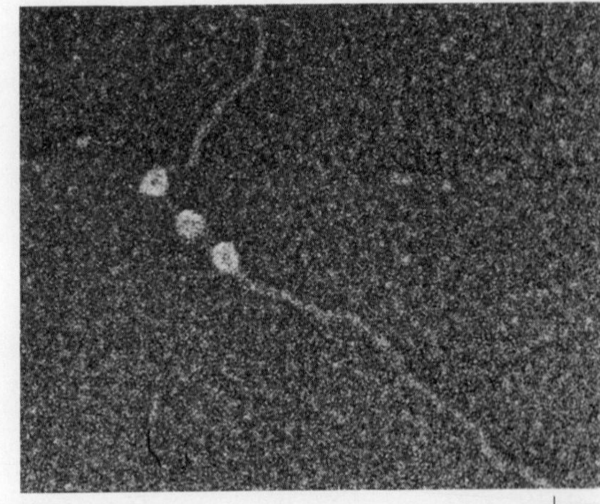

B

9.2 Transcription of a gene
(A) As the RNA polymerase complex moves along the DNA, it catalyzes transcription of only one of the two DNA strands. (B) An electron micrograph showing three polymerases transcribing a gene in *E. coli*.

[1] This use of uracil in place of thymine enables the active site of a wide variety of enzymes to distinguish DNA from RNA, and so be selective in their activity. DNA polymerase, for instance, does not attempt to replicate mRNAs, nor does DNA polymerase try to transcribe messengers, nor can enzymes in the cytoplasm that digest the DNA of invading viruses chew up the cell's own RNAs.

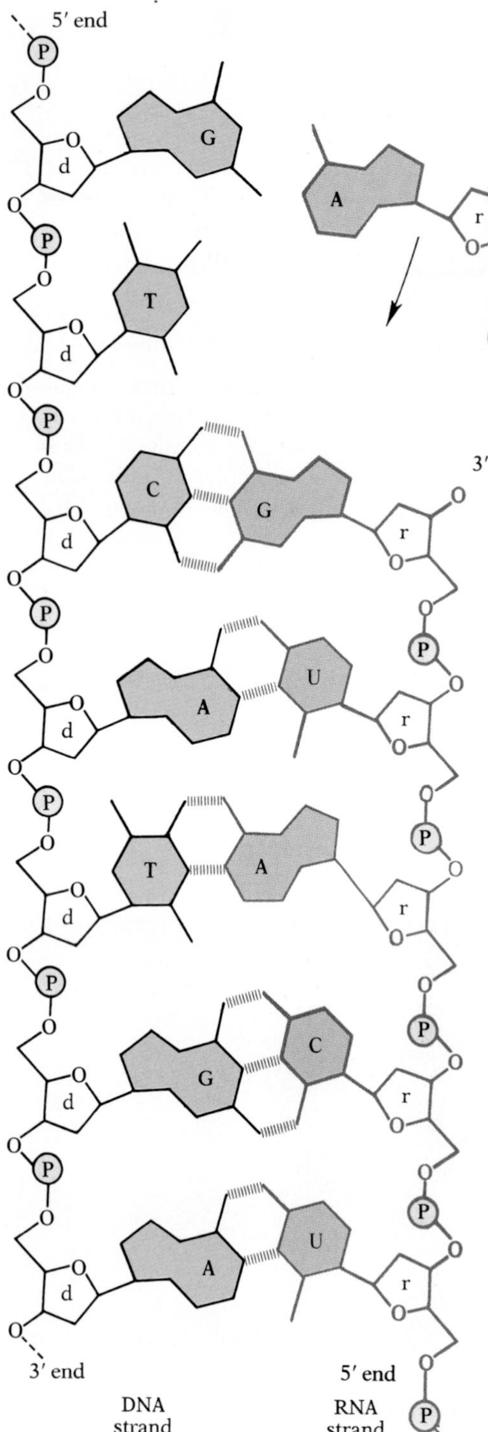

guanine (Fig. 9.3). In short, the synthesis of RNA, in a process now called ***transcription***, operates exactly like replication; the resulting strand of RNA, which is complementary to the transcribed DNA, acts as the intermediary.

We now know that there are three major types of RNA: ***messenger RNA (mRNA)***—the type just discussed—carries the information necessary to specify the sequence of amino acids in a protein to the ribosomes where proteins are synthesized; ***ribosomal RNA (rRNA)*** forms part of the ribosomes themselves; and ***transfer RNA (tRNA)*** brings amino acids to the ribosomes during protein synthesis.

MECHANISMS OF TRANSCRIPTION

Transcription of DNA into RNA is accomplished by a six-protein enzyme complex—***RNA polymerase***—which both binds to the DNA and opens up the helix. Like DNA polymerase, the transcription complex moves from the 3′ end to the 5′ end of the DNA, and synthesizes a strand of the opposite polarity (5′ to 3′), in this case a strand composed of ribonucleotides. In procaryotes, a single kind of RNA polymerase is responsible for all RNA synthesis. In eucaryotes, RNA polymerase II is responsible for the synthesis of mRNA. Two other RNA polymerases synthesize ribosomal RNA, transfer RNA, and other RNAs that serve structural and, in some cases, enzymatic roles; they work in much the same way.

Transcription in procaryotes The strategy just described raises an important question: how does RNA polymerase know where to start and stop synthesizing the messenger? The answer is that specific control sequences marking the beginning and end of a gene are embedded in the strands of DNA. The first pair of these is called the ***promoter;*** active sites of the RNA polymerase bind specifically to regions with the base sequences of the promoter. The promoter in *E. coli*, reading from 5′ to 3′ on the complementary strand, usually begins with TTGACA or a very similar sequence.[2] This is followed by a sequence of roughly 17 bases with another function; then comes TATATT or a very similar sequence (Fig. 9.4A). The polymerase recognizes these two regions in the DNA and is large enough to bind simultaneously to both. The sequences shown in Figure 9.4 are called ***consensus sequences***, since each base shown is the one most commonly found at the location in

[2] Though the RNA polymerase copies the template strand of the DNA, the custom is to refer to the sequences in question as they exist on the complementary strand. This method of reference is convenient because the RNA sequences produced in transcription are identical to those of the corresponding regions of the complementary DNA strand—except, of course, that U is substituted for T.

9.3 The synthesis of RNA by transcription of a DNA template
The sugar in RNA (ribose, *r*), is slightly different from that in DNA (deoxyribose, *d*), and uracil (U) takes the place of the thymine in DNA. Transcription is accomplished by the enzyme complex RNA polymerase (not shown).

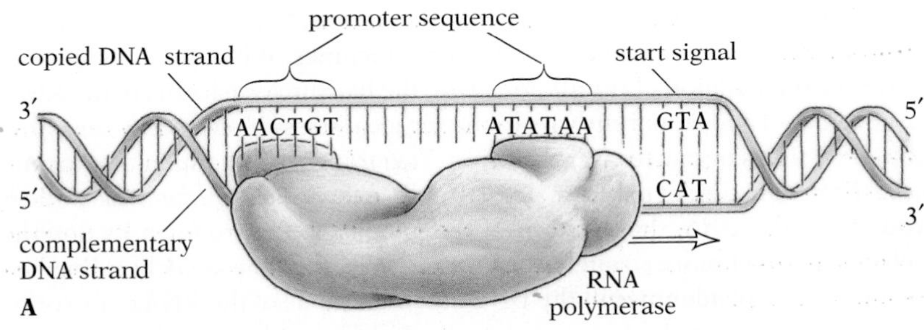

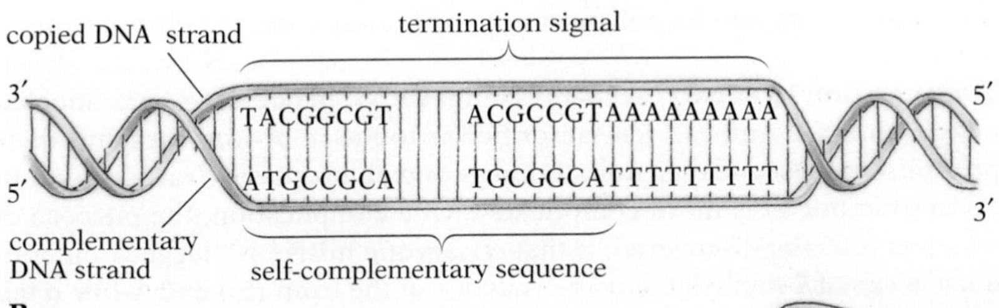

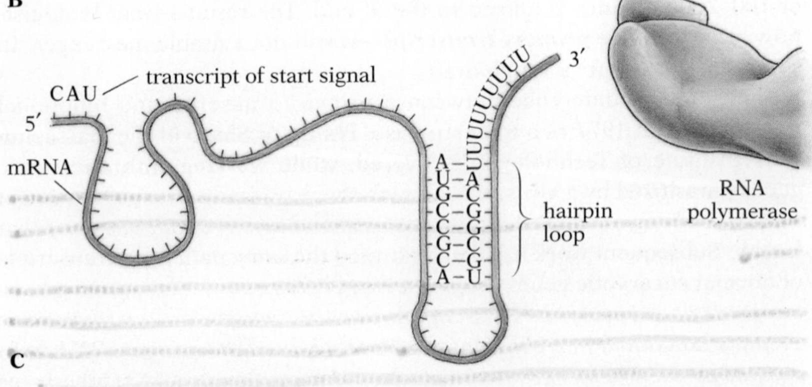

9.4 Transcription signal in *E. coli* DNA
(A) RNA polymerase binds to a region of DNA with the sequences shown. Once the polymerase is bound, transcription begins with the GTA signal (corresponding to the CAT sequence on the complementary strand) and proceeds until a termination signal is encountered. (B) The termination signal consists of a "self-complementary" sequence—a sequence that can bind to itself—and four to eight adenines (corresponding to thymines on the complementary strand). (C) At this point, the tail of the mRNA forms a short double-helix hairpin, and the polymerase stops synthesizing. Note that RNA polymerase inserts a uracil wherever DNA polymerase would add a thymine.

question in procaryotes. (Much less is known about eucaryotic promoters, but the sequence TATA appears to be important in them as well.) It should be noted that the promoter sequences of most genes differ in one way or another from the consensus sequences; more than a hundred variations have been discovered in *E. coli*, for example. Minor variations in the promoter sequences cause the polymerase to bind less strongly and less often to some of them, so some genes are transcribed less frequently than others. We will look at more sophisticated mechanisms for controlling rates of transcription in Chapter 11.

Once bound, the polymerase does not begin synthesizing mRNA immediately. Instead, it must find the start signal—often CAT—located about seven bases beyond the binding point, toward the 3′ end of the complementary strand. In most procaryotes synthesis of mRNA continues until the polymerase encounters a termination signal. The termination signal has two components. First, there is a region with a base sequence that allows the corresponding bases in the tail of the mRNA to pair off and bind together to

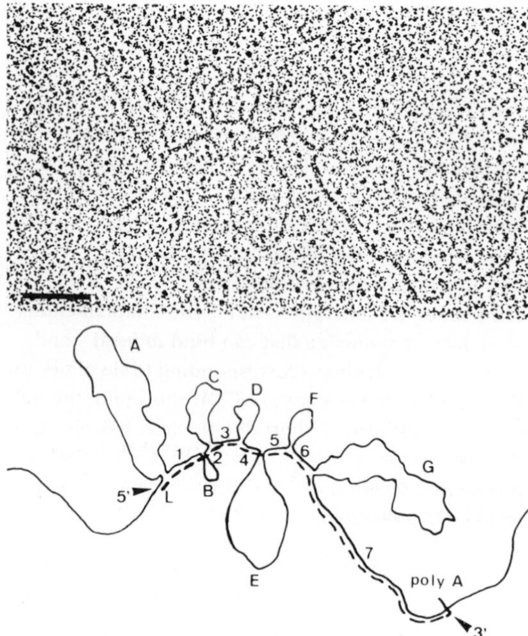

9.5 Visible evidence of introns
A gene for an egg protein, ovalbumin, was denatured to create single-stranded DNA, and then mixed with mature mRNA transcribed from this gene. Because the mRNA sequence is complementary to the DNA strand from which it is transcribed, the two molecules are able to form a double-stranded hybrid. Because the introns in the primary transcript have been removed by editing, the intron regions in the DNA form unmatched loops, revealing that this gene has six introns.

form a small loop, known as a ***hairpin loop*** (Fig. 9.4C). This is followed on the copied strand by a run of four to eight adenines. When the polymerase moves into the region of the adenine run, the hairpin loop forms in the RNA just produced and, apparently from the physical stress it puts on the enzyme complex, slows or pauses transcription. Two things now happen that terminate the production of the RNA. First, because the loop sequence has pulled away from the DNA, that portion of the gene is able to re-form its double helix, adding yet more strain to the complex. Second, the weak bonding between the run of adenines in the DNA and the uracils of the RNA copy (only two hydrogen bonds between each base pair) is unable to anchor the RNA to the gene for long; the pause and the stress together conspire to allow the transcribed copy and its polymerase to drift away from the chromosome.

Transcription in eucaryotes—messenger RNA processing The mechanism of mRNA synthesis just described is found in procaryotes and their presumptive descendants, mitochondria and chloroplasts. Transcription in eucaryotic nuclei is more complicated. One complication, the purpose of which is yet to be discovered, is that eucaryotic mRNA is "tagged" on both ends: a cap of 7-methylguanosine is added at the front (5′) end, while a tail of 100–200 adenines is affixed to the 3′ end. The result—what biologists now refer to as the ***primary transcript***—is still not a usable messenger. In some sense it is only a rough draft.

The dramatic difference between a primary transcript and functional mRNA came in 1977 as a total surprise: Phillip A. Sharp of the Massachusetts Institute of Technology discovered, while working with eucaryotic genes parasitized by a virus, that though the primary transcripts are about 6000 bases long, the mRNA actually transcribed is only about one-third that length. Subsequent work has demonstrated the same pattern in transcripts of normal eucaryotic genes: large specific regions *within* the primary transcript of most eucaryotic messengers must be removed in the nucleus to create a functional mRNA molecule. The regions of the primary RNA transcript that survive this processing and operate during protein synthesis (as well as the parts of the gene that gave rise to these sections) are called ***exons*** (because they are expressed), while the intervening sequences of the primary transcript, which are removed in the nucleus (and the corresponding regions of the gene), are referred to as ***introns*** (Fig. 9.5). Experiments have shown that despite their early removal, many introns are necessary to the functioning of RNA: most mRNA transcribed from artificially manufactured genes that lack introns fails to get into the cytoplasm, while mRNA from genes with some introns intact is often processed correctly and slips through the nuclear pores to the cytoplasm.

Intron removal is a formidable task (Fig. 9.6). Some genes have as many as 50 introns, and a mistake of even a single base in the excision process can render the mRNA useless. The precise beginning and end of introns on the primary transcripts are marked by signals so that they can be recognized and removed.[3] The intron-boundary signals are recognized by a short bit of RNA found in curious RNA/protein complexes known as the small nuclear

[3] The sequence at the boundary of the intron is AG-GUAAGU (the hyphen marks the exon-intron boundary); at the end of the intron it is CAG-G.

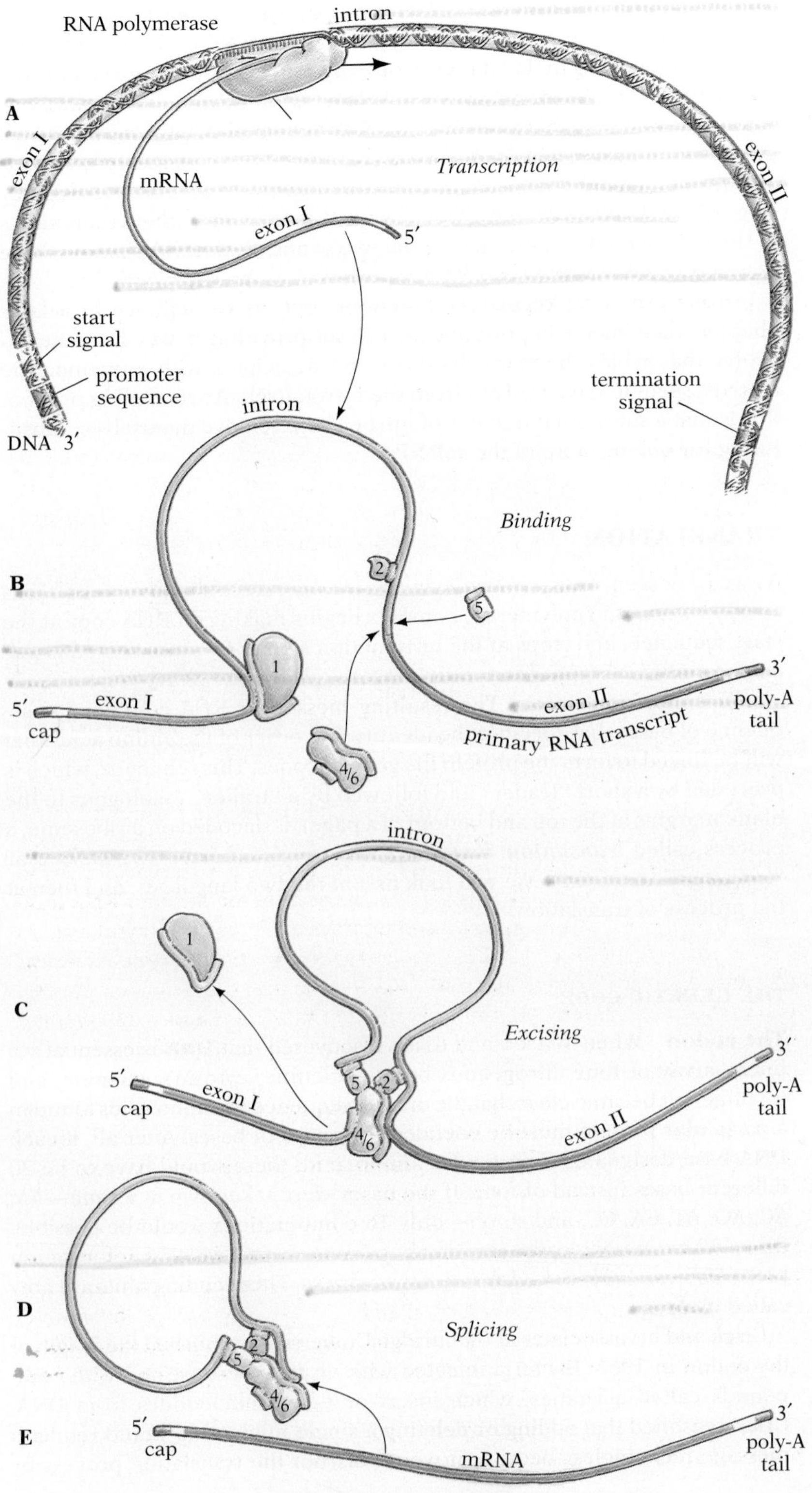

9.6 Messenger RNA processing in eucaryotes RNA polymerase binds to a promoter sequence and moves along the DNA strand, synthesizing a complementary RNA strand (A). A 7-methylguanosine cap is added at the beginning of the transcript by another enzyme. After the RNA polymerase transcribes the termination signal, the RNA between the last exon and the sequence of four to eight adenines is removed; a separate enzyme extends the poly-A tail by 100–200 bases, completing the primary RNA transcript (B). With the aid of several small nuclear ribonucleoprotein particles (snRNPs), any introns are removed, and the resulting transcript is spliced. For clarity, only one intron is shown. The process begins with recognition of the beginning of the intron by snRNP 1 (B). Next, snRNPs 2, 4/6, and 5 come together to form a complex at the end of the intron (C). The leading end of the intron is cut (D) and bound to the complex, which then cuts the trailing end of the intron and splices the two exons together (E). At this point the mature mRNA is ready for export to the cytoplasm.

9.7 An snRNP-intron complex

ribonucleoprotein particles—snRNPs or "snurps." At least four different kinds of snRNPs cooperate in most splicing. The RNA in these particles is like ribosomal RNA in that it is used directly, and has both an enzymatic and a structural role. The best known snRNP has a base sequence complementary to the boundary between the end of exons and the beginning of introns. SnRNP complexes catalyze the breakage of the bond between the primary-transcript nucleotides at each intron-exon boundary; the intron drifts away (Fig. 9.7) and is thought to be digested by other enzymes; other components of the snRNP complex then splice the two exons. Once all the introns have been removed, the mature mRNA is exported to the cytoplasm.

Introns can serve regulatory functions, but, as we will see in a later chapter, their major importance may lie in providing a way for genes to evolve that avoids the many disadvantages associated with mutation. The snRNPs seem to have evolved from the intron itself. An intriguing proof of this is that a substantial number of introns help remove themselves, substituting for one or more of the snRNPs.

TRANSLATION

As we have seen, transcription is initiated when an RNA polymerase binds to a promoter; the polymerase complex begins making an RNA copy at the start sequence, and stops at the termination signal. In eucaryotes this primary transcript is edited to remove introns, whereas in procaryotes no such modification is necessary. The resulting messenger RNA contains the sequence of bases that specifies the identity and order of the amino acids that will be linked to form the protein the gene encodes. This sequence, which is preceded by a short "leader" and followed by a "trailer" (analogous to the blank margins at the top and bottom of a page), is decoded on a ribosome, a process called ***translation:*** information is translated from one molecular "language" to another. We will look first at the two languages, and then at the process of translation.

THE GENETIC CODE

The codon When Watson and Crick discovered that DNA is essentially a linear array of four nitrogenous bases—adenine, cytosine, guanine, and thymine—it became clear that the unique sequence of amino acids found in a particular protein must be encoded by *groups* of bases. After all, if each DNA base designated a particular amino acid, there would have to be 20 different bases instead of four. If the bases were taken two at a time—AA, AC, AG, AT, CA, CC, and so on—only 16 combinations would be possible. Since 20 amino acids are commonly present in proteins, the information has to be encoded in sets of three or more bases. These coding units are now called **codons**.

Crick and his associates at Cambridge University established the length of the codon in 1961. Bacteria infected with viruses were treated with compounds called acridines, which insert or delete nucleotides from DNA. Crick reasoned that adding or deleting a single nucleotide would render a message meaningless because it would disrupt the translation process by shifting the apparent starting point in subsequent coding units. Suppose the message is

THE BIG RED ANT ATE ONE FAT BUG

Deletion of the first *E* will make it

THEB IGR EDA NTA TEO NEF ATB UGX

A single deletion redefines all subsequent codons, and if it occurs near the start of the message, the amino acid sequence of the protein is completely altered. Two deletions, Crick reasoned, would have the same effect, but if the number of deletions corresponded to the codon length, and if they all occurred near the beginning, translation would produce a mostly correct protein, an enzyme that would probably retain some activity:

THEBI GRED ANT ATE ONE FAT BUG

Crick and his associates used various concentrations of acridines to make different numbers of nucleotide deletions, and determined for each concentration whether or not active enzymes were produced. Subjecting these results to elaborate statistical analyses, they concluded that codons were three bases long.

Deciphering the code The next problem to be solved in understanding translation was to determine the exact relationships between codons and amino acids. Marshall W. Nirenberg and Heinrich Matthaei of the National Institutes of Health used an enzymatic process developed by Severo Ochoa at New York University to link nucleotides together to create synthetic RNA. When uracil was the only nucleotide available to the enzyme, for instance, a long polyuracil chain of mRNA was synthesized: UUUUUUUUUU. Similarly, when adenine alone was supplied, a polyadenine chain was formed.

When the polyuracil obtained by this method was used instead of normal mRNA for protein synthesis in an artificial environment providing all the amino acids, a polypeptide chain composed only of the amino acid phenylalanine resulted—a clear indication that the codon UUU codes for phenylalanine. Nirenberg and Matthaei also showed that AAA codes for lysine, GGG for glycine, and CCC for proline.

It was understandably more difficult to interpret codons composed of two or three different nucleotides. If, for example, uracil nucleotides and guanine nucleotides were made available to the enzyme in a ratio of 2 to 1, artificial mRNA in which the codons GUU, UGU, and UUG predominated was synthesized. When this mRNA was used as a template in polypeptide synthesis, the amino acids cysteine, valine, and leucine were the principal ones incorporated into the polypeptide, but it was impossible to determine by this technique which of the three triplets coded for which of the amino acids.

In 1964 Nirenberg and Philip Leder of the National Institutes of Health developed a technique for getting ribosomes to bind to RNA trinucleotides (three nucleotides bonded together in sequence) of known composition. A trinucleotide, acting as though it were a short piece of mRNA, would then cause the ribosome to bind a specific amino acid—an early step in translation. For example, if they used a ribosome that had bound to a trinucleotide

TABLE 9.1 *The genetic code (messenger RNA)*

First base in the codon	*Second base in the codon* U	C	A	G	*Third base in the codon*
U	Phenylalanine	Serine	Tyrosine	Cysteine	U
	Phenylalanine	Serine	Tyrosine	Cysteine	C
	Leucine	Serine	*Termination*	*Termination*	A
	Leucine	Serine	*Termination*	Tryptophan	G
C	Leucine	Proline	Histidine	Arginine	U
	Leucine	Proline	Histidine	Arginine	C
	Leucine	Proline	Glutamine	Arginine	A
	Leucine	Proline	Glutamine	Arginine	G
A	Isoleucine	Threonine	Asparagine	Serine	U
	Isoleucine	Threonine	Asparagine	Serine	C
	Isoleucine	Threonine	Lysine	Arginine	A
	Methionine[a]	Threonine	Lysine	Arginine	G
G	Valine	Alanine	Aspartic acid	Glycine	U
	Valine	Alanine	Aspartic acid	Glycine	C
	Valine	Alanine	Glutamic acid	Glycine	A
	Valine	Alanine	Glutamic acid	Glycine	G

[a] Also *Initiation* when located at leading end of mRNA.

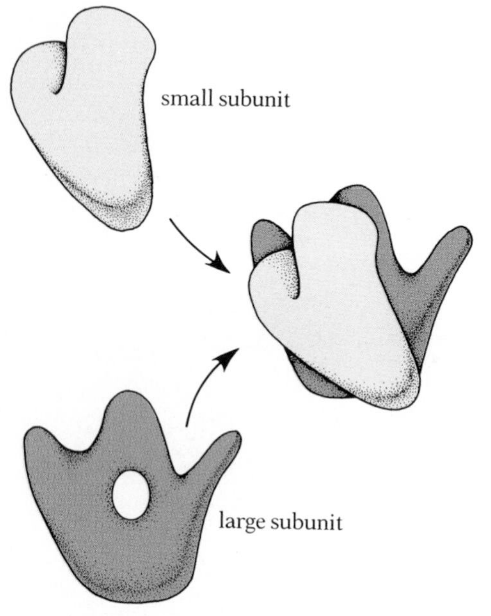

9.8 A ribosome
A functional ribosome is formed when the two kinds of independent subunits join. This joining can take place only after mRNA binds to the small subunit. The tunnel in the large subunit is about 25 Å in diameter.

composed of UUU, phenylalanine would couple with it. Since it is relatively easy to synthesize trinucleotides with a particular base sequence, each of the 64 possible triplets could be synthesized and complexed with ribosomes, and the resulting amino acid association would then be determined. A few trinucleotides were not entirely specific in their binding, however, so some ambiguity in codon interpretation remained.

Shortly after the trinucleotide-binding technique became available, procedures for producing synthetic mRNA polymers with known repeating sequences (such as AAGAAGAAG · · ·) were developed by H. G. Khorana at the University of Wisconsin, and these helped resolve the remaining ambiguities in the code. Table 9.1 summarizes the genetic code.

As the genetic dictionary of the table makes clear, all but two amino acids (methionine and tryptophan) are represented by more than one codon. The synonymous codons—those coding for the same amino acid—usually have the same first two bases, but differ in the third; thus CCU, CCC, CCA, and CCG all code for proline. In fact, U and C are always equivalent in the third position, and A and G are equivalent in 14 of 16 cases. Notice also that four codons—AUG, UAA, UAG, and UGA—serve as "punctuation," marking the beginning and end of the actual message within the mRNA.[4]

[4] Certain minor inconsistencies in the genetic code have been uncovered. For example, in some ciliates the codons UAA and UAG, which are normally termination signals, instead code for glutamine; in some bacteria and most (perhaps all) eucaryotic mitochondria, UGA, which can also be a termination signal, codes for tryptophan.

THE ROLE OF RIBOSOMES

The process of translation takes place on ribosomes. A ribosome consists of a large and a small subunit, each a complex of ribosomal RNA (rRNA), enzymes, and structural proteins (Fig. 9.8). When not carrying out protein synthesis, the ribosomal subunits exist in the cytoplasm as separate entities. The large subunit of procaryotic ribosomes is composed of two molecules of rRNA and roughly 35 proteins; the small subunit has one molecule of rRNA and approximately 20 proteins. Eucaryotic ribosomes are somewhat larger. In both eucaryotes and procaryotes, the rRNA itself is highly structured: long segments of complementary base pairs are held together in a helical chain by hydrogen bonds, creating a complex pattern of arms and loops (Fig. 9.9). Though the function of this structure is not yet known, it is probably crucial to ribosomal activity.

In procaryotes, the 5′ end of the mRNA binds to the small subunit, which is then able to bind the large subunit (Fig. 9.10). The binding of the mRNA involves base pairing between a section of the rRNA and a binding signal—

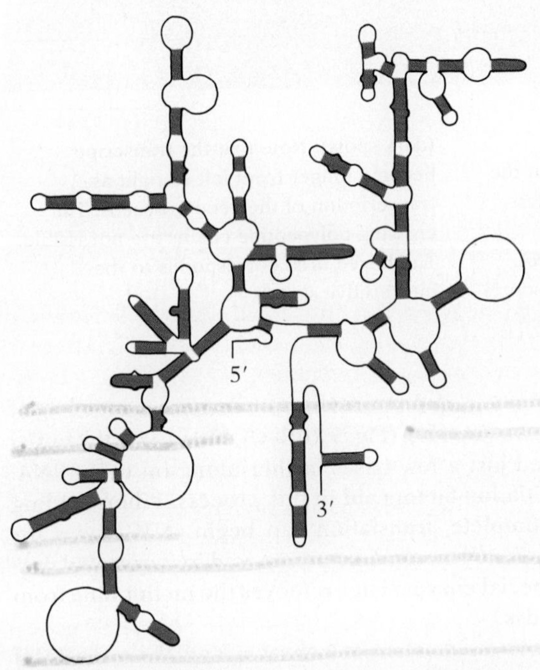

9.9 Ribosomal RNA
Ribosomal RNA has many stretches of complementary base pairs (red), which allow the chain to fold back on itself and form double-stranded helical arms. The open loops represent areas in which the bases are not complementary. This arrangement probably plays an important role in ribosomal function. The rRNA shown here is found in the small ribosomal subunit of a bacterium. Its three-dimensional structure is not yet known.

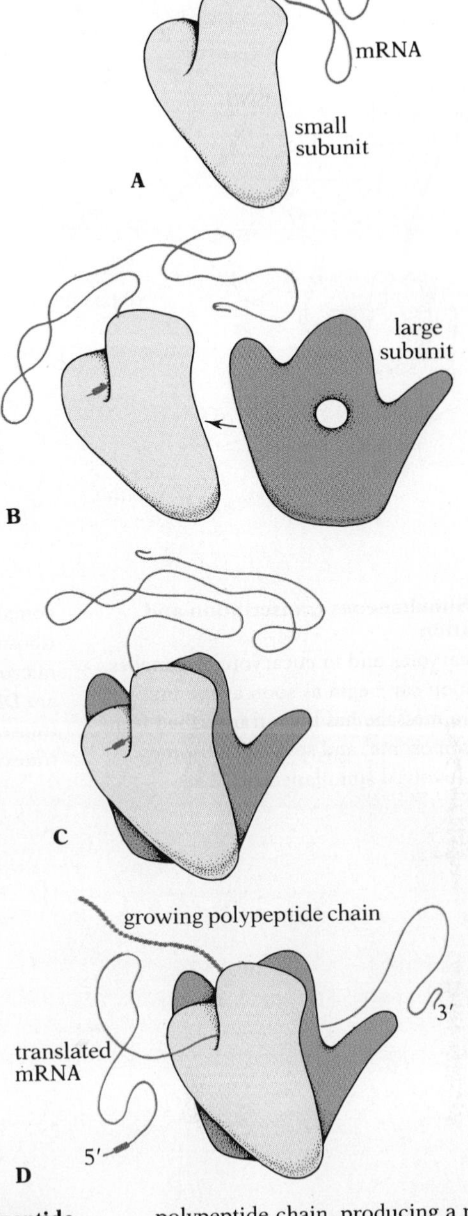

9.10 Synthesis of a polypeptide chain by a ribosome
Free ribosomes exist as two separate subunits (A–B). A signal sequence on the mRNA binds to the small subunit, causing a change that permits it to bind the large subunit (B–C). The ribosome then translates the mRNA into a polypeptide chain, producing a protein (D). The growing chain is thought to be threaded out through the tunnel in the large subunit. The mRNA is much longer than shown here, and additional ribosomes can bind and begin translation while the first ribosome is still at work.

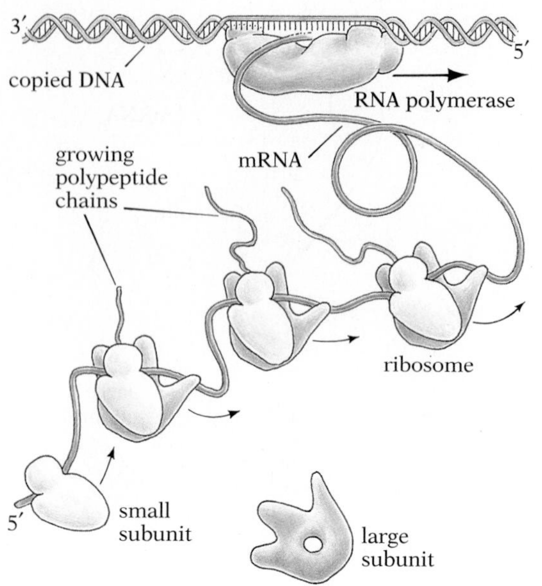

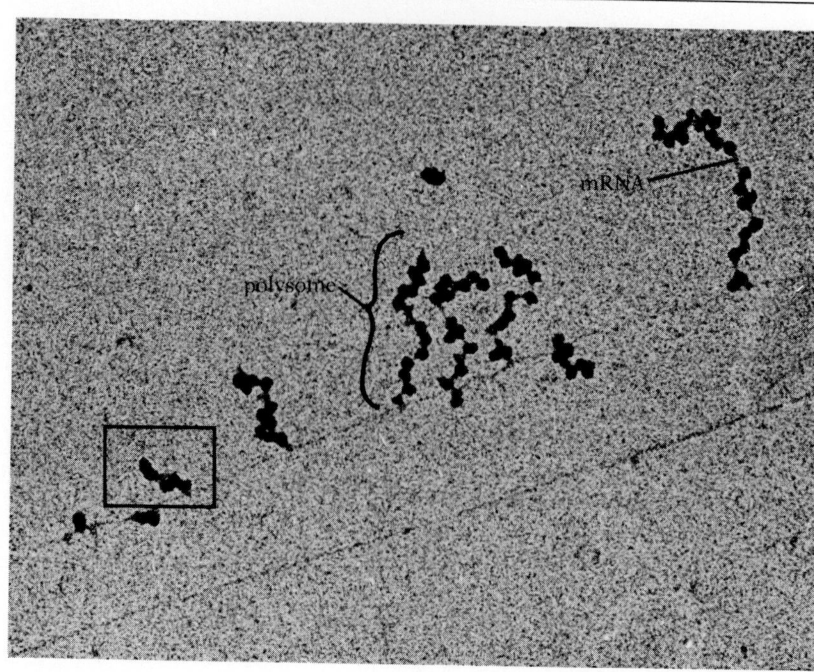

9.11 Simultaneous transcription and translation
In procaryotes and in eucaryotic organelles, translation can begin as soon as the first part of a message has been transcribed from the chromosome, and several ribosomes can be involved simultaneously. The complex of mRNA and two or more ribosomes is often called a polysome. In the micrograph, the two thin horizontal lines are DNA; the upper strand is being transcribed and, simultaneously, the RNA transcripts are being translated by ribosomes (dark spots). Note that the transcripts become longer from left to right as transcription of the gene proceeds. The growing polypeptide chains are not visible. The boxed area corresponds to the interpretive sketch.

usually AGGAGGU, near the end of the mRNA in the leader segment. This signal binds in the small subunit's cleft (Fig. 9.10B–C), along with the initiation codon—AUG—located just a few bases farther along on the mRNA. Special proteins called initiation factors aid in this process. When binding of the large subunit is complete, translation can begin. AUG codes for methionine, so methionine is always the first amino acid incorporated during translation. Another special enzyme later removes the methionine from the end of most polypeptides.

Because procaryotes have no nuclear membrane, and the process of transcription is not segregated from the ribosomes and other translational machinery in the cytoplasm, ribosomes are able to bind one end of an mRNA molecule and begin translating it into protein while the RNA polymerase is still transcribing the rest of the message from the chromosome. (The same situation occurs in translation in eucaryotic chloroplasts and mitochondria.) In both procaryotes and eucaryotic organelles, while the first ribosome to bind is translating later parts of the message, additional ribosomes can bind and begin translation (Fig. 9.11). The situation is different for mRNA produced in the nucleus of eucaryotes. Even though, as we saw in Chapter 5, important parts of eucaryotic ribosomes are synthesized

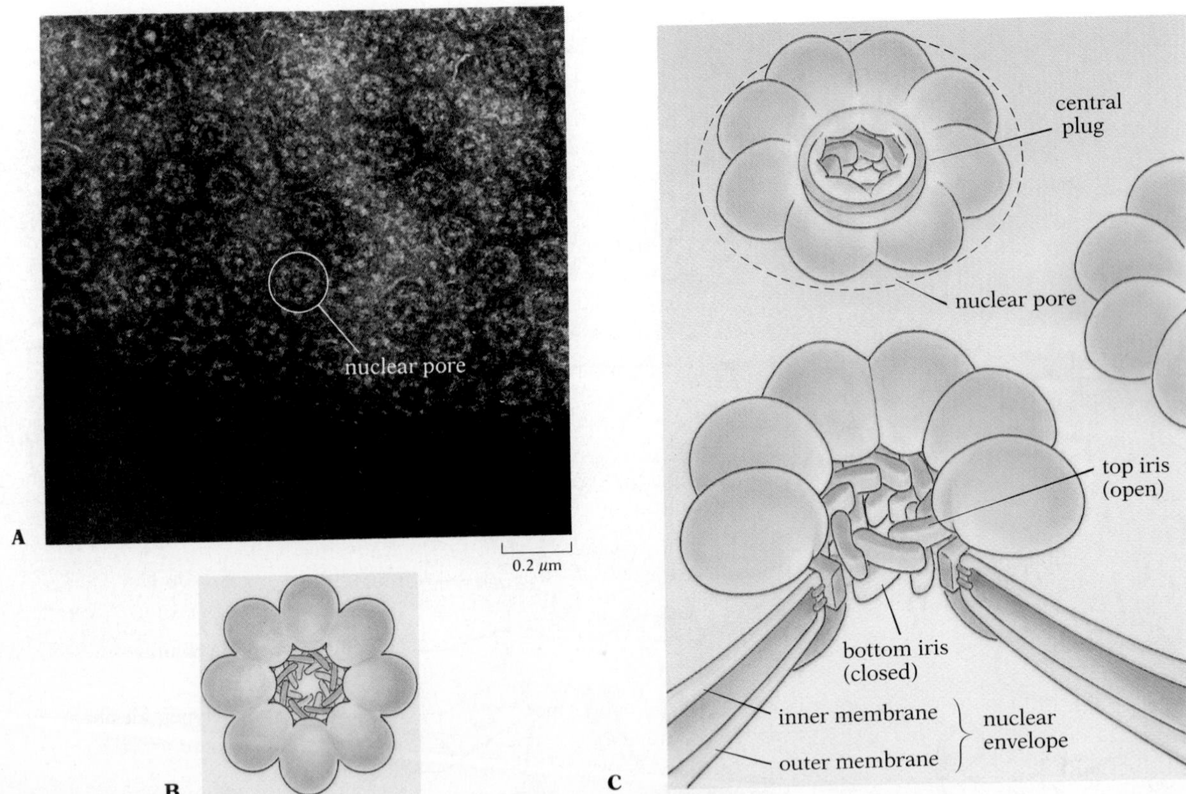

9.12 Nuclear pores
(A) The nuclear envelope contains many specialized pores, which provide a pathway from the nucleus to the cytoplasm for mRNA. The density of these portals is indicated by the micrograph of a small section of the nuclear membrane from a frog egg. (B) The probable structure of these octagonal pores includes a central plug, which appears to have two sets of eight protein arms, one on the inner face of the nuclear envelope, the other on the outer face. (C) The arms seem to act like an iris, opening to a diameter of 200 Å to admit correctly tagged macromolecules, and closing to 90 Å after they pass.

in the nucleus, translation cannot occur in the nucleus because the final steps of ribosome assembly take place in the cytoplasm.

In eucaryotes, four steps intervene between transcription and translation. We have already mentioned three of these in this chapter: modification of the transcript at the 5′ end, modification at the 3′ end, and splicing of the RNA after the removal of introns (Fig. 9.6). The fourth step is movement through the pores of the nuclear envelope (Fig. 9.12) into the cytosol, where unbound ribosomes are available to do translation. Most RNA is translated by ribosomes in the cytosol, but the mRNA for certain proteins is always translated on the rough ER. The mRNA of this type begins the process of translation in the normal way, by binding to a small ribosomal subunit in the cytosol and thereby making possible the binding of the large subunit and the start of translation. But translation ceases almost immediately; the leading end of the partially synthesized protein contains a sequence that binds to a signal-recognition complex, which is composed of its own RNA and six proteins; this binding causes translation to stop. The mRNA-recognition complex–ribosome assembly then binds to the rough ER. The partially synthesized protein is inserted into a channel of the ER, translation resumes, and the growing chain of amino acids is fed into the lumen of the ER. At the end of translation, the ribosomal subunits release the mRNA, separate, and drift free of the ER. As we saw in Chapter 5, the proteins synthesized in the rough ER are destined for noncytoplasmic roles: they may be soluble, and either drift free in the ER lumen, or be collected together in the

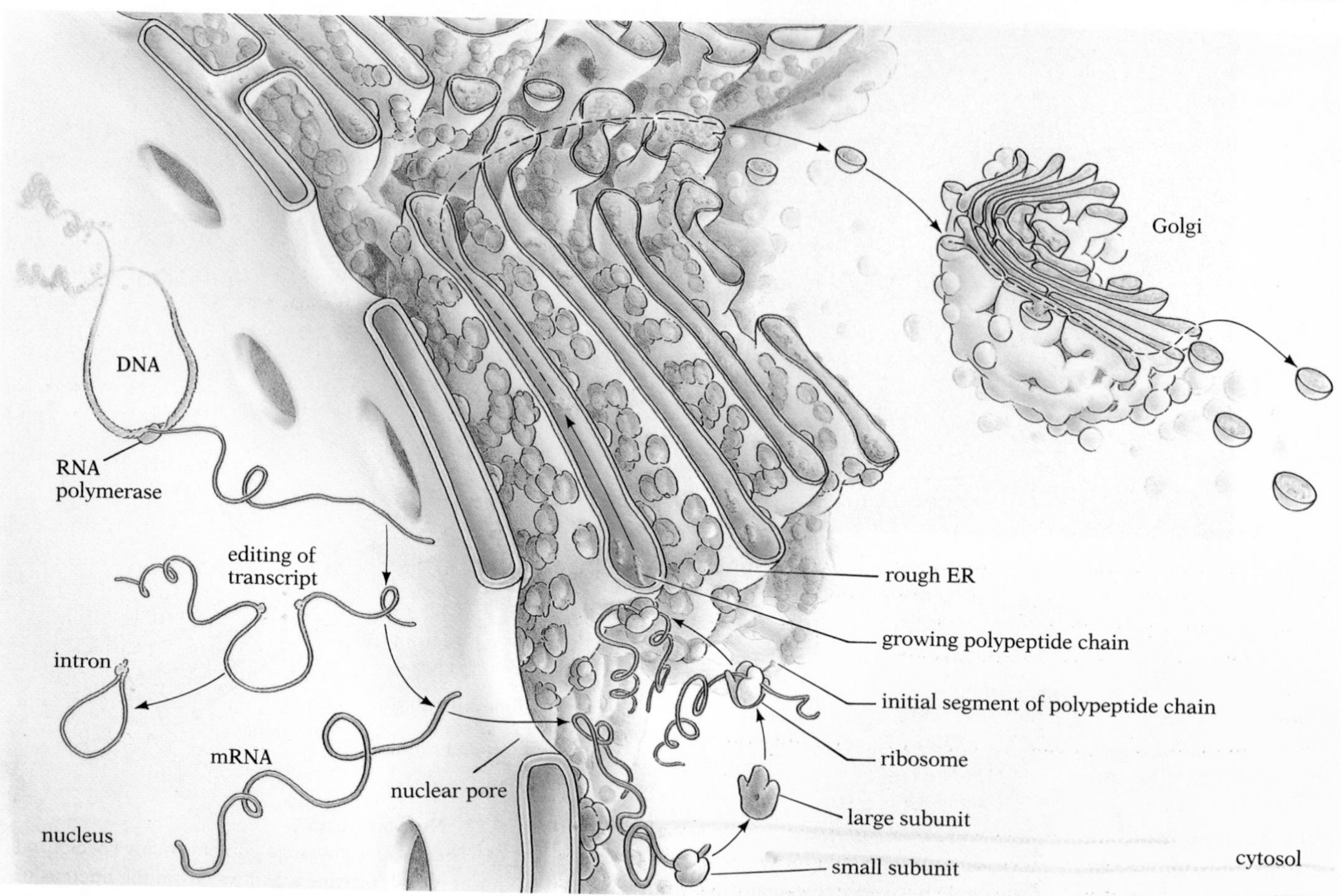

9.13 Synthesis of a hydrolytic enzyme
Most mRNA is translated freely in the cytosol, but when protein products are destined for the ER lumen, the ER membrane, or secretion, translation of mRNA occurs on the rough ER. A gene is transcribed, and the transcript is processed to form mRNA. Next, the mRNA passes through the nuclear pore (shown here without its plug) and binds to a small ribosomal subunit. A large ribosomal subunit then binds, and translation begins and continues until a signal sequence on the growing polypeptide chain binds to a signal-recognition complex which interrupts translation. The ribosome complex then binds to the rough ER and translation resumes, with the growing polypeptide chain passing into the lumen of the ER. Finally, the newly synthesized protein, in this case a hydrolytic enzyme, moves to the smooth ER, where it is packaged into a vesicle for transport to the Golgi apparatus.

smooth ER and packaged into secretory or other vesicles (Fig. 9.13); alternatively, they may become mounted in the ER membrane, either to remain there to perform their function, or to be packaged into vesicle membrane or the cell membrane itself. Most proteins synthesized on ribosomes in the cytosol, by contrast, remain in the cytosol.

TRANSFER RNA AND ITS ROLE IN TRANSLATION

We have said that in eucaryotes, mRNA is synthesized on DNA in the nucleus, moves to the ribosomes, and functions as a template for protein synthesis. Early researchers, in their attempt to understand translation, tried to ascertain whether the amino acids interact directly with the mRNA, or indirectly by way of some intermediate agent or adapter molecule.

In 1957 Mahlon Hoagland and his associates at Harvard demonstrated that each amino acid becomes attached to some form of RNA *before* it arrives at the ribosome, where it is added to the growing polypeptide. This new kind of RNA, which is neither mRNA nor rRNA, acts, they hypothe-

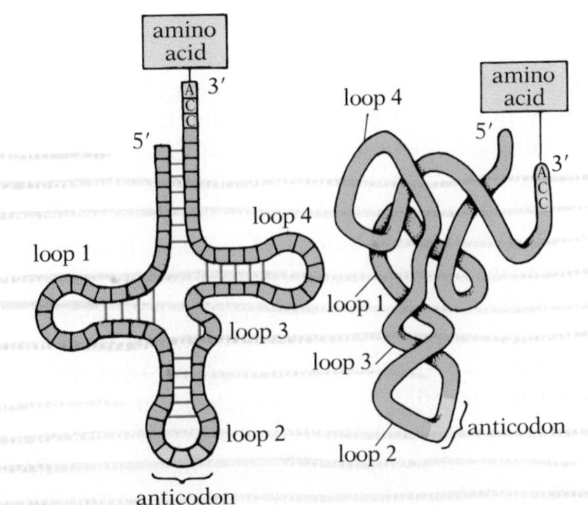

9.14 Structure of a molecule of transfer RNA The single polynucleotide chain folds back on itself, forming five regions of complementary base pairing, four loops with unpaired bases, and an unpaired terminal portion to which the amino acid can attach. The unpaired triplet that acts as the anticodon is on the second loop. Loop 4 is thought to function in binding the tRNA to the ribosome (probably by binding to rRNA), and loop 1 probably binds an activating enzyme. The molecule is shown flattened at left; its tertiary structure is diagrammed at right.

sized, to transfer amino acids to the ribosomes; they called it transfer RNA. Hoagland and his associates demonstrated conclusively that each tRNA molecule binds a single molecule of amino acid, activates it with energy from ATP, and transports it to a ribosome.

Subsequent investigation by many researchers has yielded considerable information about tRNA structure and function. We now know that at least one form of tRNA is specific for each of the 20 common amino acids; the amino acid arginine, for example, combines only with tRNA specialized to transport arginine; leucine combines only with tRNA specialized to transport leucine; and so on. But all forms of tRNA share certain structural characteristics: a length of 73 to 93 nucleotides, a structure consisting of a single chain that is folded into a cloverleaf shape, with internal base pairing (Fig. 9.14), and a CCA termination sequence. When an amino acid binds to its specific tRNA, it always binds at the CCA end.

Specific enzymes (aminoacyl tRNA synthetases) match each amino acid with its appropriate tRNA. Each enzyme is specific for a particular amino acid and its tRNA, and catalyzes a reaction that binds the two together (Fig. 9.15). The tRNA then carries the amino acid to a ribosome that has bound a strand of mRNA. There the tRNA becomes attached to the mRNA by complementary base pairing. We have seen that the codon for each amino acid is a sequence of three bases on the single-stranded mRNA. Each tRNA has, in an exposed position, an unpaired triplet of bases called an ***anticodon***, which is complementary to an mRNA codon for its particular amino acid. For example, the mRNA codon CCG codes for the amino acid proline. One type of tRNA for proline has the anticodon sequence GGC, which is complementary to CCG; similarly, the codon GUA, which codes for valine, has its complement in the tRNA for valine with the anticodon CAU. When a molecule of the proline tRNA with an attached molecule of proline approaches a molecule of mRNA, its exposed GGC triplet can bind to the mRNA only where the mRNA has a CCG triplet; similarly, the exposed CAU triplet of the valine tRNA can bind to the mRNA only at points where a GUA triplet occurs.

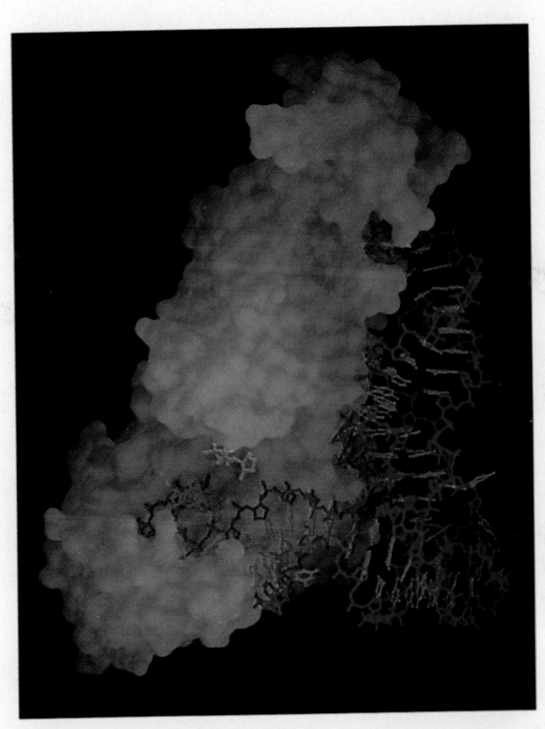

9.15 Matching of an amino acid with its tRNA The matchmaking enzyme responsible for loading a particular amino acid (glutamine in this case) onto its transfer RNA is shown in this computer-generated image in blue. It binds to the tRNA (red and yellow) at both the anticodon (bottom) and the acceptor arm (top center), the two regions of the tRNA responsible for its specificity. The green molecule is ATP.

THE TRANSLATION CYCLE

The actual binding of tRNA molecules and linkage of the amino acids they carry is managed in step-by-step fashion by the large subunit of the ribosome, but only after completion of the preliminaries: first, the tRNA with the anticodon sequence for methionine is bound to the small subunit; in eucaryotes as well as procaryotes this sequence is complementary to the initiation codon on the mRNA, which can now also bind to the small subunit; finally, the large subunit is bound to the small one. As translation then proceeds, the ribosome moves along the mRNA, with the part of the message being translated lying in the groove between the two subunits; two adjacent tRNA binding sites—the ***P-site*** and the ***A-site***—bring the appropriate nucleotide sequences together. When the cycle begins, the anticodon of the first tRNA is in register with the initiation codon of the mRNA at the P-site (Fig. 9.16A). The adjacent A-site then brings a molecule of tRNA with the appropriate anticodon together with the next codon triplet on the mRNA (Fig. 9.16B). Once the new tRNA is bound to the A-site, the enzyme peptidyl transferase moves the amino acid from the P-site tRNA and binds it to the amino acid at the A-site. The tRNA at the P-site, having relinquished its amino acid, is released (Fig. 9.16C), and the cycle is completed as the ribosome moves along the mRNA by one codon, thereby bringing the codon that was at the A-site (and its tRNA, with the growing polypeptide chain) to the P-site (Fig. 9.16D). As a result, the codon to be translated next now occupies the A-site. The translation cycle adds about 15 amino acids per second to the polypeptide chain, with an error rate of about one mistake in every 30 chains. It comes to an end when the ribosome encounters one of the termination codons and, with the aid of a protein known as the release factor, causes the completed polypeptide to be released.

The details of how ribosomes work—how they "select" the appropriate tRNA from the rich supply of different transfer molecules, for example, or how they shift the messenger along one codon at a time—are not known, though the effects of antibiotics on various steps in translation are beginning to provide some insights into the process. Tetracycline, for instance, works by binding irreversibly to the A-site on bacterial (and organelle) ribosomes; streptomycin blocks initiation, and erythromycin prevents the A → P shift, again on bacterial and organelle ribosomes. (Though reducing translation rates in pathogenic bacteria can be of great therapeutic value, it must be accomplished without using concentrations of antibiotics so high that they will kill our own mitochondria.)

The ribosome complex's potential for producing protein is enormous. Since the average protein chain is 300–500 amino acids long, a ribosome can produce a chain in about 25–35 seconds. Ribosomes outnumber mRNA by a factor of 10 in most cells, and one mRNA can be translated by many ribosomes simultaneously, so a new peptide chain can be generated from each mRNA about every 3 seconds. A typical active eucaryotic cell has on the order of 300,000 mRNA molecules in circulation to direct synthesis of the proteins necessary for self-maintenance; as a result, if sufficient raw materials are available, 100,000 proteins can be synthesized every second. And some cells—particularly those specialized for secretion or rapid growth—are far more active than this.

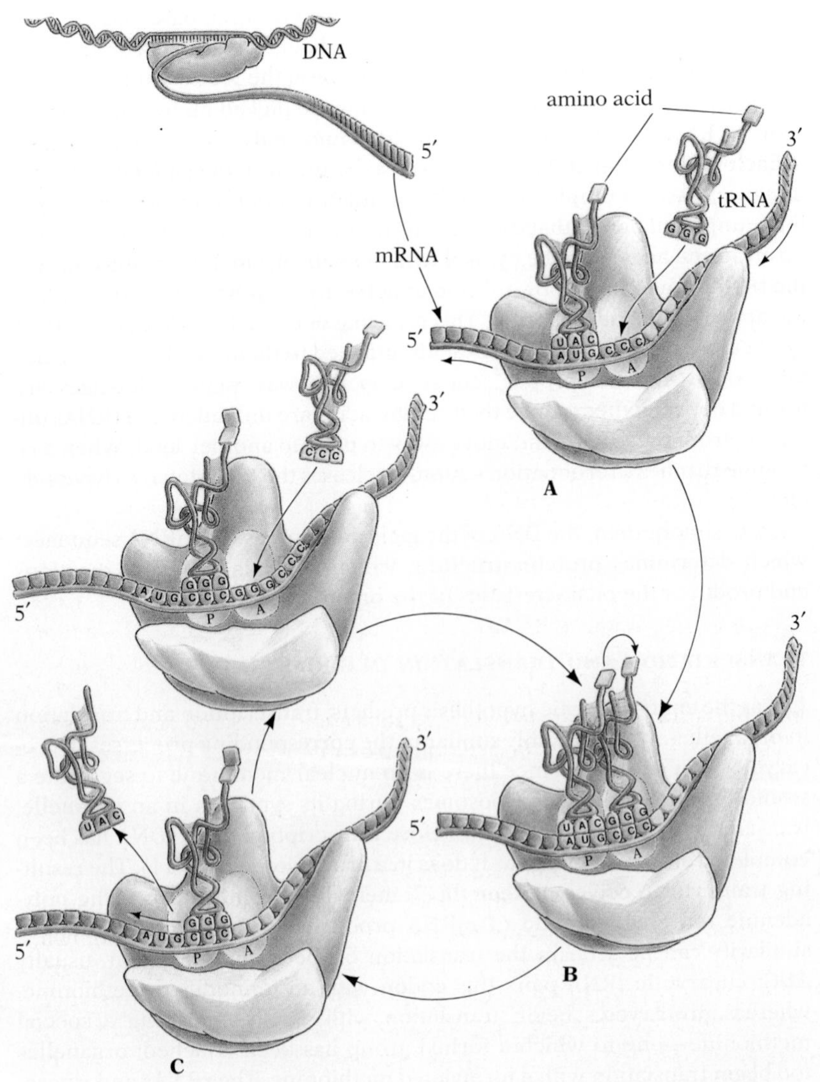

9.16 Translation cycle on the ribosome After messenger RNA created by transcription binds to a small ribosomal subunit, a large subunit can bind, and translation begins. At this point the tRNA with the anticodon for methionine (which was bound to the small subunit before the two subunits combined) is already paired with the initiation codon on the mRNA at the P-site (A). Another tRNA, bearing the appropriate anticodon, pairs with the codon at the vacant A-site (B). Next the amino acid that was attached to the tRNA at the P-site is detached and bound to the amino acid on the tRNA at the A-site. The P-site tRNA then drifts free (C). Finally, the messenger is moved so that the remaining tRNA occupies the P-site, and the next codon is brought into position at the A-site, ready to accept another complementary anticodon (D). In these drawings part of the small subunit is shown cut away to reveal the mRNA, which lies in the groove between the subunits. The size of the tRNAs is greatly exaggerated.

SUMMARY OF TRANSCRIPTION AND TRANSLATION IN EUCARYOTES

Let's now review the flow of information in a typical eucaryotic cell. First, when the double-stranded DNA that constitutes a particular gene is activated, RNA polymerase is able to recognize and bind to the promoter sequence. The DNA acts as a template for synthesis of a molecule of single-stranded messenger RNA. Transcription of the DNA begins at a start signal and ends after transcription of a termination sequence. The resulting transcript is capped with 7–methylguanosine and tagged with a polyadenine tail to form the primary transcript, and the introns are removed and the remaining segments are spliced—all steps that are unnecessary in procaryotes. The mature mRNA leaves the nucleus and moves into the cytosol, where it becomes associated with ribosomes.

The mRNA, carrying information in the form of three-base codons, acts as the template for synthesis of polypeptide chains. As the ribosomes move along the mRNA, they read the codons, starting at the 5′ end. Amino acids to be incorporated into the polypeptide chains are picked up by molecules of transfer RNA specific for each of the 20 amino acids.

Each molecule of tRNA has an exposed anticodon (a sequence of three unpaired bases) complementary to the mRNA codon (also a sequence of three unpaired bases) that codes for its particular amino acid. After picking up an amino acid from the cytosol with the help of a matchmaking enzyme, the tRNA moves to a ribosome and attaches to the mRNA at a point where the appropriate codon occurs. This ordering of the tRNAs along the mRNA molecule also orders the amino acids attached to them. As the amino acids are moved into the proper sequence in this way, peptide linkages are formed between them. Once their amino acids are unloaded, the tRNAs uncouple from the mRNA and move away to pick up another load. When a ribosome reaches a termination codon, it releases the completed polypeptide chain.

Quite simply, then, the DNA of the gene determines the mRNA sequence, which determines protein structure, which controls chemical reactions and produces the characteristics of the organism.

TRANSCRIPTION AND TRANSLATION IN ORGANELLES

Just as the endosymbiotic hypothesis predicts, transcription and translation in organelles are remarkably similar to the corresponding processes in procaryotes. For example, since there is no nuclear membrane to segregate a strand of mRNA from the ribosomes during its synthesis in an organelle, translation can begin at one end before transcription of the DNA has been completed at the other, just as it does in a procaryote (Fig. 9.11). The resulting transcript receives neither the 7–methylguanosine cap nor the polyadenine tail characteristic of mRNA produced in the nucleus. Another similarity can be seen in the translation of the initiation codon, usually AUG: eucaryotic tRNA pairs this codon with an unmodified methionine, whereas procaryotes begin translation with a tRNA carrying a special methionine—one to which a formyl group has been attached; organelles too begin translation with a formylated methionine. The rRNAs and ribosomal proteins of organelles differ in size and composition from the rRNAs and ribosomal proteins used for translation in the cytoplasm (Fig. 9.17), but are much closer in these respects to those of procaryotes. Indeed, the ribosomes of *E. coli* and chloroplasts have been shown to be essentially interchangeable.

Information flow between the nucleus and organelles As we have already noted, mitochondrial DNA usually encodes about 40 products. The entire base sequence (16,569 nucleotides) of human mitochondrial DNA is now known, as is most of the base sequence of yeast mitochondrial DNA; as a result, the products encoded by these DNAs can now be identified. Both of these kinds of mitochondrial DNA contain instructions for making two of the three rRNAs of the organelle ribosomes, all of their various tRNAs, one of the dozens of ribosomal proteins, one of the nine proteins of the F_1 com-

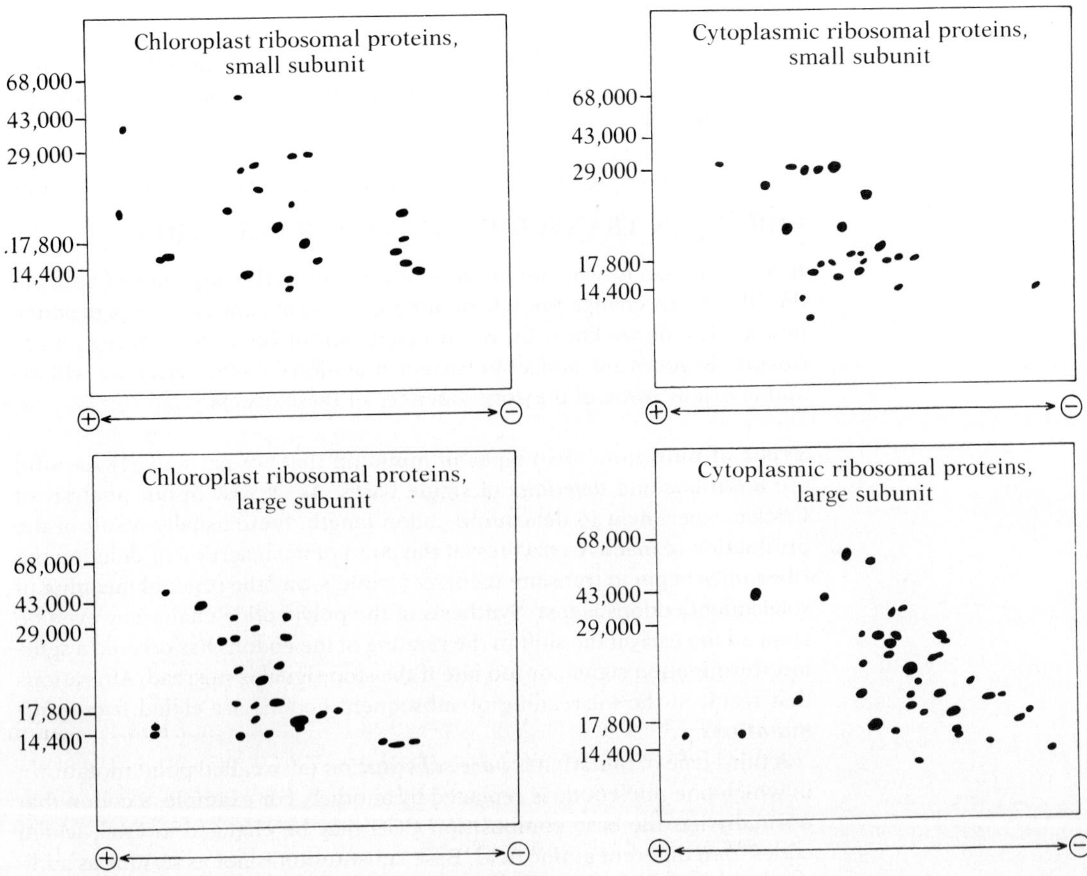

9.17 Proteins from chloroplast and cytoplasmic ribosomes of eucaryotes
The proteins from large and small subunits of organelle and cytoplasmic ribosomes have been dissociated and separated by molecular weight (vertical axis) and charge (horizontal axis). It is evident that organelle ribosomes (left) have fewer proteins than the cytoplasmic ribosomes of eucaryotes (right); furthermore, few if any of the proteins from the two sources are alike in weight and charge, an indication that they also differ in composition.

plex (which makes ATP, utilizing energy stored by the respiratory electron-transport chain), and several other organelle proteins. The other rRNA and ribosomal proteins, the replication, transcription, and translation enzymes, the electron-transport and other F_1 proteins, and so on must be encoded by nuclear genes and transported to the organelles. The nucleus, therefore, contains nearly all the genes for two different kinds of ribosomes. The abbreviated chromosome of mitochondria is probably an evolutionary consequence of a kind of intracellular competition. During early development, when a growing organism's cells are multiplying rapidly, any mitochondrion that can reproduce faster than others in the cell will contribute a greater proportion of mitochondria to the next generation of cells, including the ones destined to produce gametes in the adult. In time, natural selection favors such mitochondria, which come to dominate their slower-growing counterparts. The same argument holds for the multiple chromosomes within each mitochondrion: the one that replicates the fastest will leave more copies in daughter organelles. At either level, the key to reproductive speed lies largely in reducing the number of genes that must be replicated. Selection, therefore, has favored mitochondria that transfer as many genes as possible to the nucleus. Why then don't the organelles transfer all their remaining genes to the nucleus, and utilize the same kinds of ribosomes, tRNAs, and other enzymes as the rest of the cell? The answer to this question, as yet unknown, may reveal much about the evolution of eucaryotic life.

ERRORS IN TRANSCRIPTION AND TRANSLATION

It is now apparent why mutations—alterations in the sequence of bases in the DNA—can change the information content of genes and thus produce new alleles. As we know from our discussion of DNA repair at the end of Chapter 8, genes are subject to various mutational events. Here we will examine briefly some of the consequences of these events.

Types of mutation Two types of mutation that are particularly harmful are ***additions*** and ***deletions*** of single bases. As we saw in our analysis of Crick's experiment to determine codon length, these usually result in the production of inactive enzymes: at the point of the insertion or deletion the ribosomes begin to translate incorrect triplets, and the original meaning of subsequent codons is lost. Synthesis of the polypeptide chain may even be stopped too early if the shift in the reading of the codons has created a spurious termination signal, or too late if the stop signal is misread. Alterations that result in the misreading of subsequent codons are called ***frameshift mutations***.

A third type of mutation is ***base substitution*** (also called point mutation), in which one nucleotide is replaced by another. For example, a codon that normally has the base composition CGG may be changed to CAG, which codes for a different amino acid. Base substitution is not as serious as addition or deletion, however, because in most codons a change in the third base does not alter the meaning (Table 9.1). Roughly 30 percent of all base substitutions therefore have almost no effect.[5] Moreover, even when an amino acid is changed, the substitution usually has relatively little effect if, for example, one nonpolar amino acid replaces another. On the other hand, exchanges that occur between polar and nonpolar amino acids, or between positively charged and negatively charged amino acids, are likely to cause trouble. And alterations that create new prolines or cysteines, or remove old ones, are often serious since they may change the tertiary structure of the protein. All in all, just over half of all base substitutions are likely to affect protein function strongly.

Another very common type of mutation is ***transposition***. This phenomenon, which we shall discuss in detail in Chapter 10, results from the insertion of long stretches of DNA from one part of the genome into the middle of another. The majority of easily detectable spontaneous mutations in *Drosophila*, for example, are known to result from transposition.

[5] There is some effect of the third-base mutation because the tRNAs for some alternative codons for an amino acid can be less numerous or less active or less accurate in binding to the codon. In genes that are transcribed and translated at a high rate, selection favors the version of the codon that yields the best translation performance; for less frequently translated genes, there is little effect from a third-base substitution.

Mutagenic agents High-energy radiation, including both ultraviolet light and ionizing radiation such as X rays, cosmic rays, and emissions from radioactive materials, can cause mutations. In addition, a variety of chemicals have been found to be mutagenic. A normal spontaneous mutation rate for a single gene is one mutation in every 10^6–10^8 replications, but the rate can be greatly increased by unusual exposure to mutagenic agents.

Ionizing radiations sometimes induce simple base substitutions, but they also frequently produce large deletions of genetic material, either directly by colliding with the DNA and causing breaks to occur, or indirectly by splitting water to produce highly reactive oxygen radicals (single, unbonded, strongly electronegative oxygen atoms), which combine with and break the DNA.

As we saw in the last chapter, ultraviolet light most often exerts its mutagenic effect by causing abnormal bonding of thymine bases. When two adjacent thymine units absorb ultraviolet light and are thus energized, they often bond to each other, forming a ***thymine dimer*** (Fig. 9.18). An unrepaired dimer can inactivate a strand of DNA; not only can no mRNA be transcribed from it, but—more important for dividing cells—DNA replication cannot take place.

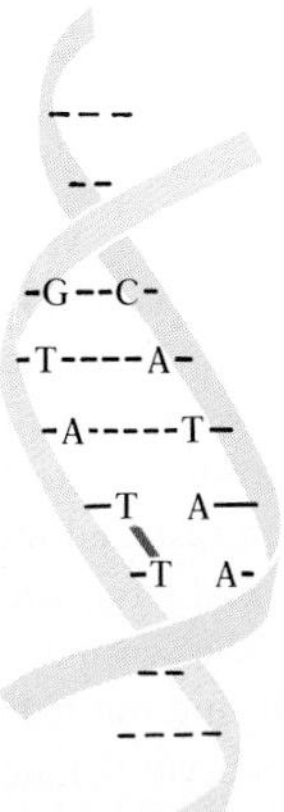

9.18 A thymine dimer within a DNA molecule Two adjacent thymine nucleotides are bonded to each other covalently, so they cannot form hydrogen bonds with the adenine nucleotides of the complementary strand. Such a mutation, often induced by ultraviolet light, will inactivate the DNA if it is not repaired. Some species that normally live in the dark (for instance, the intestinal bacterium *Salmonella*) lack the enzyme that can repair thymine dimers.

Some mutagenic chemicals produce their effects by directly converting one base into another. For example, nitrous acid (HNO_2) is a very powerful mutagen that deaminates (removes the NH_2 group from) cytosine, changing it into uracil.[6] Other chemical mutagens, of a type called base analogues, are themselves sometimes incorporated into nucleic acids in place of one of the normal bases. An example is 5-bromouracil, an analogue of thymine. When a strand of DNA contains a unit of 5-bromouracil instead of thymine, it is prone to errors of replication, because the 5-bromouracil will sometimes pair with guanine rather than with the requisite adenine. Virtually all these mutations are detected and repaired before they can exert any effect.[7]

There is now a wealth of evidence for a close correlation between the mutagenicity of a chemical and its carcinogenicity—between the potential of a chemical for producing genetic mutations and its cancer-inducing activity. This fact strongly suggests that many cancers are caused, at least in part, by mutations in somatic cells. Thus we see that alterations of the DNA are important not only in germ cells, where they may affect future offspring, but also in somatic cells, whose metabolism or growth they may disrupt, causing disease or degeneration.

The connection between mutagenicity and carcinogenicity is the basis for the Ames test,[8] in which such chemicals as environmental pollutants, reagents used in industrial processes, proposed new drugs, and food additives are screened for potential carcinogens. The compound to be tested is added

[6] Nitrous acid, a major component of cigarette smoke, also converts adenine into hypoxanthine and guanine into xanthine.

[7] By flooding a cell with 5-bromouracil, it is possible to overwhelm the repair enzymes. This strategy is often used in cancer chemotherapy to kill rapidly dividing tumor cells. An unfortunate side effect is that some highly active normal cells—those producing hair, for instance—are also killed.

[8] Named for Bruce N. Ames of the University of California, Berkeley.

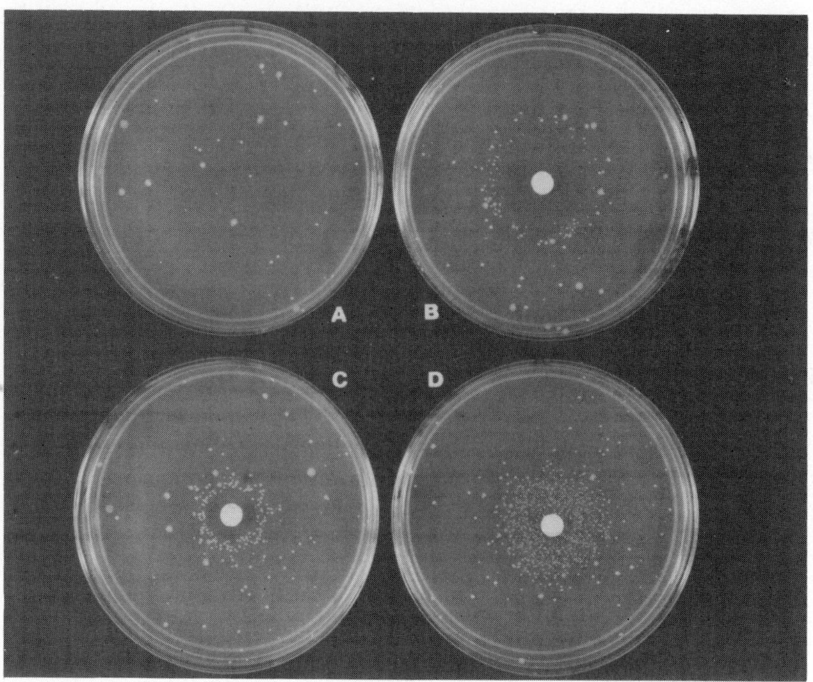

9.19 The Ames test
Each of these Petri dishes has vast numbers of histadine-requiring bacteria in a layer of agar. The agar contains everything the bacteria need to grow except histadine. The white spots in (A) indicate colonies descénded from a bacterium in which a spontaneous mutation restored the ability to synthesize histadine. The large white disks in the center of the other dishes contain a mutagenic substance that is diffusing out from the disk: furylfuramide (B), aflatoxin (C), and 2-aminofluorene. The mutagenic potential of these chemicals is indicated by the large number of colonies able to grow as a result of a genetic change that restored their capacity to synthesize histadine.

to a culture of about one billion bacteria; a special mutant strain of *Salmonella* is used, which requires the amino acid histidine as a nutrient. When the mixture of bacteria and chemical is incubated on a medium deficient in histidine, some cells undergo a mutation that is the reverse of their original mutation, and they thus regain the ability to synthesize histidine and to grow on the deficient medium. After several days a count is made of the number of so-called revertant colonies derived from these cells (Fig. 9.19). Because the reactions that damage DNA in procaryotes also damage eucaryotic genes, any increase in the mutation rate of the bacteria over the normal spontaneous level is then used to predict the likely cancer-inducing potency of the chemical in humans. (The choice of *Salmonella* with a histidine-gene mutation was arbitrary; any easily grown species with a single-base mutation blocking the synthetic pathway for a nutrient would do as well.)

In early screening tests some substances known to be potent carcinogens gave negative results in the Ames test. It was soon realized that many chemicals, though not themselves mutagens or carcinogens, are transformed in the mammalian body, especially by detoxification enzymes in the liver, into derivatives that are mutagenic and carcinogenic.[9] For this reason newer versions of the Ames test add liver homogenate to the bacterial culture, so that the bacteria will be exposed both to the original chemical and to its metabolic derivatives. We will look more closely at the genetic bases of cancer in a later chapter.

[9] As we will see in Chapter 31, the liver operates on the principle that unfamiliar chemicals are bad, and alters them with a variety of enzymes that prepare them for excretion; unfortunately, a few substances (like the pesticide contaminant dioxin) are thereby turned into carcinogens.

STUDY QUESTIONS

1. Compare and contrast replication and transcription. (pp. 223–30, 233–36)
2. Compare the several types of RNA used in eucaryotic cells. (pp. 233–34, 241–46)
3. Compare and contrast transcription and translation in procaryotes with the more complicated series of events in eucaryotes. How does the process in organelles fit into the picture? Is there any obvious reason for these differences? (pp. 234–38, 242, 247–50)
4. Make a list of all the signal sequences (and the function of each) we have encountered so far, whether in DNA, RNA, or peptides. What others are likely to exist that have not been discussed yet? (pp. 234–38, 241–46)
5. In all, it takes four high-energy phosphate bonds to catalyze the several steps necessary to add each amino acid to a growing peptide chain. Given that the average protein contains 300–500 amino acids (the lower figure is more typical of procaryotes), how expensive (in terms of glucoses metabolized) is a protein, assuming that the amino acids are freely available? (p. 182)

CONCEPTS FOR REVIEW

- RNA versus DNA
- Transcription
  - RNA polymerase
  - Signals
    - promoter
    - start sequence
    - stop sequence and hairpin loop
    - intron-exon boundary
  - Exon splicing
- Genetic Code
  - Codons
  - Degeneracy
  - Effects of mutation
  - Addition and deletion (frameshift) substitution
- Translation
  - Ribosomes
  - Signals
    - initiation sequence
    - termination sequence
    - ER binding sequence
  - Translation cycle
  - Role of tRNA
- Organelles
  - Asynchronous replication
  - Genome organization and size
  - Contrasts with "host-cell" transcription and translation

SUGGESTED READING

CECH, T. R., 1986. RNA as an enzyme, *Scientific American* 255 (5). *On the enzymatic properties of ribosomal and, especially, snRNP RNA, and the possibility that RNA originally served the functions now taken over by DNA and protein.*

CHAMBON, P., 1981. Split genes, *Scientific American* 244 (5). (Offprint 1496) *On the organization of introns and exons.*

CRICK, F. H. C., 1962. The genetic code, *Scientific American* 207 (4). (Offprint 123) *Describes Crick's demonstration that the codon is three bases long.*

DARNELL, J. E., 1983. The processing of RNA, *Scientific American* 249 (4). (Offprint 1543) *An excellent summary of the topic.*

DARNELL, J. E., 1985. RNA, *Scientific American* 253 (4). *Reviews transcription, processing, translation, and transcriptional control.*

DARNELL, J. E., H. LODISH, and D. BALTIMORE, 1990. *Molecular Cell Biology.* W. H. Freeman, New York. *Excellent detailed study of transcription and translation.*

JUDSON, H., 1980. *The Eighth Day of Creation.* Simon & Schuster, New York. *Well-written history of molecular biology.*

LAKE, J. A., 1981. The ribosome, *Scientific American* 245 (2). (Offprint 1501) *On the three-dimensional structure of the ribosome and the details of translation.*

RICH, A., AND S. H. KIM, 1978. The three-dimensional structure of transfer RNA, *Scientific American* 238 (1). (Offprint 1377) *How the three-dimensional structure of tRNA was determined, and how that structure helps explain how tRNA works.*

STEITZ, J. A., 1988. "Snurps," *Scientific American* 258 (6). *On the enzymes that remove the introns from eucaryotic primary transcripts.*

Chapter **10**

MOBILE GENES AND GENETIC ENGINEERING

ife depends on a dynamic balance between stability and change. In our discussion of information flow in the last two chapters, the emphasis has been on stability: the precise replication of DNA in preparation for cell division and reproduction, and the accurate transcription and translation of genes to produce RNA and protein. We have identified DNA proofreading and repair of mutations as mechanisms to insure the fidelity of genetic instructions. But change, too, is essential: natural selection works only if variation appears in a species, and a static genome would lead to the inevitable extinction of any species unable to evolve in the face of changing climate, resources, or competition.

Our emphasis in this chapter will be on large-scale, dynamic alterations that can occur in the genome. By "large-scale" we mean changes that involve an entire gene or major parts of it. Mutations involving single-base changes in functional genes play a comparatively small role in evolution, for the simple reason that such "tinkering" is likely to reduce a gene's ability to produce a functional product, and is therefore unlikely to survive in future generations.[1] By contrast, the consequences of large-scale chromo-

[1] As we will see in the next chapter, single-base changes in the promoter or other sequences that control gene expression probably can have a significant effect on cell physiology by altering the *amount* of enzyme produced rather than by changing its structure.

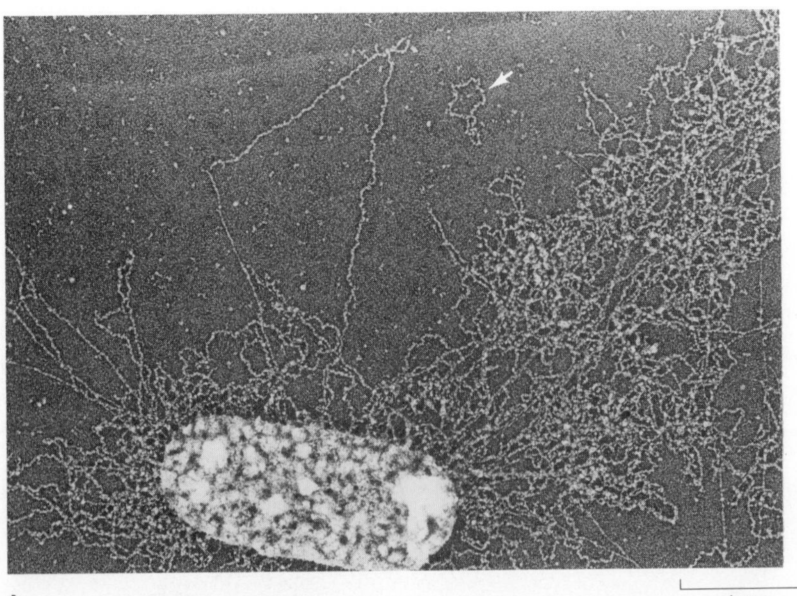

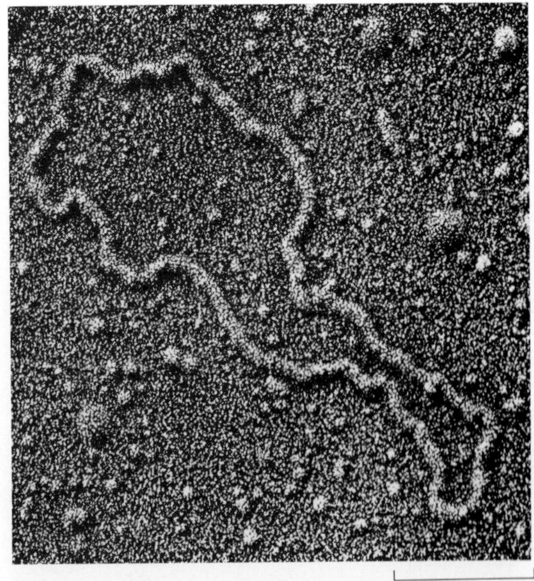

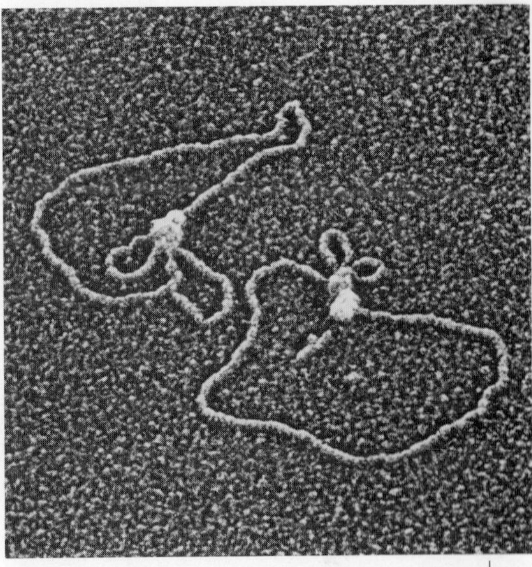

10.1 Electron micrographs of plasmids from *E. coli*

(A) A plasmid (arrow) is visible next to the main chromosome of this lysed *E. coli*. (B) The isolated plasmid seen here carries a gene encoding a product that confers resistance to the antibiotic tetracycline. (C) Plasmids are replicated independently of the main chromosome. The bacterial enzyme complex DNA polymerase can be seen copying each of these plasmids.

somal changes range from cancerous growth in the individual to the potential for major evolutionary change in the species.

Much of the work being done in genetic engineering makes use of curious "accessory" chromosomes, and it is with their discovery that we begin our discussion of gene mobility. Next we will turn our attention to viruses, whose unique reproductive strategies make them nearly ideal agents for moving altered genes into target cells. We will then look more closely at gene mobility; as we will see, the enzymes that accomplish major genomic changes also make possible the techniques of genetic engineering, with their enormous potential for improving everything from crops to human health.

PLASMIDS AND SEX FACTOR IN BACTERIA

We have said that a bacterium has one copy of its DNA instruction set on its large circular chromosome; organisms with one set of instructions are called ***haploid***. By contrast, sexually reproducing organisms are ***diploid***, having two copies in every cell except in their gametes, which fuse to restore the duplicate set prior to fertilization. In 1946 Joshua Lederberg and Edward L. Tatum, then at Yale University, demonstrated that bacterial mating (of a sort) nevertheless does take place. In so doing, they discovered ***plasmids***, small circular accessory molecules of DNA (Fig. 10.1). Plasmids have since been discovered in eucaryotes ranging from yeast to mammals.

Bacterial sex factor Lederberg and Tatum isolated two mutant strains of *E. coli*, each lacking the ability to synthesize a particular pair of nutrients. The first mutant was unable to synthesize the amino acid methionine and the vitamin biotin; the second could not make the amino acids threonine and leucine. The mutants could be grown only on a medium that supplied

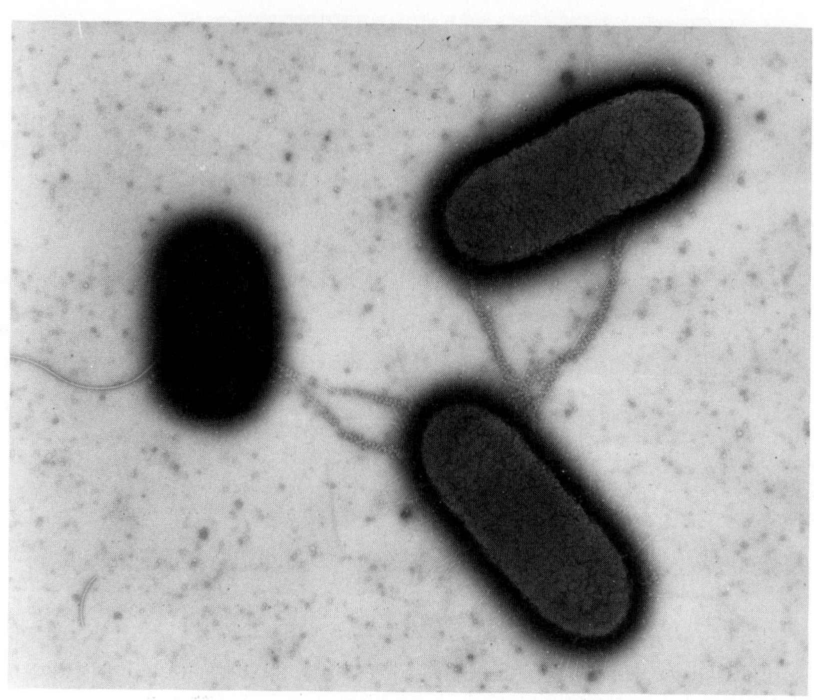

1 μm

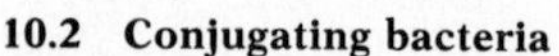

10.2 Conjugating bacteria
Long cytoplasmic bridges, or pili, connect the lower cell with two others with which it is conjugating simultaneously. The pili also provide attachment sites for certain viruses that can only infect F^+ cells.

the nutrients they could not synthesize. But when the two mutant strains were mixed on a minimal medium (one lacking all four critical nutrients), some healthy colonies nevertheless formed and these could subsist indefinitely on the minimal medium. The individuals in these colonies had somehow inherited both the first strain's ability to synthesize threonine and leucine and the second strain's ability to synthesize methionine and biotin. In short, they appeared to be the result of ***recombination*** of traits from the two original mutant strains—that is, genetic material from two different individuals had become combined in a single genome. Lederberg and Tatum demonstrated that direct contact between the cells of the two strains was necessary for this recombination to occur. Recombination occurs whenever two haploid gametes fuse to form a fertilized egg, which is thereby rendered diploid; but bacteria are never diploid, and most species possess a stiff cell wall that precludes fusion. How could recombination have come about?

A few years later several researchers showed that bacterial recombination like that found in *E. coli* is brought about through ***conjugation***, a process that is analogous to sexual mating in higher organisms. Two bacterial cells come to lie very close to each other, and a narrow cytoplasmic bridge, or pilus, extending from one connects to the other. This bridge is visible under the electron microscope (Fig. 10.2). Genetic material can pass through the bridge from one cell to the other.

Conjugation can occur only between cells of different mating types, designated F^+ and F^-. An F^- cell cannot conjugate with another F^- cell. Similarly, F^+ cannot conjugate with F^+. F^+ cells differ from F^- in containing in their cytoplasm a plasmid containing a so-called ***sex factor***, which is composed of DNA. Under most circumstances this factor can replicate within a non-dividing cell, and a copy can easily be transferred from F^+, which acts as donor and is thus analogous to a male, to F^-, which acts as recipient and is thus analogous to a female. The products of conjugation are always F^+. If this were the only type of cross possible, we might suppose all cells should eventually be F^+ and no further conjugation could take place. But conjugation is very time-consuming, which means that an F^- cell can undergo one or more fissions while an F^+ cell engages in one conjugation. Hence, the proportion of F^+ cells actually declines with conjugation.[2] In addition, the pilus provides an attachment point for certain viruses, which are thus able to infect and kill F^+ bacteria. Finally, conjugation in other types of crosses does not always convert the recipient cell from female into male, as we will see.

F^+ strains usually include a very few cells that, when they conjugate, may fail to transfer an intact sex factor to the F^- cells, but instead transfer chromosomal DNA. These are called *Hfr* cells (for high-frequency recombination; *Hfr* cells have a much lower threshold for initiating conjugation). As a result, crosses of the $F^- \times$ *Hfr* type do not usually convert the F^- cells into F^+ or *Hfr;* the female remains a female. The basis of this difference between F^+ and *Hfr* strains is the location of the sex factor in the cell; in *Hfr* strains the plasmid has become incorporated into the bacterial chromosome (Fig. 10.3). During conjugation, a copy of the main chromosome of the donor is moved through the pilus, but this delicate cytoplasmic bridge rarely remains unbroken long enough for the sex factor (part of which is at the end of the chromosome) to be transferred. The contemporary significance of the F^+/*Hfr* phenomenon is that it provided an early indication that at least some genes can move.

The sex factor is also interesting because its behavior differs from that of every other component of the cell so far discussed. At times it is apparently free in the cytoplasm as a very tiny piece of circular DNA. At other times it is incorporated into the chromosome and behaves like other chromosomal genes. A little later, we will look at closely related behavior in viruses and in movable elements in eucaryotic chromosomes called transposons.

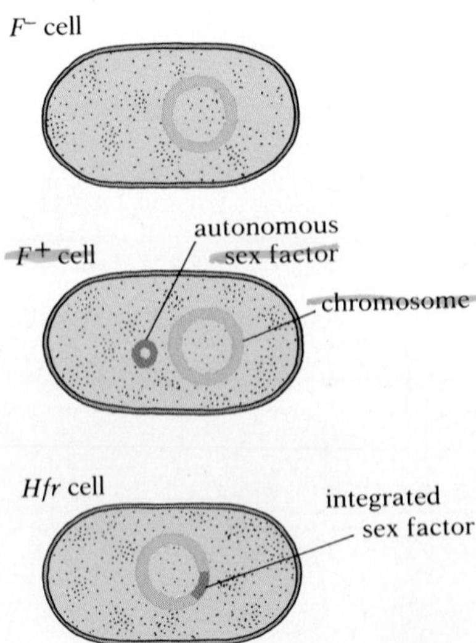

10.3 F^-, F^+, and *Hfr* cells of *E. coli* compared F^- cells lack the sex factor. F^+ cells have the sex factor as a plasmid free in the cytoplasm. *Hfr* cells have the sex factor integrated within the chromosome.

Autonomous plasmids Most bacterial cells contain some plasmids that are never integrated into the main chromosome. In other words, these plasmids exist only in autonomous form. Some such plasmids carry only one or two genes, while others may be as much as one-fifth the size of the main chromosome and carry many genes.

Especially important are plasmids with genes for resistance to various antibiotics, including streptomycin, tetracycline, and ampicillin (Fig. 10.1B). When bacteria with resistance plasmids are exposed to one of these antibiotics, the plasmids immediately begin replicating, and a cell that originally had only two or three may soon have a thousand or more. On the

[2] This reproductive disadvantage is kept to a minimum because F^+ bacteria typically initiate conjugation only when the population is growing very slowly, perhaps because of a shortage of some critical nutrient.

other hand, plasmids may carry genes that make the bacteria virulent, as in the case of the pneumococci that cause pneumonia. As we will see, because techniques now exist to insert new genes into both autonomous and integratable plasmids, which can then be made to multiply, plasmids are an invaluable tool in recombinant DNA technology.

VIRUSES AS AGENTS OF GENE MOVEMENT

Viruses are tiny obligate intracellular parasites that consist of no more than a miniature chromosome, a protective capsule, and (rarely) an enzyme; they lack even chromosomes. Their various reproductive strategies provide powerful means for both natural and artificial transfer of genes.

REPRODUCTIVE STRATEGIES OF VIRUSES

Lytic viruses In the 1940s Max Delbrück at the California Institute of Technology traced the interesting reproductive cycle of a common kind of bacteriophage. When he mixed a number of these phage with a culture of bacteria the viruses seemed to vanish, but half an hour or so later a hundred times as many phage as he had started with suddenly appeared in the culture. Delbrück hypothesized that phage of this kind enter the host cell, replicate themselves, and then cause the host cell to burst (lyse), releasing a new generation of infectious particles.

In the years since Delbrück's classic experiment, viruses have become one of the most valuable tools for studying how genes are organized, and how they work. In an earlier chapter we described the work of Hershey and Chase, which elucidated the normal life cycle of most bacteriophage. These viruses inject their DNA into the bacterial cells they attack, but leave their protein coats outside. When virulent phage (phage that kill the bacteria they invade) do this, the viral DNA promptly takes control of the bacterial cell's metabolic machinery and puts it to work manufacturing new viral DNA and new viral protein. These two components are then assembled into new infective phage. About 20–25 minutes after the initial injection of the viral DNA, the bacterial cell lyses, releasing the new phage, which may then attack other bacterial cells and start the ***lytic cycle*** over again (Fig. 10.4A). Many techniques of genetic engineering take advantage of the willingness of viral coat proteins to encapsulate nonviral DNA; such viruses can be used to deliver new genes to other cells.

Retroviruses A variety of different viral strategies have now been recognized. For example, many disease-causing viruses have an RNA chromosome, which is both replicated and transcribed (Fig. 10.4B). From the perspective of gene mobility and genetic engineering, the most important departure from the "standard" DNA- and RNA-based strategies is that employed by ***retroviruses***. These parasites have an RNA chromosome, and carry an enzyme, ***reverse transcriptase***, that catalyzes the formation of a DNA copy of the RNA (known as ***cDNA*** because it is complementary to the RNA from which it is copied); thus they reverse the usual flow of informa-

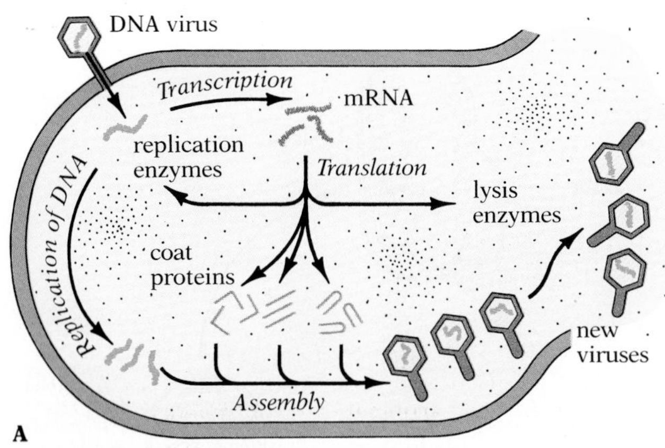

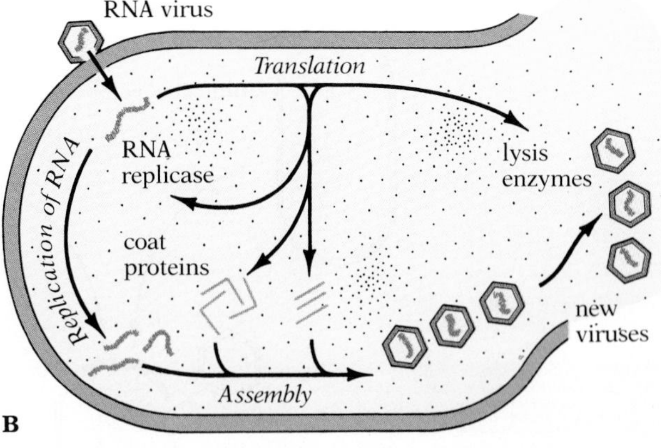

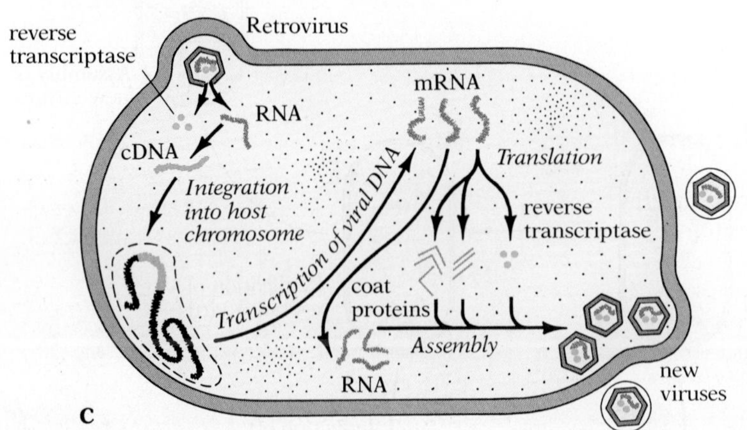

10.4 Three common strategies of viruses
Viruses exploit their hosts in a variety of ways. The most frequent strategy is for a viral genome consisting of DNA to be both replicated and transcribed, and the transcripts translated, all by means of host enzymes. Transcription and translation generate coat proteins, enzymes that modify host-cell function, and (later) enzymes that lyse the host. Other DNA viruses utilize replication enzymes produced through transcription and translation, as indicated in the drawing (A). For RNA viruses (B) the transcription step is unnecessary, but most encode a special enzyme, RNA replicase, to catalyze replication of the RNA. Some RNA viruses lyse their hosts, as shown, but many others escape from their hosts without lysing them. Retroviruses (C) are an unusual kind of RNA virus. Each retrovirus carries an enzyme, reverse transcriptase, that catalyzes the formation of a cDNA copy of the viral RNA; this copy is then incorporated into a chromosome of the host (often an animal cell). The host's enzymes then take over and accomplish replication (not shown), transcription, and translation of the viral genes. As the drawing indicates, retroviruses may enter the host cell by a kind of endocytosis, losing the protein coat only after entry; similarly, retroviruses generally do not lyse their hosts but instead leave by extrusion, a process whereby the virus is enveloped in a segment of the cell membrane as it emerges from the cell.

tion from DNA to RNA (Fig. 10.4C). Retroviral genes usually become permanent residents in the host chromosome.

Temperate viruses Some viral chromosomes act in a manner analogous to the F^+ plasmid: they can both enter and leave the host chromosome. This ability was first recognized in 1953 when André Lwoff and his colleagues at the Institut Pasteur found that if they exposed certain apparently normal strains of bacteria to ultraviolet light or X rays or various chemicals, the bacteria would lyse within an hour, releasing large numbers of infectious phage. Apparently these bacteria had been carrying viruses within their cells, but the viruses did not become active and did not usurp the cells' metabolic machinery until exposed to ultraviolet light, X rays, or chemicals. Cells that harbor inactive viruses are said to be ***lysogenic.***

The discovery by Lwoff that viruses can be present in an inactive state inside their host cells showed that some viruses must be ***temperate*** rather than virulent. Virulent phages invariably kill their hosts. Temperate phages may or may not kill their hosts, depending on a variety of conditions. When

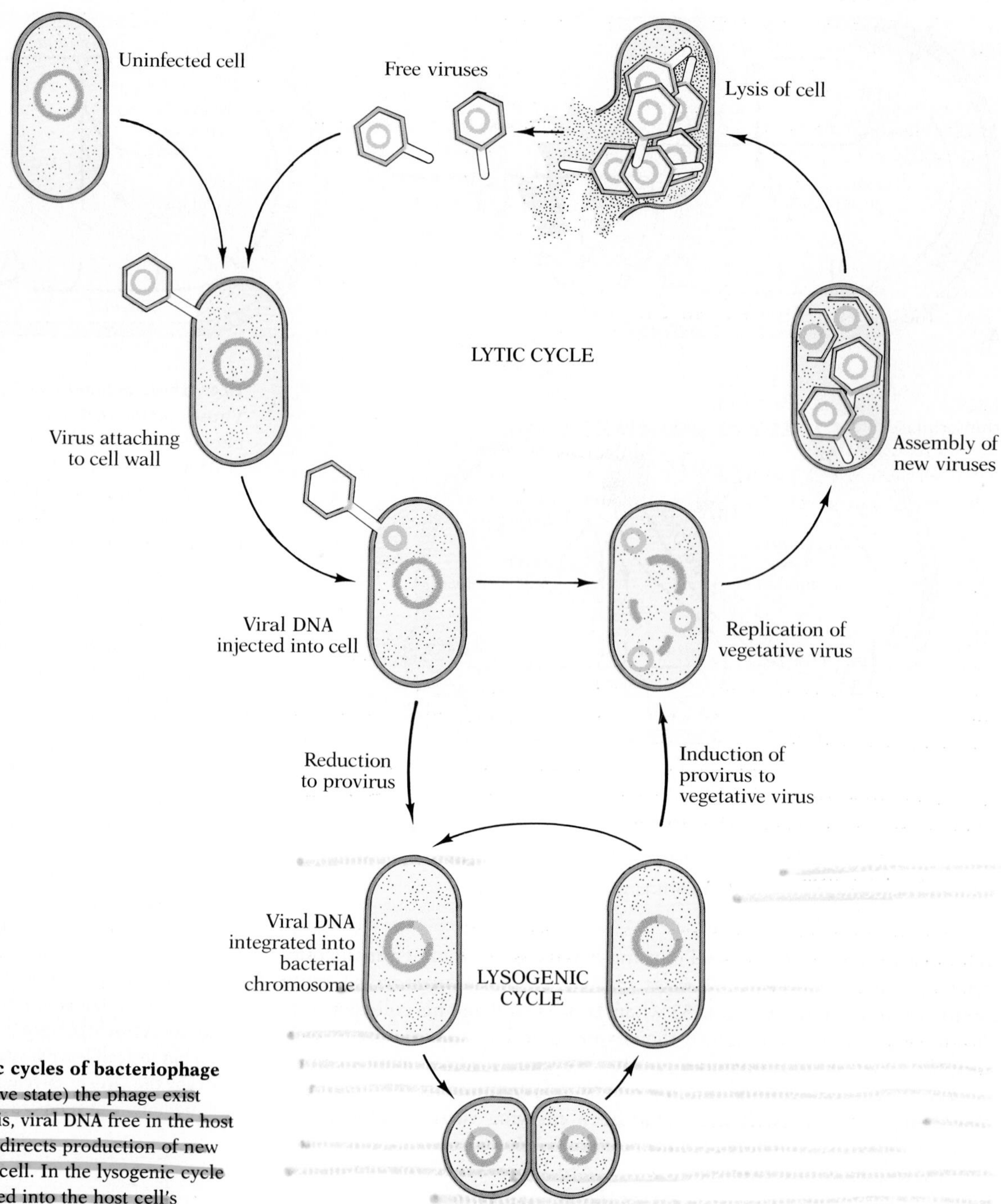

10.5 Lytic and lysogenic cycles of bacteriophage In the lytic cycle (vegetative state) the phage exist only as free viruses—that is, viral DNA free in the host cell's cytoplasm, where it directs production of new viral particles by the host cell. In the lysogenic cycle the phage DNA is integrated into the host cell's chromosome, and only on occasion is it induced to break loose and initiate viral replication. Bacteriophage are much smaller relative to their hosts than is indicated here.

they do not kill their hosts, their injected DNA usually becomes associated with the bacterial chromosome at a particular location, and the ***lysogenic cycle*** begins (Fig. 10.5). (The DNA of a few rare viruses survives as a plasmid in the host cell.) While integrated into the bacterial chromosome, the viral DNA, like the sex factor of *Hfr* strains, behaves as part of that chromosome. It is replicated with the rest of the chromosome; it can be transferred from one cell to another during conjugation; its genes can undergo recombination with bacterial genes; its genes can even produce visible effects in the host bacterium, such as modifications of colony morphology, changes in the properties of the cell wall, and changes in the production of enzymes. For example, diphtheria bacteria can produce a toxin that causes the disease only if they are carrying a specific type of viral gene. And viral genes often make the bacterium in which they reside immune to further infection by the same type of virus, a phenomenon very like the ability of the sex-factor genes to block the entry of additional F^+ DNA. We see, then, that the DNA of the temperate viruses can exist in an autonomous or vegetative state, replicating independently and eventually destroying the cell, or it can exist in the integrated state, as a ***provirus***, functioning and replicating as a portion of the chromosome. Several human ailments, including cold sores and shingles, are caused by lysogenic viruses indulging in a bout of lytic reproduction.

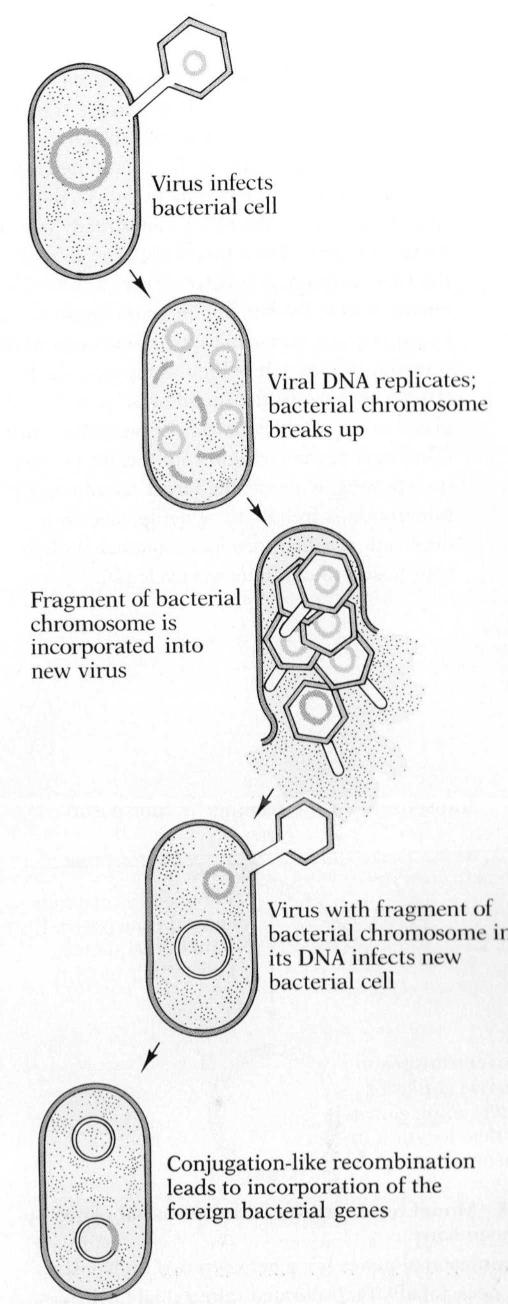
Virus infects bacterial cell
Viral DNA replicates; bacterial chromosome breaks up
Fragment of bacterial chromosome is incorporated into new virus
Virus with fragment of bacterial chromosome in its DNA infects new bacterial cell
Conjugation-like recombination leads to incorporation of the foreign bacterial genes

10.6 Model of transduction by bacteriophage

TRANSDUCTION: VIRAL TRANSFER OF NONVIRAL GENES

Not only can some viral genes move into chromosomes, but the reverse can also occur. The phenomenon is best understood in bacteria. When temperate viruses are in the vegetative state and have put the bacterial cell to work making more viruses, small fragments of the bacterial chromosome may become enclosed in the new viral coats. If a temperate virus carrying bacterial DNA in this manner infects a new host, it injects bacterial DNA into the new host. Sometimes the injected bacterial genes undergo recombination with the new host's genes. The virus has thus acted as a vehicle for transferring genes from one bacterial cell to another (Fig. 10.6). This process, called ***transduction***, was first described in 1952 by Norton D. Zinder and Joshua Lederberg, then at the University of Wisconsin.

There is good evidence that some of the genes that produce important observable effects in the human body may have been moved into human chromosomes by viruses. These genes may have been transferred from other human beings or even from other species. This can happen because some viruses (especially those responsible for influenza) can infect other hosts (swine, for example). As we will see in the next chapter, some genes transported by viruses may be involved in cancer. Viruses may also have been involved in moving organelle genes into the nucleus. In short, by moving entire genes, viruses can play a crucial role in creating the variation upon which natural selection works.

TRANSPOSITION OF GENES WITHIN THE GENOME

While we have so far emphasized the changes created by movement of genes from one organism into the chromosomes of another, equally signifi-

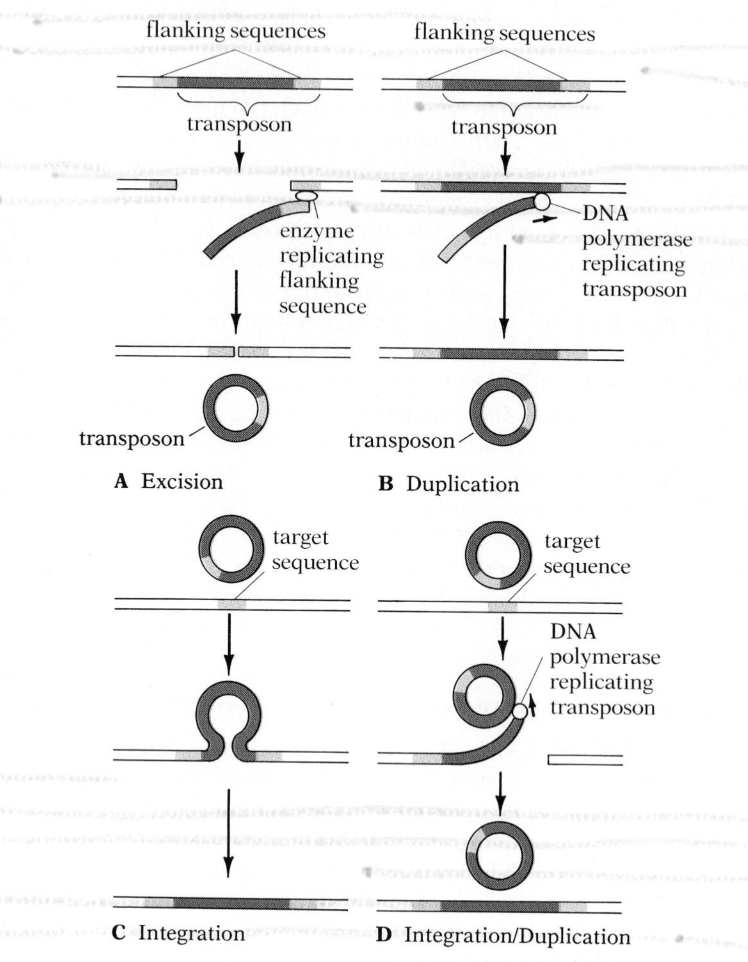

10.7 A model of transposition events
While integrated into a chromosome a transposon exists as a stretch of DNA flanked by characteristic sequences. At least one of these flanking sequences moved into the chromosome as part of the transposon, while the other may already have been there, and served as a target site. The movement of the transposon may involve either excision from the chromosome, leaving one or more target sequences behind (A), or duplication, with one copy of the transposon then left behind at the original location (B). In both cases, the mobile transposon probably exists as a circle, which can be integrated into any other part of the genome with (or, for certain transposons, without) the target sequence (C). The integration is frequently accompanied by a duplication, which produces another mobile transposon to continue the cycle (D).

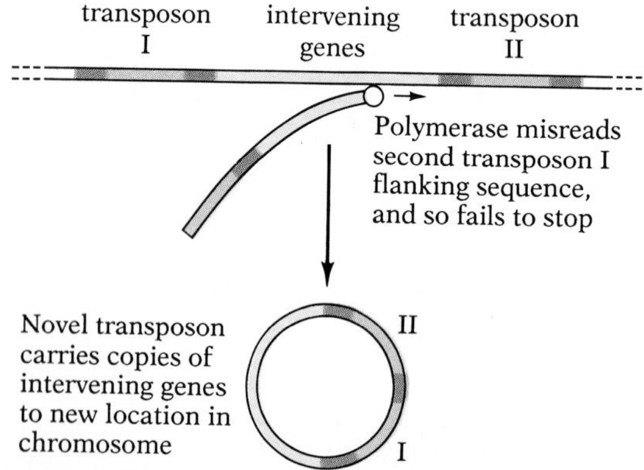

10.8 Model for capture of chromosomal genes by transposons
Chromosomal genes lying between two transposons are occasionally incorporated into a single hybrid transposon. A mutation in the second flanking sequence makes this incorporation inevitable.

cant alterations can be caused by moving genes *within* a genome. Since each gene's transcription is controlled by nearby sequences (including the promoter) that are not necessarily moved with the gene, the timing and degree of transcription can be significantly altered. Gene movement, therefore, can lead to dramatic changes.

Transposons Genetic units that move about in the genome, either by removing themselves to new locations or by duplicating themselves for insertion elsewhere, are called ***transposons***. They are found in all cells, procaryotic and eucaryotic, and can also insert themselves into plasmids. Transposition in eucaryotes was discovered more than three decades ago by Barbara McClintock of the Cold Spring Harbor Laboratory. She found that in corn certain genetic elements (which she believed to be genes) will occasionally move, particularly after cells are subjected to trauma, such as exposure to intense UV radiation. These movements produced kernels with unusual colors—colors that could not have resulted from normal recombination. Thirty years ago her results seemed so completely at odds with the prevailing concept of a static gene that the mobile genetic elements she had discovered were considered some sort of abnormality. By 1983, however, when she was awarded the Nobel Prize for her discovery, many such transposons had been discovered, and their possible role in evolution was beginning to be recognized.

We now know that transposons can consist of one or several genes, or just a control element, and can move in several ways, none of which is fully understood. Transposons within a chromosome are flanked by a pair of identical sequences, some of which are actually part of the transposon. Some transposons move from one site on the chromosome to another. A transposon that has moved may leave no trace except a telltale "scar" consisting of the flanking sequences (Fig. 10.7A), or it may move only after being copied in full by DNA polymerase, so that a complete duplicate of itself remains at its original location (Fig. 10.7B). Alternatively, it may stay put and dispatch a copy; this copy can be composed of either DNA or RNA. In the case of an RNA copy the transposon encodes reverse transcriptase. When such an RNA copy escapes to the cytoplasm it is translated; the reverse transcriptase thus produced proceeds to make a cDNA copy for insertion back into the chromosome. In many cases the host chromosome carries one or more copies of a particular sequence, which serve as targets for the transposon, and this sequence pairs with the one copy of the flanking sequence carried on the transposon. A special enzyme, usually encoded by the transposon itself, enables it to recognize and act on the target sequence in the host chromosome. The transposon can then, with the aid of another enzyme usually encoded by the transposon, incorporate itself into the host chromosome. Other transposons have enzymes able to effect insertion anywhere in the genome. Like viruses and plasmids, transposons can sometimes pick up additional genes from the main chromosome (Fig. 10.8). There is an obvious similarity between transposons, temperate viruses, and those plasmids capable of incorporating themselves into a chromosome: each generally encodes the enzymes it needs to orchestrate its movements. The likely evolutionary relationships between these genetic entities are outlined in Figure 10.9.

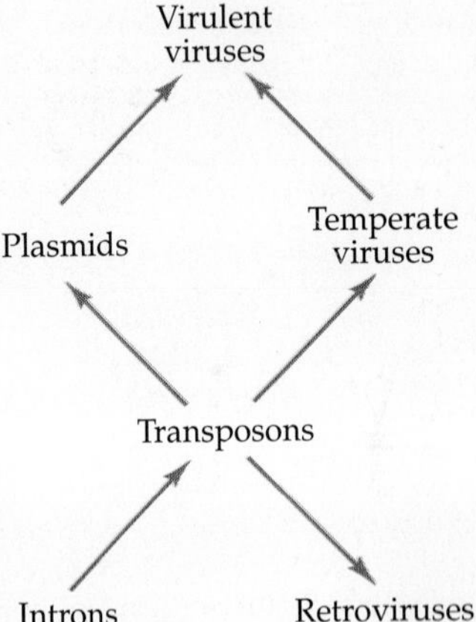

10.9 Probable evolutionary relationships between mobile genetic entities
Although it is not yet possible to trace the evolution of mobile genetic elements with certainty, it seems likely that the development of self-splicing introns may have been the first step. (The likely adaptive value of early introns will be discussed in Chapter 17) The next step would have been the evolution of a system to allow reinsertion of excised introns, producing transposons. Transposons could then have given rise to three new entities: (1) the subsequent evolution of RNA transposons would open the way for the development of retroviruses; (2) defective transposons, trapped in their independent circular state, could have been the source of plasmids; and (3) transposons that had developed the ability to move between cells could have evolved into the less benign entities we know as temperate viruses. Virulent viruses could have evolved from either plasmids or temperate viruses. Though it is not indicated here, at least some of these steps are reversible: temperate viruses, for example, could lose their ability to switch cells, and thus return to the status of transposons.

TABLE 10.1 *Sources of large-scale genetic change*

Phenomenon	*Mechanism*	*Possible consequences*
INSERTION	Transformation	New genes from a dead cell imported from surrounding medium and incorporated into chromosome
INSERTION	Transduction	New genes accidentally picked up from previous host and imported into cell by a virus
INSERTION	Plasmid insertion	Existing gene or genes become integrated into genome, and subject to novel controls
INSERTION	Lysogenic insertion	Novel genes of temperate phage inserted into host genome
INSERTION	Retroviral insertion	cDNA copy of novel genes of retrovirus inserted into genome
INSERTION	Intron insertion	Excised introns inserted into genome, mainly at exon-exon junctions in cDNA insertions
INSERTION; DUPLICATION	Retroinsertion	cDNA copies of transcribed host DNA incorporated into genome, providing duplicate copies of genes
DUPLICATION	Breakage and fusion	Part of one chromosome breaks off and fuses to the end of another during gamete formation; some gametes may obtain duplicate copies of genes on the broken fragment
DUPLICATION	Unequal crossing over	Chromosomes may be misaligned during a complex process called crossing over (to be described in Chapter 12); some gametes may obtain duplicate copies of some genes
GENE MOVEMENT	Transposition	Chromosomal DNA moved with genome, or both duplicated and moved

Other mechanisms of movement Although of less use in genetic engineering, there are other ways genes can move—ways we will refer to later in our discussions of chromosomal organization, cancer, immunology, and evolution. We will touch on them here only briefly. Perhaps the most intriguing case is the occasional instance in which an mRNA is mistakenly bound by reverse transcriptase, serving as a template for the creation of novel cDNA, which may then be incorporated back into the chromosome. Our genome has many such cDNAs, which are easily recognized from their lack of introns (missing in mRNAs after processing) and their tell-tale caps and poly-A tails (absent in normal genes, but added to mRNA during processing).

Genes can also be moved when chromosomes break and the pieces are reattached in the wrong places, or when errors occur during mitosis. The range of possible mechanisms for gene movement is summarized in Table 10.1.

GENETIC ENGINEERING

Artificial transformation The agents of gene mobility, especially reverse transcriptase and plasmids, play a major role in ***recombinant DNA*** techniques for inserting selected genes from one kind of organism into cells of another kind of organism. These techniques have great potential: specific genes can be isolated, introduced into bacteria, and used to produce large quantities of a desirable gene product—insulin, for example. Alternatively, new genes can be introduced into plants or animals that might benefit from them—genes for disease resistance in plants, let us say, or genes for growth hormones in domestic animals. However, recombinant DNA techniques have become a subject of public debate because of concern over the possibility of accidentally producing new pathogens or developing "genetic monsters." Let us take a look at how the basic technique works.

Plasmids, like DNA from bacterial chromosomes, can transform bacterial cells. In other words, if bacterial cells of certain species are placed in a medium containing free plasmids, some of the cells will pick up the plasmids, just as some of the nonvirulent pneumococci studied by Griffith picked up DNA that had been released into the medium by dead virulent pneumococci (see p. 215). Thus nonresistant bacteria can be transformed into resistant ones by exposure to a medium in which bacteria with plasmids for antibiotic resistance have been killed. Recombinant DNA technology makes use of this transforming potential of plasmids: purified plasmids are modified by the addition of foreign genetic material; when bacterial cells then pick up the modified plasmids, they acquire the foreign genes.

To follow the procedure in more detail: Bacterial cells containing plasmids are broken up and their DNA is extracted. The DNA is then centrifuged to separate the plasmids from the main bacterial chromosomes. The purified plasmids are next exposed to a restriction endonuclease, one of a class of procaryotic enzymes that cleave the DNA circle at a certain particular nucleotide sequence (Fig. 10.10). The plasmid DNA is now linear, with "sticky" ends (unpaired bases) where it was cleaved. It is next mixed with fragments of foreign DNA prepared with the same restriction endonuclease and therefore equipped with sticky ends complementary to those of the plasmid DNA. In such a mixture under the appropriate conditions of temperature, pH, and so on, the plasmid DNA and the foreign DNA spontaneously anneal by complementary base pairing, re-forming a circle in the

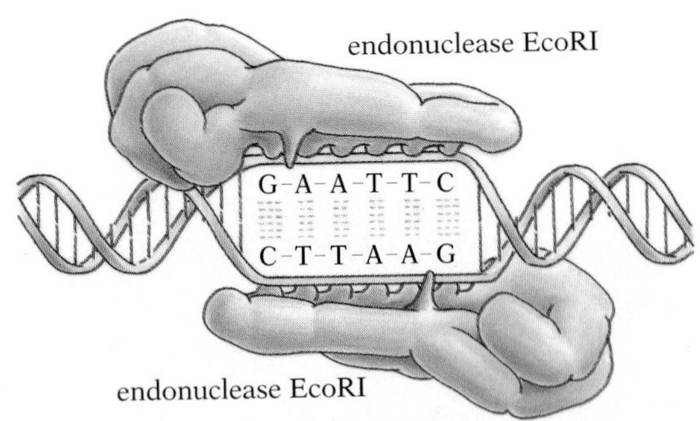

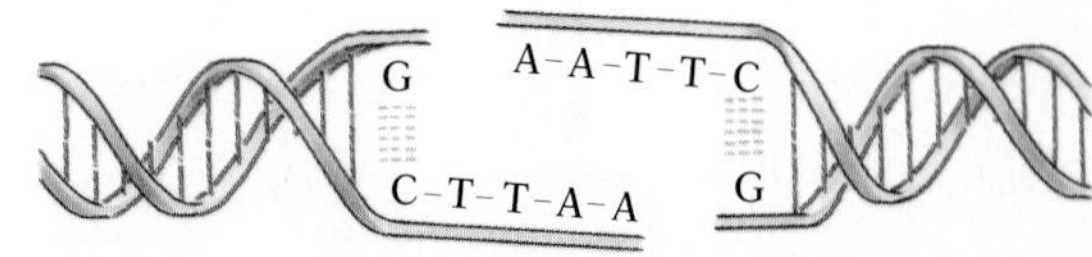

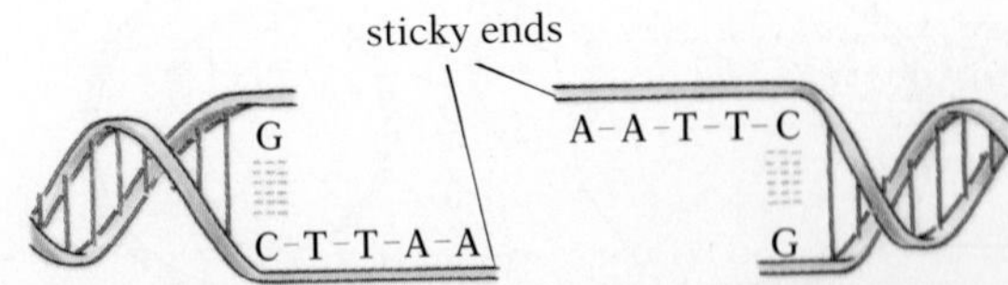

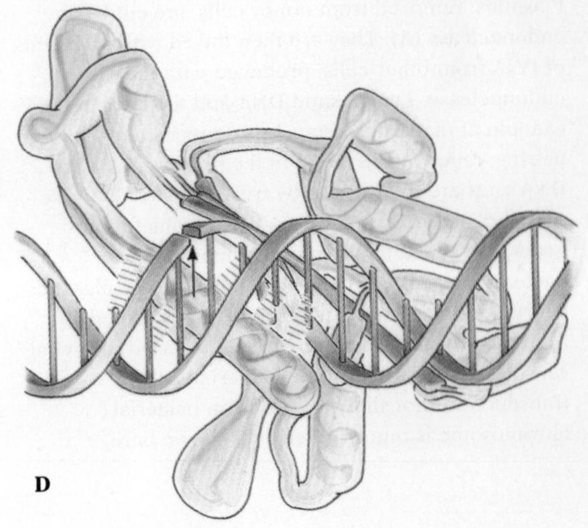

10.10 Endonuclease action

(A) Restriction endonucleases bind to a pair of target sequences—GAATTC in the case of the enzyme *Eco*RI. (B) They break a particular phosphate bond in the backbone of each strand of the DNA. (C) The resulting cut ends with unpaired bases are called sticky ends because under favorable conditions they will pair with cut ends having the complementary sequence of bases. (D) A more detailed look at a restriction endonuclease. One *Eco*RI attaches to a target sequence; the other endonuclease is omitted for clarity. The blue portion of the enzyme is important for recognizing the target sequence; the binding is indicated by dashed lines. The red portion is involved in weakening the bonds in the DNA; the arrow indicates where the helix is cut. More than 100 endonucleases are known, each with its own specific target sequence.The normal function of endonucleases is to digest infecting viral DNA.

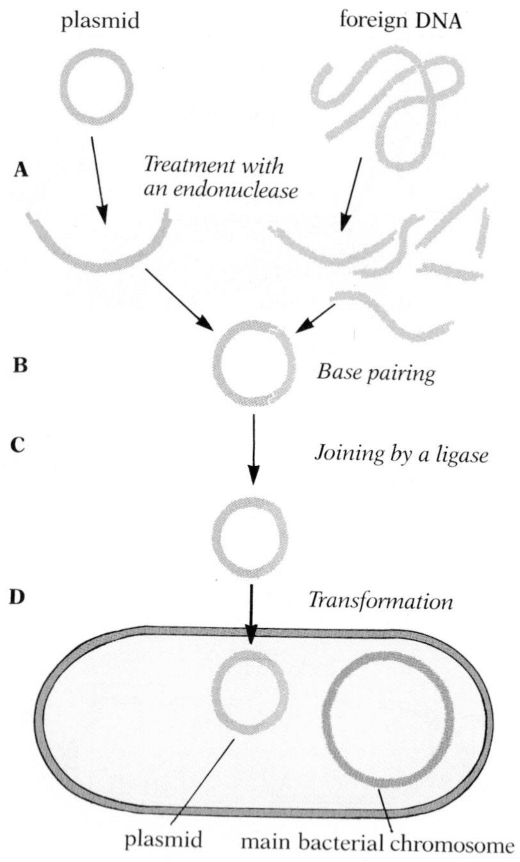

10.11 The recombinant DNA technique
Plasmids, removed from donor cells, are cut by an endonuclease (A). They are then mixed with fragments of DNA from other cells, produced with the same endonuclease. The plasmid DNA and the foreign DNA can join at their sticky ends by complementary base pairing, re-forming a circle in the process (B). The DNA ends are then sealed by treatment with a ligase (C). The modified plasmids, some bearing foreign genes, are added to a medium containing live bacterial cells. Some of the cells pick up the plasmids bearing foreign genes and are thus transformed (D). Alternatively, the plasmids can be packaged into viral capsules and inserted in the bacterial cells by transduction (not shown). The main bacterial chromosome is much larger than shown here.

process (Fig. 10.11). The backbones (chains of alternating phosphate and sugar groups) of the DNA circle can then be sealed with the enzyme DNA ligase. The end result is plasmids that contain a graft of foreign genetic material. All that remains to be done is to mix these plasmids with bacterial cells (usually treated with a calcium salt to make them more permeable). The bacterial cells will pick up the modified plasmids, which include both the original plasmid genes and fragments of the foreign genome. For more efficient transfer, transduction can be used: the hybrid plasmids are mixed with viral coat proteins, which will self-assemble around the plasmid if it is not too large; these plasmid-carrying viruses can then be used to infect target cells. In general, plasmids containing genes that confer antibiotic resistance are used in these processes; this means that if the experimenter treats the bacterial cells exposed to the recombinant plasmids with antibiotic, only the bacteria that have incorporated hybrid plasmids will survive, and bacteria lacking foreign genes will be eliminated.

One disadvantage of this approach is that the desired combination of plasmid and foreign DNA is not the sole result. After endonuclease treatment, the foreign DNA is left as a mixture of various fragments, of which only one may be of interest. Indeed, the endonuclease may well have found its target sequence within a particular gene, and so cut the gene itself. Repeating the process with a different endonuclease may yield fragments with a gene intact. Rigorous screening is often required to isolate the bacteria with plasmids that have incorporated complete copies of the desired genes. Worse yet, if the foreign DNA is from a eucaryote, as it often is in recombinant DNA research, it will contain introns, which bacteria are unable to remove before translation. This creates a serious problem for commercial applications. To avoid these and other problems, many researchers use yeast, a simple eucaryote. Others use more selective techniques to isolate and incorporate desirable genes into bacteria. One of these techniques is gene cloning.

Cloning genes The crucial step in one method of mass-producing exact copies or ***clones*** of a particular eucaryotic gene is to find cells that specialize in manufacturing that gene's product—pancreatic cells, for instance, if the desired product is insulin. The cytoplasm of such cells will have a high concentration of mRNA molecules coding for their special product, and these mRNAs will, of course, already have undergone intron removal in the nucleus. Many techniques exist to separate out the particular kinds of mRNA on the basis of physical characteristics like weight, so the mRNA found in unusual abundance in the specialist cells can be isolated. A variety of other tricks are available to identify the mRNA of genes that are only rarely transcribed. (Another method of gene cloning, the polymerase chain reaction, is described in the *Exploring Further* box on page 268).

Once the appropriate mRNA has been isolated, the next step is to produce from it the corresponding single-stranded DNA, using reverse transcriptase from a retrovirus. DNA polymerases are then used to replicate the DNA strand by complementary base pairing, a process that supplies the second strand for the double-helical structure of the transcript. The procedures already discussed for inserting foreign DNA into plasmids are then utilized: a restriction endonuclease cuts open the plasmid, creating a pair of sticky ends; the cloned DNA (equivalent to the cDNA produced by normal retroviruses), with a complementary set of sticky ends, is added, and the plasmid DNA and cDNA anneal; a ligase restores the bonds in the DNA backbone; and the plasmid is inserted into a bacterial host. This procedure has two immediate advantages: the expensive process of sorting for bacteria bearing the desired gene is eliminated, since only the appropriate cDNA is used; and introns, which cannot be removed by procaryotes, have already been eliminated in the production of the mRNA. The transformed bacterial cells grow and divide rapidly, creating limitless numbers of bacteria that may synthesize the desired product, particularly if the host plasmid has signals that enhance transcription of the cDNA. Besides insulin, this technique is used to produce large quantities of other hormones—growth hormone is particularly important—that are difficult to synthesize. Recombinant bovine growth hormone is widely used to boost milk production by 10–40 percent, and synthetic human growth hormone is now employed to prevent stunted growth in hormone-deficient children. A naturally produced but poorly understood agent called interferon, which sensitizes the immune system and so holds great promise in the treatment of various diseases, is also now widely available for research and medical applications as a result of recombinant DNA technology.

Recombinant techniques have also made possible several new approaches to vaccination. The one that seems most promising for finally defeating the common cold takes advantage of the need of viruses to locate and bind to the exposed portion of a particular membrane protein in their host's cells. Researchers can clone the portion of the gene that encodes the extracellular part of the target protein, and then mass-produce this fragment. When the body of a potential victim is flooded with these molecular decoys, the viruses attach themselves to the free-floating target segments, and so render themselves harmless. Many years of testing, however, will be needed before such treatments become widely available.

Gene cloning makes it possible not only to use the host cells as chemical factories that produce substances of medical or commercial importance, but also to study the sequencing and activity of genes from eucaryotic cells. Indeed, one of the many spin-offs from recombinant DNA technology is a method of mapping genes with enormous precision. A piece of single-stranded DNA is transcribed from an mRNA of known function in a medium containing the radioactive isotope ^{32}P. This cDNA is then mixed with a set of chromosomes treated to separate the double helix into single strands. When base pairing is again made possible in the resulting mixture, the cDNA frequently pairs with the corresponding gene on the chromosome (Fig. 10.12). Because the cDNA carries a radioactive label, the exact loca-

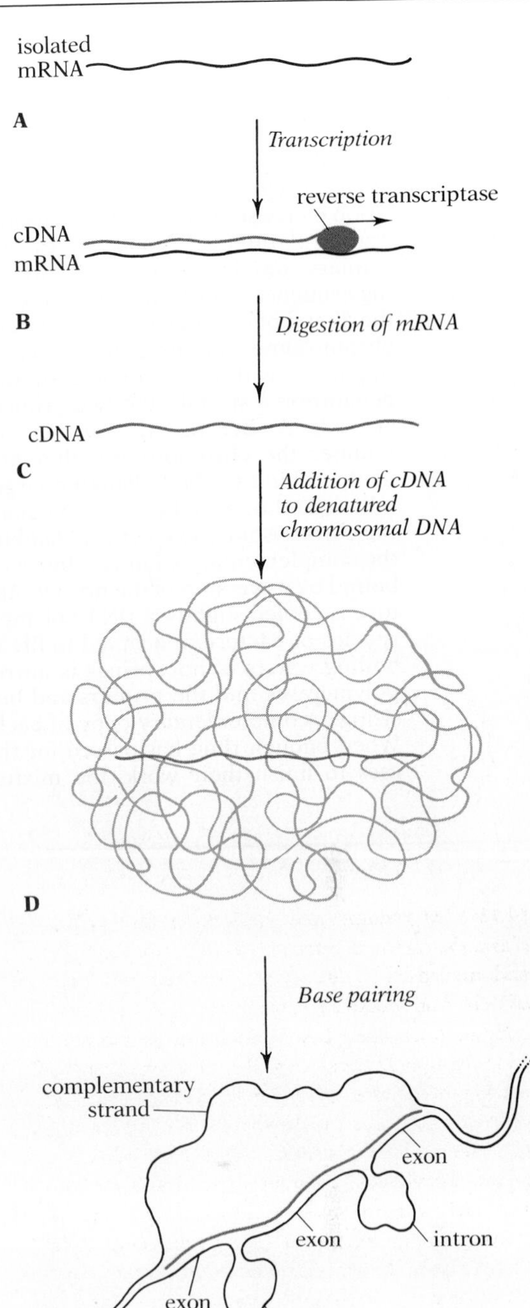

10.12 Locating a gene with a cDNA probe
A particular kind of mRNA is isolated (A), and reverse transcriptase is used to make a cDNA transcript incorporating a radioactive label (B). Then RNAse is added to digest the mRNA (C). The cDNA is next mixed with denatured chromosomal DNA (D). During base pairing, the cDNA binds to the gene coding for the mRNA, thus showing its location on the chromosome (E). Note that the intron segments loop out. They are not matched on the cDNA because they were absent from the mRNA.

Exploring Further:

THE POLYMERASE CHAIN REACTION

Another popular way of making a vast number of copies of a gene depends on knowing the composition of a base sequence flanking each end of the region of interest. Many copies of short DNA primers complementary to the two known flanking sequences must then be synthesized, one for the 3′ end of the region on each strand. After the chromosome with the gene of interest is cut into fragments with an endonuclease and heated to denature the strands, the two primer sequences are added. Because the primers greatly outnumber the chromosomes, they are far more likely to bind to the 3′ flanking target sequence on a strand than is the complementary strand of the chromosome; so too, the 3′ flanking region of the complementary strand is almost certain to be bound by its version of the primer. After the mixture is cooled slightly, a DNA polymerase from a species of bacterium adapted to life in the near-boiling waters of hot springs is introduced. The polymerases find the primers and begin synthesizing a complementary copy of each fragment. When enough time has passed for the polymerases to finish their work, the mixture is again heated to denature the two strands of each fragment. Once more, free primers will bind to the largest sequences, now present on four strands (the two original strands of the fragment with the gene of interest, plus two copies), and the mixture is again cooled to allow the polymerases to do their job. With each cycle of heating and cooling the number of copies of the original fragment doubles, enabling researchers to generate a million replicas in as little as eight hours.

This ***polymerase chain reaction*** technique (PCR) allows indefinite amplification and analysis of genes even if there is only a single copy present at the outset. In addition to the many obvious applications of the procedure to the study of genetic mechanisms, PCR has been used for a number of unusual tasks, ranging from establishing the identity of criminals based on the DNA in a single hair, spot of blood, or drop of semen, to studying the evolutionary relationships of extinct organisms using nothing more than a minute quantity of surviving mitochondrial DNA in museum specimens or fossils.

---

tion on the chromosome of the gene to which it binds is revealed (see Fig. 11.6, p. 287).

Scientists have begun the complete mapping of the human genome. Such an effort would begin with cDNA localization of most or all of the 50,000 genes, followed by base-sequence determination of the entire genome. (See *Exploring Further* box: DNA Sequencing.) There is considerable debate about whether wholesale sequencing is worth the billions of dollars it is likely to cost, especially since other projects will necessarily go unfunded to allow this initiative, and it will mean a huge redirection of talent from other genetic engineering projects—all for the rather mundane task of sequencing a set of chromosomes, 90–99 percent of whose DNA (as we will see in the next chapter) appears never to be transcribed. In the long run, however, there seems little doubt that the human genome project, if fully carried out, will make possible many kinds of therapeutic intervention, and cast considerable light on chromosomal organization and evolution.

Other spin-offs of recombinant DNA research include techniques for quickly and precisely identifying nucleotide sequences in DNA, methods of rearranging genes and their components to study gene interaction and gene control, and a technique for creating specific mutations at specific sites in genes. This last procedure makes possible precise investigation of the roles

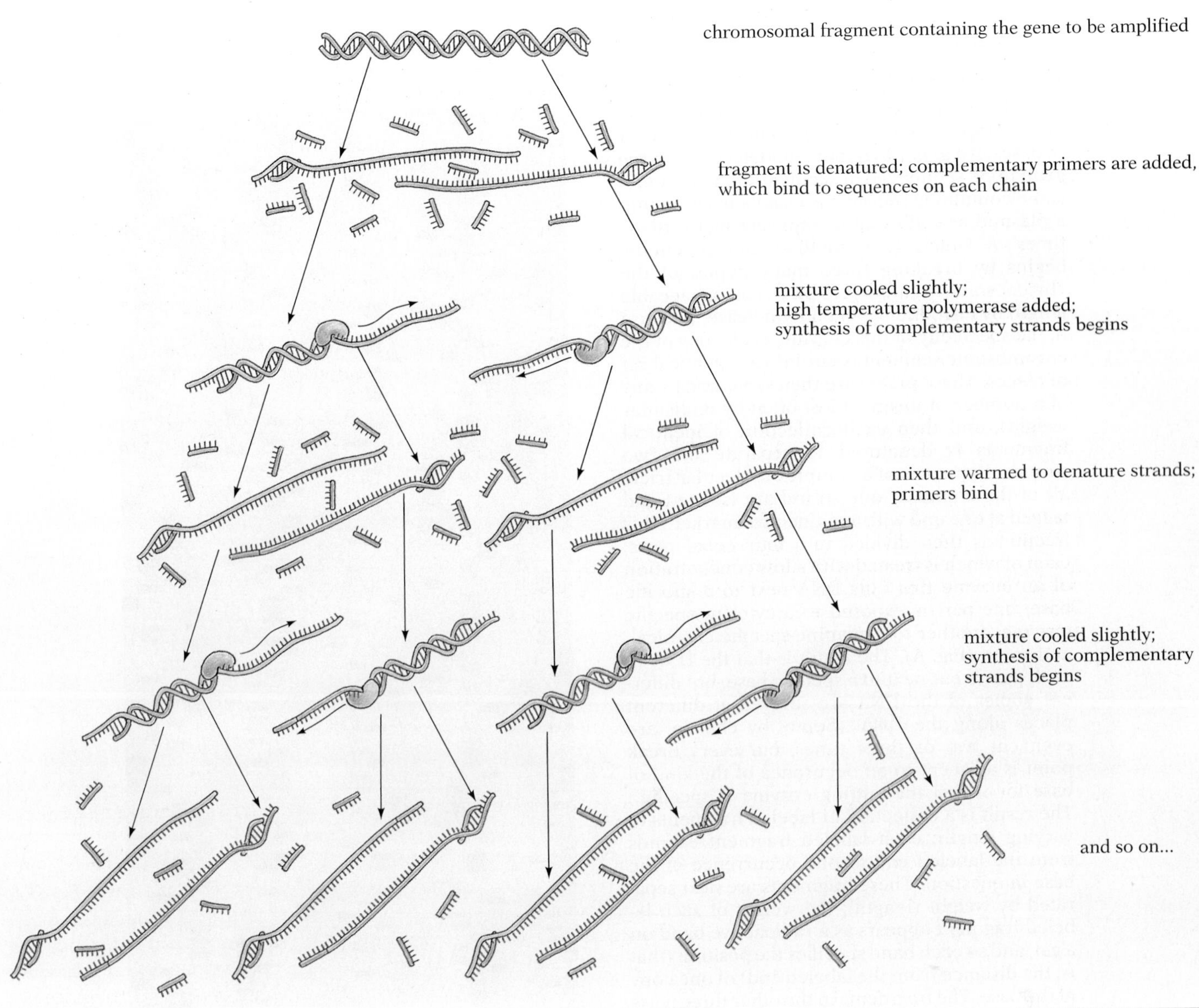
chromosomal fragment containing the gene to be amplified
fragment is denatured; complementary primers are added, which bind to sequences on each chain
mixture cooled slightly;
high temperature polymerase added;
synthesis of complementary strands begins
mixture warmed to denature strands;
primers bind
mixture cooled slightly;
synthesis of complementary
strands begins
and so on...

of signal sequences in DNA and mRNA, and also enables the researcher to alter proteins at will to explore the bases of enzyme function.

Looking into the future, some researchers have suggested using cDNA technology to transfer genes for C_4 photosynthesis to some crop plants that normally depend exclusively on C_3 photosynthesis, for the sake of greater productivity under less than optimal growing conditions. Others have proposed transferring genes for the fixation of atmospheric nitrogen to crop plants, thus eliminating the need for application of nitrogenous fertilizers. Both transfers would be formidable undertakings. A more modest but

Exploring Further:

DNA SEQUENCING

One of the cornerstones of recombinant technology is DNA sequencing—determining the order of bases in a gene. To obtain enough copies of the gene to be sequenced, a segment of the chromosome containing the gene is usually inserted into a plasmid and allowed to replicate many, many times. A widely used method for sequencing begins by breaking these many copies of the chromosome segment into pieces of manageable length by introducing an endonuclease. Because of the specificity of this enzyme, each copy of the chromosome segment is cut into an identical set of pieces. These pieces are then separated by any of a number of means (most often by molecular weight), and then each collection of identical fragments is denatured to separate the two strands. By means of a complex molecular trick, all of the copies of one strand are isolated and tagged at one end with a radioactive marker. This fraction is then divided into four equal parts, each of which is treated with a low concentration of an enzyme that cuts DNA next to a specific base; one part is exposed to a cytosine-specific enzyme, another to a guanine-specific chemical, and so on (Fig. A). The result is that the DNA in each part is cut next to a specific base, but different copies of the DNA are severed at different places along the DNA. (Some, by chance, are even cut two or three times, but every break point is adjacent to an occurence of the kind of base for which the cutting enzyme is specific.) The result is a collection of labeled fragments of varying length; each labeled fragment extends from the labeled end to one occurrence of the base in question. These fragments are next separated by weight (length); the weight of each labeled fragment appears as a radioactive band on a gel, and so each band specifies the position (that is, the distance from the labeled end) of one copy of the base. The fragments in the other three parts provide similar information about the location of the other three bases. When the resulting bands from each sample are aligned, the sequence of bases can be read out directly (photo, right).

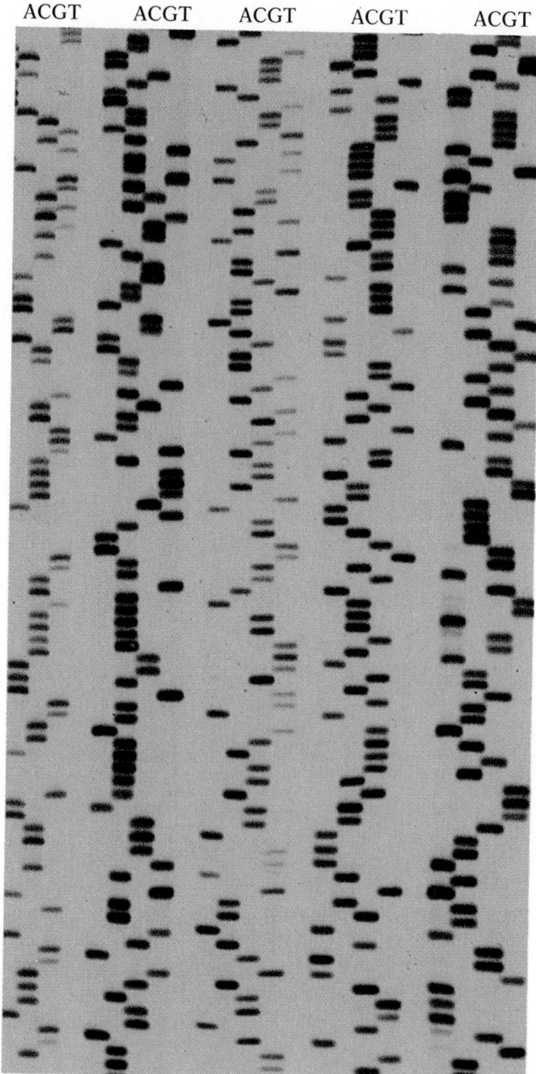

Nucleotide sequences produced by column chromotography; each sample generates a group of four vertical bands.

The same procedure must be repeated for each sample in order to sequence the entire set of chromosome pieces. At this point, however, there is still no way to put the sequences into the proper order with respect to each other. This is done by treating another collection of fragments of the same chromosome with a different endonuclease, and then sequencing those fragments (Fig. B). Because the second endonuclease will cut the DNA at different spots, the break points of the segments from the first analysis will lie within the sequences worked out in the second. Thus, by looking for the beginnings and ends of the sequences from the first digestion *within* the sequences from the second digestion, the relative order of the fragments can be established.

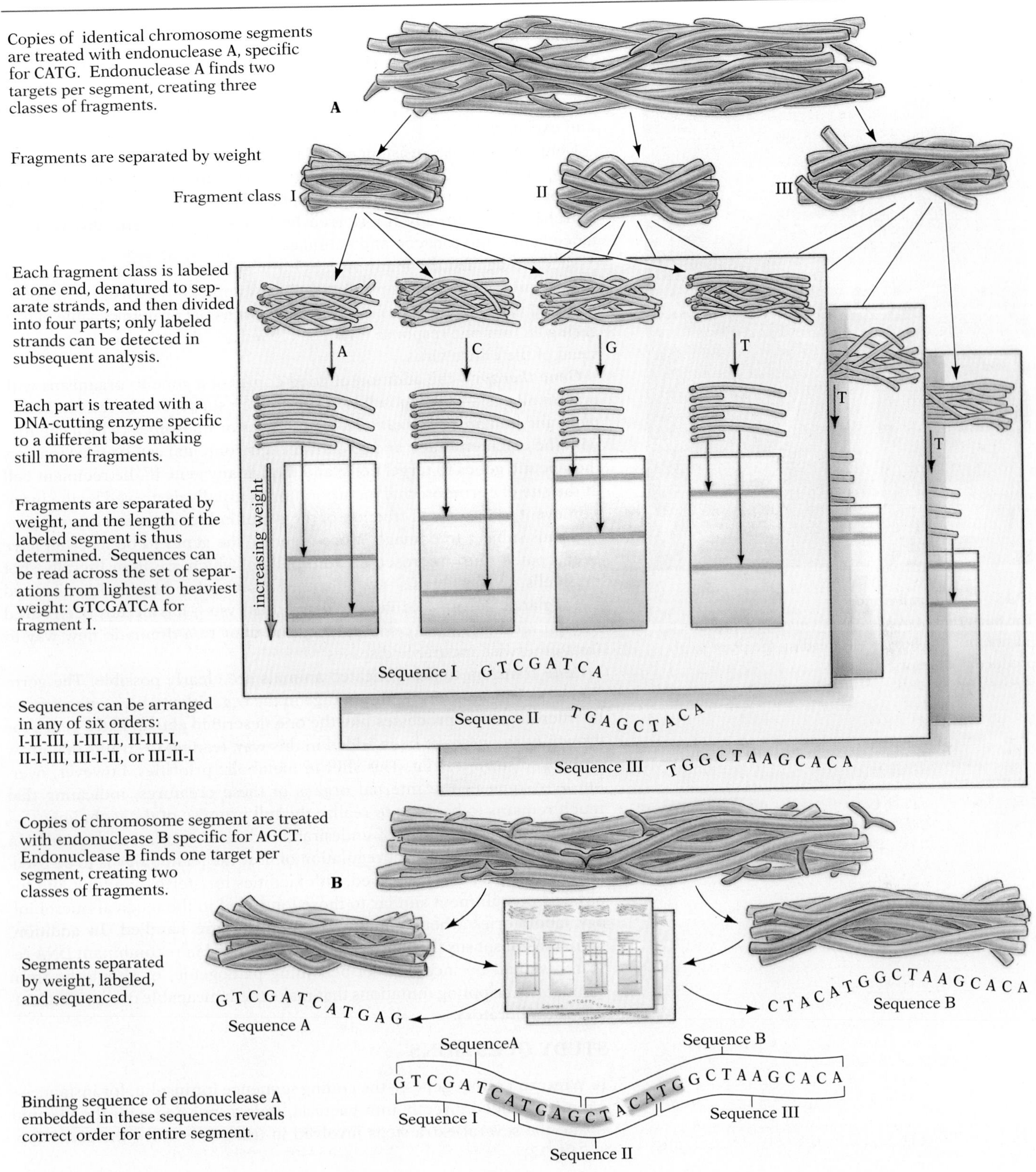
Copies of identical chromosome segments are treated with endonuclease A, specific for CATG. Endonuclease A finds two targets per segment, creating three classes of fragments.
A
Fragments are separated by weight
Fragment class I
II
III
Each fragment class is labeled at one end, denatured to separate strands, and then divided into four parts; only labeled strands can be detected in subsequent analysis.
A
C
G
T
Each part is treated with a DNA-cutting enzyme specific to a different base making still more fragments.
Fragments are separated by weight, and the length of the labeled segment is thus determined. Sequences can be read across the set of separations from lightest to heaviest weight: GTCGATCA for fragment I.
increasing weight
Sequence I GTCGATCA
Sequence II TGAGCTACA
Sequence III TGGCTAAGCACA
Sequences can be arranged in any of six orders: I-II-III, I-III-II, II-III-I, II-I-III, III-I-II, or III-II-I
Copies of chromosome segment are treated with endonuclease B specific for AGCT. Endonuclease B finds one target per segment, creating two classes of fragments.
B
Segments separated by weight, labeled, and sequenced.
GTCGATCATGAG
Sequence A
CTACATGGCTAAGCACA
Sequence B
SequenceA
Sequence B
GTCGATCATGAGCTACATGGCTAAGCACA
Sequence I
Sequence III
Sequence II
Binding sequence of endonuclease A embedded in these sequences reveals correct order for entire segment.

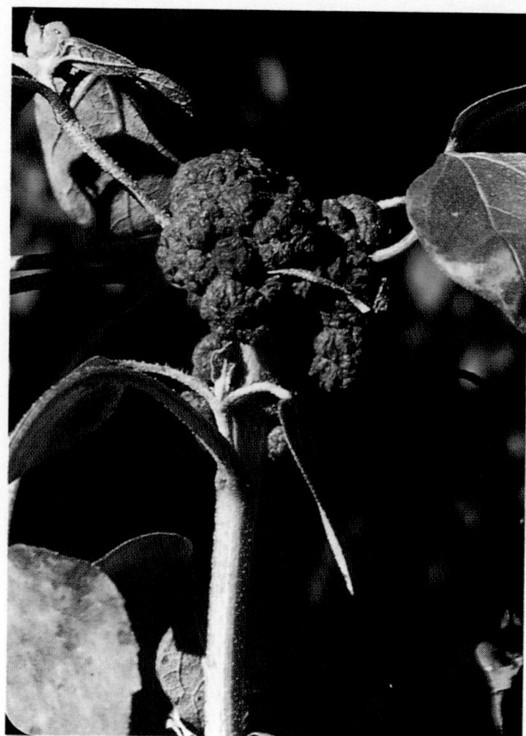

10.13 Crown gall tumor
The bacterium *Agrobacterium tumefaciens* inserts a plasmid into its host's cells, leading the plant to produce the tumor.

equally important undertaking seems almost certain to succeed: introduction of genes for proteins rich in amino acids normally rare in grains or beans, but which are essential in the human diet. The result would be a complete food, as nutritious as meat and dairy products, but far less fatty and expensive.

Genes for disease resistance might be transferred to susceptible crops. Using the bacterium responsible for crown gall disease (Fig. 10.13)—a pathogen that inserts its ten-gene plasmid into the host genome—researchers have already transferred herbicide-, insect-, and disease-resistance genes into tobacco and petunias.

Vaccines against the many viruses for which conventional vaccine techniques fail could be mass-produced. For example, if the viruses cannot be cultured in large quantities, a gene coding for a coat protein capable of triggering an immune response might be cloned and the gene product used instead of the entire virus.

Gene therapy—the addition of good copies of a gene to organisms with mutant alleles—is also possible, and has even succeeded in tests in *Drosophila* and humans. For example, retroviruses or liposomes (which can fuse with the cell membrane and so deliver their contents) can be used to carry therapeutic genes to target cells, and though any gene in the recipient cell at the site of chromosomal incorporation is usually destroyed by the insertion event, only a small fraction of the genes in the mature cell are active, and thus subject to damage. More recently, the genes to be inserted have been coated onto microscopic gold pellets, which are then literally shot into cells. Cystic fibrosis—an inherited disease that blocks digestive and sweat glands, causing painful and dangerous cysts—has already been cured in cultured laboratory cells, opening the door to a dramatic new way to treat otherwise incurable diseases.

Gene transfers to domesticated animals are clearly possible. The germ line is altered by effecting the change in the egg, either with a retrovirus or by microinjection processes like the one described above. Extra genes for growth hormone have been added in this way, leading to strains that grow larger and store less fat. This shift in metabolic priorities, however, overburdens some of the internal organs of these creatures, indicating that much remains to be done to realize the full promise of the technique.

Recognition of the risk of undesirable side effects of recombinant DNA technology has led to strict regulation of how the technology is to be used. Laboratories must be equipped with facilities for sterile handling and for physical containment similar to those long used in the medical microbiology laboratories where dangerous pathogens are handled. In addition, there is a consensus that the microorganisms used in recombinant DNA research should be incapable of becoming pathogenic, or should be from strains with crippling mutations that make them incapable of surviving outside the laboratory.

STUDY QUESTIONS

1. Why can't cDNA genes—the coding sequence for insulin, for instance—be introduced directly into bacteria, thus dispensing with the plasmid and the several extra steps involved in using it to carry the gene? (pp. 264–72)

2. Construct a plausible scenario for the evolution of a functional RNA plasmid. How does it manage to maintain itself generation after generation? (pp. 257–64)

3. Now play the role of natural selection. How would you turn the plasmid into a working RNA virus, single- or double-stranded? (pp. 258–61)

4. What sort of evidence could convince you that plasmids evolved from viruses, rather than the reverse? (pp. 258–61)

5. Based on the evidence in Chapter 5 and in Chapters 8–9, summarize the case for the endosymbiotic hypothesis. If mitochondria and chloroplasts are really tame bacteria, why do you suppose that most but not all of their genes now reside in the nucleus (including many for products used only in organelles)? Surely it would have been simpler to have left them "on site," rather than to have to transport so many special products there; or, given that most of their enzymes are being taken to the organelle anyway, why leave any genes behind? (pp. 154–56, 224–26, 248–50)

CONCEPTS FOR REVIEW

- Plasmids
  - Sequence of events in conjugation
  - Timing and consequences of conjugation
- Transposons
- Viruses
  - Alternative viral "life histories"
    - Conventional DNA viruses
    - Temperate viruses
    - Retroviruses
  - Transduction
  - Viral evolution
- Recombinant DNA Technology
  - Role of endonucleases and plasmids
  - cDNA from mRNA
  - Use of cDNA probes
  - Problems and promise of recombinant technology

SUGGESTED READING

AHARONWITZ, Y., and G. COHEN, 1981. Microbial production of pharmaceuticals, *Scientific American* 245 (3). *On how recombinant DNA techniques are used to make microbes produce antibiotics, hormones, and other drugs. There is also an explanation of how antibiotics work to destroy bacteria, which suggests how plasmid genes may confer resistance.*

BROWN, D., 1973. The isolation of genes, *Scientific American* 229 (2). (Offprint 1278) *How a particular mRNA can be used to locate and purify the gene that codes for it.*

CAMPBELL, A. M., 1976. How viruses insert their DNA into the DNA of the host cell, *Scientific American* 235 (6). (Offprint 1347)

CHILTON, M-D., 1983. A vector for introducing new genes into plants. *Scientific American* 248 (6). (Offprint 1539) *On bacteria (as opposed to viruses) that transduce host cells.*

CLOWES, R. C., 1973. The molecule of infectious drug resistance, *Scientific American* 228 (4). (Offprint 1269) *Experiments demonstrating that the bacterial genes for antibiotic resistance are carried on plasmids and can be transmitted from one bacterium to another.*

COHEN, S. N., and J. A. SHAPIRO, 1980. Transposable genetic elements, *Scientific American* 242 (2).

DEVORET, R., 1979. Bacterial tests for potential carcinogens, *Scientific American* 241 (2). (Offprint 1433) *The close relationship between mutations and cancer, and how to measure mutagenicity.*

FEDOROFF, N. V., 1984. Transposable genetic elements in maize, *Scientific American* 250 (6). *A modern interpretation of the transposition discovered by McClintock.*

GILBERT, W., and L. VILLA-KOMAROFF, 1980. Useful proteins from recombinant bacteria, *Scientific American* 242 (4). (Offprint 1466) *How recombinant methods can be used to create insulin-producing bacteria.*

GRIVELL, L. A., 1983. Mitochondrial DNA, *Scientific American* 248 (3). (Offprint 1535) *On the procaryotelike organization of mitochondrial genes, and their unique modification of the genetic code.*

HOPWOOD, D. A., 1981. The genetic programming of industrial microorganisms, *Scientific American* 245 (3). *An excellent summary of how basic recombinant DNA techniques work.*

KAPLAN, M. M., and R. G. WEBSTER, 1977. The epidemiology of influenza, *Scientific American* 237 (6). (Offprint 1375) *On how genetic recombination between human and animal strains of the influenza virus is probably responsible for the appearance of new subtypes of the virus.*

Novick, R. P., 1980. Plasmids, *Scientific American* 243 (6). (Offprint 1486)

Varmus, H., 1987. Reverse transcription, *Scientific American* 257 (3). *A detailed description of the process that reverses the usual direction of information flow.*

Verma, I. M., 1990. Gene therapy. *Scientific American*, 263 (5). *On attempts to correct genetic defects.*

Weinberg, R. A. 1985. The molecules of life, *Scientific American* 253 (4). *A good, very brief review of recombinant DNA techniques.*

Chapter **11**

CONTROL OF GENE EXPRESSION

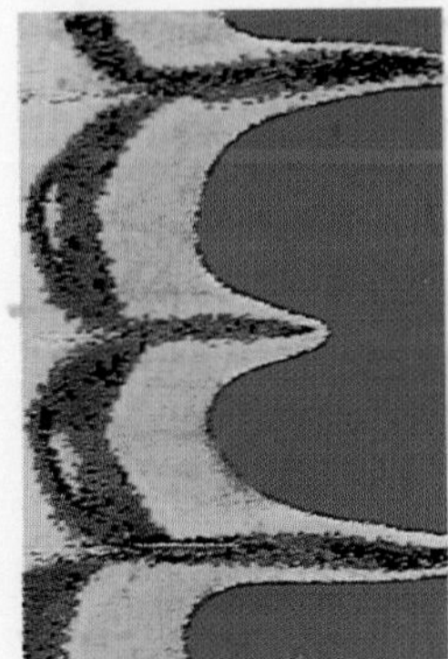

very cell in the body of a multicellular organism receives a complete set of chromosomes, a replicate of the DNA in the original cell from which the organism developed. The nucleotide sequences of this DNA carry in full the genetic information that is the cell's evolutionary endowment. But though every cell in the body receives the same set of instructions, individual cells may look entirely different from other cells, and may behave in entirely different ways. In fact, only a particular subset of all the genes a cell contains will ever generate proteins in that particular cell, and only a small percentage of this DNA is active at any one time. We have already seen that many minor, up-to-the-minute adjustments of cellular chemistry can be made by shifts in enzyme activity, such as feedback inhibition in glycolysis (*Exploring Further*, p. 177). But major adjustments calling for, say, enzymes not previously necessary (as when a new source of food becomes available), require altering the expression of a cell's genes. In this chapter we will examine the logic and mechanisms of gene control, a process we now know involves chemicals that bind directly or indirectly to DNA or mRNA.

CONTROL OF GENE EXPRESSION IN BACTERIA

Early investigators of gene expression in bacteria made several assumptions about the process they sought to understand. First, it was logical to assume that proper control of gene expression would require that only those genes whose products were needed at any given moment be expressed. Sec-

A

B

11.1 François Jacob (A) and Jacques Monod (B).

ond, since most genes code for an enzyme that controls only a single step in a biochemical pathway, the genes coding for several enzymes in the same pathway might be expected to be controlled as a group. And furthermore, since the function of a pathway is to turn a reactant into a product, the availability of the reactant in the cell might be expected to turn on transcription, while the availability of the final product might turn it off.

These assumptions have proven to be correct in many cases of transcription in bacteria. Because bacteria possess only about 3000 genes (compared to the 50,000 of humans, for instance) and can be grown rapidly in huge numbers, they have been especially useful organisms for study of the control of gene transcription. The first models of gene control emerged from research on the intestinal bacterium *Escherichia coli.*

THE JACOB-MONOD MODEL OF GENE INDUCTION

In the course of an extended investigation of enzyme synthesis in *E. coli,* beginning in the late 1940s, the French biochemists François Jacob and Jacques Monod (Fig. 11.1) formulated a powerful model of gene regulation in bacterial cells. They concentrated on the enzyme β-galactosidase, which catalyzes the breakdown of lactose to glucose and galactose, substances both used and produced by other pathways. They were awarded the Nobel Prize in 1965 for their work.

Lactose is not continuously available to *E. coli,* and so—as would be expected—the gene for β-galactosidase is normally transcribed at a very low rate. Jacob and Monod found that the further production of this digestive enzyme is triggered by the presence of a so-called ***inducer***, in this instance allolactose, a derivative of lactose automatically produced in the cell when lactose is present. Normally, then, β-galactosidase is an ***inducible enzyme***.

Jacob and Monod were eventually able to demonstrate the participation of four genes in the production of β-galactosidase and the two other enzymes involved in lactose breakdown: three so-called ***structural genes***, each specifying the amino acid sequence of one of the three enzymes, and a ***regulator gene***, which controls the activity of the structural genes. They proposed that the regulator gene, which is located at some distance from the structural genes, normally directs the synthesis of a ***repressor*** protein that inhibits transcription of the structural genes.

Jacob and Monod also discovered that a special region of DNA contiguous to the structural genes for β-galactosidase determines whether transcription of the structural genes will be initiated; they called this special region the ***operator***, and they called the combination of the operator and its three associated structural genes an ***operon***. Subsequently it was found that the operator, which does not in itself constitute a gene since it doesn't code for a specific product, is located between the ***promoter***, the region to which RNA polymerase binds, and the structural genes. Hence, when the repressor binds to the operator, RNA polymerase cannot physically bind to the promoter, and transcription is blocked (Fig. 11.2A).

If inducer is present, it will bind to the repressor, thus causing a conformational change in the repressor that forces it to dissociate from the operator; in short, the inducer inactivates the repressor (Fig. 11.2B). Now free to bind to the promoter, RNA polymerase can initiate transcription of the

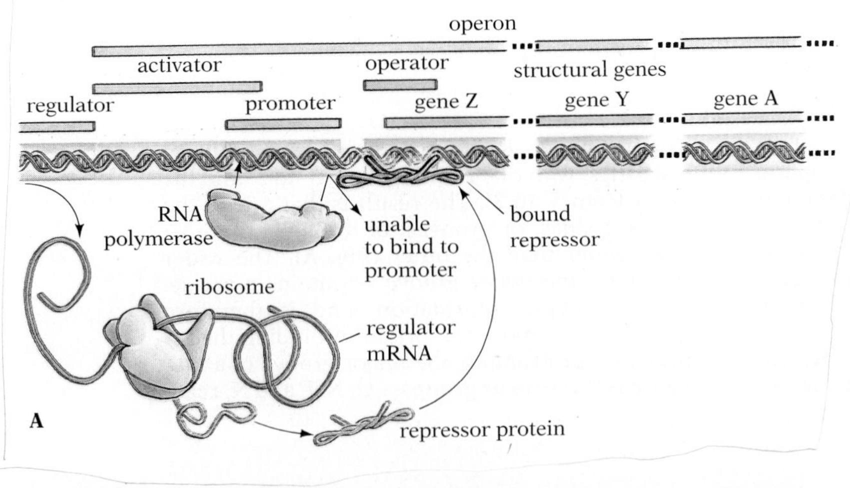

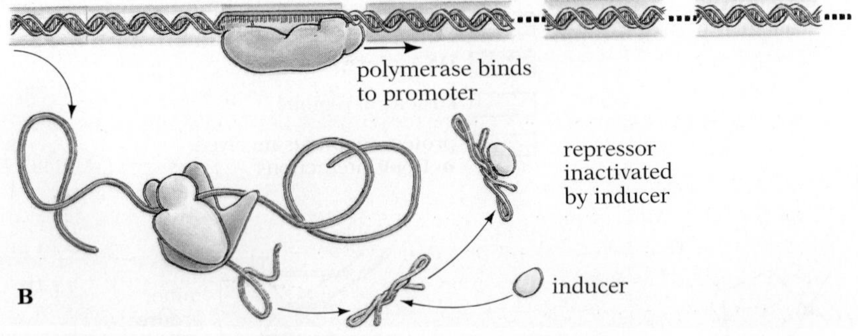

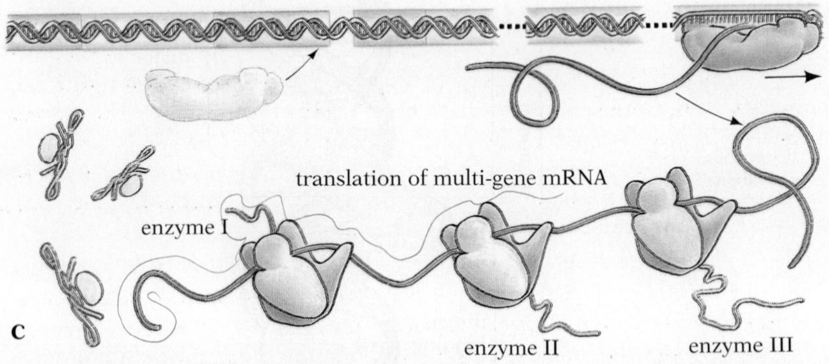

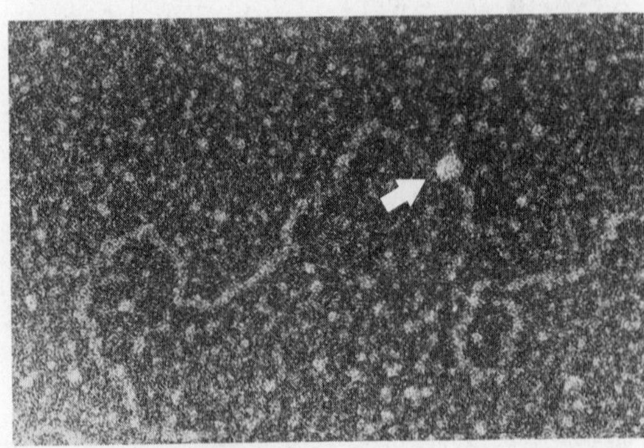

11.2 The *lac* operon as an example of an inducible operon
(A) The operon consists of a promoter/operator region and three structural genes (Z, Y, and A). The operator sequence overlaps the beginning of the structural gene. The regulator gene codes for mRNA, which is translated on the ribosomes and determines synthesis of repressor protein. When the repressor protein binds to the operator, it blocks one of the promoter's binding sites for RNA polymerase and thus prevents initiation of transcription of the structural genes. (B) Binding of inducer to the repressor inactivates the repressor, and the RNA polymerase can then bind to the promoter regions. (C) Polymerases initiate transcription of the structural genes, which are transcribed as a unit, producing an mRNA coding for three gene products. The mRNA then binds to ribosomes and is translated into three enzymes. Enzyme I is β-galactosidase; enzyme II is a permease that helps transport lactose into the cell; and enzyme III is a transacetylase, whose role in lactose utilization is not understood. The transcribed strand of the structural genes has been emphasized. The function of the activator sequence will be discussed presently. (D) Electron micrograph of the *lac* repressor protein bound to the operator region.

structural genes and the production of mRNA (Fig. 11.2C). The mRNA carries the instructions of all three structural genes. This messenger binds to ribosomes in the cytoplasm, where its information is translated and the three enzymes necessary for lactose metabolism are synthesized. The number of β-galactosidase enzymes rises to about 5000 per cell when the operon is not repressed; during repression, there are about 10 copies of the enzyme.

Exploring Further

HOW CONTROL SUBSTANCES BIND TO DNA

Heroic efforts have revealed many of the details of how a repressor protein may bind to an operator. For the repressor to bind to a particular operator, it must bear active sites that match the DNA substrate exactly. Such matching involves polar amino acids on the repressor and complementary polar groups exposed on the *sides* of the base pairs of the DNA.

The deoxyribose backbones of DNA are attached to the bases slightly off center, and as we saw in Chapter 8, the two chains have opposite polarities, with one running from 3′ to 5′ and the other from 5′ to 3′. The result is that one of the exposed sides, or grooves, of the helix is somewhat wider than the other (Fig. A). The wider opening—the ***major groove***—contains more sequence-specific information, and is therefore thought to be more important in binding the repressor. For example, the major groove of a T–A pair has three polar groups—O, NH, and N, read-

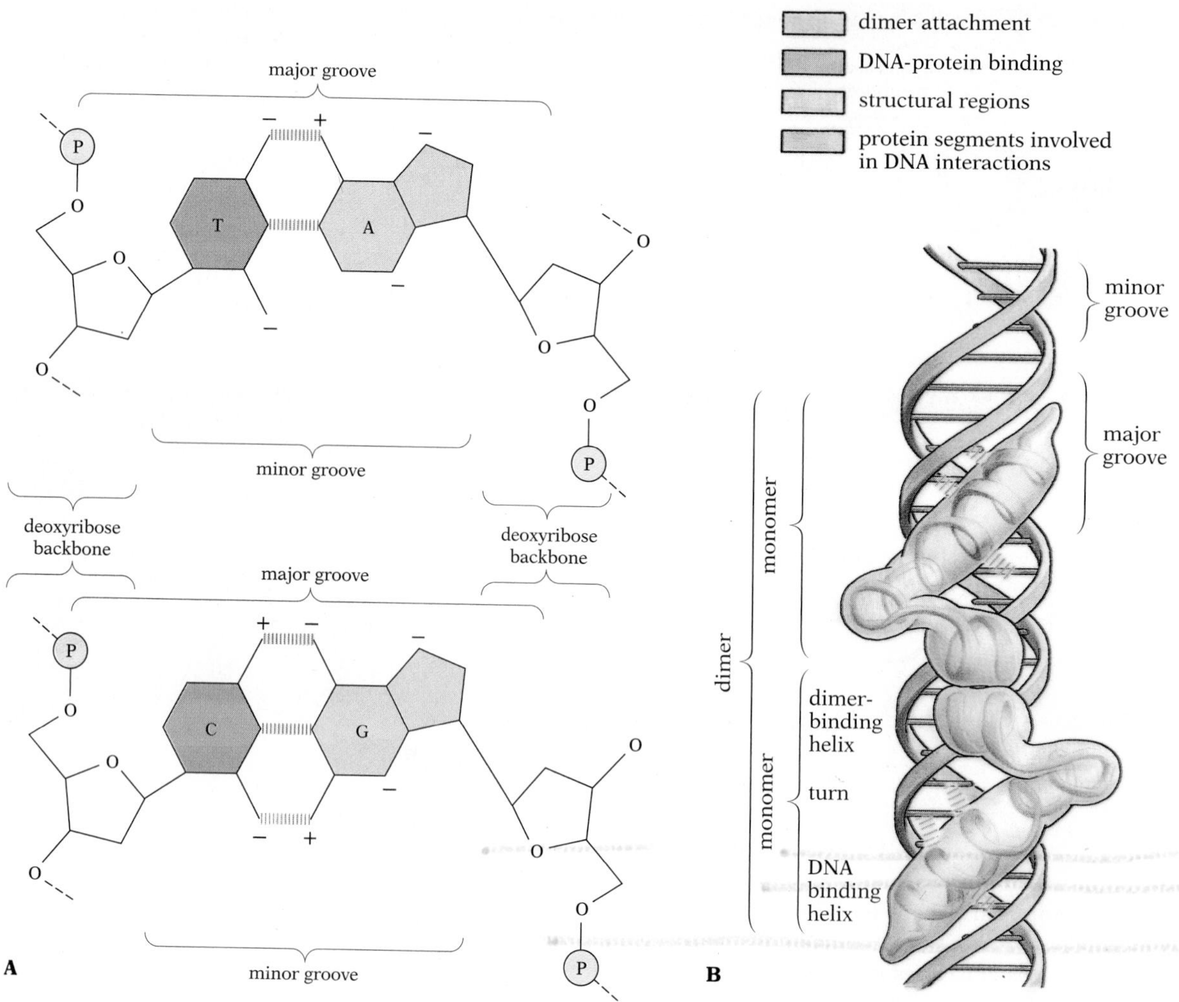

ing from T to A—their respective polarities being −, +, and −. The major groove of the C–G pair has the polar groups NH, O, and N, with polarities of +, −, and −. The ***minor groove*** is less useful: for T–A, the polar "code" is −, X, −. (where "X" means no charge), while for C–G the sequence is −, +, −. Each particular repressor has a shape corresponding to the differing physical widths of the operator's major and minor grooves, and polarity patterns complementary to those of the operator's base pairs.

Three general molecular morphologies are now recognized among DNA control substances. The "helix-turn-helix" motif involves two alpha helices, one of which binds in the major groove while the other attaches itself to a second copy of the binding chemical to create a dimer (Fig. B). The "leucine-zipper" is also a protein dimer with two helices. The shorter helix has a leucine at every seventh position, so that four of the leucine side groups project out on the same side of the short helix; these leucines interlock in a zipper-like fashion with the aligned leucine side groups of the other protein's short helix (Fig. C). The two long helices bind in the major groove of the DNA in what is called a "scissors grip." The third family of binding proteins also uses a dimer with a helix devoted to binding in the major groove. The "zinc-finger" proteins are stabilized by zinc ions, and (again) the dimers bind to each other as well as to the DNA (Fig. D).

Site-specific DNA binding is not restricted to repressor proteins. The same strategies are used by transcription-activating proteins, endonucleases, and some hormones, and it should someday be possible to create synthetic control proteins to regulate any gene, including those carried by disease-causing organisms, or those in cancerous cells that have begun to malfunction.

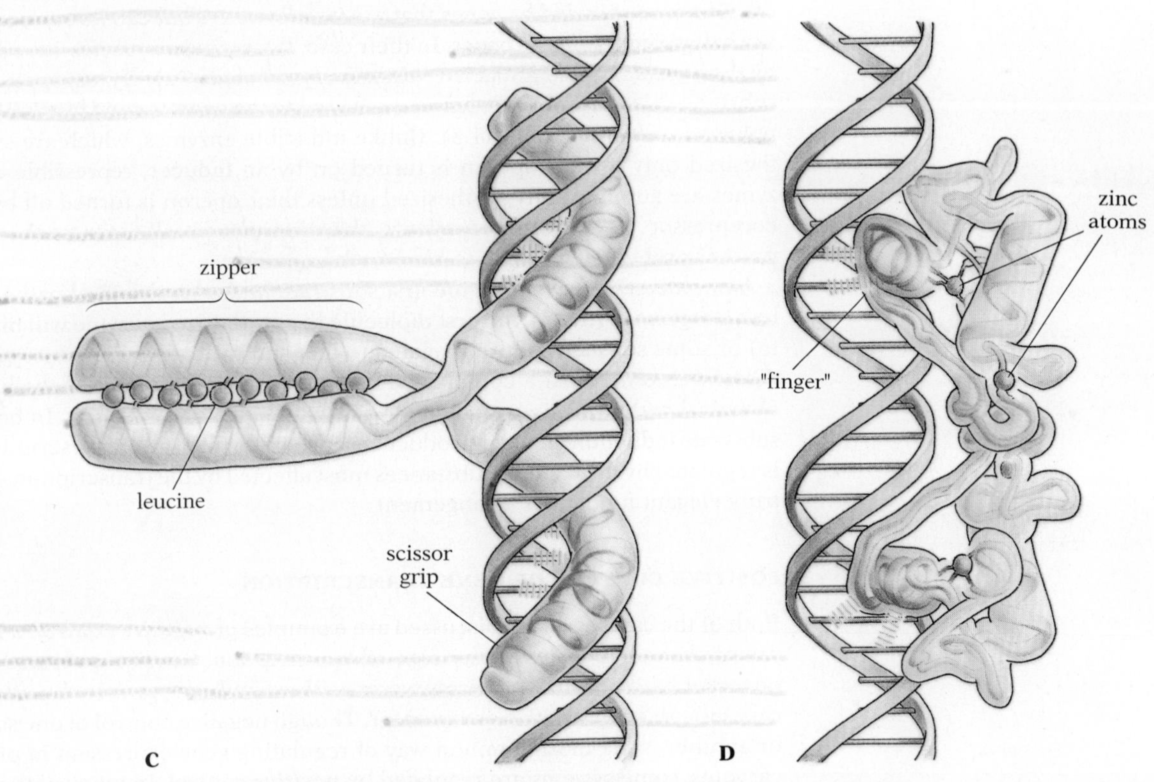

Summary of the Jacob-Monod model The condition of the operator region is one key to whether or not there will be activation of the so-called *lac* operon—the operon responsible for the synthesis of enzymes involved in the breakdown of lactose. If repressor protein is bound to the operator, there will be no transcription. If no repressor is bound to the operator (because the repressor has been inactivated by inducer), transcription can proceed freely. Note that only a few molecules of inducer are required to bind all the repressor molecules in a cell.

The three jointly controlled structural genes of the *lac* operon specify enzymes with closely related functions. It is characteristic for the structural genes of an operon to determine the enzymes of a single biochemical pathway; thus the whole pathway can be regulated as a unit.

GENE REPRESSION

In the years since the Jacob-Monod model was first proposed, it has become apparent that not all operons are regulated in the same way as the *lac* operon, which is an inducible operon—that is, one that is inactive until turned on by an inducer substance. Many operons are, instead, continuously active unless turned off by a ***corepressor*** substance. One example is the operon whose five structural genes code for the enzymes necessary to synthesize the amino acid tryptophan. This operon is normally turned on, but when *E. coli* are grown in a medium containing tryptophan, it switches off. Enzymes encoded by genes that are usually active but can be repressed are called ***repressible enzymes***. In their case, the repressor protein encoded by the regulator gene is inactive when first produced. Only if a corepressor substance binds to and activates it can it bind to the operator and block RNA polymerase binding (Fig. 11.3). Unlike inducible enzymes, which are synthesized only if their operon is turned on by an inducer, repressible enzymes are automatically synthesized unless their operon is turned off by a corepressor. In tryptophan synthesis, the tryptophan itself activates the repressor protein, enabling it to bind to the operator.

An inducer is often either the first substrate in the biochemical pathway being regulated (that is, the first molecule the synthesized enzyme will bind to) or some substance closely related to that substrate. It is not surprising, therefore, to find that a corepressor is usually the end product of the biochemical pathway being regulated, or a closely related substance. In both substrate induction and end-product corepression, then, gene transcription is regulated by the cellular substances most affected by the transcription—a truly elegant functional arrangement.

POSITIVE CONTROL OF GENE TRANSCRIPTION

Both of the cases we have discussed are examples of ***negative control***: a repressor bound to the operator turns off transcription. Control is effected in one case by deactivating the repressor with an inducer, and in the other case by activating it with a corepressor. Though negative control of one sort or another is the most common way of regulating gene expression in procaryotes, some systems are regulated by ***positive control***. In many of these cases a control protein binds directly to the DNA to activate the operon.

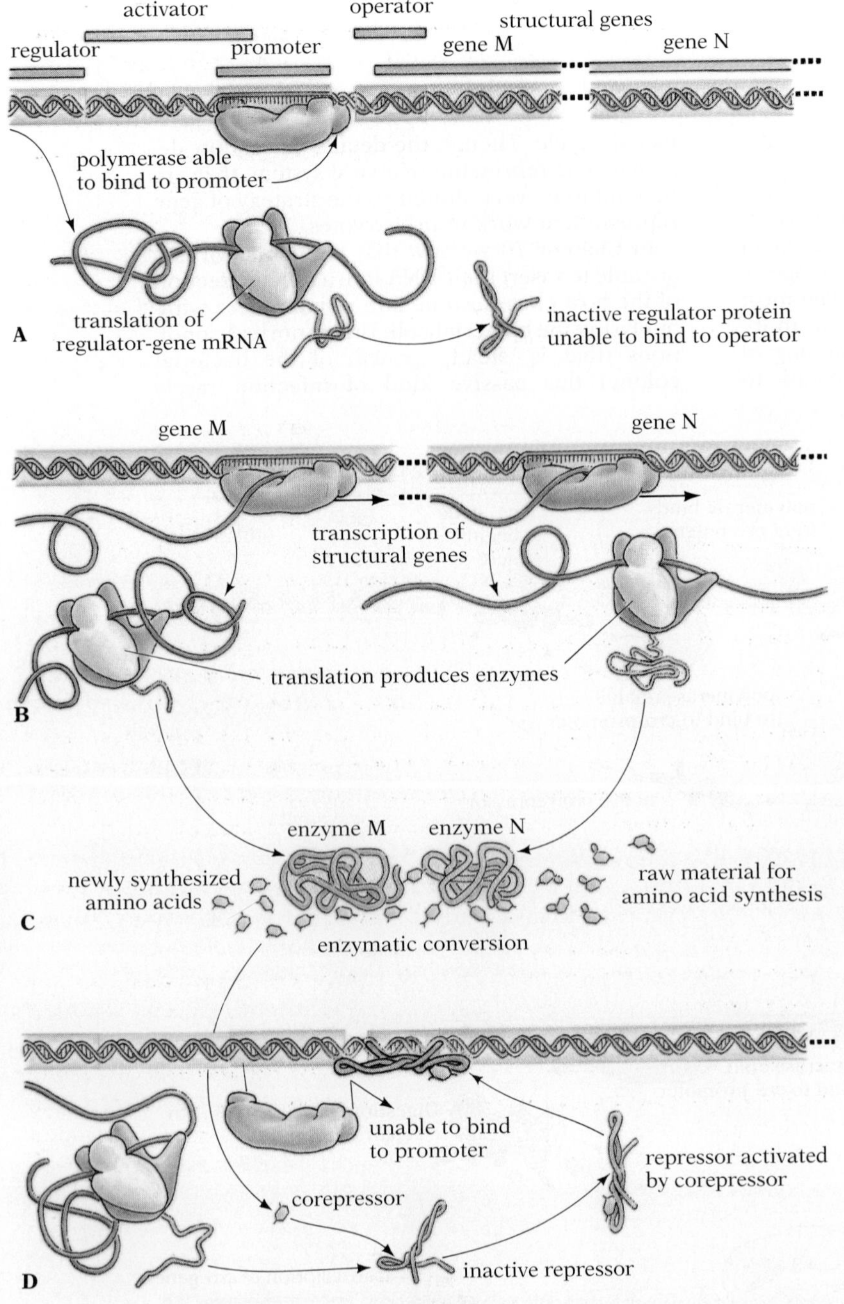

11.3 A repressible operon
The repressor protein encoded by the regulator gene is initially inactive (A). As a result, polymerases can bind to the promoter and transcribe the structural genes (B). These genes frequently encode enzymes that lead to the synthesis of an end product such as an amino acid (C). When the inactive regulator protein combines with a specific corepressor molecule (often the end product of the biochemical pathway served by the enzymes encoded by the operon), it can bind to the operator and block transcription of the structural genes (D). After the operon has been repressed, the concentration of the corepressor falls as it is used in cellular metabolism and no more is produced. When the corepressor becomes scarce, the repressor tends to lose it to metabolic enzymes. As a result, the repressor can no longer bind to the operator, the RNA polymerase binds to the promoter, and transcription resumes. The operon shown here is responsible for the synthesis of an amino acid, which is incorporated into new proteins.

Let's look at how positive control works. We have seen that RNA polymerase recognizes and binds to the promoter sequence adjacent to the operator. But few genes have promoters with exactly the same nucleotide sequence. Genes with promoters that differ significantly from the optimal

Exploring Further

THE LAMBDA SWITCH AND CONTROL OF THE LYTIC CYCLE

The expression of genes is often regulated by systems more complex than any we have described, systems that may involve the interaction of three or more control substances, two or more separate operators, and sometimes other DNA control regions in addition to the regulator genes, operators, and promoters already discussed. The most thoroughly investigated example of these more complex interactions involves the switching of the temperate virus lambda from its lysogenic to its lytic cycle. Though the details that follow describe gene repression in a viral system, they are thought to be very similar to the strategy of gene repression at work in procaryotes.

In Chapter 10 we saw that temperate viruses are able to insert their DNA into a specific section of the host chromosome and remain there without lysing the bacterial cell. Under normal conditions (that is, steady growth of the bacterial colony) this passive kind of infection rarely

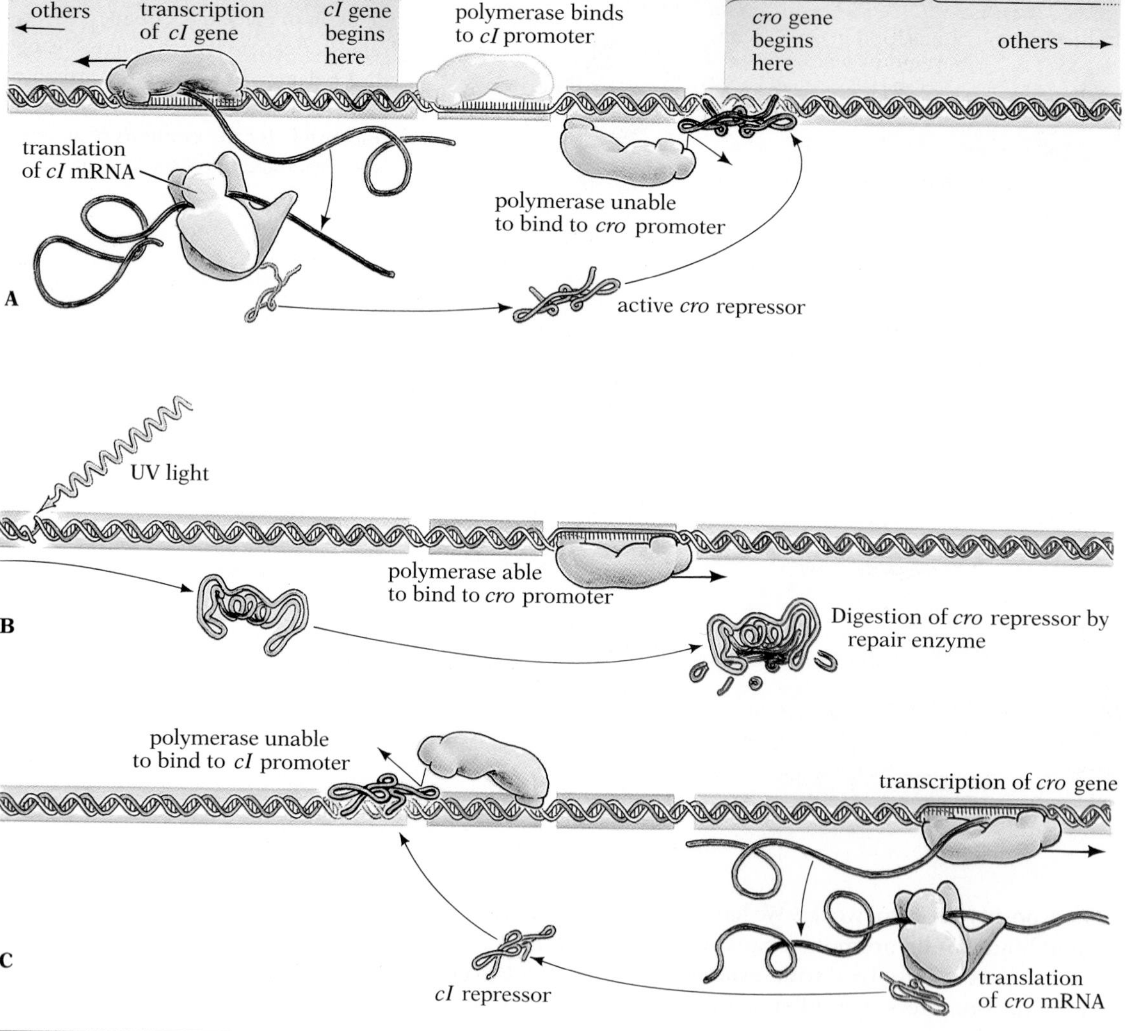

occurs, but once the virus does enter the host chromosome it can remain there, in the lysogenic cycle, indefinitely. Exposure of an infected bacterium to ultraviolet radiation, however, may incite the viral DNA to leave the host chromosome and begin reproduction. In the lambda virus, which parasitizes *E. coli,* two interacting operons on complementary strands of the DNA control this switch from a lysogenic to a lytic cycle.

The lambda system has been studied intensively. Excising the lambda DNA from the *E. coli* chromosome and initiating the destructive lytic cycle are controlled by the so-called *cro* gene and several other genes downstream from the operon. These genes are normally repressed: the *cI* gene on the complementary strand produces a helix-turn-helix repressor protein that binds to the operator region of the *cro* gene, and thereby blocks transcription (Fig. A). Repression of the *cro* and associated genes is stable until the host bacterium begins to synthesize unusually large quantities of DNA repair enzymes (Fig. B). Synthesis of these enzymes is brought about by the destructive effect of UV radiation (among other things) on the bacterial DNA. While fixing DNA damage, one of the repair enzymes digests part of the repressor synthesized by the *cI* gene, freeing the *cro* operator to begin transcription of the normally repressed *cro* and associated viral genes. The *cro* gene codes for a second repressor, which binds to the *cI* operator and turns off transcription of *cI* (Fig. C). Transcription of lytic-phase genes can now proceed (Fig. D), leading to excision of the lambda genome from the host chromosome (Fig. E), the production of replicates of the viral DNA, synthesis of capsule proteins, viral assembly, and finally host lysis.

The switch in lambda appears to be irreversible: even the disappearance of the bacterial DNA repair enzyme does not stop production of the second repressor. Apparently a survival strategy encoded in the lambda genome directs the virus to react to the first sign of trouble (UV radiation, in this case) by initiating the synthesis of new phage, which will abandon the damaged bacterium and seek out other hosts.

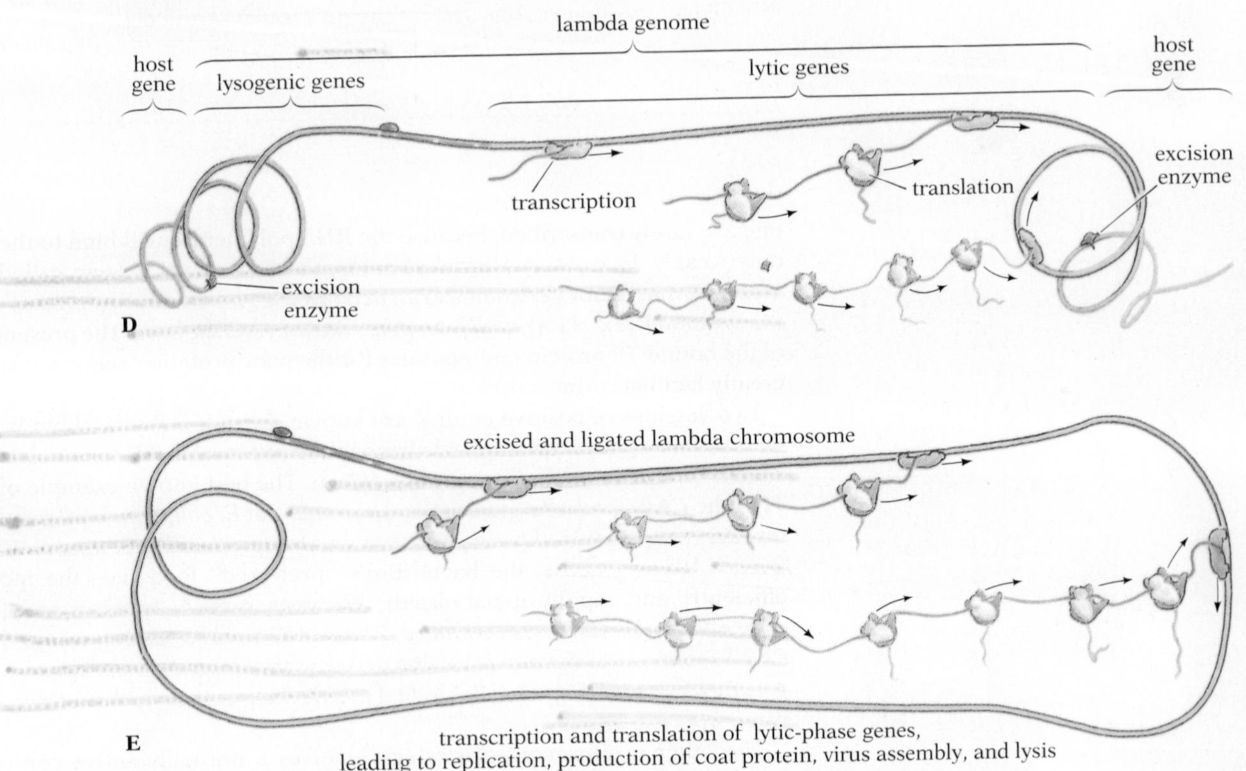

TABLE 11.1 *Summary of transcriptional control strategies in procaryotes*

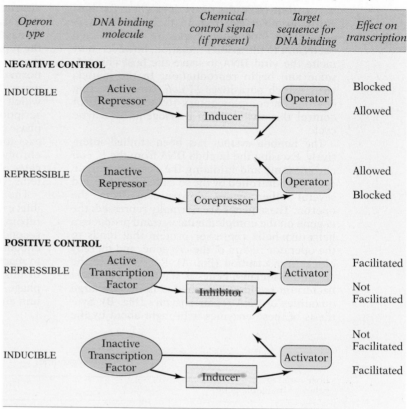

Operon type	*DNA binding molecule*	*Chemical control signal (if present)*	*Target sequence for DNA binding*	*Effect on transcription*
NEGATIVE CONTROL				
INDUCIBLE	Active Repressor	Inducer	Operator	Blocked Allowed
REPRESSIBLE	Inactive Repressor	Corepressor	Operator	Allowed Blocked
POSITIVE CONTROL				
REPRESSIBLE	Active Transcription Factor	Inhibitor	Activator	Facilitated Not Facilitated
INDUCIBLE	Inactive Transcription Factor	Inducer	Activator	Not Facilitated Facilitated

one are rarely transcribed, because the RNA polymerase will bind to them only weakly. In positive control of such genes, a control protein called a ***transcription factor (TF)*** binds to an ***activator*** region a little "upstream" of the promoter (Fig. 11.4), and then helps the polymerase bind. The presence of the bound TF protein compensates for the poor promoter sequence and greatly facilitates transcription.

Two versions of positive control are known. In one, a chemical binds to the TF, causing a conformational change that activates the TF, enabling it to bind to the DNA and facilitate transcription. The best-known example of a TF is the ***CAP*** (catabolic gene activator protein) of *E. coli,* which helps control the transcription of genes that encode enzymes that digest unusual nutrients. When glucose, the bacterium's "preferred" food (i.e., the most efficiently and rapidly metabolized), becomes rare, a "messenger substance"—cyclic AMP—is produced. The cAMP binds to and activates the CAP TF, and transcription of the alternative nutrient-metabolism operons is greatly increased (Fig. 11.4; Table 11.1). CAP, then, can affect the activity of many operons at once.

The other positive-control strategy involves a normally active control protein that is turned off by the binding of a small molecule. Once switched

TABLE 11.2 *Summary of* lac *operon control*

Environmental conditions		*Operon status*		*Transcriptional activity of lactose-metabolizing genes*
Glucose present	*Lactose present*	*cAMP-CAP TF complex bound to activator*	*Repressor bound to operator*	
yes	no	no	yes	none
yes	YES	no	NO	LOW
NO	no	YES	yes	none
NO	YES	YES	NO	**HIGH**

off, the protein cannot help the RNA polymerase bind to the poor promoter sequence, so transcription is minimized (Table 11.1). For instance the cAMP signal that, via CAP, activates the operons for alternative-nutrient-digestion genes, simultaneously *inactivates* the TF proteins that help the polymerases bind to and transcribe the genes for transporting glucose into the cell. The metabolic systems for alternative food sources are activated, and the now-useless glucose system is temporarily shut down.

The *lac* system, as it happens, is one of the most complex known procaryotic operons; it involves *both* negative control (a repressor, which can be inactivated by an inducer when lactose is present) and positive control (the CAP transcription factor, which can be induced to bind when glucose is rare). As a result, the *lac* operon is most active when lactose is present *and* glucose is absent (Table 11.2).

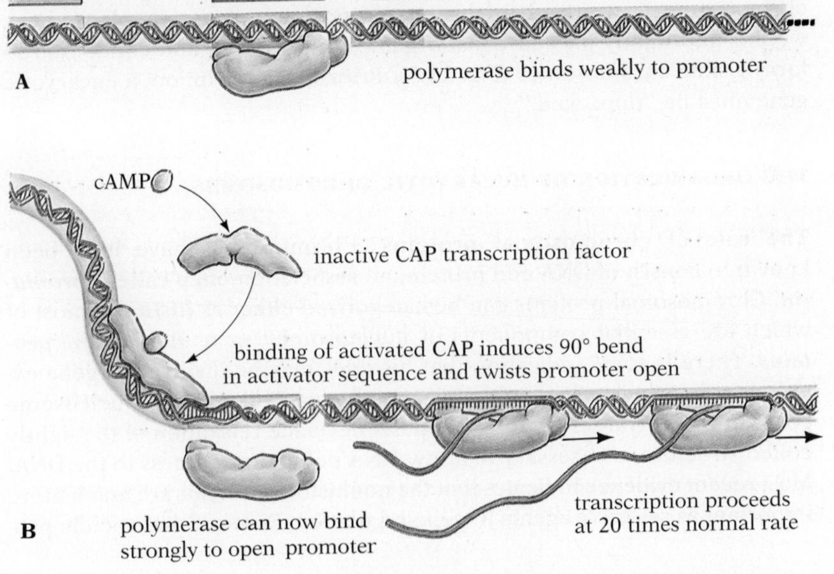

11.4 Induction of transcription by CAP
Many genes have promoters that bind polymerase at low rates (A). When a CAP transcription factor protein, itself activated when bound by a molecule of cAMP, binds to the activator sequence adjacent to the promoter, it induces a 90° bend in the DNA; the resulting twist opens the promoter, and increases the polymerase-binding rate twentyfold (B).

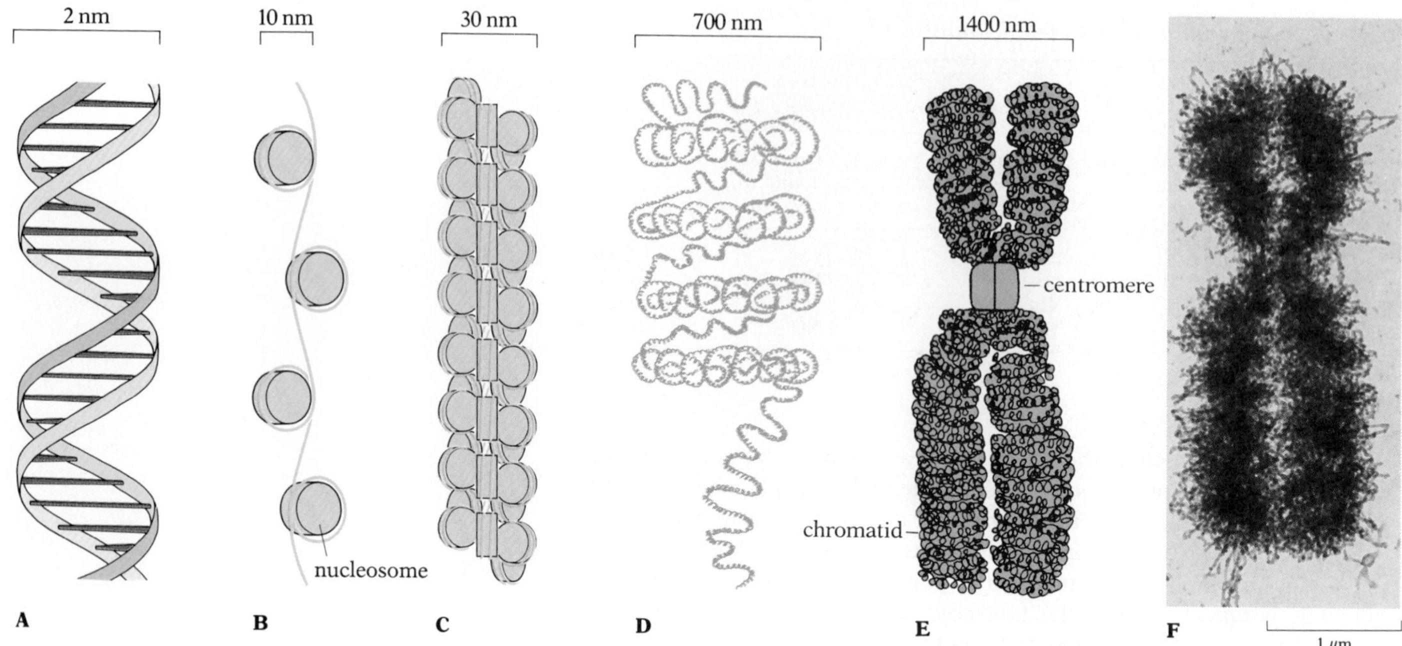

11.5 Packaging of DNA in a eucaryotic nuclear chromosome
The double helical strand of DNA (A) is wound around nucleosomes about 10 nm in diameter (B). This chain of spools is itself coiled in some way to produce a thick strand about 30 nm in diameter (C). The arrangement shown here is hypothetical. Early in cell division the thick strands are collected into a long series of loops, which are wound into a helix (D). The result is the ragged appearance of the chromosome, as seen in drawing (E) and in the electron micrograph of a human chromosome (F). Some of the loops are tightly condensed while others are not, and can be seen extending out from the chromatid cores.

CONTROL OF GENE EXPRESSION IN EUCARYOTES

The control of gene expression appears to be more complex in eucaryotes than in procaryotes. For one thing, even the simplest eucaryote can have many more functional genes to regulate than a procaryote can. And as we have seen, before translation can take place more than half of a typical eucaryotic primary RNA transcript synthesized from an active gene in the nucleus must be excised by an elaborate enzyme-mediated process. As we noted in Chapter 5, a further complication is that the DNA of the eucaryotic chromosome is wrapped on histone-protein complexes (nucleosomes), wound into tightly packed coils, organized into loops, and condensed to form visible chromosomes (Fig. 11.5). Prior to transcription, a eucaryotic gene must be "unpacked."

THE ORGANIZATION OF EUCARYOTIC CHROMOSOMES

The role of chromosomal proteins Chromosomes have long been known to consist of DNA and protein, an association often called ***chromatin***. Chromosomal proteins can be categorized either as ***histones***, most of which are essential components of nucleosomes, or as ***nonhistone proteins***. There is some indication that histones may be involved in gene expression, but their role would appear to be passive: with the nucleosome cores in place, transcription is not possible, since relaxation of the tightly coiled structure is necessary to allow RNA polymerase access to the DNA. Most recent evidence indicates that the nonhistone proteins are much more important as selective agents in gene regulation. Some of these acidic pro-

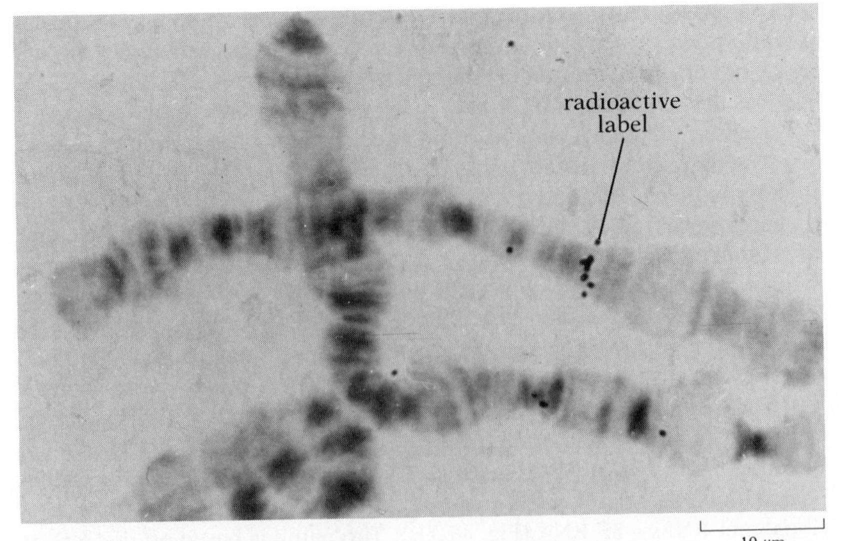

11.6 Locating a gene
In this version of the DNA probe technique, salivary-gland chromosomes from the much-studied fruit fly *Drosophila* have been treated to weaken base pairing; a radioactively labeled sample of previously identified DNA has been added and has bound to the complementary region. The dark dots reveal where radioactive decay is taking place, and therefore where this particular gene is located.

teins are bound directly to the DNA, while others are linked to the nucleosome cores. They exhibit a rich diversity, not only from organism to organism, but also from tissue to tissue within a given organism, and even within a single cell at various times, depending on its developmental stage and its current functional condition. Hence, they may have the specificity necessary for control elements. (Nucleosome cores, on the other hand, appear not to vary with gene identity or activity.) Moreover, the fact that the nonhistone proteins seem to contribute little to chromatin structure suggests that their function is regulatory. The role of at least some nonhistone proteins is to bind to specific control regions in the DNA to cause decondensation of a chromosomal loop. They also play a role in the second step of eucaryotic gene control. Since most loops are about 100,000 base pairs long, each usually contains several genes, which are often selectively activated.[1] This level of control probably includes removing one or more nucleosomes from the promoter regions of the genes destined to be switched on. As we will see, activation rather than inhibition of transcription appears to be the rule in eucaryotes.

Highly repetitive DNA The organization of eucaryotic chromosomes has been revealed in large part through a technique known as ***DNA hybridization***. The first step in this technique is to heat the chromosomal DNA under controlled conditions; this causes the two strands of each helix to separate. In one type of hybridization the next step is to add to the chromosomal DNA a labeled "probe" of a particular DNA sequence of interest, or an mRNA (or a cDNA made from it by reverse transcriptase). The mixture is then cooled very slowly. Under these conditions, the RNA or DNA will frequently bind to the complementary region of the chromosomal DNA, and because of the labeling, the gene or sequence for it can then—in theory—be located on the chromosome (Fig. 11.6). Clever variations of this basic method have pro-

[1] The length of the average eucaryotic gene is 2000 nucleotides, of which only 1200 are in exons.

duced a series of surprises that make it clear that the evolutionary history of procaryote and eucaryote chromosomes has been very different. Perhaps the most consistent contrast is that eucaryotic genomes are filled with sequences that are never transcribed, many of which exist in multiple copies. For instance, about 10 percent of most eucaryotic DNA contains base sequences that are found not once but thousands of times in the genome, constituting what is called ***highly repetitive DNA***.

There appear to be at least four classes of highly repetitive DNA. The first consists of a vast number of copies of a few kinds of short sequences located at the ***centromere*** (the specialized region that holds the two copies of a replicated chromosome together until cell division) and in large blocks in chromosome arms, especially at the ends, where they are called ***telomeres*** (Fig. 11.7A). The function of this DNA, which is never transcribed, is to facilitate particular steps in the process of cell division (which we will discuss in Chapter 12), and to maintain chromosomal stability.

Another class of highly repetitive DNA consists of long, tandemly repeated units. These areas contain genes that code for the smallest of the four ribosomal RNAs—5S RNA (Fig. 11.7B). The name is based on the rate at which this RNA sediments through a sucrose solution; the other three rRNAs—6S, 18S, and 28S—are all larger and heavier by this measure. In one well-studied species of frog, each 5S gene is associated with a nonfunctional region known as a ***pseudogene*** (so called because its sequence is almost identical to that of a functional gene) and a very long "spacer" region. Neither the pseudogene nor the spacer is ever transcribed; most of this region of the chromosome appears functionless. But though it makes up only a small fraction of the region, there is a clear reason for the large number of 5S genes: developing eggs need enormous numbers of ribosomes (perhaps 10^{12}, a trillion) to handle all the protein synthesis necessary for rapid growth. Repeated transcription of a single copy of this gene would be inadequate to meet the needs of the cell for ribosomal RNA; consequently most eucaryotes have approximately 25,000 copies of this sequence.

A third class of highly repetitive DNA, known as "long interspersed elements," consists of sequences that are several thousand bases long and are found tens of thousands of times in the chromosomes of many eucaryotes (Fig. 11.7C). Their function, if any, is not yet known. The fourth class consists of relatively short (300 base pairs) segments scattered throughout the genome. Humans have about 500,000 copies of one such sequence, that is simply a transposon (derived from an RNA polymerase promoter) that reproduced and spread almost uncontrollably at some stage in our evolution; it appears to have no function.

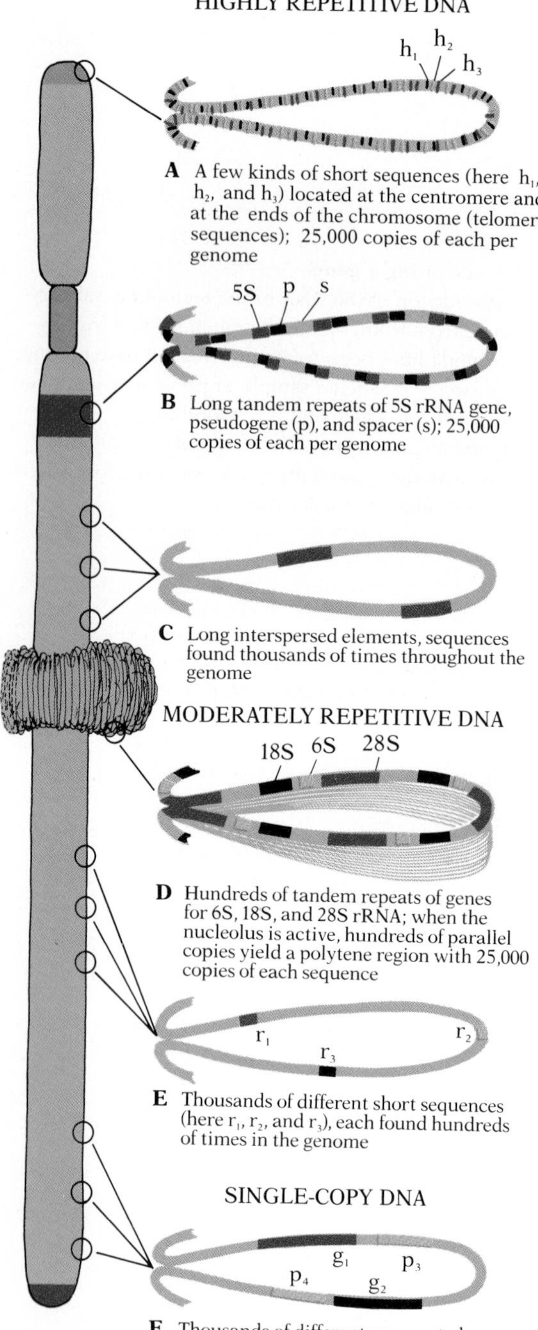

A A few kinds of short sequences (here h_1, h_2, and h_3) located at the centromere and at the ends of the chromosome (telomeric sequences); 25,000 copies of each per genome

B Long tandem repeats of 5S rRNA gene, pseudogene (p), and spacer (s); 25,000 copies of each per genome

C Long interspersed elements, sequences found thousands of times throughout the genome

D Hundreds of tandem repeats of genes for 6S, 18S, and 28S rRNA; when the nucleolus is active, hundreds of parallel copies yield a polytene region with 25,000 copies of each sequence

E Thousands of different short sequences (here r_1, r_2, and r_3), each found hundreds of times in the genome

F Thousands of different unrepeated sequences, including functional genes (here g_1 and g_2) and pseudogenes (p_3 and p_4)

11.7 Organization of eucaryotic chromosomes The short highly repetitive DNA (A), long tandem repeats (B), and polytene tandem repeats (D) are confined to different specialized regions of the chromosome. The long interspersed elements (C), moderately repetitive sequences (E), and single-copy DNA (F) are actually intermixed; for clarity, however, they are shown here in isolation on separate loops.

Exploring Further

TELOMERES

Telomeres, which are sequences of TTAGGG in all vertebrates, and are repeated at least 250 times in each chromosome, solve at least two problems. First, because nearly all eucaryotic chromosomes are linear, one replication-origin site must be located next to each end. After the RNA primer (necessary to initiate eucaryote as well as procaryotic replication) is laid down near the end on the lagging strand, it cannot be removed and replaced with DNA. This problem arises because for the primer to be removed, there must be newly added DNA abutting the RNA on the upstream side; since the site in question is the one nearest the end of the chromosome, there are none closer to the end. (As we saw in the *Exploring Further* box in Chapter 8, this problem does not arise in procaryotes because their chromosomes are circular; the leading-strand polymerase III eventually traverses the full circumference of the chromosome, and thus provides the free end of DNA needed by polymerase III to patch the primer.) Hence, at least a few bases are lost with each cell division, and the chromosomes get progressively shorter with each replication. As the chromosomal ends are padded out with telomere sequences rather than genes, this trimming is of no consequence in most cells: no cell in a developing organism ever goes through enough divisions to lose all its telomeres. But this safety measure is effective only if the chromosomes in the fertilized egg began with a sufficient supply. If gametes were to lose telomeres generation by generation, there would come a point at which the losses of chromosomal ends incurred with each replication would begin to nibble into actual genes. This problem is solved in gametes by a telomerase enzyme, which uses an RNA copy of the telomere sequence to create new DNA copies and add them to the end. This enzyme, then, is a kind of reverse transcriptase—perhaps the evolutionary source of the catalyst used by retroviruses.

The other problem that telomeres solve is that of loose ends: without special cappings, the ends of linear chromosomes would be very reactive, and could fold back and pair with sequences in the middle or at the end of other chromosomes. How telomeres keep the flexible chromosomal ends out of trouble is not fully understood, but some evidence suggests that they act to anchor these free ends to the nuclear envelope. As we would predict, the rare circular chromosomes of eucaryotes (yeast, for example) lack telomere sequences. What is less easy to understand is how the telomere sequence has managed to remain unchanged through millions of years of vertebrate evolution, and varies only slightly from that of other eucaryotes, including organisms as distantly related to us as protists. There is nothing in telomere function as now understood that would seem to require such conservatism, which almost certainly means that telomeres have some other equally critical role of which as yet we have no hint.

Moderately repetitive DNA About 20 percent of the typical eucaryotic genome consists of so-called ***moderately repetitive DNA***. Each moderately repetitive DNA sequence is found hundreds rather than thousands of times and comes in one of two varieties. The first is a tandem repeat of certain genes; in particular, genes for the other three kinds of rRNA are tandemly repeated in the order 18S rRNA, 6S rRNA, 28S rRNA, untranscribed long spacer, 18S, 6S, 28S, long spacer, 18S, 6S, 28S, long spacer, and so on (Fig. 11.7D). Since every ribosome must have one of each of the four kinds of rRNA, you may wonder how cells manage to get by with only a hundred to a thousand copies of these genes while there are thousands of copies of the 5S gene. The answer is that when new ribosomes are needed, the portion of the chromosome with the tandem repeat for 18S, 6S, and 28S rRNA can be *rep-*

licated repeatedly, *independent of the rest of the chromosome*, producing a ***polytene*** region consisting of 25–250 parallel copies of the repeat region. This process results in a total of up to 25,000 copies of each gene in active cells. The region of the chromosome with the many replicated repeating segments forms a structure we have already discussed—the nucleolus.

Moderately repetitive DNA of the other class is more mysterious, and varies widely in size and frequency between species. These sequences, of which there may be as many as 5000 different types, are only about 300–3000 bases long—as little as one-tenth the length of the average functional gene. Each type is usually scattered between functional genes throughout the chromosome in 30–500 different locations (Fig. 11.7E). Some of this DNA is clearly derived from mRNA containing, by chance, the promoter sequence for reverse transcriptase; presumably some of the DNA copies were "mistakenly" synthesized from mRNA by reverse transcriptase as an artifact of retroviral infection, and made their way into the host chromosome. One of the cell's tRNAs appears to be particularly susceptible to this process. Other cases, however, do not fit this model.

Single-copy DNA Though 70 percent of a typical eucaryotic genome consists of ***single-copy*** sequences (Fig. 11.7F), the majority of this DNA is *never* transcribed; at least in mammals, much of it exists as pseudogenes—nearly identical but untranscribed copies of functional genes. When researchers make calculations for different species, taking into account the introns that are excised after transcription, they typically find that only about 1 percent of eucaryotic DNA ever codes for mRNA that is subsequently translated. This could hardly be more dramatically different from procaryotes, in which well above 90 percent of the genome is translated at one time or another. We have suggested that the high density of information in procaryote genomes—haploid and almost devoid of introns, spacers, and pseudogenes—is probably a result of intense selection for reproductive speed. But though few eucaryotes find reproductive advantage in breakneck cellular doubling, the toleration of those countless introns, pseudogenes, and spacers, as well as transposons, retroinsertions, and other genetic debris seems incredible. Some researchers believe that eucaryote genomes simply lack efficient ways of expunging genetic junk, which therefore accumulates endlessly; others suspect that much of this vast proportion of untranscribed DNA is actively tolerated as a laboratory for evolutionary experiments in exon shuffling and mutation. The logic of eucaryotic gene organization remains one of the most interesting and elusive questions in modern biology.

VISIBLE EVIDENCE OF CONTROL OF GENE TRANSCRIPTION

Profound as the differences are in the organization of the procaryotic and eucaryotic genomes, it is not surprising that their systems of gene control differ as well. The eucaryotic genome, with its need to decondense tightly packed loops of DNA before transcribing them, must have an added level of control that procaryotes do not require. In addition, eucaryotes must orchestrate multiple constellations of genes, each specific for different tissues and different stages in their complex development from a single cell.

Patterns of activity in eucaryotic chromosomes Cytological studies of chromatin suggest that the internal structure of a chromosome is not uniform, and that this variation reflects gene activity. For example, some regions of chromosomes stain only faintly when treated with basic dyes, whereas other regions stain intensely. The nonstaining regions are called ***euchromatin*** and the staining ones ***heterochromatin***. Chromosomal mapping over the past few years has shown that euchromatic regions contain active genes whereas heterochromatic regions are inactive.

At first, some investigators thought the heterochromatic regions were simply devoid of genes, functioning merely as structural elements of the chromosome. This is probably true of the large region of heterochromatin with highly repetitive base sequences located around the centromere. But most heterochromatic regions do not lack genes; instead, the genes (often long series of genes) are simply inactive. For example, many regions completely heterochromatic in adults were euchromatic at earlier stages in the development of the organism.

Even within a region of euchromatin, only a few, widely spaced genes may be active; the clumping of a small set of related genes into an operon that is so characteristic of bacteria is rare in eucaryotes. Indeed, in many instances in which there appears to be simultaneous control of the synthesis of functionally related enzymes, the enzymes are encoded by widely separated genes. The genes that specify two polypeptide chains of a single protein (the α and β chains of hemoglobin, for example) are often on different chromosomes.

Lampbrush chromosomes and chromosomal puffs as visible evidence of gene activity The developing egg cells of many vertebrates synthesize large amounts of mRNA for later use: the mRNA is then ready during the early stages of development, after fertilization, when the chromosomes are so busy with DNA replication in support of rapid cell division that they are largely unavailable for transcription into mRNA. In vertebrate egg cells this changing activity can be observed with a phase-contrast microscope. The parts of the chromosome bearing genes that are being repeatedly transcribed—the euchromatic regions—are looped out laterally from the main chromosomal axis, while the parts bearing repressed genes are tightly compacted; chromosomes with many looped-out regions are called ***lampbrush chromosomes*** (Fig. 11.8).

The giant chromosomes of the salivary glands of larval flies (Diptera) are interpreted as lateral arrays of replicated chromosomes (polytene chromosomes) that have remained stuck together. The best estimates indicate that 10 cycles of replication are involved in the creation of a polytene chromosome, yielding 1024 parallel copies of each gene. The result of this unselective gene amplification is a capacity for rapid synthesis of large amounts of

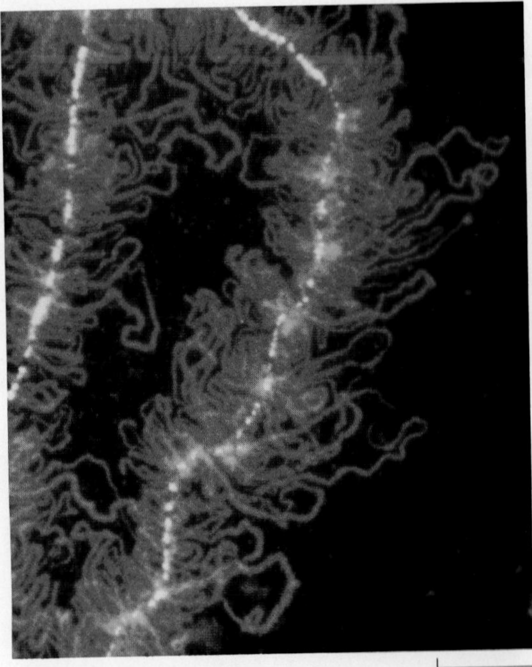

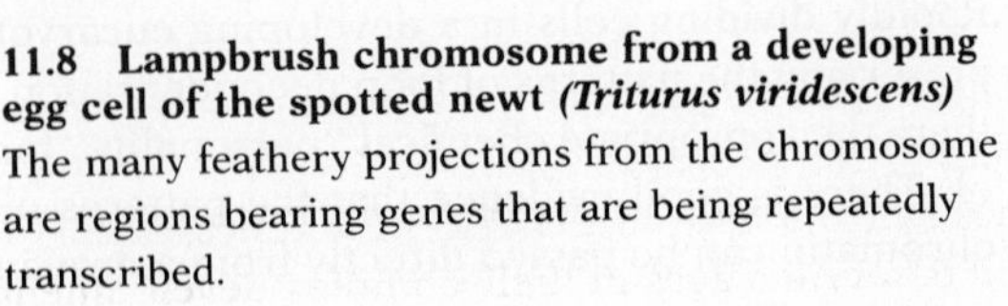

11.8 Lampbrush chromosome from a developing egg cell of the spotted newt *(Triturus viridescens)* The many feathery projections from the chromosome are regions bearing genes that are being repeatedly transcribed.

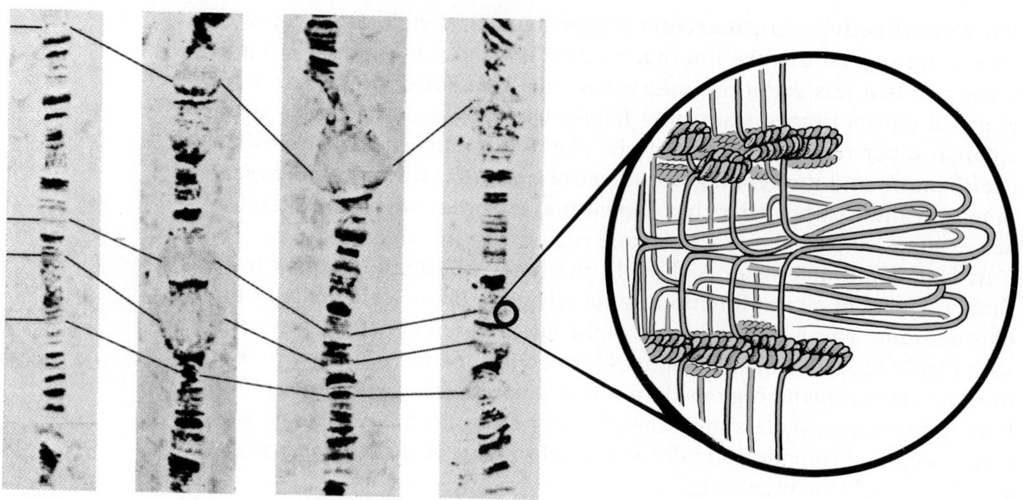

11.9 Chromosomal puffs
Puffs in polytene chromosomes consist of hundreds of parallel copies of the chromosomal DNA looped out to expose the maximum surface for synthesis of RNA. The interpretive drawing shows only a few of the loops. The series of photographs indicates how the location of puffs in a *Drosophila* chromosome changes dramatically but predictably over time; we can clearly see the varying activation of four different genes or sets of genes on chromosome 3 (each identified by a series of connecting lines) over a period of 22 hours in larval development.

RNA. When certain regions of a giant chromosome are especially active, all the parallel DNA molecules form brushlike loops in those regions (Fig. 11.9). The resulting clusters of lateral loops, called ***chromosomal puffs***, are clearly visible under the microscope. The locations of puffs are different on chromosomes in different tissues, and they are different in the same tissue at different stages of development (Fig. 11.9), though at any given time all the cells of any one type in any given tissue show the same pattern of chromosomal puffing.

The puffs indicate the location of active genes, and thus the sites of rapid RNA synthesis. As such, the puffs provide a way of determining visually whether or not changes in the extranuclear environment can alter the pattern of gene activity. For example, if ecdysone, the hormone that causes molting in insects, is injected into a fly larva, the chromosomes rapidly undergo a shift in their puffing pattern, taking on the pattern characteristically found at the time of molting in normal untreated individuals. If treatment is stopped, the puffs characteristic of molting disappear. If treatment is begun again, they reappear. Or, to give another example, if chromosomes from one type of cell are exposed to the cytoplasm of a different type of cell or of the same type of cell at a different developmental stage, they quickly lose the puffs characteristic of their original cells and develop puffs characteristic of the type and stage of the cells providing the new cytoplasm. Puffing can be prevented entirely by treatment with actinomycin, which is known to be an inhibitor of nucleic acid synthesis.

IMPRINTING: PRESERVING PATTERNS OF GENE ACTIVITY

Rapidly dividing cells in a developing eucaryote must have some way of preserving the patterns of loop decondensation and gene activity that give them the appropriate chemical "personality" for the tissue the cell is a part of. There is good evidence that the patterns of euchromatin and heterochromatin can be passed directly from a growing cell to the two daughter cells it produces when it divides through a process called ***imprinting***. One component of this imprinting system in vertebrates involves methylation of the cytosines in certain C–G sequences in inactive genes. Methylation can transmit activity patterns to daughter cells: Because only C–G sequences can be methylated by the enzyme involved, and since C–G is always paired with G–C on the other strand, the same pattern of methylation exists on both strands of the DNA:

```
              m
              |
: : :A :T :C :G :T :C :A : : :
: : :T :A :G :C :A :G :T : : :
                 |
                 m
```

(The colored "m" indicates the cytosine has been methylated by the addition of a —CH_3 group.) Replication creates a hybrid structure wherever the parental DNA had been methylated:

```
            m
            |
: : :A: T: C: G: T: C: A: : :      Parental strand
: : :T: A: G: C: A: G: T: : :      New strand
```

A special enzyme, maintenance methylase, scans the DNA for methylated cytosines in C–G sequences, methylating the corresponding cytosine on the new strand. Thus the pattern of methylation—gene inactivation—is passed on intact. If, however, both methyl groups at a site are lost by chance, cells are produced with altered gene activity. There is good evidence that at least some of the problems associated with aging arise from the progressive loss of methylation over time, and hence the activation of inappropriate genes in cells. Demethylation in gametes can even pass mistaken gene-activation patterns on to offspring, and several diseases are now ascribed to such imprinting errors.

MECHANISMS OF TRANSCRIPTIONAL CONTROL IN EUCARYOTES

Little is yet known about the selective decondensation of chromosomal loops—the necessary precursor to transcription—that creates the polytene puffs and lampbrush projections we have mentioned. Certain "master control chemicals" like ecdysone can trigger major changes in looping, but how they act is a mystery. But though decondensation is yet to be understood, knowledge about the basis of individual gene regulation on decondensed loops is growing rapidly.

As we have already said, eucaryotes rely mainly on positive control: transcription does not occur without active aid from a transcription factor, which helps the polymerase bind to the promoter. In addition, eucaryotes control TF binding in gene-specific ways from two other locations on the chromosome.

The first control locus is an ***inducer region*** just upstream of the TF se-

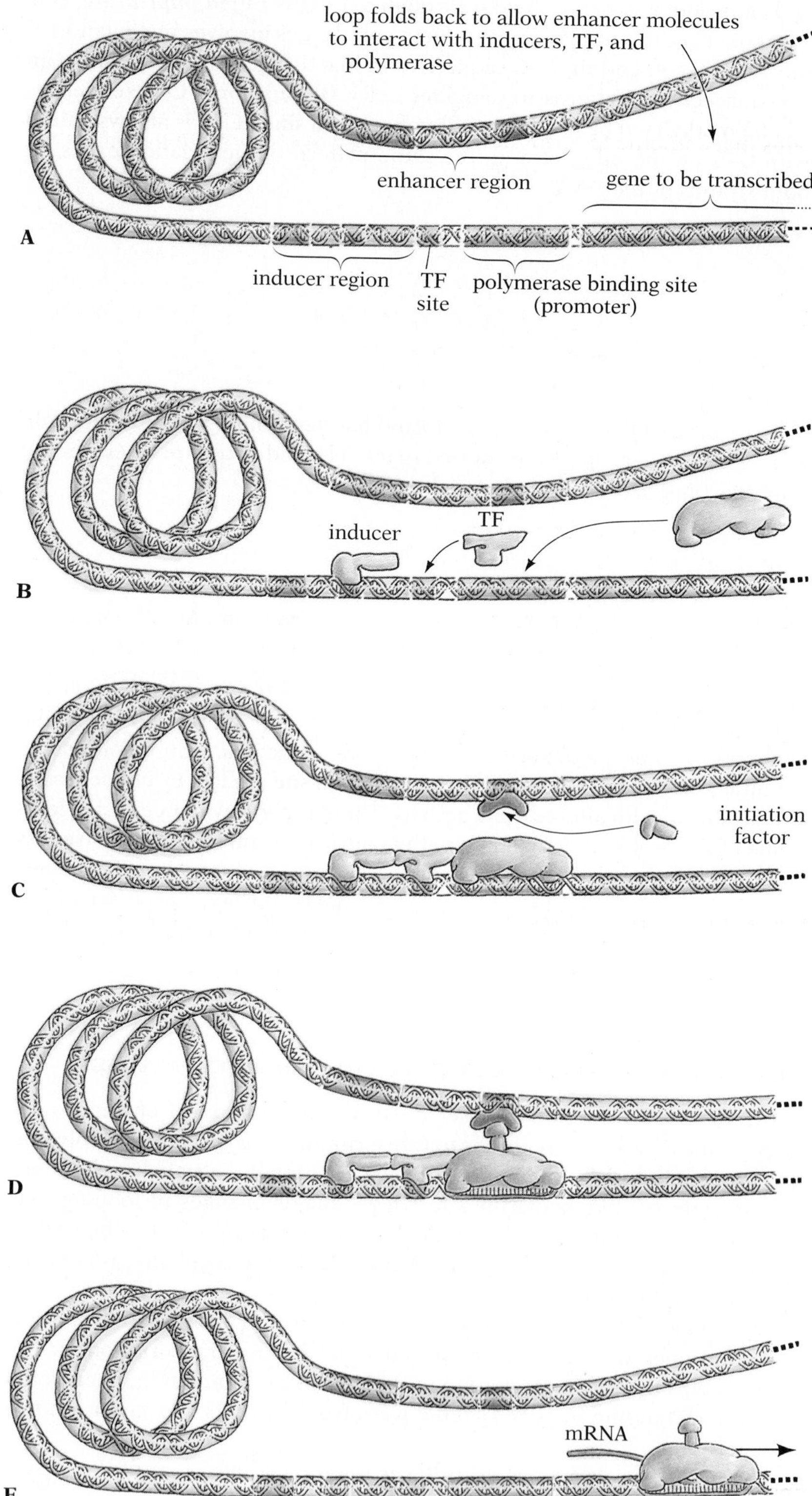

11.10 Eucaryotic gene control
For this hypothetical gene there are four inducer sites adjacent to the transcription-factor binding sequence, and four enhancer sites upstream (A). In this simplified positive control model an activating inducer binds to one site and loads the transcription factor (B), which in turn helps the polymerase to bind (C). An activating enhancer binds upstream (C) and, once the loop folds into position, loads the initiation factor onto the waiting polymerase (D), thereby initiating transcription (E). As described in the text, there are other ways inducers and enhancers can participate in starting or blocking transcription.

quence (Fig. 11.10A). It consists of one or (usually) more individual sites that specific control proteins can bind to. Some of these control molecules in turn help the TF protein bind (Fig. 11.10C), while others inhibit it. For instance, the β-globin gene (which encodes one of the two kinds of protein chains that combine to form the blood's oxygen-transport complex, hemoglobin) has seven of these control sites in its inducer region.

The other locus of action, surprisingly enough, is a few thousand nucleotides away on the same decondensed loop. It consists of a cluster of sites collectively called the ***enhancer region***. These sites exert their effects when the loop folds back on itself to bring the entire region into contact with the promoter and inducer sites (Fig. 11.10). Some enhancers help load the TF protein, which then aids in polymerase binding, as we have seen; others directly help the polymerase to bind. Still others add a protein called the ***initiation factor*** to the polymerase complex. The initiation factor greatly enhances the polymerase's efficiency in transcription. Other regulatory substances that can bind to the enhancer region actually inhibit one or another of these steps.

This action-at-a-distance system is nearly ubiquitous in eucaryotes, and a few bacterial genes have been found to be controlled in this way as well. Why do most eucaryotic genes have so many alternative ways to regulate transcription? The answer seems to be that most eucaryotes have much more complex life histories than bacteria. As we have said, eucaryotes are usually multicellular, and so must grow in a controlled and directed way to achieve a species-specific morphology. Moreover, eucaryotes usually have distinct tissues and organs whose cells must develop particular attributes to fulfill their specialized roles. In short, gene activity depends not only on immediate metabolic needs, but on a cell's location and developmental age as well. Studies of specific clusters of enhancer and inducer sites support this idea. Some sequentially active enhanced and inducer sites are found in many genes scattered throughout the chromosomes, indicating that those genes are expressed in synchrony. There are four versions of the β-globin gene, for instance: two encode hemoglobin subunits whose affinity for oxygen is appropriate for a period early in development, before the fetal circulatory system matures; another takes over for the interval during which oxygen must be obtained from the maternal placenta; another leads to the production of a subunit adapted to the high-oxygen conditions in the lungs of children and adults. Each has a regulatory sequence shared with the many other genes active at similar ages.

Other sequences are less common, and seem to promote gene transcription only in the appropriate sort of tissue; the activity of β-globin genes, for instance, in addition to being restricted to particular developmental ages, is further limited to cells in the bone marrow, along with all the other specialized genes needed to produce red blood corpuscles and maintain the marrow itself. It is only in the marrow cells that the necessary enhancers for these genes are found; the enhancers, in turn, are produced in response to earlier chemical signals from neighboring cells (signals that identify tissue type), and internal signals encoding developmental age.

Finally, there are control sequences in these same marrow cells that are unique, or are shared by only a few genes at most; at least some of these appear to adjust the concentration of hemoglobin to ensure an adequate supply. An active gene in a eucaryote, then, would probably need the coop-

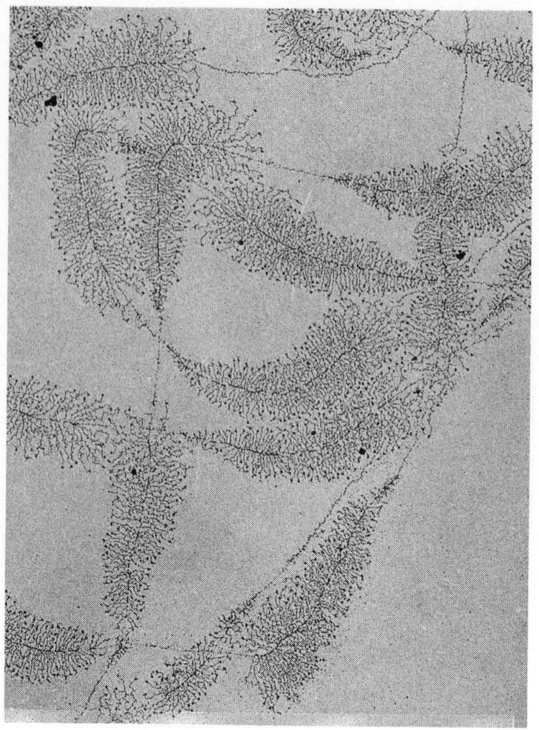

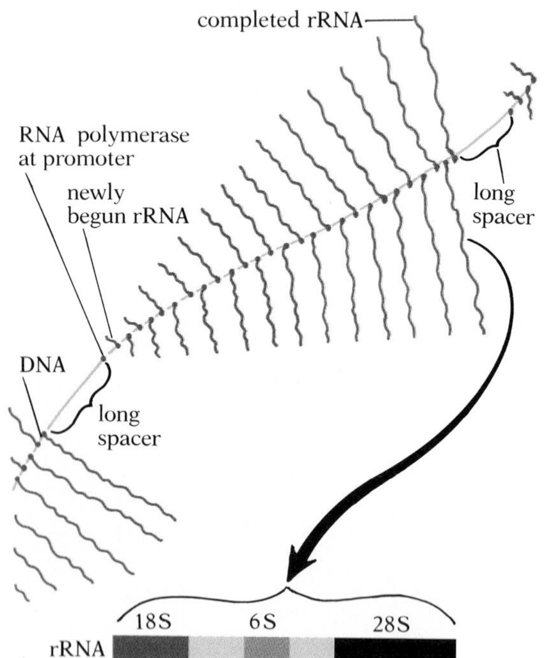

11.11 Transcription of rRNA on DNA in the nucleolus
Top: Electron micrograph of portion of the nucleolus in an egg cell of the spotted newt. This continuous strand of DNA bears multiple copies of the genes for three RNAs. Strands of rRNA at progressive stages of synthesis feather out from each set of genes. Because the successive gene sets are separated by regions of intergenic DNA spacers where no rRNA is being synthesized, we can see more or less exactly where on the DNA molecule each set of genes begins and ends. Bottom: Diagram of rRNA synthesis on one gene set. Many molecules consisting of the three kinds of rRNA joined together in a single strand (red) are being synthesized, as RNA polymerase molecules (red circles) specialized for transcribing rRNA genes move along the DNA. Strands of rRNA attached to the DNA near the beginning of the gene set are still short, their synthesis having just begun.

TABLE 11.3 *Comparison of procaryotic and eucaryotic gene control*

	Procaryotes	*Eucaryotes*
Control of DNA polymerase binding	Usually negative, blocked by a repressor	Usually positive, possible only with aid of transcription factor (TF)
Direct control of TF binding: inducer sites	Allosteric change after binding of TF to activator (inducer) site	Inducer must bind to DNA before helping TF bind; inhibitors can bind to DNA and block TF or inducers
Indirect control: enhancer sites	Very rare; exerts negative effect	Very common: can exert positive or negative effect
Levels of control	Usually one; rarely two	Usually at least three; very often four or more

eration of at least three control substances bound to the inducer and enhancer regions, as well as the absence of any inhibitors, in order to be transcribed. The main features of eucaryotic gene regulation are summarized in Table 11.3.

GENE AMPLIFICATION AS A CONTROL MECHANISM

Ordinarily a single copy of each gene per chromosome is sufficient to meet the needs of a cell for the substance coded by the gene. The rate of transcription of a single gene is such that in four days it may produce 100,000 messenger RNA molecules, which can lead to the synthesis of as many as 10 billion (10^{10}) protein molecules. But in some cases the demand for a product is so great that a single gene cannot meet it. For example, we've already seen that so much ribosomal RNA is required by a eucaryotic cell for construction of ribosomes that the 5S rRNA gene is repeated in tandem 25,000 times (Fig. 11.7B). The other three rRNA genes are repeated only 100–1,000 times in tandem, but when large numbers of ribosomes are needed, many parallel copies of this region of the chromosome are synthesized (Fig.

11.7D). This region usually loops out from the main axis of one of the chromosomes to form the nucleolus, and rRNA is produced there at a high rate (Fig. 11.11). The production of these extra copies of the rRNA genes from moderately repetitive DNA is one type of ***gene amplification***.

When amplification of the rRNA genes was first discovered, the possibility was immediately suggested that a similar amplification might occur for other sets of genes at various stages of embryonic development. But contrary to expectation, no additional instances of localized gene amplification in normal cells have been positively demonstrated; as we will see presently, however, gene amplification is common in many kinds of cancer.

POST-TRANSCRIPTIONAL CONTROL

Our discussion of cellular control mechanisms has so far focused on regulation of the amount of mRNA synthesis. But there are numerous other points in the flow of information within the cell where control can be exerted. Several involve mRNA. For example, some primary transcripts can be processed in more than one way, yielding slightly different products, depending on the cell's needs. In *Drosophila*, for example, the way one gene's mRNA is spliced controls the expression of a constellation of other genes that help orchestrate the development of either a male or female fly (Fig. 11.12). The critical event occurs when the splicing complex prepares to join the end of exon 3 to the beginning of the next exon. If a particular gene (*transformer*, or *tra*) is fully expressed, it causes exon 4 to be added, but blocks further splicing; if active *tra* gene product is not available, exon 4 is skipped and all the remaining exons are spliced normally. (The *tra* protein mimics the portion of the splicing complex that binds to the leading end of

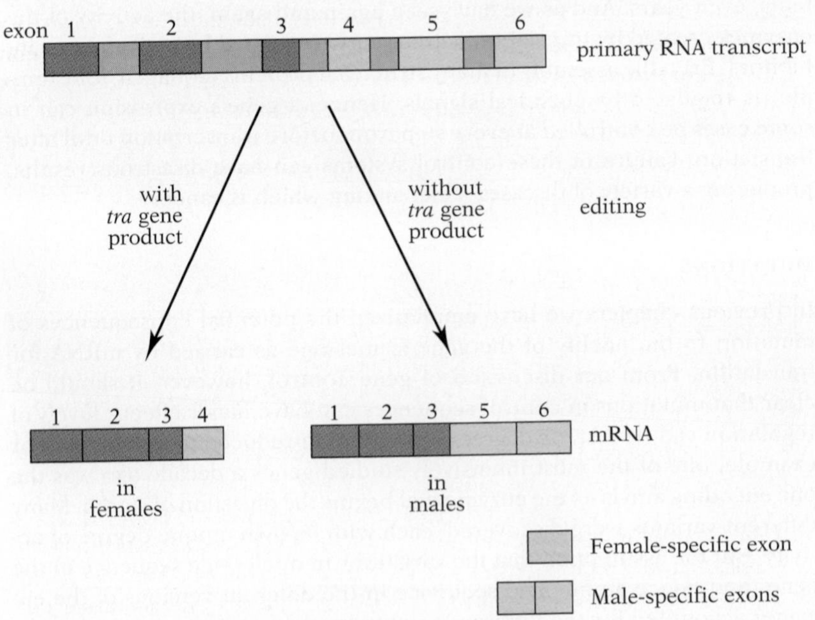

11.12 Alternative splicing in *Drosophila* The primary transcript of the "double-sex" gene in *Drosophila* can be spliced in two ways, depending on whether the *tra* gene product is present to intervene in the process. If the exon sequence is 1-2-3-4, the protein translated from the mRNA leads to the expression of female-specific genes, presumably by binding to the appropriate places in the DNA. If the sequence is 1-2-3-5-6, the mRNA encodes a product that leads to the development of a male fly.

exons, but how it alters splicing is not known.) The *tra* product itself is generated in response to higher-level signals in the nucleus that reflect whether the developing organism is a male or a female fly.

Even after splicing, almost half of the mature mRNA produced in the nuclei of most cells never reaches the cytoplasm for translation. There must be a system that can selectively block the export of specific mRNA molecules, probably in response to chemical signals that reflect the cell's needs. And even when mRNA reaches the cytoplasm, it may not be translated. Inhibitor substances can bind to specific messengers in the cytoplasm and block either ribosomal binding or complete translation. This strategy can reduce the cell's response lag to changing conditions by maintaining a pool of mRNAs ready to begin or complete translation when an appropriate chemical signal binds to the inhibitor and causes it to disassociate from the messenger. In most cases, the translation inhibitor is a protein, but in a few instances a segment of RNA with a sequence complementary to the mRNA binds though base pairing. This binding of ***antisense RNA*** is the mechanism by which herpes viruses are able to remain dormant in cells for extended periods. How antisense inhibitors come to be removed is not yet understood. Other control chemicals facilitate rather than inhibit translation: these bind to specific mRNAs and help them to bind ribosomes.

Finally, the expression of mRNA can be regulated by the rate at which the messenger is broken down: some sorts of mRNA have long life-spans, while others are destroyed within minutes. An mRNA's degree of resistance to digestion by RNAse is encoded near the trailing end of the messenger: a sequence with mostly adenines and uracils targets the mRNA for early destruction, whereas a paucity of these nucleotides prevents efficient digestion. Control can also be exerted after translation: protein longevity (i.e., resistance to proteases), for example, is written into the sequence of subunits, and ranges from a few minutes to days or, for some structural proteins, even years. And as we have seen again and again, the activity of the enzymes created by translation is frequently regulated by activators or inhibitors. Even the assembly of many structural proteins (collagen, for example) is regulated by chemical signals. Hence, a gene's expression can in some cases be controlled at every step from before transcription until after translation. Failure of these control systems can have disastrous results, producing a variety of diseases, chief among which is cancer.

MUTATIONS

In previous chapters we have emphasized the potential consequences of mutation to the fidelity of the genetic message as carried by mRNA for translation. From our discussion of gene control, however, it should be clear that mutations in control sequences can have major effects: levels of regulation can be lost, or degrees of regulation reduced or enhanced. For example, one of the most intensively studied genes a decade ago was the one encoding amylase, the enzyme that begins the digestion of starch. Many different variants were discovered, each with its own unique degree of activity. But the assumption that the variations in nucleotide sequence in the gene (and hence amino acid sequence in the different versions of the enzyme) accounted for the different activity levels proved largely incorrect:

the critical alterations are in the control regions regulating the rate of gene transcription; the variations in amino acid sequence have relatively little effect on amylase activity. This observation has had a major impact on the way scientists now think about evolution.

Alterations of regulatory regions can also be enormously important to human health. Consider, for example, the two steps involved in detoxifying alcohol in humans:

$$\text{ethyl alcohol} \xrightarrow[\text{dehydrogenase}]{\text{alcohol}} \text{acetyl aldehyde} \xrightarrow[\text{dehydrogenase}]{\text{aldehyde}} \text{acetic acid}$$

In many individuals, the transcriptional activity of the aldehyde dehydrogenase gene associated with the secretion of this enzyme into the stomach is unusually low, leading to a buildup of the toxic intermediate acetyl aldehyde; entire races can be affected by this genetic factor, leading to a low tolerance for alcohol. Individuals of Asian descent and American Indians have characteristically low aldehyde dehydrogenase levels. Studies of alcoholics, on the other hand, show that they almost invariably have elevated levels of aldehyde dehydrogenase transcription; presumably, they differ from most individuals in the population in that one of the control sequences regulating transcription of this gene is unusually active. Such patterns suggest a genetic basis for susceptibility to at least some diseases or behaviors. The most dramatic cases of serious effects of altered gene regulation, however, come from studies of the many forms of vertebrate cancer.

CANCER: A FAILURE OF NORMAL CELLULAR CONTROLS

Biologists hope to learn more about how normal cellular controls operate by studying how they fail—by investigating how the normal controls can become so deranged as to permit cancerous growth. The most distinctive feature of cancer cells is their unrestrained proliferation, which results in formation of malignant tumors and often in the spread of the cells from the original site of growth to many other parts of the body, a process known as ***metastasis***. Studies of these cells are leading to better ways of preventing or treating cancer.

Study of cells in culture One of the most profitable ways of studying the properties of specific kinds of cells is to grow them in tissue culture in the laboratory. Embryonic cells and tumor cells grow best, while fully differentiated cells are poor candidates for culturing. Among nontumor cells, embryonic fibroblasts outgrow all other cells in culture. Fibroblasts—literally, fibrous germinating cells—serve many regenerative functions such as producing scar tissue. The usual lab procedure is to seed a large number of cells onto (or into) a sterile culture medium to which a variety of nutrients and growth-stimulating factors have been added. The nutrients and other factors are frequently supplied by serum—the liquid portion of blood. Serum contains, for example, the platelet-derived growth factor that signals fibroblasts to initiate the growth of scar tissue; in the body, the growth factor escapes from the circulatory system and binds to receptors on fibroblasts whenever a wound breaks blood vessels. While a cell culture, no matter

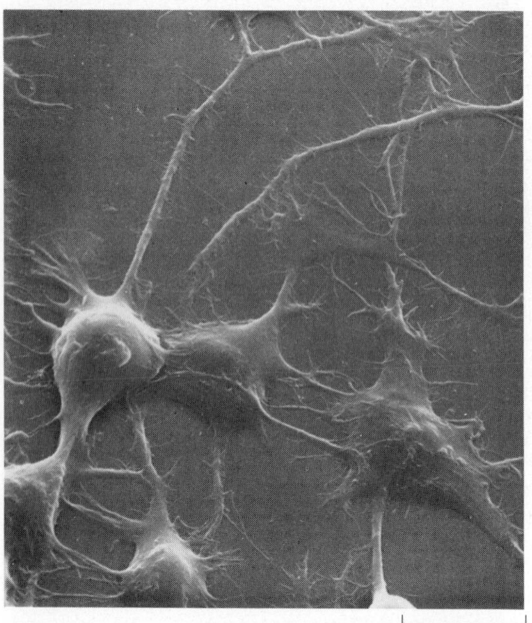

11.13 Phase-contrast photomicrograph of cultured neuroblastoma cells from a mouse These cells, from a malignant tumor of the nervous tissue (neuroblastoma), are growing on a solid surface.

how elaborately set up and controlled, is far from the normal environment of the cells, some of the cells may nevertheless survive and engage in many of their usual developmental and functional activities, including cell division (Fig. 11.13). Most cultured lineages of animal cells stop dividing when they become crowded or, if crowding is prevented, after a certain number of cellular generations; the number of generations tends to be specific for the species and tissue of origin.

These limitations on the growth of cultured cells exemplify two mechanisms that prevent unlimited cell growth in the organism as well. One is the cell's reaction to the effect of crowding, which we called ***contact inhibition*** in Chapter 4; when receptors in the cell membrane recognize markers in the cell coats (glycocalyces) of adjacent cells of the same type, the receptors signal the nucleus and further cell division is inhibited. This inhibition is not absolute, however: adding high concentrations of growth factors, for example, can often stimulate a further round of cell division even in a crowded culture. The second mechanism, the automatic cessation of cell division after a set number of divisions, is an independent system for limiting cell proliferation; in most tissues it comes into play only if contact inhibition fails to work normally. Cultured cells that continue to divide indefinitely fall into one of two categories: those that have lost the fixed-number-of-divisions control, and will go on growing as long as they are not crowded, and those that have lost both kinds of control. The cells of each category are potentially "immortal" in culture, but the former resemble ***benign*** (that is, self-limiting) tumors, while the latter resemble cancers: they will grow indefinitely regardless of crowding to create an ever-larger mass of tissue. The study of cultured cells that have lost both levels of control has been vigorously pursued in recent years.

A recently developed technique called ***somatic cell hybridization*** has made possible exciting new approaches to the study of cellular control systems. Two cultured cells, often of different origins, are fused together, with the result that the cell has a hybrid group of chromosomes, one set from each "parental" cell. In time, the nuclei of fused cells usually fuse as well. The initial result is that each daughter cell receives a single nucleus containing two sets of chromosomes, which may be from different species or from different tissues from the same species. In this way, hybrid cells containing human chromosomes in combination with chromosomes from a variety of other animal species have been produced. With each round of cell division, however, chromosomes of one species are progressively lost; in mouse/human hybrids, for instance, the human chromosomes are lost, while in mouse/hamster fusions, it is the mouse's genetic contribution that slowly disappears. (The selective nature of this loss is not well understood.)

The cell-fusion technique has been especially valuable in cancer research, because study of fused cancerous and noncancerous cells, and of fused cancerous cells of two different types, helps reveal how the altered control system or cytoplasmic milieu of one type of cell influences the control system of the other type. It is also useful in chromosomal mapping: by choosing a mouse cell line with a specific mutation that makes it unable to grow in culture, and then growing the mouse/human hybrids until only one human chromosome is left, the human gene that supplies the gene product the mouse line lacks can be confidently assigned to the surviving chromosome. Mapping greatly facilitates gene sequencing by identifying the chromosome to be isolated and treated with endonuclease; mapping also helps identify clusters of genes that are expressed as a unit. As we will see in Chapter 15, the cell fusion technique has also proved invaluable in producing monoclonal antibodies, molecules that can bind to other molecules with high specificity, and so localize that substance in cells or tissues (see *Exploring Further*, p. 404).

The characteristics of cancer cells The loss of normal control systems can produce a variety of changes in a cell's anatomy, chemistry, and general behavior. Studies of these changes tell us quite a bit about how cancer cells differ from normal ones, and sometimes suggest potential avenues of treatment. We will mention a few of the characteristics of cancer cells that these studies have revealed.

One property of cultured cancer cells is that they almost always have an abnormal set of chromosomes. For example, the cells of the human cancer line called HeLa,[2] by far the most widely studied line of cultured human cells, typically have 70–80 chromosomes instead of the normal 46 (Fig. 11.14). Interestingly enough, noncancerous cell lines that pass the "crisis stage" in culture and become potentially immortal also have extra chromosomes; hence it seems likely that possession of extra chromosomes somehow frees cultured cells from some of the normal constraints on proliferation. In cancerous tissue in organisms, however, such an increase in chromosome number is not typical. Instead, there are frequently numerous self-perpetuating chromosomal segments called "minute chromosomes," or specific rearrangements of the normal chromosomes. For example, in 90 percent of patients with Burkitt's lymphoma (a cancer of cells in the immune system) the tip of one specific chromosome (8) has been moved to the end of another (14). The insertion point in chromosome 14 is next to a gene coding for an immune-system polypeptide (see Fig. 12.4, p. 312), while the translocated region contains a gene that is frequently involved in cancers. The other 10 percent of patients with Burkitt's lymphoma have other translocations from chromosome 8. The implication is that the critical site in chromosome 8 has a control region that can contribute to cancer by stimulating the transcription of specific genes. As we will see, predictable translocations are also found in the cultured cells of several other sorts of cancer.

Besides differences within the nucleus, cancer cells and normal cells display a host of significant differences pertaining to cell shape and to the nature of the cell surface. Cultured cancer cells, for instance, tend to have a rather spherical shape, one seen in normal cells only during a short period immediately following cell division. This peculiarity of shape is probably a result of the abnormally small number of functional structure-stabilizing microfilaments, which also makes these cells more mobile than normal cells. This characteristic is also related to a striking feature of cultured cancer cells known as anchorage independence: most cells must "cling" to

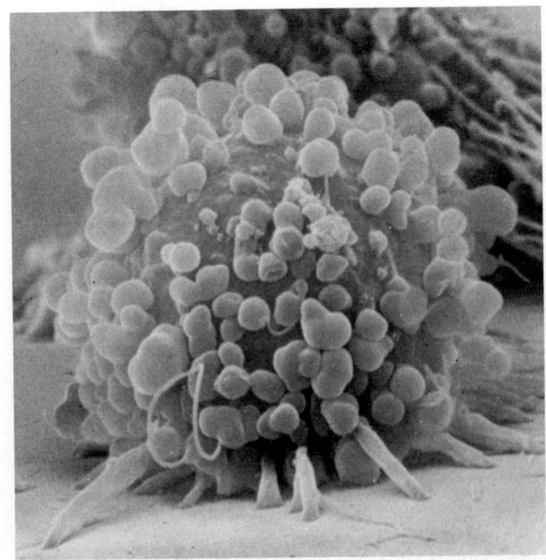

11.14 A typical cancer cell
This cell, from the HeLa line, is growing in tissue culture. The cell is rather spherical, and is covered with many small "blisters," called blebs, whose significance is not understood.

[2] The HeLa cell line is derived from a carcinoma of the cervix of Henrietta Lacks, who died of her cancer in 1951. This was the first stable, vigorously growing line of cultured human cells used in cancer research.

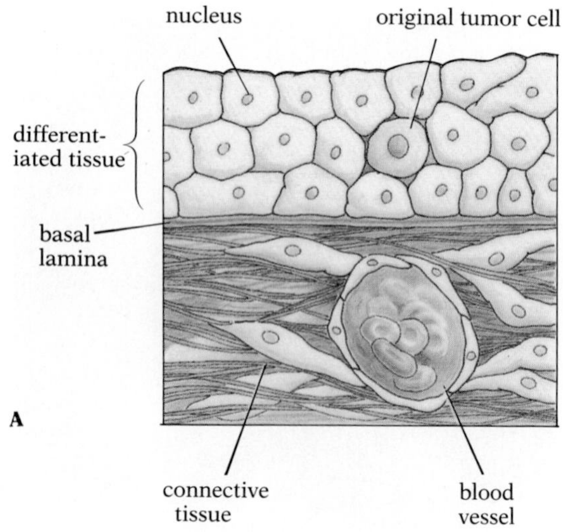

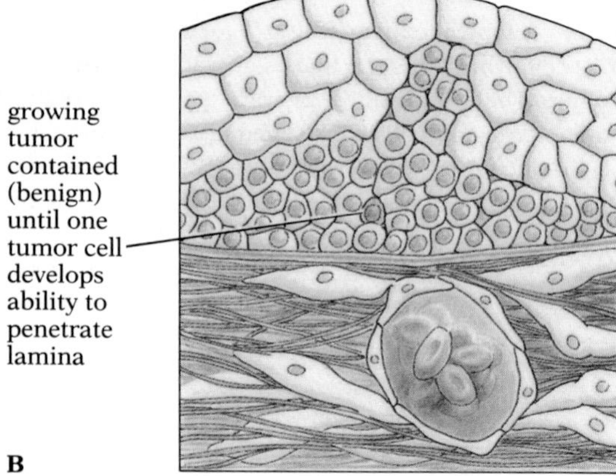

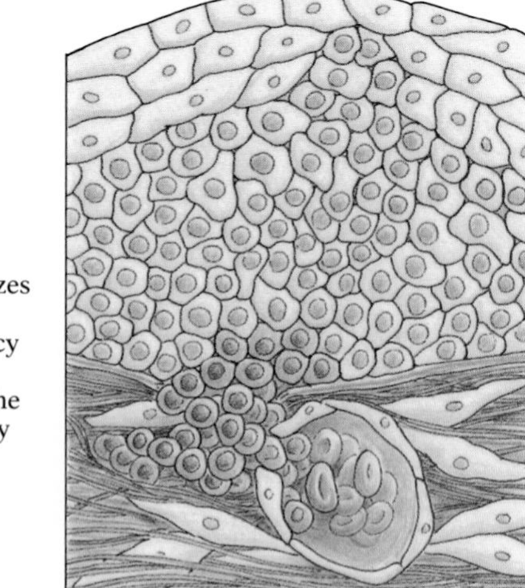

a solid surface in order to grow—a useful constraint, since they will require support later in order to function properly. Cancer cells are not constrained in this way, and so they can grow in liquids or on soft surfaces.

In order to metastasize in an individual, and so escape from its tissue or organ of origin, a cancerous cell must usually be able to produce and transport to its own membrane a receptor called laminin, which allows cancer cells to bind to the basal lamina (a tough layer that underlies or surrounds many tissues and organs, including the vessels of the circulatory system). It must then be able to secrete collagenase, which digests the lamina and allows the cell to cross the barriers that otherwise contain the growth of tumors (Fig. 11.15). In particular, the capacity to break through the lamina surrounding capillaries allows cancer cells to spread throughout the body. Since most normal cells can neither produce laminin nor secrete collagenase, the majority of tumors are contained at their site of origin, and so are benign.

Cancer cells, with their abnormal surfaces, appear to suffer from an inability to react appropriately to environmental changes. Cancer cells typically have fewer and different glycolipids and glycoproteins in their cell coats; these differences are probably correlated both with the absence of normal contact inhibition during cell division, and with the apparent inability of cancer cells to recognize other cells of their own tissue type. Whereas normal cells from two different tissues (such as liver and kidney) mixed in culture tend to sort themselves out and reaggregate according to tissue type, cancer cells do not do so. This absence of normal cellular affinities is probably one of the reasons why malignant cells of many cancers can spread.

The multistep hypothesis Our current understanding of cancer suggests that the conversion of a normal cell into a cancer cell involves several changes: loss of fixed-number-of-division control, loss or reduction of contact inhibition, loss of anchorage dependence, and, sometimes, tissue-specific cell-surface changes. But in many tissues even these changes result only in benign tumors that grow slowly and then stop, apparently because the cells fail to produce the chemical signal or signals that cause vascularization (the establishment of a capillary system that supplies oxygen and nu-

11.15 Development of a malignant tumor
Normal tissue has differentiated cells with small, quiescent nuclei, separated from the circulatory system and loosely packed connective tissue by a tough basal lamina (A). When a cell incurs enough changes to begin proliferating, its nucleus enlarges, its shape changes, and it begins to replicate its DNA. As the cell reproduces, the clone it forms begins to push other cells back. If it lacks contact inhibition, it will grow regardless of crowding, and if it is not limited to a fixed number of divisions, the clone will grow as long as nutrients (blood supply) permit, but may still be contained by the lamina (B). If a cell in the clone develops the ability to penetrate the basal lamina, it will create a breach and its descendants will move across and proliferate. The tumor will metastasize widely when it penetrates the lamina protecting blood vessels (C).

trients to growing tissue). Even when it can recruit capillaries, the tumor cannot metastasize without being able to bind to and destroy the lamina. Do all these changes take place at once as a result of a single genetic event, or is each change the result of one or more separate events?

Health statistics suggested the answer to this question before the actual mechanisms began to be elucidated. As Alfred Knudson of the Institute for Cancer Research in Philadelphia pointed out, if just a single genetic event were required, the probability of an individual's having contracted a particular type of cancer should increase proportionally with age. For example, if the chance of contracting skin cancer were 1 percent per year, then the chance of having contracted it would be 2 percent for a two-year-old, 3 percent for a three-year-old, and so on. In fact, however, the incidence of cancers in the population as a whole does not rise in this linear fashion (Fig. 11.16). Instead, cancer is normally a disease of old age. Knudson concluded that to account for the exponential rise of the incidence of cancer, several independent genetic events are required. The curve would be the product of the individual probabilities of the separate events. If we knew that the different genetic events had exactly equal probabilities of occurring, we could calculate with great precision the number of events required to induce a particular kind of cancer. But since this is an unlikely assumption, and since the curves for different kinds of cancer are not identical, it is not possible to be exact. Nevertheless, it seems almost certain that some cancers (particularly those, like leukemia and immune-system cancers, involving tissues in which tissue-type affinity, vascularization, crossing of laminae, and a solid surface for growth are not necessary for the normal cells) require only two or three genetic events, while others require four to seven. This is in reasonably good agreement with the two to seven levels of control thought to be involved—a fixed number of cell divisions, contact inhibition, anchorage dependence, tissue-type affinity, and the need for laminin-binding proteins, collagen digestion enzymes, and vascularization.

Additional evidence for a multistep model in at least some cancers comes from Knudson's studies of individuals who inherit a proclivity for certain cancers (retinoblastoma, a cancer of the eye, is a well-documented example). Unlike most people, these individuals usually contract cancer when young, and the cumulative incidence curve for their families, whose members share an inborn tendency for some particular form of the disease, shows a linear rise with age, rather than the more common exponential one. The linear curve suggests that in such individuals—those with a predisposition for retinoblastoma, for example—only a single event is necessary to trigger the disease. Because the inheritance of the genetic susceptibility in this kind of cancer appears to behave as a single-gene characteristic, contracting it probably requires a total of two transforming events—just one in addition to the one already built into the genome.

These statistical studies alone, however, cannot tell us what constitutes a genetic event leading to cancer (among the possibilities are single-base changes, deletions, insertions, or chromosomal rearrangements); nor do the studies tell us the precise effects of the events. Recent evidence, however, strongly suggests that most steps are directly linked to the loss of or escape from a particular level of control. The best data on this come from studies of oncogenes.

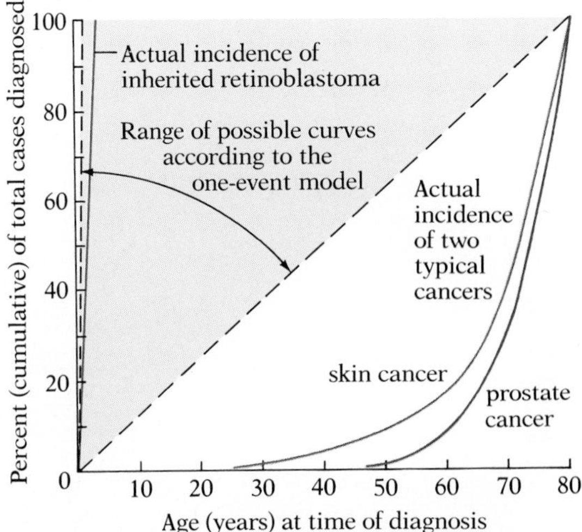

11.16 Incidence of representative cancers
If each type of cancer were caused by a single event, the probability of contracting a particular cancer would increase linearly with age; the range of possible curves is indicated. Actually, the incidence of most cancers increases exponentially, indicating that several contributing events are usually involved. The incidence curve for skin cancer closely approximates a three-event curve, while the curve for prostate cancer approximates a five-event curve. By contrast, the curve for inherited retinoblastoma is linear, approximating a single-event curve.

Oncogenes The discovery of ***oncogenes***—genes that cause cancer—has a complex history. In 1910 Peyton Rous, working at the Rockefeller Institute in New York, found that a viruslike entity, now known as Rous sarcoma virus, could cause cancer in chickens. Rous's conclusions were not widely accepted for decades, and only in 1966 was he awarded a Nobel Prize for his pioneering work. We know now that the causative agent in this particular case is a retrovirus, which, like other retroviruses, produces a cDNA copy of its RNA that is incorporated into the host chromosome (see p. 258). Some strains of Rous sarcoma virus cause cancer quickly, and at a high rate, because the cDNA brings an oncogene (known as *v-src*—the *v* identifies it as a viral gene) with it. The expression of this gene in the host chromosome, which does not depend on the site of incorporation, makes the cell cancerous. Ordinary retroviruses—those not containing oncogenes—can also cause cancer, but only at a very low rate, and even then after a long latency. Incorporation of cDNA into the host genome is apparently random, and cancer from ordinary retroviruses occurs only when the incorporation takes place near one of a few particular genes in the host, called ***proto-oncogenes***. Presumably an active control region in cDNA alters the expression of one or more of these genes. Oncogenes, then, are genes that inevitably cause cancer; proto-oncogenes, on the other hand, have the potential to cause cancer, but some change is required to convert them into oncogenes.

Perhaps the most surprising discovery about the oncogene *v-src* is that it is virtually identical to a normal gene in chickens; indeed, this normal gene (designated *c-src* to indicate that it is cellular in origin) is a proto-oncogene: incorporation of ordinary cDNA near *c-src* can (eventually) cause cancer. Since 1970 more than nearly two dozen rapidly acting, cancer-inducing retroviruses infecting a variety of birds and mammals have been discovered, each carrying an oncogene almost identical to a normal host proto-oncogene.

The viral cancers are convenient to study, but as yet few forms of cancer in humans have been found to be definitely viral in origin. However, studies of cultured human and nonviral animal tumors have revealed strong similarities to the observed behavior of retroviruses: some of the genetic changes that are responsible for making the cells in these cultures cancerous involve the same oncogenes also found in cancer-inducing retroviruses. Some of the nonviral oncogenes appear to have arisen through a mutation in a proto-oncogene that altered the base sequence through a base

TABLE 11.4 *Events that can create oncogenes*

I. Altered gene, resulting in altered product
1. Incorporation of retroviral oncogene
2. Mutation of normal proto-oncogene to create an oncogene
II. Altered gene expression, in which a gene is transcribed at the wrong time or at too high a rate
1. Insertion next to a proto-oncogene of retroviral cDNA containing an active control region
2. Translocation of a proto-oncogene to a position next to an active control region
3. Translocation of an active control region to a position next to a proto-oncogene
4. Mutation of a control region next to a proto-oncogene
III. Loss of anti-oncogene activity

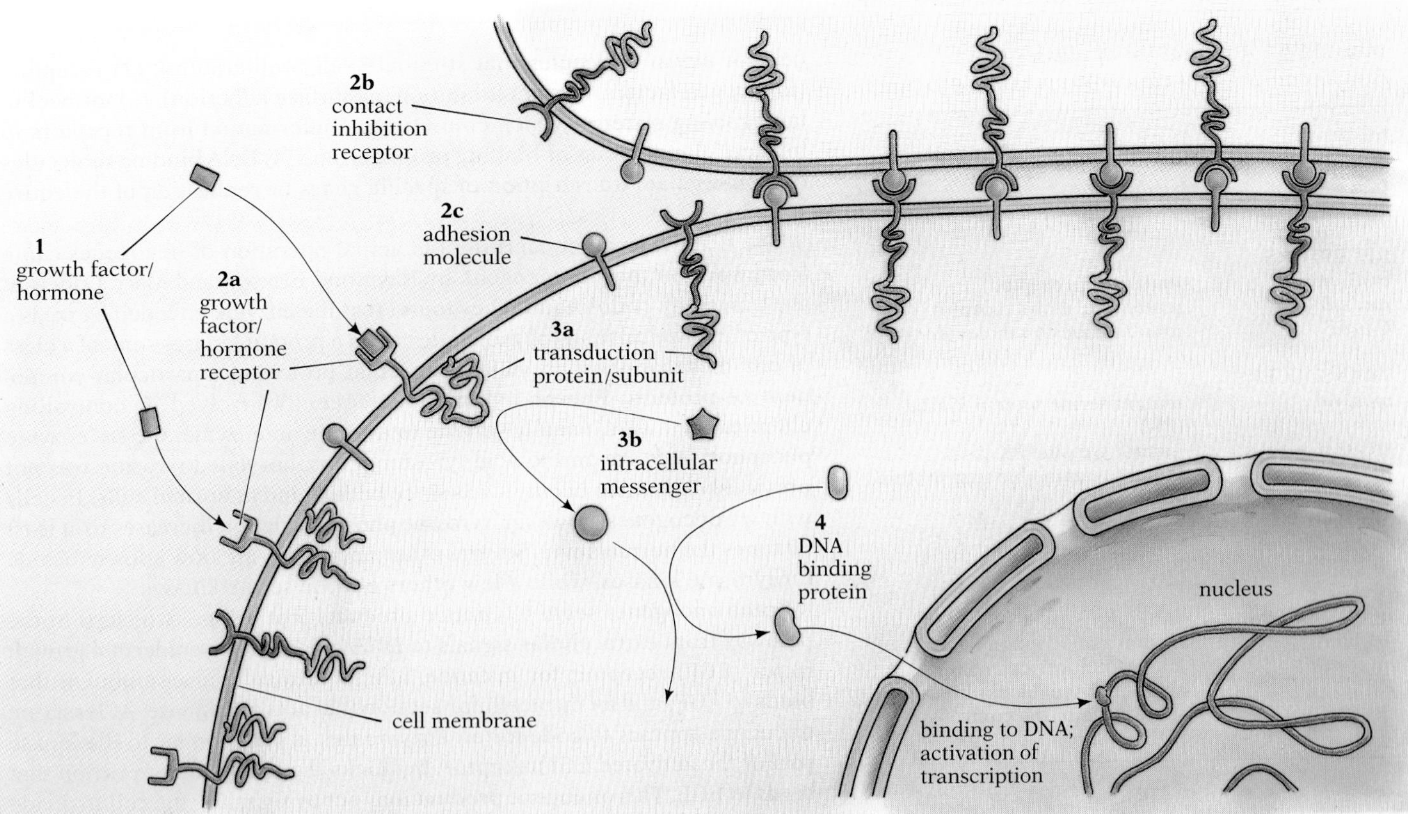

11.17 General sites of oncogene action Oncogenes can exert their effect at any point in the flow of information from the extracellular environment to the chromosomes. Some act to create high levels of growth factor (1) or GFs with unusual properties. Others alter the sensitivity of membrane receptors for GF or other relevant signals, like those involved in adhesion or contact inhibition (2). Still others alter the communication between the receptors and the intracellular messenger chemicals (3); some receptors incorporate domains that alter intracellular messengers, while others interact with an intermediate enzyme that makes the appropriate change. Finally, some oncogenes alter the DNA-binding proteins that are activated or repressed by the intracellular messengers (4).

change, insertion, or deletion. Others appear to be proto-oncogenes that have been moved (translocated) from their usual locations and so perhaps released from their normal controls, to become oncogenes. Conversely, either mutation or translocation involving a control region next to a proto-oncogene can convert a proto-oncogene into an oncogene. Indeed, when grown in culture, cells of several leukemias, one ovarian cancer, and several other kinds of cancer display predictable chromosomal translocations or losses. The translocations seem to involve moving a structural gene and its control region, which is responsible for its transcription in the tissue at a high rate, to a position next to a proto-oncogene, or moving the proto-oncogene next to a structural gene and its control region. Finally, there are ***anti-oncogenes***—genes that encode proteins that deactivate oncogene products. Failure of such an inhibitor gene is equivalent to activating a proto-oncogene. The gene involved in retinoblastoma is an anti-oncogene.

The various mechanisms that may result in the presence of an oncogene in the genome are summarized in Table 11.4.

We have seen that the conversion of a normal cell into a cancer cell requires two or more genetic changes specific to that kind of cell, and at least some of these changes involve oncogenes. We have also seen that some changes that characteristically give rise to cancer cells involve a loss of one or more levels of control. Do oncogenes, then, affect the control of cell division, contact inhibition, anchorage dependence, tissue-type affinities, and vascularization, as well as the expression of the genes needed to breach the basal lamina?

Current research indicates that most oncogenes encode products that fall into one of four general categories (Fig. 11.17): (1) growth factors (extra-

TABLE 11.5 *A partial list of classes of cancer-causing and cancer-suppressing genes*

Name	*Type*
EXTRACELLULAR SIGNALS	
sis	mutant growth factor (GF)
RECEPTORS	
erbB, neu	mutant GF receptor
ros, erbA	mutant hormone receptor
C-myc	mutant adhesion molecule
TRANSDUCERS	
mos, raf	mutant serine protein kinase (PK)
src, met	mutant tyrosine PK
ras	mutant guanine binding protein
SECOND MESSENGERS	
(none known)	
DNA BINDING	
fos, jun	mutant transcription factor
myc, myb	mutant inducer or enhancer
RB	anti-oncogene; blocks binding at some stage in the control pathway

cellular signal molecules that stimulate cell proliferation), (2) receptors (for growth factors, contact inhibition, or surface adhesion), (3) intracellular signaling systems (which communicate information from receptors to intracellular enzymes or binding proteins), and (4) DNA binding molecules (which regulate transcription of specific genes or replication of the entire genome).

The first evidence concerning the actual operation of oncogenes came from work on the *src* oncogene by Raymond Erikson and Marc Collett at the University of Colorado. They found that the enzyme encoded by *src* is a type of intracellular signal molecule called a protein kinase—one of a class of enzymes that phosphorylate a particular protein or a particular component of proteins. Phosphorylation is frequently involved in controlling chemical pathways, usually serving to activate an enzyme; the *src* enzyme phosphorylates the amino acid tyrosine. Phosphorylated tyrosine was not previously known to occur. It has since been found in normal cells. In cells with *src* oncogenes, however, tyrosine phosphorylation increases to at least 10 times the normal level. Several other oncogenes are now known to code for tyrosine kinases, while a few others encode serine kinases.

Some oncogenes seem to encode products that bridge two steps in the pathway from extracellular signals to DNA binding. The epidermal growth factor (EGF) receptor, for instance, has an extracellular component that binds to EGF, and an intracellular section that acts as a kinase. At least one oncogene appears to code for an enzyme that is very similar to the kinase part of the activated EGF receptor, but lacks the extracellular portion that binds to EGF. This oncogene product may act by signaling the cell to divide continuously, whether or not EGF is present.

The pattern that is emerging suggests that understanding cancer in the near future may be a realistic hope: there appear to be only a limited number of ways cancer is triggered, only a limited number of proto-oncogenes in any cell (fewer than a hundred in humans), and an even smaller number of functions for oncogene products (Table 11.5). Understanding oncogenes will probably prove the most important step in finding effective treatments for cancer.

Environmental causes of cancer As we have seen, mutations, translocations, and retroviruses can each play a role in triggering cancer. We also know that many, if not all, mutations and translocations are caused by external agents—radiation and mutagenic chemicals. But linking particular chemicals or radiation sources to cancer in humans and evaluating the relative danger of each has proven difficult. Not every mutation occurs in a cancer-causing location, and even a mutation that is in such a location will trigger only one of the steps necessary for development of a cancer. We have seen that two to seven specific independent genetic events are required; the final step may occur years after the first significant exposure to a mutagenic chemical or radiation. Only in the case of colon cancer are researchers even close to understanding the steps involved (Fig. 11.18).

The absence of strict cause-and-effect relationships between particular mutagenic agents and particular cancers has made the identification and evaluation of causative agents difficult, even in the most obvious cases. For example, though health statistics have shown that cigarette smoking is probably responsible for about 150,000 fatal cancers annually (and for 25 percent of fatal heart disease), critics in the tobacco industry deny the link. While granting that smokers develop lung cancer and heart disease far more often than nonsmokers, the critics point out that many smokers never develop lung cancer, and that some nonsmokers do. An informed understanding of how oncogenes work and of the probabilistic nature of their development, however, reveals the weaknesses in these arguments: the genetic events necessary for inducing lung cancer must be greatly facilitated by the highly mutagenic "tar" in cigarette smoke (a deadly combination of chemicals that kills two-pack-per-day smokers eight years early on average), but the changes must occur in the correct locations (including a specific deletion on chromosome 3) in at least one cell, so even with regular exposure to mutagenic smoke, some smokers will escape unscathed. At the same time, other sources of mutation, such as radiation, are also present, and though less potent, they will sometimes cause the changes necessary to trigger lung cancer even in nonsmokers.

Another way to evaluate the carcinogenic potential of chemicals and radiation is to expose animals—usually mice—to measured doses of these agents. This procedure is expensive and time-consuming, and in addition we must assume that what causes cancer in mice is equally carcinogenic in humans. Another assay is the Ames test, described on page 251, which involves exposing bacteria to potential carcinogens to see if mutations result. The assumption is that a mutagen for *E. coli* is a carcinogen for humans. The Ames test has identified many substances as potential carcinogens. Others having no initial effect are converted into mutagens when liver homogenate is added to the bacterial culture, and are presumably also thus converted in the liver of an organism. The coal-tar components of many hair dyes and cosmetics, as well as hexachlorophene soaps, flame-retardant chemicals in children's sleepwear, the seared protein of grilled meat, the smoke from wood fires, and several chemicals in certain vegetables and spices are just a few of the substances identified by the Ames test as possible carcinogens. When tested on animals each of these has in fact proven carcinogenic. It seems only sensible to assume that all mutagens are carcinogenic, but given the ubiquity of environmental mutagens, avoiding them all is impossible. The only sensible course, then, is to eliminate our exposure to the most potent—especially cigarette smoke and coal tars—and, weighing the costs and benefits (just as we do each time we decide to take the risk of driving a car), to minimize contact with the rest.

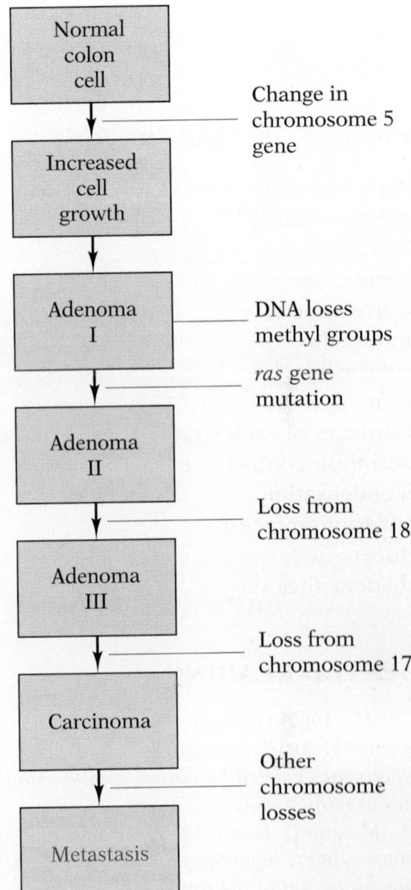

11.18 Chromosomal changes leading to colon cancer Colon cancer usually begins with small growths called adenomas, which are categorized according to morphology. The cancer is unusual because some of the chromosomal changes involved are cytologically visible and must occur in more or less one particular order; these peculiarities have made it possible to outline the steps involved. The *ras* oncogene encodes an intracellular signaling enzyme, which couples the response of growth-factor receptors to DNA-binding protein, which in turn regulates part of the cell-division control system.

STUDY QUESTIONS

1. Are there good reasons why eucaryotes should rely so much more than procaryotes on positive gene control? (pp. 286–99)
2. Why do you suppose that skin cancer is so often controllable? (pp. 299–302)
3. How is it that the eucaryotic genome has not been strangled by bureaucracy, with its 50,000 genes, each managed by one or more control genes, with these in turn orchestrated by another hierarchy of thousands of middle-level executive genes, themselves responding to hundreds of upper-management sequences, and so on?
4. Why bother keeping inactive regions of chromosomes condensed?
5. What sorts of mutations could affect the *lac* operon, and what effects would they have? What would happen if a cell had two of these mutations at once? (pp. 276–77, 280–81, 284–85)

CONCEPTS FOR REVIEW

- Transcription control in bacteria
  - Negative control
    - inducible
    - repressible
  - Positive control
    - inducible
    - repressible
  - The *lac* operon
  - Advantages of each strategy for particular sorts of genes
- Transcription control in eucaryotes
  - Decondensation
  - Transcription factor
  - Inducer sites
  - Enhancer sites
  - Need for multilevel control
- Chromosomal organization
  - Gene amplification
  - Repetitive DNA: sources and possible uses
- Post-transcriptional control
  - Alternative splicing
  - Delayed or interrupted translation
  - Variable lifespans of mRNA and protein
- Cancer
  - Changes necessary for malignancy
  - Multistep hypothesis
  - Oncogenes
    - sites of action in cellular information flow
    - role of mutation and viruses

SUGGESTED READING

BISHOP, J. M., 1982. Oncogenes, *Scientific American* 246 (3). (Offprint 1513) *An illuminating look at the relationship between cancer genes carried by viruses and the similar, noncancerous genes in normal cells.*

CROCE, C. M., and G. KLEIN, 1985. Chromosome translocations and human cancer, *Scientific American* 252 (3). (Offprint 1558) *A clear discussion of the translocations involved in Burkitt's lymphoma.*

FELDMAN, M., and L. EISENBACH, 1988. What makes a tumor cell metastatic? *Scientific American* 259 (5). *About the oncogenes that permit tumor cells to stop adhering to other cells or structures, and so spread to other parts of the body.*

FELSENFELD, G., 1985. DNA, *Scientific American* 253 (4). *The role of DNA structure in the regulation of gene expression.*

HUNTER, T., 1984. The proteins of oncogenes, *Scientific American* 251 (2). (Offprint 1553)

MCKNIGHT, S. L., 1991. Molecular zippers in gene regulation, *Scientific American* 264 (4). *On the operation of the leucine zipper.*

MOSES, P. B., and N.-H. CHUA, 1988. Light switches for plant genes, *Scientific American* 258 (4). *How light energy is used to activate the genes involved in photosynthesis.*

NICOLSON, G. L., 1979. Cancer metastasis, *Scientific American* 240 (3). (Offprint 1422)

NOMURA, M., 1984. The control of ribosome synthesis, *Scientific American* 250 (1). (Offprint 1546)

PTASHNE, M., 1989. How gene activators work, *Scientific American* 260 (1). *A very up-to-date summary of how promoters work in bacteria and yeast.*

PTASHNE, M., A. D. JOHNSON, and C. O. PABO, 1982. A genetic switch in a bacterial virus, *Scientific American* 247 (5). (Offprint 1526) *An excellent description of the details of the lytic/lysogenic switch of lambda virus.*

ROSS, J., 1989. The turnover of messenger RNA, *Scientific American* 260 (4). *On how the rate of degradation of different messengers is controlled.*

RUDDLE, F. H., and R. S. KUCHERLAPATI, 1974. Hybrid cells and human genes, *Scientic American* 231 (1). (Offprint 1300) *The technique of fusing human cells with cells from other mammals in order to map human genes and study their regulation.*

SAPIENZA, C., 1990. Parental imprinting of genes, *Scientific American* 263 (4). *On the inheritance of gene switches bound to the DNA of gametes.*

TIOLLAS, P., and M. A. BUENDIA (1991). Hepatitus B virus, *Scientific American* 264 (4). *On the life history of a cancer-promoting virus.*

WEINBERG, R. A., 1983. A molecular basis of cancer, *Scientific American* 249 (5). (Offprint 1544) *A good discussion of oncogenes and the discovery that a single-base change can transform a prepared cell into a cancer cell.*

WEINBERG, R. A., 1988. Finding the anti-oncogene, *Scientific American* 259 (3). *On the genes that restrain cell growth, focusing on retinoblastoma.*

WEINTRAUB, H., 1990. Antisense RNA and DNA, *Scientific American* 262 (1). *On the use of antisense RNA to regulate mRNA activity.*

Chapter **12**

CELLULAR REPRODUCTION

n the last four chapters, we have traced the flow of genetic information within individual cells, as well as ways in which that flow and the information itself are changed—the control of gene expression and gene evolution, respectively. Now we will turn our attention to how cells themselves reproduce, passing their genetic endowment from cell to cell, and from parent to offspring. In the next two chapters, we will follow this pathway still further to see how the genome organizes the development and cellular specializations typical of complex, multicellular organisms.

A

B

C

D

E

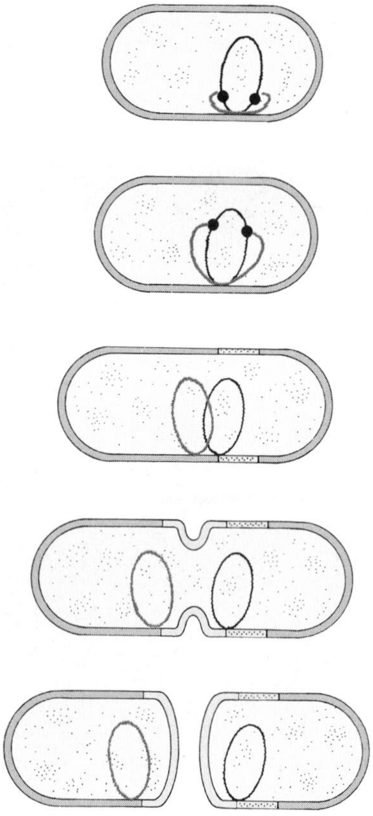

12.1 Binary fission of a procaryotic cell
(A) The circular chromosome of a cell is attached to the plasma membrane near one end; it has already begun replication (the partially formed second chromosome is shown in red). (B) Replication is about 80 percent complete. (C) Chromosomal replication is finished, and the second chromosome now has an independent point of attachment to the membrane. During replication additional membrane and wall (stippled) has formed. (D) More new membrane and wall (tan) has formed between the points of attachment of the two chromosomes. Part of this growth forms invaginations that will give rise to a septum cutting the cell in two. (E) Fission is complete, and two daughter cells have formed. The chromosomes, which are actually so long that they must be looped and tangled to fit into a cell, are depicted here as small open circles. At the scale of this drawing, the actual circumference of each chromosome would be about 30 m.

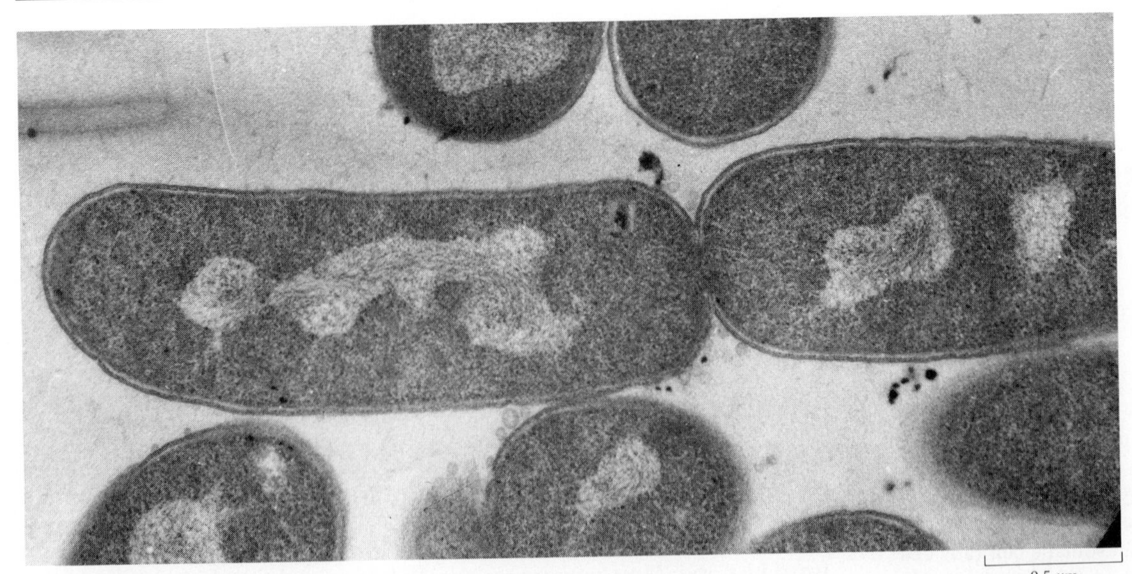

12.2 Electron micrograph of a dividing bacterial cell
New wall is growing inward, cutting this *E. coli* in two. The nucleoid (white areas inside cells) has already divided.

THE TRANSFER OF GENETIC INFORMATION

Before a cell can divide to produce two viable new cells, it must first replicate all the genetic information in its nucleus and then make sure that a full set of information is given to each daughter cell. We examined the process of DNA replication in detail in Chapter 8; our emphasis here will be on the larger-scale organizational task of duplicating the entire genome and accurately segregating the copies into two progeny cells.

Procaryotic chromosomes As we saw in Chapter 8, cell division is less complex in procaryotes than in eucaryotes. As the procaryotic cell elongates sufficiently to form two independent daughter cells, the single circular chromosome replicates and the resulting second chromosome attaches to a different point from the first one on the expanding plasma membrane (Fig. 12.1). Next, new plasma membrane and wall material form near the midpoint of the parental cell and grow slowly inward, cutting through both the cytoplasm and the nucleoid (Fig. 12.2). Thus each new daughter cell receives a complete chromosome, a full complement of genetic information encoded by perhaps a million base pairs. This process is known as transverse fission, or ***binary fission***.

Eucaryotic chromosomes As we saw in Chapter 11, the chromosomes of eucaryotes differ from those of procaryotes in several ways. First, except in mitochondria and other organelles, eucaryotic chromosomes are linear rather than circular. Second, eucaryotic chromosomal DNA in the nucleus is wound on nucleosome cores. Third, each chromosome is organized into loops, all of which are in a highly condensed state during cell division, probably to prevent tangling (Fig. 12.3).

The final major difference between eucaryotic and procaryotic chromosomes arises because most eucaryotes receive chromosomes from two parental cells—from a sperm cell and an egg cell, for instance—rather than

12.3 Condensed chromosome showing highly condensed loops

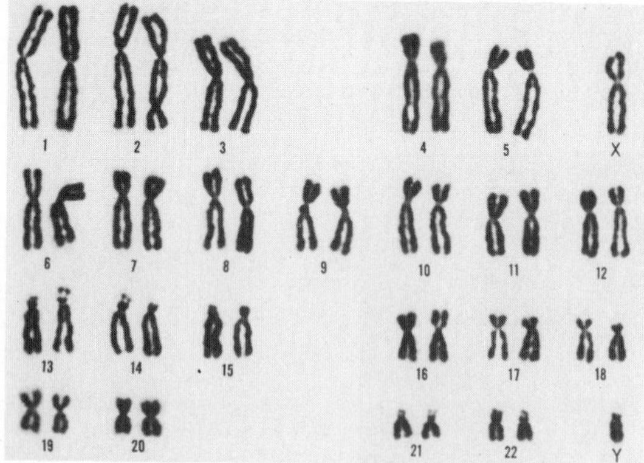

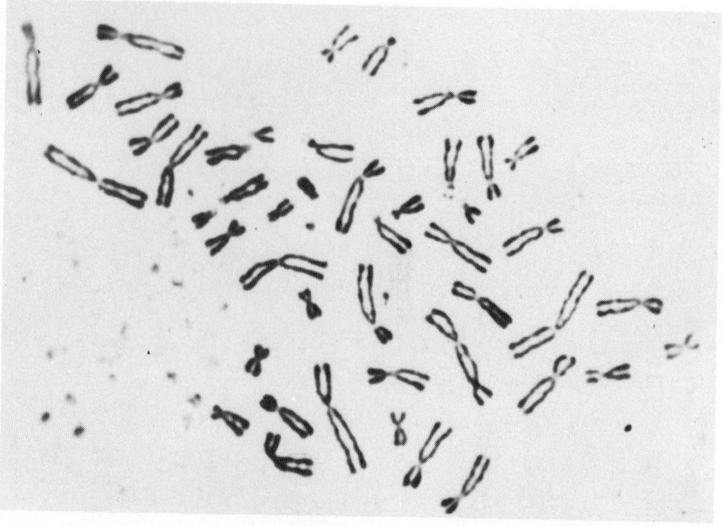

12.4 Photograph of chromosomes of a human male
At left, the chromosomes have been arranged as homologous pairs and numbered according to accepted convention. A human somatic cell (any cell except the egg or sperm cells) contains 23 pairs of chromosomes, including a pair of sex chromosomes (for a male, X and Y).

from just one. Most eucaryotes are diploid, and their chromosomes thus occur in ***homologous pairs***, each consisting of one chromosome from each parent bearing basically the same genes in the same order (Fig. 12.4). Procaryotes, as we have seen, are haploid: their chromosome is unpaired. Though chromosomal number may vary enormously between species—cats, for instance, have 19 pairs, fruit flies 4 pairs, onions 8 pairs—the chromosomal number within a species is normally uniform. The 23 pairs of human chromosomes have been studied extensively, and their distinctive "personalities" are beginning to be familiar to researchers interested in the bases of human genetic diseases and developmental abnormalities.

When eucaryotic chromosomes condense and become visible early in cell division, each chromosome has already been replicated to produce compact twins, connected by a centromere (Fig. 12.5). Each member of this bound pair is referred to as a ***chromatid***.

We mentioned earlier two basic kinds of cell division in eucaryotes: division as part of the growth of an organism, also known as somatic or mitotic cell division, and division to produce the gametes that give rise to new individuals, also known as meiotic cell division. We will describe each of these in turn, in separate sections.

MITOTIC CELL DIVISION

Cell division in eucaryotic cells involves two fairly distinct processes that often, but not always, occur together: division of the nucleus and division of the cytoplasm. The process by which the nucleus divides to produce two new nuclei, each with the same number of chromosomes as the parental nucleus, is called ***mitosis*** (from the Greek *mitos*, "thread"). The process of division of the cytoplasm is called ***cytokinesis***. We will discuss mitosis and cytokinesis separately.

MITOSIS

For convenience, it is customary to separate each mitotic cycle, from one cell division to the next, into a series of stages, each designated by a special name. Though each stage will be discussed separately here, the entire process is a continuum rather than a series of discrete occurrences.

Interphase The nondividing cell is said to be in the interphase state. The nucleus is clearly visible as a distinct membrane-bounded organelle, and one or more nucleoli are usually prominent. But chromosomes, as ordinarily pictured, are not visible in the nucleus; there are none of the distinct rodlike bodies that microscopes and sophisticated staining techniques enable us to see in a dividing cell (Fig. 12.6). Interphase chromosomes are so thin and tangled that they cannot be recognized as separate entities. They appear only as an irregular granular-looking mass of chromatin.

In interphase animal cells there is a special region of cytoplasm just outside the nucleus that contains two small cylindrical bodies oriented at right angles to each other. These are the ***centrioles*** (see Fig. 5.32, p. 148, and Fig. 5.33B, p. 148), which will duplicate themselves, move apart, and become associated with the poles of the mitotic apparatus of the dividing cell. Because they form the basis for producing flagella and cilia, the cells' centrioles must be distributed to daughter cells as reliably and accurately as the chromosomes themselves. In many animal cells this separation of the centrioles occurs just before the onset of mitosis, but in some cells it occurs during interphase, long before mitosis begins. No centrioles have yet been detected in the cells of most seed plants, which is not too surprising since these species lack cilia and flagella, but centrioles do occur in some algae, fungi, bryophytes (mosses, liverworts, etc.), and ferns in association with the production of motile sperm. Despite their high visibility, centrioles are not essential for cell division even in animal cells; they remain close to the spindle organizing centers during mitosis, but do not participate in spindle formation.

In the past, interphase cells were commonly called resting cells, but that terminology has been rejected as grossly inappropriate. An interphase cell is definitely not resting; it is carrying out all the innumerable activities of a living, functioning cell—respiration, protein synthesis, growth, differentiation, and so forth. During interphase, furthermore, the genetic material is

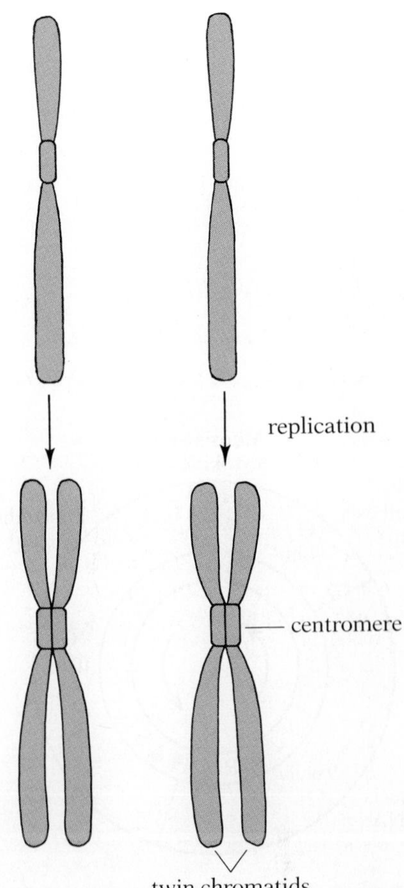

12.5 Production of chromatids through replication
At some time prior to cell division the genetic material of each chromosome duplicates itself. As a result, twin chromatids take form upon condensation, joined to each other by a centromere. For clarity, the unduplicated chromosomes are shown in condensed form, even though condensation normally takes place only after duplication.

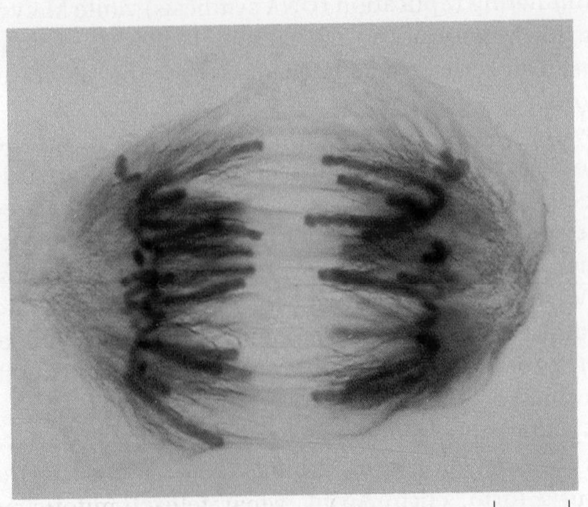

12.6 Chromosomes in a dividing cell of the African blood lily
The separating chromosomes for each of the new nuclei are easily distinguishable.

replicated and all the cellular machinery is duplicated in preparation for the next division sequence.

The replication of the genetic material does not, however, begin immediately after completion of the last division sequence. There is a gap in time, designated the G_1 stage, before genetic replication. This is when ribosomes and organelles begin to be duplicated. Next comes the S stage, during which the synthesis of new DNA takes place, along with the further duplication of organelles. Another period, designated G_2, separates the end of replication from the onset of mitosis proper. During this time the cell prepares for mitosis. These three subdivisions of interphase, along with mitosis and cytokinesis (together referred to as the M stage), constitute what is called the ***cell cycle*** (Fig. 12.7). Cells in tissue culture show striking morphological changes as they pass through the cell cycle, but the significance of these changes, especially in the intact organism, is not well understood.

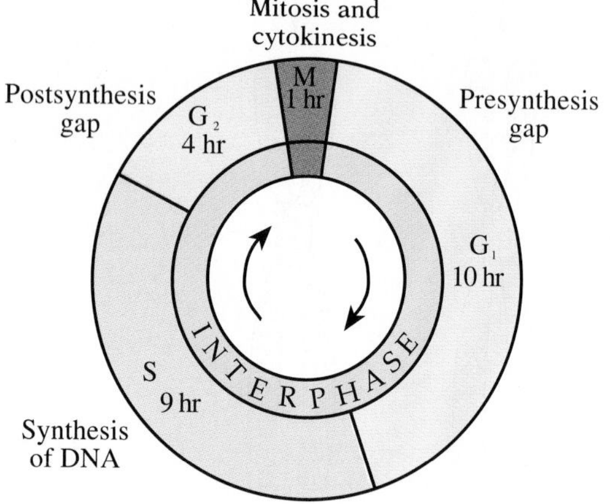

12.7 The cell cycle
This particular cycle assumes a period of 24 hours, but some cells complete the cycle in less than an hour and others take many days. Similarly, the ratios of the four stages of the cycle vary, with G_1 exhibiting the most variation.

The duration of the complete cell cycle can vary greatly. Though usually lasting 10–30 hours in plants and 18–24 hours in animals, it may be as short as 20 minutes in some organisms or as long as several days or even weeks. All the stages can vary in duration to some degree, but by far the greatest variation occurs in the G_1 stage. At one extreme very rapidly dividing embryonic cells may pass so quickly through the G_1 stage that it can hardly be said to exist at all, whereas at the other extreme some cell types become arrested in the G_1 stage; differentiated skeletal-muscle cells and nerve cells, for example, are suspended in the G_1 stage and normally never divide again. There are a few cases in which cells are arrested in the G_2 stage: they have replicated their DNA, but do not divide. The heart-muscle cells of human adults are an example of cells in G_2 arrest, a phenomenon for which there is, at present, no clear explanation.

Both G_1 and G_2 arrest result from a failure to produce an essential control chemical. If the nucleus from a cell in G_1 arrest is transplanted into a cell that is just entering the S stage, the transplanted nucleus will promptly be activated and itself enter the S stage, apparently because it has been stimulated by a control substance present in the cytoplasm of the host cell. Similarly, when a cell in G_2 arrest is fused with a mitotic cell, its chromosomes soon begin to condense and the cell enters mitosis.

The chemicals that control the cell cycle are called ***cyclins***. S-cyclin is involved in stimulating replication (DNA synthesis) while M-cyclin helps trigger mitosis. Each can bind to the same cell-division-cycle protein (***cdc***), which in turn can activate a cellular messenger specific for either replication or mitosis.

To see how the system works, let's follow a complete cycle beginning with the G_1 phase (Fig. 12.8). As G_1 begins, the cdc protein is in its inactive form, cdc_i. If the cell is to divide again, S-cyclin is steadily synthesized; the faster it is made, the sooner the S phase will begin. As the S-cyclin concentration rises, it starts to bind to cdc_i, converting it into its synthesis-promoting form, cdc_s. When a threshold is reached, cdc_s becomes activated and begins phosphorylating a messenger compound (Fig. 12.8); thus activated, the messenger triggers replication, and other enzymes quickly destroy the S-cyclin, restoring cdc_s to its inactive cdc_i form.

During G_2, the mitosis-inducing cyclin—M-cyclin—is steadily synthesized. As its concentration rises, it binds to cdc_i and converts it into a mitosis-promoting form, cdc_m. Again, when a threshold is passed, cdc_m is

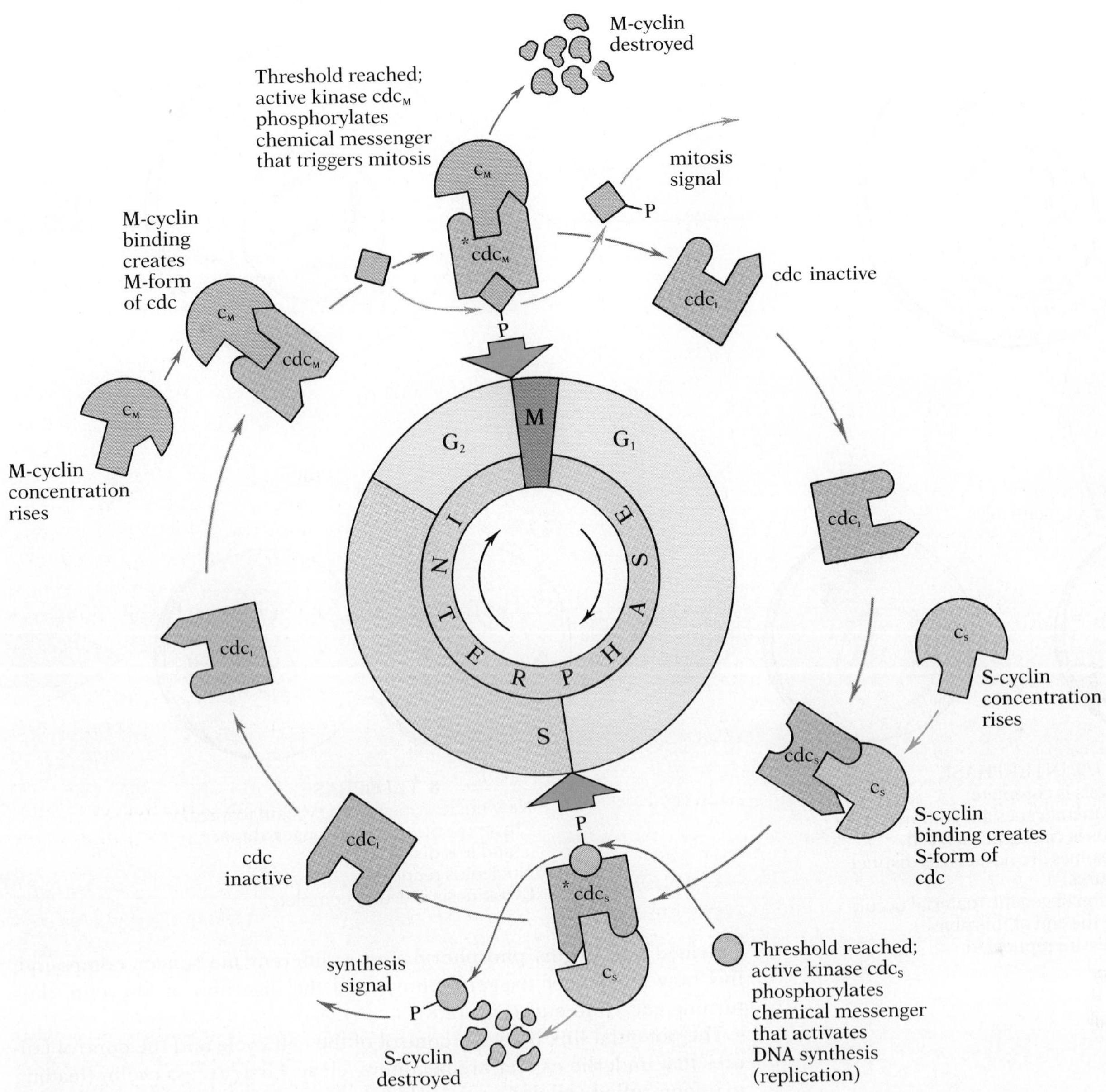

12.8 Control of cell division
A single protein kinase, cdc, is responsible for signaling the cell to replicate or undergo mitosis. It adopts the replication-promoting form as the concentration of S-cyclin rises in G_1 (lower right), becoming activated at a certain threshold (the activation is indicated here by an asterisk); it then phosphorylates a messenger that triggers replication and the destruction of S-cyclin. Similarly, as the M-cyclin concentration rises in G_2 (upper left), it binds to cdc, which adopts a mitosis-promoting form; once the threshold has been reached, the cdc becomes activated and catalyzes the phosphorylation of a messenger that triggers mitosis and the destruction of M-cyclin.

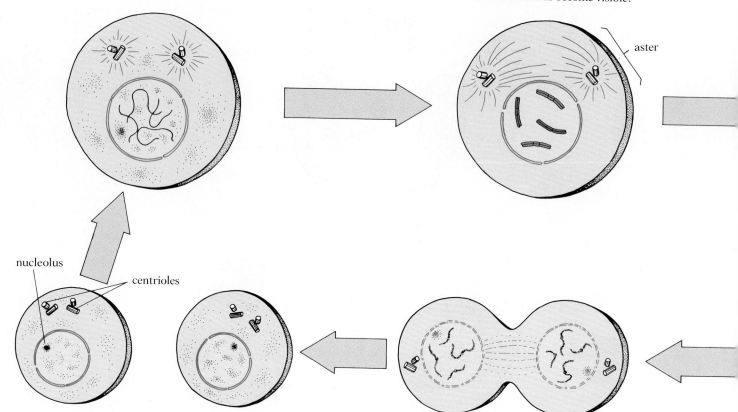

2. EARLY PROPHASE
Centrioles begin to move apart.
Chromosomes appear as long thin threads.
Nucleolus becomes less distinct.

3. MIDDLE PROPHASE
Centrioles move farther apart.
Asters begin to form.
Twin chromatids become visible.

aster

nucleolus

centrioles

1/9. INTERPHASE
Cytokinesis is complete.
Nuclear membranes are complete.
Nucleolus is visible in each cell.
Chromosomes are not seen as distinct structures.
Replication of genetic material occurs before the end of this phase.
Centrioles are replicated.

8. TELOPHASE
New nuclear membranes begin to form.
Chromosomes become longer, thinner, and less distinct.
Nucleolus reappears.
Cytokinesis is nearly complete.

activated and begins phosphorylating a different messenger compound; this new messenger triggers mitosis and the digestion of M-cyclin, thus turning cdc_m into cdc_i (Fig. 12.8).

The potential link between control of the cell cycle and the control failures that underlie cancer are becoming clear. First, excess cyclin (leading to uncontrolled cell division) could be produced, either through alteration of cyclin transcriptional control, or (as in the case of several oncogenes mentioned in the last chapter) through overstimulation of growth-factor receptors. Alternatively, a mutant form of cdc could remain constantly active, or be impossible to inactivate; the *RB* proto-oncogene of retinoblastoma, for instance, is thought to be involved in keeping cdc turned off. With the same effect, the messenger compound itself may be made permanently active. Finally, the DNA target sequence of a messenger could be altered, throwing the gene it regulates into a transcriptionally active form. It is widely hoped that studies of the control of replication and mitosis will lead to new techniques for restoring quiescence to tumor cells.

4. LATE PROPHASE
Centrioles nearly reach opposite sides of nucleus.
Spindle begins to form and kinetochore microtubules project from centromeres toward spindle poles.
Nuclear membrane is disappearing.
Nucleolus is no longer visible.

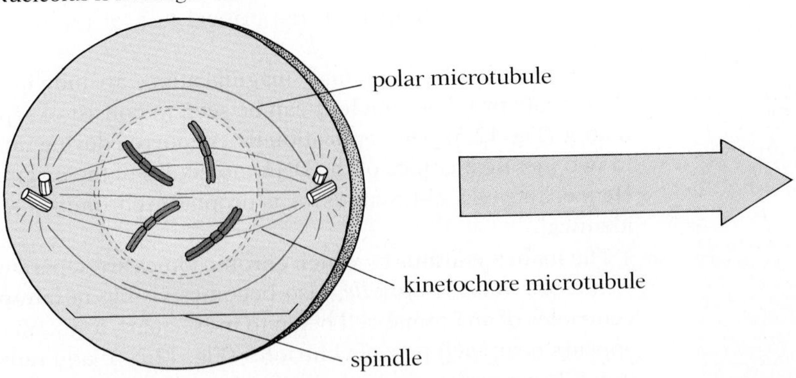

5. METAPHASE
Nuclear membrane has disappeared.
Kinetochore microtubules move each twin-chromatid chromosome to the midline; other spindle microtubules interact with spindle tubules from opposite pole.

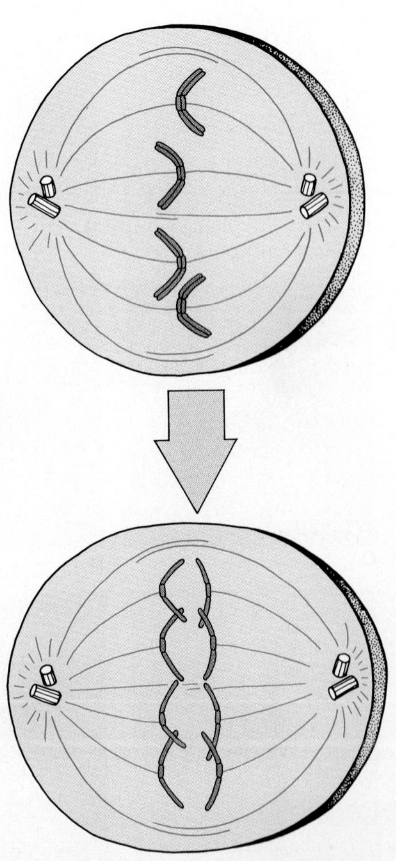

6. EARLY ANAPHASE
Centromeres have split and begun moving toward opposite poles of spindle.
Spindle microtubules from opposite poles force the poles apart.

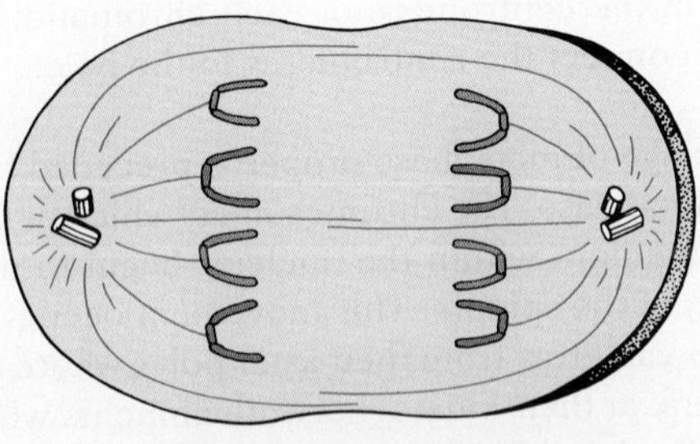

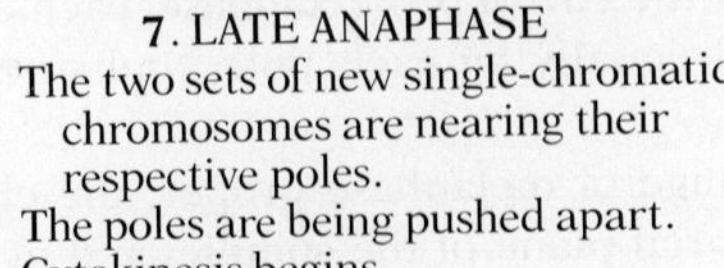

7. LATE ANAPHASE
The two sets of new single-chromatid chromosomes are nearing their respective poles.
The poles are being pushed apart.
Cytokinesis begins.

12.9 Mitosis and cytokinesis in an animal cell
In these drawings, purple and green are used to distinguish the two pairs of homologous chromosomes.

Let's assume now that the cell we are examining has passed through the G_1, S, and G_2 stages of interphase and is entering mitosis proper—itself a process so complex that it is customarily divided into four stages.

Prophase During prophase, the cell readies the nucleus for the crucial separation of two complete sets of chromosomes into two daughter nuclei. As the two centrioles of an animal cell move toward opposite sides of the nucleus, the initially indistinct chromosomes begin to condense into visible threads, which become progressively shorter and thicker and more easily stainable with dyes. When the chromosomes first become visible during early prophase, they appear as long, thin, intertwined filaments, but by late prophase the individual chromosomes can be clearly discerned as much shorter rodlike structures. As the chromosomes become more distinct, the nucleoli become less distinct, often disappearing altogether by the end of prophase (Fig. 12.9).

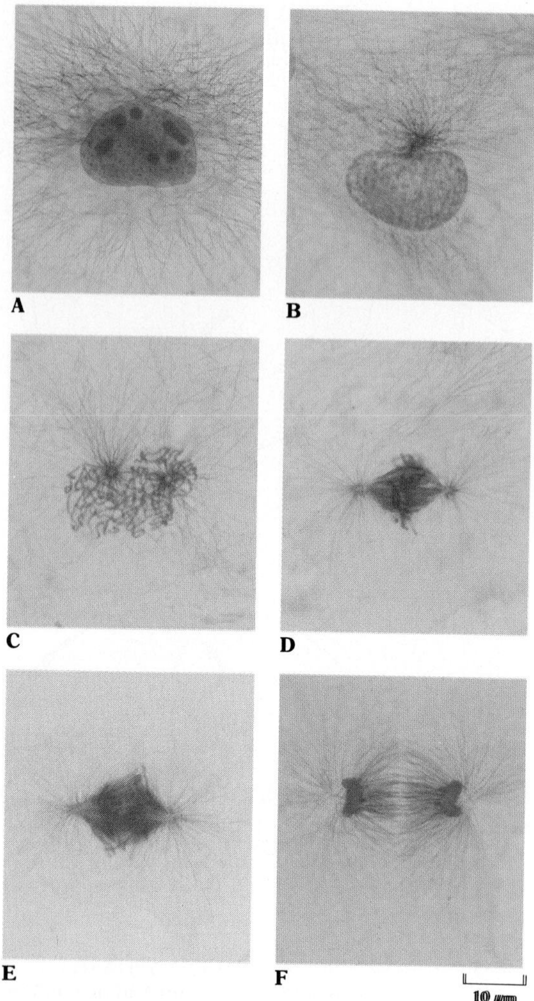

12.10 Mitosis in a cell from the endothelium of a frog's heart

(A) Interphase: a diffuse network of microtubules (stained red) is visible. (B) Early prophase: the chromosomes begin to condense; the centrioles have not yet separated. (C) Late prophase: prophase ends as the nuclear envelope disappears; the centrioles have moved apart, and the two asters are prominent. (D) Metaphase: the chromosomes are lined up along the equator of the spindle, midway between the two asters. (E) Middle anaphase: the two groups of chromosomes have begun to move apart from each other, toward their respective poles of the spindle. (F) Early telophase: the two groups of chromosomes are being organized into new nuclei; cytokinesis has begun, and the line dividing the two daughter cells is clearly visible.

The shortening of the chromosomes during cell division has one obvious advantage. In their shorter form, they can be moved about freely without becoming hopelessly tangled. But only in their long uncoiled form can the chromosomes participate in the critical task of replication during interphase.

Examined under very high magnification, an individual chromosome from a late-prophase nucleus can be seen to consist of separate twin chromatids (Fig. 12.5). The replication that occurred during interphase resulted in two identical copies of the DNA molecules of the original chromosome. Hence, the two chromatids of a prophase chromosome are genetically identical.

The main apparatus by which chromosomes are separated during cell division, the mitotic ***spindle***, also becomes visible in late prophase. As the centrioles of an animal cell begin to move apart, a system of microtubules[1] appears near each pair of centrioles (Fig. 12:2–3) and radiates in all directions. These series of blind-ended microtubules are called the ***asters*** (Fig. 12.10:C–D). As a pair of centrioles approaches a pole, some microtubules attach to microtubules from the opposite pair of centrioles forming ***polar microtubules***; asters and polar microtubules together form the basket-like spindle (12.9:4). In late prophase, the nuclear membrane gradually disappears and some aster microtubules bind to protein plates called ***kinetochores*** that form on the centromere of each chromatid; the kinetochore microtubules thus connect the centromeres to the poles.

Metaphase The stage of metaphase proper is preceded by a short period known as the prometaphase. The chromosomes, which at first were distributed essentially at random within the nucleus, begin to move toward the equator, or middle, of the spindle. This movement occurs because kinetochore microtubules can grow from their aster poles where tubulin subunits are added, and shrink at their kinetochore attachments where tubulin subunits are digested. By the end of prometaphase, each chromosome has apparently been pulled equally by each pole, and so is positioned on the midline.

During the brief stage of metaphase proper, the chromosomes are arranged on the equatorial plane of the spindle, and in side view appear to form a line across the middle of the spindle (Fig. 12.10D). Metaphase ends when the centromeres of each pair of twin chromatids split apart. Each chromatid then becomes an independent chromosome with its own centromere. By the end of metaphase the total amount of genetic material is unchanged though the number of independent chromosomes in the nucleus has doubled.

Anaphase The preceding stages of mitosis prepare the nucleus for the critical event that occurs in anaphase: the separation of two complete sets of chromosomes. At the beginning of anaphase the centromeres that hitherto held the twin chromatids together have just broken apart. The two new sets of single-chromatid chromosomes now begin to move away from each

[1] Until recently, the hollow microtubules were thought to be solid fiber. Hence the references to "spindle fibers" and "astral fibers" in the literature.

other, one going toward one pole of the spindle and the other going toward the opposite pole. This movement toward the respective poles is accomplished in two ways. The same processes that lined the chromosomes up in a kind of cellular tug of war can now pull the centromeres, with their attached chromosomes, to the poles. Meanwhile, the polar microtubules from opposite ends of the dividing cell form cross bridges between themselves (similar to the dynein arms that are thought to enable cilia to move) and push the poles apart (Figs. 12.9:6 and 12.10E), becoming longer (by the addition of molecular subunits to their ends) as they do so. By late anaphase the cell contains two groups of chromosomes that are widely separated, the two clusters having almost reached their respective poles of the spindle (Fig. 12.9:7). Cytokinesis often begins during late anaphase.

Telophase Telophase (Fig. 12.10F) is essentially a reversal of prophase. The two sets of chromosomes, having reached their respective poles, become enclosed in new nuclear membranes as the spindle disappears. Then the chromosomes begin to uncoil and to resume their interphase form, while the nucleoli slowly reappear (Fig. 12.9:8). Cytokinesis is often completed during telophase. Telophase ends when the new nuclei have fully assumed the characteristics of interphase, thus bringing to a close the complete mitotic process. What was a single nucleus containing one set of twin-chromatid chromosomes in prophase is now two nuclei, each with one set of single-chromatid chromosomes.

ALTERNATIVE PATTERNS OF NUCLEAR DIVISION

The preceding description of mitotic division fits most but not all eucaryotes. Most of the deviations from the pattern illustrated in Figure 12.9 are seen in primitive eucaryotes, and they may indicate how the typical eucaryotic pattern of nuclear division evolved from the sequence of cell division in procaryotes. Some primitive eucaryotic dinoflagellates, for example, even though they have numerous chromosomes packaged in a membrane-bounded nucleus, still divide by a process remarkably like the binary fission of procaryotes. The chromosomes, which are attached to the inner surface of the nuclear membrane by very short kinetochore microtubules running from their centromeres, are first replicated, and then separated into two groups by growth of the membrane between the points of attachment of the original chromosomes and their replicates. As a result, two new nuclei are organized.

These dinoflagellates rely on microtubules in their cytoplasm to determine the direction of nuclear division. Bundles of these tubules run in parallel from each end of the cell toward the other, meeting in "tunnels" in the nucleus (Fig. 12.11A), and the two new nuclei move away from each other along this scaffolding. There is, however, no physical connection between the microtubules and the nuclear membrane, and the means by which the daughter nuclei are pulled apart is not known.

Advanced dinoflagellates and some fungi and protozoans illustrate a possible further evolutionary progression in their method of cell division, while still relying on cytoplasmic microtubules to orient the process. As in the primitive dinoflagellates, their chromosomes are attached to the nu-

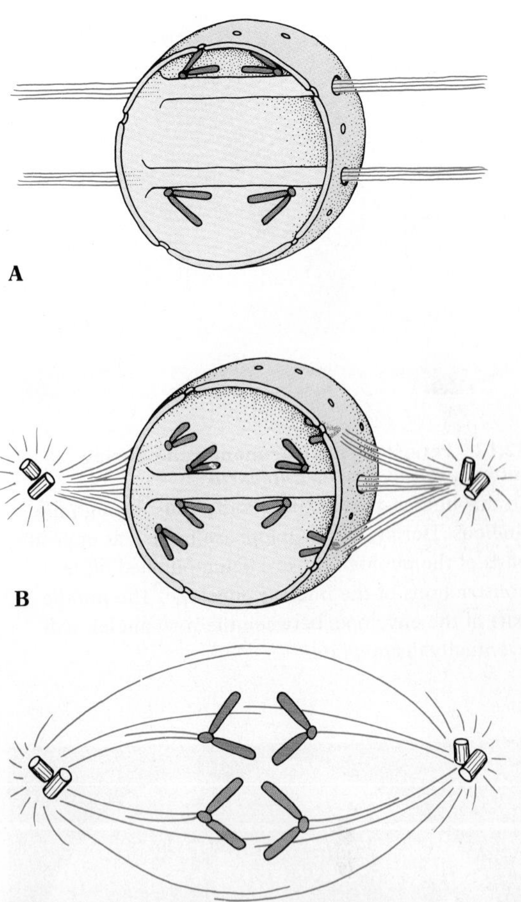

12.11 Evolution of nuclear division
(A) In primitive dinoflagellates, parallel bundles of microtubules run from each end toward the other, meeting and interacting in tunnels in the nucleus. The chromosomes are attached to the nuclear membrane and separate as the nucleus divides, and the daughter nuclei slide along the microtubules. (B) In advanced dinoflagellates there is only a single microtubule tunnel through the nucleus. Additional microtubules run from the poles to the nuclear membrane and pull the daughter nuclei apart. (C) In higher eucaryotes the nuclear membrane disappears during division and certain of the polar microtubules interact with the kinetochore microtubules to pull the individual chromosomes toward the poles.

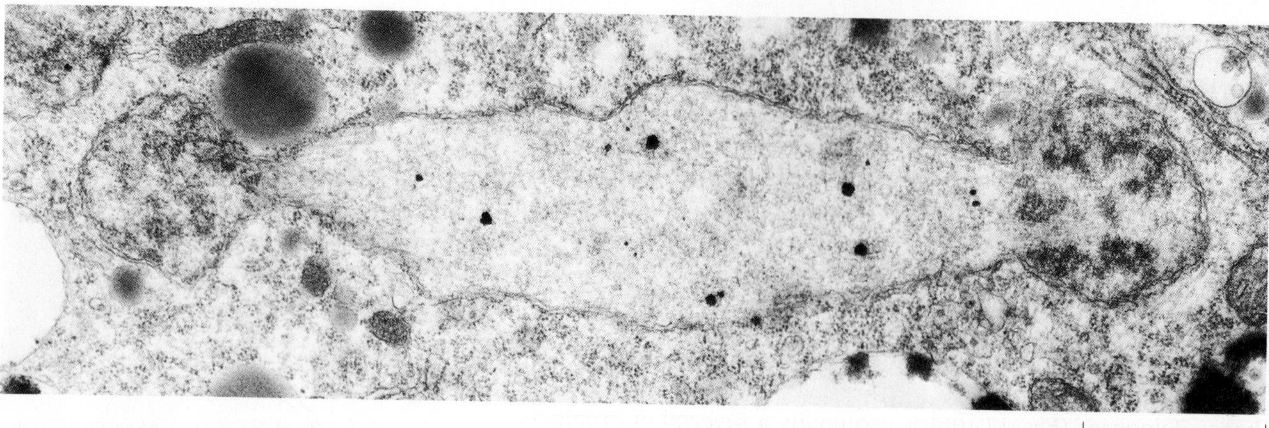

12.12 Persistent nuclear membrane during mitosis in the fungus *Catenaria*

In this fungus, part of the spindle runs through the nucleus. Here two new telophase nuclei, at opposite ends of the photograph, are being pinched off by constrictions of the nuclear envelope. The middle part of the envelope, between the new nuclei, will eventually disintegrate.

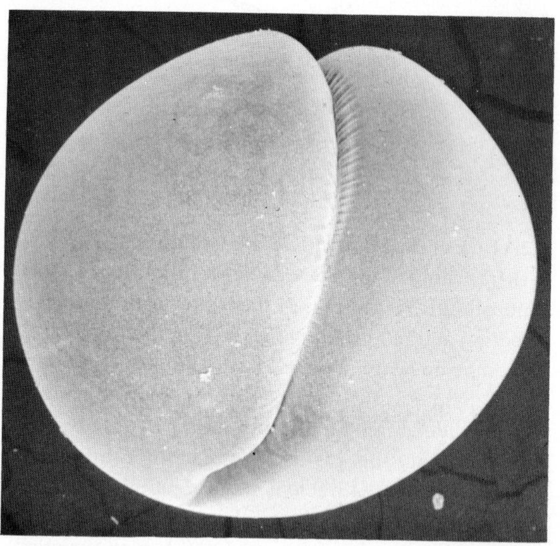

12.13 Scanning electron micrograph of a dividing frog egg

The cleavage furrow is not yet complete (see bottom of cell). Note the puckered stress lines in the furrow.

clear membrane by very short kinetochore microtubules from the centromeres, and bundles of long spindle microtubules that run through a nuclear tunnel determine the direction of nuclear division. In these organisms, however, there is just a single tunnel. Additional spindle microtubules, originating at the same poles as the transnuclear bundles, run to the nuclear membrane and physically pull the two newly formed nuclei apart (Figs. 12.11B and 12.12).

As we have seen, in higher eucaryotes the nuclear membrane disappears before division, and there is no apparent bundling of microtubules. Instead of securing the chromosomes to the nuclear envelope, as they do in some dinoflagellates, microtubules connect to polar microtubules or to the poles themselves. These tubule complexes exert a direct pull on the chromosomes, rather than on the daughter nuclei (Fig. 12.11C).

The cell-division cycle of higher eucaryotes requires the intricate nuclear membrane to be disassembled and reassembled. Partial disassembly is also necessary in some eucaryotes with persistent nuclear envelope, as seen in Figure 12.12. The disassembly process is not fully understood. Phosphorylation of one of the three proteins that form the nuclear lamina appears to be the cue that triggers disassembly of the membrane. One of these three proteins remains embedded in the fragments of nuclear membrane as they are incorporated temporarily into the endoplasmic reticulum of the cytoplasm, perhaps serving to mark these specific bits of membrane for recovery at the end of telophase. Other pieces apparently remain attached to chromosomes. The fate of the proteinaceous pores of the nucleus is unknown, though some evidence suggests that these specialized channels may become temporarily associated with the chromosomes.

CYTOKINESIS

We said earlier that division of the cytoplasm often accompanies division of the nucleus and that it often begins in late anaphase, reaching completion during telophase. But this is not always the case. Mitosis without cytokinesis is common in some algae and fungi, producing ***coenocytic*** plant bodies (bodies with many nuclei, but no, or few, cellular partitions). It regularly

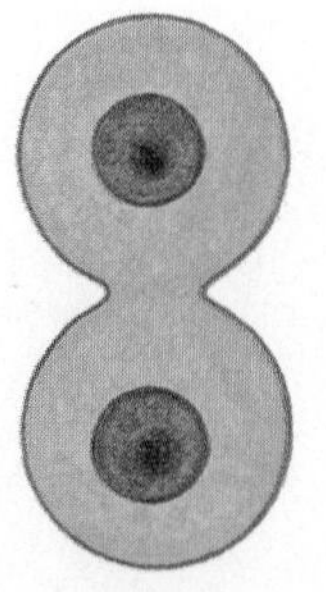
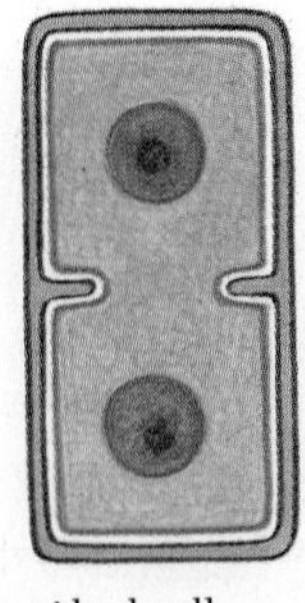

12.14 Three mechanisms of cytokinesis
(A) Cytokinesis in animal cells typically occurs by a pinching-in of the plasma membrane. (B) In many algal cells cytokinesis occurs by an inward growth of new wall and membrane. (C) In higher plants cytokinesis typically begins in the middle and proceeds toward the periphery, as membranous vesicles fuse to form the cell plate.

occurs during certain phases of reproduction in seed plants and certain other vascular plants. It is also common in a few lower invertebrate animals with coenocytic bodies. During the early development of insect eggs, mitosis without cytokinesis produces hundreds of nuclei in a limited amount of cytoplasm; later, cytokinesis cuts up this cytoplasm to produce many new cells in a very short time.

Cytokinesis in animal cells Division of an animal cell normally begins with the formation of a ***cleavage furrow*** running around the cell (Fig. 12.13). When cytokinesis occurs during mitosis, the location of the furrow is ordinarily determined by the orientation of the spindle, in whose equatorial region the furrow forms (Fig. 12.9:7). The furrow becomes progressively deeper, until it cuts completely through the cell (and its spindle), producing two new cells.

Very little is known about how the cleavage furrow forms. Since the location of the furrow is usually related to the position of the centrioles and spindle, an early hypothesis was that some of the astral microtubules of the mitotic apparatus were attached to the cell surface and pulled the surface inward to form the cleavage furrow. But cytokinesis is known to occur in many instances long after mitosis is complete and the spindle has disappeared; indeed, removal of the entire mitotic apparatus from a sea-urchin egg does not inhibit furrow formation. Recent research suggests that a dense belt of actin and myosin microfilaments at the site of the cleavage furrow is probably responsible. This view is supported by the finding that drugs like cytochalasin, which block the activity of microfilaments, stop cytokinesis. Since agents that bind actin and myosin also stop the cleavage process, the sort of actin-myosin interaction known to be involved in cell movement, as discussed in Chapter 5, is probably responsible for cytokinesis as well.

Cytokinesis in plant cells The relatively rigid cell walls of plants, essential for their support, required the evolution of a different kind of cytokinesis. Since plant cells cannot develop cleavage furrows, it is not surprising that cytokinesis is different in plant cells. In many fungi and algae, new plasma membrane and wall grow inward around the wall midline until the growing edges meet and completely separate the daughter cells (Fig. 12.14). In higher plants a special membrane, called the ***cell plate***, forms

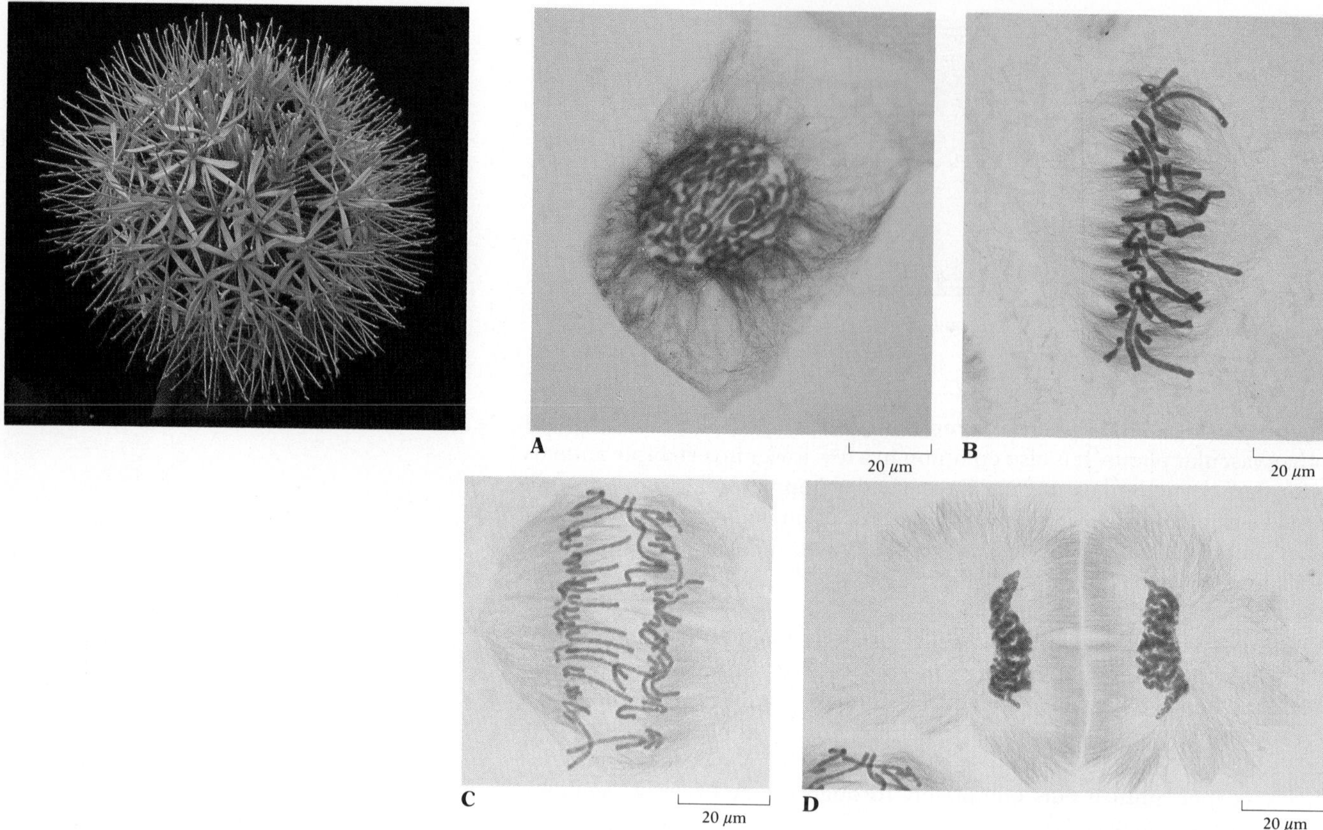

12.15 Cell division in a plant, the African blood lily

(A) Prophase: the chromosomes have condensed; microtubules are visible. (B) Metaphase.
(C) Anaphase: the two groups of chromosomes are moving to opposite poles of the cell. (D) Telophase: a cell plate has begun to form.

halfway between the two nuclei at the equator of the spindle if cytokinesis accompanies mitosis (Fig. 12.15D). The cell plate begins to form in the center of the cytoplasm and slowly becomes larger until its edges reach the outer surface of the cell and the cell's contents are cut in two. We see, then, that cytokinesis of higher-plant cells progresses from the middle to the periphery, whereas cytokinesis in animal cells progresses from the periphery to the middle.

The cell plate forms from membranous vesicles that are carried to the site of plate formation by the microtubules that remain after mitosis. There they first line up and then unite (Fig. 12.16). The vesicles are derived mainly from the Golgi and, to a lesser extent, from the ER. As the vesicle membranes fuse with one another and, peripherally, with the old plasma membrane, they constitute the partitioning membranes of the two newly formed daughter cells. The contents of the vesicles, trapped between the daughter cells, give rise to the middle lamella and to the beginnings of primary cell walls.

MEIOTIC CELL DIVISION

As we have seen, mitosis serves to maintain a constant number of chromosomes in somatic cells. What would be its effect in reproductive cells? As we

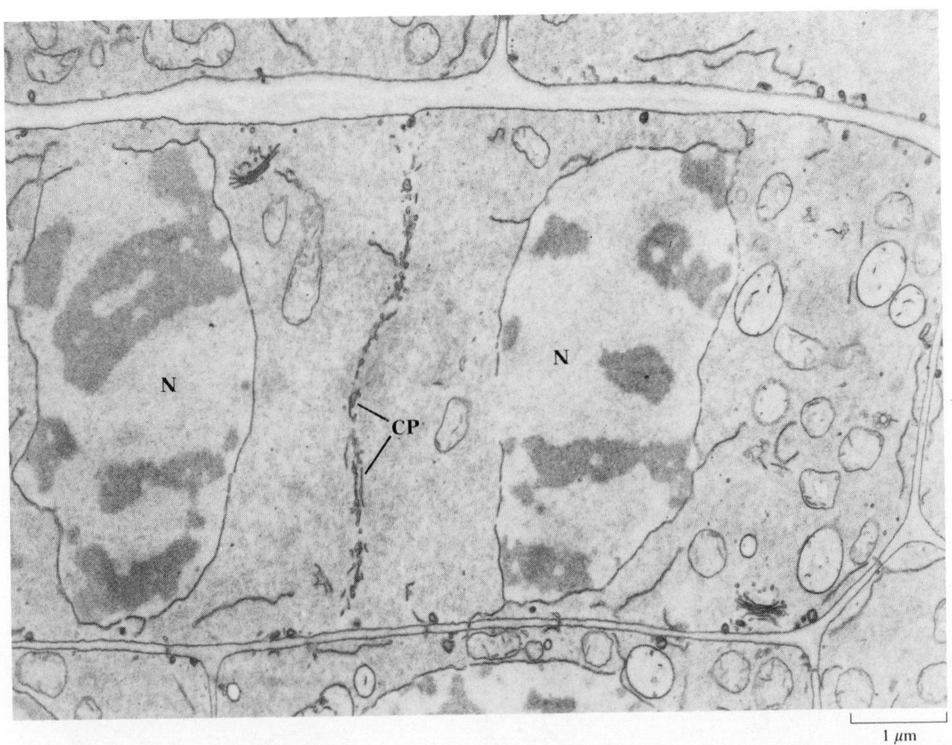

12.16 Electron micrograph of a late-telophase cell in corn root, showing formation of cell plate
Mitosis has been completed, and the two new nuclei (N) are being formed; the chromosomes (dark areas in the nuclei) are no longer visible as distinct structures, but the nuclear membranes are not yet complete. A cell plate (CP) is being assembled from numerous small vesicular structures. At the lower end of the nucleus on the right, a length of endoplasmic reticulum can be seen that appears to run from the nuclear envelope to the cell wall, through the wall, and into the adjacent cell.

know, in sexual reproduction two gametes (an egg and a sperm, for example) unite to form the first cell (called a ***zygote***) of the new individual. If those two gametes were produced by normal mitosis in human beings or in the hypothetical four-chromosome organism of Figure 12.9, the zygote produced by their union would have double the normal number of chromosomes, and at each successive generation the number would again double, until the total chromosome number per cell approached infinity. This does not happen: the chromosome number normally remains constant within a species. At some point, therefore, a different kind of cell division must take place, a division that reduces the number of chromosomes by half, so that when the egg and sperm unite in fertilization the normal diploid number is restored. This special process of reduction division is called ***meiosis*** (from the Greek word for "diminution"). In all multicellular animals meiosis occurs at the time of gamete production. Consequently each gamete possesses only half the species-typical number of chromosomes. It is important to note that in the reduction division of meiosis the chromosomes of the parental cell are not simply separated into two random halves; the diploid nucleus contains two of each type of chromosome, and meiosis partitions these chromosome pairs so that each gamete contains one of the two homologues. Such a cell, with only one of each type of chromosome, is therefore haploid. When two haploid gametes unite in fertilization, the resulting zygote is diploid, having received one of each chromosome type from the sperm of the male parent and one of each type from the egg of the female parent.

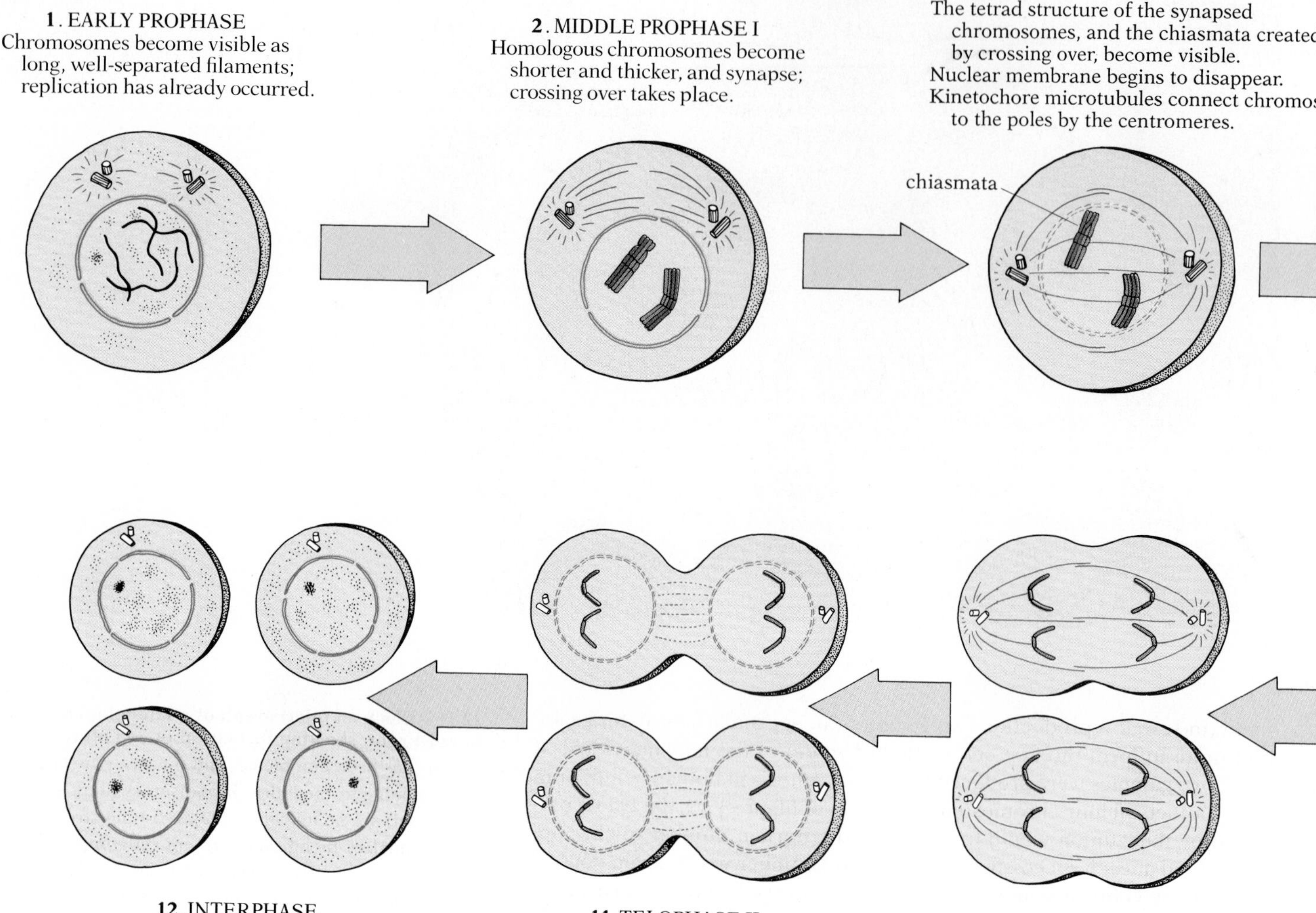

1. EARLY PROPHASE
Chromosomes become visible as long, well-separated filaments; replication has already occurred.
2. MIDDLE PROPHASE I
Homologous chromosomes become shorter and thicker, and synapse; crossing over takes place.
3. LATE PROPHASE I
The tetrad structure of the synapsed chromosomes, and the chiasmata created by crossing over, become visible.
Nuclear membrane begins to disappear.
Kinetochore microtubules connect chromoso
to the poles by the centromeres.
chiasmata
12. INTERPHASE
11. TELOPHASE II
10. ANAPHASE II

12.17 Meiosis in an animal cell

In these diagrams, the members of each pair of homologous chromosomes are shown in different colors to aid in visualizing the results of crossing over.

THE PROCESS OF MEIOSIS

Production of two haploid gametes from a diploid cell could be accomplished by a single division; instead, complete meiosis (Fig. 12.17) involves two successive division sequences, which result in four new haploid cells. As in mitosis, meiosis is preceded by replication, which creates chromatid pairs. This step seems unnecessary, and may exist solely to make possible the process called crossing over; we will discuss the logic of crossing over presently. In any case, the result is that the gamete-producing cell now has four times as much DNA as the gametes require, and meiosis proceeds to correct this excess. The first division sequence accomplishes the reduction in the number of chromosomes; the second separates chromatids. The

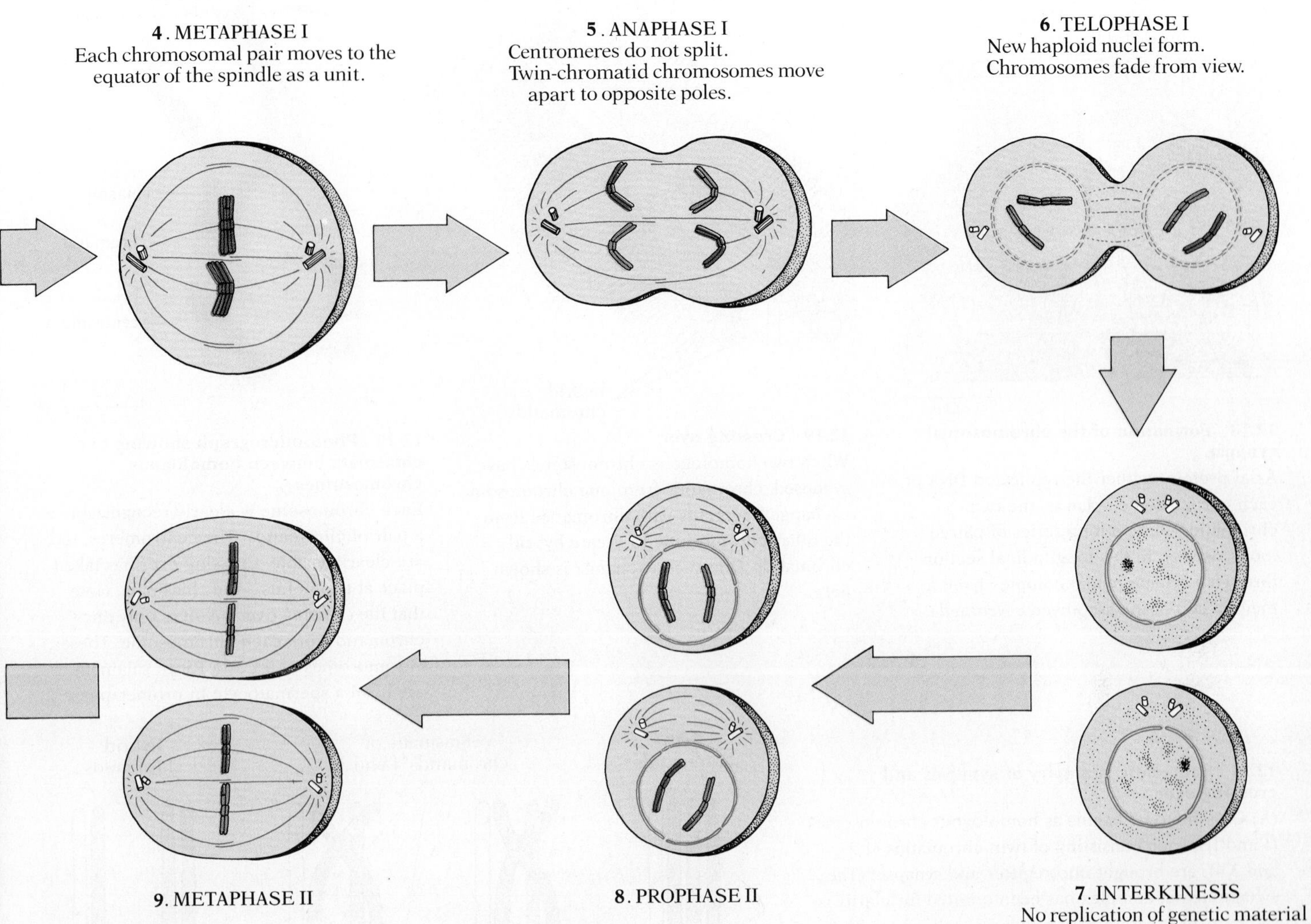
4. METAPHASE I
Each chromosomal pair moves to the equator of the spindle as a unit.
5. ANAPHASE I
Centromeres do not split.
Twin-chromatid chromosomes move apart to opposite poles.
6. TELOPHASE I
New haploid nuclei form.
Chromosomes fade from view.
7. INTERKINESIS
No replication of genetic material occurs.
8. PROPHASE II
9. METAPHASE II

same four stages as in mitosis—namely prophase, metaphase, anaphase, and telophase—are recognized in each division sequence.

Prophase I Many of the events in prophase I of meiosis superficially resemble those in the prophase of mitosis. The individual chromosomes come slowly into view as they coil and become shorter, thicker, and more easily stainable. The nucleoli slowly fade from view, and, finally, the nuclear envelope disappears and the spindle is organized. Radioactive tracer studies show that the replication of the genetic material occurs during the interphase that precedes prophase I, as in mitosis, but the twin chromatids are not yet distinguishable. There are, however, important differences between the prophase of meiosis and that of mitosis.

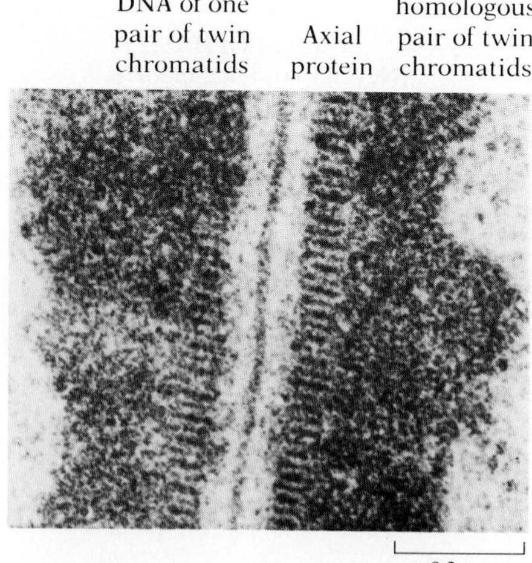

12.18 Formation of the chromosomal synapse
Axial proteins gather the replicated DNA of each chromosome—that is, the twin chromatids—into a long series of paired loops, as seen in this longitudinal section through a synaptonemal complex from a meiotic cell of the ascomycete *Neottiella*.

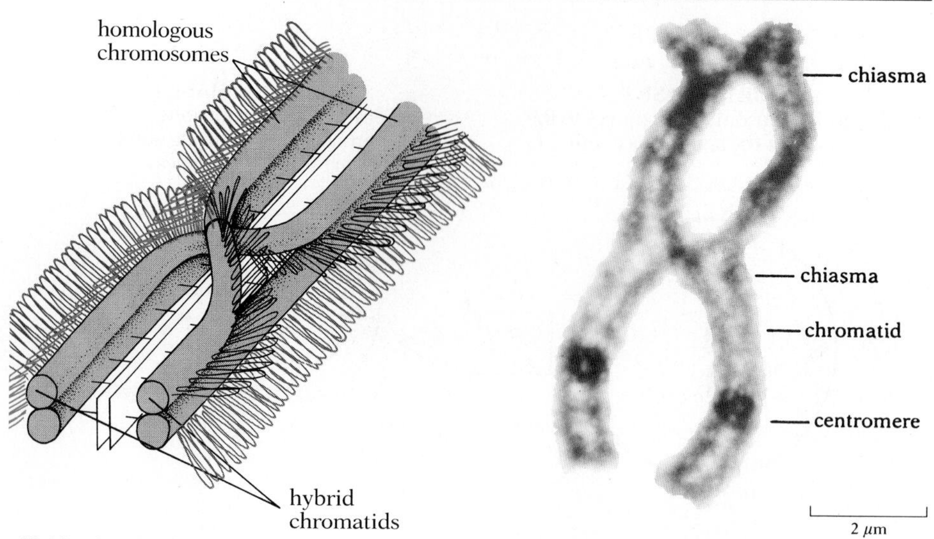

12.19 Crossing over
When two homologous chromosomes have synapsed, chromatids from one chromosome exchange fragments with chromatids from the other chromosome to create hybrid chromatids. One such exchange is shown here.

12.20 Photomicrograph showing two chiasmata between homologous chromosomes
Each chromosome is clearly recognizable as a pair of chromatids. The centromeres, too, are clearly visible. Crossing over has taken place at two points—the chiasmata. Note that the crossing over involves only one chromatid from each chromosome. These chromosomes of a Costa Rican salamander are from a spermatocyte in prometaphase I.

12.21 Schematic summary of synapsis and crossing over
(A) Crossing over begins as homologous chromosomes (I and II), each consisting of twin chromatids (1,2 and 3,4), are brought into register and synapse. (The synaptonemal complex has been omitted for clarity.) (B) Parts of separate chromatids are spliced together, while the protein axis (not shown) keeps each segment firmly attached to its twin. (C) When the synaptonemal complex breaks up, the homologous chromosomes begin to drift apart, but the chiasmata of spliced hybrid chromatids prevent them from separating. (D) The homologous chromosomes separate fully during anaphase.

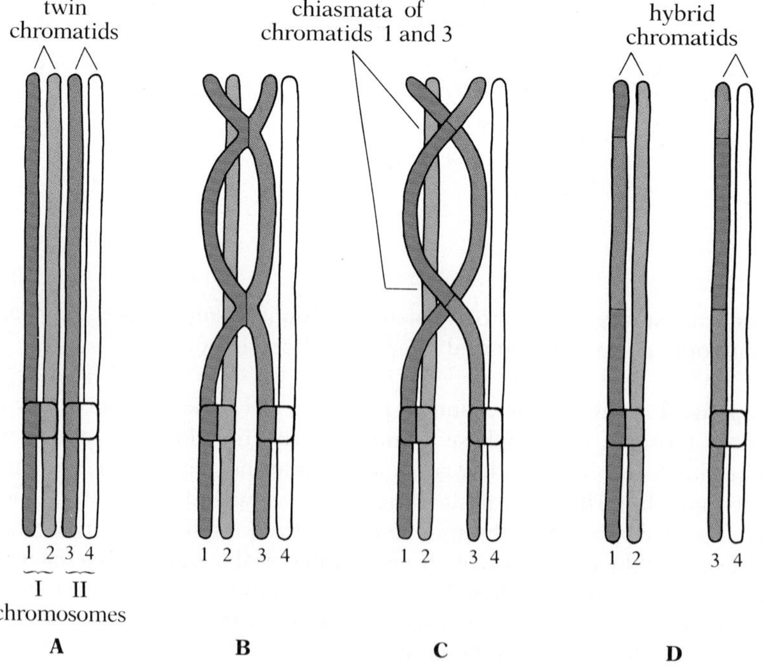

The chief difference is that in meiosis the members of each pair of homologous chromosomes move together and lie side by side (Fig. 12.17:2). Also, in addition to being fastened at their centromeres, as in mitosis, the twin chromatids of each meiotic chromosome are also held together by a special pair of long, thin protein axes that run their entire length. The DNA is gathered into loops as part of the condensation process. The protein axes of the two homologous chromosomes now join by means of protein cross bridges to form an intricate compound structure known as a ***synaptonemal complex***, which lines up the four chromatids (often referred to as a tetrad) in perfect register (Fig. 12.18). This process is known as ***synapsis***.

Now the important process known as ***crossing over*** begins. Large protein complexes called recombination nodules appear along the ladderlike cross bridges. These nodules probably determine the locations at which genetic material will be exchanged between homologous chromosomes. The number of nodules depends on the species of organism and the length of the chromosome; each human chromosome has an average of three. In many species there is also a sexual difference in crossing over, with fewer crossover events in male cells. Next, each nodule begins a process by which two of the chromatids, one from each of the homologous chromosomes, are clipped open at precisely the same place, and the resulting fragments are spliced to each other (Fig. 12.19). The all-important result of this splicing is that the chromatids of the tetrad no longer form two sets of twins; the recombined chromatids are now hybrids, containing genetic material descended from both the mother's and the father's homologous chromosomes.

When the synaptonemal complex begins to break up in late prophase, the points at which crossing over has taken place begin to become visible. The hybrid chromatids produced by crossing over link the two homologous chromosomes at these points, called ***chiasmata***. Each chiasma represents one crossover event (Fig. 12.20). Each crossover event can involve a different pair of chromatids. The process is summarized in Figure 12.21. Crossing over is not rare or accidental: it is a frequent and highly organized mechanism, and probably has significant adaptive value, as we will see presently.

As prophase ends most chromosomes now consist of two hybrid chromatids rather than twin chromatids, since genetic material has been exchanged between the chromatids of homologous chromosomes. Spindle microtubules appear, radiating from the two poles of the cell, and kinetochore microtubules attach to the centromeres. But there is a distinct difference between the activity of the centromeric microtubules in mitosis and in meiosis. In meiosis microtubules from one centromere of each homologous pair attach to a pole so the two pairs are pulled toward opposite poles (Fig. 12.22).

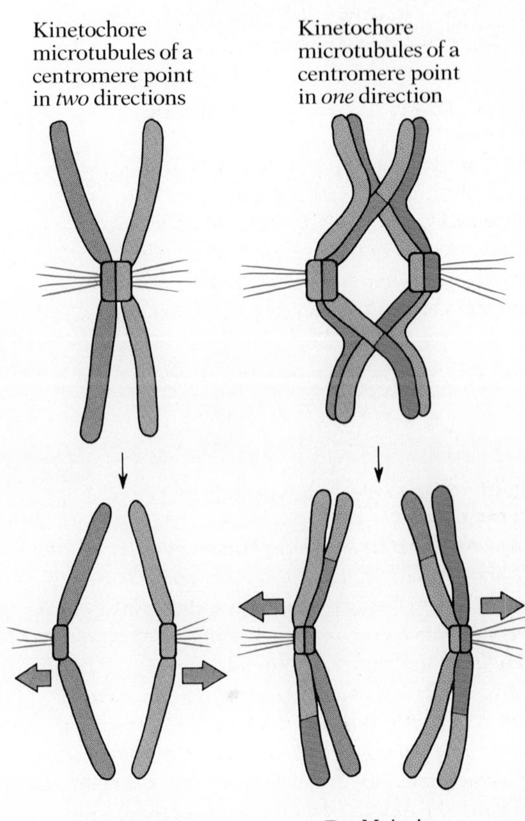
Kinetochore microtubules of a centromere point in *two* directions
Kinetochore microtubules of a centromere point in *one* direction
A Mitosis
B Meiosis

12.22 Kinetochore microtubules in mitosis and meiosis

(A) In mitosis, each centromere is attached by kinetochore microtubules to both poles. As a consequence, the two chromatids are pulled apart and wind up at opposite ends of the cell during anaphase. (B) In meiosis I, however, one centromere in each pair is attached to one or the other of the two poles. The result is that in anaphase I the chromatids remain joined, and homologous two-chromatid chromosomes wind up in separate cells.

Metaphase I In mitosis each chromosome consists of a pair of twin chromatids, and moves independently to the midline of the cell. In meiosis each chromosome typically consists of two hybrid chromatids, and homologous chromosomes move as a unit to the midline in preparation for anaphase. As a result, in meiosis the number of independent units waiting at the midline for cell division is only half the number of mitosis (Fig. 12.17:4).

Anaphase I In mitosis, metaphase ends and anaphase begins when the centromere of each twin chromatid splits, and the two independent single-chromatid chromosomes thus formed move away from each other toward opposite poles of the spindle. But in the first division of meiosis, the separation occurs instead between the two chromosomes that have been joined since the middle of prophase (Fig. 12.22). Since each chromosome has its own centromere, there is no splitting of centromeres. With their respective centromeres attached by microtubules to only one pole, the homologous chromosomes move away from each other toward opposite poles during anaphase. Thus in our hypothetical organism, only two chromosomes, each with two hybrid chromatids, move to each pole (Fig. 12.17:5), in contrast to the four single-chromatid chromosomes that move to each pole in mitosis. Because the synaptic pairing was not random, but involved the two homologous chromosomes of each type, the two daughter nuclei get not just any two chromosomes, but rather one of each type.

Telophase I Telophase of mitosis and meiosis are essentially the same, except that each of the two new nuclei formed in mitosis has the same number of chromosomes as the parental nucleus, whereas each of the new nuclei formed in meiosis has half the chromosomes present in the parental nucleus (Fig. 12.17:6). At the end of telophase of mitosis, the chromosomes are single when they fade from view; at the end of telophase I of meiosis, the chromosomes are composed of two chromatids each when they fade from view.

Interkinesis Following telophase I of meiosis, there is a short period called interkinesis, which is similar to an interphase between two mitotic division sequences except that no replication of the genetic material occurs and hence no new chromatids are formed (replication is unnecessary, since each chromosome already has two chromatids when interkinesis begins).

Second division sequence of meiosis The second division sequence of meiosis, which follows interkinesis, is essentially mitotic from the standpoint of mechanics, though the functional result is different (Fig. 12.17C). The chromosomes do not synapse; they cannot, since the cell contains no homologous chromosomes. Each two-chromatid chromosome moves to the midline independently, and its centromere sends kinetochore microtubules toward each pole. At the end of metaphase II each centromere splits, and during anaphase II the single-chromatid chromosomes thus formed move away from each other toward opposite poles of the spindle. The new nuclei formed during telophase II are therefore haploid (Fig. 12.23).

In summary, then, the first meiotic division produces two haploid cells containing double-chromatid chromosomes. Each of these cells divides

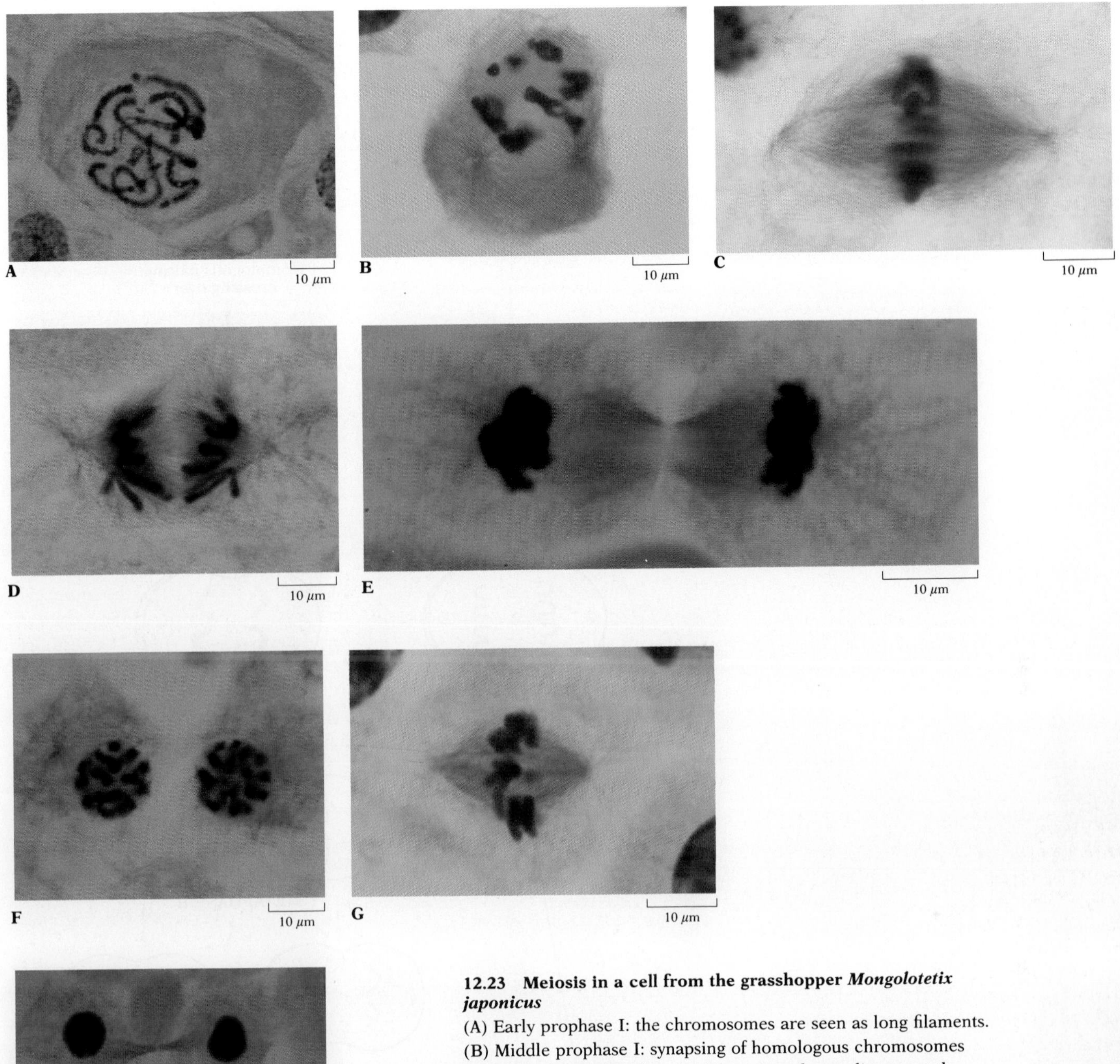

12.23 Meiosis in a cell from the grasshopper *Mongolotetix japonicus*
(A) Early prophase I: the chromosomes are seen as long filaments. (B) Middle prophase I: synapsing of homologous chromosomes takes place. (C) Metaphase I: chromosomal pairs line up at the equator. (D) Anaphase I: homologous chromosomes have separated and are moving to opposite poles. (E) Telophase I: division into two haploid nuclei has begun. (F) Prophase II: early in this stage, separate chromosomes are barely distinguishable. (G) Metaphase II: chromosomes in each cell are at the equator, and the spindles are evident. (H) Telophase II: four new haploid nuclei can be seen.

12.24 Comparison of mitosis and meiosis
The first step, interphase (during which replication occurs), is omitted, as are the stages between anaphase I and metaphase II of meiosis.

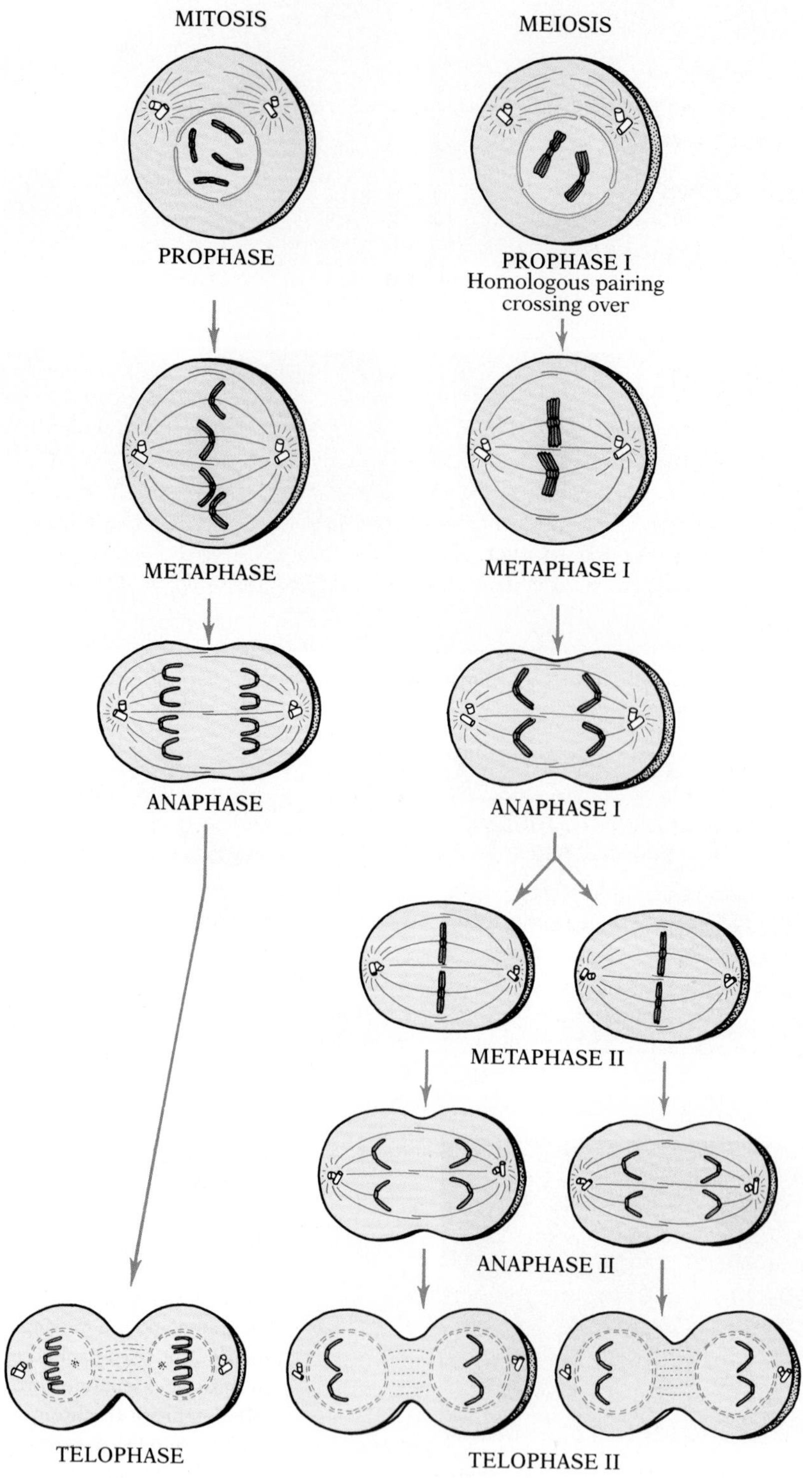

again in the second meiotic division, producing a total of four new haploid cells containing single-chromatid chromosomes. Mitosis and meiosis are compared in Figure 12.24.

THE ADAPTIVE SIGNIFICANCE OF RECOMBINATION AND CROSSING OVER

As we have said, life thrives on an optimal balance between stability and change—maintaining the fidelity of the genetic message while at the same time providing enough variation to permit selection for improved genes, particularly in the face of unpredictable changes in competition, predation, habitat, and climate. Sexual organisms go to great lengths to produce haploid gametes for mating (***sexual recombination***): asexual cloning of diploid offspring is metabolically less expensive[2], and gamete-producing cells are put to considerable trouble to make crossing over possible. Most researchers believe that these two processes serve the goal of either maintaining stability or generating change, but disagree about which is the object. We will consider the adaptive value of sex in some detail in Part III; our intention here is to describe the mechanisms and consequences of recombination and crossing over, and their potential for both creating and limiting variation. We will look first at how recombination generates change.

Variation The most important consequence of sexual reproduction is that it creates offspring with new combinations of characteristics. Each member of a homologous pair of chromosomes comes from a different parent, and contains genes coding for the same kinds of RNA, structural proteins, and enzymes, but the exact base sequences in two homologous genes are not necessarily identical. Instead, the copy of a gene inherited from an organism's father is often at least slightly different from the copy inherited from the mother. Different versions of the same gene are called ***alleles***, and the enzymes or structural proteins they produce are likely to have slightly (or even very) different activities. For example, blue eyes develop in people when both copies of their eye-pigment gene code for a defective colorless screening pigment; people with brown eyes have at least one copy of the allele that codes for a functional pigment. Since an individual organism is a mix of alleles from two different parents, its chromosomes are almost certain to contain an ensemble of alleles different from that of either parent, and many aspects of its morphology, physiology, and behavior will be correspondingly different.

To understand why variation arises in sexual recombination, let's consider a hypothetical six-chromosome organism. Three chromosomes came originally from the gamete of the male parent (the product of meiosis), and three from the gamete of the female parent, to form the three pairs of homologous chromosomes in this diploid organism. Each time meiosis

[2] Cloning identical copies of an organism is a realistic alternative even in multicellular diploids. A variety of species, ranging from certain lizards and fish to the ubiquitous dandelion, always reproduce asexually. Many more species can reproduce either sexually or asexually; this group includes the many plants that can spread by runners, as well as tens of thousands of invertebrates (the most familiar of which are the aphids).

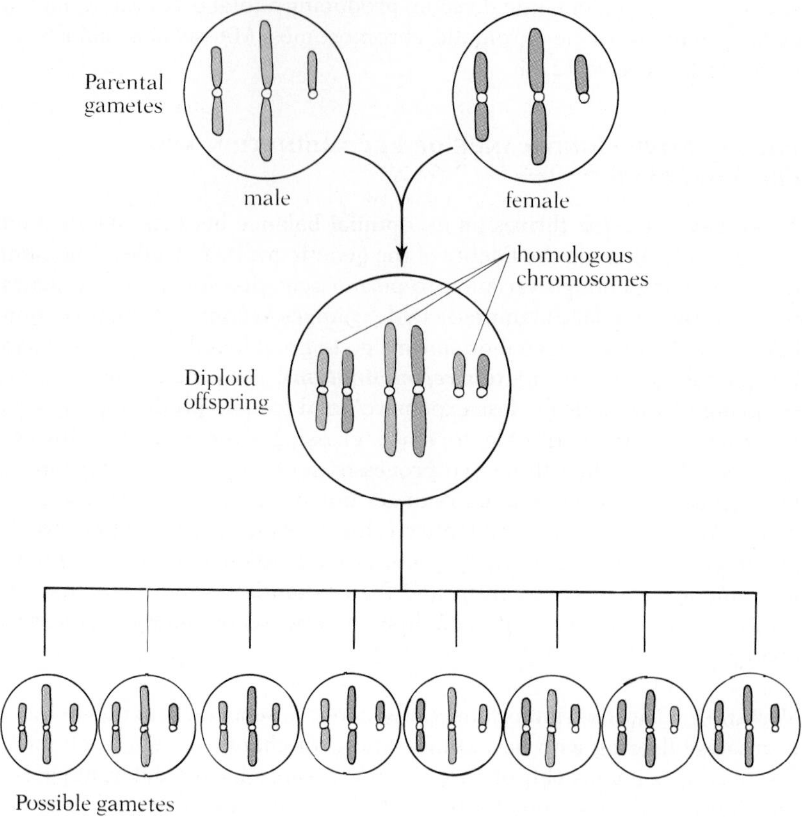

12.25 Variation in gametes
Even without crossing over, a diploid organism produces 2^n different kinds of gametes, where n is the number of pairs of homologous chromosomes. In this hypothetical organism with an n of 3, there are eight (2^3) possible kinds of gametes.

occurs in an individual producing gametes for reproduction, all of that organism's maternal chromosomes *might* go into one gamete and all the paternal ones into another, but many other combinations are equally likely. If an organism has three pairs of chromosomes, eight different combinations may be found in the gametes (Fig. 12.25). In the zygote formed by two gametes in such a species, 64 (8×8) different combinations of chromosomes are then possible. For humans, with our unusually large number of chromosomes, the number of different chromosome combinations possible in the offspring of the same two parents is about 7×10^{13}.

Even without crossing over, simple sexual recombination produces a substantial amount of variation, and this would be obtained even if the elaborate two-stage process of meiosis were replaced by a much simpler sequence: the single-chromatid homologous chromosomes with which the cell ended the previous telophase of mitosis could in theory segregate into two haploid cells. The replication of DNA before meiosis, and the whole second division sequence, would then be unnecessary. But this shortcut, though economical, would rob the organism of the benefits of crossing over, the second way in which new combinations are generated.

The highly ordered process of crossing over increases the variety of possible gametes astronomically, since crossover points are essentially random, and virtually every hybrid chromatid is spliced together at a unique set of points. If we narrow our focus to an example using only two of a chromosome's thousands of genes, we can get an idea of what happens to the genetic information during meiosis. In this case, let's say one gene at one end of a chromosome controls the animal's fur color, and a gene at the other end of the same one codes for body size.

Let's assume that the father contributes a gene coding for black fur (which we can call *F*) and a gene coding for a large body (*S*), while the alleles from the mother are different, coding for gray fur (*f*) and for a small body (*s*). Though the *F* and *S* of the chromosome contributed by the father remain together (or ***linked***) throughout the *mitotic* life of a cell, as do the *f* and *s* on the maternal chromosome, crossing over in *meiosis* can rearrange things (Fig. 12.26). As a result, the chromosome in the offspring's gamete is likely to have a new combination—*F* and *s*, or *f* and *S*—rather than one of the parental combinations. In fact, since crossing over usually occurs at several locations and can involve any two homologous chromatids (see Fig. 12.21), the chance of a parental chromosome's surviving meiosis unshuffled is quite a bit lower than the 50 percent of Figure 12.26. Furthermore, crossing over can occur even within a gene, so if the copies of a particular gene are different alleles, two entirely novel alleles can also be created.

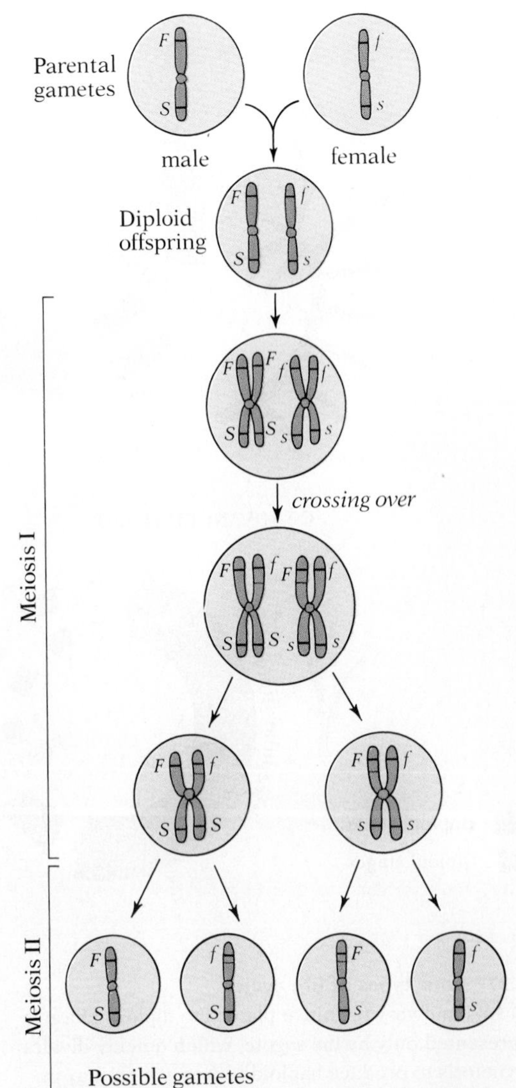

12.26 Variation in gametes from crossing over In this example, the parents contribute chromosomes with two different alleles of genes for fur color (*F* and *f*) and body size (*S* and *s*); these chromosomes form a homologous pair in the diploid offspring. At meiosis, crossing over breaks up the paternal and maternal combinations, so some of the resulting gametes have hybrid chromosomes, unlike those of either parent.

Stability So much for how recombination and crossing over can create genetic novelty. What about the alternative: maintaining stability? An evolutionary argument suggests that unrepaired mutations slowly accumulate in a genome until both copies of some genes are damaged or inactivated; as a result, an organism's clonal offspring become weaker as generations pass and yet more mutations are incurred. A good way to obtain working copies of such genes is to engage in recombination: a fertilized egg probably has at least one good copy of each gene since no two gametes from unrelated individuals are likely to have the same mutations. Sex, then, might exist to maintain stability.

The argument for a conservative role for crossing over is somewhat similar. When, as sometimes happens, both strands of the DNA of a chromosome are damaged, repair enzymes cannot fix the problem—there is no good strand to use as a template. But there is an intact copy of the gene on the homologous chromosome. Some researchers believe that crossing over occurs near sites of double-stranded damage, and allows repair enzymes to fix the damaged chromosome by making a copy of the same region on the homologous chromosome.

Finally, it is entirely possible that recombination and crossing over evolved to serve opposite purposes, or to do both at once—to repair or compensate for serious mutations *and* generate large amounts of diversity at the same time. This important issue, which is one of many that bring molecular and evolutionary biologists together, is yet to be resolved.

THE TIMING OF MEIOSIS IN THE LIFE CYCLE

Meiosis in the life cycle of plants Meiosis in plants usually produces haploid reproductive cells called ***spores***, which often divide mitotically to develop into haploid multicellular plant bodies. At one extreme in plant

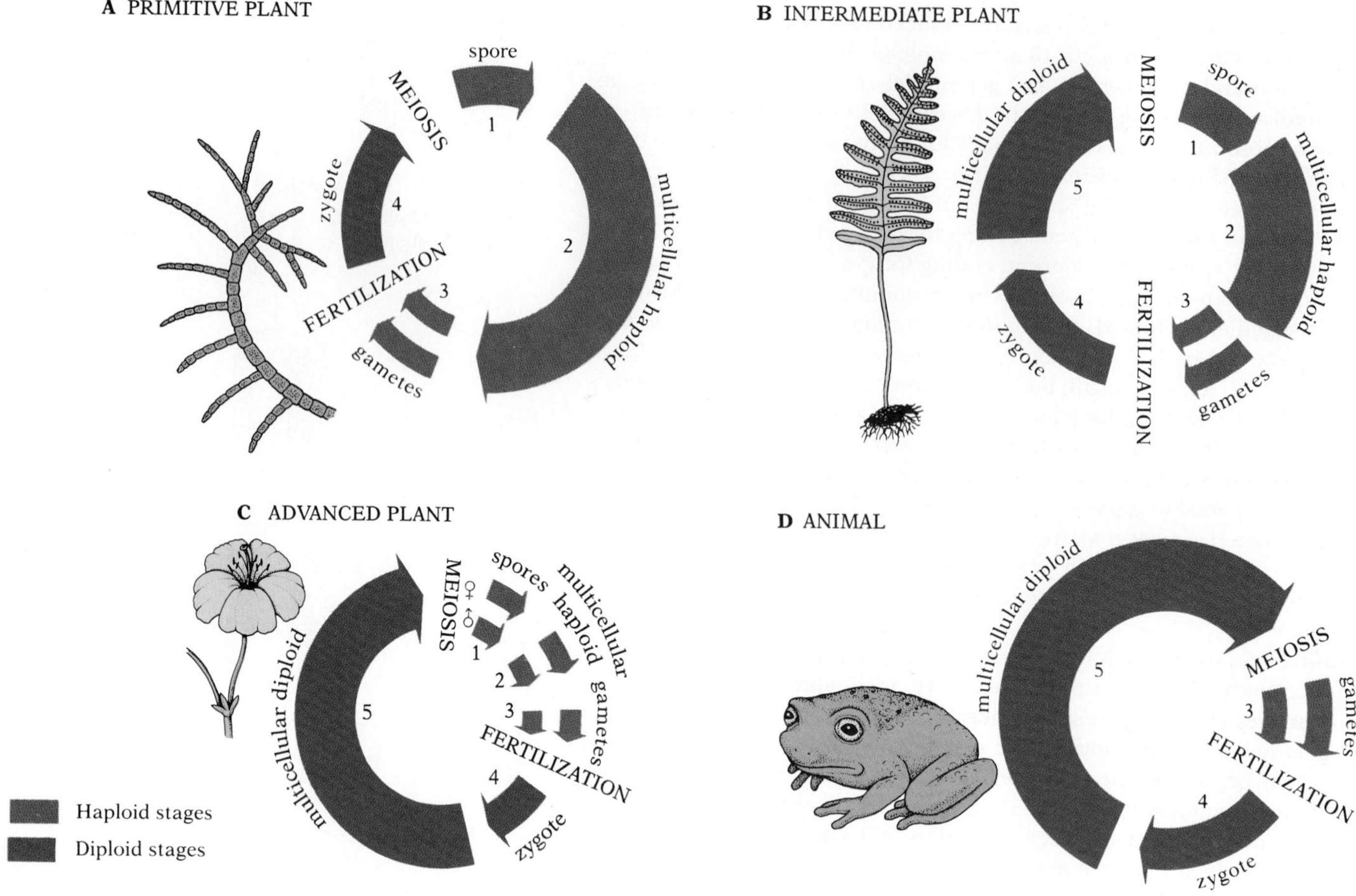

12.27 Four types of life cycles

(A) In some very primitive plants the diploid phase is represented only by the zygote, which quickly divides by meiosis to produce haploid spores, which may in turn divide mitotically to produce a multicellular haploid plant. (B) In most multicellular plants there are two multicellular stages, one haploid and one diploid (stages 2 and 5); the relative importance of these two stages varies greatly from one plant group to another (the cycle shown here is an intermediate one in which stages 2 and 5 are nearly equal). (C) In flowering plants the multicellular diploid stage (stage 5) is the major one, and the multicellular haploid stage (stage 2) is much reduced, being represented by a tiny organism with very few cells. (D) Animals and a very few plants have a life cycle in which meiosis produces gametes directly—the spore stage (stage 1) and the multicellular haploid stage (stage 2) being absent.

reproductive patterns are some primitive plants like algae (Fig. 12.27). The haploid spore cells (stage 1) divide mitotically and develop into a rapidly growing, poorly differentiated, haploid, multicellular stage (stage 2) in which the organism passes most of its life. This multicellular plant eventually produces, by mitosis, cells specialized as gametes (stage 3), which unite to form the diploid zygote (stage 4); the zygote promptly undergoes meiosis to produce four haploid spores, thus beginning the cycle again. In such an organism, then, the haploid phase of the cycle (particularly stage 2) is dominant, and the only diploid stage—the zygote—is very transitory.

Most plants devote more of their life cycle to the diploid state. Many ferns, for instance, divide their time about evenly between haploid and diploid phases (Fig. 12.27). While the fern is in the diploid portion of its life cycle, certain cells in its reproductive organs divide by meiosis to produce haploid spores (stage 1). These spores divide mitotically and develop into haploid multicellular plants (stage 2). The haploid multicellular plant eventually produces cells specialized as gametes (stage 3). As with algae, the gametes are produced by mitosis, not meiosis, because the cells that divide to produce the gametes are already haploid. Two of these gametes unite in fertil-

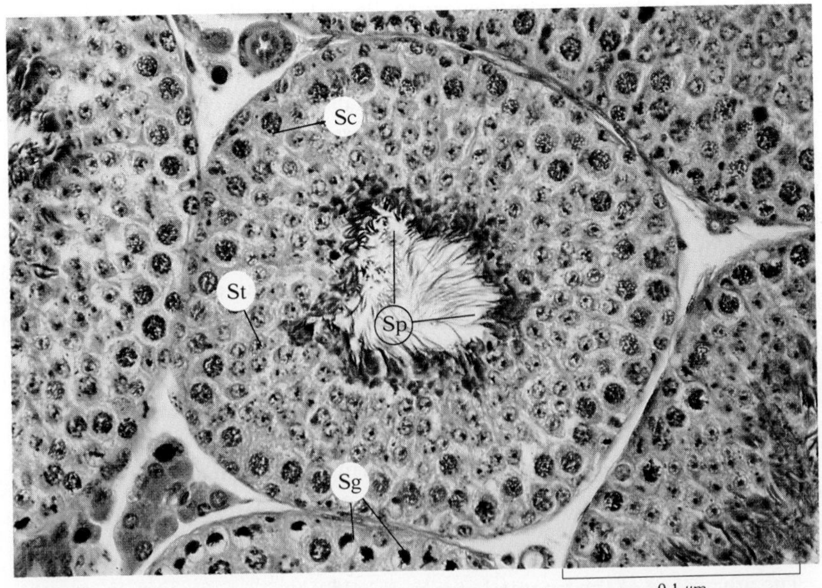

12.28 Photograph of cross section of rat seminiferous tubule, showing spermatogenesis The dark-stained outermost cells in the wall of the tubule are spermatogonia (Sg), which divide mitotically, producing new cells that move inward. These cells enlarge and differentiate into primary spermatocytes (Sc), which divide meiotically to produce secondary spermatocytes and then spermatids (St). The spermatids differentiate into mature sperm cells, or spermatozoa (Sp), whose long flagella can be seen in the lumen of the tubule in this photograph.

ization to form the diploid zygote (stage 4), which divides mitotically and develops into a diploid multicellular plant (stage 5). In time, this plant produces spores and the cycle starts over again.

As we will see in some detail in a later chapter, the various groups of plants vary greatly in the relative importance of the diploid and haploid phases in their life cycles. A very few plants have cycles almost like the animal life cycle shown in Figure 12.27D; stages 1 and 2 are absent, and the haploid phase of the cycle is represented only by the gametes. In the flowering plants (Fig. 12.27C), stages 1 and 2 have not been abandoned altogether, but stage 2 has been reduced to a tiny three-to-eight-cell entity that is not free-living, and the plant spends most of its life as a multicellular diploid organism (stage 5).

We will consider the costs and benefits of the differing reproductive strategies of plants in Chapter 22; for now, let us emphasize the characteristics that will form the basis for the discussion: haploid-dominated plants tend to be short-lived, to reproduce rapidly and prolifically, and to be poorly differentiated; diploid-dominated plants are at the other end of the scale, and more like animals in each regard.

Meiosis in the life cycle of animals With rare exceptions, higher animals exist as diploid multicellular organisms through most of their life cycle. At the time of reproduction, meiosis produces haploid gametes, which, when their nuclei unite in fertilization, give rise to the diploid zygote. The zygote then divides mitotically to produce the new diploid multicellular individual. The gametes—sperm and egg cells—are thus the only haploid stage in the animal life cycle (see Fig. 12.27D).

In male animals, sperm cells (spermatozoa) are produced by the germinal epithelium lining the seminiferous tubules of the testes (Fig. 12.28). When one of the epithelial cells undergoes meiosis, the four haploid cells that result are all quite small, but approximately equal in size (Fig. 12.29).

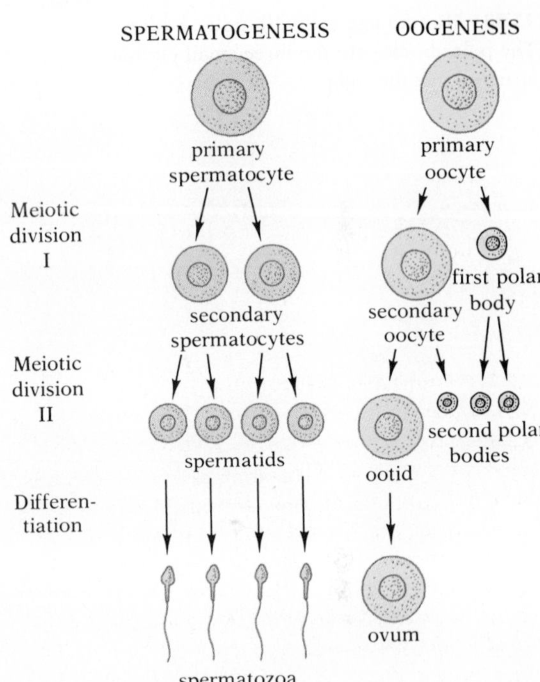

12.29 Schematic illustration of spermatogenesis and oogenesis in an animal
In some animals the first polar body does not divide.

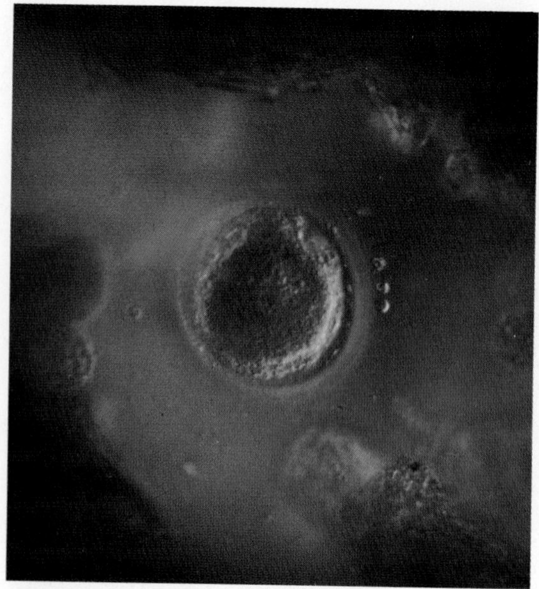

12.30 Human egg cell with polar bodies
The polar bodies are the three small circular structures at the right.

All four soon differentiate into sperm cells with long flagella, but with very little cytoplasm in the head, which consists primarily of the nucleus. This process of sperm production is called ***spermatogenesis***. Meiosis in human males begins at puberty. In female animals the egg cells are produced within the follicles of the ovaries by a process called ***oogenesis***. When a cell in the ovary undergoes meiosis, the haploid cells that result are very unequal in size. The first meiotic division produces one relatively large cell and a tiny one called a first ***polar body***. The second meiotic division of the larger of these two cells (secondary oocyte) produces a tiny second polar body and a large cell that soon differentiates into the egg cell (or ovum). The first polar body may or may not go through the second meiotic division. If it does redivide, there are three polar bodies altogether (Fig. 12.30). Thus, when a diploid cell in the ovary undergoes complete meiosis, only one mature ovum is produced (Fig. 12.29); the polar bodies are essentially nonfunctional. By contrast, a diploid cell undergoing complete meiosis in the testis gives rise to four functional sperm cells. In human females, the oocytes complete the first meiotic prophase in the fetal ovaries, and then complete meiosis upon ovulation (when a mature egg is released from an ovary). The interval between these two events can exceed 40 years.

The advantage of the unequal cytokinesis of oogenesis is obvious. By this mechanism an unusually large supply of cytoplasm and stored food is allotted to the nonmotile ovum for use by the embryo that will develop from it. In fact, the ovum provides almost all the cytoplasm and initial food supply for the embryo. The tiny, highly motile sperm cell contributes, essentially, only its genetic material.

STUDY QUESTIONS

1. It has been said that the only possible function of the odd process of meiosis is to make crossing over possible. Compare mitosis and meiosis with this suggestion in mind. (pp. 312–19, 322–33)

2. Crossing over is less frequent in males than in females in many species —male *Drosophila*, for instance, never undergo crossing over in meiosis. Even in hermaphroditic species, in which every organism can have both male and female reproductive organs, the same pattern holds. What might be the evolutionary logic behind this curious asymmetry? (pp. 331–33)

3. Interpret the following observations and formulate a master hypothesis about what sorts of chemical signals must appear in the cell at various points, and what effect they have:

   a) When a cell just entering S phase is fused with a G_1-phase cell arrested in that state, a process that mixes their cytoplasms but leaves the nuclei separate, both nuclei enter S phase.
   b) When a mid-S-phase cell is fused with a G_1-phase cell, the S nucleus remains frozen until the G_1 nucleus catches up.
   c) When a G_1-phase cell and a G_2-phase cell are fused, the G_2 nucleus remains frozen until the G_1 nucleus catches up.
   d) When an M-phase cell is fused with an S-phase or either sort of G-

phase cell, the M nucleus is unaffected, but the other nucleus loses its nuclear envelope and its chromosomes condense. (pp. 313–17)

4. Do animals with larger numbers of chromosomes but similar crossover rates have more, less, or the same general amount of variation in their progeny? (pp. 331–33)

CONCEPTS FOR REVIEW

- Cell division in procaryotes
- Terminology
  - Chromosomes versus chromatids
  - Homologous versus nonhomologous chromosomes
- Mitosis
  - Events
    - interphase
      - G_1 and G_2 gaps
      - S phase
    - prophase
    - metaphase
    - anaphase
    - telophase
  - Control of cell cycle: roles of cyclins and cdc protein
  - Role of microtubules, centrioles, and centromeres
  - Cytokinesis in animals versus plants
- Meiosis
  - Events of meiosis I
  - Process and consequences of crossing over
  - Events of meiosis II
  - Possible roles of crossing over and recombination
- Life cycles: haploid versus diploid phases in algae, plants, and animals
- Gamete formation in animals

SUGGESTED READING

ALBERTS, B., et al., 1989. *Molecular Biology of the Cell* 2nd ed. Garland, New York. *Contains a brief but up-to-date discussion of cell division in molecular terms.*

GOULD, J. L., and C. G. GOULD, 1989. *Sexual Selection.* Scientific American Library, New York. *A wide-ranging, nontechnical account of the many theories that seek to account for the evolution of sex and gender.*

MCINTOSH, J. R., and K. L. MCDONALD, 1989. The mitotic spindle, *Scientific American* 261 (4).

MAZIA, D., 1974. The cell cycle, *Scientific American* 230 (1). (Offprint 1288) *The stages of interphase and mitosis proper; experiments conducted to ascertain the characteristics of these stages and the controls governing them.*

MURRAY, A. W., and M. W. KIRCHNER, 1991. What controls the cell cycle, *Scientific American* 264 (3). *On the biochemical control of cell division, with emphasis on cyclin and cdc.*

STAHL, F. W., 1987. Genetic recombination, *Scientific American* 256 (2).

Chapter **13**

THE COURSE OF ANIMAL DEVELOPMENT

o biologists and nonbiologists alike, probably no aspect of biology is more amazing than the development of a complete new organism from one cell. The process of development is so precisely controlled that the entire intricate organization of cells, tissues, organs, and organ systems that will characterize the functioning adult comes into being, each element in just the right place with respect to the others, with rarely a flaw. Moreover, the millionth of a billionth of a gram of DNA in a mouse egg inevitably produces a mouse, while the same minute quantity of DNA in an oak seed produces an oak. In other words, the chromosomes of a fertilized seed or egg contain all the information necessary to direct the precisely ordered construction of an organism that may ultimately consist of more than 10^{12} cooperating cells.

The process of development requires a precisely programmed and coordinated series of changes in gene expression, in cell-surface proteins and the recognition and adhesion properties they display, in cell shape, and in cell motility, as cells possessed of all the necessary information interact to create a characteristic species morphology, and give rise to the lines of specialist cells that constitute the various tissues and organs. A complete understanding of development requires familiarity with nearly the entire range of biological processes. This chapter will describe the major physiological events of development in animals, from fertilization through embryonic development, birth, and postnatal development; the next chapter will examine the molecular and biochemical mechanisms that control and coordinate development.

We will concentrate in these chapters on the patterns and mechanisms of animal development, and will consider development in plants and organisms of the other kingdoms in Parts IV and V. There are a number of reasons for this conceptual scheme, most of them related to the very different requirements of plants and animals. The vast majority of plants are autotrophic, and are adapted to remain rooted during a life cycle to a specific locale and to collect sunlight. Since plants are not adapted for locomotion, they can enjoy the structural advantages of more rigid cell walls—a specialization that fixes each cell in place. The intricate migration of cells that takes place during the growth of the embryo in animal development is therefore impossible in plants.

Furthermore, because few plants feed on other organisms, they do not need nervous systems and muscles to guide and power the capture or harvesting of food; nor do they need mouths, stomachs, or digestive systems, or the more than 300 specialized cell types that comprise kidneys, bladders, rectums, and other specialized organs and tissues of animals. Because plants compete with each other by growing taller or broader to gather more light, and by extending their roots to obtain more water and minerals, they produce new organs like leaves and flowers whenever and wherever they are needed throughout their life cycle. As we will see, animals usually generate their entire array of tissues and organs early in development, and devote their later energies to the maintenance and growth of these organs and tissues. We will return to the subject of plant development in the context of the hormonal control of day-to-day plant growth in Chapter 32.

FERTILIZATION

In the ***ovaries***, or egg-producing organs, of a female animal certain cells are set aside early as egg primordia. These cells grow to an unusually large size; when they undergo meiosis the divisions are unequal, and almost all the cytoplasm is retained in the ripe ovum. The other haploid cells are the tiny polar bodies, and soon deteriorate (see Fig. 12.30, p. 336). The sperm, on the other hand, are unusually small cells with very little cytoplasm. The ovum thus furnishes most of the initial cytoplasm for the ***embryo***—the entity that develops from the fertilized egg and orchestrates its own conversion into a new individual. Moreover, even in eggs with relatively little stored food, or ***yolk***, the high density of yolk causes it to accumulate at the bottom, creating a polarity of cytoplasmic materials in the egg. As we will see in the next chapter, this polarization helps to provide direction in development. Even before fertilization, then, the ovum provides many of the initial cues that help to give order and direction to early development.

The fusion of the sperm with the egg cell creates the ***zygote*** and provides a stimulus that initiates its development into an embryo. The triggering, however, depends on the contact of the two cell membranes, not the fusion of the sperm nucleus with the egg nucleus to form a diploid nucleus. Apparently ***fertilization***—the joining of two haploid sets of chromosomes—is not necessary to induce embryonic development in many animals, even ani-

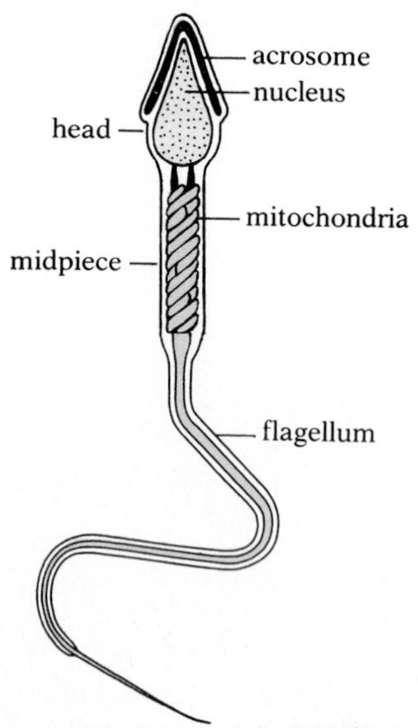

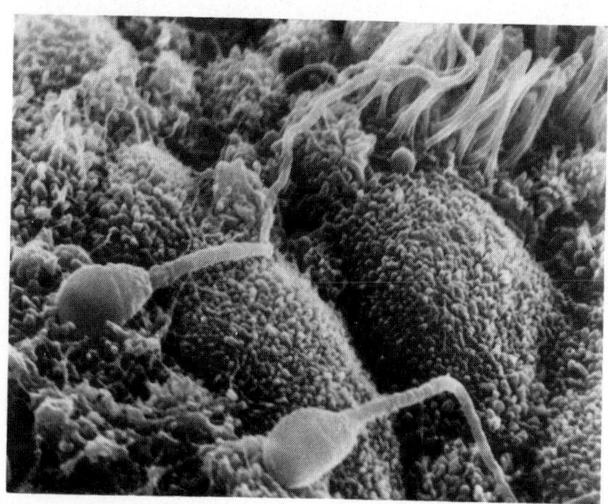

13.1 Human sperm
Three sections are easily distinguishable: the head, containing the nucleus; the midpiece, tightly packed with mitochondria, which produce the ATP that fuels the sperm's swimming; and the long flagellum. The chromosomes in the nucleus are supercondensed into an almost crystalline, genetically inactive form. In the apex of the head, immediately in front of the nucleus, is the acrosome, a membrane-bounded vesicle derived from the Golgi apparatus. There is only a tiny amount of cytosol in the cell. The photograph shows human sperm that have not yet reached the egg.

mals that do not normally reproduce parthenogenetically (from unfertilized eggs). It is easy to induce unfertilized frog eggs, for example, to begin development in the laboratory by pricking them with a fine needle dipped in blood. A few such eggs will develop into viable, apparently normal tadpoles.[1] Adult rabbits have been produced from unfertilized eggs by similar procedures. Unfertilized eggs can even be stimulated to begin developing by a mild electric shock, or by a change in the salt concentration in the surrounding fluid, or by a physical jolt, but in most species the development aborts after only a few divisions.

Let us look more closely at the process of fertilization of an egg cell by a sperm cell. In many mammalian species the egg is initially enclosed in a thin protective layer of cells called the ***follicle***. This layer, a barrier to the sperm, is loosened by hyaluronidase, an enzyme released by a vesicle called the ***acrosome***, which is located at the apex of the sperm in many species (Fig. 13.1); similar enzymes are used by bacteria in penetrating host tissues. But even after the follicle cell barrier has been loosened, the sperm, attracted by chemicals secreted by the egg, now encounters a thick coat, called in mammals the ***zona pellucida***. Receptors in the head of the sperm must bind to a long, filamentous glycoprotein embedded in the coat. Species-specific markers on the glycoprotein prevent the sperm of other species from attaching.

At this point, the acrosome fuses with the plasma membrane of the sperm cell and releases enzymes that act on the jelly coat and cell membrane of the egg cell (Fig. 13.2). A tubular filament derived from the acrosome membrane penetrates the coat and then fuses with a microvillus of the egg cell; the fusion is probably facilitated by enzymes derived from the acrosome. Once this fusion of sperm and egg membranes has occurred, the electrical potential of the egg-cell membrane changes dramatically in many species (thus making fusion with the other sperm difficult for the next minute or so), and the sperm nucleus moves into the cytoplasm of the egg. At the same time, the structure of the glycoprotein responsible for sperm binding is altered to prevent the attachment of additional sperm even after the membrane potential returns to normal.

As the sperm nucleus, now known as a pronucleus, moves into the egg, many changes begin in the egg. First, calcium-containing vesicles in the sperm release their contents, triggering release of even more calcium from vesicles in the egg, initiating the earliest steps in post-fertilization development. (The tricks for inducing development in unfertilized eggs, mentioned earlier, probably succeed by triggering calcium release.) Another process initiated immediately after fusion is completion of the oogenic meiosis if that process is not already complete. Among the many other changes regularly seen in the egg after fusion are striking alterations in the permeability of the plasma membrane (especially a much enhanced permeability to inorganic phosphate) and an increased rate of oxygen consumption.

[1] Most frog embryos developed from unfertilized eggs are haploids, and die after reaching, at most, the tadpole stage. Such embryos as survive to reach adulthood are the few that undergo spontaneous chromosomal doubling to become diploid (though the two sets of chromosomes are identical). A few embryos even become $4n$, $6n$, or $8n$; these too may survive.

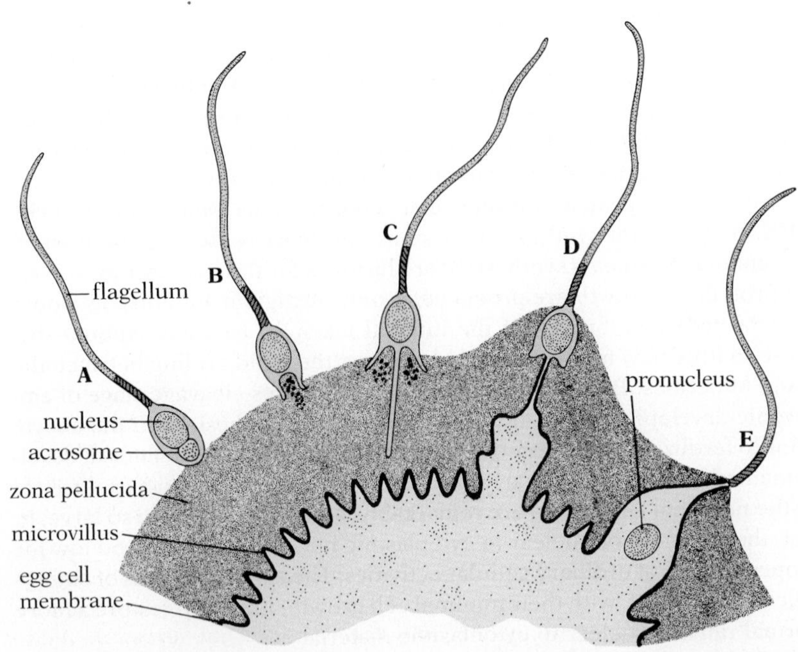

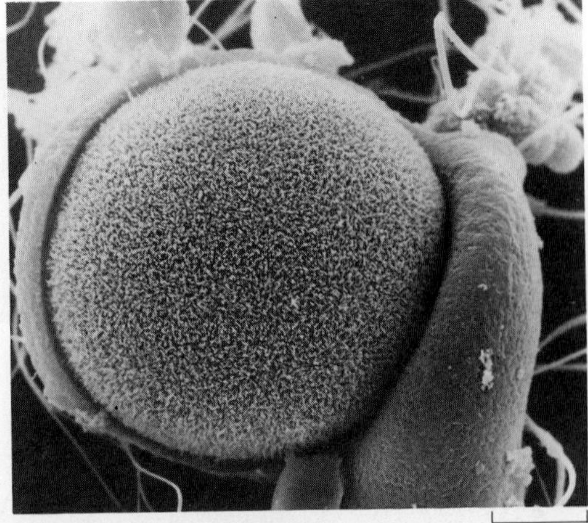

13.2 The fertilization process
(A) After the follicle cell coat has been broken, a sperm cell comes into contact with the zona pellucida surrounding the egg cell membrane. (B) If the recognition of species-specific markers on the two gametes is successful, the membrane of the acrosome fuses with the plasma membrane of the sperm and ruptures, releasing the acrosomal contents, which include enzymes that act on the jelly coat and on the membrane of the egg cell. (C) A tubular filament derived from the acrosome membrane pushes through the zona pellucida. (D) The tube fuses with an enlarged microvillus of the egg cell; there is now no membranous barrier between the contents of the sperm cell and the egg cell. (E) The sperm pronucleus moves into the egg cell. (F) SEM of a hamster cell showing the zona pellucida pulled back and the exposed microvilli of the egg cell membrane.

True fertilization—the union of the two gamete nuclei—depends on some attraction of the sperm pronucleus by the egg pronucleus, the nature of which is still unknown. If sperm are induced to penetrate immature sea-urchin eggs in which the chromosomes are still condensed, no attraction of the sperm pronucleus occurs, nor is there any tendency for the supercondensed chromosomes of the sperm pronucleus to undergo decondensation. In short, the mature egg pronucleus must be responsible both for attracting the sperm pronucleus and for inducing the decondensation of its chromosomes. Decondensation of both sperm and egg chromosomes is necessary if the zygote nucleus produced by fusion of the two gamete pronuclei is to carry out DNA replication preparatory to the first mitotic division of embryonic development.

EMBRYONIC DEVELOPMENT

EARLY CLEAVAGE AND MORPHOGENETIC STAGES

In normal development the zygote begins a rapid series of mitotic divisions immediately after fertilization has taken place. In many animals these early cleavages are not accompanied by cytoplasmic growth. They produce a cluster of cells that is no larger than the single egg cell from which it is de-

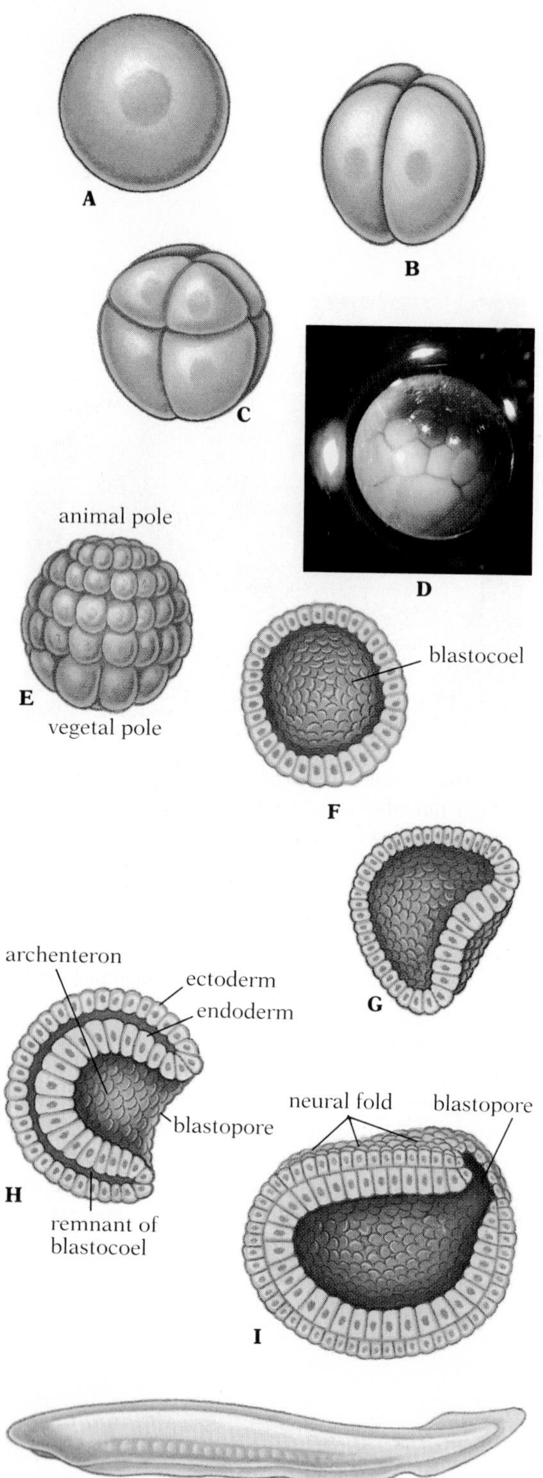

rived (Fig. 13.3C). The cytoplasm of the one large cell is simply partitioned into many new cells that are much smaller. But in some animals, notably reptiles and birds, some cytoplasmic growth does occur, as nutrients from the yolk (stored food) are consumed.

During this early cleavage stage of development, the nuclei cycle very rapidly between chromosomal replication (the S period of the cell cycle) and mitosis (the M period); the G_1 and G_2 periods are practically absent. Such rapid cycling, in which G_1 and G_2 are skipped, is possible because the ovum already contains huge quantities of the DNA polymerase necessary for catalyzing repeated chromosomal replication, as well as most of the mRNA required for synthesis of proteins during early cleavage (at least 50 percent of the proteins synthesized are histones for the new chromosomes; the proteins of growth are produced in only negligible amounts, as would be expected). The brevity of the interval taken up by transcription—because so little new mRNA is needed—allows the rapid cycling between the S and M periods. But note that, since control of this cleavage stage of embryonic development depends largely on mRNA synthesized in the oocyte prior to fertilization, the paternal genes have little input until later in development; the course of the early cleavages is determined almost exclusively by the maternal genes. Ova are remarkably large cells. They are so large, in fact, that the ratio of nuclear to cytoplasmic material would be too low for proper control of ordinary cellular activities. The early cleavages of embryonic development, with their minimal cell growth, thus help restore a more normal ratio of nuclear to cytoplasmic material.

As cleavage continues, the newly formed cells (blastomeres) of many species begin to pump sodium ions into the center of the mass of cells. As a result, water diffuses in and the blastomeres come to be arranged in a sphere surrounding a fluid-filled cavity called a ***blastocoel*** (Fig. 13.3E). An embryo at this stage is termed a ***blastula***.

Next begins a series of complex movements important in establishing the definitive shape and pattern of the developing embryo. The establishment of shape and pattern in all organisms is called ***morphogenesis*** (meaning "the genesis of form"). Morphogenetic movements of cells in large masses always occur during the early developmental stages of animals.

The mechanism of these movements is still very poorly understood.

13.3 Early embryology of amphioxus
(A) Zygote. (B–E) Early cleavage stages, culminating in formation of a blastula (E). (D) Photograph of an early cleavage showing that the cells in the vegetal hemisphere are enlarged with yolk (yellow). (F) Longitudinal section through a blastula, showing the blastocoel. (G–H) Longitudinal sections through an early and a late gastrula. Notice that the invagination is at the vegetal pole of the embryo, where the cells are largest. (I) The blastopore becomes the anus of the gastrula, and a neural fold begins to form. Subsequent steps are traced in Figure 13.5. An adult amphioxus is also shown.

There are often changes in the shapes of the cells, probably effected by interactions between actin microfilaments and myosin microtubules. The changes in shape may be relatively small (Fig. 13.4), or they may be extensive. Important in some of the movements are changes in the adhesive affinities of the cells for neighboring cells or, in the case of epithelial (surface-covering) cells, for the basement membrane on which these cells sit. It may be relatively easy for a group of cells that adhere tightly to each other to slide as a group over the surface of an underlying layer to which their affinity has been at least temporarily reduced.

Since the pattern of cleavages and cell movement is greatly influenced by the amount of yolk in the egg, we will examine, first, the pattern in an animal whose eggs have little yolk, then that in animals whose eggs have more.

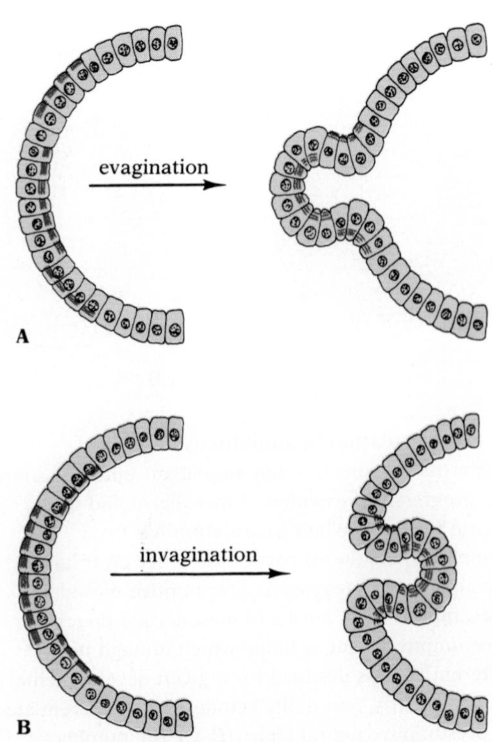

13.4 The mechanism of some morphogenetic movements in cells
Contraction of microfilament-microtubular complexes (red), asymmetrically positioned in the cells, may change the shapes of the cells and produce evaginations (A), invaginations (B), or other alterations of the arrangement of cells in a developing organ. The contractile interactions are of the same type as those involved in amoeboid movement and cytokinesis.

Development in amphioxus In amphioxus (Fig. 13.3), a tiny marine animal whose egg has very little yolk, the movements that occur after formation of the blastula (when it is composed of about 500 cells) convert it into a two-layered structure called a ***gastrula***. The process of ***gastrulation*** begins when a broad depression, or invagination, starts to form at a point on the surface of the blastula where the cells are somewhat larger than those on the opposite side (Fig. 13.3F). The differences in cell size are not very great in amphioxus embryos; they are more pronounced in many other animals. The smaller cells make up the ***animal hemisphere*** of the embryo. The larger cells make up the ***vegetal hemisphere***. It is at the pole of the vegetal hemisphere that the invagination of gastrulation occurs in amphioxus. As gastrulation proceeds, the invaginated layer bends farther and farther inward, until eventually it comes to lie against the inside of the outer layer, nearly obliterating the old blastocoel (Fig. 13.3G). In the sea urchin, which has a transparent blastula that permits detailed observation of gastrulation, pseudopods extend from the inner membranes of the cells where invagination began. The pseudopods adhere to the inner membranes of other blastula cells; then they shorten, pulling the invaginated layer farther inside, and new pseudopods form and continue the process. The embryos of other animals probably gastrulate in a similar way. The resulting gastrula is a two-layered cup, with a new cavity that opens to the outside via the ***blastopore***. The new cavity, called the ***archenteron***, will become the cavity of the digestive tract, and the blastopore will become the anus. A very similar process occurs in nearly all animals, though with a few fundamental differences—for instance, in the vast majority of animals (but not in the group called the chordates, to which amphioxus and human belong) the blastopore becomes the mouth rather than the anus. Because they are so basic to organisms, comparison of developmental programs is a critical tool in reconstructing the evolutionary relationships of major groups, as we will see in more detail in Chapter 24.

Gastrulation, as it occurs in amphioxus, first produces an embryo with two primary cell layers, an outer ***ectoderm*** and an inner layer. The latter subsequently separates into two layers, the ***endoderm*** and the ***mesoderm;*** the mesoderm lies dorsally (toward the top side) between the ectoderm and the endoderm. In amphioxus the mesoderm originates as pouches flanking a supportive central rod, the ***notochord;*** these all pinch off from the inner

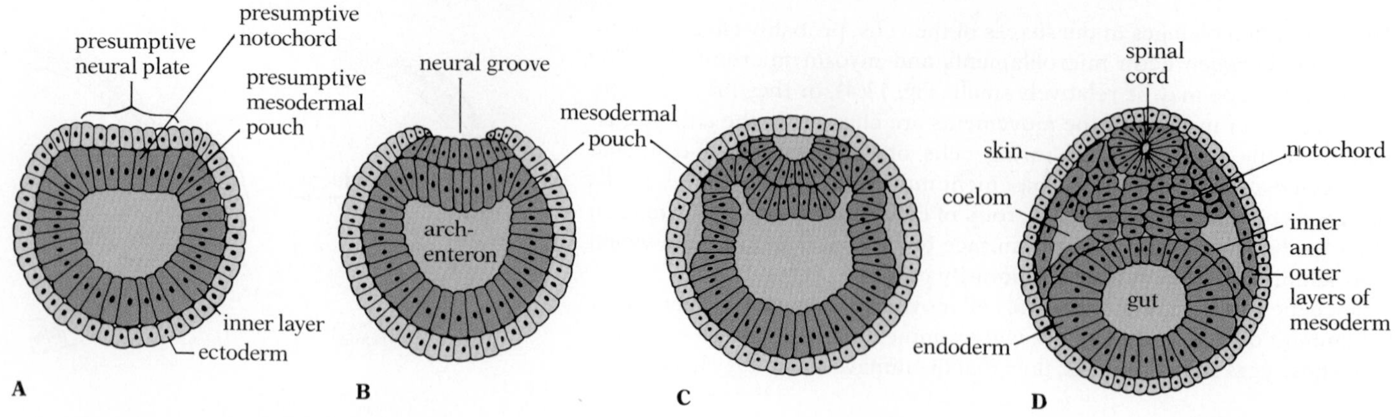

13.5 Neurulation in amphioxus
The cross sections through amphioxus embryos show the progressive formation of mesoderm and the neural tube. (A) When gastrulation has been completed, the dorsal part of the inner layer has already been segregated as presumptive mesoderm—presumptive notochord and mesodermal pouches. (Presumptive tissue is tissue which, though not yet differentiated, is destined for a given developmental fate.) Similarly, part of the ectoderm has differentiated as presumptive neural plate. (B–C) The notochord and mesodermal pouches form as evaginations from the inner layer, and the neural plate invaginates from the ectoderm as it begins to form the spinal cord. (D) In this later embryo, called the neurula, both the spinal cord and the mesoderm are taking their definitive form. Notice that there is a cavity (the coelom) in the mesoderm.

layer (Fig. 13.5). The remaining part, the endoderm, pinches together to form a tube that becomes the digestive tract. Figure 13.5 shows the further development of the embryo to the ***neurula*** stage, so called because it incorporates the beginnings of the nervous system; the details will be discussed below. All these changes, of course, are orchestrated by well-timed changes in gene activity, which we will explore in the next chapter.

In the amphioxus egg, where the distinction between animal and vegetal hemispheres is only slight, owing to the small amount of yolk in the vegetal hemisphere, the early cleavages are nearly equal. The new cells are thus of nearly the same size, and gastrulation occurs in a direct and uncomplicated manner. But the eggs of many organisms have far more yolk in their vegetal hemisphere, and this deposit of stored food imposes complications and limitations on such processes as cleavage and gastrulation. Generally, the more yolk an egg contains and the more eccentric its cytoplasmic distribution, the more cleavage tends to be restricted to the animal hemisphere and the more gastrulation departs from the pattern in amphioxus.

Development in frogs The frog egg, which contains far more yolk than that of amphioxus but much less than that of a bird, serves as an example of eggs with an intermediate amount of yolk. The first two cleavages, which are perpendicular to each other, cut through both the animal and vegetal poles, producing cells of roughly the same size (Fig. 13.6B). But the next cleavage is equatorial (parallel to the egg's equator) and located decidedly nearer the animal pole (Fig. 13.6C); hence the four cells produced at the animal end of the egg are considerably smaller than the four at the vegetal end. From this stage onward, more cleavages occur in the animal hemisphere of the embryo than in the vegetal hemisphere as the blastula develops. As in amphioxus, there is no increase in total mass during these early cleavage stages (Fig. 13.7).

After the blastula has been formed, the frog embryo begins gastrulation. Simple invagination of the vegetal hemisphere is not mechanically feasible, because of the large mass of inert yolk. Instead, portions of the cell layer of the animal hemisphere move down around the yolk mass and then turn in at the edge of the yolk. This involution begins at what will be the dorsal side

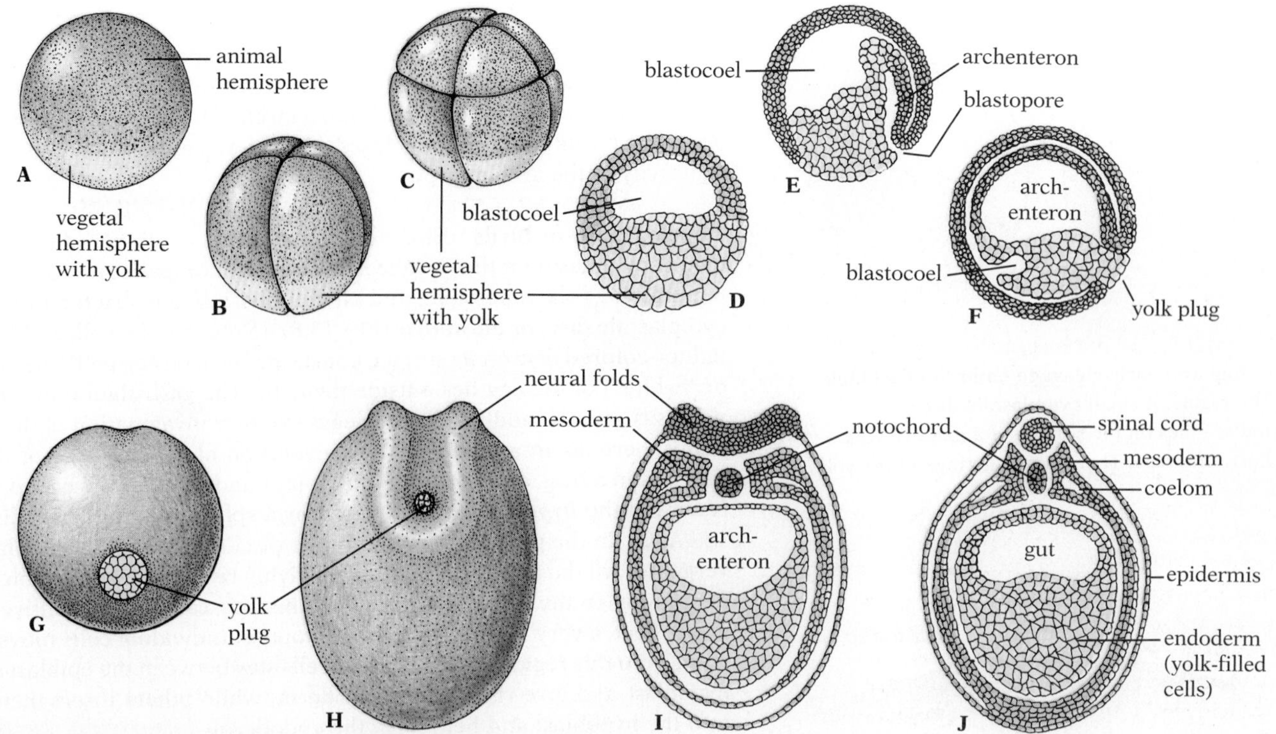

13.6 Early embryology of a frog The large amount of yolk in the frog egg causes its pattern of gastrulation to differ from that in amphioxus. (A) Zygote. (B–C) Early cleavage stages. Note that because the first horizontal cleavage is nearer the animal pole, the cells at the vegetal pole are much larger. (D) Longitudinal section of a blastula. (E–F) Longitudinal sections of two late gastrula stages. (G) End view of late gastrula with first hint of developing neural fold at top. (H) End view of an early neurula, showing the neural folds and neural groove. (I) Cross section of a neurula after formation of mesoderm. (J) Cross section of a later embryo, showing definitive spinal cord; this stage is reached about 24 hours after fertilization.

13.7 Scanning electron micrographs of frog egg and some early cleavage stages Left: Unfertilized egg. Middle: eight-cell stage. Right: 32- to 64-cell stage. All three micrographs are at the same magnification: × 46. Note that there has been no overall growth in size during these cleavage stages—the 32- to 64-cell embryo is no larger than the egg cell.

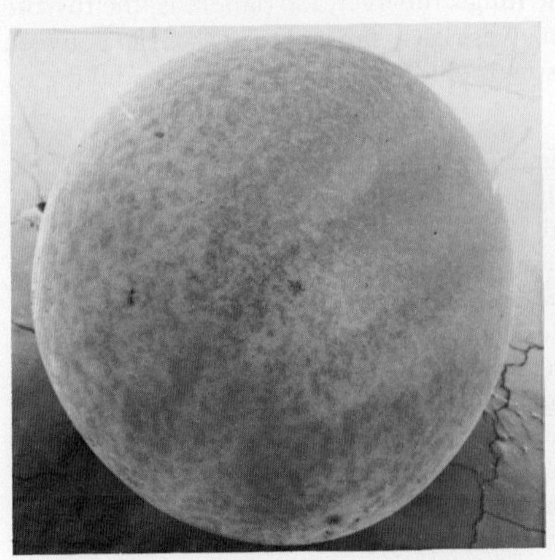

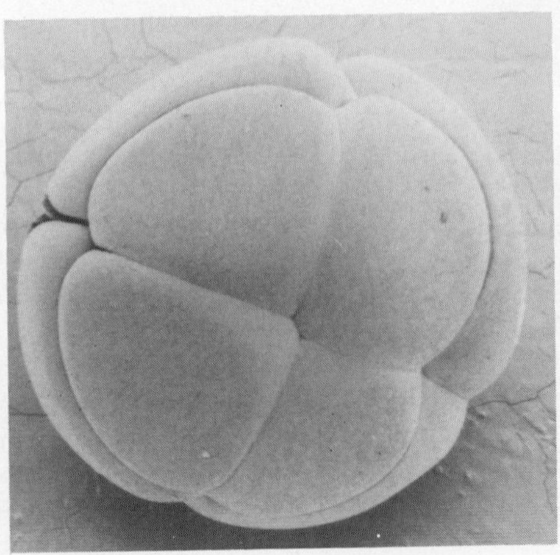

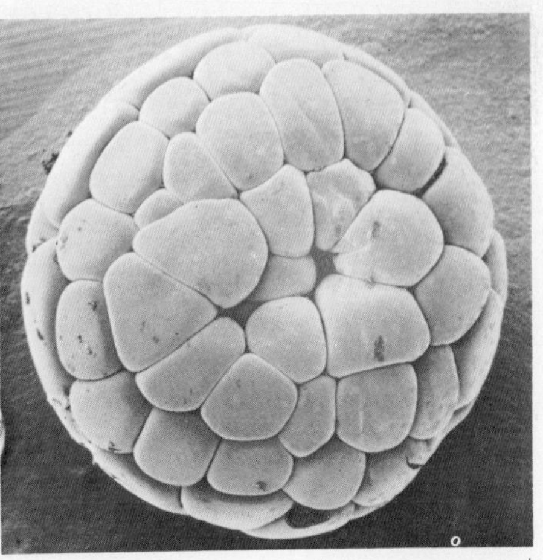

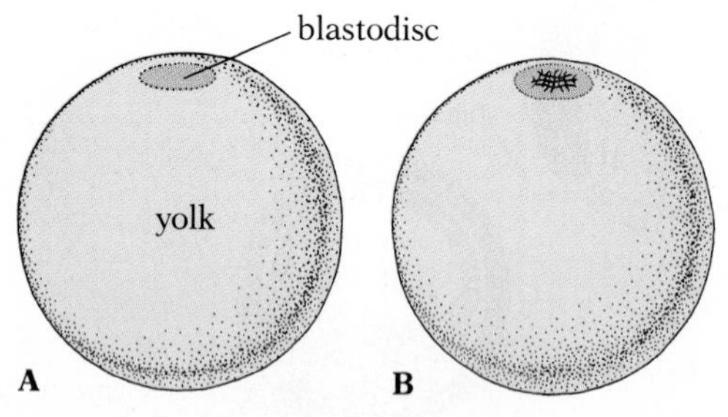

13.8 Egg and early-cleavage embryo of a chick (A) The zygote. A small cytoplasmic disc—the blastodisc—lies on the surface of a massive yolk. (B) Early cleavage. There is no cleavage of the yolk.

of the yolk mass, forming initially a crescent-shaped blastopore at the edge of the yolk. The infolding slowly spreads to all sides of the yolk, so that the crescent blastopore is converted into a circle. Movement of the other cells around the yolk eventually encloses this material almost completely within the cavity of the archenteron.

Development in birds Birds' eggs contain so much yolk that the small disc of cytoplasm on the surface is dwarfed by comparison. No cleavage of the massive yolk is possible, and all cell division is restricted to the small cytoplasmic disc, or ***blastodisc*** (Fig. 13.8). (Note that the yolk and the small lighter-colored disc on its surface constitute the true egg cell. The albumin, or "white," of the egg lies outside the cell.) The gastrulation process is of necessity greatly modified in such eggs. Neither invagination of the vegetal hemisphere, as in amphioxus, nor involution along the edge of the yolk mass, as in a frog, can occur. Instead, outer and inner layers called the ***epiblast*** and the ***hypoblast*** are produced by a splitting of the blastodisc (Fig. 13.9A–C). In the posterior portion of the disc, the cells of the epiblast converge toward the longitudinal midline, giving rise to a clearly visible line or depression on the epiblast (Fig. 13.9D); this line, called the primitive streak, is, in effect, a very elongate closed blastopore. Individual cells move downward from this region. Some of these cells stay between the epiblast and the hypoblast and give rise to the mesoderm, while others insert themselves into the hypoblast and help form the endoderm.

Developmental fates The fates of cells in different parts of the three primary layers of vertebrates have been determined by staining them with dyes of different colors, or by marking them with carbon or other particles, and then following their movements. As you might expect, the ectoderm eventually gives rise to the outermost layer of the body—the top layer of the skin, called the ***epidermis***—and to structures derived from the epidermis, such as hair, nails, the eye lens, the pituitary gland, and the epithelium of the nasal cavity, mouth, and anal canal. As you might also expect, the endoderm gives rise to the innermost layer of the body—the epithelial lining of the digestive tract and of other structures derived from the digestive tract, such as the respiratory passages and the lungs, the liver, the pancreas, the thyroid, and the bladder. The mesoderm gives rise to most of the tissues in between, such as muscle, connective tissue (including blood and bone), and the no-

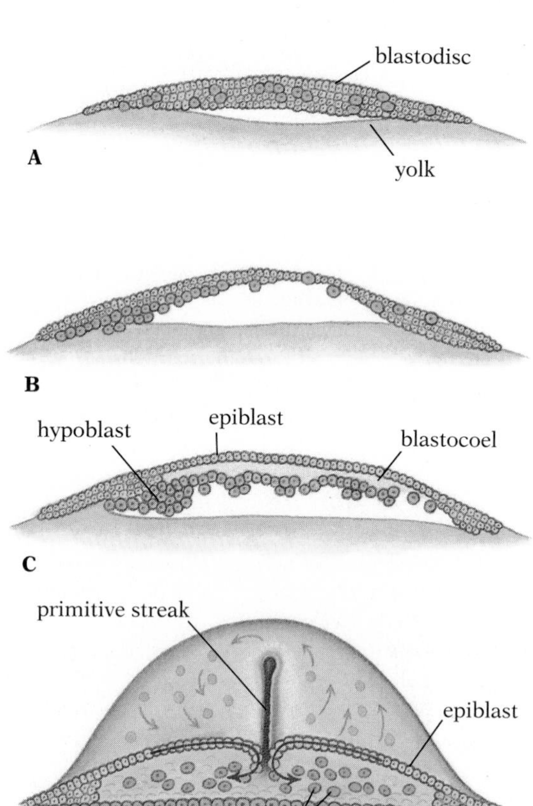

13.9 Gastrulation in the chick embryo (A) Longitudinal section through the midline of a blastula. Larger yolk-laden cells are intermixed with smaller cells. (B) The larger cells begin to accumulate on the lower surface of the cell mass. (C) The layer of larger cells separates from the layer of smaller cells to become the hypoblast; the cavity between the two layers is the blastocoel. (D) Surface view and section across the midline of a gastrula. Involution of cells along the midline of the embryo during gastrulation (red arrows) produces a clearly visible primitive streak, which is essentially a very elongate blastopore. Some epiblast cells of the primitive streak move downward to form the mesoderm; others combine with the hypoblast to help form the endoderm.

tochord (the dorsally located supportive rod found in all chordates, at least in the embryonic stages). The origins of various body organs and tissues are summarized in Table 13.1.

One major tissue located topographically between the skin and the gut does not develop from the mesoderm. This is the nervous tissue, which, curiously enough, is derived from the ectoderm. Soon after gastrulation, the ectoderm becomes divided into two components, the epidermis and the neural plate. A sheet of ectodermal cells lying along the midline of the embryo dorsal to the newly formed digestive tract and developing notochord bends inward in a process called ***neurulation***, and forms a long groove extending most of the length of the embryo (Figs. 13.5A–C and 13.6H–I). The dorsal folds that border this groove then move toward each other and fuse, converting the groove into a long tube lying beneath the surface of the back. This neural tube becomes detached from the epidermis dorsal to it, and in time differentiates into the spinal cord and brain (Figs. 13.5D and 13.6J).

We see, then, that the morphogenetic movements of gastrulation and neurulation give shape and form to the embryo, and bring masses of cells into the proper position for their later differentiation into the principal tissues of the adult body. In effect, the movements mold the embryonic mass into the structural configuration on which differentiation will superimpose the finer detail of the finished organism.

TABLE 13.1 *Origins of certain organs and tissues*

Primary layer	*Tissue or organ*
Ectoderm	Epidermis Hair Nails Eye lens Nervous system Lining of nose and mouth
Mesoderm	Muscle Bone Blood Notochord
Endoderm	Lungs Liver Pancreas Lining of stomach Bladder

LATER EMBRYONIC DEVELOPMENT

Gastrulation and neurulation provide the organization that shapes the development of the early embryo; the late embryo must be converted from this promising beginning into a fully developed young animal ready for birth. The individual tissues and organs must be formed, and an efficient circulatory system must quickly come into being (Fig. 13.10). In a verte-

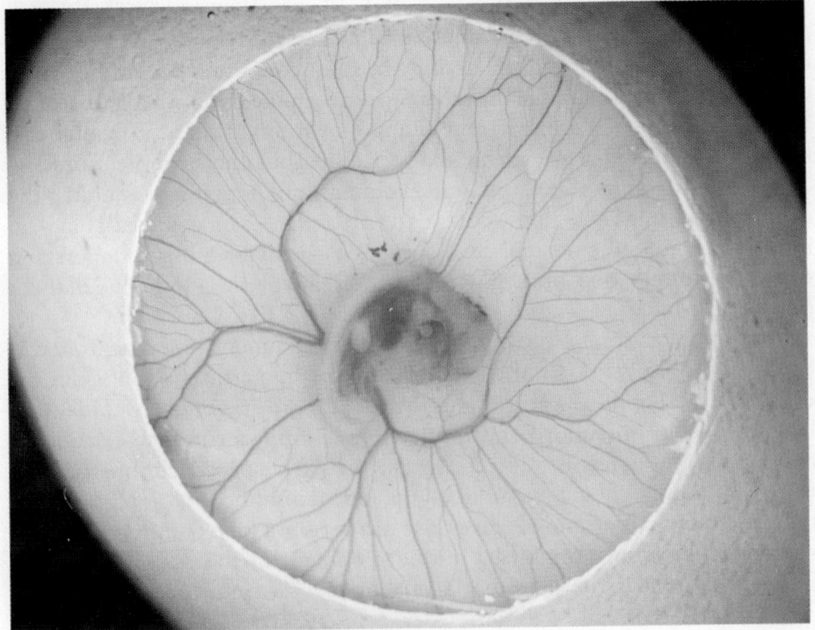

13.10 Chick embryo after four days of incubation The tiny embryo lies on the surface of the yolk. It has a functional circulatory system, including a beating heart, even at this early stage of its development. Note the long branching blood vessels that run out of the embryo into the yolk; they transport nutrients back to the embryo.

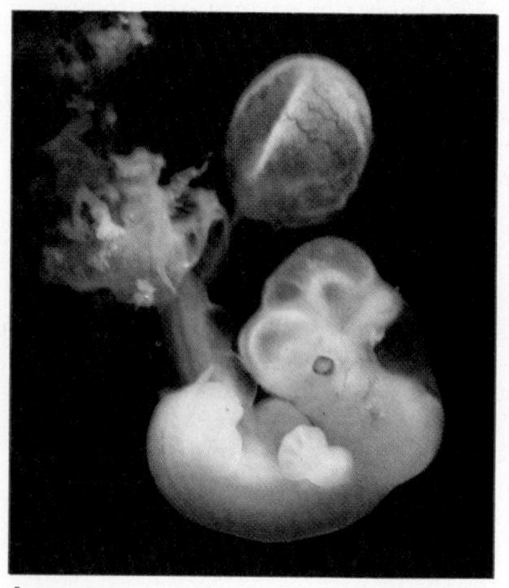
A

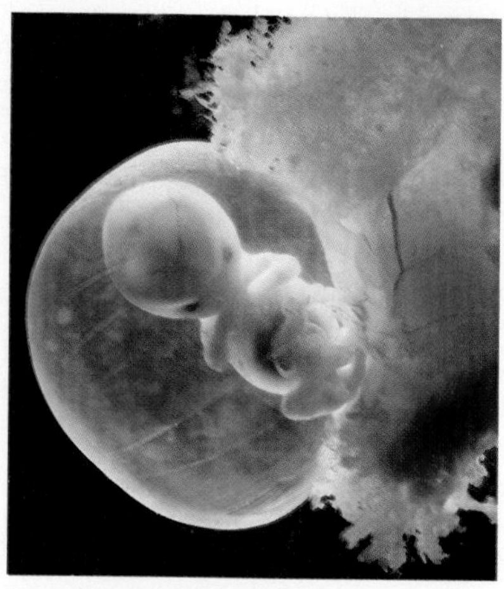
B

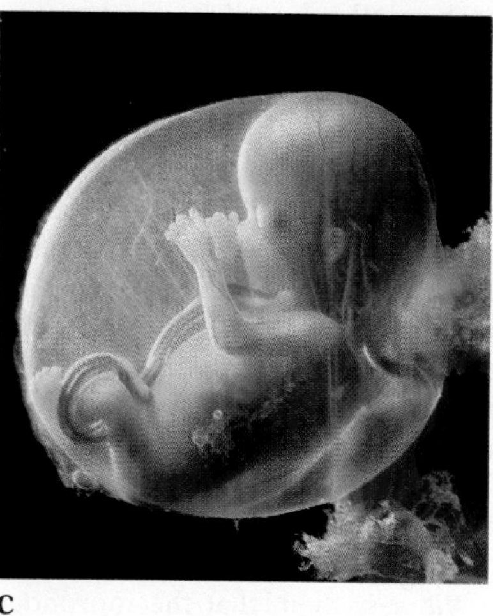
C

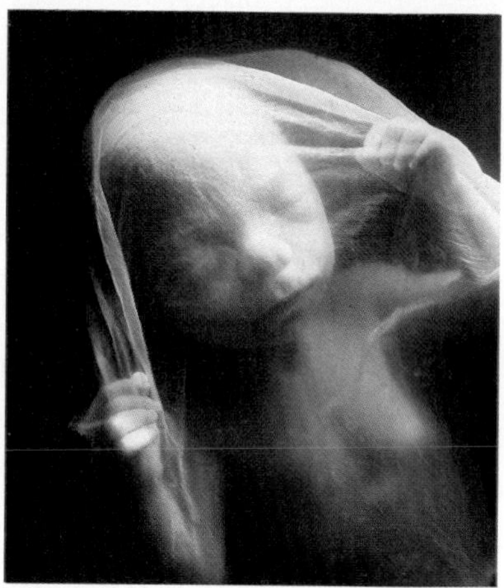
D

13.11 Human embryos at successive stages of development
(A) The five-week embryo, 1 cm long, shows the beginnings of eyes but no distinctive face; note its mittenlike hands and feet, with no separation between the digits. (B) By contrast, the seven-week embryo, 2 cm long, has a distinct face, and there is separation between its fingers. (C) At 13 weeks the fetus is over 7 cm long and weighs about 30 gm, 15 times more than at seven weeks. (D) By 17 weeks the fetus is over 15 cm long and seven times heavier than it was a month earlier; all the internal organs have formed.

brate the four limbs must develop, the elaborate system of nervous control must be established, and so on. The complexity and precision characterizing these developmental changes are staggering to contemplate. To give but one example, approximately 43 muscles, 29 bones, and many hundreds of nervous pathways must form in each human arm and hand. To function properly, all these components must be precisely correlated. Each muscle must have exactly the right attachments; each bone must be jointed to the next bone beyond it in a certain way; each nerve fiber must have all the proper synaptic connections with the central nervous system and must terminate on the right effector cells. Incredibly sensitive mechanisms of developmental control must operate for such an intricate structure to arise from a mass of initially undifferentiated cells. Yet the developmental processes that produce all these later embryonic changes are the same ones we have seen at work in the early embryo—cell division, cell growth, cell ***differentiation*** (as cells take on increasingly specialized roles), and morphogenetic movements. Bursts of mitotic activity in some areas and cessation of cell division in other areas alter the balance between the parts. Special patterns of cell growth produce important changes in size and shape. Through differentiation cells may lose particular capacities, but may become more efficient at performing other functions. Foldings and pouchings establish the primordia of lungs and glands, of eyes and bladder. Even cell death plays an important role in the normal development of the living animal: fingers and toes, for example, become separated by the death of the cells between them (Fig. 13.11B).

One simplification in the organization of development—a pattern we will explore more fully both in the next chapter and in our discussion of animal evolution in Chapters 24 and 25—is the repetitive organization of older embryos. Once the notochord or primitive streak is fully formed (which occurs about a day after fertilization in birds), regularly spaced clumps of cells called ***somites*** begin to appear along the dorsal midline (Fig. 13.12). In vertebrates, each pair of somites gives rise to a vertebra (one of the bones in the spinal column), and organizes the development of the nerves, muscles, bones, and other structures that a vertebra serves. In most invertebrates, each somite directs the development of one segment of the adult. Hence, the developmental program of most animals is compartmentalized. As we will see in the next chapter, specialized characteristics of segments (like those of somites that produce limbs) arise as these organizing centers "read" their anterior-posterior position in the embryo and then activate the appropriate genes.

It is beyond the scope of this book to discuss in detail the many events that occur during later embryonic development. Yet these are the events that mold morphologically similar gastrulas into a fish in one instance, a rabbit in another, and a human being in still another, depending on the genetic endowment of the gastrula in question; the developmental events are programmed differently for each species. An understanding of how such different programs arise and how they are carried out is one of the important goals of developmental biologists.

One interesting aspect of the differences in the developmental programs of different species should be mentioned here—namely that the early embryos of most vertebrates closely resemble one another. For example, the early human embryo, with its well-developed tail and a series of pouches in the pharyngeal region (where the neck will appear), looks very much like an early fish embryo (Fig. 13.13), and it looks even more like an early rabbit embryo. Only as development proceeds do the distinctive traits of each kind

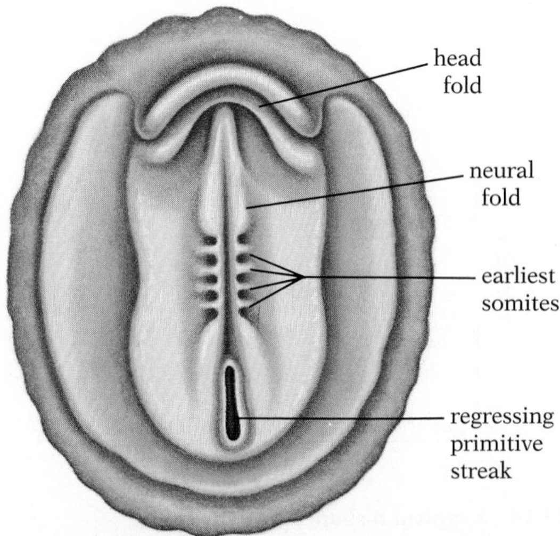

13.12 Formation of somites in the chick embryo After the primitive streak forms, it develops into a neural groove, beginning at the anterior end and proceeding posteriorly. Somites begin to appear, each of which organizes the development of a separate segment of the body.

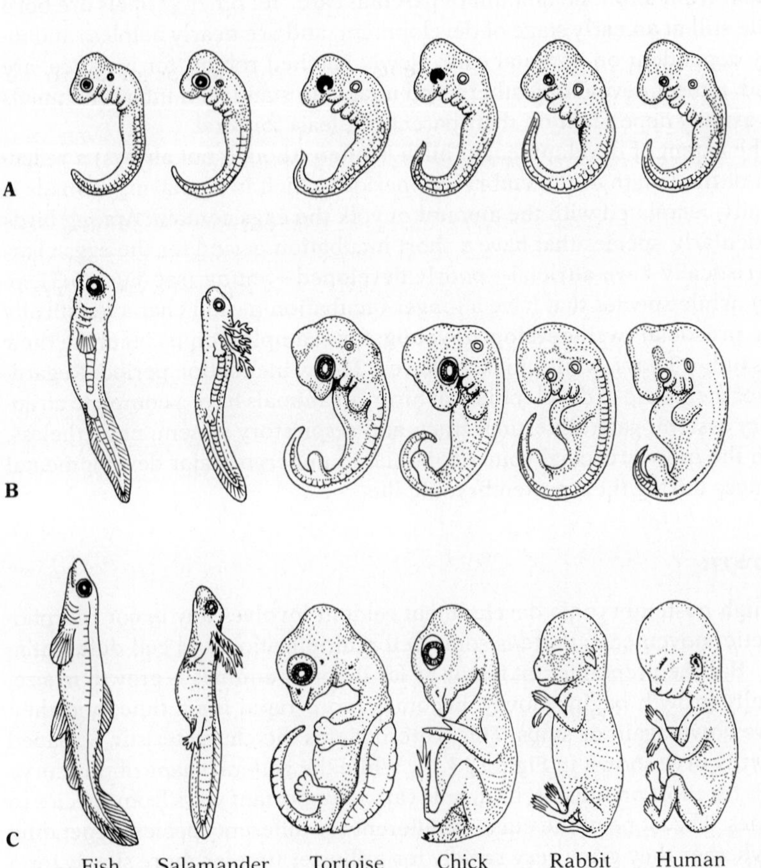

13.13 Vertebrate embryos compared at three stages of development At stage A, all the embryos—whether fish, amphibian, reptile, bird, or mammal—strongly resemble one another. Later, at stage B, the fish and salamander are noticeably different, but the other embryos are still very similar; note the pharyngeal pouches in the neck region and the prominent tail. By stage C, each embryo has taken on many of the features distinctive of its own species.

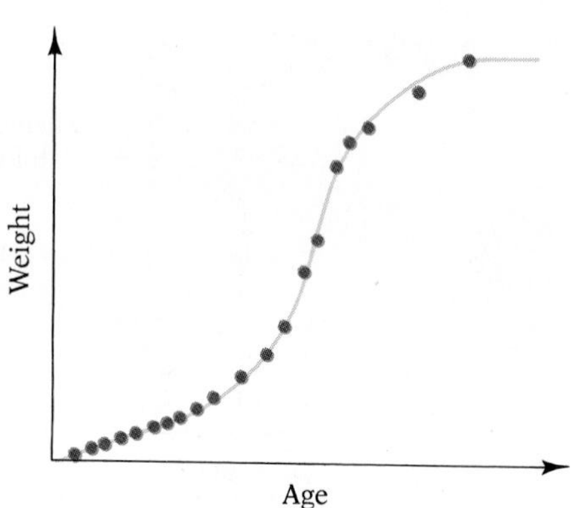

13.14 A typical S-shaped growth curve

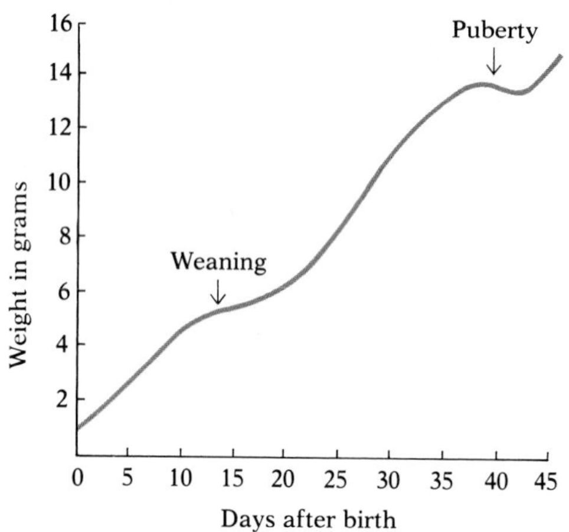

13.15 Growth in weight of a mouse
The rate of growth is slower at weaning and at puberty.

of vertebrate become apparent. About a hundred years ago, the German scientist Ernst Haeckel took this observation beyond its general interpretation as vague evidence of common descent, and proposed that the development of each organism retraces in detail its evolutionary history—that is, that "ontogeny recapitulates phylogeny." According to this hypothesis, early human embryos resemble fish because mammals are the evolutionary descendants of fish. While it is certainly true that the suites of genes working together to control developmental patterns are conserved by natural selection, changing little over time as compared with superficial morphology, we now know that the general developmental pattern of a species can skip the steps of ancestors, or create new structures *de novo*. There is even evidence to suggest that novel groups of organisms can arise through major and relatively rapid changes in developmental programs.

POSTEMBRYONIC DEVELOPMENT

The extent to which an animal has developed by the time of birth varies greatly among different species. Some young animals, such as baby guppies, are entirely self-sufficient from the time they are born and neither need nor receive parental care. Others—baby chicks and ducks, for example—can run about and feed themselves as soon as they are born, but still benefit from a limited amount of parental care. Yet other animals are born while still at an early stage of development, and are nearly helpless and totally dependent on parental care. Newly hatched robins, for instance, are blind, almost devoid of feathers, and unable to stand, and infant mammals are usually dependent on their parents, at least for food.

The extent of development at birth is often (though not always) a reflection of the length of the embryonic period, which in egg-laying animals is usually correlated with the amount of yolk the eggs contain. Among birds particularly, species that have a short incubation period for the eggs characteristically have altricial—poorly developed—young (see Fig. 25.31, p. 725), while species that have a longer incubation period characteristically have precocial (well-developed) young; for example, robins incubate their eggs only 13 days, while chickens have a 21-day incubation period. Regardless of their state of development at birth, all animals have a complete circulatory system, gastrointestinal tract, and respiratory system; nevertheless, even the most precocial young continue to undergo major developmental changes during their postembryonic life.

GROWTH

Though postembryonic development seldom involves any major morphogenetic movements, there is some cell multiplication and cell differentiation. But the preponderant factor by far in many animals is growth in size. Usually growth begins slowly, becomes more rapid for a time, and then slows down again or stops. This pattern yields the characteristic S-shaped growth curve shown in Figure 13.14. While the general shape of this curve holds for most organisms, its details vary in important ways from species to species. The slope of the curve is different for different species, depending on whether they grow very rapidly for a shorter time or more slowly for a longer time. (Compare the rate of increase in weight of, say, a calf and a child.) The shape of the curve is seldom as smooth as it appears in a generalized growth curve because many factors can affect the rate of growth. For example, in most mammals growth slows down for a while immediately after weaning, and it often varies greatly during puberty; such irregularities are reflected in bumps and dips in the curve (Fig. 13.15). An especially marked departure from the smooth generalized curve is seen in the growth of arthropods, a group that includes, among other creatures, the many kinds of insects. These animals can undergo only limited growth between molts, because the hard exoskeleton that encases their bodies can be stretched only slightly. However, at each molt there is a sharp burst of growth during the short period after the old exoskeleton has been shed and before the new one has hardened. The resulting growth curve shows a steplike pattern (Fig. 13.16).

Growth does not occur at the same rate and at the same time in all parts of the body. It is obvious to anyone that the differences between a baby chick and an adult hen or rooster, or between a newborn baby and an adult human being, are differences not only in overall size but also in body proportions. The head of a young child is far larger in relation to the rest of its body than that of an adult, while a child's legs are much shorter in relation to its trunk than those of an adult. If the child's body were simply to grow as large as an adult's while maintaining the same proportions, the result would be a most unadultlike individual. Normal adult proportions arise because the various parts of the body grow at quite different rates or stop growing at different times (Fig. 13.17).

Two closely related species that differ in size are frequently also quite different in body proportions, not because of any basic difference in the growth patterns of the two species, but simply because a slight increase in overall body size automatically results in disproportionate increases in some parts of the body. In elk, for example, the size of the antlers increases much faster than the overall body size. Consequently, if species A grows slightly larger than species B and natural selection does not intervene, then A will have proportionately much larger antlers.

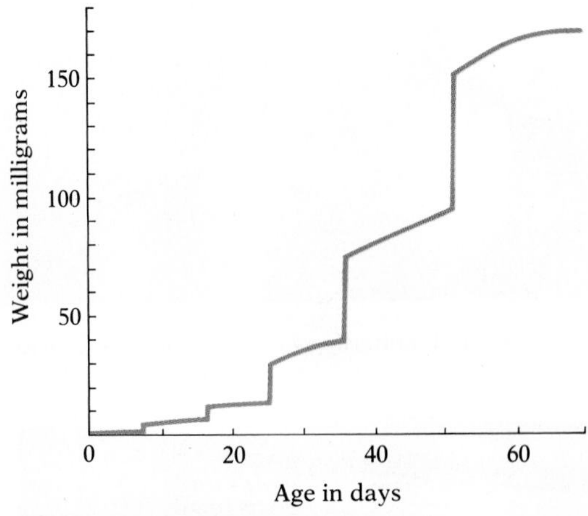

13.16 Growth in weight of an insect
The growth spurts of the water boatman, an aquatic insect, occur at the time of molt, when the old exoskeleton has been shed and the new one has not yet fully hardened.

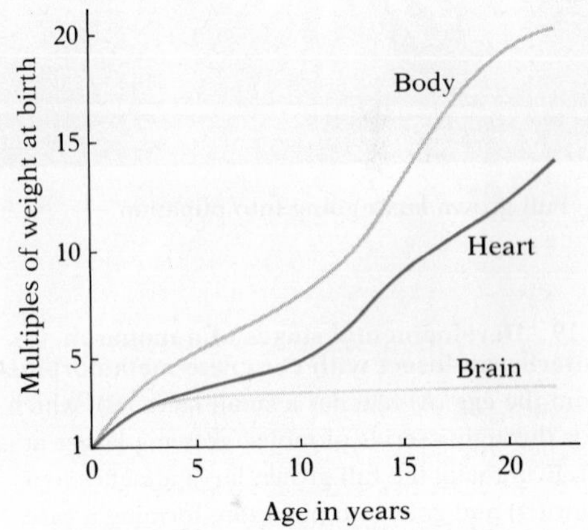

13.17 Graph showing differences in relative growth of body, heart, and brain of a human

LARVAL DEVELOPMENT AND METAMORPHOSIS

Growth in size is not always the principal mechanism of postembryonic development. Many aquatic animals, particularly those leading sessile (nonmobile) lives as adults, go through ***larval*** stages that bear little resemblance to the adult (Fig. 13.18). The series of sometimes drastic developmental

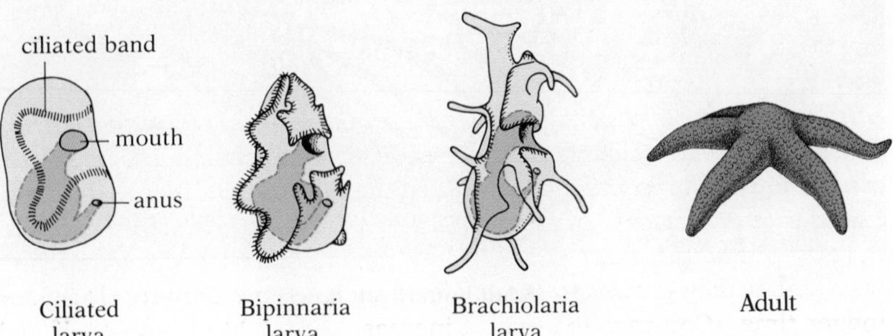

13.18 Three larval stages and adult of the sea star *Asterias vulgaris*
The gastrula develops into the ciliated larva, which changes into the bipinnaria larva, which changes into the brachiolaria larva, which metamorphoses into the characteristically shaped sea star, with five arms.

A Monarch butterfly egg

B First-stage larva eating its own egg shell

C Later-stage larva eating its own molted skin

D Full-grown larva going into pupation

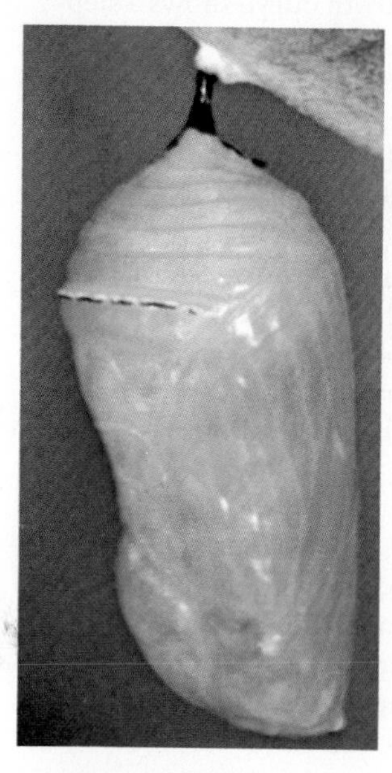

E Early chrysalis

F Late-stage chrysalis

G Inflation of wings by newly emerged butterfly

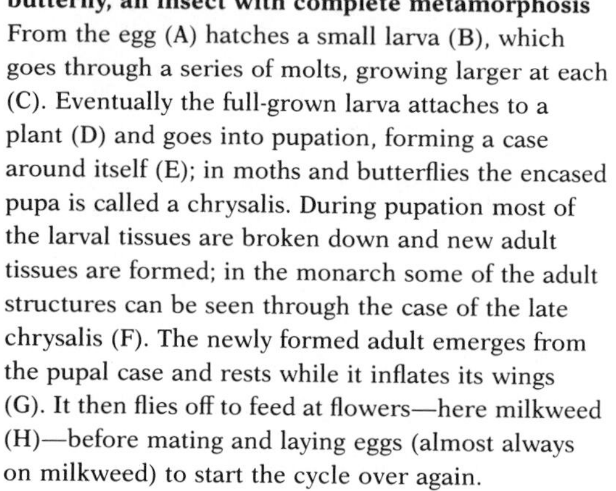

13.19 Developmental stages of a monarch butterfly, an insect with complete metamorphosis From the egg (A) hatches a small larva (B), which goes through a series of molts, growing larger at each (C). Eventually the full-grown larva attaches to a plant (D) and goes into pupation, forming a case around itself (E); in moths and butterflies the encased pupa is called a chrysalis. During pupation most of the larval tissues are broken down and new adult tissues are formed; in the monarch some of the adult structures can be seen through the case of the late chrysalis (F). The newly formed adult emerges from the pupal case and rests while it inflates its wings (G). It then flies off to feed at flowers—here milkweed (H)—before mating and laying eggs (almost always on milkweed) to start the cycle over again.

H Adult butterfly at flower

changes that convert an immature animal into the adult form is called ***metamorphosis***. It often involves extensive cell division and differentiation, and sometimes even morphogenetic movement; growth alone could not accomplish the transformation of a larva into an adult.

In many aquatic animals dispersal of the species depends on the larval stage; the tiny larvae either swim or are carried passively by currents to new locations, where they settle down and undergo metamorphosis into sedentary adults. In other species, such as frogs, where the adult is not sedentary, the adaptive significance of the larval stage (the tadpole in the case of frogs) seems less a matter of dispersal than of exploiting alternative food sources (tadpoles feed primarily on microscopic plant material, while adult frogs are carnivorous and take fairly large prey).

Though a larval stage occurs in the life history of many aquatic animals, probably the most familiar larvae are those of certain groups of terrestrial insects, including flies, beetles, wasps, butterflies, and moths. The young fly, wasp, or beetle is a grub that bears no resemblance to the adult. The young butterfly or moth is a caterpillar. In the course of their larval lives, these insects molt several times and grow much larger, but this growth does not bring them any closer to adult appearance; they simply become larger larvae (Fig. 13.19B–D). Finally, after they have completed their larval development, they enter an inactive stage called the ***pupa***, during which they are usually enclosed in a case or cocoon. During the pupal stage most of the old larval tissues are destroyed, and new tissues and organs develop from small groups of cells called ***imaginal discs*** that were present in the larva but never underwent much development. The adult that emerges from the pupa is therefore radically different from the larva; it is the product of an entirely different developmental program—almost a new organism built from the raw materials of the larval body (Fig. 13.19H).

Insects with a pupal stage and the type of development just described are said to undergo ***complete metamorphosis***. The sharp distinction between the larval and adult stages in such insects has meant evolution in two markedly different directions; in general, the larva is more specialized for feeding and growth, while the adult is more specialized for active dispersal and reproduction.

Complete metamorphosis is not characteristic of all insects. Many, such as grasshoppers, cockroaches, bugs, and lice, undergo ***gradual metamorphosis*** (Fig. 13.20). The young of such insects resemble the adults, except that their body proportions are different (the wings and reproductive organs, especially, are poorly developed). They go through a series of molts during which their form gradually changes and becomes more and more like that of the adult, largely as a result of differential growth of the various body parts. They have no pupal stage and experience no wholesale destruction of the immature tissues.

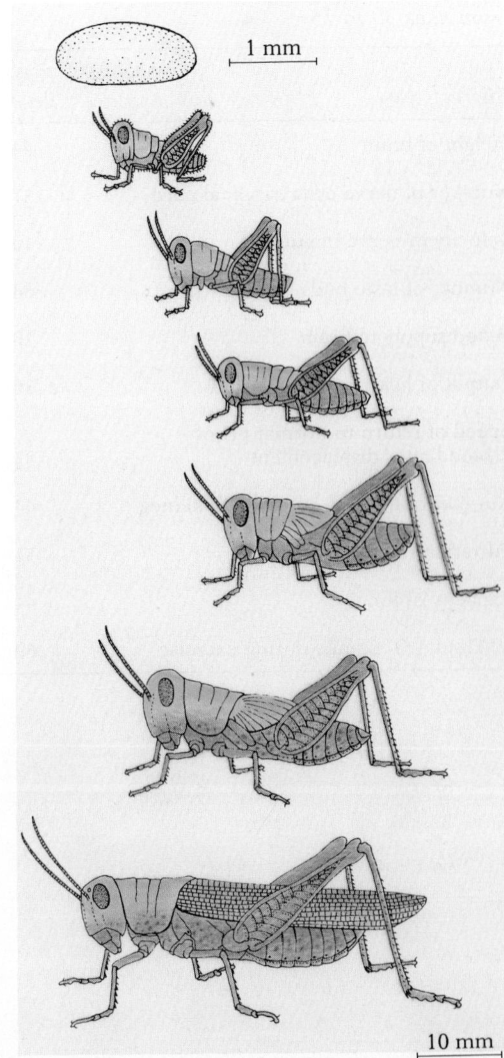

13.20 Gradual metamorphosis of a grasshopper The insect that emerges from the egg (top) goes through several nymphal stages that bring it gradually closer to the adult form (bottom).

AGING AND DEATH

Discussions of development often stop with the completely mature adult. But development in its full biological sense does not cease then. The adult organism is not a static entity; it continues to change, and hence to develop, until death brings the developmental process to an end.

TABLE 13.2 *Average decline in a human male from ages 30 to 75*

Characteristic	*Percent decline*
Weight of brain	44
Number of nerve cells in spinal cord	37
Velocity of nerve impulse	10
Number of taste buds	64
Blood supply to brain	20
Output of heart at rest	30
Speed of return to normal pH of blood after displacement	83
Number of filtering subunits in kidney	44
Filtration rate of kidney	31
Capacity of lungs	44
Maximum O_2 uptake during exercise	60

The term "aging" is applied to the complex of developmental changes that lead, with the passage of time, to the deterioration of the mature organism (Table 13.2) and ultimately to its death. For many years, little research was devoted to aging, but now it has become a major field of investigation. Modern scientific progress and improved medical techniques have greatly increased our ability to protect ourselves against disease, starvation, and the destructive forces of the physical environment. More and more people are living to an advanced age. And as the life expectancy of humans increases and the proportion of the population in the upper age brackets rises, the changes associated with aging become more obvious and more important to all of us.

Little is known about the factors involved in aging. The process seems to be correlated with specialization of cells for one or a few highly specific functions. Cells that remain relatively unspecialized and continue to divide do not age as rapidly (if at all) as cells that have lost the capacity to divide. Cancer cells, of course, divide continually and are essentially immortal. Bacteria and some other unicellular organisms cannot be said to age, for any cell that is not destroyed eventually divides to produce two young cells; division is thus a process of rejuvenation. Within the body of a multicellular animal, tissues like muscle and nerve, in which cell division has normally ceased, slowly deteriorate, whereas tissues like those of the liver and pancreas, in which active cell division continues, age much more slowly. Furthermore, animals such as the tortoise that grow as long as they live seem to show fewer symptoms of aging than, for example, mammals and birds, which cease growing soon after they reach maturity. Within a single species, individuals whose period of growth and development is slowed and extended by a very limited diet are usually older than normal before they begin to show signs of aging.

Clearly, then, the aging and death of individual cells and the aging and death of the multicellular organism as a whole are two rather different things. Paradoxical as it may seem at first, cell death is an essential part of life: the death of individual cells, as noted previously, plays an essential role in the development of an animal embryo and in the complete metamorphosis of some insects. And early death of individual red blood corpuscles and epidermal cells is entirely normal even in a young healthy mammal. Aging of the whole organism, therefore, is a matter not simply of the death of its cells, but of the deterioration and death of those cells and tissues that cannot be replaced.

What makes irreplaceable tissues age? We don't know. However, we do know some of the factors that contribute to aging:

1. The replacement of damaged tissue by connective tissue places an increased burden on the remaining cells in that tissue. When cells in most types of tissue die as a result of disease or injury, no new cells are formed to replace the ones lost. Wound healing in such cases involves growth of connective (scar) tissue, which serves as a patching material but cannot function like the original cells. As more and more irreplaceable cells die, the increased burden placed on the remaining cells of that type may contribute to their aging.

2. Changing hormonal balance—caused by a drop in the level of sex hor-

mones, for example—may disturb the function of a variety of tissues and perhaps cause them to function less well.

3. As cells become older, they tend to accumulate some metabolic wastes that they apparently cannot expel, and these wastes—particularly highly reactive products like hydrogen peroxide—may contribute to the eventual deterioration of the cells. At the same time, aging cells produce ever smaller amounts of the antioxidant molecules needed to detoxify these dangerous chemicals.

Though all these factors are doubtless involved in aging, they are not really explanations but symptoms. The real question is why these changes occur, and to this question scientists cannot as yet give a satisfying answer. Some investigators have suggested that somatic cells slowly cease to function and eventually die as a result of damage by radiation (particularly X rays and cosmic rays). However, all laboratory experiments indicate that radiation damage is greatest to actively dividing cells—the cells that age most slowly. Furthermore, the amount of radiation damage would be proportional to the chronological age of the cells, but we know that aging is a function of physiological age, not of chronological age—a five-year-old rat is physiologically very old indeed, and its tissues show pronounced symptoms of aging, whereas five-year-old tissues in a human being are not yet even mature.

Other investigators put major emphasis on intrinsic rather than extrinsic factors. They believe that the changes characteristic of aging are programmed in the genes just like the earlier developmental changes, and that, though extrinsic environmental factors doubtless influence aging, they do so only by speeding up or slowing down processes that would occur anyway. These processes may involve a decline in the production of important enzymes or an altered chemical balance or physical structure, with an ensuing loss of ability to perform certain functions; or they may involve development of autoimmune reactions (allergies against parts of the organism's own body) that result in destruction of essential tissues; or they may involve increased rupture of lysosomes and release of destructive hydrolytic enzymes within the cells.

A particularly interesting proposal is a variation of two already mentioned: that aging is a developmentally programmed termination of life that results from an organism's failure to repair somatic-cell mutations or other damage. According to this hypothesis, the different rates of aging in various species reflect different inherited capacities for DNA repair and the production of antioxidant molecules. It is known that long-lived species are the ones with high levels of DNA repair enzymes, and this pattern holds true for the concentration of antioxidant chemicals as well. Conversely, species with short life spans and early aging invest little in either DNA repair or cellular detoxification. If, as seems indicated by such data, aging is a genetically programmed trait, it must have originated and been maintained because of a selective advantage for a shorter rather than a longer average lifespan. As yet, however, no general explanation for why allowing itself to age and die should be to an individual's advantage has been formulated. The opportunity to synthesize concepts and data from molecular biology and evolution is one of the most exciting challenges of modern biology.

STUDY QUESTIONS

1. What evolutionary pressures and aspects of life history are likely to favor eggs with small versus large quantities of yolk? (p. 399)
2. What would be the consequence of having several sperm fertilize an egg? (pp. 399–400)
3. Why do you suppose sperm are small and streamlined? Why should selection not have favored male gametes that bring large quantities of cytoplasmic resources for the zygote to use? (pp. 399–400)
4. Why might the nervous system have evolved from ectodermal tissue? What does this suggest about the course of that evolution? (pp. 346–47)
5. Summarize the case for programmed aging. Explain what might be the adaptive value of such a system in a species whose life history you know well enough to be able to speculate on the value of early death. (pp. 353–55)

CONCEPTS FOR REVIEW

- Steps in fertilization of egg by sperm
- Pattern of early cell division and blastula formation
- Pattern of gastrula formation
- Pattern of neurula formation
- Role of somites
- Variations related to different amounts of yolk
- Patterns of development in insects: imaginal discs and metamorphosis
- Characteristics and theories of aging

SUGGESTED READING

Epel, D., 1977. The program of fertilization, *Scientific American* 237 (5). (Offprint 1372) *The numerous changes that occur in an egg cell as soon as a sperm cell reaches it.*

Gordon, R., and A. G. Jacobson, 1978. The shaping of tissues in embryos, *Scientific American* 238 (6). (Offprint 1391)

Hayflick, L., 1980. The cell biology of human aging, *Scientific American* 242 (1). (Offprint 1457)

Wassarman, P. M., 1988. Fertilization in mammals, *Scientific American* 259 (6). *On how eggs manage to be fertilized by only one sperm.*

Chapter 14

MECHANISMS OF ANIMAL DEVELOPMENT

single fertilized egg cell contains all the genetic information necessary to orchestrate the precise series of changes associated with the stages of development discussed in the last chapter, from a blastula, to a gastrula, to a neurula, to a mature multicellular organism, through aging, and even to death. It is now clear that the intricacies of eucaryotic gene control, which we discussed in Chapter 11, can account for the vast changes in gene expression that successful development must require. Throughout development, cells need to know their location in the organism, what their immediate neighbors are doing, and the organism's developmental age. The major sources of all this information, as we will see, are chemical signals that modify each cell's pattern of gene expression.

Our goal in this chapter is to explore how genes and the products they encode control the many characteristics of development evident during an animal's life: the differentiation of cells into ever more specific classes, the movement of cells at particular times, the formation of morphological patterns (morphogenesis) to create organs and limbs, and the repeating cycles of growth. Much remains to be learned about development, but the basic principles now seem clear, and the essential molecular mechanisms are being uncovered at an ever-increasing rate. In particular, the process of ***induction***, by which chemical signals alter a cell's DNA to specify or determine its developmental fate, is beginning to yield to analysis. We will see in broad terms how, in the process of ***cell differentiation***, each cell type follows a different avenue of determination leading to a distinctive cellular biochemistry and morphology. The principles that emerge from the molecular biology of development have much to tell us about the genetic bases of cancer and birth defects.

THE POLARITY OF EGGS, ZYGOTES, AND BLASTOMERES

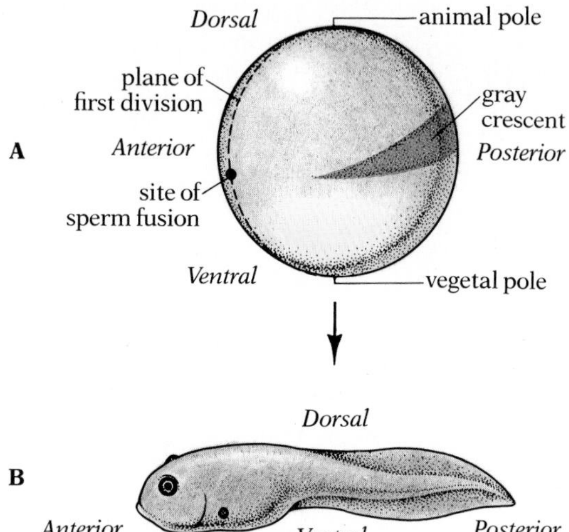

14.1 Polarity of the fertilized frog egg
After a sperm fuses with the egg, a gray crescent develops opposite the point of fusion (A). The crescent, in turn, helps define the anterior/posterior and dorsal/ventral axes of the tadpole (B). The first cleavage divides the embryo's left side from the right. Not all parts of the tadpole arise from corresponding sites on the egg: the anterior half of the egg contributes all of the ectoderm; cells developing from the posterior half fold back into the anterior and give rise to mesoderm.

The early steps in the development of an embryo establish the basic body plan, a kind of coordinate system by which the blastula will orient during gastrulation and further development will proceed. The egg, as we have seen, is often polarized into animal and vegetal hemispheres by the concentration of yolk—dense stored food—at the vegetal (lower) end. There are also crucial differences in the concentration of proteins and mRNAs between the two hemispheres that will, in time, lead the animal hemisphere in most species to become the dorsal (top) part of the organism, as opposed to the ventral (bottom) part; thus, the animal–vegetal polarity of the egg usually defines a dorsal–ventral axis. The second axis, the anterior–posterior axis, is often defined by the point at which the sperm penetrates the egg to create the zygote. As we will see, the cells formed in the subsequent early stages of cell division—the blastomeres—also play a role in the polarity of cytoplasmic substances.

Let us examine an actual example. As soon as the egg cell of a leopard frog fuses with a sperm cell, some of the contents of the egg shift position and a crescent-shaped grayish area appears on the egg opposite the point where the sperm entered (Fig. 14.1); these two points define the embryo's anterior–posterior axis. The material in this so-called ***gray crescent*** plays a very prominent role throughout embryonic development. The first cleavage of the frog zygote normally passes through the gray crescent, so that each daughter cell receives half (Fig. 14.2A). If these two cells are separated, each will develop into a normal tadpole, since each cell retains the information specifying both axes. But if the plane of the first cleavage is experimentally made to pass to the side of the gray crescent, the result of separating the daughter cells will be very different; the cell that contains the gray crescent will develop into a normal tadpole, but the other cell will form only an unorganized mass of cells (Fig. 14.2B). In other words, the way in which the material of the egg, especially the gray crescent, is distributed is of utmost importance to the developmental potential of the cells.

As the above experiment confirms, after the first cleavage of the frog zygote, both of the new cells are ***totipotent:*** they have the full developmental potential of the original zygote (since the polarized cytoplasmic substances are equally distributed in each) and all pathways of differentiation remain open to them. In this respect their development is characteristic of sea stars and many of their relatives, and of most vertebrates, including humans. Mammalian blastomeres generally remain totipotent until at least the eight-cell stage. In fact, eight-cell-stage embryos of two different strains of mice can be mixed to create a 16-cell ***chimera*** (after the Chimera, a monster in Greek mythology composed of parts from several different animals) that develops normally. This technique enables researchers to study how cells from one strain interact with those of another during development. That

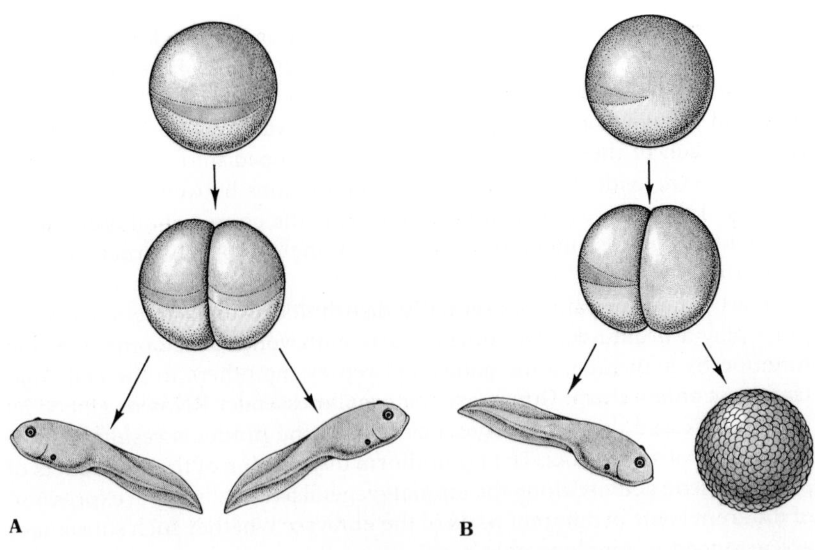

14.2 Importance of the gray crescent in the early development of a frog embryo

(A) If the two cells produced by a normal first cleavage, which passes through the gray crescent, are separated, each develops into a normal tadpole.
(B) If the first cleavage is experimentally oriented so that it does not pass through the gray crescent, and the two daughter cells are separated, the cell with the crescent develops normally, but the other cell develops into an unorganized cellular mass.

early human blastomeres should contain totipotent cells was expected, of course, since only animals with this kind of development can give birth to identical twins.[1]

In some other groups of animals, such as molluscs and segmented worms, the normal first cleavage partitions critical cytoplasmic constituents asymmetrically; therefore, when the daughter cells are separated, they do not have equivalent developmental potentialities. For example, in some molluscs, such as mussels and sea snails, a protuberance called the polar lobe develops on the fertilized egg cell just before the first cleavage occurs. The plane of cleavage is oriented in such a way that one of the two daughter cells receives the entire polar lobe (Fig. 14.3). If the two daughter cells are separated, the one with the lobe material (which was drawn back into the main body of the cell soon after division was accomplished) will form a normal embryo possessing two prominent structures—a so-called apical organ and the posttrochal bristles—that are seen on normal larvae; the one with no lobe material forms an aberrant embryo lacking these structures. Something in the polar-lobe material must be essential for formation of the apical organ and the bristles.

Sea urchins, which are related to sea stars, provide another example. If the animal and vegetal halves of the embryo are separated (along the plane of the third cleavage), the animal half will give rise to an abnormal blastula-like larva with overdeveloped cilia, and the vegetal half will give rise to a different type of abnormal larva with an overdeveloped digestive cavity

[1] Identical twins develop from the same zygote, usually as a result of a double gastrulation event. Nonidentical (fraternal) twins develop from separate zygotes when two egg cells are released from the ovaries at the same time and are fertilized by different sperm cells. Consequently, identical twins are genetically identical, whereas fraternal twins are no more alike genetically than any two siblings. Fraternal twins are much more common than identical twins.

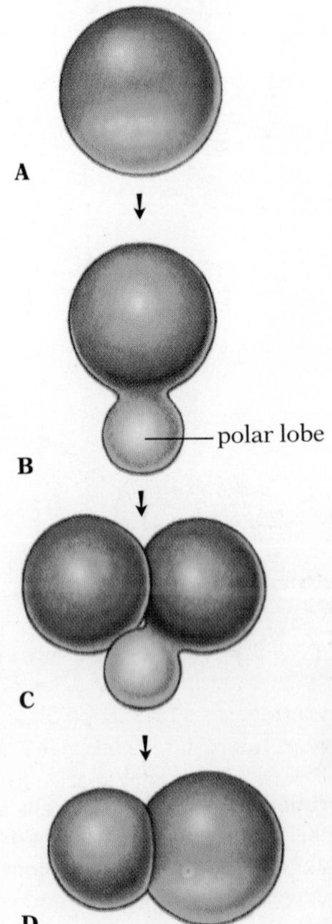

14.3 Cleavage of the fertilized egg of the sea snail *Ilyanassa*

(A) At one pole of the zygote is a region of clear cytoplasm (yellow). (B) Just before the first cleavage, this polar cytoplasm moves into a large protuberance, the polar lobe. The first cleavage partitions the zygote in such a way that the entire polar lobe goes to one of the daughter cells (C). The lobe recedes during interphase (D), but it will form again prior to the next division sequence. Only the cell that receives the polar-lobe material can give rise to a specific set of external structures that are seen in normal larvae.

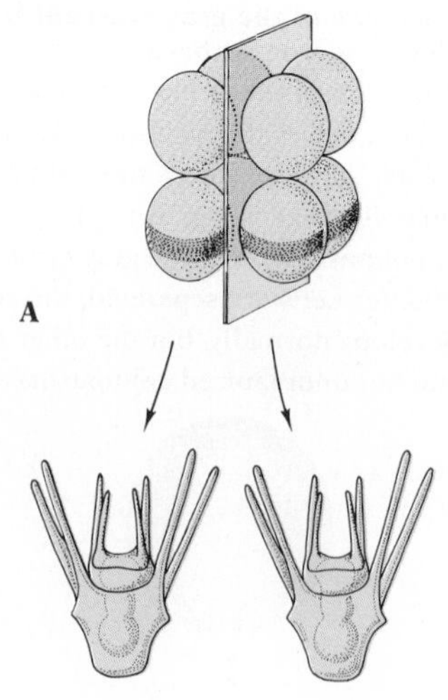

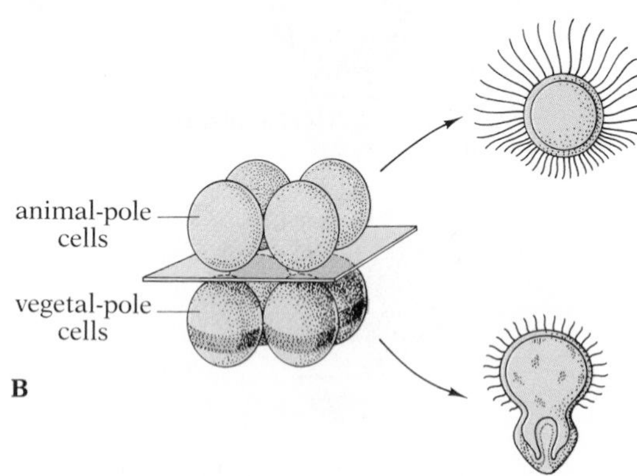

14.4 Experimental separation of cells after the third cleavage in the embryo of a sea urchin
(A) If the embryo is partitioned meridionally, so that each half receives both animal-pole and vegetal-pole cells, each half develops into a normal larva. (B) If the embryo is partitioned equatorially, so that one half receives only animal-pole cells and the other half only vegetal-pole cells, the animal half develops into an abnormal blastulalike larva with overdeveloped cilia, and the vegetal half develops into an abnormal larva with an overdeveloped digestive cavity.

(Fig. 14.4B). Cytoplasmic determinants promoting various kinds of differentiation are distributed along the animal–vegetal axis of the embryo. By contrast, if the eight-cell embryo is divided along its animal–vegetal axis, each half will develop into a small but normal larva (Fig. 14.4A). In an intact embryo each of these halves would have developed into only half of the normal larva; evidently, as in the frog, interactions between the different regions of the embryo normally act to modify the course of differentiation of neighboring cells, integrating them into a single, properly structured and proportioned larva.

Clearly, then, certain differentially distributed cytoplasmic substances must play a prominent role during early embryonic development. Some function by activating some genes and repressing others in the cells that come to contain them. Others are maternal messenger RNAs produced by genes expressed during oogenesis as well as the products resulting from translation of such RNAs. The nonuniform distribution of these products of previous gene activity along the animal–vegetal axis leads to the expression of different traits in different parts of the embryo. Whether such substances are arranged in a pole-to-pole gradient (as they apparently are in sea urchins) or in a less symmetrical pattern (as they are in fertilized frog and mollusc eggs, with their gray crescents and polar lobes), cells with different cytoplasmic components are produced at oogenesis, and further polarized early in embryonic development (at the first cleavage in some organisms, at the third cleavage or later in others).

INDUCTION IN EMBRYOGENESIS

INDUCTION AND THE DEVELOPMENTAL CLOCK

As we have seen, the first major event that absolutely requires an established polarity is gastrulation, in which one end of the alimentary canal (the anus in vertebrates, the mouth in insects) folds into the rest of the blastula. But there is another factor that needs to be taken into account: time. The embryo needs to know when the appropriate moment for this critical morphological alteration has arrived. Since, as we saw in Chapter 11, most cells stop growing after a fixed number of division cycles, cells must be able to "count"; thus it could be that the blastula somehow keeps track of the number of cell divisions until a particular number corresponding to the right developmental age has been attained. However, most evidence to date suggests instead that cell division proceeds in response to changing concentrations of chemical signals: developmental events are cued directly by the concentration of specific molecules, a kind of molecular developmental clock. Support for this idea comes from a variety of so-called heterochronic mutations that are known to cause developing cells to misread the molecular timer and produce structures at the wrong times.

Once the time has arrived for gastrulation in a frog embryo, cells in the region of the gray crescent, which runs horizontally across the posterior pole of the blastula at the animal–vegetal boundary, begin the process of involution. In the early gastrula the cells derived from the gray-crescent portion of the egg become the ***dorsal lip*** of the blastopore. These cells soon move inward and form the ***chordamesoderm***, which is at first located in the roof of the newly forming archenteron (see Fig. 13.6E, p. 345), but soon detaches from the roof of the archenteron to form the notochord and other mesodermal structures (see Fig. 13.6I, J). The chordamesoderm also exerts a very important influence on the ectodermal tissue lying over it; this tissue folds inward during neurulation, forming the neural tube, which differentiates into the brain and spinal cord. Neurulation does not occur if the chordamesoderm is missing. In fact, mesodermal tissue seems to play a dominant role throughout the embryo, migrating to new locations and inducing adjacent endoderm and ectoderm to differentiate.

In a series of classic experiments performed in 1924, Hans Spemann of the University of Freiburg, Germany, who had previously demonstrated the importance of the gray crescent in early cleavage, and his student Hilde Mangold turned their attention to the dorsal lip of the blastopore in the early gastrula stage of a salamander embryo. They transplanted the dorsal lip from its normal position on a light-colored embryo to the belly region of a darker-colored embryo. After the operation, gastrulation occurred in two places on the recipient embryo—at the site of its own blastoporal lip and at the site of the implanted lip. Although the original chemical polarity of the egg was characterized by a gentle gradient, only the dorsal lip sensed a signal strong enough to induce it into its role as the initiator of gastrulation; adjacent cells lacked this potential. Eventually two nervous systems were formed, and sometimes even two nearly complete embryos developed, joined together ventrally (Fig. 14.5). Most of the tissue in both embryos was dark-colored, an indication that the transplanted blastoporal lip had altered the course of development of cells derived from the host. Similar transplants of tissues from other regions of embryos failed to produce comparable results. The dorsal lip of the blastopore plays a crucial role even after it takes the lead in gastrulation. Signals from those cells as they make their way anteriorly under the dorsal ectoderm induce the formation of the neural plate, which in turn is important in establishing the longitudinal axis of the embryo and in inducing formation of other structures.

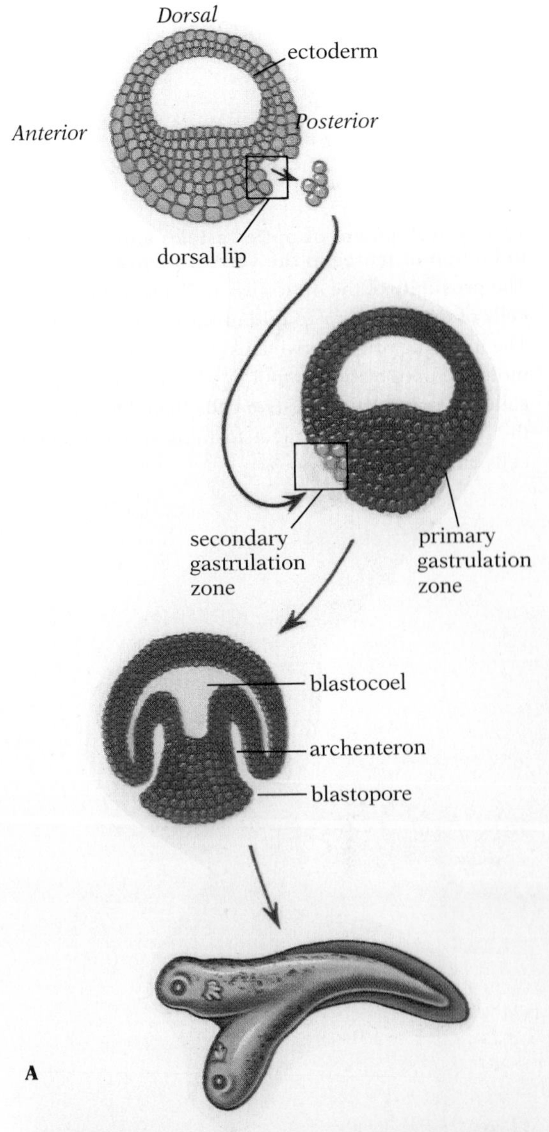

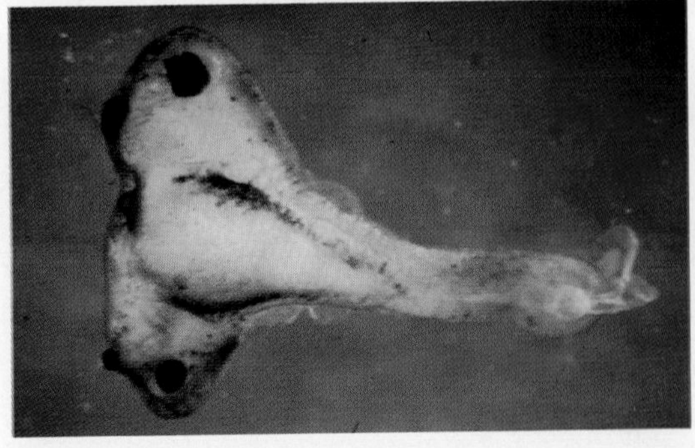

14.5 Spemann's experimental transplantation of the gray crescent
(A) When Mangold and Spemann transplanted dorsal-lip material from a light-colored salamander embryo to a dark-colored one, the blastula with two dorsal lips proceeded to form an embryo with two gastrulation zones—one at the original dorsal lip of the blastopore, and one at the transplanted dorsal lip. A double larva of mostly dark-colored tissue was the result. (B) A similar transplantation experiment in the frog *Xenopus laevis* produced an embryo with two body axes, including two heads and a second spinal cord.

INDUCTION OF ORGANS

Some of the first definitive studies on embryonic induction in an animal were performed in 1905 by Warren H. Lewis of Johns Hopkins University. Lewis worked on the development of the eye lens in frogs. In normal devel-

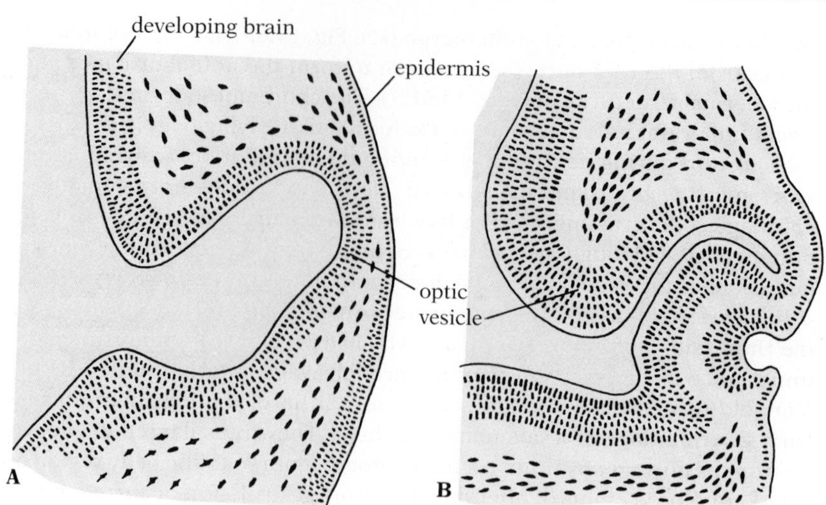

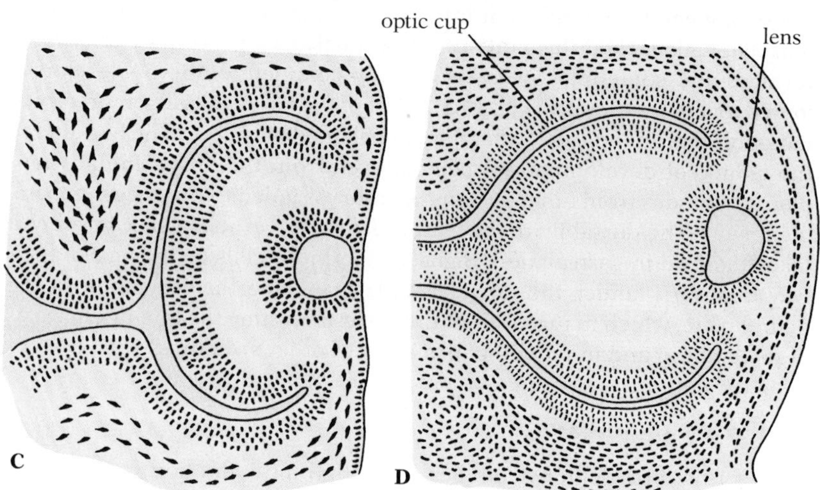

14.6 Development of optic vesicles and their induction of lenses in the vertebrate eye
The proximity of the optic vesicle induces the nearby cells of the epidermis to fold inward and form a lens. The infolding presumptive lens tissue, in turn, helps mold the optic vesicle into a two-layered structure called the optic cup. The cup cells then differentiate, the layer adjacent to the lens forming visual receptor cells and nerve cells.

opment the eyes form as lateral outpockets (or optic vesicles, as they are called) from developing brain tissue. When one of these outpockets comes into contact with the epidermis on the side of the head, the contacted epidermal cells promptly undergo a series of changes and form a thick plate of cells that sinks inward, becomes detached from the epidermis, and eventually differentiates into the eye lens (Figs. 14.6 and 14.7). Lewis cut the connection between one of the optic vesicles and the brain before the vesicle came into contact with the epidermis. He then moved the vesicle posteriorly into the trunk region of the embryo. Despite its lack of connection to the brain, the optic vesicle continued to develop, and when it came into contact with the epidermis of the trunk, that epidermis differentiated into a lens. The epidermis on the head that would normally have formed a lens failed to do so. Clearly, the differentiation of epidermal tissue into lens tissue depends on some inductive stimulus from the underlying optic vesicle. Later experiments have shown that if a barrier is inserted between the vesicle and the epidermis no lens develops.

Other experiments have shown that the regulation is not all one-way; the lens, once it begins to form, also influences the further development of the optic vesicle. If epidermis from a species with normally small eyes is transplanted to the sides of the head of a species with large eyes, the eyes that are formed do not have a large optic cup (formed from the optic vesicle) and a small lens, as might be expected. Instead, both the optic cup and the lens are intermediate in size and correctly proportioned to each other. Obviously, each influences the other as they develop together.

Of the many chemicals that can act as intercellular inducers in embryonic development, some play a so-called instructive role, others play a more permissive role. The ***instructive inducers*** restrict the developmental potential of the target cell and thus help determine the course of differentiation. The ***permissive inducers*** amplify potentialities already expressed. For example, a rudimentary organ may form fully committed cells during embryology, but will not complete its development and become functional until acted on by a permissive inducer.

We must stress that the so-called instructive inducers do not give their target cells specific instructions about the design of the tissues or organs they are to form; they instruct the cells only in the sense that, through repression of some genes and induction or derepression of others, they tell the cells what part of their genetic endowment they are to use. A dramatic example of this principle is the experiment in which Spemann and Oscar E. Schotté transplanted ectoderm from the flank of a frog embryo to the mouth region of a salamander embryo. They found that, though induced by salamander endoderm, the transplanted tissue formed the typical horny jaws of a frog: the constellation of genes for producing a jaw was activated by instructive inducers from the adjacent tissue in the salamander, but the induced genes, being from frog ectoderm, encoded the morphology of a frog jaw.

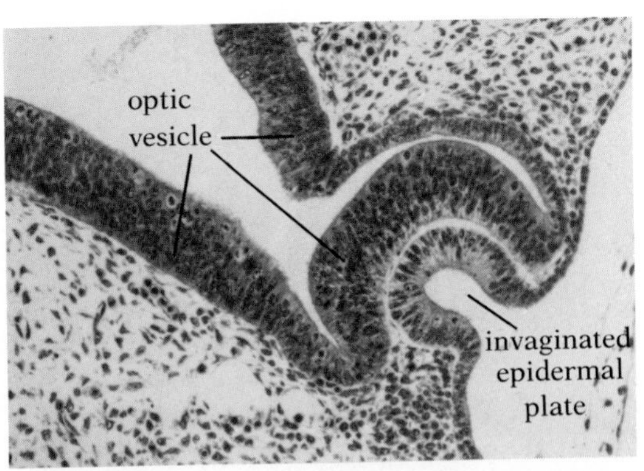

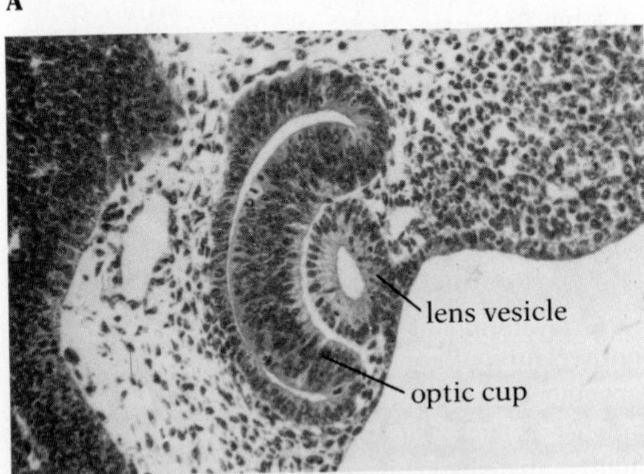

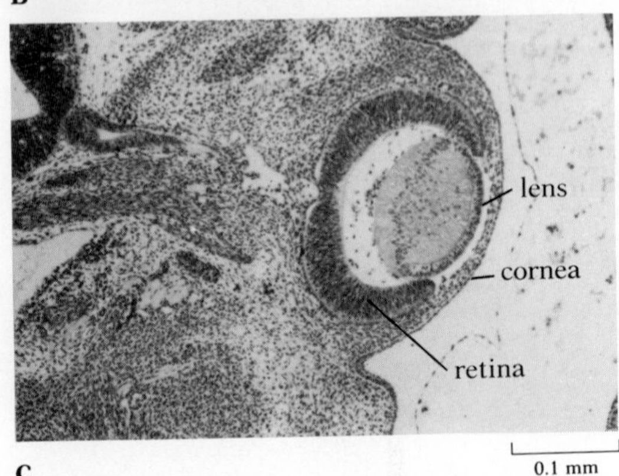

14.7 Development of a mammalian eye
Photographs of mammalian eye development over a four-day period: The contact with the tip of the optic vesicle causes the epidermis to invaginate (A). The epidermal region differentiates to form the lens vesicle, while the optic vesicle develops into the optic cup (B), which ultimately becomes the retina—the tissue at the rear of the eye that contains light receptors (C).

HORMONES AS INDUCERS

The body uses many internal chemical signals, called ***hormones***, to pass messages over relatively long distances, often through the circulatory system; testosterone (the male sex hormone) and adrenalin (an alerting hormone) are just two examples, both of which, along with several others, will be discussed in detail in Chapter 33.

In vertebrates hormones play a predominantly permissive role in development, helping differentiated tissue assume its full character, but they are not, on that account, any less important than instructive inducers. What good would it do an organism if cells were set aside as a presumptive tissue or organ (as a result of determination in response to instructive inducers) and their developmental potential could never be expressed for lack of a necessary permissive hormone? The essential interplay between instructive inducers and permissive hormones can be studied in the gonads of amphibians. The ***gonads*** are the primary sex organs—the sperm-producing ***testes***

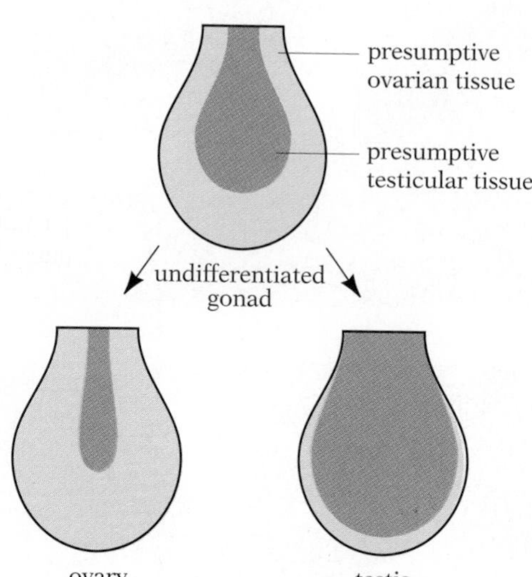

14.8 Development of ovary or testis from the undifferentiated gonad of an amphibian
Depending on the hormonal condition of the embryo, one or the other of the two types of tissue in the undifferentiated gonad gains developmental ascendancy and the other type is repressed.

of the male, and the egg-producing ovaries of the female. Two distinct kinds of cells are set aside early to become gonads in amphibians: peripherally located cortical cells and centrally located medullary cells (Fig. 14.8). The cortical cells have the potential for forming ovarian tissue, and the medullary cells for forming testicular tissue. Only when sex hormone acts on the sexually uncommitted embryonic gonad does one or the other of its potentialities gain ascendancy.

Similarly, the same embryonic primordia give rise to the accessory sexual organs of both sexes in humans (Fig. 14.9). Whether these primordia form male or female structures depends on whether or not male sex hormone is present in the embryo at the critical stages of embryonic development. In birds, the situation is reversed; the sexual organs will follow a masculine developmental course unless female hormone is present at the critical embryonic stages.

INDUCTION OF CELL MIGRATION

One of the most remarkable phenomena of development is the organized movement of cells and the folding of tissues into new shapes (morphogenesis). Dorsal lip cells crawl in the right direction (dragging other parts of the blastula behind them) and stop at the correct spot; cells of the optic vesicle "know" when to begin migrating outward toward the ectoderm, when they have reached their target, and then, how to create a suitable cup. We know that the blastula and the later embryo are chemically polarized so as to define morphologically important axes, but what molecules provide directional information to migrating cells, and how are the coordinates "read"?

The first hint of how cells sort themselves out has come from early tissue-disassociation experiments: if cells from two different developing organs—the liver and kidney, say—are separated from one another and the two groups mixed together, they slowly but accurately sort themselves out into two clumps. Under the microscope we can see the cells each extending several pseudopods, touching and adhering to neighboring cells, and then pulling themselves toward some and not others.

At the molecular level, we now know that each cell membrane is richly supplied with one or (usually) several kinds of ***cell-adhesion molecules*** (CAMs). Unlike the typical binder-receptor systems we have dealt with up to now, in which two different molecules—glucose and the glucose receptor, for instance—bind to one another, each kind of CAM attaches itself exclusively to other molecules of the same sort of CAM (Fig. 14.10), a phenomenon called ***homophilic binding***. The proportion of the several different classes of CAMs on the membranes of various liver cells is fairly similar, but does not match the corresponding ratio on spleen cells. As a result, liver cells stick slightly to spleen cells, but more firmly to their own kind. When a migrating liver cell attempts to withdraw its pseudopods, it drags itself toward the cells it is most strongly bonded to, and loses its tenuous molecular grip on spleen cells. It is this differential affinity that permits the two classes of organ cells to sort themselves out.

A very similar process may well occur when the dorsal lip cells begin to move. Their developmental clock (probably a rising concentration of a spe-

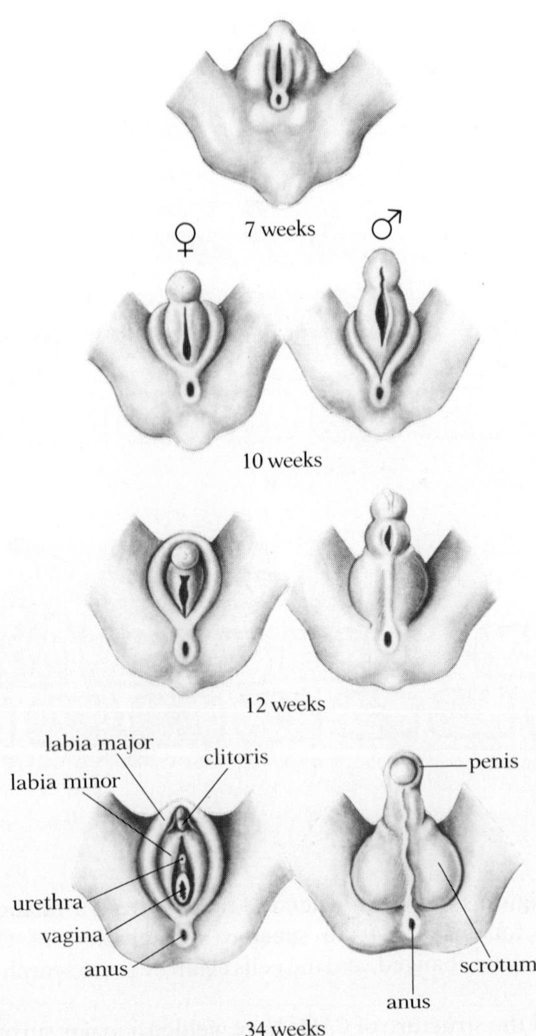

14.9 Development of the external genitalia of human beings
At seven weeks the genitalia of male and female fetuses are virtually identical. At 10 weeks the penis of the male is slightly larger than the clitoris and labia minor, which form from the same primordium in the female. At 12 weeks these differences are more pronounced, and the male scrotum has formed from the tissue that becomes the labia major in the female. At 34 weeks the distinctive features of the genitalia of the two sexes are fully apparent. It is largely the concentration of male sex hormones like testosterone that determines which of these developmental pathways will be followed.

cific chemical) tells them it is time for gastrulation, and causes a new set of CAMs to be mounted in the membrane. General chemical gradients may provide some initial orientation, and the presence of other cells or structures to attach to constrains the choice of routes available to migrating cells, but CAMs appear to be the decisive factor in determining among a cell's alternative routes once it begins to grow. Pseudopods begin to form and seek attachments, and find the strongest interactions in the anterior-dorsal direction. Pulled along by the discovery of even better matches farther forward, the dorsal lip cells drag themselves (and the posterior surface of the blastula) forward until their CAMs find the best matches available; when no direction offers the prospect of any stronger attachments, migra-

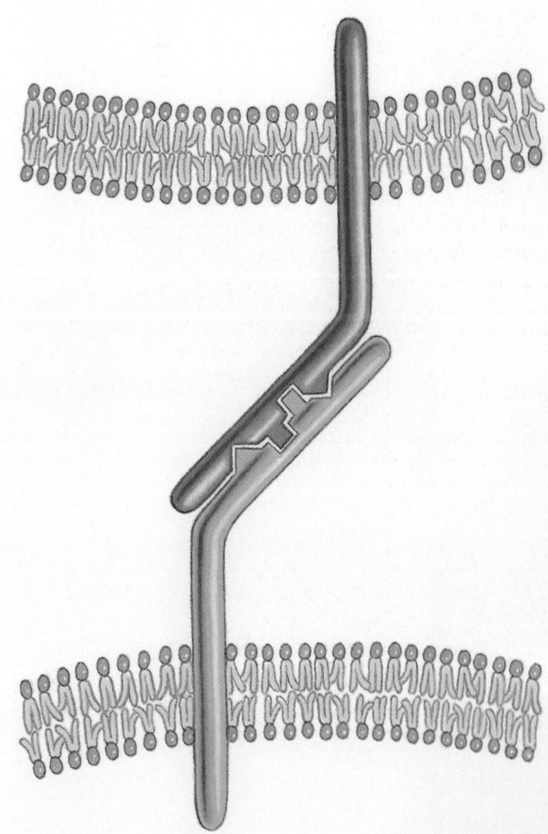

14.10 Homophilic binding of cell-adhesion molecules
The complementary structure of the arms of CAMs allows each type to bind to others of the same class. Several dozen kinds of CAM are thought to exist, though fewer than 10 have yet been characterized.

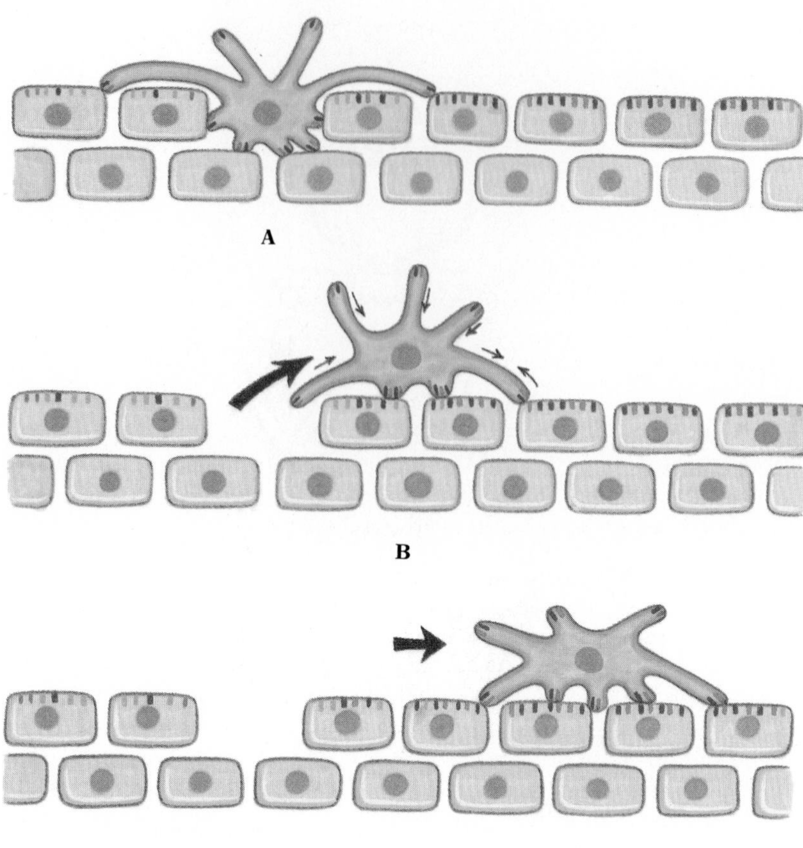

14.11 A model of cell migration
(A) A cell induced to begin searching for a better CAM match extends pseudopods and attaches more or less strongly to other cells. (B) When the pseudopods are periodically retracted, the cell is pulled in the direction of the best match. After many such steps (C), the cell reaches a point from which there is no available improvement in CAM matching.

tion stops (Figure 14.11). But when the time comes for further changes—neurulation, for instance—the specific constellation of CAMs in the membrane may be changed, and the cells again begin to search for compatible tissues.

Analyses of the structure of CAMs have yielded a major surprise: they are closely related to specialized proteins of the immune system. Since CAMs are found in the fruit fly *Drosophila* and in lower vertebrates, which lack an immune system, it seems likely that this developmental recognition system provided the basis for the evolution of the molecules of the immune system of mammals (described in the next chapter).

PATTERN FORMATION

Once the embryo has undergone gastrulation and neurulation, different organs begin to appear up and down its length. The strategy of further steps in development, from insects to humans, is one of subdivision of the embryo into a series of domains, followed by largely independent development of each domain. We will look first at how the domains are established, and then at the more local process of organ and limb formation.

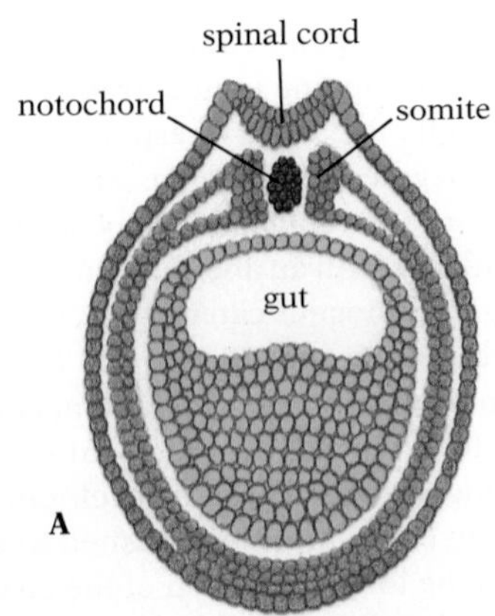

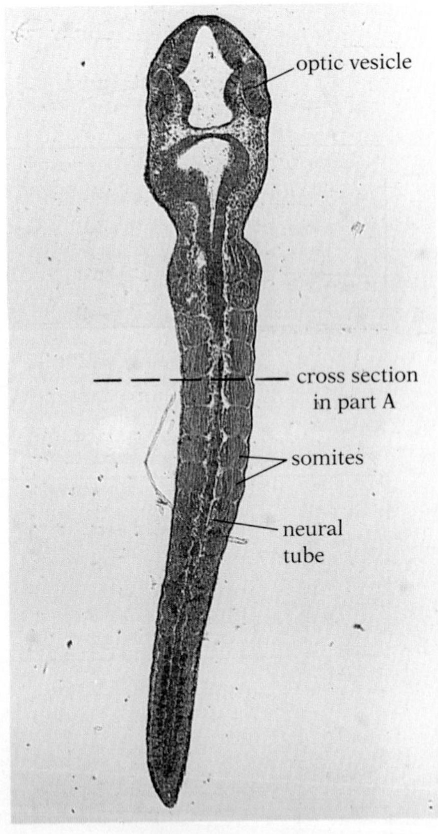

14.12 Somite formation
The dorsal portion of embryonic mesoderm (A) differentiates into a series of cell blocks, called somites, which give rise to vertebrae and dermis, as well as (depending on the somite) ribs and muscles of the limbs. (B) A photograph of an amphibian larva with somites clearly visible.

LONGITUDINAL SEGMENTATION OF THE EMBRYO

Vertebrate somites Once the spinal cord has fully formed in a vertebrate embryo, the most dorsal region of mesoderm begins to differentiate into blocks of tissue called ***somites*** (Fig. 14.12). This reorganization of the dorsal mesoderm is based both on an anterior-posterior chemical gradient, which triggers formation of the somites, and local interactions between somites, which assure proper grouping and segregation.

Each somite goes on to produce a vertebra, the ribs (if any) associated with it, the muscles unique to that vertebra (most notably, those serving the limbs), and the dermis (the layer of cells just below the epidermis). Each somite "knows" which set of bones, nerves, and muscles to construct on the basis of its anterior–posterior location, and becomes ***determined***—that is, committed to that fate. In general, positions along an axis are interpreted by the genes of somites and other units by measuring the concentration of one or more chemicals called ***morphogens***, which are secreted from a specific point in the embryo and (normally) diffuse away. (Thus the yolk in the animal–vegetal yolk gradient, though it provides a concentration axis, is not usually called a morphogen.)

Morphogens and segmentation in *Drosophila* The genetic and molecular bases of subdivision, and the use of morphogens for position finding, are best understood in *Drosophila*, where the segmentation in both the larva and adult is clearly visible externally. (In the adult, there are three segments in the head, three in the thorax—the middle section, to which the legs and wings are attached—and eight in the abdomen.) Two chemical axes are established in the blastoderm—one anterior–posterior, the other dorsal–ventral—which determine the overall organization of the larva and adult. The dorsal–ventral axis is established and locations on it are interpreted by a group of about 20 genes; another 30 or so genes are involved in the construction and reading of the anterior–posterior axis, and it is this longitudinal axis that generates segmentation.

Segmentation depends on gradients of two protein morphogens already being produced in the egg; one diffuses from the front toward the rear,

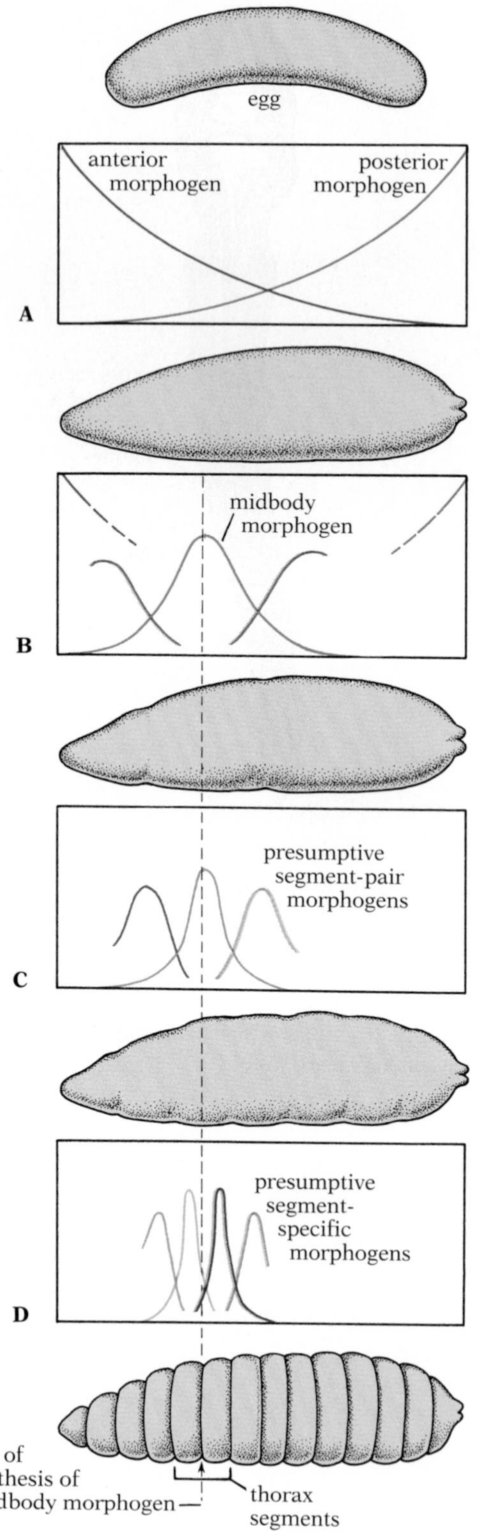

while the other (or perhaps a pair) emanates from the posterior end forward. The result is two overlapping concentration gradients (Fig. 14.13A). The genes of blastula cells apparently respond to the ratio of these two chemicals. Even before segmentation begins, if cytoplasm is removed from the anterior end of the egg and replaced with posterior cytoplasm, the larva develops without a head, but with an abdomen at each end. Mutants lacking one of these polarizing morphogens can be "rescued" through an injection of the absent chemical into the appropriate end of the egg.

Though the overlapping concentration gradients inform cells of their approximate location in the embryo, this information is not precise enough on its own to direct complete and accurate development. To aid in this goal, at least one intermediate landmark is established to provide a closer point of reference for cells in the long midregion of the larva. For example, a tiny band of cells about a third of the way back along the larva, stimulated by the appropriate ratio of anterior-to-posterior morphogens, begins synthesizing and releasing a different morphogen that will help organize the middle of the embryo (Fig.14.13B); other morphogens may be produced by bands elsewhere along the axis to aid development in those areas. The site of synthesis of the middle morphogen becomes the eventual location of the junction between the second and third thoracic segments, and this morphogen's localizing effects extend from the last head segment well into the anterior part of the abdomen. Once this band of cells begins synthesizing its morphogen, it inhibits neighboring cells from taking on the same task, thereby insuring that the resulting gradient will be well focused.

The next step in organizing the embryo requires the genes in the various cells to use the anterior and posterior morphogen gradients and the local morphogen to determine which of seven segment pairs the cell is located in (Figure 14.13C). (There are 14 segments in the mature larva.) Presumably this triggers the release of further chemicals, which enable cells to localize themselves to the anterior or posterior member of their segment pair (Figure 14.13D). Finally, a set of 10 or so genes sets to work polarizing each segment individually. The sequential and hierarchical organization of whole-body morphogens, followed by ever-more-local morphogens, produces a gradual, step-by-step determination of cells toward increasingly specific fates.

14.13 Model of morphogen concentration gradients responsible for segmentation in *Drosophila*

(A) Two morphogens important in establishing segmentation are produced in *Drosophila* eggs. One, encoded by the gene *bicoid*, is synthesized near the anterior end and diffuses toward the rear; the other, the product of the gene *oskar*, is generated at the posterior end and diffuses toward the front. (A second posterior morphogen may also exist.) The result is a pair of overlapping concentration gradients. (The gradient pictured for the posterior morphogen may not be this regular; there is some evidence that it is actively transported anteriorly before beginning to diffuse.) Additional gradients are created later in development. One, based on a morphogen encoded by the gene *Krüppel*, helps organize the middle of the larva (B). These more local morphogens allow the larva to divide itself into seven segments (C), which then give rise to the final complement of 14 (D), each of which then organizes itself internally.

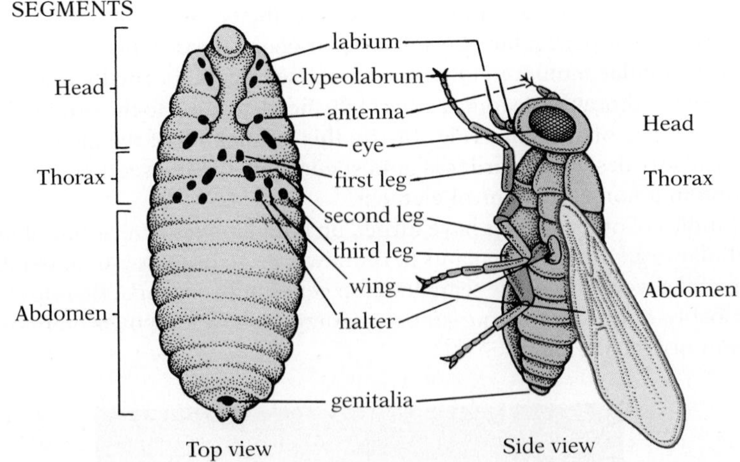

14.14 Segmentation of a *Drosophila* larva Induced by morphogens, homeotic genes in each segment give rise to imaginal discs in the larva that control the development of particular structures in the adult.

The roles of homeotic genes At this point, as segment identity is fixed and the polarizing coordinates established, particular members of a family of about 20 ***homeotic genes*** (so called because the mutant alleles of these genes alter a segment or structure so that it resembles another) are activated according to which segment they are in. The products of homeotic genes, like the morphogens that induced them, are generally control substances that bind to DNA and orchestrate the operation of many other genes; in this way, they cause each segment to express its unique character. Like the other control genes already mentioned, homeotic genes make permanent changes in a cell's genome that are passed on to its daughters when it divides—that is, they induce cellular determination. Unbound homeotic gene products have long since disappeared when the cells begin to differentiate, and their effects thus become visible.

The most dramatic consequence of the action of homeotic genes in insects is the creation of imaginal discs—those platelike clumps of cells we mentioned in the last chapter that differentiate during the pupal stage to form adult structures (Fig. 14.14). But though differentiation occurs in the pupal form, determination took place much earlier, in the larval stage, soon after segmentation, when homeotic genes first became active: if a presumptive leg disc is exchanged with an antenna disc, the resulting adult will have a leg on its head and an antenna on its thorax.

The homeotic genes are organized on the chromosome in two tight groups, the so-called bithorax and antennapedia complexes. A gene's position within a complex correlates with its site of action on the embryo: beginning with those in the antennapedia complex that operate at the extreme anterior end of the embryo, genes further along the chromosome operate on ever-more-posterior structures, and the pattern holds for the bithorax group. The logic of this organization is not yet understood.

Another intriguing discovery is the existence of a 180-nucleotide sequence called the ***homeobox*** in each of the homeotic genes, as well as in many other developmental-control genes. Although the exact sequence varies from gene to gene, a strong similarity is clear. Homeobox gene prod- ucts are transcription factors, which suggests that the variation in basic ho- meobox sequence accounts for the specificity of homeotic genes i activating the particular constellation of genes needed in each segment One particular mutation, for example, causes a gene normally transcribe only in the thorax to become active in the head, leading to the production o legs in place of antennae (Fig. 14.15); this mimicking of the effects of dis transplants described earlier is almost certainly the consequence of a mu tation in a homeotic control element.

Studies of other species have turned up homeoboxes in all kinds of multi cellular organisms, from plants to roundworms to humans. An understand ing of the workings of the segmentation genes in *Drosophila*, therefore, wil probably contribute to our understanding of how our own embryos g about organizing themselves.

A

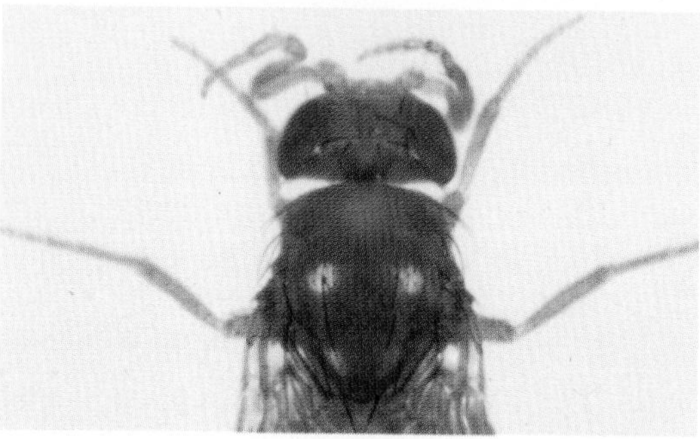

B

14.15 (A) Bithorax and (B) antennapedia mutations

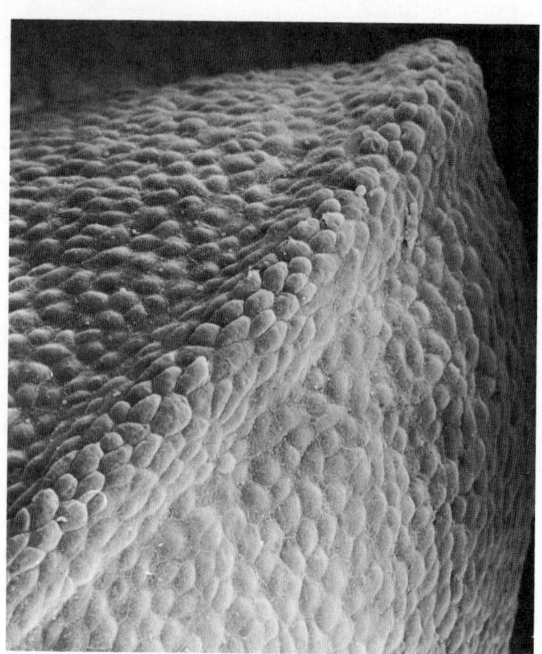

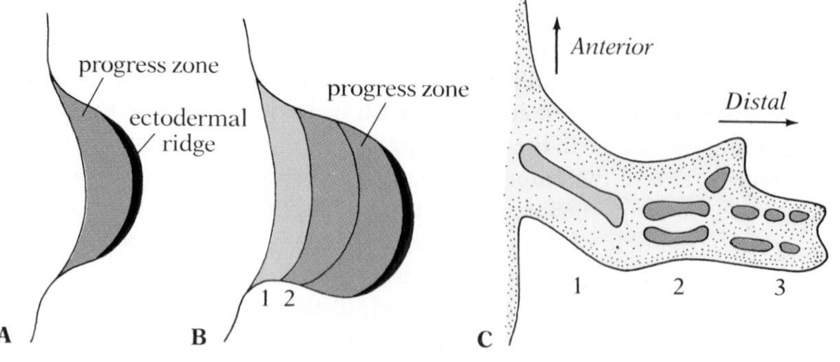

14.16 Pattern formation in the development of the chick wing
Above right: Scanning electron micrograph of the ectodermal ridge of a chick wing bud. Right: Diagram of three stages of development of the wing bud, according to Wolpert's model. (A) Just behind the ectodermal ridge is a progress zone, where new cells are produced. (B–C) The first band of cells derived from the progress zone will become the humerus section of the wing; the second band will become the section containing the radius and ulna; and a third band, derived from the progress zone late in the development of the wing bud, will become the distal part of the wing.

LIMB FORMATION

Once a particular somite or imaginal disc is told to help produce a limb or an organ, there comes the major developmental task of orchestrating its construction. Lewis Wolpert of Middlesex Hospital Medical School, London, has suggested a model to account for limb growth and patterning, based on his investigations of wing development in chicks. His model assumes a bicoordinate system: a proximodistal axis (from the body to the end of the extremity) and an anterioposterior (front-to-rear) axis.

The proximodistal coordinate Wolpert proposes is linked with the progress zone for development of the wing, an area associated with an ectodermal ridge running across the tip of the limb bud (Fig. 14.16A). New cells are produced in the progress zone and left behind as the area is pushed farther and farther away from the body. A cell's proximodistal positional values might be determined by the time it spent in this progress zone (Fig. 14.16B). Cells left behind by the progress zone very early in development of the wing would have a low positional value, which might cause them to develop into the basal part of the wing (the humerus portion; see Fig. 1.16); cells left behind a bit later would have an intermediate positional value, appropriate to formation of the middle part of the wing (the radius and ulna portion); and cells left behind late in the development, having spent a long time in the progress zone, would have a high positional value, appropriate to formation of the distal part of the wing. This information is somehow stored in the cells as they are left behind, enabling them to regenerate new limbs from any point of amputation. As to the anterioposterior coordinate, Wolpert proposes that a diffusion gradient of a morphogen (now widely thought to be retinoic acid[1]) secreted by a small group of cells at the rear margin of the wing bud polarizes the bud. With these two coordinates, the cells could "read" their position with sufficient accuracy to ensure their differentiation into an appropriate structure.

To test the first part of this model, Wolpert grafted the progress zone (the ectodermal ridge and the associated area of actively dividing cells) from an early wing bud onto the end of an older bud whose own progress zone had been removed. The result was development of a wing with two humerus sections and two radius-ulna sections (Fig. 14.17). As Wolpert interprets this result, the cells of the graft had no way of telling they were so far out on the wing that they should form only its distal parts; because they had spent so little time in the progress zone, some of these cells were left behind, programmed to read their position as being near the wing base, and developed into structures appropriate to such a position. The converse experiment of grafting the progress zone from an older bud onto an early bud produces a wing with only the distal parts, the phalanges: the transplanted tissue has no indication that the humerus, radius, and ulna have not yet developed. Having spent a long time in the progress zone, it develops structures appropriate to the wing tip.

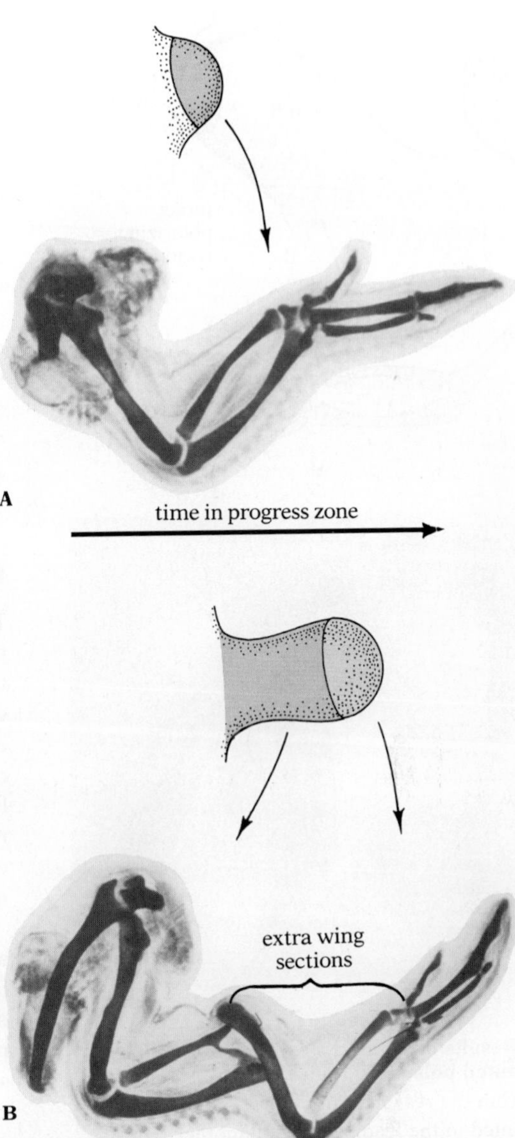

14.17 Results of Wolpert's experiment with a grafted progress zone of a chick wing bud (A) A normal wing developed from an intact wing bud. The tan area in the bud is the progress zone. (B) The wing that developed when an early wing bud (lighter tan) was grafted onto the original bud (darker tan) after the basal part of the wing had already begun to develop. The wing has extra humerus and radius-ulna sections.

[1] There is some evidence that retinoic acid is not the actual morphogen, but instead that it triggers the release of the true morphogen from the cells it enters.

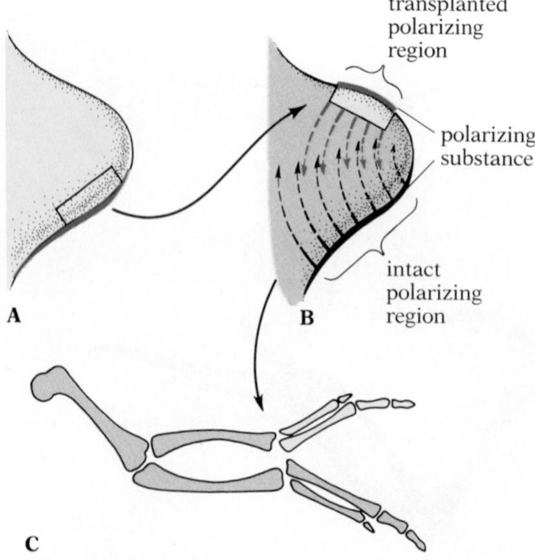

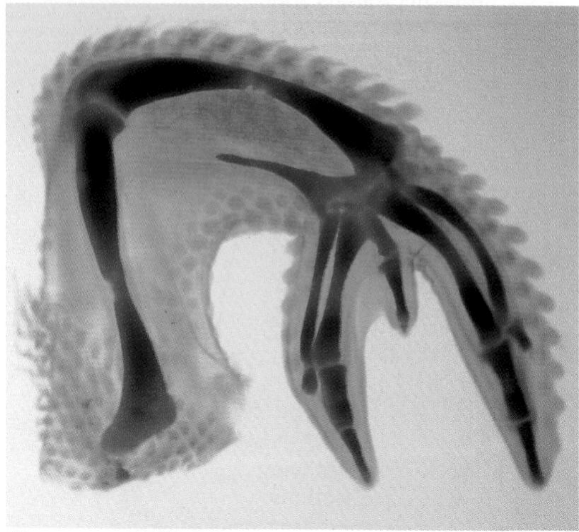

14.18 Results of Wolpert's experiment with a transplanted polarizing region
Tissue from the rear margin of one wing bud (A) is transplanted to the leading edge of another (B). There the morphogen diffusing from the transplant causes cells to differentiate into a second set of phalanges (C). The same effect can be achieved by treating the leading edge of a limb bud with retinoic acid (D).

To test the gradient part of the hypothesis, Wolpert transplanted part of the supposed polarizing region of the bud from the rear (posterior) edge to the front (Fig. 14.18). The results were dramatic: a partial set of mirror-image phalanges developed on the front edge. Their orientation to the transplanted part of the polarizing region was identical to the orientation of the normal set of bones to the polarizing tissue at the rear of the wing; evidently because the morphogen from the transplanted region polarized the front half of the bud, diffusing from front to rear, just as the morphogen from the intact polarizing region on the rear edge polarized the rear half of the bud, diffusing from rear to front. Note that because of the elevated total concentration of morphogen, the level was not low enough anywhere for the normal front digit to develop. Further experiments strongly support the gradient model: if the transplant is large, nearly a full set of reversed digits develops, but if only a small portion of tissue is moved, only a single digit is induced. Implants of retinoic acid mimic these transplants, and retinoic-acid receptors in wing-bud cells, once bound, are known to move to the nucleus and bind to the DNA, just as we would expect from a gene-control substance (Fig. 14.18D). Moreover, which homeotic gene is expressed correlates with the concentration of retinoic acid.[2]

DEDIFFERENTIATION AND REGENERATION

The usual sequence of development, as we have seen, is for a morphogen or other chemical (often from an adjacent cell) to act as an inducer, altering the pattern of gene expression of a cell. This cell has now become to some degree determined—that is, its range of potential specializations has become restricted. Subsequent experience with additional morphogens or other control substances further focuses a cell's fate, until it may become fully determined. At the same time, a cell's morphology and chemistry are responding to the activity of the genes in its nucleus, so that the cell differentiates (though often after considerable delay). Determination and differentiation, therefore, are usually gradual and proceed in that order.

But though differentiation usually follows determination, certain cells are designed to become differentiated *before* determination fully fixes their fate. When such cells are transplanted to a new location, they can lose their differentiated appearance and take on a morphology appropriate for their new location. In mammals and birds, dedifferentiation of cells is usually restricted to embryos. In amphibians, however, this ability to maintain incomplete determination in at least some cells into adulthood leads to remarkable powers of regeneration. For example, if the leg of a salamander is amputated, cells near the wound begin to dedifferentiate under the epidermis at the tip of the stump. Gradually, as mitotic activity and cellular redifferentiation take place within this area, the mound comes to look more and more like the normal limb bud of an embryonic salamander (Fig.

[2] Retinoic acid and its chemical relatives are widely used to treat acne. They work by stimulating rapid cell division, preventing to a large extent the formation of pimples, cysts, and scars. On the other hand, these chemicals leave treated cells highly sensitive to the carcinogenic effects of UV light, and can disrupt fetal development in pregnant users.

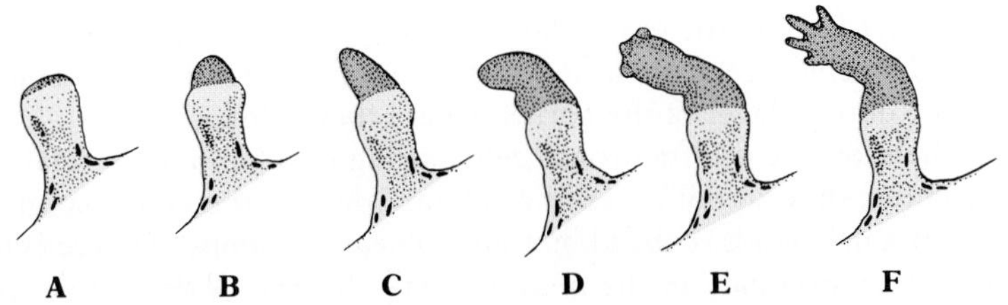

14.19 Regeneration of a salamander arm

14.19). It slowly elongates, and after several weeks a distinct elbow and digits appear, complete with muscle, tendon, bone, connective tissue, etc.

Dedifferentiation—which is to say, incomplete determination—is most common in animals that grow throughout their life cycles: species like many fish, reptiles, and amphibians that have no typical adult size, but instead simply grow larger and larger until they die. With the need to be able to grow indefinitely, it may be equally necessary to keep the developmental options of at least some cells open forever.

Although dedifferentiation is possible for some cells in some species, can determination itself be reversed? In fact, we have already seen that it can: cancerous cells suffer mutations that remove some of the determining constraints, and return a cell to what often appears to be an earlier, less differentiated, rapidly growing stage. What is possible through mutation is, in theory, also possible through selective removal or exchange of the DNA-binding control substances and transcription factors that are responsible for a cell's determined status in the first place. However, there are as yet few cases to suggest that a determined state can be altered. In one of the most promising experiments, J. B. Gurdon of Oxford University found that about 2 percent of nuclei taken from the intestinal cells of swimming tadpoles and injected into frog eggs from which the original nucleus had been removed could, in spite of their relatively determined state, direct the development of normal frogs. Possibly these few nuclei had suffered mutations that had dedetermined them, or perhaps there really is a small probability of spontaneous loss of determination, which in the normal course of events, with the cell surrounded by correctly determined cells, is self-correcting.

THE ORGANIZATION OF NEURAL DEVELOPMENT

Most aspects of development are illustrated with particular clarity in the development of the nervous system. Nerve cells, or ***neurons***, are "born," migrate to their proper places, send out fibers called axons to specific target locations, and so come to form a highly integrated functional network that is more complex than any other system in the body.

MIGRATION OF NERVE CELLS

The first step in the life of a newly formed presumptive neuron, like that of many other cells in a developing organism, is usually movement from where it was formed to where it is supposed to be. The cells that give rise to the retina, for instance, must grow out from the developing brain to where the eyes are to be located, and then form the optic vesicle, while the cells of

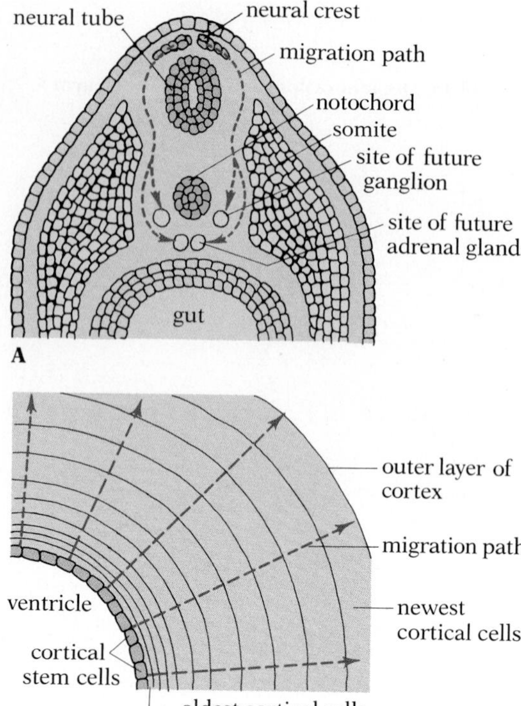

14.20 Migration of nerve cells
Before nerve cells join together in a network, many must move from their place of origin to their final location. (A) Cells from the neural crest move down and along glycoprotein filaments to positions at the future sites of spinal ganglia and the adrenal glands. (Some neural-crest cells encounter another set of filaments, leading to positions just under the ectoderm; cells following this pathway become pigment cells in the skin.) (B) New cells of the cerebral cortex are generated by a basal layer of stem cells lining each fluid-filled cavity, or ventricle, and migrate out along filaments to form a layer on the outside of the cortex. Cells produced later will move through this layer to positions still farther out.

the outer layer of the brain, the cerebral cortex, must move through layers of older cells to get from the core of the brain, where they were engendered, to the outside layer of the cortex, where they belong.

The movement of neuronal cells during development is relatively well oriented; they do not move at random. Consider the cells that give rise to the adrenal glands of the kidney and to nearby clumps of nerve cells (ganglia). They originate in the neural crest—the region just above the spinal cord—and migrate down to specific spots near the notochord (Fig. 14.20A). They move in the proper general direction from the outset, follow highly predictable pathways, and stop at precise spots. Similarly, cells of the cerebral cortex are generated in a basal layer of tissue and then migrate outward through other, older cortical cells until they reach the outer layer of the cortex (Fig. 14.20B). The three stages the neural-crest and cortical cells pass through—determining the initial direction, following a path, and determining the stopping point—are characteristic of the migration of developing cells. At least three mechanisms appear to be involved, utilizing diffusing chemicals, cell-adhesion molecules, and tactile cues.

The gradients of the diffusing chemicals help guide neuronal cells by causing them to move in an amoeboid fashion toward the source of the chemical, leading neural-crest cells "down," cortical cells "outward," and other cells in an anterior or posterior, or dorsal or ventral, direction. But most path-finding appears to involve the other two mechanisms—CAMs and tactile cues. Neural-crest cells (guided by filaments of glycoproteins that lead around the spinal cord and past the notochord) and cortical cells (following a radial array of filaments up through the cortex) apparently recognize their respective pathways by means of the particular ratios of CAMs they encounter; they partially envelop the guide cells, and then move along them, maintaining intimate tactile contact, finding ever-better matches between self- and substrate-borne CAMs.

When a migrating cell encounters the optimum CAM correspondence on the cells it has touched, it stops moving and proceeds to form the cell-to-cell attachments that will anchor it in place. The molecular specificity of nerve-cell surface markers is so precise that each class of neurons (and as few as two cells in an entire animal can constitute a class) has distinctive molecules in its membrane; this may mean that surface chemicals more specific than CAMs are involved in the final stages of neuronal wiring.

FORMATION OF AXONS AND SYNAPSES

Once a neuronal cell has reached its permanent place in the nervous system, it must send axons (the long, thin processes specialized for transmitting information) to specific target cells. Here again, both chemical and tactile information seem to play a role. The leading edge of the developing axon displays an unusual type of structure known as a ***growth cone*** (Fig. 14.21), first described almost a century ago by the great Spanish histologist Santiago Ramon y Cajal. The growth cone continually extends and retracts spikelike pseudopods called filopodia that probably sample the environment for specific chemicals and for the actual presence of certain guide cells. If the cone encounters the chemical or tactile stimulus of such a guide cell (usually the axon of another nerve), it partially envelops it and grows along it.

The evidence for a chemical "stepping-stone" strategy, first proposed by Roger Sperry of the California Institute of Technology, is especially compelling in the growth of axons. Many neuronal cells, once they have reached their destinations, send axons by circuitous but predictable routes to target cells. Rearrangement of the cellular "terrain" through which the axons must travel can cause equally predictable rerouting. For example, in mice, the axons from the visual area of one part of the brain normally extend to layer 4 of the visual cortex (Fig. 14.22A). We might suppose that the axons simply grow radially through the lower layers of cortex until they encounter the specially marked cells of layer 4, but this is not what happens. In a strain of mice affected by the so-called reeler mutation, in which the cortical layers are inverted, the axons grow up through the inverted cortex, continue past their targets in layer 4 until they encounter some essential chemical marker in layer 6, and then turn and migrate back through the cortex to layer 4 (Fig. 14.22B). This suggests that these axons are programmed to find layer 6 of the cortex first, and then to move to layer 4.

In vertebrates, as many as a million axons may project information from the body to one of several large arrays of target cells like those found in the visual, auditory, and tactile areas of the cortex. These axon-target cell connections are spatially organized and genetically predetermined. It seems unlikely that 10^6 specific molecular labels exist in the visual system to assure that all axons find their correct target cells. Nor does it seem plausible to imagine that a simple bicoordinate morphogen gradient could be sufficiently precise. Most researchers believe that the initial wiring is actually not very exact, and that the final pattern is determined dynamically through a process called neural competition.

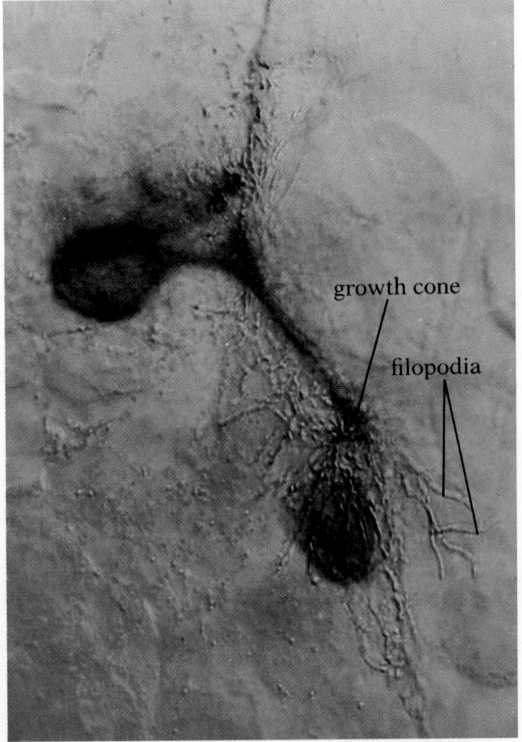

14.21 Growth cone of a developing axon in a grasshopper
The filopodia have made contact with another neuron (lower right).

CELL DEATH AND NEURAL COMPETITION

In most animal nervous systems, many more cells are born than are actually put to use. In vertebrates, for example, identical ganglia containing vast numbers of cells develop next to each of the vertebrae. And yet only the ganglia serving the many muscles and sensory receptors of the arms and legs require so many cells; in the other ganglia, the extra cells die. Apparently it is easier or more efficient in some important way for the develop-

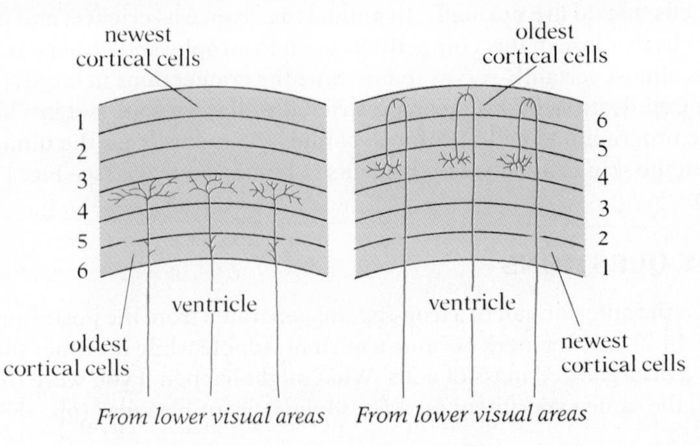

14.22 Connections to the visual cortex
(A) In normal mice the axons extend directly to their targets in layer 4. (B) In reeler mice, whose cortical layers are inverted, the axons travel out to layer 6 and then return to layer 4. This pattern suggests that layer 6 is a necessary intermediate landmark for the development of these axons.

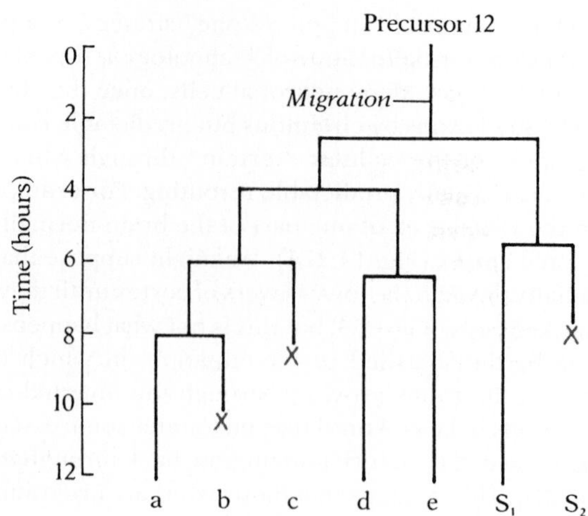

14.23 Cell death during the formation of nematode ganglion 12
This family tree of the cells in the last ganglion of a nematode suggests that the developmental program for the neurons involves a fixed series of divisions regardless of which neurons are needed by the ganglion. For example, the ganglion at the tip of the tail (illustrated here) does not need cells b and S_2, which in other ganglia connect to the next ganglion to the rear. Cell c is also unnecessary except in the middle ganglia of males, where it controls some of the animal's reproductive behavior. Nevertheless, apparently cell a can be produced only if cells b and c are produced as well.

mental program to build all segments alike initially, and then to allow functionless cells to die.

In nematodes (roundworms), which are the subject of extensive developmental research, a similar pattern can be readily observed. Their simple nervous system consists of a long ventral nerve cord containing 12 ganglia. Each arises during development from one of 12 precursor cells, each part of a clump of embryonic cells. They mature to control rhythmic swimming movements and mating. The pattern of cell division is the same in all 12 ganglia even though the eventual organization of the ganglia may be very different. Cells that become neurons in only some (or even just one) of the ganglia are nevertheless formed in all 12, and those not needed are then allowed to atrophy (Fig. 14.23).

An analogous phenomenon occurs in many animals at the level of neuron–neuron connections: many more of the axonal endings specialized to communicate neural activity from one cell to another—structures called synapses—are formed than survive. When cells wired to a particular target are prevented from responding to external sensory stimuli (either by experimental manipulation or because the cells are misplaced or defective), their synapses seem to be crowded out on the target neuron by synapses from other cells that do fire normally. In animals as diverse as crickets and mammals, cells that lose in this competition seem to atrophy and disappear. This process almost certainly serves to fine-tune the connections in large spatial arrays. Cell death is by no means restricted to the nervous system: highly specific programmed cell attrition separates the radius from the ulna, and culls out the skin cells between the digits in human embryos (see Fig. 13.11, p. 348).

STUDY QUESTIONS

1. When the anterior half of a frog egg was separated from the posterior half (Fig. 14.2), one segment became a normal tadpole while the other turned into a disorganized mass of cells. What might happen if you were to perform the same experiment on a *Drosophila* egg or blastula? (pp. 358–60, 367–68)

2. How does the overlapping two-gradient strategy of *Drosophila* allow for greater accuracy in reading anterior-posterior location than the use of a single morphogen? Is there any evidence that *Drosophila* need this degree of precision? (pp. 367–68)

3. Can you think of any reason why it might make sense to arrange the homeotic genes in anatomical order? (pp. 369–70)

4. Think of five different ways unnatural chemicals in the diet of a pregnant mammal might disrupt the development of an embryo.

5. Why might incomplete determination be essential to organisms with indefinite growth? Why might incomplete determination be a disadvantage to species with fixed growth—species that, most often, live longer and grow larger? (pp. 372–73)

CONCEPTS FOR REVIEW

- Induction versus determination versus differentiation
- Determinants of polarity in eggs
- Determinants and visual signs of polarity in zygotes
- Induction in gastrulation and neurulation
- CAMs and cell migration
- Morphogens and the formation of somites or segments
- Morphogens and the formation of limbs
- Dedifferentiation as a mechanism for regeneration
- Role of cell death in developmental programs

SUGGESTED READING

BRYANT, P. J., S. V. BRYANT, and V. FRENCH, 1977. Biological regeneration and pattern formation, *Scientific American* 237 (1). (Offprint 1363) *On basic principles of the organization and growth of complex structures in animals.*

COOKE, J., 1988. The early embryo and the formation of body pattern, *American Scientist* 76, 35–41. *A wide-ranging review looking for common mechanisms.*

COWAN, W. M., 1979. The development of the brain, *Scientific American* 241 (3). (Offprint 1440) *At the peak of brain growth, hundreds of thousands of neurons are added each minute, and yet they are wired together correctly.*

DEROBERTS, E. M., G. OLIVER, and C.V.E. WRIGHT, 1990. Homeobox genes and the vertebrate body plan, *Scientific American* 263 (1). *Application of the principles from* Drosophila *development to vertebrates.*

EDELMAN, G. M., 1984. Cell-adhesion molecules: A molecular basis for animal form, *Scientific American* 250 (4). (Offprint 1549) *On the likely molecular basis of cell-to-cell adhesion and changes in adhesion during embryonic development.*

EDELMAN, G. M., 1987. Topobiology, *Scientific American* 260 (5). *More on cell-adhesion molecules, and their evolutionary connection with the immune system.*

GARCIA-BELLIDO, A., P. A. LAWRENCE, and G. MORATA, 1979. Compartments in animal development, *Scientific American* 241 (1). (Offprint 1432) *An excellent discussion of imaginal discs and insect development.*

GEHRING, W. J., 1985. The molecular basis of development, *Scientific American* 253 (4). *Focuses exclusively on* Drosophila *development, with a nice discussion of homeotic mutations.*

GIERER, A., 1974. Hydra as a model for the development of biological form, *Scientific American* 231 (6). (Offprint 1309) *On the physiochemical basis of pattern development.*

GOODMAN, C. S., and M. J. BASTIANI, 1984. How embryonic nerve cells recognize one another. *Scientific American* 251 (6). (Offprint 1556) *An excellent description of how axons of invertebrates employ the stepping-stone strategy, following first gradients and then one preexisting axon after another to reach their targets.*

GURDON, J. B., 1968. Transplanted nuclei and cell differentiation, *Scientific American*, 219 (6). (Offprint 1128)

LEVI-MONTALCINI, R., and P. CALISSANO, 1979. The nerve-growth factor, *Scientific American* 240 (6). (Offprint 1430) *About the best-understood molecule important in creating chemical gradients for axon growth and development.*

STENT, G. S., and D. A. WEISBLAT, 1982. The development of a simple nervous system, *Scientific American* 246 (1). (Offprint 1508) *On how the nervous system of the leech is organized and wired up during development.*

WOLPERT, L., 1978. Pattern formation in biological development, *Scientific American* 239 (4). (Offprint 1409) *Wolpert's model for pattern development, based on studies of the chick wing.*

Chapter **15**

IMMUNOLOGY

e all know that once we have had diseases like measles and chicken pox, we cannot contract them in the same form a second time. This immunity develops because the complex set of cells that fights the disease while it is present has acquired the power to recognize and destroy the disease-causing entity far more rapidly in the future. The number of foreign cells and unfamiliar chemicals the immune system can "learn" to recognize—including many that cause disease—is essentially infinite, yet only a few genes are actually involved in encoding the vast number of different proteins, each of which binds specifically to a particular target. How does such a remarkable line of molecular defense work? How could it have evolved, and how can its incredible specificity be generated by only a few genes? The answers, as we will see, take advantage of the cell-adhesion and cell-surface marker molecules we encountered in the last chapter, and introduce one of the most powerful mechanisms of genetic variation—a theme to be treated at length in Chapters 16, 17, and 18. We will also see how immunological disorders like AIDS produce their effects, and will look briefly at current approaches to treating this disease.

THE IMMUNE RESPONSE

Nearly all animals have phagocytic cells that ingest bacteria and dead cells. These ***macrophages*** are attracted by chemicals released by damaged tissue and many foreign cells; they create a localized inflammation where they are active, as well as a liquid mass of dead cells and other debris (commonly called pus). This slow and relatively unselective line of defense has been elaborated in vertebrates into a highly specific immune system that is probably an adaptation to larger body size and longer life. As we will see, many receptors involved in the immune response clearly evolved from the receptor proteins used to guide the movement of cells during development and to mediate selective cell adhesion in tissues. The best-known receptor molecules of the immune system are the ***antibodies***, which bind to molecules that are foreign to the organism. These foreign substances, collectively called ***antigens*** (from "antibody generating"), are almost always large molecules (usually proteins or polysaccharides). The antigens may be free in solution, as are the toxins secreted by some microorganisms, or built into the outer surfaces of viruses or foreign cells like bacteria and pollen. The antigens stimulate certain cells in the immune system to produce highly specific antibodies—proteins that bind to these antigens exactly as enzymes bind to reactants. The binding of antibodies to toxins inactivates them, often by simply covering the toxin's active site, and therefore altering their toxicity. Antibodies inactivate viruses by binding to the receptors by which they recognize their hosts. When antibodies bind to microorganisms, however, they target these pathogens for subsequent destruction by one of several mechanisms to be described below. The human immune system recognizes as foreign not only nonhuman cells but also nearly all cells from other individuals; this explains why successful organ transplants are so difficult. The vast number of potential antigens means that literally millions or perhaps even billions of different antibodies must exist for the immune system to function properly.

The first vaccines against disease Immunologic reactions have been the subject of much study since the English physician Edward Jenner discovered in 1796 that people develop immunity to smallpox if they are injected with material that induces a very mild form of the disease (actually cowpox). This was the first vaccine (the term is from the Latin *vacca*, "cow"). Further dramatic demonstrations of the immune reaction were made by Louis Pasteur in France during the latter half of the nineteenth century.

The object of one of Pasteur's many investigations was a disease of cattle and sheep called anthrax, which was ravaging the herds of Europe at that time. Persuaded that a certain type of bacterium caused the disease, he exposed bacteria of this type to temperatures that were high enough to weaken but not kill them. When he injected these weakened bacteria into healthy sheep, the sheep became slightly ill, but thereafter they exhibited immunity to further infection by this disease. To convince the skeptics of his day, Pasteur arranged a demonstration attended by his most influential contemporaries. With these as witnesses, he injected weakened bacteria into 25 sheep, leaving 25 others uninjected as controls. Several weeks later, with the witnesses again assembled, he gave all 50 sheep a massive injection of fully active bacteria—more than enough to kill any normal healthy sheep. A few days later, all 25 control sheep were dead, while all 25 of the treated sheep were alive and healthy.

We know now that Pasteur's vaccinated sheep reacted to the antigens of the weakened anthrax bacteria by producing antibodies against them. The

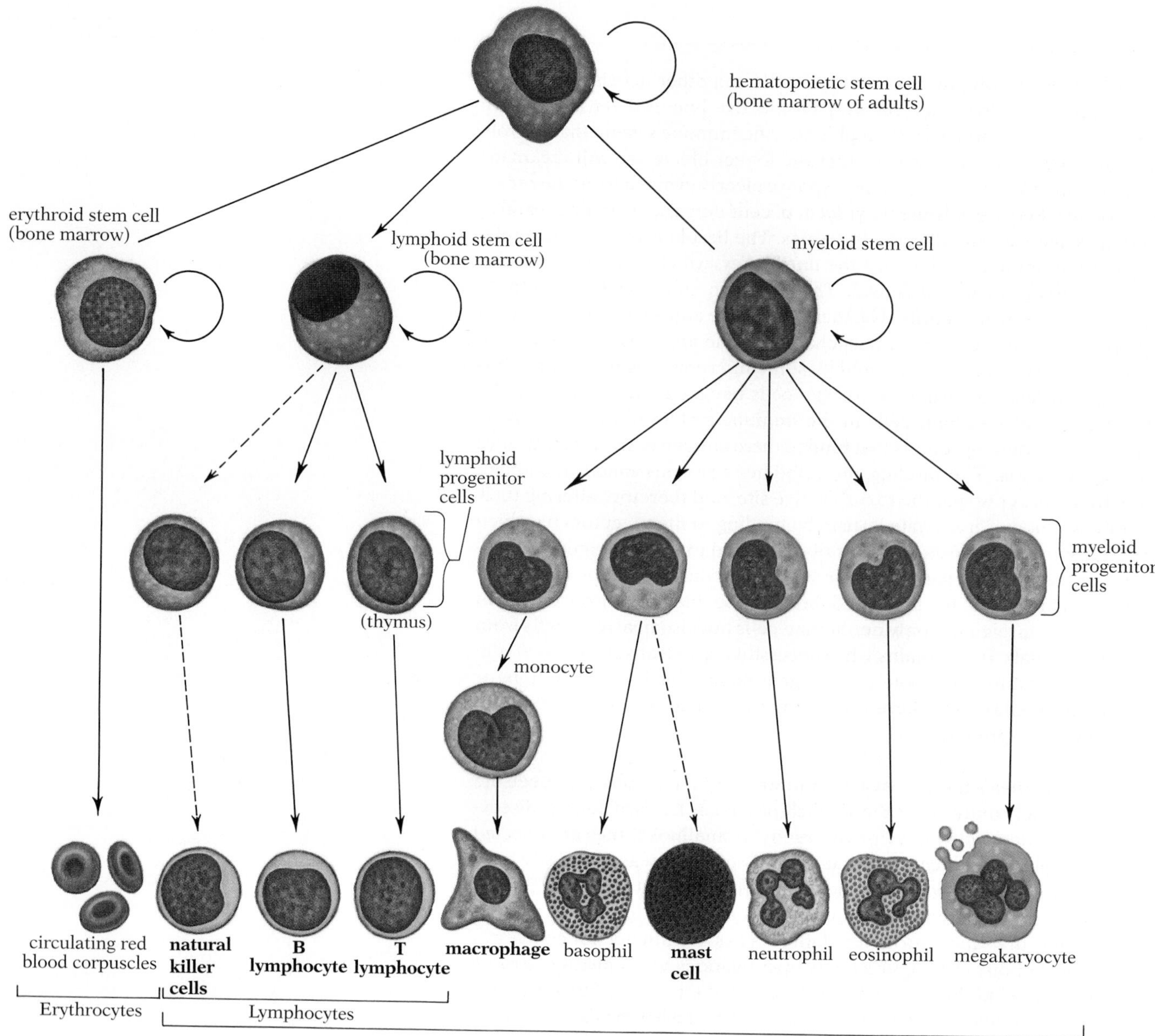

15.1 Hematopoiesis: the origin of blood cells
Hematopoietic stem cells give rise to at least three kinds of more highly determined stem cells: erythroid stem cells, lymphoid stem cells, and myeloid cells. All four of these stem cells have the ability to regenerate themselves. Each of the three more highly determined stem cells in turn gives rise to more specialized cells: erythroid stem cells produce red blood corpuscles; lymphoid stem cells give rise to three yet more highly determined progenitor cell types, which in turn produce the natural killer cells,B lymphocytes, and T lymphocytes that we will discuss in this chapter; myeloid stem cells produce at least five classes of more highly determined cells, which themselves go on to generate the more specialized monocytes (which mature into the cell-scavenging macrophage discussed later), as well as basophils and (perhaps) mast cells. (Myeloid stem cells also give rise to neutrophils and eosinophils, which attack bacteria and larger parasites, as well as megakaryocytes, which produce substances crucial to clotting, discussed in Chapter 30.) The antigen-presenting cells (which we will encounter later in this chapter) are derived from hematopoietic stem cells along an unknown pathway. The text discussion will focus on the interactions of the best-understood immunocytes, designated here with boldface type.

animals' immune systems had thus acquired the ability to recognize and destroy anthrax bacteria. But how does exposure to an antigen stimulate an organism to make antibodies, and how does the immune system "remember"? The immune system of vertebrates is so richly complex that it has taken nearly 200 years to arrive at satisfactory solutions to these questions. We will begin to answer them by first looking at the specialized cellular components of the immune system.

CELLS AND ORGANS OF THE IMMUNE SYSTEM

Origin of immune system cells In most vertebrates, all ***immunocytes***—cells with immunological function—derive from precursors that form in the yolk sac of the early embryo. These so-called hematopoietic stem cells (so named because they have become developmentally determined as progenitors of blood cells) migrate to specific tissues and organs, where they give rise to the red and white blood cells (Fig. 15.1). Red cells carry oxygen in the blood (see Chapter 30), and have no immunological role. A class of white cells that come from a gland called the thymus and from bone marrow give rise to ***lymphocytes;*** it is these cells that respond to the presence of foreign antigens or kill microorganisms tagged by antibodies. Lymphocytes derive their name from the high proportion of their time they spend in the lymphatic system, where dead cells and debris tend to collect, and into which toxins and infectious organisms usually find their way.

The lymphatic system Blood consists of cells and a liquid portion called plasma. As the blood flows through any of the many fine capillary beds in the body, a portion of the plasma leaks out between capillary cell walls into surrounding tissue. This fluid, supplemented by liquid transported endocytotically or lost osmotically, serves to bring nutrients to tissue cells outside the bloodstream, and to pick up waste products. As blood leaves the capillaries for larger vessels, most but not all of the lost plasma and metabolic wastes of tissues is reabsorbed. (We will examine the dynamics of this process in Chapter 30.) The rest is collected into a parallel system of lymphatic veins and funneled into two large thoracic ducts in the chest, from which it rejoins the blood (Fig. 15.2). Fluid and unanchored cells (generally dead or foreign) from body tissues can also enter the lymphatic system. The transport of this fluid, called ***lymph***, is passive: body movements squeeze lymph past one-way check valves on its slow journey to the chest. When feet swell after resting for an extended period, the accumulation of stationary lymph is responsible.

Lymphatic veins from capillary beds in the chest contain one or more **lymph nodes**, which are regions with a fine tissue mesh that filters dead

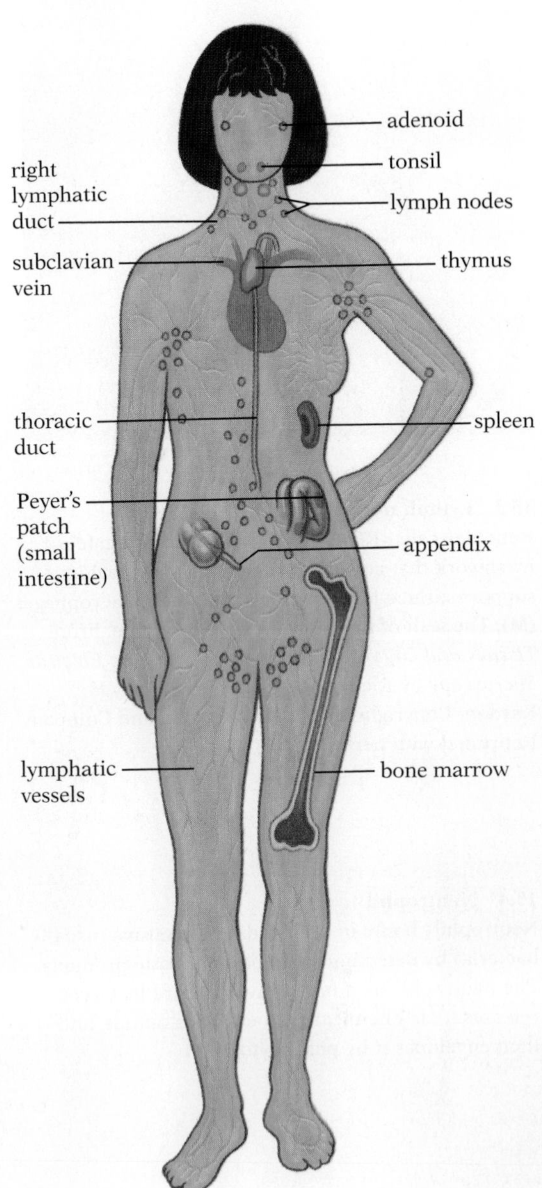

15.2 The human lymphatic system
The primary organs of the lymphatic system, the thymus and bone marrow (only one bone is shown), produce lymphocytes that circulate from extremely fine blood vessels, called capillaries, into the lymphatic system. Other tissues and organs with connections to the lymphatic system are shown.

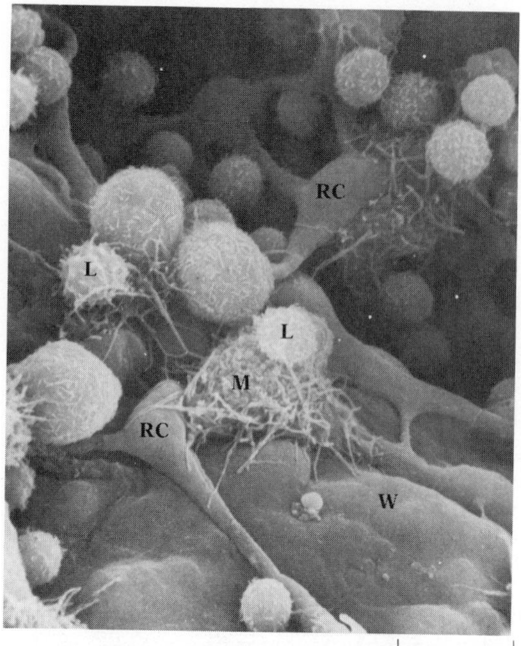

15.3 Lymph node
Reticular cells (RC) and their extensions create a meshwork that acts as a crude filter and provides a support surface for lymphocytes (L) and macrophages (M). The wall of the node (W) is also visible. From *Tissues and Organs: A Text-Atlas of Scanning Electron Microscopy* by Richard G. Kessel and Randy H. Kardon. Copyright 1979 W.H. Freeman and Company. Reprinted with permission.

cells and other large fragments from the lymph (Fig. 15.3). The nodes are home to many phagocytic cells (which consume the material trapped in the mesh) and lymphocytes. Like the spleen, through which all of the body's blood is frequently filtered, the nodes provide a convenient place for immunocytes to monitor the plasma for foreign antigens, and immune responses are frequently localized in the nodes, particularly those in the tonsils.

The diversity of immunocytes The two main types of lymphocytes are B cells (which mature in the bone marrow) and T cells (which mature in the thymus). B cells manufacture and secrete antibodies, and are responsible for the **humoral immune response**, that is, the response triggered at least in part by antibodies circulating in the body fluids, blood, and lymph. Because their antibodies can recognize and bind to cell-surface markers, B cells are particularly effective against bacteria, fungi, parasitic protozoans, and viruses, as well as toxins free in the blood and plasma. The T cells are responsible for the **cell-mediated response**, which kills infected cells. T cells can also bind to and modulate the activity of B cells.

The major phagocytic cells of vertebrates are the macrophages, mentioned earlier. As we will see, the macrophages of vertebrates can be far more selective than those of invertebrates because they look specifically for cells to which antibodies have been bound, thus identifying them as foreign. Another kind of white cell, the natural killer (NK) cell, is also found in invertebrates, but like the macrophages, has evolved in vertebrates to home in on and attack antibody-tagged cells. The discussion that follows will emphasize the best-understood actors in immune reactions: macrophages, B cells, T cells, NK cells, and mast cells. Other white blood cells with immunological functions include neutrophils (Fig. 15.4) and eosinophils, which attack bacteria and larger parasites, megakaryocytes (which, as we will see in Chapter 30, produce platelets crucial to clot formation), and basophils, which, like mast cells, release a chemical (histamine) that signals injury or

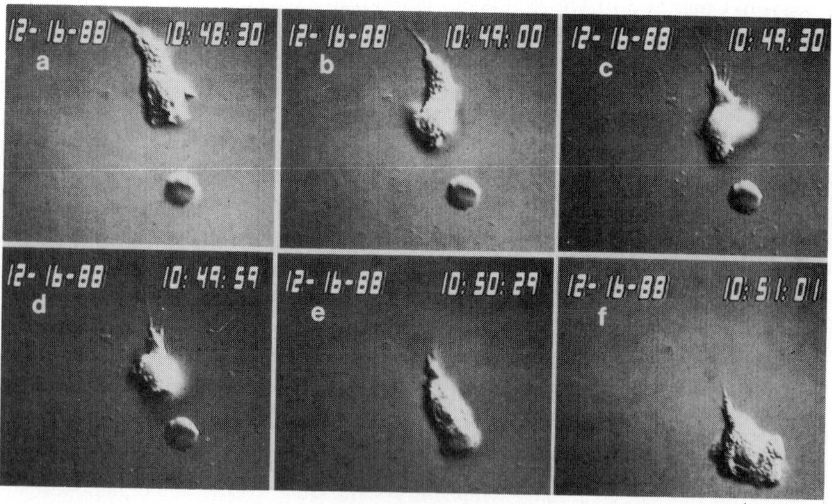

15.4 Neutrophil
Neutrophils home in on invading organisms (usually bacteria) by detecting some of their waste products. The neutrophil seen here crawls toward its target, releases toxic chemicals and enzymes onto it, and then envelopes it by phagocytosis.

infection, and thus create inflammation; basophils are specialized for circulating in the blood, while mast cells reside in tissues. Antigen-presenting cells (of which we will have more to say presently) are derived from hematopoietic stem cells along an unknown pathway.

THE HUMORAL IMMUNE RESPONSE

Most B-cell lymphocytes migrate back and forth from the blood to the lymph, taking up temporary residence in the spleen and lymph nodes where, using their membrane-mounted antibodies, they are well placed to monitor the body's fluids or dispense free antibodies; when needed, they can disperse to the circulation and tissues.

The B-cell antibody molecule Each antibody molecule consists of four polypeptide chains—two identical "heavy" chains and two identical shorter "light" chains, linked by disulfide bonds (Fig. 15.5). There are two general classes of light chains; these display no clear functional difference, but are encoded by separate genes. There are five classes of heavy chains—A, D, E, G, and M; these differ in the amino acid sequence at the COOH (tail) ends. The tail regions play no role in antigen specificity, but are involved in determining which reaction of the humoral antibody response will take place. For example, after an antigen has been bound, heavy chains with a so-called G tail (by far the most common class of antibodies in higher vertebrates) undergo an allosteric change in conformation that allows them to be recognized by macrophages, which can then ingest the antibody along with whatever the bound antigen is attached to—a virus, for instance. Antibodies with other heavy chains are used to activate other parts of the immunological reaction: antibodies with E tails become mounted on the membranes of mast cells, signaling these early-warning cells to begin releasing histamine when an antigen is encountered. Others are specialized for tasks to be described presently, like conferring maternal immunity on newborns, or activating a series of enzyme reactions known as the complement system. Together, all the antibodies of all classes are called ***immunoglobulins (Ig)***, and the individual classes are often referred to as IgG, IgE, and so on.

Regardless of which class of heavy or light chain is incorporated into an antibody, most of each chain has a constant amino acid sequence and structure; the variability crucial to antigen specificity lies mostly at the free amino ends. The binding sites for antigens (two identical sites on each antibody molecule) are at the ends of the variable portions. Each binding site is a pocket or cleft bounded partly by the heavy chain and partly by the light chain (Fig. 15.5). This region can bind to approximately six amino acids or carbohydrate units of an antigen in just the way enzymes bind to substrates.

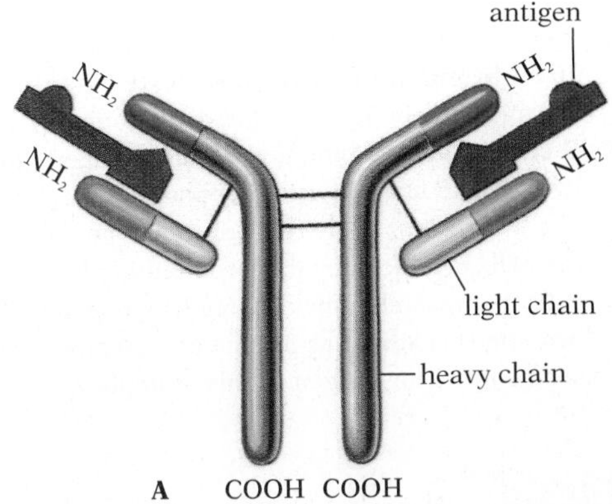

15.5 The B-cell antibody molecule
(A) B-cell antibodies consist of two identical pairs of polypeptides; each pair has a heavy chain and a light chain, which is readily seen in this schematic representation. The sections shown in gray have relatively constant sequences, while the colored portions vary greatly from one B cell to another. Antigens bind in the cleft between the heavy and light chain of each pair. B-cell antibodies can exist as free circulating molecules (as shown here) or mounted in the membrane of B lymphocytes or mast cells. (B) A space-filling model showing the three-dimensional shape of antibodies. Each sphere represents one amino acid. The antibody shown here is IgG.

Development of the humoral response Before an organism has encountered a particular antigen, the B lymphocytes that can recognize the antigens are small, metabolically quiescent ***virgin cells*** that can move freely between the blood and the lymphatic tissues by squeezing between the cells that line the blood vessels. Each of the millions of virgin cells generated during embryonic development has thousands of identical antibodies

15.6 Stimulation of B lymphocyte by antigen
When the membrane-mounted antibodies of a virgin B cell are able to bind a particular antigen, the lymphocyte first grows larger and then begins a series of cell divisions (only two are shown here). Some of the cells produced by this proliferation are memory cells that resemble the original lymphocyte; others become specialized as plasma cells, which secrete antibodies. The antigen in this example is a toxin.

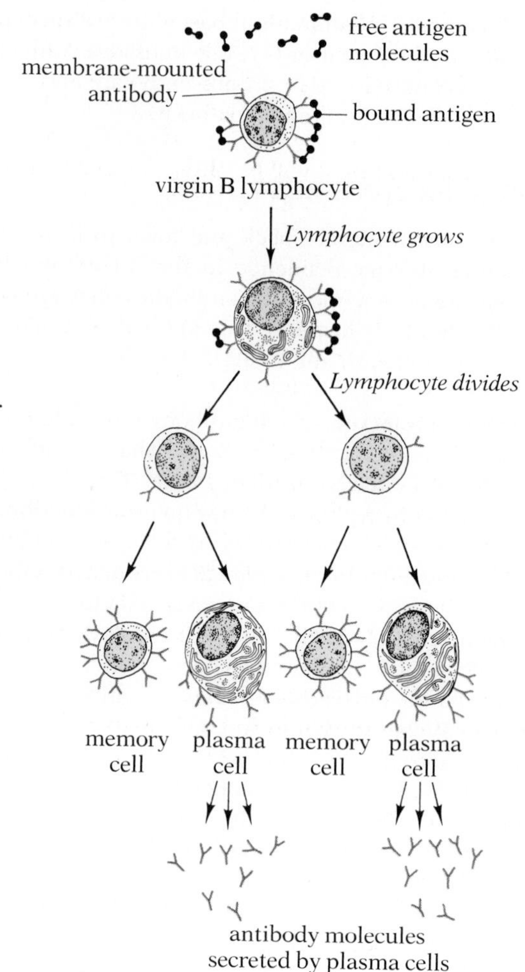

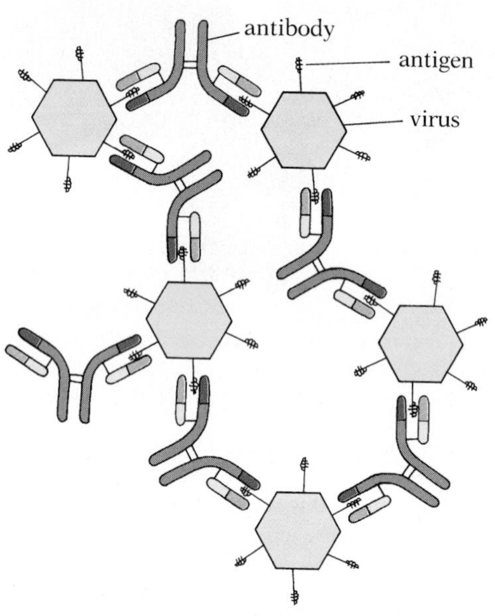

15.7 Agglutination by antibodies bound to antigens
Each antibody molecule can bind to two antigen molecules; hence the microorganisms or viruses bearing the antigens can be held together in large clumps. Here the antigens are surface proteins or carbohydrates on invading viruses. (For clarity, the antibodies and antigens have been enlarged. They are in fact much smaller than viruses.) This agglutination aids in the destruction of antibody-bearing pathogens by macrophages, killer cells, or complement-system proteins.

mounted in the membrane, but no two virgin cells are likely to display antibodies with the same antigenic specificity. When a cell's antibodies begin binding antigen, these small lymphocytes start to grow larger and divide (Fig. 15.6). Let us assume, for the moment, that the lymphocyte we are following begins as a virgin ***B lymphocyte***. Cell division by the stimulated B lymphocyte gives rise over a period of several days to numerous ***plasma cells***, and it is primarily these cells that secrete antibody molecules. A stimulated B lymphocyte also gives rise to other lymphocytes like itself, which serve as ***memory cells*** and make possible (in a manner to be described later) the more rapid response that will occur if the same antigen should be encountered again. It is the rapidity of this subsequent response that confers immunity.

Each of the millions of kinds of antibodies has a different set of active sites, so each binds to one or more different antigens or antigen regions. Some antibodies are so well matched to an antigen that this bonding is rapid and strong; other antibodies bind with a lower affinity. Because each antibody molecule can bind to two antigen molecules, the antibodies tend to agglutinate, or lump together, the antigens and any microorganisms or viruses bearing them (Fig. 15.7), which helps neutralize these pathogens directly. Even if a virus is not part of an agglutinated clump, if its surface is covered by bound antibodies, it will be physically unable to bind to the appropriate cell-surface marker on its host cell.

The agglutination in turn can trigger three reactions. First, large phagocytic macrophages (see Fig. 4.26, p. 113) in the lymph recognize antigen-bound antibodies, and engulf them and their targets (Fig. 15.8A). This reaction is effective against toxins, viruses, and most bacteria. Second, lymphocytes of several different kinds—the natural killer cells mentioned earlier—recognize the bound antibodies, bind to them, and destroy any foreign eucaryotic cells they have marked (Fig. 15.8B). The two mechanisms by which these lymphocytes kill are not yet fully understood, but one makes holes in the membrane of the target cell; osmotic entry of extracellular water kills the cell. Finally, the bound antibodies can trigger the ***complement system***, a cascade reaction involving more than 20 plasma proteins—many of which exist as inactive zymogens. In the cascade reaction, each protein in turn catalyzes the activation of another. Four kinds of complement-system proteins then assemble into an 18-unit membrane channel in the invading cell; the channel allows osmosis of water into the microorganism, which swells and bursts (Fig. 15.8C). Some membrane-enclosed viruses are neutralized in this way.[1] This system operates almost exactly like the hole-punching strategy of natural killer cells. Having detected an invader, the immune system clearly takes no chances; several redundant systems work simultaneously to attack anything foreign.

These three reactions, which depend on circulating antibodies secreted by B lymphocytes, are each part of the humoral immune response, but they are not the end of the B-lymphocyte story. In addition to binding directly to antigens and thus targeting them for destruction, some circulating antibodies can become attached by their bases to ***mast cells***. The antibodies thus mounted on the cell membrane of the mast cell have their antigen-binding region free and exposed to the surrounding medium. When an antigen is bound by a mast-cell-mounted antibody, the mast cell is induced to release histamine and other chemicals. Histamine in turn causes nearby blood vessels to dilate and become "leaky," allowing the plasma, rich in antibodies and complement-system proteins, to reach the site of the histamine release. Lymphocytes and macrophages are also attracted. The mast-cell/antibody system, then, acts as a sort of cellular burglar alarm to recruit the other elements of the immune system to concentrations of antigens. It is particularly important in the response to roundworms and other parasites that are found embedded in tissue rather than moving through the blood or lymph.

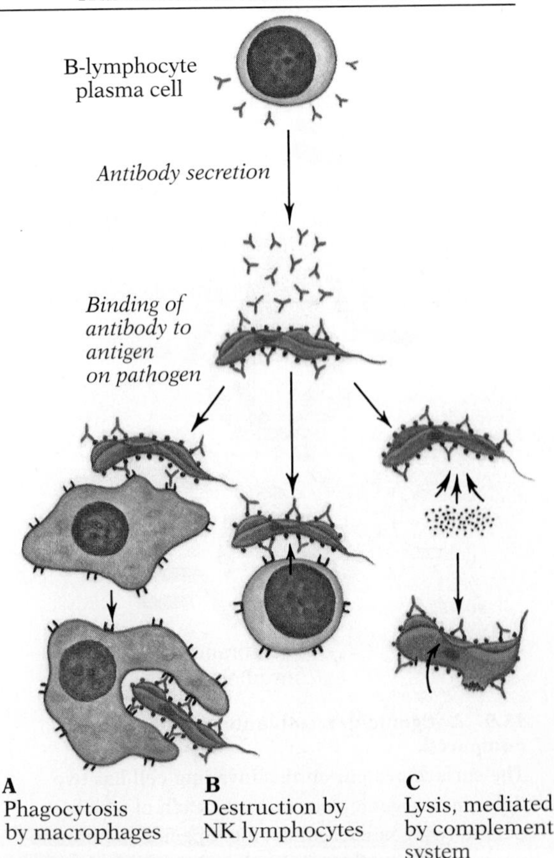

15.8 How humoral antibodies facilitate the destruction of pathogens
Once bound to antigens, the antibodies secreted by B lymphocytes can trigger three sorts of reactions. (A) Bound antibodies, and the pathogen to which they are attached, are ingested by phagocytic macrophages; this is also the primary mechanism by which agglutinated toxins and viruses are eliminated. (B) Bound antibodies are recognized by NK lymphocytes, which destroy the marked pathogen by unknown mechanisms. (C) Bound antibodies trigger a cascade reaction by which zymogens of the complement system are activated and catalyze the construction of a membrane channel in the invading microorganism; water moves into the invader through this channel by osmosis, causing the cell to swell and burst. The action of circulating antibodies on mast cells is not shown in this drawing.

[1] Some viruses seem to have evolved defenses against the complement system. Herpes and Epstein-Barr viruses, for instance, bear receptors that bind and inactivate the third protein in the cascade, bringing this (to them) dangerous reaction sequence to a halt; vaccina virus (one of the pox viruses) protects itself by binding the fourth protein.

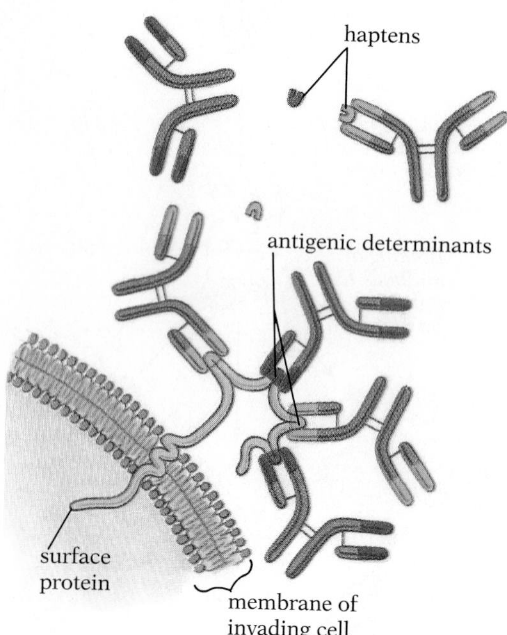

15.9 Antigenic determinants and haptens compared
The surface protein on this invading cell has two different antigenic determinants, each of which occurs twice in each molecule. Specific membrane-mounted antibodies can bind to these regions and trigger an immune response, which includes the production of free antibodies (shown). The surface protein and its antigenic determinants may be unique to this kind of invading cell, or they may be shared by cells of other kinds. Isolated antigenic determinants, or haptens, can also be bound by antibodies and stimulate an immune response if the immune system has been previously exposed to a large molecule bearing that particular determinant.

The mechanism of antigen stimulation An antigen is almost always a large molecule—usually a protein, polysaccharide, glycoprotein, or glycolipid. Not all of the antigen molecule stimulates lymphocytes to begin the immune response. Rather, certain regions (probably, in the case of proteins, about six amino acids long) serve as so-called ***antigenic determinants***—sites of interaction with the receptors on lymphocytes. A single large antigen molecule may include several different kinds of antigenic determinants, which are bound by a corresponding number of different antibodies (Fig. 15.9). Conversely, different antigen molecules may, by chance, have one or more antigenic determinants in common, and so "share" antibodies. The initial reaction to an antigen is triggered only if the antigenic determinant is part of a large molecule, but subsequent reactions can be initiated by the isolated antigenic determinant. Such an isolated determinant is called a ***hapten*** (Fig. 15.9).

Proper functioning of the immune system is predicated upon the availability of an enormous number of slightly different lymphocytes, each specific for a particular antigenic determinant. Estimates of the number of different kinds of lymphocytes that exist in any individual run as high as 10 billion (10^{10}) or more. We will see in a later section how this diversity is generated from only a few genes. Each antigen reacts only with those very few lymphocytes that have antibodies capable of binding to part of it, and this binding is necessary to induce proliferation of the appropriate lymphocyte types (Fig. 15.10). In Pasteur's demonstration, the sheep of both groups had lymphocytes specific to regions of the anthrax antigens, but only in the 25 that had been previously exposed to anthrax had the lymphocytes proliferated enough to win the race against the invading bacteria.

Each stimulated lymphocyte gives rise to a clone of cells (a group of genetically identical cells descended from a common virgin cell). Hence the proliferation of the particular lymphocytes that react with a specific antigen is called ***clonal selection***. Each of the plasma cells to which a given B lymphocyte gives rise when stimulated may transcribe as many as 20,000 mRNA molecules from its genes for antibody, enabling each of the plasma cells to secrete 5,000,000 identical antibody molecules per hour.[2] But the immune system is careful not to trigger this massive response by accident. Recall that we said that the binding of antigens by lymphocyte receptors allows the bound receptors to cross-link with each other. This cross-linking of receptors (which, by requiring at least two simultaneous antigen-binding events, serves to reduce false alarms, and probably explains why haptens cannot stimulate virgin cells) appears to be the event that induces the lymphocytes to begin dividing (Fig. 15.10).

[2] A partial but rapid immune response can also be triggered in individuals directly inoculated with the appropriate antibodies. Though there are no memory cells to begin supplying new antibodies, the antibody molecules introduced from outside can deactivate pathogens, or mark them for phagocytosis or complement-mediated destruction. Breast feeding is one means of creating this "passive immunity": maternal antibodies specialized to cross the intestinal wall (by virtue of a special heavy-chain tail) and enter the fetal bloodstream provide a significant degree of temporary protection until the infant's own immune system gains experience. Another source of antibodies is an injection of gamma globulin, a protein extract from the blood containing antibodies from individuals immune to a particular disease, though the extract contains many other kinds of antibodies as well. Gamma globulin treatment may reduce the severity of symptoms in someone who has been exposed to certain diseases, and can be useful in the absence of an effective vaccine, or when there is no time for an immunization to "take."

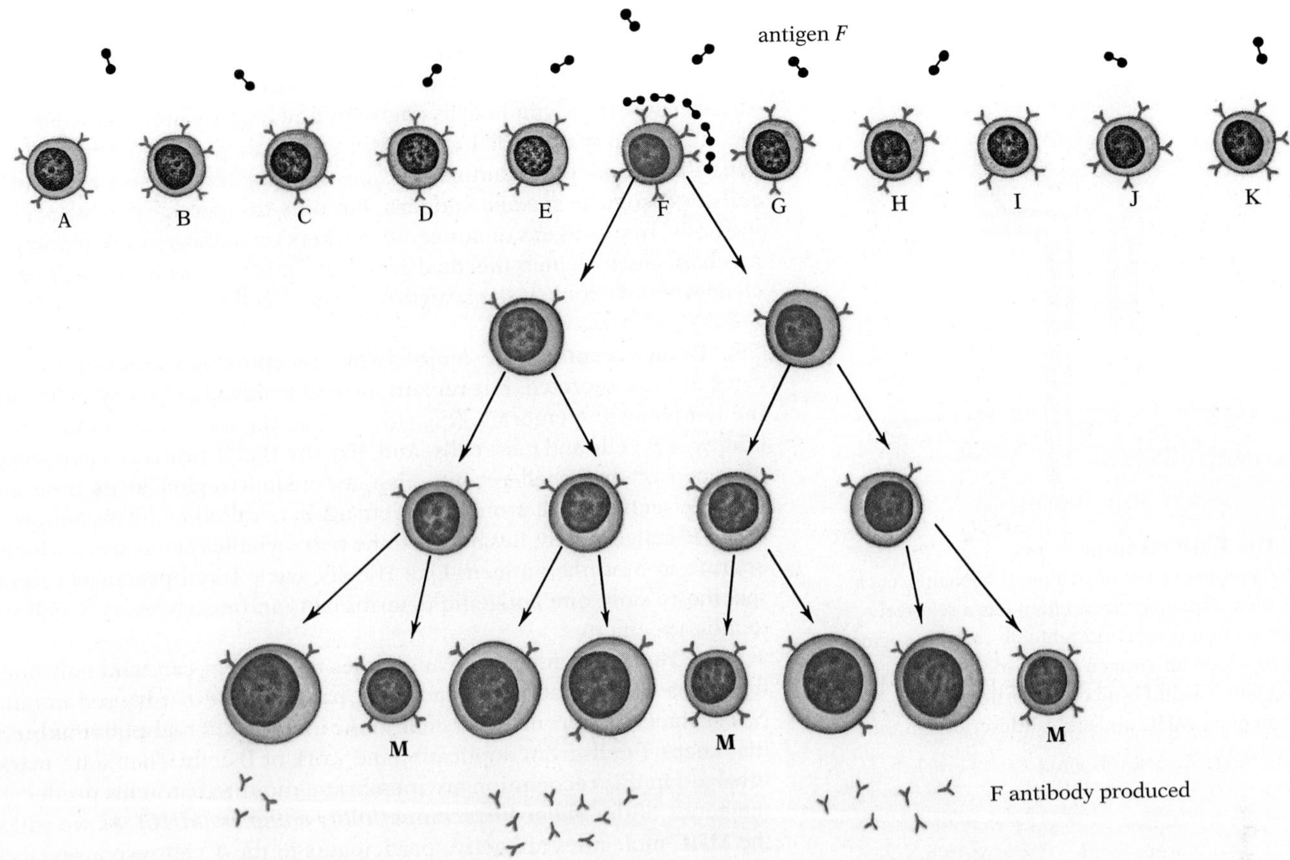

15.10 Clonal selection
Antigens are able to react with the membrane-mounted antibodies of only one or a few very specific lymphocytes from among the billions of kinds of lymphocytes in the organism's body. In this example antigen *F* can be bound only by the B lymphocyte of type F; it does not affect the other lymphocytes (top row). Lymphocyte F, stimulated by the binding of the antigen, proliferates to form a clone of genetically identical cells. Some cells of the clone are memory cells **(M)**; others are cells that actively secrete F antibody. The cross-linking of the antibodies on lymphocyte F greatly facilitates the proliferation of the F clone.

TABLE 15.1 *Simplified outline of immune-system cells*

Function	*B-cell system (humoral response)*	*T-cell system (cell-mediated response)*
Early-warning	Mast cells	Presentation cells
Diversity generation	Virgin B cell	Virgin T cells
Modulation		Helper T cells Suppressor T cells
Effector systems	Complement cascade Natural killer cells Macrophages	Cytotoxic T cells

THE CELL-MEDIATED IMMUNE RESPONSE

Another set of immune reactions is mediated by ***T lymphocytes***, which mature in the thymus. One kind—the ***cytotoxic T cell***—kills infected cells of an individual's own body before the pathogen (usually a virus) can spread further. Cytotoxic cells, sometimes referred to as "killer" T cells, are sometimes confused with NK cells; the distinction between the two is made clear in Table 15.1, which is an outline of the classes of immunocytes discussed in this chapter, organized roughly by system and action. The B and T systems are not, however, as independent as Table 15.1 suggests. For example, two

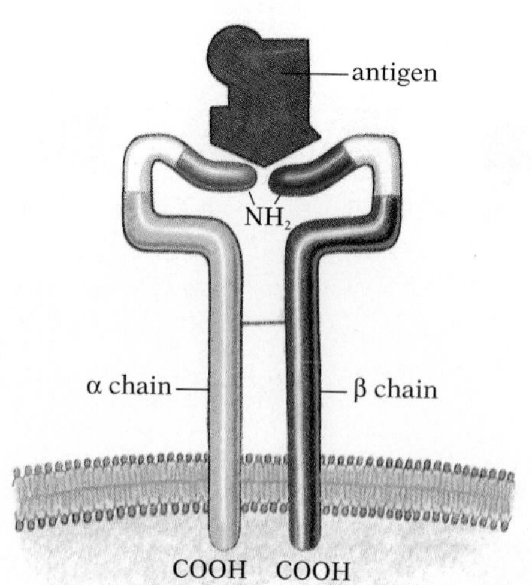

15.11 The T-cell receptor
T-cell receptors consist of two peptide chains, each with its own sequence. Each chain has a relatively constant section (gray) and a highly variable region (color) to which an antigen can bind. Another section in each chain (white) binds to a portion of complementary MHC molecules (described in the next section).

other sorts of T cells modulate the activity of the immune system—both humoral and cell-mediated—accelerating the response at the outset, and preventing it from getting out of hand. Controlling the immune reaction involves tuning a lymphocyte's activity, and is antigen-specific. For example, a particular clone of T cells will manage the activity of the clone of B cells that reacts to the antigen in question. Each of these jobs requires T cells to recognize specific antigens, but only to interact with other cells of the body; free antigens or antigenic markers on pathogens are ignored. The mechanisms by which this dual recognition is accomplished will become clear when we look at the structure of the T-cell receptor.

The T-cell receptor The antibody-like receptor molecules of T lymphocytes are not secreted, but remain instead firmly attached by their tails to the lymphocyte membrane (Fig. 15.11), like the membrane-mounted antibodies of B cells and mast cells. And, like the B-cell antibody, each polypeptide arm of the T-cell receptor has a constant region at its base and a variable section further out. The cleft, again paralleling the organization of of the B-cell antibody, lies between the two variable-region arms, which cooperate to bind the antigen. Like B cells, each T cell produces receptors specific to only one antigenic determinant, and nearly every T cell has a unique specificity.

T-cell receptors differ from antibodies in that they can bind only one antigen at a time. In addition, each arm has a region that is used to bind to cell-surface markers on other cells in the individual's body. It is this binding that keeps T cells from duplicating the work of B cells. The "self" markers involved in this recognition are membrane-mounted proteins produced by the genes of the ***major histocompatibility complex (MHC)***. As we will see, the MHC molecules are active participants in the T-cell response.

The MHC system There are two general types of MHC molecules: MHC-II proteins are found on the membranes of B cells, cytotoxic T cells, and certain immune-system cells specialized for antigen presentation and located in the tissues; MHC-I proteins are found on all other cells in the body. This dichotomy reflects the dual role of T cells as modulators of B-cell activity and assassins of disease-infected cells: cells bearing MHC-II proteins participate in regulation of the immune system, while those with MHC-I molecules can be killed by cytotoxic T cells.

MHC molecules, like the immune-system antibodies and receptors we have already discussed, are composed of two chains, and involve both constant and variable regions (Fig. 15.12A,B); it is likely that T-cell receptors evolved from MHCs. MHC molecules bind antigens and then "present" them on the surface of the cell for binding by appropriate T-cell receptors (Fig. 15.12C). This binding is then stabilized by a specific class of glycoprotein called a cluster determinant: CD8 in the case of MHC-I, and CD4 when MHC-II is involved. (The CD4 molecule will become important when we discuss AIDS.)

Development of the cell-mediated response Like their humoral counterparts, T lymphocytes begin as antigen-specific virgin cells. The first step in their activation occurs when MHC-I molecules in an infected cell begin

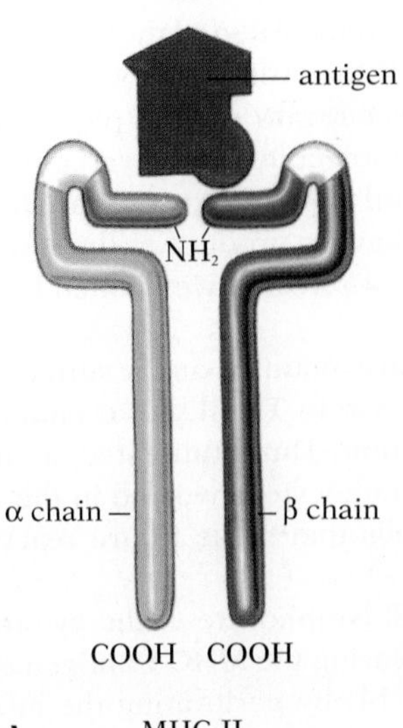

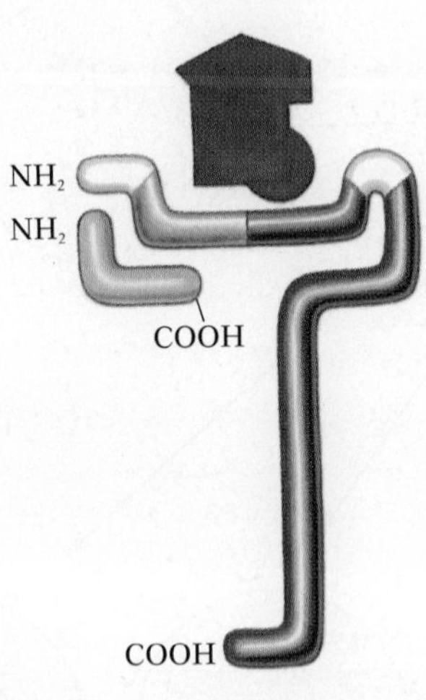

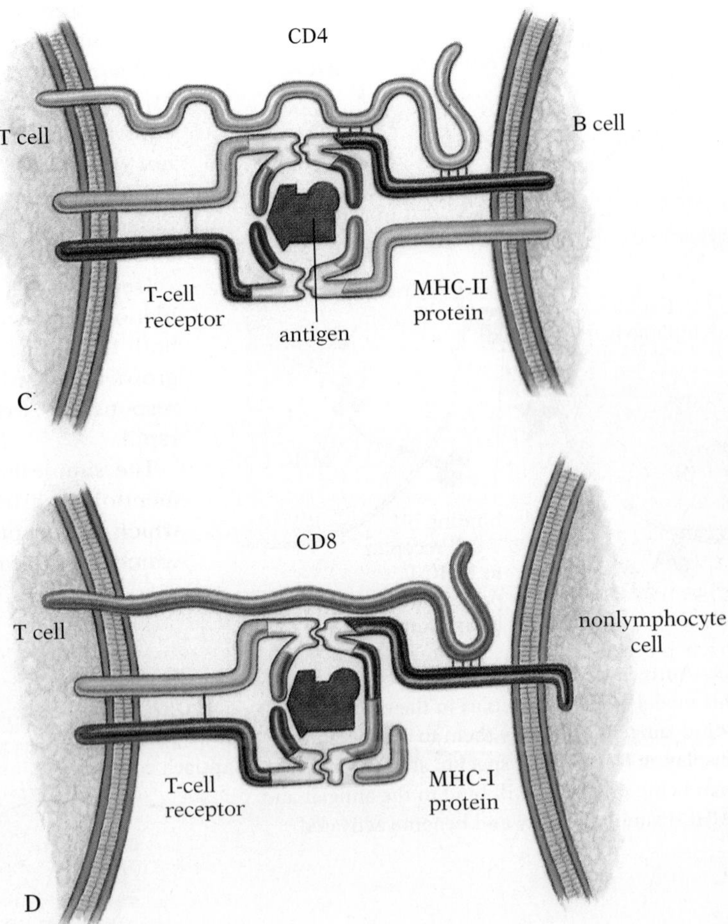

15.12 Structure and function of MHC molecules (A) MHC-II molecules are similar to T-cell receptors, having a pair of distinct chains, each with a constant region (gray), a variable section that binds an antigen (color), and a portion (white) that is complementary to a corresponding T-cell receptor. (B) MHC-I molecules are functionally almost identical to MHC-II, though structurally dissimilar. They too bind antigen (though the two variable domains are on the same chain) and a matching T-cell receptor. When either MHC molecule binds to a T-cell receptor, the antigen-binding portions of the T-cell receptor and MHC protein each interact with different parts of the antigen they hold in common. The binding is stabilized by a CD protein. (C, D)

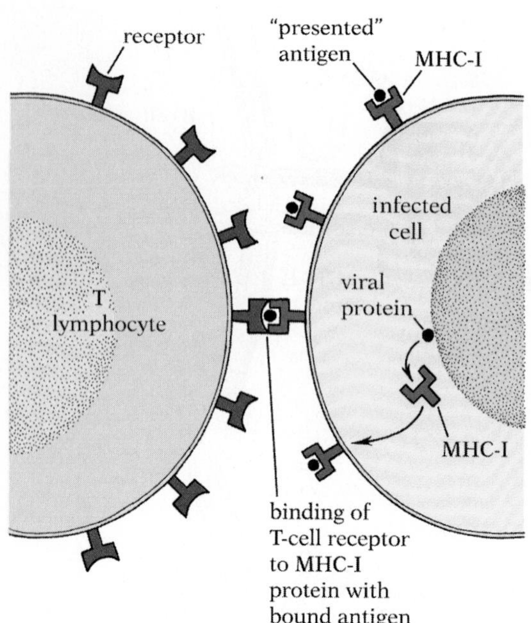

15.13 Antigen display by MHC-I
In this model MHC-I molecules in the cytoplasm of a cell bind antigens and carry them to the membrane for display. A T cell with a specific affinity for the antigen being presented will bind to the antigen and the MHC-I simultaneously and become activated.

presenting pathogenic antigens. The antigens in question are usually viral-coat proteins synthesized by the infecting organism in preparation for its further spread in the host. MHC-I molecules probably intercept these foreign compounds in the cytoplasm or on the ER, and then carry them to the cell surface for display (Fig. 15.13). If the antigen is too large to fit in the MHC pocket, it is "processed" (digested) into smaller pieces functionally equivalent to haptens. Though MHC molecules have a variable region, it is not very specific: a given MHC can bind 10–20 percent of the antigens it encounters. Otherwise each cell would have to produce millions or billions of different MHCs to match the degree of selectivity exhibited by antibodies and T-cell receptors.

Once the MHC/antigen complexes are mounted on the surface of the population of infected cells, a matching virgin T cell will eventually bind to both the antigen and the MHC-I protein. Thus stimulated, a virgin T cell grows and divides to produce the lymphocytes involved in the immediate response, as well as the memory cells that make future reactions more rapid.

The simplest variety of activated T lymphocyte is the cytotoxic T cell mentioned earlier. It destroys cells bearing the MHC-I/antigen complex to which its receptor can bind (Fig. 15.14) by perforating the infected cell, which allows water to enter and burst the target. The other classes of spe-

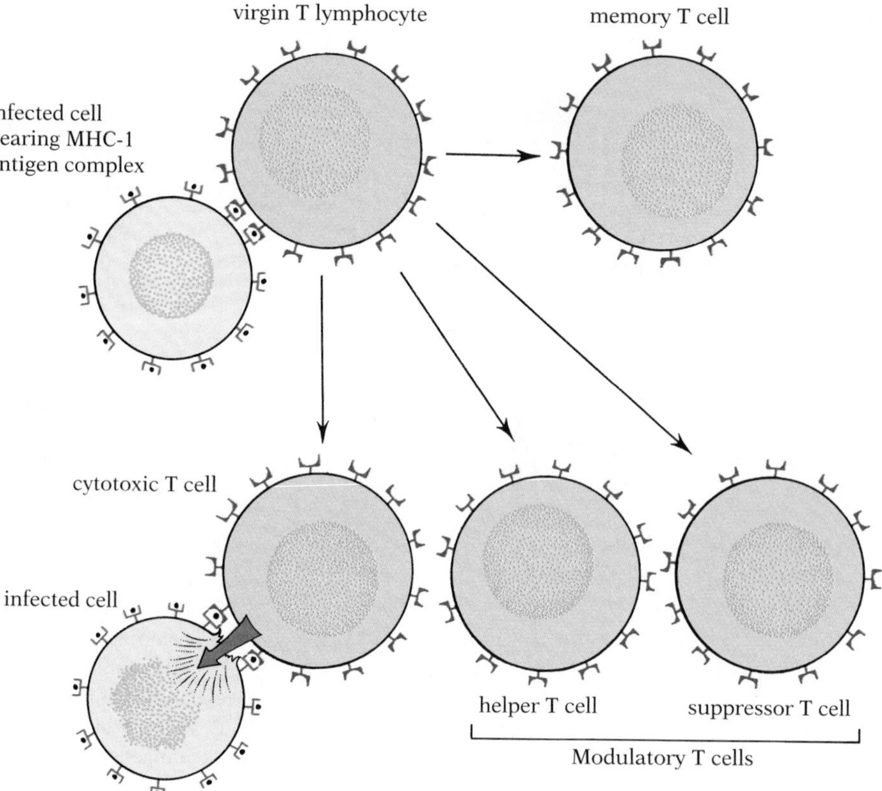

15.14 The cell-mediated immune response
When a virgin T lymphocyte binds to an infected antigen-presenting body cell, it becomes activated, grows, and begins dividing to generate memory cells (which permit faster responses when the antigen is encountered again), cytotoxic T cells (which kill infected cells), and modulatory T cells (which regulate the responses of B lymphocytes, cytotoxic T cells, macrophages, and other elements of the immune system to the same antigen).

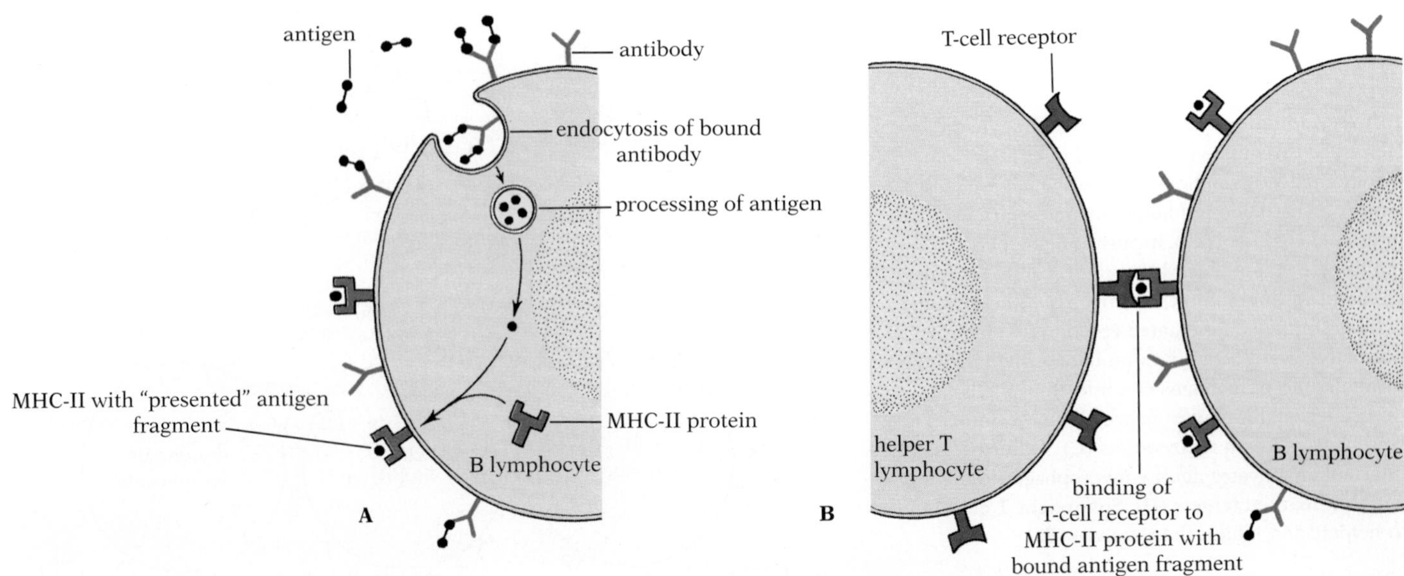

15.15 Presentation of antigen by MHC molecules of a B cell
MHC-II molecules are found in B lymphocytes, cytotoxic T cells, and presentation cells. (A) When the membrane-mounted antibodies of a B cell (shown here) or the receptors of a cytotoxic T cell bind their particular antigen, the antibodies are taken in through endocytosis. Antigens are removed in special proteolytic (protein-digesting) vesicles and, if necessary, broken into convenient lengths, and these fragments are then bound by MHC-II molecules in the cytoplasm and displayed on the cell surface. (B) When a helper T cell specific for the antigen is encountered, the two lymphocytes bind and the T cell becomes active. Activation by a presentation cell (not shown) involves nonselective endocytosis of intercellular material, processing, and display on its MHC-II molecule. This system of indirect activation is essential if the immune reaction is to be accurately modulated.

cialized T lymphocytes—helper and suppressor cells—react to MHC-II/antigen complexes, and are involved in modulating the immune response.

THE MODULATORY ROLE OF T CELLS

While MHC-I complexes collect antigens found in the cytoplasm, the immune-system cells with MHC-II proteins actively import antigens, process them if necessary, and then display them on the MHC-II complex. B cells use their membrane-mounted antibodies to collect specific antigens (Fig. 15.15A); T cells use the T-cell receptor. ***Presentation cells***, whose structure we have not yet explored, have neither receptors nor antibodies to capture antigens. These curious cells disperse from the bone marrow to the tissues, differentiate in tissue-specific ways, and take up their role as antigen presenters—a kind of cellular burglar alarm. They continually and unselectively endocytose material from the surrounding tissue fluid, process it, and then display the fragments on their MHC-II molecules.

Helper T cells begin their modulation of the immune response by binding to MHC-II proteins displaying the appropriate processed antigens on the cell surface of other immune-system cells (Fig. 15.15B). But helper T cells do not attack the antigen or its source directly; instead, once bound to B cells, cytotoxic T cells, or presentation cells displaying the appropriate antigenic determinant, they help orchestrate the activation of these elements of the system.

Helper cells become aware of the presence of an antigen when it is exposed in the binding cleft of an MHC-II protein. Whether the antigen is dis-

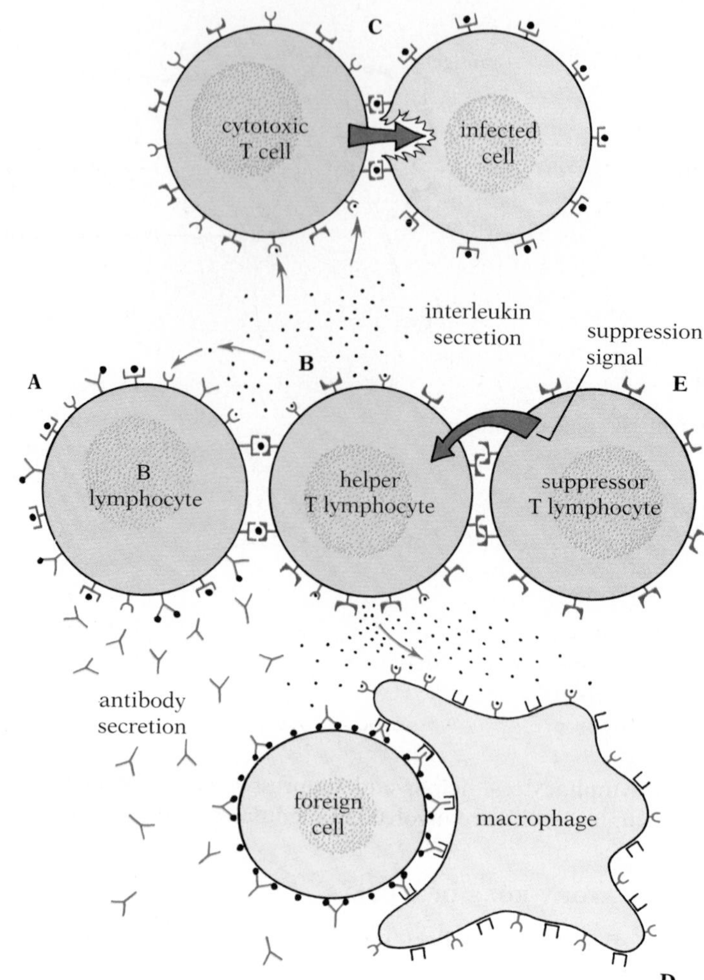

15.16 Modulatory action of T lymphocytes Helper cells regulate the humoral immune response by binding to B lymphocytes displaying a unique antigen (A). The bound helper secretes various interleukins, one of which causes the helper to multiply (B); another induces the B cell to secrete antibodies, which bind the antigens on the invading cell (A). In the cell-mediated response (which, normally, would be active simultaneously only if the invading microorganisms can infect host cells), a third interleukin induces nearby bound cytotoxic T lymphocytes to kill their targets (C). A fourth interleukin activates nearby macrophages to ingest antibody-marked targets (D). Suppressor T cells bind to helpers and inhibit their activity (E).

played by a B cell (Fig. 15.16A), cytotoxic T cell, or presentation cell, only a helper cell specific for the antigenic determinant *and* the MHC-II molecule holding it can bind and become activated. An activated helper can advance the campaign against the antigen and its source in several ways. Its first, rather circuitous action is to install receptors for a chemical-signal molecule, ***interleukin***, in its own membrane, and then to begin secreting interleukin. The binding of interleukin to its own receptors causes the helper cell to proliferate (Fig. 15.16B). The secreted interleukin also induces multiplication of any activated cytotoxic T cells nearby that have recently encountered their specific antigens; more often than not, the helper and the cytotoxic neighbor will be responding to the same pathogen.

The helper also binds to cytotoxic T cells that display the appropriate antigenic determinant, inducing them (by an unknown mechanism) to far more vigorous attacks on infected cells; local interleukin secretion has a similar effect (Fig. 15.16C). When bound to stimulated B lymphocytes, helper T cells produce a second kind of interleukin that encourages the B cells to secrete antibodies (Fig. 15.16A). Yet another kind of interleukin energizes nearby macrophages (Fig. 15.16D).

This two-step activation mechanism, by which B lymphocytes and cytotoxic T cells must both bind an antigen *and* be induced by a helper T cell specific for the same antigen, is probably a safety feature designed to prevent the potent cell-destruction capacity of the immune system from making mistakes. In particular, this double-checking strategy prevents the immune system from erroneously attacking the organism's own proteins, thereby bringing about an ***autoimmune response***, which can lead to a slow and sometimes fatal process of self-digestion. We will look at the problem of recognizing and ignoring friendly molecules in the next section.

The role of the ***suppressor T cell*** is to prevent the immune system from overreacting. Suppressor cells act by binding to helper cells in a poorly understood but antigen-specific manner. The helpers are then inhibited by an unknown mechanism (Fig. 15.16E). An antigen-specific class of helper cells begins proliferating when there is an increase in the number of MHC-II/antigen complexes to which it can bind, but the corresponding suppressors must wait for a rise in the number of helper cells to begin to proliferate and act; as a result, the suppression response lags behind the helping efforts so long as the antigen-encounter rate is on the rise. Once the immune system begins getting the upper hand, the suppressors start to catch up and turn the response down and finally off. One consequence of a failure to end the immune response promptly would be an overabundance of antigen-specific mast cells, and therefore a hypersensitivity to the antigen in question (Fig. 15.17). Even a slight exposure to the antigen under these conditions could lead to a massive release of histamine, triggering excessive loss of fluid from the blood; the result is an allergic reaction that, if extreme, can cause anaphylaxis: loss of consciousness as blood pressure falls, and even asphyxiation as fluid-induced swelling in the throat flattens the trachea (windpipe). There is good evidence that most or all allergic reactions result from a failure of antigen-specific suppressor T cells.

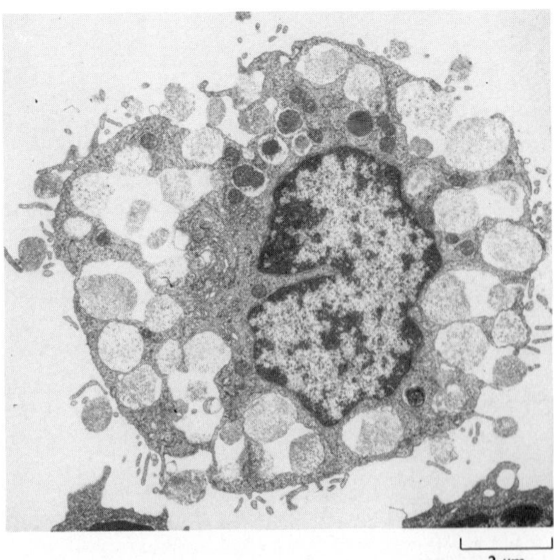

15.17 Mast cell
Massive exocytosis of histamine by a mast cell from a rat.

RECOGNITION OF SELF

While the immune system of an organism is able to recognize and destroy almost any foreign antigen or antigen-bearing cell, it must not confuse the cell-surface proteins on its own cells, "self" antigens, with foreign antigens. Tissue transplantation experiments demonstrate that this recognition of self, or self-tolerance, is learned early in the embryonic development of the immune system: if a piece of tissue from one organism is transplanted into or grafted onto another adult animal, the recipient's immune system will almost always reject the tissues by mounting an immune reaction against the donor's cell-surface proteins. But foreign tissue transplanted prenatally, while the recipient's immune system lacks experience, will be accepted and transplants from that same donor will be accepted later, even when the recipient is an adult.

"Self" antigens are generally membrane-anchored glycoproteins of the glycocalyx (see pp. 119–120). Some are specific to particular tissues, and seem to play a role in cell-type recognition and cell-to-cell adhesion. Others are encoded by the MHC, which devotes at least seven genes to the task, each of which can exist in as many as 100 versions. The diversity of MHC proteins is important if they are to bind and display the full range of processed antigens; it also has important consequences for cell recognition.

Inactivation of virgin cells specific for "self" antigens Self-tolerance originates during fetal development when a process of selective lymphocyte inactivation takes place. Millions of virgin B cells are released into the prenatal blood at a time when no foreign antigens are present. Those cells whose antibodies do bind during this stage of development must therefore be attaching themselves to normal proteins, and so are inactivated; this still leaves vast numbers of virgin cells specific for "non-self" antigens capable of proliferating and mounting an immune response should the need arise. The mechanism of this inactivation is not yet understood.

The inactivation of T cells is somewhat more complicated. A T cell must be repressed not only if the receptor binds to a normal protein present during fetal development, but also if it *fails* to bind to one of the organism's MHC complexes. Some such T cells are actually killed, a process known as ***clonal deletion***. Inactivation or deletion is essential if T cells are to perform their normal role of killing infected cells and modulating the activity of other immune-system cells.

Inactivation is an active, ongoing process that depends on the continued presence of each "self" antigen. Even with continued exposure, however, inactivation may not be permanent: should control over inactivation be loosened, as by infection with certain bacteria or viruses, an autoimmune disease can result. One example is myasthenia gravis, a neurological disorder involving the loss of tolerance of the body's billions of receptors for acetylcholine, a chemical that is used for communicating between nerve cells and muscles. The result is a completely debilitating deterioration of muscle control. Multiple sclerosis, rheumatoid arthritis, and Type I diabetes are among the other, more familiar autoimmune diseases. How and why self-tolerance is overcome in such ailments is not yet well understood.

Cell recognition and transplants When a foreign cell is introduced into an organism as part of an organ transplant, it bears several different MHC-I proteins on its surface. The diversity of MHC proteins plays an important role in cell recognition, as we have said. Most of the MHC variants on the cells of the transplant will be novel to the host's immune system, and so B-cell antibodies will bind to them and trigger a humoral response. At the same time, the overwhelming odds are that the foreign cell and the host will share at least one MHC-I variable region; as a result, at least one class of T-cell receptor will be able to bind to one region of this one kind of MHC protein. Unfortunately, the rest of that foreign MHC molecule will almost certainly contain novel regions, and so be seen as an antigen by the T cell; as a result, a cell-mediated response will also be triggered. The resulting two-pronged immune reaction will lead to transplant rejection. This attack on transplanted tissue can be suppressed with certain drugs (like cyclosporin, which blocks an enhancer of interleukin genes), but not without cost: the recipient, whose entire immune system will be depressed, will be at risk even from common colds. Somehow, however, a recipient's lymphocytes

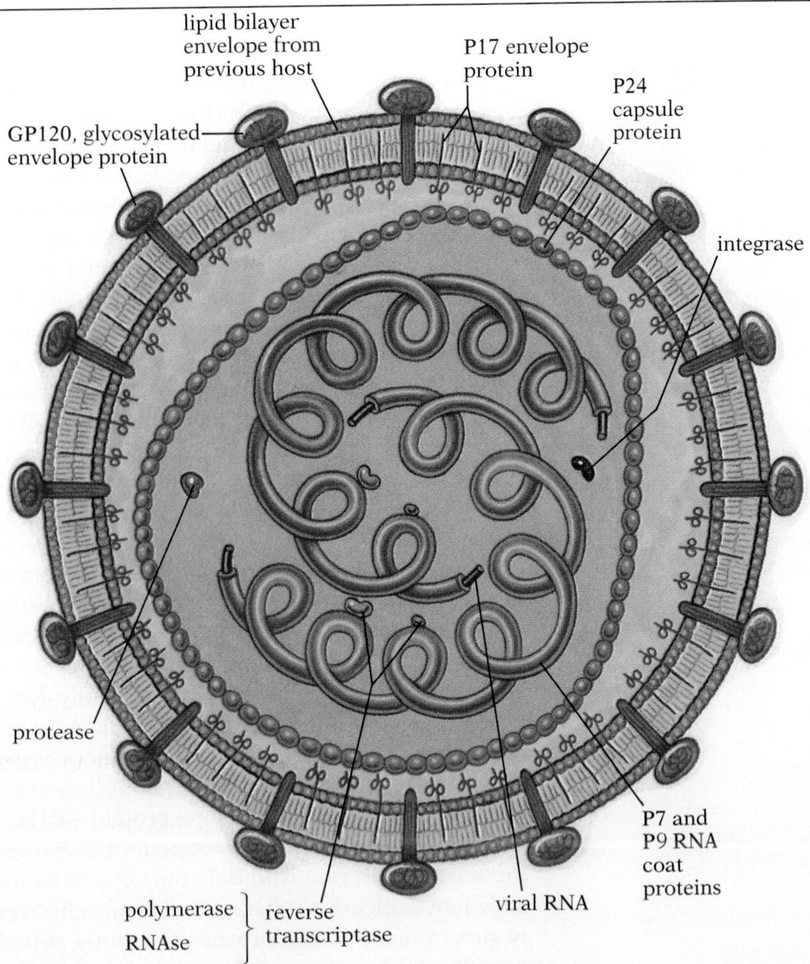

15.18 Structure of HIV
The two copies of the viral RNA are protected by a protein coat and enclosed in a capsid, which also contains the reverse transcriptase enzymes (which convert the single-stranded RNAs into double-stranded DNA copies), the integrase (which inserts the DNA version into the host genome), and the protease (which catalyzes the budding off of the virus). The capsid is itself enclosed in a bilayer membrane obtained from the previous host cell, in which are mounted the remains of a protein involved in budding off and GP120, the glycoprotein that binds the helper T-cell receptor and so enables the virus to gain entry into its host.

may, with the passage of time, come to learn to ignore the foreign antigens, and then immunosuppression therapy can be relaxed or discontinued.

ACQUIRED IMMUNE DEFICIENCY SYNDROME

The adaptive value of the vertebrate immune system is graphically illustrated by acquired immune deficiency syndrome, a disease that destroys most of the immune response by eliminating the all-important helper T cells.

The AIDS virus The virus that causes AIDS—HIV-1 (human immunodeficiency virus, type 1)—is an unusually complex retrovirus (Fig. 15.18). It consists of two copies of its RNA genome, each coated by a pair of proteins

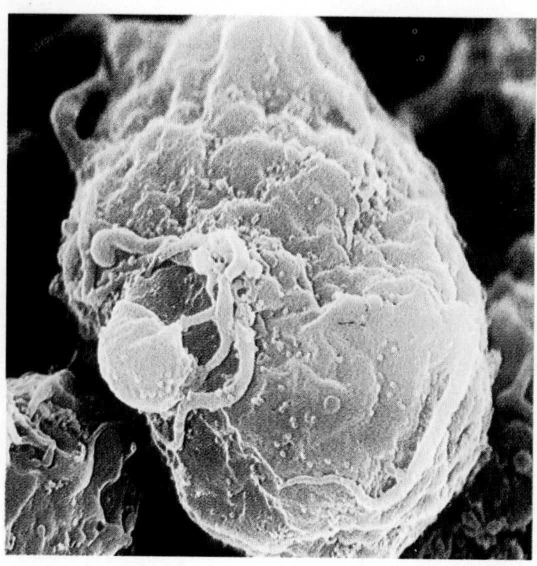

15.19 HIV viruses budding from a cultured lymphocyte

(called P7 and P9), and all enclosed by two very different layers. The inner layer is more or less a conventional viral capsule, a self-assembling structure of one kind of protein (P24). The outer envelope is more complex: its primary constituent is a lipid bilayer obtained from a host cell as the virus buds off (Fig. 15.19), but embedded in this vesicle are one sort of protein facing inward (P17) and a glycoprotein (GP120) directed outward. HIV locates and enters its host by means of the GP120 glycoprotein, which binds specifically to the CD4 protein of helper T cells.

The HIV-1 retrovirus carries four enzymes. The first and second work together as the reverse transcriptase: a polymerase synthesizes a complementary DNA copy of the single-stranded RNA genome, and an RNAse separates the viral template from the complementary DNA allowing the polymerase to complete the job of making the HIV DNA double-stranded for insertion into the host genome. The third enzyme, an integrase, cuts the host DNA and inserts the parasite's own instructions into the genetic library. The role of the fourth enzymatic passenger, a protease, will become clear presently.

The viral genome contains several regulatory sequences whose operation is not yet fully understood. Apparently the virus can exist in a low impact, wholly lysogenic or semi-lytic phase; in the semi-lytic state it will produce a slow stream of progeny but still not overwhelm its host. As with more conventional viruses, induction into the fully lytic state seems to require a shock or threat to the host cell, including, for example, its participation in an immune response. The virus reproduces both by replicating its entire genome and by producing various mRNAs, some polygenic, that subsequently direct synthesis of the envelope protein GP120, the "P" proteins, and the integrase and protease. The proteins, polyproteins, and RNA aggregate in the host membrane, with only the GP120s facing out (Fig. 15.20). When a sufficient collection of membrane proteins accumulate, the membrane begins to bud and the protease evidently cuts the polyproteins into their constituent elements, producing the enzymes destined to be packaged with the virus, as well as the proteins of the coat, or ***capsid***. Once the capsid is formed, the virus buds off and is free to bind another host cell.

The attack on the immune system Initially, the immune system responds normally to HIV, with B cells releasing suitable antibodies, helper T cells amplifying the response, and macrophages consuming free viruses. But the virus lives on, hidden in helper T cells and, undigested, in macrophages. Though the total population of T cells remains normal for (on average) about a year, infected cells slowly begin to be recruited into the lytic or semilytic phase. As GP120 antigens begin to appear on the surface of T cells, these hosts are destroyed by cytotoxic T cells and natural killer cells, but this is not of any real benefit. Not only is the body's arsenal of helper T cells being culled (dropping, on average, to 25 percent of normal three years after infection), but GP120s are released into the blood and lymph, where they bind to other helper cells, which are then, though not actually infected, attacked and killed. Worse, an infected cell with GP120 on its membrane will bind to other T cells and, just as the viral envelope fuses with the host, the membrane of the infected cell will fuse with the bound cell. Dozens or hundreds of T cells will be drawn into this fusion event, with the result that all die together, either at the hands of killer or cytotoxic cells, or because the resulting multinucleate cell becomes too large to function.

AIDS also affects the nervous system. Macrophages can cross from blood vessels into the nervous system where, as the viruses they harbor escape, HIV infects the special cells that insulate neurons (the glia, which, as it happens, have receptors that bind GP120). As the insulating cells die, conduction becomes slower and less efficient, and the accuracy with which new neurons are routed to their targets declines; the result can be actual "miswiring."

Things go from bad to worse. Soon the number of helper cells is down to 5 percent of normal or below, and any infection that comes along faces only a minor immune response. AIDS patients generally die five to ten years after infection, often of diseases that pose no particular threat to humans with healthy immune systems.

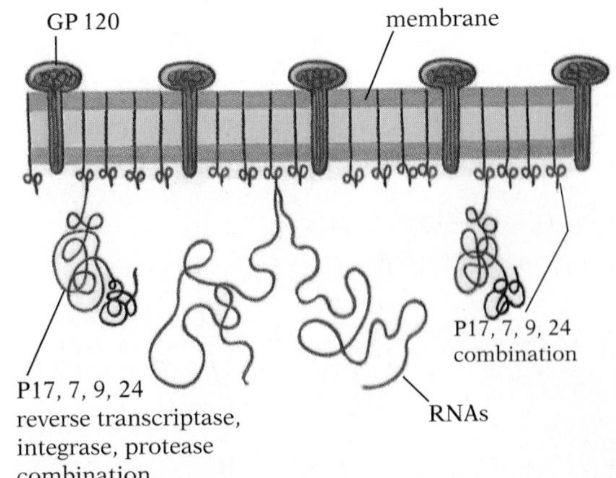

15.20 Assembly of a new HIV virus
Assembly occurs on the host-cell membrane where viral RNA, GP120, the P-protein precursor, and the long P-protein/enzyme precursors aggregate. When the collection of parts is sufficient, the piece of membrane begins to bud outward. Then a protease is freed and quickly cuts out the many enzymes and P-proteins, allowing coating of the RNA and capsid formation, steps that lead to complete budding off.

The epidemiology of AIDS Studies of how AIDS spreads indicate that HIV transmission is almost always from the blood or seminal fluid of one individual to the circulatory system of another. Hence the disease spreads readily through transfusion of infected blood, reuse of hypodermic needles, and anal intercourse. The virus can also be transmitted through vaginal intercourse, though the rate of infection is lower.

The spread of AIDS (or of any other epidemic) depends on the average incubation period of the infective agent, the length of the infectious period, the rate of contact (most often, in this case, the number of sexual partners or shared needles), and the efficiency of transferring the infection with each encounter. Most of this information can now be estimated for the United States. It appears that the average time from infection to death is about eight years (the mortality rate as of 1992 was essentially 100 percent) and carriers are infectious throughout this period, though more so early and late in the cycle.

Among American homosexuals and intravenous drug users, the infection rates are as high as 70 percent in urban areas like San Francisco and New York City, where contact rates are very high; on average, each infected individual communicates the disease to several others. In less urban regions, where the contact rate is lower and the susceptible pool of potential victims smaller, the spread is slower. Overall rates of transmission in the United States are declining simply because a large portion of the high-risk group is already infected; several million Americans are thought to carry the virus.

Among heterosexuals, both transmission rates and contact rates are lower. At present it is not clear whether there will be an epidemic among heterosexuals in Europe or the Americas: the statistic that defines an epidemic is whether the typical carrier will infect, on average, more than one other individual. In Africa, HIV transmission is primarily heterosexual, and risk correlates with the number of sexual partners. The World Health Organization estimates that nearly 60 percent of all AIDS deaths, which totalled 2.5 million through 1992 (and will rise beyond 20 million by the end of the decade), have occurred in Africa, where millions more are infected. Clearly a heterosexual epidemic *is* possible. There is some indication that

15.21 AZT
AZT is a DNA thymidine mimic. The base (T) is normal, but the deoxyribose unit (R) differs from the sugar in that the hydroxyl group to which the next nucleotide is attached is replaced by N_3, which cannot be used by DNA polymerase.

the predominant strain of HIV in Africa is more readily transmitted than the other varieties, but if the transmission rate is even slightly above 1.0, an epidemic is inevitable.

Prospects for treatment Early attempts to combat AIDS suffered from a lack of knowledge about the mode of transmission of this rapidly spreading disease, its remarkably long latency, and an almost complete ignorance of its elaborate life history. Now that more is known, one thing is immediately obvious: though nearly all conventional vaccines use a surface antigen as the agent to teach the immune system to recognize a particular disease, this approach will not work with AIDS since it merely prepares the body to destroy its own immune-system cells. Instead, some step in the life cycle must be blocked. Preventing binding by masking the CD4 molecules is also clearly useless, since this would inactivate the immune response. Blocking the GP120 active site is a possibility—unless, of course, it turns out to be an analogue of the MHC molecule to which the T-cell receptors normally bind, which seems all too likely.

Preventing a step inside the infected cell is more promising, though that step must be unique to the virus or the treatment will be toxic to the host. A good example of this tradeoff is seen with 3′-azido-2′, 3′-dideoxythymidine (AZT, Figure 15.21). This thymidine analogue is actually preferred by HIV's DNA polymerase over thymidine, whereas host-cell polymerases prefer conventional thymidine. When AZT is incorporated into a growing DNA, replication stops because the host-cell polymerase is unable to add the next nucleotide. At suitably low doses of AZT, host cells replicate fairly normally (though there is substantial attrition in the bone marrow, where rapid production of blood cells is always under way, and anemia is a frequent side effect), but HIV only rarely reproduces successfully in the presence of this drug. AZT can delay the death of patients with clear symptoms by about a year, and may be more effective among those more recently infected.

One big risk with AZT is that the HIV polymerase gene, which has an unusually high error rate (one mispairing per 2000 bases in a genome only about 10,000 nucleotides long) will mutate to discriminate against AZT. Fortunately, this has not yet been observed. The drug is also expensive to synthesize and must be administered several times a day indefinitely. The search for a more practical and effective treatment is being intensively pursued. An anti-sense RNA complementary to part of the HIV genome, for instance, would be a better treatment because it would be completely specific

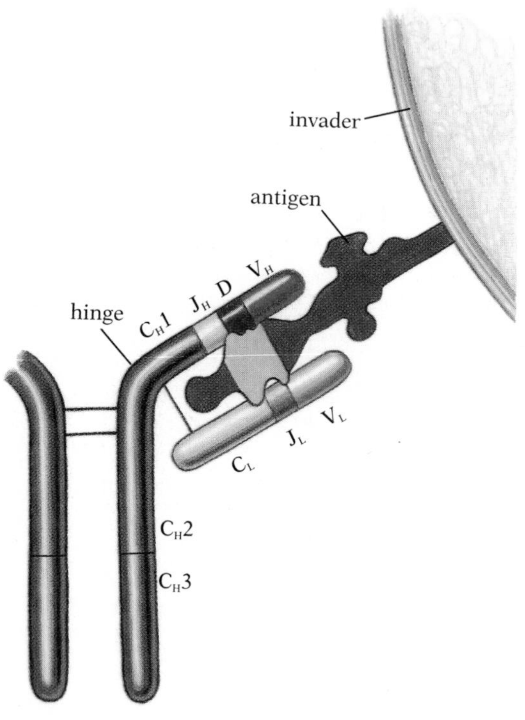

15.22 The antigen-binding site of an antibody molecule
The binding site is a pocket formed by the interaction of a heavy chain and a light chain. The sectors labeled C_H3, C_H2, hinge, C_H1, J_H, D, V_H, C_L, J_L, and V_L on this schematic representation are encoded by separate exons in the genes. The five different classes of heavy chains have different C_H2 and C_H3 regions; class E and class G chains have a fourth C_H region. Only one of the antigenic determinants of the antigen (shaded) is bound by this antibody. Completely different antibodies may bind to separate sites on the same kind of antigen. (The left half of the antibody beyond the hinge is omitted.)

and nontoxic. To make it work, however, means must be found to deliver it efficiently to target cells and make it resistant to RNAse. Injecting the portion of CD4 to which the virus binds holds promise as well. At the moment, though, prevention through behavioral changes among those at risk seems far and away the most practical alternative.

THE GENETIC BASIS OF ANTIBODY DIVERSITY

THE ROLE OF EXON RECOMBINATION

How is the vertebrate genome able to produce an almost infinite variety of antibodies and T-cell receptors without employing millions or billions of different genes, one for each potential antigen? The answer to this question involves a remarkable strategy of exon selection, which operates initially during the fetal development of the immune system, and leads to the vast diversity of lymphocytes available to the organism during its lifetime. In B cells, the process involves selection from a pool of exons in only three genes—one for the heavy chains and one for each of the two classes of light chains. The strategy used for T cells is identical in most details. We will discuss only B cells.

Let's look at the chromosomes of a developing human B lymphocyte in the bone marrow before the cell specializes to produce a single type of antibody—that is, before it becomes a virgin B lymphocyte. In the human genome, the part of the DNA coding for the heavy chain is found on chromosome 14 (the full complement of human chromosomes, with their assigned numbers, is shown in Fig. 12.4, p. 312); the segment that codes for the constant regions of this chain is composed of 22 exons arranged in five sets, one corresponding to each class of heavy chain—A, D, E, G, and M. Each set contains exons for regions C_H1, hinge, C_H2, C_H3, and, in the sets that code for heavy chains of classes E and G, C_H4 (Fig. 15.22).

This arrangement of exons in the heavy-chain gene seems at first glance to be counterproductive, since so much has to be transcribed that is ultimately not used: thus, even if the virgin lymphocyte will produce only antibodies with class A chains, exons for the other classes would be transcribed as well. Toward the 5′ end of the coding DNA strand, the organization seems even more counterproductive: there are four different exons for the J_H (joiner) region, even though each heavy chain has only one J_H segment. Still farther toward the 5′ end, beyond the exons for the joiner region, are approximately 12 different exons for the D region. Again, each antibody produced by a virgin cell has only one D segment. Still farther toward the 5′ end lies a string of roughly 200 different V_H exons, only one of which is eventually translated.

How is it, then, that the roughly 240 exons in the original heavy-chain antibody gene give rise to the protein product encoded by only seven or eight exons? The removal of superfluous exons is not yet completely understood, but it seem to be accomplished in two stages. In the first step, which takes place during fetal development, large regions of the gene are cut out of chromosome 14, and the remaining ends are spliced together by specialized enzymes. One such excision, for example, extends a random distance from the D region into the V_H exons (Fig. 15.23). Any number of V_H exons

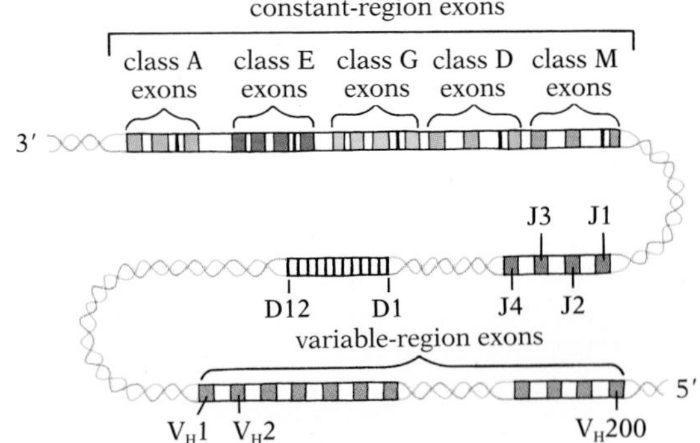

A Immature gene, before excision of any exons

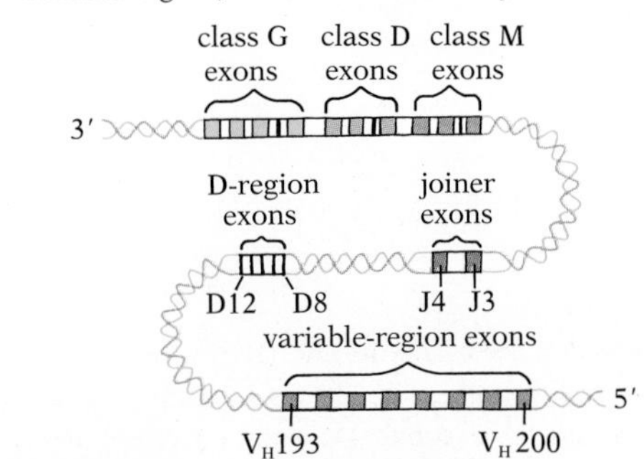

B Mature gene, as it will be transcribed

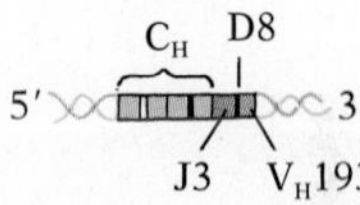

C mRNA, after processing of the primary transcript

15.23 Organization of immature and fully mature heavy-chain antibody gene, and corresponding mRNA
The immature heavy-chain gene (A) originally contains five sets of constant-region exons (one set for each class of heavy chain), four different joiner exons, approximately 12 different D-region exons, and about 200 different variable-region exons. When the gene is fully mature (B), many of the exons have been removed. The remaining exons and introns are transcribed, but only the exon closest to the 5′ end of the primary RNA transcript of each exon group survives processing and appears in the mature mRNA (C). The two steps of exon removal enable a B lymphocyte to produce a unique kind of antibody from millions of possible alternatives.

from 0 to 199 may be removed, but excisions of a relatively small number of exons are more common. The first V_H exon not excised is the one ultimately translated and expressed. During transcription, RNA polymerase copies all the remaining V_H exons on the 5′ side, but these superfluous transcribed exons are removed during RNA processing to yield an mRNA with only the first V_H exon that survived excision. As a result, only the V_H exon closest to the D-region exon will be translated (Fig. 15.23). All but one transcribed joiner exon and D-region exon are removed by means of two similar steps. A corresponding process is seen in the gene for the light chain, which has one C_L exon, four alternative joiner exons, and about 300 alternative V_L exons.

By analogy with the V_H, D, and joiner exons, you might guess that some C_H exons would be excised from the chromosomes, and all but one of the surviving set of exons removed from the primary transcript. There is debate on this point, but it is clear that something like this must be happening in at least some cells: many fully mature B lymphocytes remain able to switch between C_H classes, or even to produce two kinds of heavy chains simultaneously. Apparently the initial RNA transcript can contain the exons for more than one of the five C_H classes, and all but one set are removed along with introns in the course of RNA processing.

The diversity made possible by an active but random assembly of exons like the one just described is vast. For instance, since the gene for the heavy chain has four alternative joiner exons and 12 alternative D-region exons, there are 48 (4×12) possible joiner/D-region combinations. And since there are roughly 200 V_H alternative exons, a lymphocyte can produce antibodies with one of 9600 (48×200) forms of the variable region of the heavy chain. Similarly, the light chain, with its four alternative joiner exons and 300 V_L exons, can be produced in 1200 different forms. Since each antibody consists of one kind of randomly "selected" heavy chain and one kind of light chain, these different forms will give rise to about 12 million (1200×9600) different antibodies during the development of the immune system. Additional variation is generated by a process of active mutation in the joiner and D-region exons and parts of the V_L and V_H exons. Since the mutations take place in developing B cells, which are somatic cells, they are unique to each cell and are not perpetuated in the germ line (gametes).

HYPERMUTATION

As an immune response gets under way, the affinity of the B-cell antibodies produced by some cells in the activated clone actually increases—that is, the specificity of the antibodies involved appears to be fine-tuned in some way. How is this possible? The answer reveals yet another level of immunological "learning." Just as localized mutations can create variety beyond the 12 million types of B-cell antibodies possible from simple exon recombination, mistakes are permitted at a million times the normal rate when the exons for the variable regions of the antibody genes are replicated after a virgin cell is stimulated by an antigen. This error rate means that roughly half of the daughter cells will have slightly modified versions of the original antibody. Because the further cloning of cells from this first generation of daughters depends on how well each lymphocyte's version of the antibody binds the antigen on the cell membrane, a mutant with a better match will outreproduce its clone mates, with the result that the cells producing the better-binding antibody will begin to dominate the response. Here is a kind of natural selection in miniature. As it happens, the production of unique antibody genes by means of exon recombination provides a powerful model for how novel genes can be (and certainly have been) generated over the course of evolution. We will look at this hypothesis in detail in Chapter 17.

EVOLUTION OF THE IMMUNE SYSTEM

We said at the outset of this chapter that the molecules of the immune system are related to the cell-adhesion system we examined in the last chapter. Of course, differential cell adhesion is critical for immune-system cells: not only must they attach themselves to their targets, but also lymphocytes must be able to move to and from the blood, lymph, and any tissue that is releasing the chemical signals that draw these cells to the site of trouble. Every change of location involves turning off at least one specific set of cell-adhesion molecules and activating at least one other. But the similarity is more than descriptive: the molecules upon which the lymphocytes rely clearly evolved from the proteins that animals have been using for hundreds of millions of years—long before the first immune systems appeared. For example, some vertebrate immunoglobulins are virtually identical to CAMs discovered in *Drosophila,* where they are involved in axonal guidance. As we will see in Chapter 17, the process has been one of repeated gene duplication, followed by the independent evolution of the different copies.

Let's look a bit closer at a partial cast of these molecular actors (Fig. 15.24); the representation used here emphasizes structural similarities, but the likeness extends to nucleotide sequence as well. One of the most telling similarities is between the light chains of antibody molecules, the α chain of the T-cell receptor, and the cell-adhesion molecule ICAM-2. This particular CAM is expressed on cells in inflamed regions (in response to locally secreted chemicals like the interleukins) and thus encourages T-cell adhesion. ICAM-2 is also very similar to the MHC molecules, CD2 (which helps T cells find antigen concentration in lymph nodes), LFA-3 (which is the molecule to which CD2 binds), and the β chain of the T-cell receptor. Except that they are missing a transmembrane anchor, the light chains of antibody molecules also strongly resemble the members of this ancient cell-adhesion family.

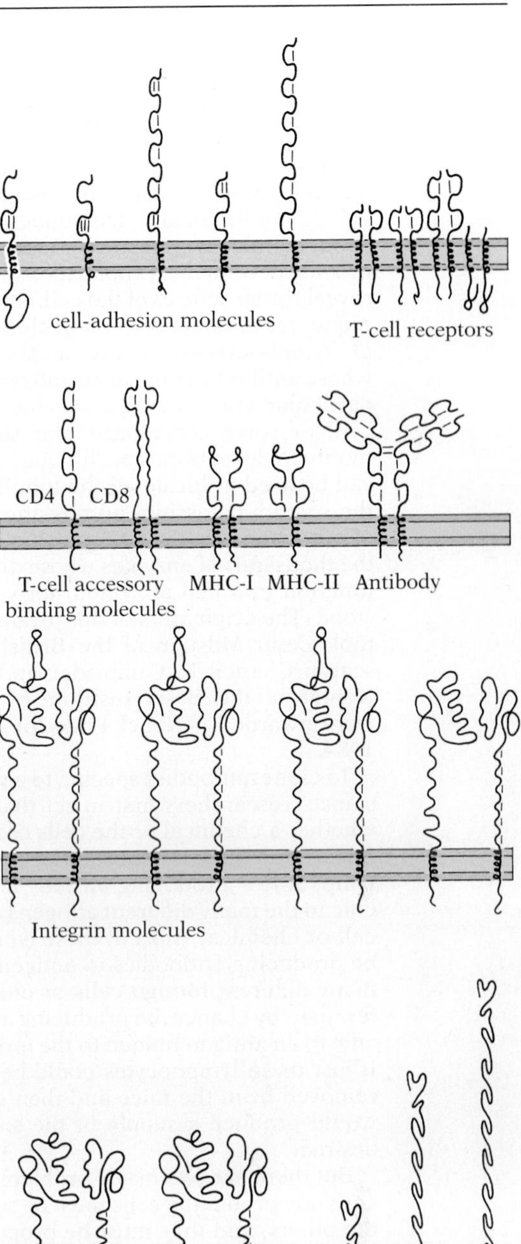

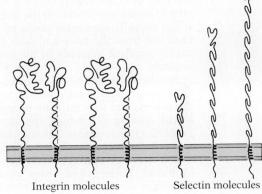

15.24 Some molecules used by the immune system
Most of the proteins and protein complexes used by immunocytes for cell binding fall into three molecular "families," each of which is part of a larger family of developmental-control substances. The similarities are obvious at a glance: compare the structure of antibodies to MHC complexes, T-cell receptors, and CAMs. Other immune-system-specific members of developmental-control families are shown in the lower half of the figure.

Exploring Further

MONOCLONAL ANTIBODIES

A growing understanding of the workings of the immune system has enabled researchers to develop revolutionary techniques for learning about cellular architecture—techniques that can provide new medical applications while they help reveal the structures of the cell. In one such technique, researchers select and clone a single type of lymphocyte—typically a B lymphocyte—whose antibodies bind to an antigenic region on a particular kind of cell, a specific structure in a cell, or some other particular substance. This monoclonal antibody technique, as it is called, can be used to locate all the tubulin in a cell, or the sodium-potassium pumps, the F_1 complexes of mitochondria, the RNA polymerases, or any of the thousands of enzymes whose distribution and function in a cell are often not yet well understood. The originators of this invaluable research tool, Cesar Milstein of the British Medical Research Council in Cambridge and George J. F. Kohler of the Basel Institute of Immunology, were awarded a Nobel Prize for their work in 1984.

To clone antibodies specific to a particular substance, researchers first inject that substance—whether a chemical or the cells of interest—into mice. Soon there is a substantial increase in the B lymphocytes producing various antibodies specific to the many different antigens on the foreign cell or chemical. Most of these lymphocytes will be producing antibodies to antigens common to many different foreign cells or chemicals, but a few may, by chance, be producing antibodies specific to an antigen unique to the foreign material. If just these lymphocytes could be selected and removed from the mice and then cultured, they would produce a supply of the single antibody desired.

But there are two major problems: the specific antibody-producing cells must be separated from the others, and they must be propagated. In the actual cloning procedure, the second of these problems is solved first. You may recall that normal cells have a set number of cell divisions, after which they age and die, while cancer cells are immortal. If the lymphocytes were cancerous, then, they would multiply indefinitely. Hence the next step is to force cells of the mixed collection of lymphocytes to fuse with cells from a cancerous line of B lymphocytes. Those that fuse become immortal; they are then called hybridomas.

This still leaves the problem of selection. It is

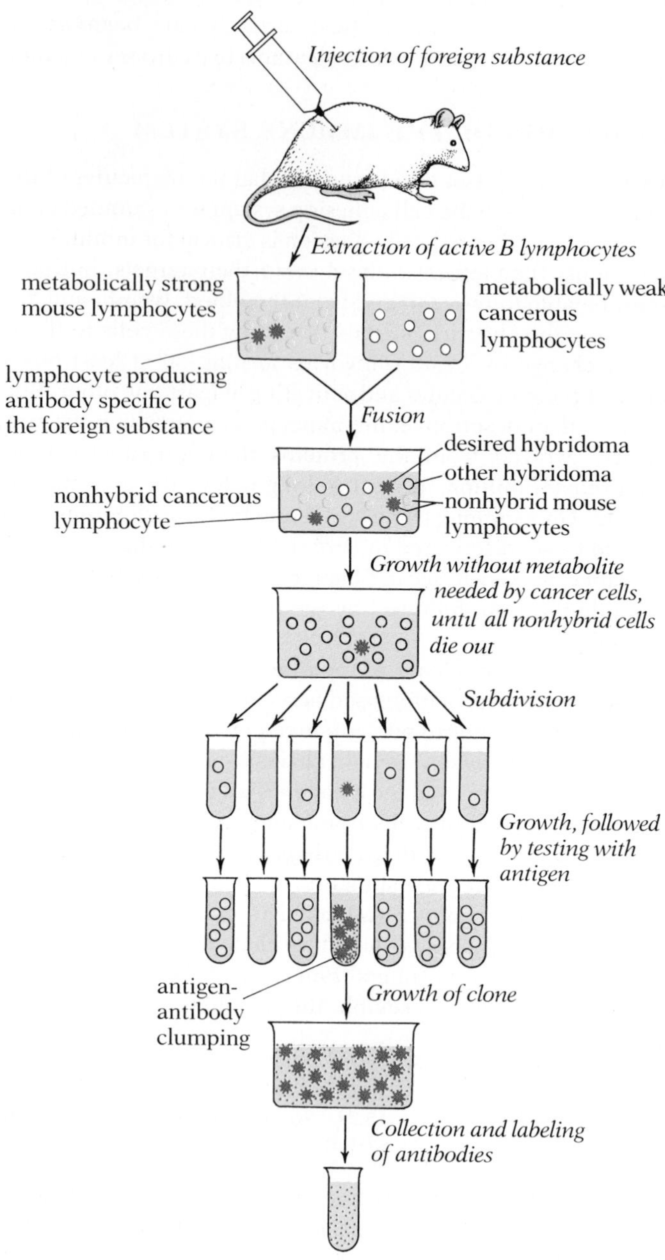

necessary to eliminate all but the hydridoma cells, and then to select those producing the desired antibody. The elimination is relatively easy. The nonhybrid lymphocytes from the mouse just die off after completing their set number of cell divisions. Milstein and Kohler solved the more difficult problem of sorting out the nonhybrid cancerous lymphocytes by using a line lacking a crucial synthetic pathway. When these cancer cells are cultured in a medium without the substance they cannot make, they die. Only the hybridomas—immortal by virtue of genes from the cancer cells, and able to synthesize the missing nutrient because of genes from the mouse lymphocytes—will survive.

The next step is to separate the individual hybridoma cells as much as possible, to culture each one separately, and then to test each culture by adding the antigen. If the cells of a particular group are producing the desired antibody, the test will result in antigen-antibody clumping. These cells can then be cultured further, to form a clone of immortal cells producing a continuous supply of a specific antibody. Recently researchers have accomplished a real *tour de force* by transplanting the antibody genes from certain hybridoma cells into plants, and (via the phage lambda) into *E. coli*, where these immunoglobins are produced far more rapidly and cheaply.

Researchers use monoclonal antibodies to tag specific cells, cellular structures, or chemicals by labeling the antibodies with fluorescent dye, the electron-dense substance ferritin, or a radioactive marker. Once a labeled antibody has reached its target, the location of the antigen can be found with a light microscope, an EM, or any of a variety of radioisotope analyses. A labeled antibody can even be used to mark certain kinds of cancer cells selectively for destruction; ovarian cancer cells, for example, can be bound by ferritin-linked antibodies and then destroyed by ferritin-specific T cells. The monoclonal antibody technique, then, has enormous potential both for basic research and for medical technology.

Three other components of the T-cell receptor (ε, γ, and δ, which communicate the fact of T-cell receptor binding to the interior of the T cell) are just truncated versions of ICAM-2, while CD8 is very like ICAM-2 with a long insertion. CD4, the heavy chains of antibody molecules, and two other immune-system CAMs are essentially ICAM-2s with a doubled outer segment.

Two other families of developmental adhesion molecules are also involved in the immune response. The immune system's ***integrins*** help lymphocytes home in on infection sites, and modulate T-cell proliferation. ***Selectins***, for their part, regulate lymphocyte binding to tissue at sites of inflammation. In short, lymphocytes have evolved to use the selective adhesion molecules of morphogenesis (suitably modified) to track down infections, where they bind both normal cell receptors and unfamiliar antigens prior to exerting their specific effects. In a way, many lymphocytes are like determined but undifferentiated stem cells in early development, migrating to their proper location by means of adhesion molecules and chemical gradients, except that the cells of the immune system come equipped with dozens of CAMs and other morphogenic binding molecules, and respond to any of a large number of gradients; hence, they have the potential to go anywhere in the body in search of cells broadcasting chemical cries for help. Selection has acted to transform the existing genetic systems of organismal development into a self-guided internal security system. As we will see in Chapter 17, many important genetic changes leading to the creation of new species occur in developmental-control genes, and the practice of remodeling an existing genetic system is far more common than building a new one from scratch.

Exploring Further

SLEEPING SICKNESS: HOW THE TRYPANOSOME CHANGES ITS COAT

Sleeping sickness is one of the most debilitating diseases in the world, and one of the most difficult to treat effectively. Its cause is a protozoan known as a trypanosome, which bloodsucking tsetse flies carry from an infected host to a potential one. The trypanosomes multiply in the blood of the host, and later spread into the nervous system. Symptoms begin with fever and fatigue, and progress to drowsiness during the day and insomnia at night. The victim ultimately becomes too sleepy to eat, then comatose, and finally dies.

Since the trypanosomes are in the bloodstream of the host for weeks before causing death, how is it that the immune system, with its millions of antibodies generated by exon shuffling, fails to react effectively? The answer is that during these weeks the trypanosomes are shuffling their genes as well.

Like all organisms, trypanosomes have proteins exposed on their cell membranes. But unlike most organisms, an individual trypanosome has only one kind of surface protein. Still, the immune system has one or more antibodies that can bind to the antigen of any particular trypanosome surface protein, so in just a few days a full-scale antibody response is under way, with macrophages, killer cells, and complement-system channels working in concert to destroy the invading organisms.

But by the fifth day after infection, all the surviving trypanosomes have changed their surface proteins. The trypanosome genome has hundreds—perhaps thousands—of surface-protein genes, and a new one is transposed into an active site in the genome roughly every five days. On each trypanosome the new surface protein may be any of many, so the number of trypanosome antigens may increase by the fifth day from the few present at the time of infection to a great many. As a result, the existing highly specific immune reaction becomes irrelevant, and a new one, responsive to a larger set of antigens, must begin essentially from scratch: a new, different set of lymphocytes must now bind to the parasites, to begin again the process in which virgin B and T lymphocytes are activated and divide, antibodies are produced, and so on.

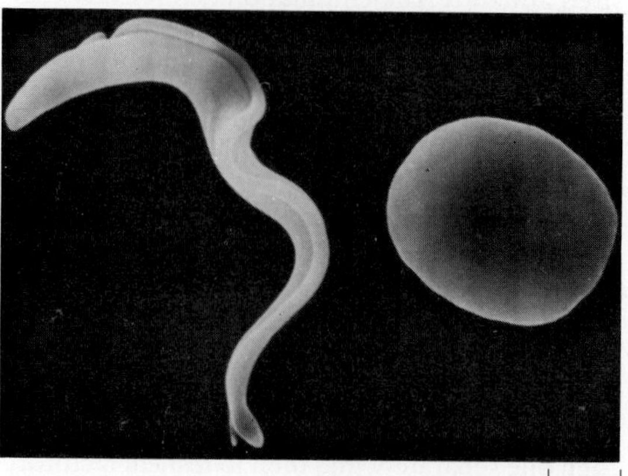

Trypanosome and a red blood corpuscle

But again, even if the battle is being won, the surface-protein genes are shuffled. New surface proteins appear, and the number of new antigens multiplies further. Eventually, the body's immune system is simply overwhelmed. Transposition in trypanosomes, then, seems to be a precisely controlled event. Transposons may be more common and more important in the history of disease than most researchers used to think.

STUDY QUESTIONS

1. Compare and contrast clonal selection with clonal deletion. (pp. 386–87, 394)

2. What would be the physiological effect on an individual born with a defect in either histamine release or reception? (p. 385)

3. What happens when, by chance, the virgin B cell for an antigen is never generated (that is, the appropriate antibody gene is not created by recombination and hypermutation) but the helper and suppressor cells

are present? What if it's only the helper that is missing? Or just the suppressor? (pp. 383–93)

4. What would be the consequences of having a B-cell system that produced hybrid antibodies—that is, every cell produced antibodies, one arm of which bound to one antigen while the other bound to another? (pp. 383–86)
5. What happens to a cell if it suffers a mutation in its MHC gene? (Consider both immunocytes and other cell types.) (pp. 388–95)
6. How could the MHC system be exploited by the body to avoid inbreeding (mating with close kin)? (pp. 393–94)

CONCEPTS FOR REVIEW

- Role of the lymphatic system
- Humoral immune response
  - Role of cells involved
    - virgin B cells
    - memory B cells
    - plasma (antibody-secreting) B cells
    - macrophages
    - natural killer cells
    - mast cells
  - Role of molecules involved
    - antibody types
    - complement cascade
    - histamine
    - interleukin
  - Structure and function of antibodies
  - Clonal selection
- Cell-mediated immune response
  - Structure and function of the T-cell receptor
  - Cytotoxic T cells
    - structure and function of MHC-I
    - operation of cytotoxic cells
  - Helper T cells
    - presentation mechanism
    - structure and function of MHC-II
    - operation of helpers
    - operation of suppressor T cells
    - role of chemical signals in T-cell activity
- Development of self-tolerance
  - B-cell system
  - T-cell system
  - Transplant (graft) rejection
  - Autoimmunity
- Basis of AIDS-mediated destruction of the immune system
- Generation of antibody diversity

SUGGESTED READING

ANDERSON, R. M., and R. M. MAY, 1992. Understanding the AIDS pandemic, *Scientific American* 266 (6). *On the epidemiology of the AIDS epidemic from an ecological and evolutionary point of view.*

BEAUCHAMP, G. K., K. YAMAZAKI, and E. A. BOYSE, 1985. The chemosensory recognition of genetic individuality, *Scientific American* 253 (1). *On how MHC complex is also responsible for individual-specific odors some animals use for recognizing one another.*

BUISSERET, P. D., 1982. Allergy, *Scientific American* 247 (2). (Offprint 1522) *A clear explanation of how the immune system's overreaction to harmless antigens can create annoying and even dangerous allergies.*

CAPRA, J. D., and A. B. EDMUNDSON, 1977. The antibody combining site, *Scientific American* 236 (1). (Offprint 1350) *On the evolution of antibodies and how they react with antigens.*

COHEN, I. R., 1988. The self, the world, and autoimmunity, *Scientific American* 258 (4). *About the rare failures in the immune system's ability to "remember" and ignore self-antigens. It argues for a speculative hypothesis that assumes that the variable region of each kind of antibody is recognized, bound, and deactivated by another kind of antibody. Out of these complex interactions immunological responses are regulated, or autoimmune attacks generated.*

COLLIER, R. J., and D. A. KAPLAN, 1984. Immunotoxins, *Scientific American* 251 (1). (Offprint 1552) *On attempts to bind toxins to monoclonal antibodies specific for tumor cells.*

COOPER, M. D., and A. R. LAWTON. 1974. The development of the immune system, *Scientific American* 231 (5). (Offprint 1306)

CUNNINGHAM, B. A., 1977. The structure and function of histocompatibility antigens, *Scientific American* 237 (4). (Offprint 1369) *On the role of the MHC antigens on the surface of normal cells, including transplant rejection on the one hand and defense against cancer and infection on the other.*

DONELSON, J. E., AND M. J. TURNER, 1985. How the trypanosome changes its coat, *Scientific American* 252 (2). (Offprint 1557) *A discussion of how the parasite evades the immune system.*

EDELSON, R. L., and J. M. FINK, 1985. The immunologic function of skin, *Scientific American* 252 (6). *On how cells in the skin inter-*

act with T cells.

GOLDE, D. W., and J. C. GASSON, 1988. Hormones that stimulate the growth of blood cells, *Scientific American* 259 (1). *On the differentiation of blood cells and its hormonal control.*

GOSDON, G. N., 1985. Molecular approaches to malaria vaccines, *Scientific American* 252 (5). *An example of how haptens are used in designing vaccines.*

GREY, H. M., A. SETTE, and S. BUUS, 1989. How *T* cells see antigen, *Scientific American* 261 (5). *A clear discussion of how processed antigens are "presented" by MHC proteins, and recognized by T-cell antibodies.*

HASSELTINE, W. A., and F. WONG-STAAL, 1988. The molecular biology of the AIDS virus, *Scientific American* 259 (4). *An up-to-date description of how HIV-1 works, emphasizing gene regulation. This same issue has nine related articles on various aspects of the AIDS epidemic.*

HENLE, W., G. HENLE, and E. T. LENNETTE, 1979. The Epstein-Barr virus, *Scientific American* 241 (1). (Offprint 1431) *A wide-ranging discussion that pulls together immunology, research with monoclonal antibodies, viral biochemistry, and disease statistics to link Epstein-Barr virus with Burkitt's lymphoma.*

LEDER, P., 1982. The genetics of antibody diversity, *Scientific American* 246 (5). (Offprint 1518) *On how the exons of antibody genes are combined to create enormous diversity.*

LERNER, R. A., 1983. Synthetic vaccines, *Scientific American* 248 (2). (Offprint 1533) *On how a knowledge of antigen–antibody interaction enables researchers to synthesize a single one of the antigenic determinants of a virus or bacterium and use the resulting harmless chemical as an effective and safe vaccine.*

MARRACK, P., and J. KAPPLER, 1986. The *T* cell and its receptor, *Scientific American* 254 (2). *An excellent summary of how T cells interact with the other elements of the immune system.*

MAYER, M. M., 1973. The complement system, *Scientific American* 229 (5). (Offprint 1283) *On the way an intricate set of enzymes works with antibodies to make novel channels in the membranes of foreign cells, thereby destroying them.*

MILLS, J., and H. MASUR, 1990. AIDS-related infections, *Scientific American* 263 (2).

MILSTEIN, C., 1980. Monoclonal antibodies, *Scientific American* 243 (4). (Offprint 1479)

OLD, L. J., 1977. Cancer immunology, *Scientific American* 236 (5). (Offprint 1358) *On the distinctive antigens on the surfaces of cancer cells and the problem of mobilizing the immune system to combat cancer.*

OLD, L. J., 1988. Tumor necrosis factor, *Scientic American* 258 (5). *On the many messenger chemicals immune-system cells use to regulate each other's activity.*

REDFIELD, R. R., and D. S. BURKE, 1988. HIV infection: the clinical picture, *Scientific American* 259 (4). *On the progression from infection to death, detailing how the immune system holds the disease at bay for so long.*

ROSE, N. R., 1981. Autoimmune diseases, *Scientific American* 244 (2). (Offprint 1491) *What happens when the immune system attacks an organism's own cells.*

SNYDER, S. H., and D. S. BREDT, 1992. Biological roles of nitric oxide, *Scientific American* 266 (6). *On the many roles of this newly discovered transmitter, local chemical mediator, and toxic weapon.*

TONEGAWA, S., 1985. The molecules of the immune system, *Scientific American* 253 (4). *An excellent review of antibody structure, antigen binding, and B- and T-cell function. Does not discuss the interactions between T cells, B cells, macrophages, and the other elements of the immune system.*

YOUNG, J. D.-E., and Z. A. COHN, 1988. How killer cells kill, *Scientific American* 258 (1). *Compares the action of the complement system with the analogous strategy of killer cells in perforating the membranes of target cells.*

Chapter **16**

INHERITANCE

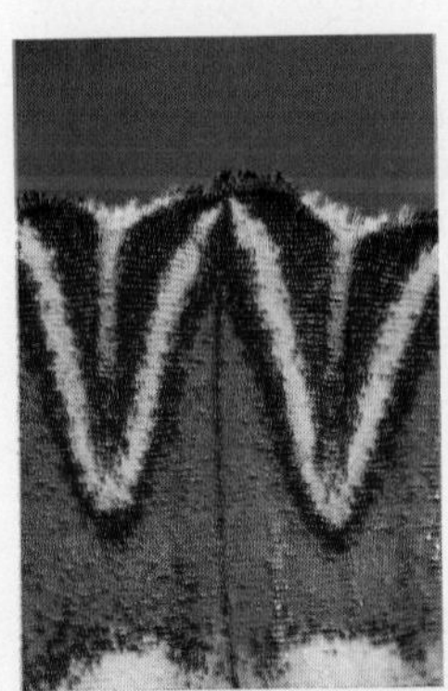

e have examined the nature of the genetic material in the preceding chapters—how it is replicated, and the ways in which elaborate control systems influence the expression of genes and orchestrate the development of multicellular organisms. We have had relatively little to say so far about the characteristics, or traits, that these genes produce, or how different alleles of a gene can interact, or how recombination can affect the traits expressed in the progeny. In short, what are the practical, visible effects of the complex genetic mechanisms we have so far studied?

For at least as long as history holds records of the subject, people have known that many observable traits, ranging from morphological (such as, in dogs, height and build) to behavioral (such as the strong breed propensities toward herding or hunting or retrieving) can be inherited. Hundreds and even thousands of years before the discovery of DNA this knowledge was used in the development of better grains, vegetables, fruits, and domesticated animals. At the same time, many puzzling observations emphasized how much remained to be understood. For instance, breeders noticed that when plants produce runners, thus reproducing asexually by cloning, the progeny were exactly like the parents. When plants reproduced sexually, however, the progeny could be very different not only from the parents but from each other as well. The most common hypothesis until the beginning of this century, that of ***blending inheritance***, predicted that progeny should always embody a recognizable blend of parental traits. Nearly everyone knew, however, that this hypothesis did not hold true: among Caucasians, for example, a small proportion of the children of two brown-eyed parents would have blue eyes; and in any group, traits absent for generations would sometimes mysteriously reappear, while, rarely, altogether new ones would be generated.

The discovery of the laws of inheritance not only made sense of these seeming contradictions, opening the way to the discovery of their genetic bases, it also made possible a far deeper understanding of how selection works to create evolution, which will be the topic of the next two chapters.

MONOHYBRID INHERITANCE

The first person known to have made sense of the conflicting phenomena of inheritance was the Austrian monk Gregor Mendel. Working at first to breed an industrious yet docile honey bee, Mendel began his experiments in inheritance by crossing a hardworking race of German bee, *Apis mellifera carnica*, with the gentler and more attractive Italian race, *A. m. ligustica*. The result of this cross was neither industrious nor tractable. Here was a perfect example of the problem with the hypothesis of blending inheritance.

In 1856, Mendel wisely shifted his attention to garden peas, an organism he found considerably more manageable. The experiments he performed over the next twelve years laid the groundwork for what is now called the ***chromosomal theory of inheritance***. First published in 1866, Mendel's results are all the more remarkable in that they represent the only known line of lucid genetic research at the time. A combination of insight and careful methodology—Mendel meticulously counted all the different types of progeny produced in his experiments—led him to conclusions reached in his day by no other researchers on inheritance. Unlike many other great discoveries, Mendel's seem to have been the achievement of an isolated genius.

EXPERIMENTS BY MENDEL

Mendel began with several dozen strains of peas, mostly purchased from commercial sources. He raised each variety for several years to discover which strains had recognizable morphological variations that bred true (i.e., reappeared consistently in each generation).

Mendel's results In the end, Mendel reported his work on seven of the numerous characteristics of garden peas that he studied. He noticed that each of these seven characteristics occurred in two contrasting forms: the seeds were either round or wrinkled, the flowers were red or white, the pods were green or yellow, and so on (Table 16.1). When Mendel bred plants with contrasting forms of just one of these characteristics, he found that all the offspring (usually referred to as the F_1, or first filial, generation) were alike and resembled only one of the two parents (the ***P***, or parental, generation). When these offspring were crossed among themselves, however, some of their offspring (the F_2, or second filial, generation) showed one of the original contrasting traits and some showed the other (Table 16.1). In other words, a trait that had been present in one of the parents, but not in any of

TABLE 16.1 *Mendel's results from crosses involving single character differences*

P characters	F_1	F_2	F_2 *ratio*
1. Round × wrinkled seeds	All round	5474 round : 1850 wrinkled	2.96 : 1
2. Yellow × green seeds	All yellow	6022 yellow : 2001 green	3.01 : 1
3. Red × white flowers	All red	705 red : 224 white	3.15 : 1
4. Inflated × constricted pods	All inflated	882 inflated : 299 constricted	2.95 : 1
5. Green × yellow pods	All green	428 green : 152 yellow	2.82 : 1
6. Axial × terminal flowers	All axial	651 axial : 207 terminal	3.14 : 1
7. Long × short stems	All long	787 long : 277 short	2.84 : 1

their children, reappeared in the next generation, just as blue eyes can reappear after a generation or more of brown eyes.

Mendel's cross of red-flowered with white-flowered plants, for example, produced red flowers in all plants of the F_1 generation (Fig. 16.1). Similarly, when he crossed plants having round seeds with plants having wrinkled seeds, all the offspring had round seeds. Apparently one form of each characteristic had taken precedence over the other: red color had taken precedence over white in the flowers, and round form had taken precedence over wrinkled in the seeds. Mendel referred to the traits that appear in the F_1 offspring of such crosses as ***dominant characters***, and the traits that are latent in the F_1 generation (in these examples, white flowers and wrinkled seeds) as ***recessive characters***.

When Mendel allowed the F_1 peas from the cross involving flower color, all of which were red, to breed freely among themselves, their offspring (the F_2 generation) were of two types; there were 705 plants with red flowers and 224 with white flowers. The recessive character had reappeared in approximately one-fourth of the F_2 plants. Similarly, when F_1 peas from the cross involving seed form, all of which had round seeds, were allowed to breed freely among themselves, their F_2 offspring were of two types; 5474 had round seeds and 1850 had wrinkled seeds. Again, the recessive character had reappeared in approximately one-fourth of the F_2 plants. The same thing was true of crosses involving the other five characters that Mendel reported on in his classic paper (Table 16.1); in each case the recessive character disappeared in the F_1 generation, but reappeared in approximately one-fourth of the plants in the F_2 generation. We can summarize the results of the experiment involving flower color as follows:

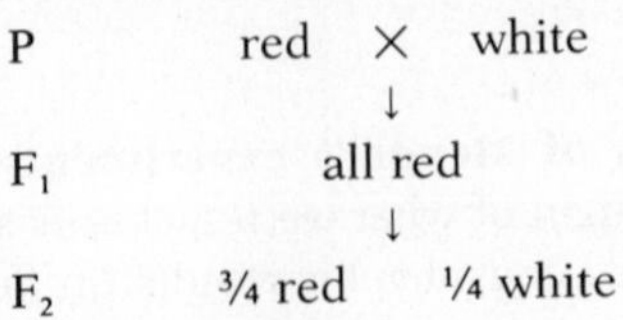

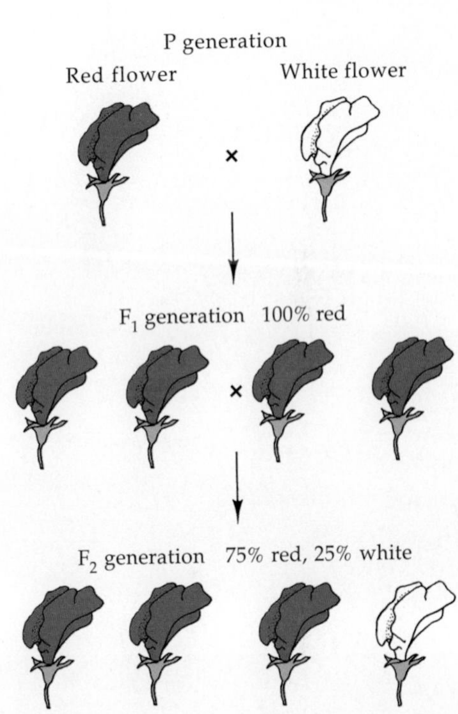

16.1 Results of Mendel's cross of red-flowered and white-flowered garden peas

Mendel's conclusions Experiments of this type led Mendel to formulate and test a series of hypotheses. Finally, he drew the revolutionary conclusion that each pea plant possesses two hereditary "factors" for each character, and that when gametes are formed the two factors segregate and pass into separate gametes. Each gamete, then, possesses only one factor for each character. Each new plant thus receives one factor for each character from its male parent and one for each character from its female parent. That two contrasting parental traits, such as red and white flowers, can both appear in normal form in the F_2 offspring indicates that the hereditary factors must exist as separate entities in the cell; they do not blend or fuse with each other. Thus the cells of an F_1 pea plant from the cross involving flower color contain, according to Mendel, one factor for red color and one factor for white color, the factor for red being dominant and the factor for white being recessive. But their coexistence in the same nucleus does not change them; the red and white do not alter each other. They remain distinct, and segregate unchanged when new gametes are formed. Mendel referred to this pattern of inheritance as the ***Principle of Segregation***.

Mendel's conclusions are consistent with what we now know about the chromosomes and their behavior in meiosis: the diploid nucleus contains two of each type of chromosome, one homologous chromosome from each parent. Each of the two chromosomes in any given pair bears genes for the same characters; hence the diploid cell contains two copies of each type of gene; these are the two hereditary factors for each character that Mendel described. Since homologous chromosomes segregate during meiosis, gametes contain only one chromosome of each type and hence only one copy of each gene, just as Mendel deduced.

Mendel's theories seem rather obvious in the light of the events of cell division. But he did his work before the details of cell division had been learned—before, in fact, the significance of chromosomes for heredity had been discovered. Mendel arrived at his conclusions purely by reasoning from the patterns of inheritance he detected in his experiments, without any reference to the structural components of the cell or its nucleus.

Oddly enough, Mendel's well-controlled experiments and brilliant deductions had little effect on the scientific world of his day. Many historians argue that the scientific "establishment" was not ready for the radical notion that pairs of nonblending "factors" underlie inheritance, but Mendel himself may also have underestimated the value of his conclusions. Some of his unpublished results could not be reconciled with his model, for, as we will see, not all genetic characters can be categorized as simply dominant or recessive, while others do not segregate independently. Mendel abandoned his effort to comprehend the rest of the puzzle in 1868. In 1900, after cell division and chromosome segregation had been observed and described in detail, three biologists—Hugo De Vries in Holland, Carl Correns in Germany, and Erich von Tschermak-Seysenegg in Austria—independently rediscovered the phenomenon of segregation and, almost immediately, Mendel's original paper.

A modern interpretation of Mendel's experiments Mendel's results make perfect sense in the light of what we now know about chromosome function and anatomy. Let's consider his results on flower color in peas from a modern perspective. In the cells of a pea plant, a gene for flower color exists at the same location, or ***locus***, on two homologous chromosomes (Fig. 16.2). The gene can exist in many different forms, or ***alleles***, but no individual can have more than two: one on each member of the pair of homologous chromosomes involved. It is customary to designate genes by letters, using capital letters for dominant alleles and small letters for recessive alleles. Flower color results from the presence of pigment molecules, whose structure is determined by the genetic information in the chromosomes. Mendel's red allele (*C*) encodes a functional pigment that absorbs all wavelengths of light except the red that we see reflected, while the white allele (*c*) either encodes a nonfunctional pigment molecule that reflects all wavelengths of light and so appears white, or fails to produce any pigment at all. Since a diploid cell contains two copies of each gene, it may have two copies of the same allele or one copy of one allele and one copy of another allele (Fig. 16.2). Thus the cells of a pea plant may contain two copies of the allele for red flowers (*C*/*C*), or two copies of the allele for white flowers (*c*/*c*), or one copy of the allele for red and one copy of the allele for white (*C*/*c*). Cells with two copies of the same allele (*C*/*C* or *c*/*c*) are said to be ***homozygous*** for that trait. Those with one each of two different alleles (*C*/*c*) are said to be ***heterozygous***. (The slash between letters indicates that the two copies of a gene are on separate chromosomes.)

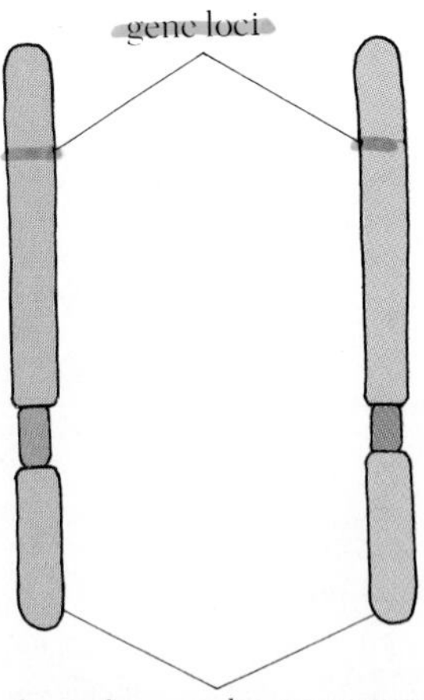

16.2 Anatomy of segregation
Somatic cells have pairs of homologous chromosomes, a pair consisting of one chromosome from each parent. Each gene is found in two copies, one on each chromosome of the homologous pair, at corresponding loci. When the genes at the corresponding loci are different, they are called alleles. In meiosis, each gamete receives a copy of only one chromosome from each homologous pair, and hence only one of the two alleles.

Unfortunately, we cannot tell by visual inspection whether a given pea plant is homozygous dominant (*C*/*C*) or heterozygous (*C*/*c*), because the two types of plants look alike: both have red flowers. In other words, where one allele is dominant over another, the dominant allele takes full precedence over the recessive allele, and a heterozygous organism exhibits the trait determined by that dominant allele; one copy of the dominant allele is as effective as two copies in determining the character trait. For this reason, there is often no one-to-one correspondence between the different possible genetic combinations, or ***genotypes***, and the possible appearances, or ***phenotypes***, of the organisms. In the example of flower color in peas discussed here, there are three possible genotypes, *C*/*C*, *C*/*c*, and *c*/*c*, but only two possible phenotypes, red and white. The heterozygote (*C*/*c*) produces a red phenotype because the red allele is dominant; the functional red pigment it specifies masks the product, if any, of the white allele. Just as we often cannot tell an organism's genotype—that is, whether an individual is heterozygous or homozygous for a dominant allele—natural selection cannot "see" or act on genotypes directly. Hidden recessives are immune to selection; it is an organism's phenotype—the sum of its expressed traits—that matters in the struggle to survive and reproduce. Only when recombination produces a homozygous recessive, and the trait it encodes is therefore expressed, can selection operate directly on the allele.

The functional/nonfunctional dichotomy is the most common basis of dominant and recessive phenotypes. Many traits result from alleles that code for nonfunctional structural proteins or inactive enzymes; in the presence of an active enzyme, or functional structural protein, the existence of the nonfunctional recessive version is hidden. Wrinkled peas, for instance, involve a gene encoding an enzyme that polymerizes sugar to make starch. The recessive allele has an insertion, and thus encodes a defective version of the enzyme. In the homozygous condition, this leads to a buildup of sugar, which draws extra water in osmotically; the developing seed thus becomes abnormally large. Water is removed from seeds when they mature; because these swollen seeds have more water and wall area than normal, wrinkles develop when the water is removed.

Using the modern conception of gene function, we can rewrite the summary of Mendel's pea cross as follows:

P	C/C red	×	c/c white	
		↓		
F_1	C/c red	×	C/c red	
		↓		
F_2	C/C red	C/c red	c/C red	c/c white

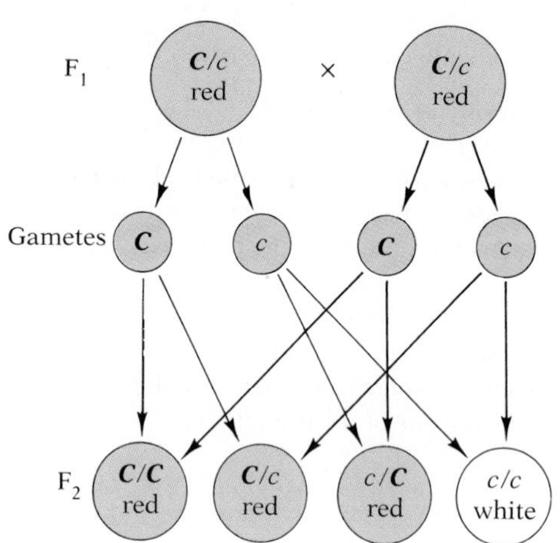

16.3 Gametes formed by F_1 individuals in Mendel's cross for flower color, and their possible combinations in the F_2

Here we see both the genotypes and the phenotypes of the plants in all three generations. Mendel began with a cross in the parental generation between a plant with a homozygous dominant genotype (red phenotype) and a plant with a homozygous recessive genotype (white phenotype). All of the F_1 progeny had red phenotypes, because all of them were heterozygous, having received a dominant allele for red (C) from the homozygous dominant parent and a recessive allele for white (c) from the homozygous recessive parent. But when the F_1 individuals were allowed to cross freely among themselves, the F_2 progeny were of three genotypes and two phenotypes; one-fourth were homozygous dominant and showed red phenotypes, two-fourths were heterozygous and showed red phenotypes, and one-fourth were homozygous recessive and showed white phenotypes. Thus the ratio of genotypes in the F_2 was 1:2:1, and the ratio of the phenotypes was 3:1.

How do we figure out the possible genotypic combinations in the F_2? This is an easy matter in a monohybrid cross (a cross involving only one character) such as this. All individuals in the F_1 generation are heterozygous (C/c); that is, they have one of each of the two types of alleles. Each of these two alleles is located on a different one of the two chromosomes of a homologous pair. In meiosis these two chromosomes synapse, move onto the spindle as a unit, and then separate, moving to opposite poles, so that the chromosome bearing the C allele is incorporated into one new haploid nucleus and the chromosome bearing the c allele is incorporated into the other new haploid nucleus. This means that half the gametes produced by such a heterozygous individual will contain the C allele and half the c allele. When two such individuals are crossed (Fig. 16.3), there are four possible combinations of their gametes:

C from male parent, C from female parent
C from male parent, c from female parent
c from male parent, C from female parent
c from male parent, c from female parent

The first of these four possible combinations produces homozygous dominant offspring (red); the second and third produce heterozygous offspring (also red); and the fourth produces homozygous recessive offspring (white).

Since each of these four combinations is equally probable, we would expect, if a large number of F_2 progeny are produced, a genotypic ratio close to 1:2:1 and a phenotypic ratio close to 3:1, just as Mendel found.

An easy way to figure out the possible genotypes produced in the F_2 is to construct a so-called Punnett square, named after the Cambridge geneticist R. C. Punnett. Along a horizontal line, write all the possible kinds of gametes the male parent can produce; in a vertical column to the left, write all the possible kinds of gametes the female parent can produce; then draw squares for each possible combination of these, as follows:

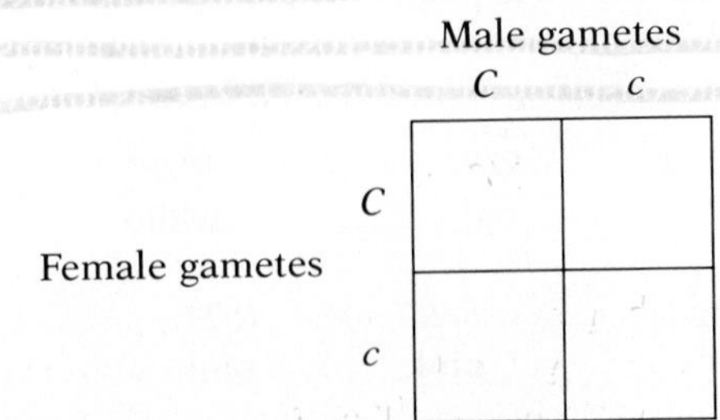

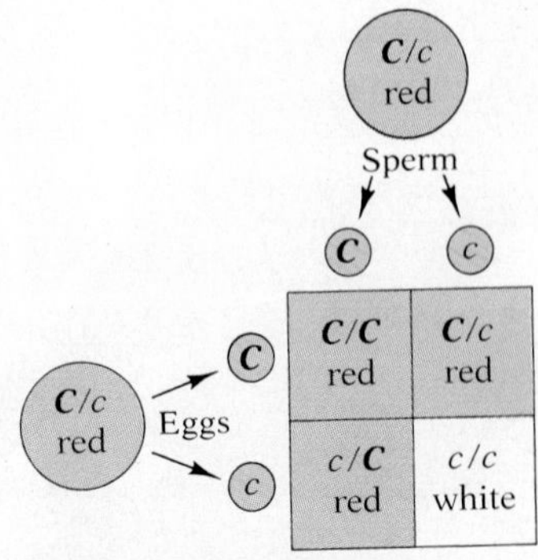

16.4 Punnett-square representation of the information shown in Figure 16.3

Next, write in each box, first, the symbol for the female gamete and, second, the symbol for the male gamete. Each box will then contain the symbols for the genotype of one possible zygote combination from the cross in question. A glance at the completed Punnett square in Figure 16.4 shows that the cross yields the expected 1:2:1 genotypic ratio and, since dominance is present, the expected 3:1 phenotypic ratio.

Extensive investigation of a vast array of plant and animal species by thousands of scientists has demonstrated conclusively that the results Mendel obtained from his monohybrid crosses, and the interpretations he placed on them, are not limited to garden peas but are of general validity. Whenever a monohybrid cross is made between two contrasting homozygous individuals, regardless of the character involved, the expected genotypic ratio in the F_2 is 1:2:1. And whenever there is dominance, the expected phenotypic ratio is 3:1. If these ratios are not obtained in large samples, some complicating condition must be present.

Sometimes, when monohybrid crosses have failed to yield the expected results, that failure has been a clue leading to the discovery of genetic phenomena that are critical to the understanding of evolution. Two of these phenomena—linkage and sex-specific effects—will be discussed presently. Others include strict maternal inheritance, which led to the discovery of organelle genes; genetic "imprinting," by which an inherited pattern of methylation acts to modify the expression of certain genes (see p. 292–293); and alleles that eliminate any heterozygous allele on the homologous chromosome, thereby assuring their own transmission in all gametes. These exceptions provide insights into the workings of inheritance.

PARTIAL DOMINANCE

The seven characters of peas that Mendel used in the experiments reported in his paper were all of the kind in which one allele, known as a full dominant, shows complete dominance over the other. Many characteristics in a variety of organisms show this mode of inheritance. But many others do not, including some that Mendel himself studied but did not discuss in his paper. When a heterozygous individual clearly shows effects of both alleles, the alleles are referred to as partial, or incomplete, dominants.

In many cases of partial dominance, heterozygous individuals have a phenotype that is actually intermediate between the phenotype of individuals homozygous for one allele and the phenotype of individuals homozygous for the other allele; here, finally, is a kind of blending inheritance. For example, crosses between homozygous red snapdragons or sweet peas and homozygous white varieties yield plants producing pink blossoms (Fig. 16.5). When these pink-flowering plants are crossed among themselves, they yield red, pink, and white offspring in a ratio of 1:2:1, as follows:

P	R/R red	×	R'/R' white	
		↓		
F_1	R/R' pink	×	R/R' pink	
		↓		
F_2	R/R red	R/R' pink	R'/R pink	R'/R' white

16.5 Partial dominance in sweet peas
When plants homozygous for red flowers are crossed with plants homozygous for white blossoms, the heterozygous F_1 sweet-pea plants have pink flowers.

Notice that when there is partial dominance both alleles are designated by a capital letter, and one is distinguished from the other by a prime, as here, or by a superscript (C^r for red, C^w for white). Note also that since both alleles affect the phenotype, each is subject to selection.

The chemical events underlying the production of pink flowers are not fully understood. Heterozygous pea flowers, as we have seen, appear red. One likely explanation is that the red pigment of pea flowers is so intense that it masks the white pigment totally, while the less saturated red pigment of snapdragons does not. It is instructive to see how little pure red pigment must be mixed into a gallon of white base paint to produce a can of bright red paint.

In other cases of partial dominance, the heterozygous phenotype is not so obviously intermediate between the two homozygous phenotypes. For example, in a certain strain of chickens a mating between a black chicken and a so-called splashed white chicken produces offspring all of which have a distinctive appearance called blue Andalusian. A cross between two blue Andalusians produces black, blue Andalusian, and splashed white offspring in a ratio of 1:2:1, as follows:

P	black	×	white	
		↓		
F_1	blue	×	blue	
		↓		
F_2	black	blue	blue	white

Here the gene products of two alleles, one for black and one for white, are interacting to produce a phenotype different from the gray that might be expected as the intermediate between black and white. As we will see, the presence in heterozygotes of traits unlike those of homozygotes of either of the alleles involved can lead to unusual consequences when those traits are subjected to selection.

To summarize, the inheritance pattern of partial dominance differs in the following ways from that in complete dominance: (1) the F_1 offspring of a monohybrid cross between parents each homozygous for a different allele have a phenotype different from that of either parent, and (2) the F_2 phenotypic ratio is 1:2:1 (just like the genotypic ratio) rather than 3:1.

MULTIHYBRID AND MULTIGENIC INHERITANCE

We have limited our discussion so far to monohybrid crosses, those in which a single gene controls a single character (a monogenic trait). But most characters are controlled by several different genes at once, and so involve multigenic inheritance; in addition, organisms have many different characters, and so all crosses are actually multihybrid crosses. Trihybrid and trigenic crosses mark the practical limits of Mendelian analysis.

THE BASIC DIHYBRID RATIO

Mendel's experiments on garden peas were not limited to single characters, but sometimes involved two or more of the characters listed in Table 16.1. For example, he crossed plants bearing round yellow seeds with plants producing green wrinkled seeds. The resulting F_1 plants all had round yellow seeds. That is, all showed dominant phenotypes, as we have come to expect. When these plants were crossed among themselves, the resulting F_2 progeny showed four different phenotypes:

315 had round yellow seeds
101 had wrinkled yellow seeds
108 had round green seeds
32 had wrinkled green seeds

These numbers represent a ratio of about 9:3:3:1 for the four phenotypes.

This experiment produced a new and interesting result: even if full dominance is present, a dihybrid cross can produce new plants phenotypically different from either of the original parental plants; here the new phenotypes were wrinkled yellow and round green (Fig. 16.6). It demonstrated, in other words, that during meiosis the genes for seed color and the genes for seed form do not necessarily remain paired as they were in the parental generation, but can separate and reassemble in different allelic combinations in the gametes. From what we know of meiosis, this result implies that the genes for seed color are on the chromosomes of one homologous pair while the genes for seed form are on the chromosomes of another pair. Consequently, the genes for the two characters segregate independently during meiosis, sometimes producing new phenotypes on which selection can operate.

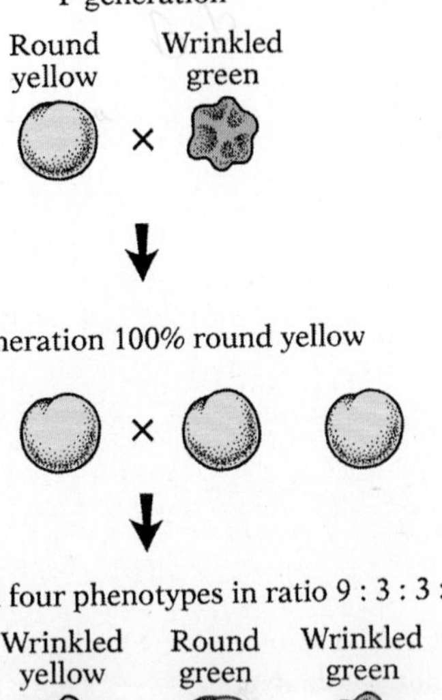

16.6 Expression of novel phenotypes
Crosses between plants homozygous for round yellow seeds and plants homozygous for wrinkled green seeds produce two novel phenotypes in the F_2 generation: wrinkled yellow seeds and round green seeds.

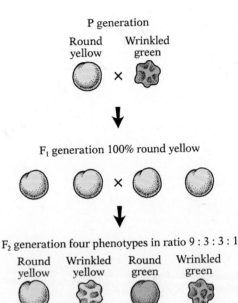

The 9:3:3:1 phenotypic ratio is characteristic of the F_2 generation of a dihybrid cross (with dominance) in which the genes for the two characters are ***independent*** (located on nonhomologous chromosomes). Each independent gene behaves in a dihybrid cross exactly as in a monohybrid cross. If we view Mendel's F_2 results as the product of a monohybrid cross for seed color (ignoring seed form), we find that there were 416 yellow seeds (315 + 101) and 140 green seeds (108 + 32), which closely approximates the 3:1 F_2 ratio expected in a monohybrid cross. Similarly, if we treat the

experiment as a monohybrid cross for seed form and ignore seed color, the F_2 results also show a phenotypic ratio of approximately 3:1. The dihybrid F_2 ratio of 9:3:3:1 is thus simply the product of two separate and independent 3:1 ratios.

Let us examine in somewhat more detail a cross of this type, using the symbols *R* for the allele for round seed and *r* for the allele for wrinkled seed, and the symbols *G* for the allele for yellow seed and *g* for the allele for green seed. In the summary of Mendel's cross below, the dash means that it does not matter phenotypically whether the dominant or the recessive allele occurs in the spot indicated.

P		$R/R\ \ G/G$ round yellow	×	$r/r\ \ g/g$ wrinkled green	
			↓		
F_1		$R/r\ \ G/g$ round yellow	×	$R/r\ \ G/g$ round yellow	
			↓		
F_2	9 $R/–\ \ G/–$ round yellow	3 $r/r\ \ G/–$ wrinkled yellow		3 $R/–\ \ g/g$ round green	1 $r/r\ \ g/g$ wrinkled green

The round yellow parent could produce gametes of only one genotype, *RG*. The wrinkled green parent could produce only *rg* gametes. When *RG* gametes from the one parent united with *rg* gametes from the other parent in the process of fertilization, all the resulting F_1 offspring were heterozygous for both characters (R/*r G*/*g*) and showed the phenotype of the dominant parent (round yellow). Each of these F_1 individuals could produce four different types of gametes, *RG*, *Rg*, *rG*, and *rg*. When two such individuals were crossed, there were 16 possible combinations of gametes (4 × 4). As shown in Figure 16.7, these 16 combinations included nine genotypes, which determined four phenotypes in the ratio of 9:3:3:1.

As Figure 16.7 indicates, one way to determine the genotypic and phenotypic ratios in a dihybrid cross is to construct a Punnett square and then count the number of boxes representing each genotype and each phenotype. This method, though satisfactory in a monohybrid cross and only moderately laborious in a dihybrid cross, becomes prohibitively tedious in a trihybrid cross or a cross involving more than three characters. There is an alternative procedure that is much easier. It is based on the principle that *the chance that a number of independent events will occur together is equal to the product of the chances that each event will occur separately*, a principle known as the ***Product Law***. It is best explained by an example.

Suppose we want to know how many of the 16 combinations in Mendel's cross will produce the wrinkled yellow phenotype. We know that wrinkled is recessive; hence it is expected in $1/4$ of the F_2 individuals in a monohybrid cross. We know that yellow is dominant; hence it is expected in $3/4$ of the F_2 individuals. Multiplying these two separate values ($1/4 \times 3/4$) gives us $3/16$; three of the 16 possible combinations will produce a wrinkled yellow phenotype. Similarly, if we want to know how many of the combinations will produce a round yellow phenotype, we multiply the separate expectancies for two dominant characters ($3/4 \times 3/4$) to get $9/16$.

The Product Law applies equally well to more complex examples. Sup-

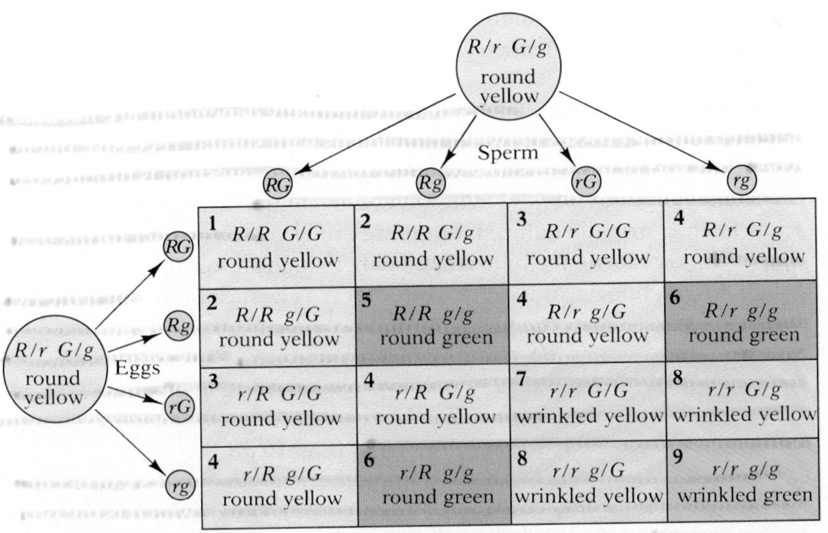

16.7 Punnett-square representation of an F_1 dihybrid cross
When two individuals heterozygous for both seed color and shape were crossed, Mendel obtained roughly nine plants that produced round yellow seeds, three that produced round green seeds, and three that produced wrinkled yellow seeds for every one that produced wrinkled green seeds. Underlying these results were nine different genotypes, as indicated by the numbers in the boxes. Combinations of alleles that represent identical genotypes are shown with the same number; combinations of alleles that give rise to identical phenotypes have the same shading.

pose we want to find out, for a trihybrid cross involving the two seed characters and flower color, what fraction of the F_2 individuals (produced by allowing *C/c R/r G/g* individuals to cross among themselves) will exhibit a phenotype combining red flowers, wrinkled seeds, and yellow seeds. The separate probability for red flowers in a monohybrid cross is ³⁄₄, that for wrinkled seeds is ¹⁄₄, and that for yellow seeds is ³⁄₄. Multiplying these three values ($3/4 \times 1/4 \times 3/4$) gives us $9/64$. This tells us that of the 64 possible combinations in a trihybrid cross, nine will produce the phenotype here specified. Now try determining the proportion of offspring that will exhibit a phenotype combining white flowers, round seeds, and green seeds.

GENE INTERACTIONS

We have already seen how different alleles of the same gene can interact when both are partially dominant. Separate genes can also interact. Indeed, most phenotypic characteristics, whether they are morphological, like flower color, or more subtle chemical characteristics, like enzyme pathways, are controlled by several genes. You may recall that glycolysis, for instance, is a dozen-step process. Each step requires a different enzyme, and each enzyme is produced by the action of a specific gene. Most of these enzymes depend on one or more others operating earlier in the chemical chain reaction to provide a substrate with which they can interact, and so in turn provide a substrate suitable for the next enzyme in the series.

Glycolysis is only one of the many life processes that require the "cooperation" of many genes. But tracking the influence each gene brings to bear in such complicated interactions can be difficult: an individual unable to perform glycolysis at a normal rate, for instance, may be homozygous recessive for any of twelve separate genes—homozygous recessive because a heterozygous individual produces at least some functional enzyme. How can we sort out gene functions when genes depend on each other for expressing their true phenotype? We will look briefly at ways of solving these problems,

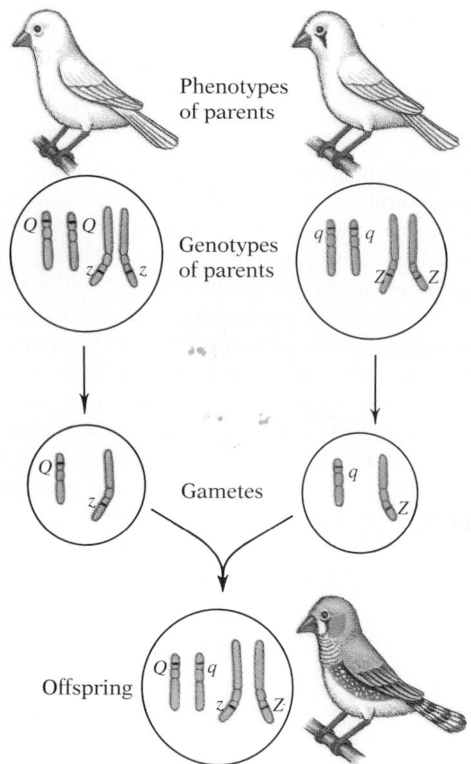

16.8 Complementation test with zebra finches There are two varieties of albino zebra finch. When these are crossed, the offspring have the slate-gray bodies and colorful markings of normal wild zebra finches. The genetic basis of this phenomenon is shown here. Finches of one strain (right) are homozygous for the recessive allele at the *Q* locus (genotype *q*/*q*), and therefore fail to produce functional pigment. Those of the other strain are homozygous recessive at the *Z* locus (genotype *z*/*z*), and lack normal color because of flaws in the mechanisms controlling gene activity. The offspring of the cross are heterozygous at both loci, and so can produce the normal amount of pigment. In the example shown here, both parents are homozygous for both genes. If one or both of the parents were heterozygous for one gene (with the genotype *Q*/*q* *z*/*z* or *q*/*q* *Z*/*z*), we would expect some of the offspring to be normal and some to be albino.

and then at how the problems themselves provide useful clues about the operation of cellular chemistry.

Complementary genes Genes that are mutually dependent are normally detected during crosses between individuals with similar recessive phenotypes. This sort of test can show whether the trait in question is produced by the same gene in both, or by two separate genes. For example, a normal zebra finch is brightly colored, but there are also two distinct forms of albino finch. One form of albino zebra finch, though almost totally white, has a slight brown tinge near its eyes. The other is a classic albino. But if the two forms are crossed in a ***complementation test***, the offspring are normal wild types (the most common naturally occurring form). This result indicates that each form of albinism is produced by a separate gene, at a different locus on the chromosome; the offspring of the cross, being heterozygous in both genes, display the normal phenotype (Fig. 16.8).

A failure of the normal pigmentation system in zebra finches can occur at two known points: the pigment protein can be defective, which is the problem in strain B; or the control mechanisms that turn on pigment production and control the amount made can be defective, which is the problem in strain A. A finch that is homozygous recessive for either element of the pigment system will be an albino. If two white individuals with the *same* defect are crossed, all the progeny will be white, but if two white individuals with different defects are crossed, some or all of the offspring will be normal (Fig. 16.8). This reversion to the normal phenotype in offspring heterozygous at both loci exemplifies the phenomenon known as complementarity, and the two mutually dependent genes are said to ***complement*** each other.

Complementarity can lead to some unusual phenotypic ratios in crosses. Consider the following cross between zebra finches:

P		*Q*/*Q* *Z*/*Z* normal	× ↓	*q*/*q* *z*/*z* albino	
F_1		*Q*/*q* *Z*/*z* normal	× ↓	*Q*/*q* *Z*/*z* normal	
F_2	9 *Q*/– *Z*/– normal	3 *Q*/– *z*/*z* albino		3 *q*/*q* *Z*/– albino	1 *q*/*q* *z*/*z* albino

Were this a dihybrid cross, involving two loci and two traits, we would have expected four phenotypes in a ratio of 9:3:3:1; instead, this monohybrid cross involves two loci but only *one* trait. In this example, the most common F_2 combination (*Q*/– *Z*/–) yields the normal phenotype; the last three genotype combinations yield the albino phenotype.

Epistasis As we have just seen, complementary genes find expression in the dominant phenotype if and only if the organism is heterozygous or homozygous dominant at both loci. Each gene in some sense has veto power over the other, because their products must cooperate. In ***epistasis***, by contrast, these two genetic "votes" do not have equal power: because of the biochemistry of the interaction of their products, only one of the genes can be vetoed by the other. When onegene has the effect of suppressing the phenotypic expression of another gene but not vice versa, the first gene is said to be epistatic to the second. (The Greek root of *epistasis*, logically enough, means "standing upon.") In guinea pigs, for example, a gene for the production of the skin pigment melanin is epistatic to one for the *deposition* of melanin. The first gene has two alleles: *C*, which causes pigment to be produced, and *c*, which causes no pigment production; hence a homozygous recessive individual, *c/c*, is an albino. The second gene has an allele *B* that causes deposition of much melanin, which gives the guinea pig a black coat, and an allele *b* that causes deposition of only a moderate amount of melanin, which gives the guinea pig a brown coat. Neither *B* nor *b* can cause deposition of melanin if *C* is not present to make the melanin. We can summarize a cross involving these two genes as follows:

P		*C/C B/B* black	× *c/c b/b* albino	
		↓		
F_1		*C/c B/b* black	× *C/c B/b* black	
		↓		
F_2	9 *C/– B/–* black	3 *C/– b/b* brown	3 *c/c B/–* albino	1 *c/c b/b* albino

Instead of an F_2 phenotypic ratio of 9:3:3:1, this monohybrid cross has yielded a ratio of 9:3:4. The last two genotypic groups of the 9:3:3:1 ratio produce the same phenotype.

It is important to distinguish epistasis from dominance, which it superficially resembles. Dominance is the phenotypic expression of one member of a pair of alleles at the expense of the other. Epistasis is the suppression by one gene of the phenotypic effect of another entirely different gene; it prevents selection from acting on either copy of the second gene. Dominance refers to interaction between alleles, epistasis to interaction between nonallelic genes.

Collaboration Sometimes two genes influencing the same character interact to produce a novel phenotype that neither gene could produce independently. Such collaborative interaction is seen in the control of the form of the comb in chickens (Fig. 16.9). One gene, *R*, produces rose comb, while its recessive allele, *r*, produces single comb. Another gene, *P*, produces pea comb, while its recessive allele, *p*, also produces single comb. When *R* and *P* occur together, they collaborate to produce walnut comb, a type of comb that neither could produce alone. Rose comb is characteristic of Wyandotte chickens, and pea comb is characteristic of Brahma chickens. A cross between a Wyandotte and a Brahma could be summarized as follows:

P		*R/R p/p* rose	× *r/r P/P* pea	
		↓		
F_1		*R/r P/p* walnut	× *R/r P/p* walnut	
		↓		
F_2	9 *R/– P/–* walnut	3 *R/– p/p* rose	3 *r/r P/–* pea	1 *r/r p/p* single

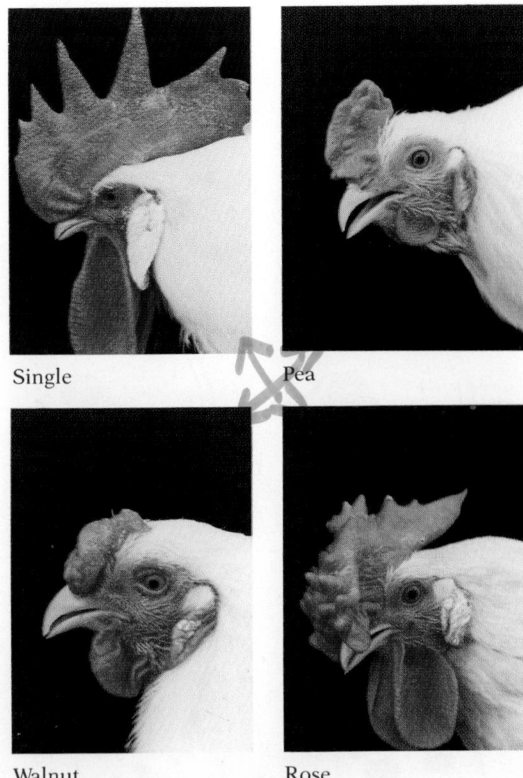

16.9 Comb types in chickens

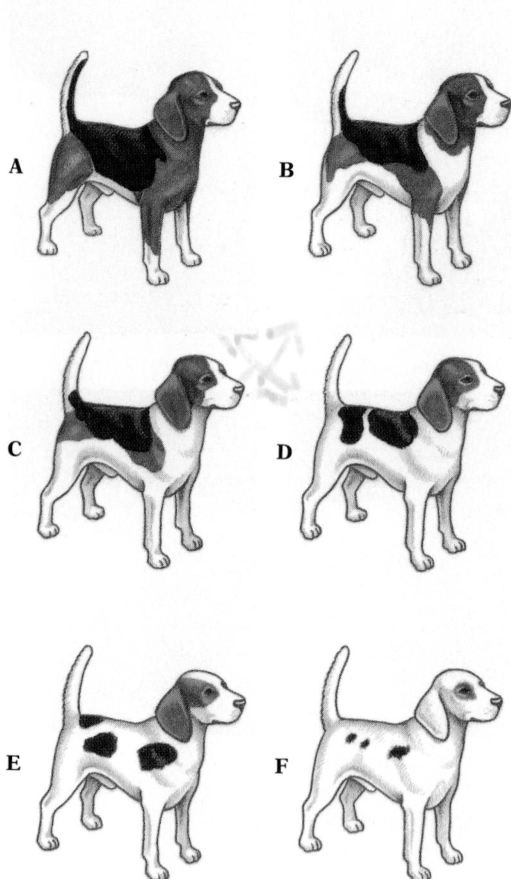

16.10 Variation in spotting of beagle dogs as a result of the action of many modifier genes

Notice that a cross of the more usual sort—between a homozygous dominant walnut, $R/R\ P/P$, and a homozygous recessive single, $r/r\ p/p$—would yield the same F_1 and F_2 pattern. Furthermore, the 9:3:3:1 ratio obtained in either version of this monohybrid cross is the same that we observe in conventional dihybrid crosses—Mendel's cross between peas that produced round yellow seeds and peas that produced wrinkled green seeds, for example. The difference, then, is descriptive: when collaboration occurs, it is not possible to identify the action of each gene on the basis of its phenotypic effects in the clear-cut way Mendel could with a cross involving seed shape and color. The action of selection on any one allele, then, may depend on the alleles present at another locus.

Modifier genes Probably no inherited characteristic is controlled exclusively by one gene pair. Even when only one principal gene is involved, its expression is influenced to some extent by countless other genes with individual effects often so slight that they are very difficult to locate and analyze. An example is eye color in human beings.

Human eye color is largely controlled by one gene with two alleles—a dominant allele, *B*, for brown eyes, and a recessive allele, *b*, for blue eyes. Brown-eyed people (*B/B* or *B/b*) have branching pigment cells containing melanin in the front layer of the iris. Blue-eyed people (*b/b*) lack melanin in the front layer; the blue is an effect of the black pigment on the back of the iris seen faintly through the semiopaque front layer.

This description of the inheritance of eye color on the basis of a single-gene system assumes only two phenotypes, brown and blue. And it is, in fact, possible to assign most people to one or the other of these two phenotypic classes. But we all know that eyes exhibit endless variations in hue; everyday terminology recognizes eyes as green, gray (both genetically forms of blue), hazel, and black (both forms of brown), to name the most familiar variations. Obviously, then, an explanation of eye color in terms of a single-gene system is an oversimplification. Many modifier genes are also involved, some affecting the amount of pigment in the iris, some the tone of the pigment (which may be light yellow, dark brown, etc.), some its distribution (uniform over the whole iris, or in scattered spots, or in a ring around the outer edge, etc.). In fact, in rare cases two blue-eyed people can have a brown-eyed child, because one of them, in whom the lack of pigmentation is a consequence of the action of modifier genes, actually carries the genotype *B/b* instead of *b/b*.

Another example of the action of modifier genes is seen in the size of the spots on beagle dogs (Fig. 16.10). A great many modifier genes are involved; no one of them by itself produces marked effects, but in combination they can radically alter the dogs' appearance.

Multiple-gene inheritance So far we have discussed characteristics that have only a limited number of relatively distinct phenotypes. Pea flowers are either red or white, pea seeds either yellow or green, chicken combs either walnut or rose or pea or single. Modifiers may blur the boundaries of the classes, as in human eye color or spotting in dogs, but a fairly limited number of major phenotypes—blue versus brown or spotted versus unspotted—can be meaningfully discussed. Many traits, however, show much

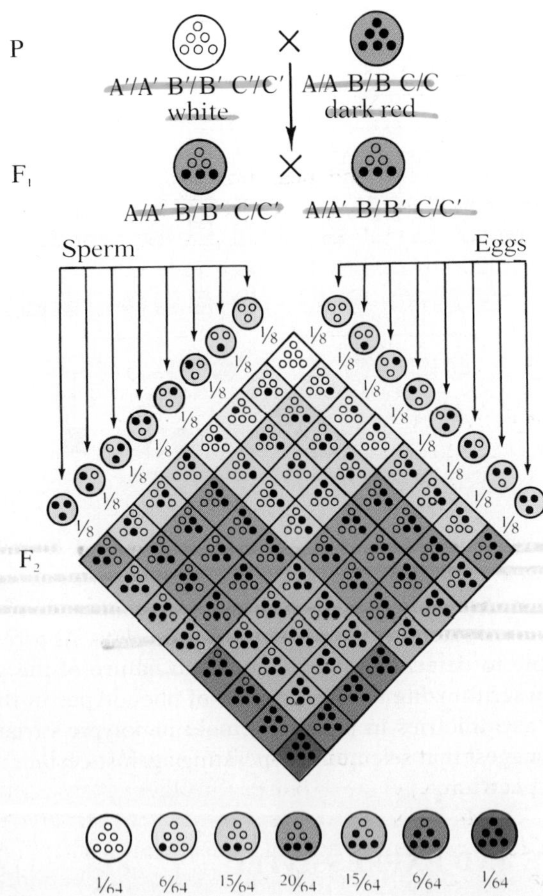

16.11 Nilsson-Ehle's trigenic cross
A line of wheat with pure white kernels and a line with dark red kernels were crossed to produce hybrid medium red F_1 offspring, which were then crossed to produce offspring with the distribution of phenotypes shown here. Nilsson-Ehle concluded from his observations that (1) there are three genes for kernel color in wheat, each with a partially dominant allele for red (*A*, *B*, *C*) and a partially dominant allele for white (*A′*, *B′*, *C′*); (2) that the genotype of the dark red line is $A/A\ \ B/B\ \ C/C$, while that of the white line is $A'/A'\ \ B'/B'\ \ C'/C'$; and (3) that the F_1 hybrids therefore have the genotype $A/A'\ \ B/B'\ \ C/C'$. A gradation of phenotypes is seen in the offspring because all the alleles are additive. Eight different kinds of gametes occur because each of the genes for color is on a different chromosome. The existence of the seven distinct phenotypes of F_2 offspring in the ratio 1:6:15:20:15:6:1 is predicted by this Punnett square of a trigenic cross postulating partial dominance in kernel color.

greater phenotypic variation, with less distinct boundaries. Human height, skin pigmentation, and IQ are just three of many possible examples. What can be said about the genetic basis of characteristics such as these?

One explanation of these more complex genetic phenomena is that two or more separate genes can affect the same character in the same way, in an additive fashion. This kind of inheritance is called ***multigenic*** or ***polygenic***. The first clear demonstration of this kind of interaction came in 1909, when the Swedish geneticist Herman Nilsson-Ehle showed that the color of wheat kernels, which can vary from white through various shades of pink and red to a very dark red, results from the interactions of three genes. Each gene has two alleles, a partially dominant allele for red and a partially dominant allele for white. Dark red kernels are homozygous for the red allele in all three genes, while pure white kernels are homozygous for the white allele in all three genes. All the phenotypes in between result from different heterozygous mixtures of the alleles (Fig. 16.11).

If we graph the probable frequency of colors in the F_2, we obtain a jagged approximation of the bell-shaped curve that represents the so-called normal distribution of most continuously varying traits in a population, traits like skin color, height, IQ, and the like (Fig. 16.12). The greater the number of

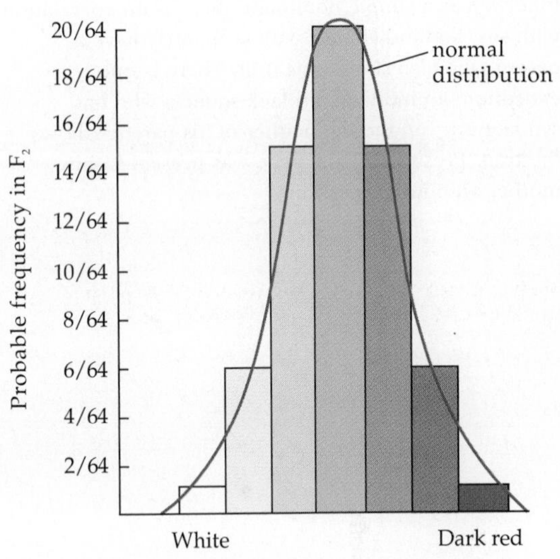

16.12 Frequency distribution of kernel color in wheat
As Nilsson-Ehle showed, a cross of heterozygous wheat kernels results in offspring with seven classes of kernel color, their relative frequencies producing a so-called normal distribution centered around light red.

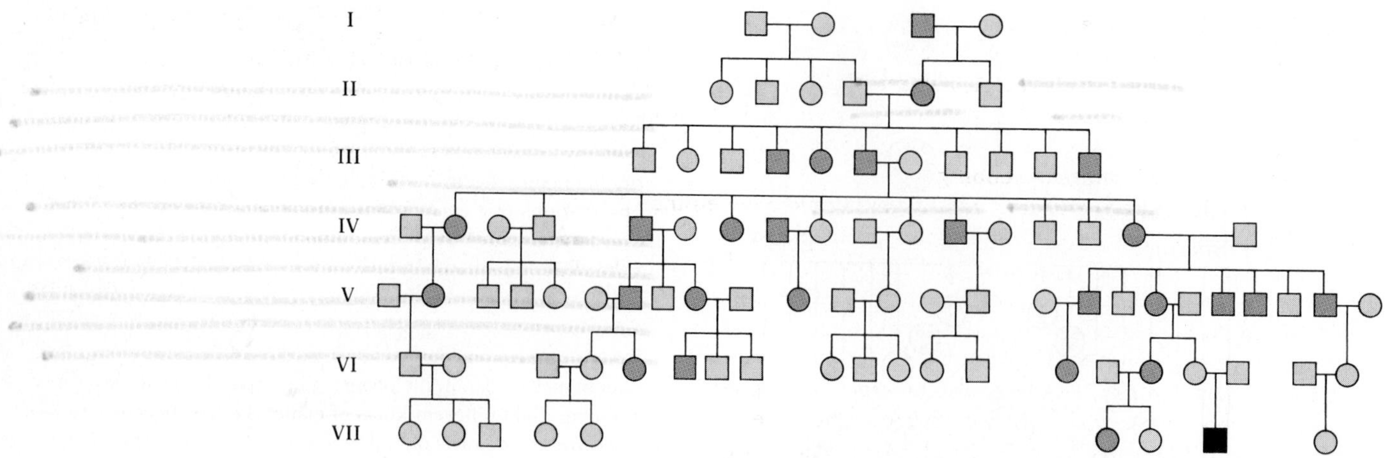

16.13 Pedigree of syndactyly in seven generations of one family

Squares represent males, and circles represent females. Red indicates syndactyly and gray indicates normal phenotype. The character appears to be inherited as a simple dominant: there is no correlation with sex, and individuals with syndactyly have a parent who also shows this trait. There is one exception: an individual (black square) who has syndactyly even though neither of his parents shows the trait. Presumably the gene was present in his mother without expression.

phenotypes, the less jagged the distribution should appear. In most cases, however, the genes involved in multiple-gene effects do not all contribute equally to the phenotype, and the effects of modifier genes and of the environment also tend to complicate the analysis of the data. As a result, it is usually impossible to determine the number and nature of the genes involved simply by scrutinizing the distribution of phenotypes in the F_2. On the other hand, asymmetries in plots of actual phenotypic variation in a population can suggest that selection is operating against certain ranges of the phenotypic spectrum.

PENETRANCE AND EXPRESSIVITY

Because genes interact, an individual can carry a dominant allele that is not expressed phenotypically. Complementary or epistatic genes, for example, or a combination of modifier genes, may prevent the expression of a dominant allele. And even when an allele is expressed, the effects of other genes may modify the degree of intensity of phenotypic expression. We can speak, therefore, of the incomplete penetrance and variable expressivity of certain genes. ***Penetrance*** refers to the percentage of individuals that, carrying a given gene, actually express that gene's phenotype; in the rest, the alleles contributing to the expected phenotype are hidden from selection. ***Expressivity*** denotes the manner in which the phenotype is expressed.

Incomplete penetrance and variable expressivity are illustrated by a gene in human beings that causes blue sclera, a condition in which the whites of the eyes appear bluish. This gene usually behaves as a simple dominant, and anyone who possesses the gene, whether in heterozygous or homozygous form, might be expected to show the blue-sclera phenotype. In fact, however, only about 9 out of 10 people who have the gene actually show the phenotype. We can say, therefore, that the penetrance of this gene is about 90 percent (or 0.9). Among those that do show the phenotype, the expressivity is variable, with the intensity of the bluish coloration ranging from very pale whitish blue to very dark blackish blue.

Figure 16.13 shows a portion of the pedigree of the Hancock family of

Virginia, which for generations has had many members with syndactyly of the ring and little fingers—the two fingers are joined by a web of muscle and skin (Fig. 16.14). This character, like blue sclera, exhibits variable expressivity: a few individuals have three fingers webbed; most have two fingers fully webbed; some have two fingers partially webbed; and a few have a crooked finger with no webbing.

Like blue sclera, too, syndactyly exhibits incomplete penetrance. Of the more than 50 individuals in this family known to have had syndactyly, all but one had a parent who also had syndactyly. This pattern of inheritance is very close to the one expected for a dominant character; as a rule, only individuals with a parent showing the character will show the character themselves, because anyone carrying the dominant gene ordinarily shows its phenotype. (By contrast, a recessive character often appears in a person neither of whose parents showed the character, because they were heterozygous.) The pedigree in Figure 16.13 indicates that syndactyly in this family is inherited as a simple dominant, and that the abnormality in the one man having two parents with normal fingers probably indicates incomplete penetrance of the gene for syndactyly: the man's mother must have been carrying the gene, even though she did not exhibit its phenotype.

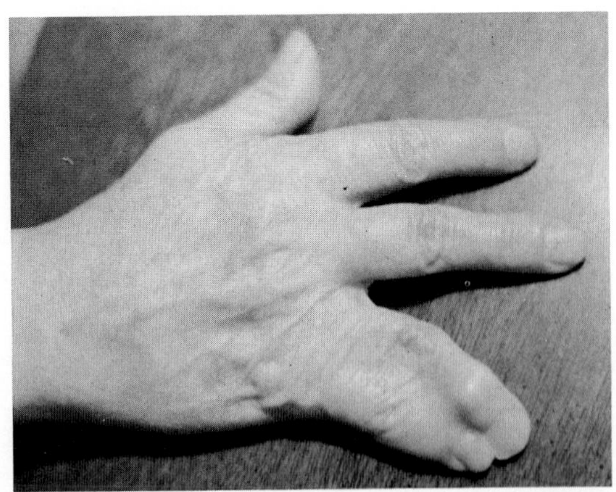

16.14 Syndactyly

Both the examples cited make it clear that penetrance and expressivity are aspects of the same phenomenon. Lack of penetrance of the gene for blue sclera produces white color, which is simply one extreme of the expressivity gradient from white through very pale blue to dark blue. Lack of penetrance of the gene for syndactyly is simply one extreme of the expressivity gradient whose other extreme is the full webbing of three fingers.

As we have indicated, incomplete penetrance and variable expressivity may result from peculiarities of the genetic background against which the gene in question must act. This explanation points up a general truth: *The action of any gene can be fully understood only in terms of the overall genetic makeup of the individual organism in which it occurs.*

Penetrance and expressivity are often also affected by the environment. For example, *Drosophila* homozygous for the gene for vestigial wings have wings that are only tiny stumps if they are reared at normal room temperatures (about 20°C), but if they are reared at temperatures as high as 31°C, their wings grow almost as long as normal wings. Himalayan rabbits are normally white with black ears, nose, feet, and tail (Fig. 16.15), but if the fur on a patch on the back is plucked and an ice pack is kept on the patch, the new fur that grows there will be black: the gene for black color can express itself only if the temperature is low, which it normally is only at the body extremities. The same phenomenon underlies the coloring of Siamese cats. It is thought to occur because in this breed an enzyme required for pigment synthesis retains its normal active conformation only at low temperatures,

16.15 Effect of temperature on expression of a gene for coat color in the Himalayan rabbit
(A) Normally, only the feet, tail, ears, and nose are black. (B) Fur is plucked from a patch on the back, and an ice pack is applied to the area. (C) The new fur grown under the artificially low temperatures is black. Himalayan rabbits are normally homozygous for the gene that controls synthesis of the black pigment, but the enzyme encoded by the gene is active only at low temperatures (below about 33°C).

and denatures—changes to an inactive conformation—when heated to body temperature. Such a susceptibility to heat might result from an inherited change in the DNA coding for one of the amino acids in the enzyme. The consequent change in the amino acid could reduce the number of internal hydrogen or disulfide bonds in the enzyme, making it unstable and hence inactive at higher temperatures.

We see, then, that the expression of a gene can depend both on the other genes present (the genetic environment) and on the physical environment (temperature, sunlight, humidity, diet, etc.). We don't inherit characters. We inherit only genes, only potentialities; other factors govern whether or not the potentialities are realized. All organisms are inevitable products of both their inheritance and their environment.

MULTIPLE ALLELES

Mendel simplified his analysis by concentrating on just two alleles at each locus. But, as we know, genes may exist in any number of allelic forms in a population. Of course, under normal circumstances, the maximum number of alleles for each gene that any diploid organism can possess is two, because the organism has only two copies of each gene. But many other alleles may be present in the population to which it belongs, and so multiple alleles can play a major role in evolution.

Eye color in *Drosophila* One of the first examples of multiple alleles was discovered in the tiny fruit fly, *Drosophila melanogaster*—a species whose genetics have been extensively studied. Though normally red-eyed, fruit flies may have eyes of other colors—white, eosin (a brightly fluorescing red), wine, apricot, ivory, or cherry. Each of these eye colors is controlled by a different allele of the same gene; about two dozen such alleles have been discovered, and others may well exist. The allele for the wild-type eye (red) is dominant over all the rest—that is, the normal red pigment masks the pigment produced by any other allele. When two of the other alleles occur together in a heterozygous fly, however, they produce an intermediate eye-color phenotype.

Human A-B-O blood types A well-known example of multiple alleles in human beings—and a relatively simple one, since only a few alleles are involved—is that of the A-B-O blood series, in which four blood types are generally recognized: A, B, AB, and O. The erythrocytes in type A blood bear antigen A on their surface; in type B, antigen B; in type O, neither A nor B; in type AB, both A and B.

An antigen, as you may recall, is a chemical capable of triggering an immune reaction by which antibodies are produced that bind to and help destroy that particular antigen and the cell that bears it. Because the immune system soon becomes insensitive to antigens present from birth, an individual whose red corpuscles bear antigen A has no antibodies—called anti-A—to that antigen. Similarly, an individual with antigen B has no anti-B antibodies. The person with type AB blood, therefore, having corpuscles that bear both A and B antigens, will have neither anti-A nor anti-B in the blood plasma; the person with type O blood, on the other hand, having corpuscles that bear neither antigen, will have both anti-A and anti-B in the blood plasma (see Table 16.2). In short, a person's plasma contains antibodies corresponding to any antigens his own corpuscles do not bear.

The presence of these antigens and antibodies in the blood has important implications for blood transfusions. Because the antibodies present in the plasma of blood of one type tend to react with the antigens on the erythrocytes of other blood types and cause clumping, it is always best, when transfusions are to be given, to obtain a donor who has the same blood type as the patient. When such a donor is not available blood of another type may be used, provided that the *plasma of the patient* and the *erythrocytes of the donor* are compatible; in other words, doctors can usually ignore the erythrocytes of the patient and the plasma of the donor. The reason is that, unless the transfusion is to be a massive one or is to be made very rapidly, the donor's plasma is sufficiently diluted during transfusion so that little or no agglutination occurs. This means that type O blood can be given to anyone, because its erythrocytes have no antigens and hence are obviously compatible with the plasma of any patient; type O blood is sometimes called the universal donor. But type O patients can receive transfusions only from type O donors, because their plasma contains both anti-A and anti-B and hence is obviously not compatible with the erythrocytes of any other class of donor. Luckily for those of us with type O blood, it is the most common variety in most parts of the world. Conversely, people with the rare type AB, whose plasma contains no anti-A or anti-B antibodies, are universal recipients, but cannot act as donors for any except type AB patients. Table 16.3 summarizes these transfusion relationships.

At first glance, you might suppose that two independent genes are involved in the A-B-O system, one determining whether the A antigen is present and another whether the B antigen is present. But actually the inheritance of the A-B-O groups is for the most part controlled by three alleles of the same gene, here designated I^A, I^B, and i. Both I^A and I^B are dominant over i, but neither I^A nor I^B is dominant over the other. Accordingly the four blood-type phenotypes correspond to the genotypes indicated in Table 16.4. From our discussion of the molecular basis of dominance in the early part of this chapter, you can probably guess that I^A and I^B are alleles that code for different functional proteins, while i codes for a nonfunctional protein. In fact, the protein is now known to be an enzyme that modifies the carbohydrate groups on a membrane component of the glycocalyx.

Blood typing is often used as a source of evidence in paternity cases in court. For example, a man with type O blood could not possibly be the father of a child with type A blood whose mother is type B. The child's true father must be either type A or type AB, because the child must have received its I^A allele from its father; an O man has no such allele. Similarly, a man with type AB blood could not possibly be the father of a type O child, because the child must have received an i allele from each parent, but an AB man has no such allele. Of course, blood-type analysis can only determine who could *not* be the father. Genetic "fingerprinting" techniques, which usually measure the number of copies of each type of repetitive DNA (a pattern which is similar between related individuals, but unique in detail for each person) can be more reliable, and are now commonly used to determine paternity among animals in an effort to decipher the operation of various social systems.

TABLE 16.2 *Antigen and antibody content of the blood types of the A-B-O series*

Blood type	*Blood contains*	
	Cellular antigens	Plasma antibodies
O	None	anti-A and anti-B
A	A	anti-B
B	B	anti-A
AB	A and B	None

TABLE 16.3 *Transfusion relationships of the A-B-O blood groups*

Blood group	*Can donate blood to*	*Can receive blood from*
O	O, A, B, AB	O
A	A, AB	O, A
B	B, AB	O, B
AB	AB	O, A, B, AB

TABLE 16.4 *Genotypes of the A-B-O blood types*

Blood type	*Genotype*
O	i/i
A	I^A/I^A or I^A/i
B	I^B/I^B or I^B/i
AB	I^A/I^B

TABLE 16.5 *Frequencies of A-B-O blood groups in selected populations*

Population	O	A	B	AB
United States whites	45%	41%	10%	4%
United States blacks	47	28	20	5
African Pygmies	31	30	29	10
African Bushmen	56	34	8	2
Australian aborigines	34	66	0	0
Pure Peruvian Indians	100	0	0	0
Tuamotuans of Polynesia	48	52	0	0

As indicated in Table 16.5, the frequencies of the various A-B-O blood types vary in populations of different ancestral extraction. Anthropologists have found data on the frequencies of blood types useful in tracing the prehistoric movements and derivations of the various subgroups of the human species. Analogous studies help delimit natural breeding populations of animals. Since the most frequent phenotype in most human populations is type O, which corresponds to the homozygous recessive phenotype, the i allele is more common than the I^A or I^B alleles. Here we have a good illustration of an important fact: whether an allele is dominant or recessive does not determine whether it will be common or rare in the population. Many people have the mistaken impression that dominant alleles are always the common ones and recessive alleles the rarer ones. "Dominant" and "recessive" describe the way the alleles interact when they occur together in a heterozygous individual; they do not indicate which allele determines the more advantageous phenotype. Natural selection normally leads to an increase in the frequency of the allele that determines the more adaptive phenotype, whether that allele is dominant or recessive, and a decrease in the frequency of the allele that determines the less adaptive phenotype; it is the relative adaptiveness of the phenotype that determines which allele is the more common.

Rh factors The A and B antigens are not the only surface proteins on red blood corpuscles. You have probably heard of Rh factors, the antigens, first identified in rhesus monkeys, that are produced by alleles of the Rh gene. Individuals whose two copies of the Rh gene code for nonfunctional products are said to be Rh-negative (Rh^-); theirs is a situation analogous to that of type O individuals, in whom both copies of the blood-type gene code for an enzyme that produces nonfunctional surface proteins. Individuals with at least one functional Rh allele (and there are many, including four common ones) are said to be Rh^+. Another gene for surface proteins on red corpuscles has two alleles, M and N, giving rise to the genotypes M/M, M/N, and N/N. Each of these groups of blood antigens can create immunological problems during blood transfusions.

MUTATIONS AND DELETERIOUS ALLELES

As we saw in Chapters 8, 9, and 11, a variety of influences can cause slight changes—mutations—in the chemical structure of a gene. Because cells have elaborate DNA-repair mechanisms, the rate at which any particular gene undergoes mutation is ordinarily extremely low. But every individual organism has a large number of different genes, and the total number of genes in all the individuals of a species is vast indeed. Hence mutations are constantly occurring within a species; pure chance determines in which individual any given mutation will occur.

Every living organism is the product of billions of years of evolution and is a finely tuned, smoothly running, astoundingly intricate mechanism, in which the function of every part in some way influences the function of every other part. By comparison, the most complex computer is simple indeed. If you were to take such a machine and make some random change in its parts, the chances are great that you would make it run worse rather than better. A random change in any delicate and intricate mechanism is far more likely to damage it than to improve it. Genes are different in that most mutations do not greatly alter the gene product and therefore have little or no phenotypic effect. However, since mutational changes in genes occur at random, it is easy to understand why the vast majority of the mutations that do have obvious phenotypic effects are deleterious. Only very rarely is a mutation beneficial. Nevertheless, mutations are a major source of the variation upon which natural selection operates. (The others include crossing over and sexual recombination, each of which creates new combinations of pre-existing genes.)

Heterozygous versus homozygous effects When a deleterious allele arises by mutation, natural selection can act against it only if it causes some change in the organism's phenotype. Because dominant deleterious mutations will be expressed phenotypically, they can be eliminated from the population rapidly by natural selection, and most mutations that persist in a population will therefore be recessive to the normal alleles. Recessive mutations are also relatively abundant because new gene products are likely to be less active than normal ones. Since the probability of the same mutation's occurring twice in the same diploid individual is vanishingly slight, most new alleles appear in combination with a normal, usually dominant allele. The individual is then heterozygous for that particular trait, the new mutant allele generally masked, and its deleterious effects not fully expressed. As a result, natural selection cannot eliminate it from the population very rapidly. Deleterious alleles that are not dominant may be retained in the population in heterozygous condition for a long time. Clearly, organisms that are diploid throughout their lives are much less sensitive to mutation than those that have extended haploid stages. This is probably the main reason diploidy predominates among long-lived organisms.

16.16 A creeper hen
These birds have very short legs and cannot walk normally. Such a hen is heterozygous for an allele that is lethal to homozygotes.

When a mating occurs between two diploid individuals carrying the same deleterious recessive allele in heterozygous condition, about one-fourth of the progeny will be homozygous for the deleterious allele, and these homozygous offspring will have the harmful phenotype. Selection (death before reproduction) culls out such organisms, thereby reducing the frequency of the deleterious allele in the population. A recessive allele whose phenotype is fatal is called ***lethal***. The occurrence of lethals can modify the phenotypic ratios obtained in the progeny of some crosses, as the following example shows:

In chickens one allele of a certain gene, when it occurs in heterozygous condition with the normal allele, causes the chicken to be a "creeper," with short crooked legs (Figure 16.16). When two creeper chickens are crossed, their offspring are of two phenotypes, normal and creeper, in a ratio of approximately 1 : 2. This ratio, which is different from any we have previously encountered, occurs because about one-fourth of the incubated eggs fail to hatch. If the embryos that die before hatching are regarded as a third phenotypic class, the cross can be said to have produced a phenotypic ratio of 1 : 2 : 1, the typical one for the F_2 generation of a monohybrid cross where dominance is lacking. The ratio of 1 : 2 seen in the live chicks is the result of the lethality of the creeper allele when it occurs in homozygous condition.

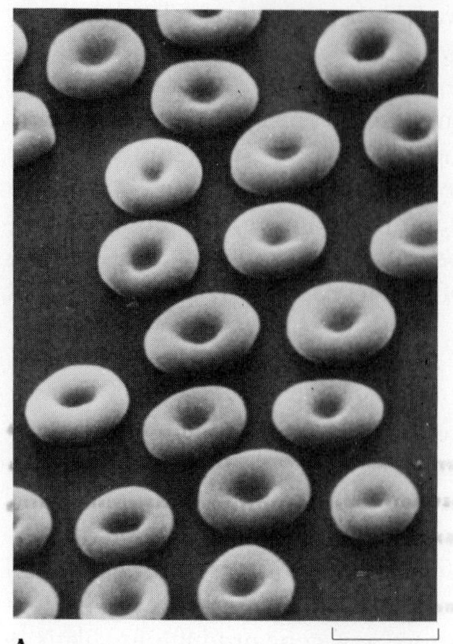

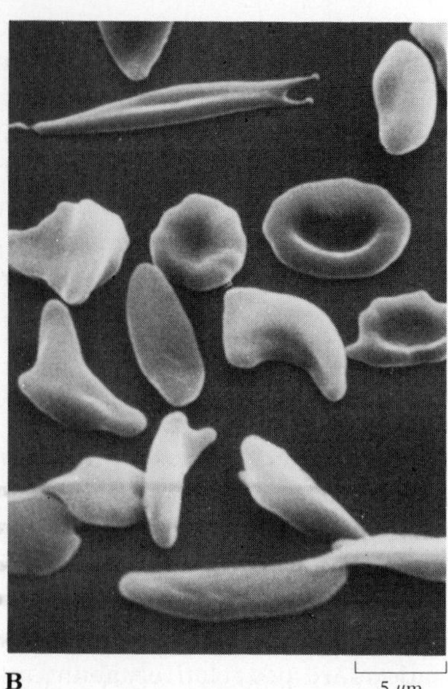

16.17 Scanning electron micrographs of normal and sickled erythrocytes
Normal erythrocytes (A), which are biconcave discs, look dramatically different from sickled cells (B). Some of the sickled cells seen here bear the filamentous processes that may cause clogging of the body's smaller blood vessels.

In numerous instances alleles harmful or even lethal when homozygous are actually beneficial when heterozygous. For example, in England there is a breed of cattle called Dexter, a good beef producer, for which it is impossible to establish a pure-breeding herd because some of its most desirable characteristics are caused by the heterozygous expression of an allele that is lethal when homozygous.

An example in human beings is the allele for ***sickle-cell anemia***. In an individual homozygous for the sickle-cell allele, the mutant hemoglobin tends to crystallize under acidic conditions (as when blood levels of CO_2 rise during exercise). When the hemoglobin in a corpuscle crystallizes, the corpuscles become curved like a sickle and bear long filamentous processes (Fig. 16.17). These abnormal corpuscles tend to form clumps and to clog the smaller blood vessels. The resulting impairment of the circulation leads to severe pains in the abdomen, back, head, and extremities, and to enlargement of the heart and atrophy of brain cells. In addition, the tendency of the deformed corpuscles to rupture easily brings about severe anemia. As might be expected, victims of sickle-cell anemia usually suffer an early death. Individuals heterozygous for the sickle-cell allele sometimes show mild symptoms of the disease, but the condition is usually not serious.

You might suppose that natural selection would operate against the propagation of any allele so obviously harmful and that such an allele would be held at very low frequency in the population. But the allele is surprisingly common in many parts of Africa, being carried by as much as 20 percent of the black population. What is the explanation? A. C. Allison of Oxford showed that individuals heterozygous for this allele have an unusually high resistance to malaria. Since malaria is very common in many parts of Africa, the sickle-cell allele must be regarded as beneficial when heterozygous. Hence, in Africa there is selection for the allele because of its heterozygous effect on malarial resistance and selection against it because of its homozygous production of sickle-cell anemia. In other parts of the world, however, malaria is infrequent, and the benefits of the sickle-cell allele are outweighed by its costs. The balance between the two opposing selection pressures determines the frequency of the allele in any population, and it comes as no surprise that the frequency of the sickle-cell allele in the descendants of African blacks living in the United States has steadily declined, a clear instance of evolution.

The case of sickle-cell anemia is a dramatic example of an allele that has more than one effect. Such an allele is said to be ***pleiotropic***. Pleiotropy is, in fact, the rule rather than the exception. All genes probably have many effects on the organism. Even when a gene produces only one perceptible phenotypic effect, it doubtless has numerous physiological effects more difficult to detect. For example, the temperature-sensitive coat-color allele of Siamese cats mentioned earlier also causes a mysterious misrouting during development of the axons carrying visual information from the eye to the brain. Many Siamese cats are cross-eyed, thus compensating for their visual miswiring.

The effect of inbreeding The conditions under which genes cause deleterious phenotypes explain the danger of matings between closely related human beings. Everyone probably carries in heterozygous combination many alleles that would cause harmful effects if present in homozygous combination, including some lethals. But because most of these deleterious alleles originated as rare mutations, and are limited to a tiny percentage of the population, the chances are slight that two unrelated individuals will be carrying the same deleterious recessive alleles and produce homozygous offspring that show the harmful phenotype. The chances are much greater that two closely related persons will be carrying the same harmful recessives, having received them from common ancestors, and that, if they mate, they will have children homozygous for the deleterious traits. In short, close inbreeding increases the percentage of homozygosity, as Figure 16.18 shows. You can see from the graph that brother-sister matings and matings between double first cousins cause rapid increases in homozygosity, and that matings between first cousins cause slight increases. As we will see, many species have evolved specific behavioral or (particularly in plants) physiological mechanisms that prevent close relatives from mating.

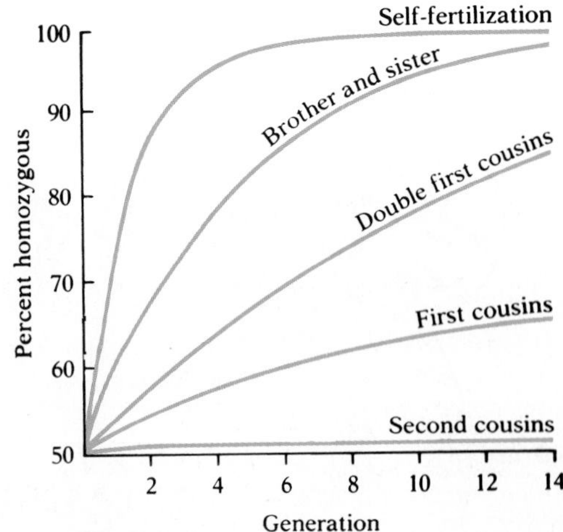

16.18 Graph showing percentage of homozygotes in successive generations under different degrees of inbreeding

It is assumed in this graph that the initial condition is two alleles of equal frequency (hence the 50 percent homozygosity at the start). If a similar graph were drawn for very rare recessive alleles, the initial rises in homozygosity would be much steeper, particularly in the three curves for cousin-to-cousin matings. Thus matings between first or second cousins, though of little effect in the situation graphed here, may greatly increase the percentage of homozygosity of rare, perhaps deleterious recessive alleles. (Double first cousins result when siblings of one family marry siblings of another family—when two brothers marry two sisters, for example. The offspring of such marriages are first cousins through *both* of their parents rather than through just one of them.)

SEX AND INHERITANCE

SEX DETERMINATION

The sex chromosomes We have said repeatedly that a diploid individual has two of each type of chromosome, identical in size and shape, and hence two copies of each gene in every cell. We must now qualify that statement: in most higher organisms where the sexes are separate (that is, where males and females are separate individuals), the chromosomal endowments of males and females are slightly different. In general, one of the two sexes has one pair of chromosomes in which the members differ markedly from each other in size and shape. These are the ***sex chromosomes***, which play a fundamental role in determining the sex of the individual; they exhibit their sex-specific effects primarily in the brain, certain hormone-producing

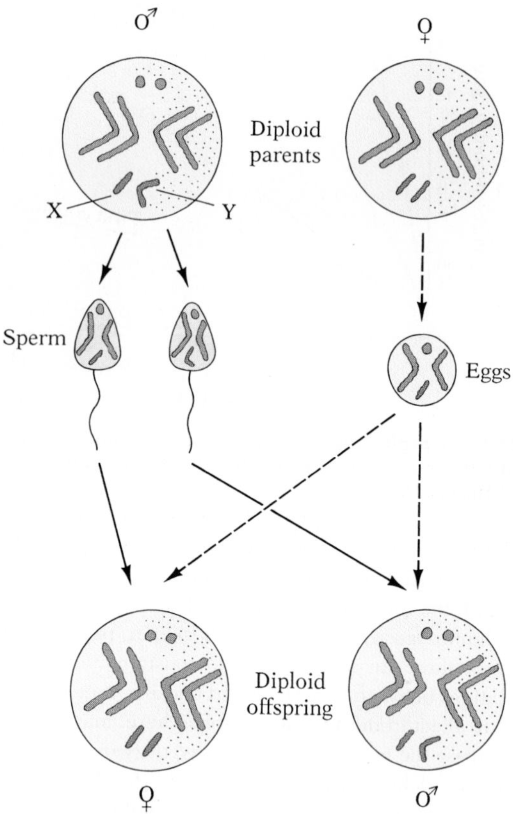

16.19 Chromosomes of male and female *Drosophila melanogaster*
There are three pairs of autosomes and one pair of sex chromosomes. Males (symbolized by ♂) have one X chromosome and one Y chromosome; since these separate at meiosis, half the sperm carry an X and half carry a Y. Since females (♀) have two X chromosomes, all eggs have an X. The sex of the offspring depends on which type of sperm fertilizes the egg.

organs, cells in sexually dimorphic portions of the external and internal anatomy, and the gonads. All other chromosomes are called ***autosomes***.

Let's look first at the chromosomes of *Drosophila* and of human beings. In each case the sex chromosomes are of two sorts: one, bearing many genes, is conventionally designated the ***X chromosome***, and one of a different shape and bearing only a few genes is designated the ***Y chromosome***. In both humans and fruit flies normal females have two X chromosomes, while males have one X and one Y. The diploid number in *Drosophila* is eight (four pairs); a female therefore has three pairs of autosomes and one pair of X chromosomes, but a male has three pairs of autosomes and a pair of sex chromosomes consisting of one X and one Y (Fig. 16.19). The diploid number in human beings is 46 (23 pairs); a female therefore has 22 pairs of autosomes and one pair of X chromosomes, while a male has 22 pairs of autosomes plus one X and one Y (see Fig. 12.4, p. 312).

When a female produces egg cells by meiosis, all the eggs receive one of each type of autosome plus one X chromosome. When a male produces sperm cells by meiosis, half the sperm cells receive one of each type of autosome plus one X chromosome, while the other half receive one of each autosome plus one Y chromosome. In short, all the egg cells are alike in chromosomal content, but the sperm cells are of two types, occurring in equal numbers (Fig. 16.19). When fertilization takes place, the chances are approximately equal that the egg will be fertilized by a sperm carrying an X chromosome or by a sperm carrying a Y chromosome. If fertilization is by an X-bearing sperm, the resulting zygote will be XX and will develop into a female. If fertilization is by a Y-bearing sperm, the resulting zygote will be XY and will develop into a male. We see, therefore, that the sex of an individual is normally determined at the moment of fertilization and depends on which of the two types of sperm fertilizes the egg.

This XY-male system, though characteristic of many plants and animals (including all mammals), is by no means universal. Birds, butterflies and moths, and a few other animals have just the opposite system, where XX is male and XY is female. (To distinguish this from the usual XY system, the symbols Z and W are often substituted, ZZ being male and ZW female.) In still other species—many reptiles and marine fish, for example—sex is environmentally determined: the sex of alligators depends on the temperature of the eggs during development, while some coral-reef fish reverse sex as adults to take advantage of variations in food and competition. One of the most surprising discoveries in recent years is that many species of animals (ranging from insects to mammals) have the ability to manipulate the sex ratio of their offspring. As we will see in later chapters, this capacity allows them to exploit changes in habitat conditions, competition, or individual social status that maximize the reproductive potential of their progeny.

The role of the Y chromosome in sex determination Occasionally the members of a homologous pair of chromosomes fail to separate properly in meiosis and both move to the same pole. The effect of such ***nondisjunction*** is that one of the daughter cells receives one too many chromosomes and the other daughter cell receives one too few. If a human gamete carrying an extra chromosome is involved in fertilization, the zygote produced has 47 chromosomes (the normal two of most types plus three of one type) instead of the normal 46. Sometimes it is the sex chromosomes that fail to separate in meiosis, and XXY individuals may result. Since XXY individuals in *Drosophila* are essentially normal females, it seemed reasonable to suppose that only the X chromosomes function in sex determination, with one producing a male and two a female, regardless of the presence or absence of a Y chromosome. And indeed in *Drosophila*, sex is determined by the ratio of certain gene products produced by the X chromosomes to certain products produced by the autosomes. A low ratio results in a male, while a high ratio produces a female. The Y chromosome in *Drosophila* does not determine the sex of the offspring, though it does contain one or more genes essential for normal male fertility.

The idea that the number of X chromosomes is the one crucial variable for determining sex in most animals was also consistent with the observation that grasshoppers and certain other animals lack the Y chromosome entirely—females have two X chromosomes and males a single unpaired X. As a result, females have one more chromosome than males. Such a system of sex determination is known as the XO system, with O denoting the absence of a chromosome.

We now know that sex determination of this kind, in which the Y chromosome is virtually irrelevant, is not universal. The human Y chromosome, for example, bears a gene with strong male-determining properties. Its product is a DNA-binding protein which, in the sixth or seventh week of fetal life, binds to numerous sites on the autosomes, initiating the developmental program that will produce a male. In the absence of this gene—even in XY individuals in whom this gene is deleted—a female develops.

SEX-LINKED CHARACTERS

Many genes occur on the X chromosome and not on the Y chromosome. Such genes are said to be sex-linked. The inheritance patterns for the characteristics controlled by sex-linked genes are completely different from those for characteristics controlled by autosomal genes, for obvious reasons. Females have two copies of each sex-linked gene, one from each parent, but males have only one copy of each sex-linked gene, and that one copy always comes from the mother, since the father contributes a Y chromosome instead of an X to his sons. Hence, in the male, all sex-linked characteristics are inherited from the mother only. Furthermore, since the male has only one copy of each sex-linked gene, recessive alleles cannot be masked, so that selection is more rigorous against deleterious recessives on the X chromosome. Recessive sex-linked phenotypes, such as red-green color blindness, occur much more often in males than in females.

Sex linkage was discovered in 1910 by the great American geneticist Thomas Hunt Morgan of Columbia University. It was Morgan who began the systematic use of *Drosophila* in genetic studies. This little fruit fly made it possible to perform in a few months experiments that would have taken Mendel years to perform on peas. *Drosophila* can be easily and economically cultured in large numbers in the laboratory. They can produce a new generation every 10 or 12 days, and are subject to a remarkable number of easily detectable genetic variations. Most of the modern knowledge of eucaryotic genetics derives from work on this tiny insect.

The first sex-linked trait discovered by Morgan was white eye color in *Drosophila*. This mutation arose spontaneously from true-breeding red-eyed stock; it is controlled by a recessive allele *r*. The normal red eye color is controlled by a dominant allele *R*. If homozygous red-eyed females are crossed with white-eyed males, all the F_1 offspring, regardless of sex, have red eyes, since they receive from their mother an X chromosome bearing an allele for red. In addition, the F_1 females receive from their father an X chromosome bearing an allele for white eyes, but the allele for red, being dominant, masks its presence. The F_1 males, like the females, receive from their mother an X chromosome bearing an allele for red eyes. But unlike the females, they receive no gene for eye color from their father, who contributes a Y chromosome instead of an X (in writing the genotype of a male for a sex-linked character, the Y is customarily shown, in order to indicate clearly that no second X chromosome is present and hence there is no second copy of the sex-linked gene). We can summarize this cross as follows (♀ denotes females, ♂ males):

P		*R/R* red-eyed ♀ ×	*r/Y* white-eyed ♂	
		↓		
F_1		*R/r* red-eyed ♀ ×	*R/Y* red-eyed ♂	
		↓		
F_2	*R/R* red-eyed ♀	*r/R* red-eyed ♀	*R/Y* red-eyed ♂	*r/Y* white-eyed ♂

Notice that when the F_1 flies of this cross are allowed to mate among themselves, the F_2 flies show the customary 3:1 phenotypic ratio of a monohybrid cross where dominance is present. But notice also that this 3:1 ratio is rather different from the 3:1 ratio obtained in a cross involving autosomal genes. In an autosomal cross there is no correlation of phenotype with sex, but in this cross all F_2 individuals showing the recessive phenotype are males. In other words, an autosomal cross gives a 3:1 F_2 ratio for both females and males, but this cross yields females of a single phenotype and males with a 1:1 phenotypic ratio. This asymmetry illustrates the potential for selection to operate differentially on the two sexes.

Two well-known examples of recessive sex-linked traits in human beings are red-green color blindness and hemophilia (a failure to form blood clots, which can lead to uncontrolled bleeding). Color blindness occurs in about 8 percent of white males in the United States and in about 4 percent of black males. It occurs in only about 1 percent of white females and about 0.8 percent of black females. It is expected, of course, that more men than women will show such a trait. A man needs only one copy of the allele to show the phenotype, and he can inherit this one copy from a heterozygous mother who is not herself color-blind. But for a woman to be color-blind, she must have two copies of the allele, and so be homozygous; not only must her father be color-blind, but her mother must be either color-blind or a heterozygous carrier of the allele. Since the allele is not very common in the population, it is not likely that two such people will marry; hence the low number of color-blind women.

The statement that females have two copies of each sex-linked gene whereas males have only one, though technically correct, requires qualification. In most interphase somatic cells of females, one of the X chromosomes condenses into a tiny dark object called a ***Barr body*** (Fig. 16.20). Most of the genes on this condensed X chromosome are inactive. Hence, a normally functioning female cell contains only one active copy of most sex-linked genes. (The specific genes escaping inactivation include the one whose product—probably an RNA—binds to a DNA sequence in the genes on the chromosome destined to be turned off.) Why, then, are sex-linked recessive traits expressed in females only when they are homozygous? And why, if the cells of both males and females contain only one active copy of each sex-linked gene, are the patterns of inheritance in the two sexes markedly different?

The explanation is that it is not always the same X chromosome that condenses into a Barr body in the different somatic cells of a given individual. Let us call the two X chromosomes X_1 and X_2. Characteristically, about half the cells in any given female show active X_1 chromosomes, the other half active X_2 chromosomes, with no discernible pattern as to which chromosome, X_1 or X_2, is active in which cell. For example, consider the distribution of alleles of the sex-linked gene for the enzyme glucose-6-phosphate dehydrogenase; one of the alleles codes for an active version of the enzyme, the other for a defective version. When the erythrocytes of women heterozygous for these two alleles were examined, roughly half the erythrocytes in each individual revealed normal enzyme activity, the other half a complete lack of it. Apparently female cells differ in their effective genetic makeup as far as sex-linked traits are concerned; women are, in a sense, genetic mosaics for sex-linked traits. For some of these traits, the mosaic pattern finds phenotypic expression; an example is the coat color of tricolor (tortoiseshell and calico) cats (Fig. 16.21). For others, it does not. It seems, for instance, that as long as half the cells are normal in a woman heterozygous for red-green blindness or for hemophilia, she will be phenotypically normal.

The phenomenon of X-chromosome inactivation explains why the sex of mammals depends on the presence or absence of the Y chromosome rather

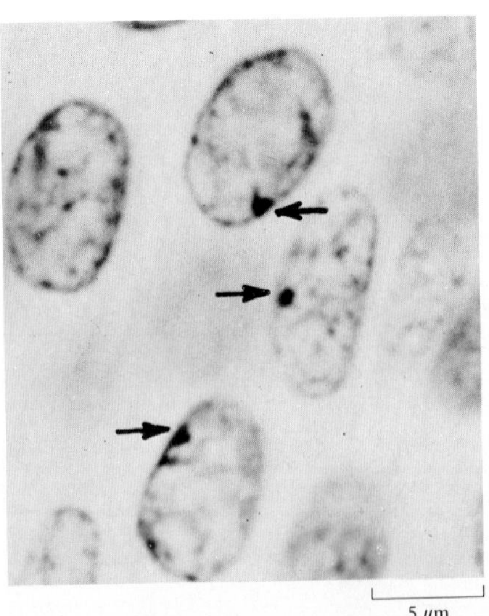

16.20 Nuclei from epidermal cells of a human female

The arrows indicate the Barr bodies. Since Barr bodies are present in the cells of female fetuses, the sex of an unborn child can be ascertained by examination of the nuclei of cells sloughed off into the fluid of the mother's womb.

16.21 X-chromosome inactivation in cats

One of the most common natural demonstrations of X-chromosome inactivation is the coat color of tortoiseshell and calico cats. A gene for color found on the cat's X chromosome has two common alleles—black and yellow (or orange). Males can have only one allele, so (in the absence of modifier genes) they are either yellow or black, usually with various white markings. Females, however, can have both alleles, and after inactivation some cells will express the black allele and others the yellow allele. Large patches of cells with one allele or the other develop, and the result is a mosaic of yellow, black, and (usually) white patches. Except for rare XXY individuals, all cats with both yellow and black fur are females.

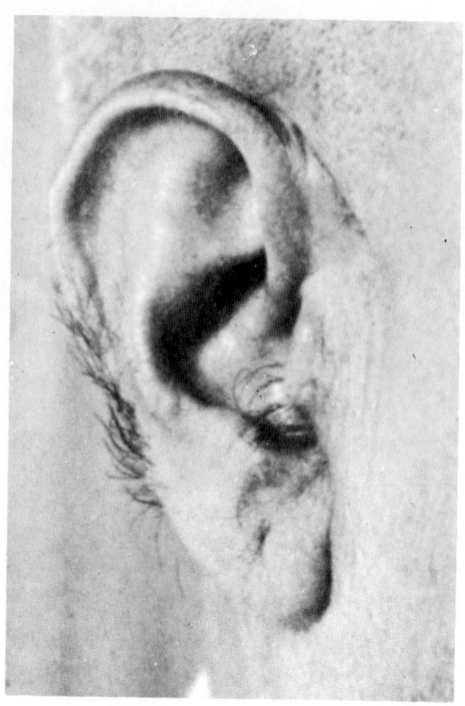

16.22 Hairy pinna
This trait is thought to be determined by a gene on the Y chromosome.

than on the ratio of X chromosomes to autosomes, as in *Drosophila:* because of inactivation each mammalian somatic cell has only one functional X chromosome, and so is either XO (female) or XY (male); both sexes have the same number of X-chromosome genes available for transcription. In *Drosophila,* on the other hand, females have twice the number males have. As we have seen, the resulting difference in the amount of certain gene products determines the sex of the individual. But for many other products, various mechanisms have evolved to achieve ***dosage compensation***—to equalize the amount resulting from transcription of the two X chromosomes in females with the amount resulting from transcription of the single chromosome in males. One of the most common methods of dosage compensation is a feedback system by which the transcription rate is automatically adjusted to match a cell's requirements. On average, then, genes on the single X chromosome of *Drosophila* males do double duty: they are transcribed twice as often as the corresponding genes on any one female X chromosome.

GENES ON THE Y CHROMOSOME

Genes unique to the Y chromosome are termed ***holandric***. The phenotypic traits they control appear, of course, only in males (Fig. 16.22). There are apparently very few genes on the human Y chromosome; we have mentioned the maleness determiners that are thought to be on the Y chromosome in humans.

In some species a few genes occur on both the X and the Y chromosome; their inheritance patterns are the same as for autosomal genes. No such genes have been demonstrated conclusively in humans.

SEX-INFLUENCED CHARACTERS

As we have seen, sex-linked genes may control characters not customarily regarded as "sexual." Furthermore, not all genes commonly associated with sex are sex-linked; many genes that control "sexual" characters are located on the autosomes. For example, a number of genes that control growth and development of the sexual organs, such as the penis, the vagina, the uterus, and the oviducts, or that control distribution of body hair, size of breasts, pitch of voice, or other secondary sexual characteristics, are autosomal and are present in individuals of both sexes. That their phenotypic expression is different in the two sexes indicates that they are sex-limited, not that they are sex-linked. Apparently the sex chromosomes determine what hormones are to be synthesized in each sex, and these hormones influence the activity of the sex-limited autosomal genes secondarily, either inhibiting or stimulating them.

We have already encountered two other instances of sex-correlated inheritance. One involves the inheritance of mitochondria: in most higher organisms, sperm and pollen only rarely contribute mitochondria to the zygote, so mitochondrial genes are essentially always inherited from the female parent. The other case is "imprinting": as we saw in Chapter 11, progeny can (rarely) inherit genes with different patterns of methylation; if the gene in question is permanently inactivated in the gamete of one parent, the phenotype of the zygote is determined by the allele contributed by the gamete from the other parent.

LINKAGE

Recall from the first part of this chapter that Mendel's observations led to two generalizations. The first was that each individual carries two copies of every hereditary factor—every gene—and these copies segregate during gamete formation; this Principle of Segregation is often called Mendel's first law. The second generalization was that when several genes are involved in a cross (as in a dihybrid cross), they sort out into the gametes independently of one another; this ***Principle of Independent Assortment*** is frequently referred to as Mendel's second law. All seven of the traits Mendel reported on from garden peas did indeed assort independently, and were free of the complications of modifiers, partial dominance, sex linkage, multiple alleles, and so on. But independent assortment describes only the simplest genetic interactions—the behavior of genes on separate chromosomes. Mendel almost certainly knew that some pairs of factors do not assort independently to produce offspring in neat 9:3:3:1 ratios, and that some do not show a simple pattern of recessiveness and dominance. He probably chose to ignore such anomalies, and the theory he published nevertheless represented a tremendous step forward in understanding inheritance. But these nagging problems—particularly the failure of many crosses to exhibit independent assortment—were well known to geneticists and students of evolution. Not until the discovery and description of chromosomes could such major obstacles to Mendel's theory be overcome.

The chromosomal basis of linkage Cytology, the study of cells with microscopes, was in its infancy in Mendel's day. Though good microscopes existed, chromosomes had not yet been discovered because appropriate stains had not yet been developed. A hint that something in the nucleus might be special came in 1869, three years after Mendel's paper, when a young Swiss student named Friedrich Miescher discovered in cell nuclei large amounts of a strange acid he called nuclein. The actual carriers of genetic information were not seen until 1882, when Walther Flemming successfully stained mitotic chromosomes. With the observation and description of mitosis, the stage was set for the reevaluation of Mendel's work. In 1900, shortly after the rediscovery of Mendel's paper, W. S. Sutton of Columbia University pointed out the striking accord between Mendel's conclusion that hereditary factors (genes) occur in pairs in somatic cells, and separate in gametogenesis, and the recent cytological evidence that somatic cells contain two of each kind of chromosome and that these chromosomes segregate in meiosis. Sutton interpreted this agreement as powerful evidence that the chromosomes are the bearers of the genes.

Now that Mendel's ideas could be directly related to observations made with the aid of a microscope, scientists could focus on what appeared to be anomalies. In time it became apparent, for example, that traits assort independently, as specified by Mendel's second law, only when their respective genes occur on two different chromosomes. Genes that occur on the same chromosome cannot assort independently during meiosis unless separated by crossing over. Such genes are said to be ***linked***. One of the first examples of linkage was reported in 1906 by R. C. Punnett and William Bateson of Cambridge University. They crossed sweet peas that had purple flowers and long pollen with ones that had red flowers and round pollen. All the F_1 plants had purple flowers and long pollen, as expected (it was already known that purple was dominant over red and that long was dominant over round). The F_2 plants from this cross did not show the expected 9:3:3:1 ratio, however, but a highly anomalous one. Next, Bateson and Punnett crossed the F_1 plants back to homozygous recessive plants (with red flowers and round pollen). Their results were equally anomalous. Using the symbols *B* for purple, *b* for red, *L* for long, and *l* for round, we can summarize them as follows:

BbLl		×	*bbll*
purple long			red round
		↓	
7 *BbLl*	1 *Bbll*	1 *bbLl*	7 *bbll*
purple long	purple round	red long	red round

According to the Principle of Independent Assortment, the heterozygous purple long parent should have produced four kinds of gametes *(BL, Bl, bL,* and *bl)* in equal numbers. When united with the *bl* gametes from the homozygous recessive parents, *BL* gametes should have given rise to purple long offspring, *Bl* gametes to purple round, *bL* to red long, and *bl* to red round, and these four phenotypes should have occurred in equal numbers, in a 1:1:1:1 ratio. But the result Bateson and Punnett actually obtained—a ratio of 7:1:1:7—makes it appear that the heterozygous parent produced far more *BL* and *bl* gametes than *Bl* and *bL* gametes.

Only in 1910 did Thomas Hunt Morgan, who had obtained similar results from *Drosophila* crosses, provide the explanation accepted today. He postulated that the anomalous ratios were caused by linkage. Hence we should write the genotypes of the parents in Bateson and Punnett's second cross *BL*/*bl* and *bl*/*bl* to show, by the positions of the slashes, that *B* and *L* are on one chromosome and *b* and *l* on the other (we would have written these genotypes *B*/*b L*/*l* and *b*/*b l*/*l* if the genes were not linked).

Now, if in Bateson and Punnett's cross the genes for purple and long and the genes for red and round were linked, we might expect the *BL*/*bl* parent in the second cross to have produced only two kinds of gametes, *BL* and *bl*, and the second cross to have yielded offspring of only two phenotypes, purple long and red round, in equal numbers. Yet the cross also yielded some purple round and red long offspring. How could the *BL*/*bl* parent have produced *Bl* and *bL* gametes? Morgan suggested that some mechanism occasionally breaks the original linkages between purple and long and between red and round and establishes in a few individuals new linkages between purple and round and between red and long, making possible the production of *Bl* and *bL* gametes. The mechanism of this recombination is, of course, crossing over.

From an evolutionary perspective, crossing over is important because it increases the number of genetic combinations a cross can produce, and therefore the number of phenotypes upon which selection can operate.

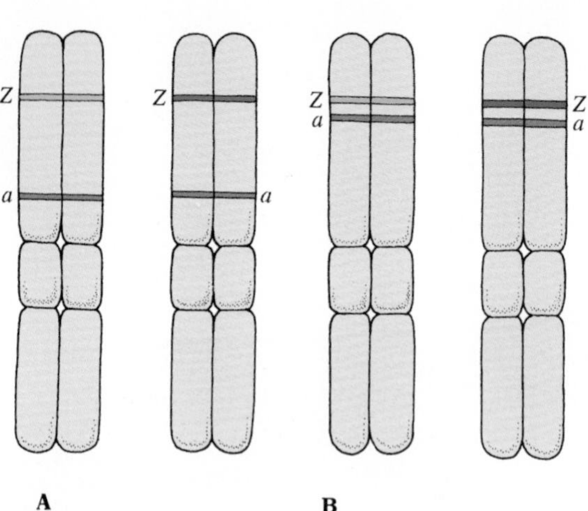

16.23 Effects of linkage on crossing over
When alleles *Z* and *a* are located far apart on a chromosome (A), the odds of a crossover event occurring between the two loci (and thus separating the alleles) are greater than when the two loci lie close together (B). In the latter case, selection operating on the allele at one locus can strongly affect the allele at neighboring loci.

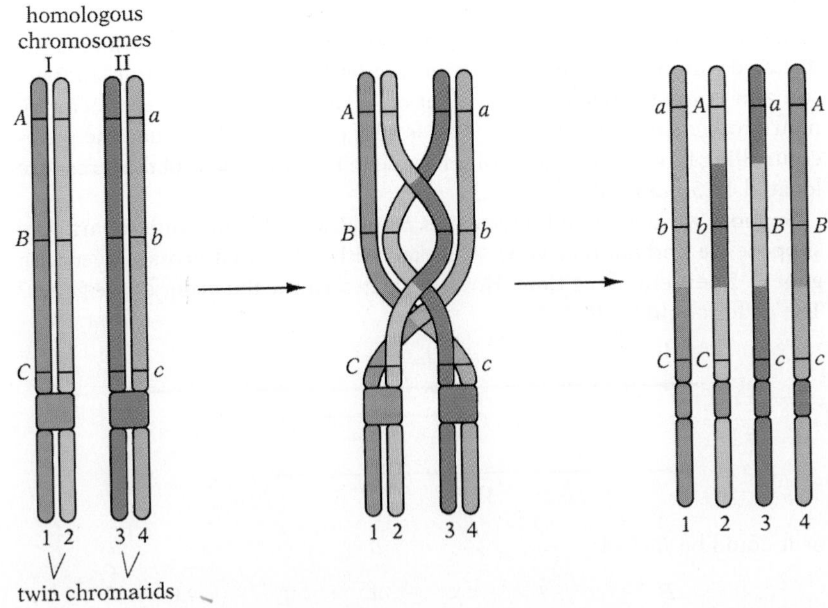

16.24 Linkage and recombination
After this typical round of crossing over, the three genes originally linked as *ABC* and *abc* are combined in four ways: *abC*, *AbC*, *aBc*, and *ABc*. Notice that the farther apart two genes are, the more likely it is that a crossover event will take place between them. In this example, crossing over between *A* and *C* takes place in every instance, whereas crossing over between *A* and *B* occurs only half as often. Notice also that crossing over between two genes does not always recombine them: because chromatids 2 and 3 have two compensating crossover events, *A* and *C* remain together, as do *a* and *c*, even though segments between the two genes have been interchanged. As a result, recombination frequency is always lower than crossover frequency.

Linkage can also have a major impact, though the reason is not as obvious. Imagine that there were no crossing over. In such a case, an allele *a* on a chromosome containing an allele *Z* producing a deleterious phenotype would be culled by selection operating against *Z*, even though *a* may be innocuous or beneficial. Crossing over allows *a* to escape from its association with *Z*, but the more tightly linked two genes are—that is, the nearer they are to one another on the chromosome—the less often escape can occur (Fig. 16.23). Two unrelated genes located next to each other may be so tightly linked that selection for or against one is, in practice, selection for or against the other. Thus, it is possible for the frequency of one allele to be strongly affected by selection on an unrelated locus.

Chromosomal mapping If we assume, as did the geneticist Alfred H. Sturtevant—then an undergraduate working in Morgan's lab—that the probability of crossing over is approximately equal at any point along the length of a chromosome, then the greater the distance between two linked genes, the greater the frequency with which they will be separated. Or, to be more precise, the frequency of recombination between any two linked genes will be proportional to the distance between them. Sturtevant postulated that the percentage of recombination can serve as a tool for mapping the location of genes on chromosomes. We speak in terms of the percentage of recombination rather than the number of crossover events because most chromosomes cross over at more than one place; therefore two crossover events can cancel each other and so go undetected (Fig. 16.24).

The percentage of recombination gives us no information about the absolute distances between genes, nor even relative distance (Fig. 16.25). Crossover events occur primarily at special DNA sequences, and selection can act to increase or decrease the number of such sites in any part of the chromosome if unusual crossover rates between particular genes are adap-

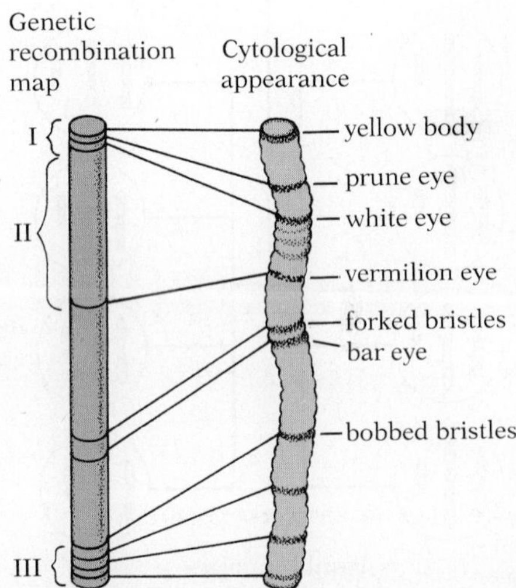

16.25 Comparison of a genetic-recombination map and cytological appearance of a portion of the X chromosome in *Drosophila melanogaster*
Staining and photographic techniques now exist to localize the genes on chromosomes directly. When these cytological results are compared with crossover frequencies (linkage maps), the effects of variations in crossover rates in different parts of the chromosome become apparent. In this case, crossing over is suppressed within two groups of genes (I and II), and greatly enhanced between the loci for white eyes and vermilion eyes (III). Suppression results from a relative paucity of crossover sequences; enhancement is a consequence of an overabundance.

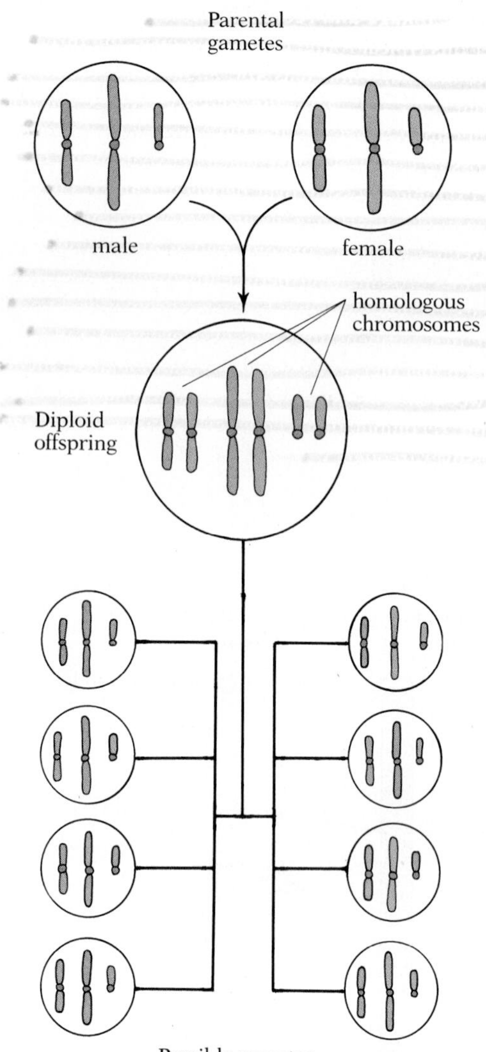

16.26 Gamete diversity in the absence of crossing over
Even in the absence of crossing over, each individual is capable of generating 2^n different kinds of gametes (where n is the number of chromosomes). Hence this hypothetical three-chromosome species produces eight classes of gametes, while our own species, with 23 chromosomes, can generate more than 8 million.

tive. On the other hand, crossover rates do give us information about gene order. By convention, one unit of map distance on a chromosome is the distance within which recombination occurs 1 percent of the time. In Bateson and Punnett's test cross, two out of 16 of the offspring were recombinant products of crossing over. Two is 12.5 percent of 16; hence the genes controlling flower color and pollen shape in the sweet peas of this cross are located 12.5 map units apart.

Suppose we know that linked genes *B* and *L* are 12.5 map units apart. And suppose we find another gene, *A*, linked with these, that crosses over with gene *L* 5 percent of the time. How do we determine the order of the genes? The order could be *B–A–L:*

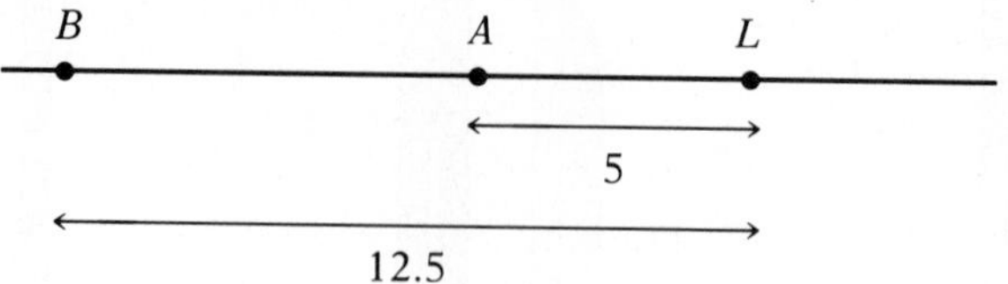

or it could be *B–L–A:*

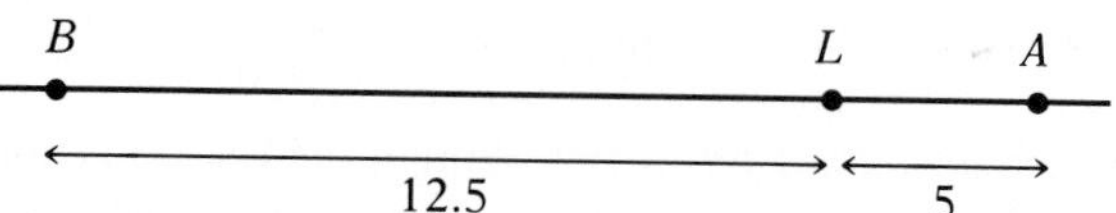

Obviously, the way to decide between these two alternatives is to determine the frequency of recombination between *A* and *B*. If this frequency is 7.5 percent (12.5 − 5), then we know that the first alternative is correct; if it is 17.5 percent (12.5 + 5), then we know that the second alternative is correct. In this way, by determining the frequency of recombination between each gene and at least two other known genes, it is possible to build up a map showing the arrangement of many different genes on a chromosome. Such maps help us understand which traits are so closely linked that selection appears to act on them as a unit (Fig. 16.25).

Linkage and variation As we said at the outset of this chapter, genetics is critical to understanding evolution because it is the major basis of heritability and variation, two crucial requirements for natural selection. As we saw in Chapter 12, the recombination of chromosomes during gamete formation and fertilization generates enormous diversity (Fig. 16.26). In our species, for instance, even without crossing over each individual can produce more than 8 million different kinds of gametes, and thus, when two unrelated individuals pair, 70 trillion different zygotes. Crossing over increases these values enormously. The average human gamete, for instance, carries a set of chromosomes that has undergone an average of 30 crossover events during meiosis, so that most chromosomes have been recombined internally at least once. The result is that any individual can generate an essentially infinite number of distinct gametes. Recombination and crossing over provide far more variation for natural selection to operate on than does mutation.

CHROMOSOMAL ALTERATIONS

A number of genetic alterations that can have important consequences for selection and the evolution of new species—mutation and transposon movement, for instance—were discussed in Chapters 8 through 11. In this section we will look at a few of the other chromosomal events that affect variation, linkage, and the emergence of new species.

Structural alterations In one form of alteration, called ***translocation***, portions of two nonhomologous chromosomes are interchanged, with a consequent modification in linkage groups. Suppose, for example, that a pair of bar-shaped chromosomes in a certain species bear the genes *ABCDEFG* and that a pair of J-shaped chromosomes bear the genes *lmnopqrst*. If the *EFG* end of one bar chromosome and the *st* end of one J chromosome were interchanged, the result would be a shorter bar chromosome bearing only the genes *ABCDst* and a longer J chromosome bearing the genes *lmnopqrEFG*. By changing the linkage relationships of genes, translocations can have important effects on phenotypes and selection.

Sometimes a piece breaks off one chromosome and fuses onto the end of the homologous chromosome. Such an alteration is a kind of ***duplication***. An example would be loss of the *ABC* portion from a chromosome bearing the genes *ABCDEFGH* and fusion of this portion onto the homologous chromosome. The chromosome in which the loss occurred would thus bear only the genes *DEFGH*, while the chromosome undergoing the duplication would bear the genes *ABCABCDEFGH*. Translocation and duplication are most often caused by X-ray-induced damage to chromosomes.

Since translocations and duplications do not result in a net loss of genes in a cell, they usually have little effect on a somatic cell. But these chromosomal rearrangements create problems in meiosis, when homologous chromosomes first synapse and then separate in the formation of gametes: If one chromosome of a pair has lost genes in a duplication event, half the gametes produced will lack these genes. If one chromosome of a pair has undergone a translocation exchange with a nonhomologous chromosome, synapsis with its homologue may be difficult, and meiosis may be blocked altogether. When meiosis does occur, the two nonhomologous chromosomes that have exchanged genes in translocation (losing some and gaining others) will segregate together only half the time. When they do not, half the resulting gametes will have nonhomologous duplications, while the other half will have two chromosomes with genes missing. A gamete that has lost essential genes may be inviable. Any zygote that does result from fusion involving such a gamete will lack the second copy of the genes in question; as a result, deleterious recessive alleles carried by the other gamete may be exposed.

Changes in chromosome number As we saw earlier, the separation of chromosomes in cell division does not always proceed normally, and chromosomes that should have moved away from each other to opposite poles of the spindle may move instead to the same pole and become incorporated into the same daughter nucleus. The result of this nondisjunction may be production of an organism with an extra chromosome (or sometimes two,

Exploring Further

TRISOMY IN HUMANS

In humans most trisomies are lethal. Trisomy-18 (Edwards's syndrome) and trisomy-13 (Patau's syndrome), for example, produce physical malformations and mental and developmental retardation so severe that most afflicted infants die within a few weeks after birth. Because trisomies of most other autosomes result in spontaneous abortion, they are not found in live births. Two kinds of trisomy—trisomy-21 (Down's syndrome) and trisomies of the sex chromosomes—are exceptional in that their victims may survive.

Down's syndrome (formerly often called mongolian idiocy), in which three chromosomes of type 21 occur in the individual's cells, was the first clinical condition ever linked to a chromosomal abnormality. It is associated with a variety of characteristic physical features (broad head, rounded face, perceptible folds in the eyelids, a flattened bridge of the nose, protruding tongue, small irregular teeth, short stature) and also mental retardation (the most common IQ is about 42). The incidence of Down's syndrome is often related to the age of the mother. It occurs in fewer than one out of 1000 births to women under 20; it is more than seven times more common in births to women 35–39 years old, more than twenty times more common when the mother is 40–44, and more than fifty times more common when she is 45 or older. A similar association with the age of the mother is seen in Edwards's and Patau's syndromes.

Trisomy of the sex chromosomes can take several forms. In one, called Klinefelter's syndrome, the chromosomal makeup is XXY, and the individuals are males. Though the symptoms of the condition are variable, and some of those affected are nearly normal, most show a variety of physical abnormalities and mental retardation; furthermore, they often suffer from thyroid dysfunction, chronic pulmonary distress, and diabetes.

Males with a second type of sex-chromosome trisomy, the XYY syndrome, generally show fewer and less severe abnormalities, though they often have subnormal intelligence. Because the incidence of XYY individuals is often significantly higher in penal institutions than in the general population, some investigators have suggested that men with the XYY condition are predisposed to aggressive behavior, but the evidence for this conclusion is weak.

Women with triple-X syndrome (XXX) usually have underdeveloped sexual characteristics and often subnormal intelligence, but since their abnormalities are not debilitating, most live a relatively normal life.

These trisomic conditions, as well as many other kinds of genetic or chromosomal diseases, can be detected during embryonic development by the process of ***amniocentesis***, in which fluid containing sloughed-off epidermal cells from the fetus is withdrawn from the uterus with a long needle inserted through the mother's abdominal wall. The fetal cells are then cultured and examined for abnormalities. A new technique, in which embryonic tissue cells are obtained directly through the maternal vagina, promises earlier and safer detection of these and other devastating handicaps.

three, or more extra chromosomes). The presence of three chromosomes of one type in an otherwise diploid individual is called ***trisomy***.

Occasionally cell division may be so aberrant that all the chromosomes move to the same pole, giving rise to a daughter cell with twice the normal number of chromosomes. If this happens during meiosis, the gamete produced is diploid instead of haploid. If such a gamete unites at fertilization with a normal haploid gamete, a triploid zygote results; if it unites with another diploid gamete, also produced by aberrant meiosis, a tetraploid zygote results. Cells or organisms that have more than two complete sets of chromosomes—that are triploid, tetraploid, hexaploid, etc.—are said to be ***polyploid***.

Polyploidy is fairly common in plants. It has sometimes given rise to new species that are adaptively superior to the original diploid species under certain environmental conditions. Polyploidy can be stimulated in the laboratory by treating plants with certain chemicals that cause nondisjunction during cell division. This procedure has been used in the production of many of the new strains of cultivated plants developed in the last few decades (Fig. 16.27). Polyploidy is very rare in animals and has probably not been an important factor in the origin of new animal species; there are, however, at least two exceptions: Several species of triploid lizards exist, which are of special interest because they are asexual. In addition, polyploidy may have played a role in the evolution of bees and related insects, among which the various species usually have 4, 8, 12, or 16 chromosomes; a similar pattern exists in certain groups of fish and some frogs.

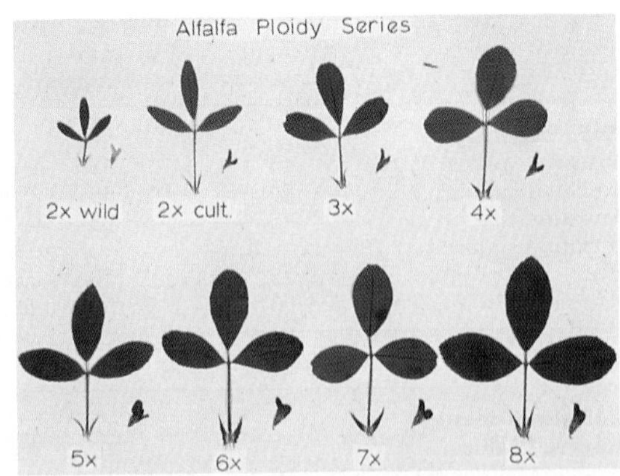

16.27 Induced polyploidy in alfalfa
Tetraploidy and octoploidy in alfalfa and other commercially valuable crops can be induced experimentally. Tetraploid alfalfa (4x) is the most stress-tolerant, and is the type cultivated by farmers; cultivated alfalfa is a naturally occurring tetraploid. Octoploid alfalfa (8x) grows well in the greenhouse, but is sensitive to the stress induced by lack of water in the field. (Diploid alfalfa, 2x, is shown at the upper left.) The polyploids shown here were obtained by a process known as sexual polyploidization, in which gametes with unreduced numbers of chromosomes are produced during meiosis. The union of such gametes results in tetraploid plants; repetition of the process in the next generation yields octoploids. Sexual polyploidization is the principal form of polyploidization of plants in nature.

THE EVALUATION OF EXPERIMENTAL RESULTS

The need for statistical analysis We have mentioned ratios such as 1 : 2 : 1 or 3 : 1 or 1 : 1 or 9 : 3 : 3 : 1, expected in the results of various types of crosses. Geneticists frequently use the phenotypic ratios obtained in breeding experiments to deduce the underlying genetic phenomena. Evolutionary biologists, on the other hand, look for evidence of evolution—that is, changes in gene frequencies—in the form of departures from the expected Mendelian ratios. For example, when selection is operating against a homozygous dihybrid recessive, the size of the fourth class of the 9 : 3 : 3 : 1 distribution is reduced. But chance deviations from a predicted distribution are common: although four coin tosses have an expected distribution of two heads and two tails, a 3 : 1 ratio is hardly surprising. But if the four-toss sequence were repeated 100 times, yielding 300 heads and 100 tails, most of us would suspect that the coin was biased in some way. To judge whether a phenotypic distribution indicates the operation of selection (or of some unexpected genetic process), we therefore must consider both the degree of discrepancy from the expected ratio *and* the sample size.

Scientists in all fields of research constantly encounter the same fundamental question—whether the deviations they observe in their experimental results are significant or not. They cannot rely simply on a guess. They cannot say, "That looks pretty close to what I predicted," or "That looks odd." To help them arrive at a decision, they can refer to a system of standards based on the mathematical probability that any observed deviation in their sample could have occurred by chance alone. This type of mathematical treatment of data is known as statistical analysis. Statisticians have devised many mathematical tests for evaluating experimental or observational data. Though these tests differ in their form and in the sorts of data to which they can validly be applied, all are simply ways of calculating the probability that the deviations of the observed values might be due to chance alone.

The chi-square test One test of statistical significance, devised by Karl Pearson of the University of London in 1900, represented a fundamental breakthrough for evaluating the results of experimental science. Pearson's so-called chi-square (χ^2) test is particularly applicable to many genetic ex-

TABLE 16.6 *Chi-square analysis of two crosses*

	First phenotype	*Second phenotype*
45:55 experiment:		
Observed values	45	55
Expected values (e)	50	50
Deviation (d)	−5	+5
Deviation squared (d^2)	25	25
d^2/e	25/50 = 0.5	25/50 = 0.5
$\chi^2 = \Sigma(d^2/e) = 0.5 + 0.5 = \mathbf{1.0}$		
5:15 experiment:		
Observed values	5	15
Expected values (e)	10	10
Deviation (d)	−5	+5
Deviation squared (d^2)	25	25
d^2/e	25/10 = 2.5	25/10 = 2.5
$\chi^2 = \Sigma(d^2/e) = 2.5 + 2.5 = \mathbf{5.0}$		

periments. This test measures whether any deviation from the predicted norm that occurs in experimental results exceeds the deviation that might occur by chance. The formula for chi-square is

$$\chi^2 = \Sigma(d^2/e)$$

where d is the deviation from the expected value, e is the expected value, and Σ means "the sum of."

Consider two hypothetical crosses in which we expect that the phenotypic ratio should be 1:1 in the absence of selection for or against one of the phenotypes. In one cross we actually got values of 45 and 55 instead of 50 and 50, and in the other we got values of 5 and 15 instead of 10 and 10. We want to know in each case whether the deviation of the observed from the expected values can reasonably be attributed to chance, or implies that selection is at work.

First we must determine the chi-square value for the two crosses (Table 16.6). Notice that in each of these experiments the absolute deviations of the observed values from the expected values are the same: a deviation of 5 in each phenotype. But notice also that the chi-squares obtained in the two crosses are very different—the one based on a sample of 20 being five times as large as the one based on a sample of 100. This illustrates well how sensitive chi-square is to sample size: the difference in sample size alone has made the great difference in the two chi-square values. To interpret the values, however, we need to know a little more.

Each of these crosses involves only two classes, in this case two different phenotypes. Hence their chi-square values were calculated on the basis of only two squared deviations. But suppose we had been analyzing a cross involving three different phenotypes. Then the chi-square would have been calculated on the basis of three squared deviations, and it is only reasonable to expect that the chi-square value obtained would have been higher than one based on only two. It is clear, then, that in evaluating chi-square values we must also take into account the number of classes on which they are based. By convention, the number of independent classes in a chi-square test is termed the ***degree of freedom***. The number of independent classes is usually one fewer than the total number of classes in the cross. Thus, in our crosses involving two phenotypes, there is only one independent class (and so one degree of freedom), while in a cross involving three phenotypes there would be two independent classes and two degrees of freedom. A moment's thought will tell you why this is so. In our cross based on a sample of 100, once we know that 45 offspring show the first phenotype, we automatically know that 55 must show the other phenotype. Since we know the total, the number in one class automatically tells us the number in the other class. In other words, the number in the second class is dependent upon the number in the first class. Therefore, only the first class is an independent class. The same reasoning applies if we perform a cross involving three different phenotypes, and the total number of observations in our sample is 100; once we know the number showing the first and second phenotypes, we automatically know the number showing the third phenotype, because the number in the third class is dependent upon the number in the first two classes.

We now know the chi-square values (1.0 and 5.0) and the degrees of free-

TABLE 16.7 *Probabilities for certain values of chi-square*[a]

Degrees of freedom	$P = 0.20$ (1 in 5)	$P = 0.10$ (1 in 10)	$P = 0.05$ (1 in 20)	$P = 0.01$ (1 in 100)
1	1.64	2.71	3.84	6.64
2	3.22	4.60	5.99	9.21
3	4.64	6.25	7.82	11.34
4	5.99	7.78	9.49	13.28
5	7.29	9.24	11.07	15.09
6	8.56	10.64	12.59	16.81
7	9.80	12.02	14.07	18.48
8	11.03	13.36	15.51	20.09
9	12.24	14.68	16.92	21.67
10	13.44	15.99	18.31	23.21
15	19.31	22.31	25.00	30.58
20	25.04	28.41	31.41	37.57
30	36.25	40.26	43.77	50.89

[a] Based on a larger table in R. A. Fisher, *Statistical Methods for Research Workers*, 10th ed., Oliver & Boyd, 1946.

dom (one for each experiment) for our two hypothetical crosses. The next step is to consult a table of chi-square values. Table 16.7 gives four different chi-square values for each of a series of different degrees of freedom, and gives the probability (P) that a deviation as great as or greater than that represented by each chi-square value would occur simply by chance.

Now let us evaluate the results obtained in the first of our hypothetical crosses. Here the deviation of our results from those expected was such as to yield a chi-square value of 1.0. The cross had one degree of freedom. According to the table, a value as high as or higher than 1.64 has a chance probability of 0.20 (20 percent); that is, deviation from the expected as great as or greater than that represented by 1.64 will occur about once in five trials by chance alone. Our chi-square is less than 1.64; hence the deviation in the experiment can be expected to occur by chance even more often than once in five trials. Most biologists agree that deviations having a chance probability as great as or greater than 0.05 (5 percent, or 1 in 20) will not be considered statistically significant. Since the deviation in our experiment has a chance probability much greater than 5 percent, it is not regarded as statistically significant, and is presumed to be a chance deviation, which can be disregarded.

In our second experiment, the chi-square value representing the deviation from the expected results turned out to be 5.0. Again there was one degree of freedom. Looking at the listings in the table for one degree of freedom, we find that the value of 5.0 is greater than 3.84, which has a probability of 0.05 (5 percent), but less than 6.64, which has a probability of 0.01 (1 percent). Hence the probability that the deviation in this cross resulted purely from chance is less than 5 percent but greater than 1 percent.

According to biological convention, then, the deviation from the expected results in the second cross is significant: some factor other than chance was involved in producing the disagreement between result and prediction. At this point, a geneticist would begin the search for a reasonable explanation: the original observations are always open to scrutiny; selection may have acted against one of the phenotypes, so that some of those individuals died, thus leading to fewer representatives of this class than were expected; or perhaps the assumptions concerning the genetics involved in this cross need modification. In this particular case, one of the first things to do is to perform a similar experiment using a larger sample to minimize chance error. After all, as we saw when we calculated the outcome of a dihybrid cross, the probability of two events happening together is the product of their individual probabilities of happening alone. The probability of this deviation occurring twice by chance is 0.05×0.05, or only 0.25 percent.

STUDY QUESTIONS

1. In squash an allele for white color (W) is dominant over the allele for yellow color (w). Give the genotypic and phenotypic ratios for the results of each of the following crosses:

$$W/W \times w/w$$
$$W/w \times w/w$$
$$W/w \times W/w$$

2. A heterozygous white-fruited squash plant is crossed with a yellow-fruited plant, yielding 200 seeds. Of these, 110 produce white-fruited plants, while only 90 produce yellow-fruited plants. Using the chi-square test, would you conclude that this deviation is the result of chance, or that it probably represents some complicating factor? What if there were 2000 seeds, and 1100 produced white-fruited plants while 900 produced yellow-fruited individuals?

3. In human beings, brown eyes are dominant over blue eyes. Suppose a blue-eyed man marries a brown-eyed woman whose father was blue-eyed. What proportion of their children would you predict will have blue eyes?

4. If a brown-eyed man marries a blue-eyed woman and they have ten children, all brown-eyed, can you be certain that the man is homozygous? If the eleventh child has blue eyes, what will that show about the father's genotype?

5. The litter resulting from the mating of two short-tailed cats contains three kittens without tails, two with long tails, and six with short tails. What would be the simplest way of explaining the inheritance of tail length in these cats? Show genotypes.

6. When Mexican hairless dogs are crossed with normal-haired dogs, about half the pups are hairless and half have hair. When, however, two Mexican hairless dogs are mated, about a third of the pups produced have hair, about two-thirds are hairless, and some deformed puppies are born dead. Explain these results.

7. In peas an allele for tall plants (T) is dominant over the allele for short plants (t). An allele of another independent gene produces smooth peas (S) and is dominant over the allele for wrinkled peas (s). Calculate both

phenotypic ratios and genotypic ratios for the results of each of the following crosses:

$$T/t\ S/s \times T/t\ S/s$$
$$T/t\ s/s \times t/t\ s/s$$
$$t/t\ S/s \times T/t\ s/s$$
$$T/T\ s/s \times t/t\ S/S$$

8. In some breeds of dogs a dominant allele controls the characteristic of barking while trailing. In these dogs an allele of another independent gene produces erect ears; it is dominant over the allele for drooping ears. Suppose a dog breeder wants to produce a pure-breeding strain of droop-eared barkers, but he knows that the genes for silent trailing and erect ears are present in his kennels. How should he proceed?

9. In Leghorn chickens colored feathers are produced by a dominant allele, C; white feathers are produced by the recessive allele, c. The dominant allele, I, of another independent gene inhibits expression of color in birds with genotypes C/C or C/c. Consequently both $C/–I/–$ and $c/c–/–$ are white. A colored cock is mated with a white hen and produces many offspring, all colored. Give the genotypes of both parents and offspring.

10. If the dominant allele K is necessary for hearing, and the dominant allele M of another independent gene results in deafness no matter what other genes are present, what percentage of the offspring produced by the cross $k/k\ M/m \times K/k\ m/m$ will be deaf?

11. What fraction of the offspring of parents each with the genotype $K/k\ L/l\ M/m$ will be $k/k\ l/l\ m/m$?

12. Suppose that an allele, b, of a sex-linked gene is recessive and lethal. A man marries a woman who is heterozygous for this gene. If this couple had a large number of normal children, what would be the predicted sex ratio of these children?

13. Red-green color blindness is inherited as a sex-linked recessive. If a color-blind woman marries a man who has normal vision, what would be the expected phenotypes of their children with reference to this character?

14. In cats short hair is dominant over long hair; the gene involved is autosomal. An allele, B^1, of another gene, which is sex-linked, produces yellow coat color; the allele B^2 produces black coat color; and the heterozygous combination B^1/B^2 produces tortoiseshell and calico coat color. If a long-haired black male is mated with a tortoiseshell female homozygous for short hair, what kind of kittens will be produced in the F_1? If the F_1 cats are allowed to interbreed freely, what are the chances of obtaining a long-haired yellow male?

15. In *Drosophila melanogaster* there is a dominant allele for gray body color and a dominant allele of another gene for normal wings. The recessive alleles of these two genes result in black body color and vestigial wings respectively. Flies homozygous for gray body and normal wings were crossed with flies that had black bodies and vestigial wings. The F_1 progeny were then crossed, with the following results:

Gray body, normal wings	236
Black body, vestigial wings	253
Gray body, vestigial wings	50
Black body, normal wings	61

Would you say that these two genes are linked? If so, how many units apart are they on the linkage map?

16. The recombination frequency between linked genes *A* and *B* is 40 percent; between *B* and *C*, 20 percent; between *C* and *D*, 10 percent; between *C* and *A*, 20 percent; between *D* and *B*, 10 percent. What is the sequence of the genes on the chromosome?

CONCEPTS FOR REVIEW

- Genotype versus phenotype
- Dominance, recessiveness, and partial dominance
  - Phenotypic effects
  - Genetic and chemical bases
- Gene, locus, and allele
- Heterozygous versus homozygous
- Construction and use of Punnett squares
- Patterns of inheritance
  - Monohybrid cross
  - Dihybrid cross
  - Trihybrid cross
  - Product Law
  - Polygenic inheritance
  - Inheritance with multiple alleles
- Gene interactions
  - Complementation tests
  - Epistasis
  - Collaboration and modifiers
  - Penetrance and expressivity
- Inheritance and selection
  - Heterozygote advantages
  - Pleiotropic alleles
  - Selection as the sum of positive and negative effects
  - Inbreeding and homozygous recessives

SUGGESTED READING

CROW, J. F., 1979. Genes that violate Mendel's rules, *Scientific American* 240 (2). *On alleles that manipulate the genome to enhance the probability of their own transmission.*

GOODENOUGH, U., 1978. *Genetics*, 2nd ed. Holt, Rinehart & Winston, New York, NY. *One of several very good introductory genetics texts currently available.*

HOLLIDAY, R., 1989. A different kind of genetic inheritance, *Scientific American* 260 (6). *On genetic "imprinting."*

STERN, C., and E. R. SHERWOOD, 1966. *The Origin of Genetics: A Mendel Sourcebook.* W. H. Freeman, San Francisco, CA. *Provides translations of Mendel's papers and letters, and other early papers in genetics from Mendel's time through the rediscovery of his work. Also includes modern analyses of why Mendel reported on only a few of the strains he tested.*

WHITE, R., and J. M. LALOUEL, 1988. Chromosome mapping with DNA markers, *Scientific American* 258 (2). *On modern methods for mapping the human genome.*

PART **III**

EVOLUTIONARY BIOLOGY

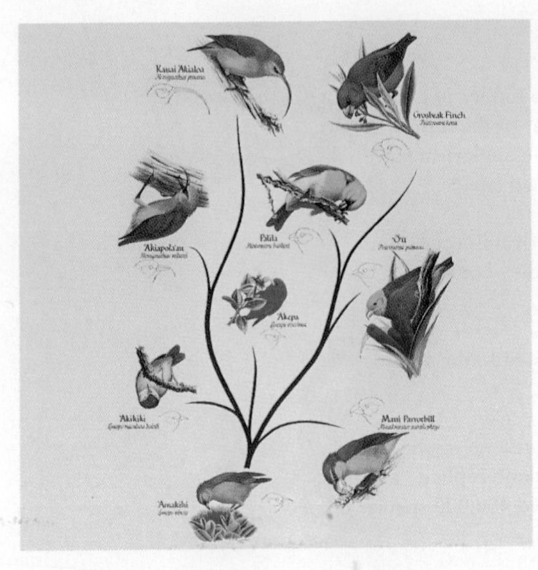

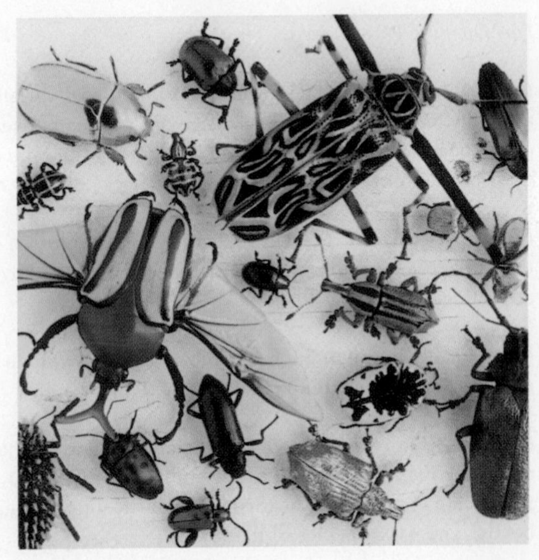

Part III

Part Opening Photographs:

(1) Female (and male) anglerfish. Morphological variation is the fodder of evolution and may lead, in time, to major size differences between mates. In the murky depths of the ocean floor, finding a mate is problematic. Miniscule male anglerfish solve their problem by hitching a ride on the backs of their much larger mates.

(2) A male peacock. Darwin recognized that evolution selects or reinforces some traits, like the peacock's ornate tail, because of the advantages they confer in attracting mates.

(3) Colorful Hawaiian finches exhibit their bills, which are adapted to the varied ecological niches within the islands.

(4) Beetles. In response to a theologian who asked what inference one might draw about the nature of God from the study of his works, J.B.S. Haldane replied "an inordinate fondness for beetles." The earth supports an estimated 400,000 species of beetles and 4100 species of mammals.

VARIATION, SELECTION, AND ADAPTATION

The study of inheritance is concerned with the transmission of alleles from parent to offspring. We saw in the last chapter that in most species every organism carries two alleles for a given genotype, and that the phenotype is usually determined by just one of the two alleles. An individual, of course, cannot evolve: its genetic inheritance is fixed at birth and so is the developmental potential of its cells. But a group of interbreeding individuals—a population—can evolve over time. Although individuals carry only two alleles for any character, one on each homologous chromosome, populations can contain many distinct alleles for any character.

In Chapter 1, we were careful to distinguish between the phenomenon of evolution—genetic change—and Darwin's astute hypothesis to explain the direction and speed of that change: natural selection. In this chapter, we will study changes in allelic frequencies in populations, the precise definition of evolution. At its simplest, evolution can occur through the emigration or immigration of members of a population. We will concentrate here, however, on the roles played by chance and natural selection in evolution.

VARIATION AND SELECTION

Fundamental to the modern theory of evolution by natural selection are five concepts: (1) that more offspring are produced than can survive to re-

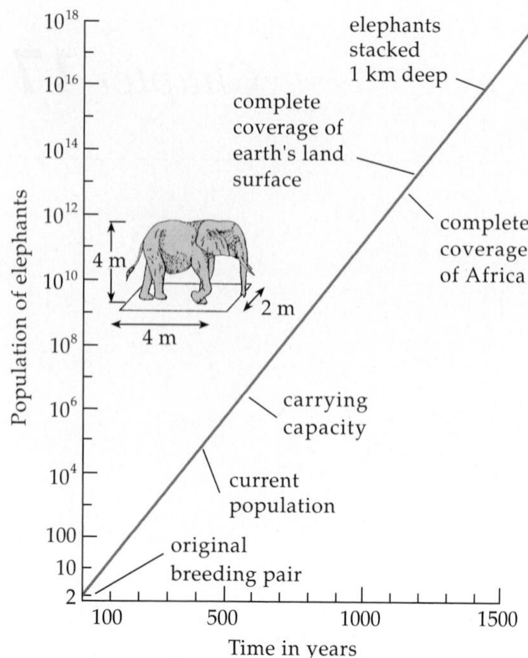

17.1 Unrestrained growth of a hypothetical elephant population
When Darwin's calculations for the slowest reproducing animal known are graphed and extrapolated, his point that not all offspring can survive becomes clear. (Darwin overestimated the reproductive lifespan and the age of sexual maturity in elephants, but these minor errors cancel one another out.)

produce (excess progeny); (2) that the characteristics of living things differ between individuals of the same species (variability); (3) that many differences are the result of heritable, genetic differences (heritability); (4) that some differences affect how well adapted an organism is (differential adaptedness); and (5) that some differences in adaptedness are reflected in the number of offspring successfully reared (differential reproduction).

The excess reproductive capacity of organisms (concept 1) is self-evident. Darwin made the point incisively:

> The elephant is reckoned the slowest breeder of all known animals, and I have taken some pains to estimate its probable minimum rate of natural increase; it will be safest to assume that it begins breeding when 30 years old and goes on breeding until 90 years old; if this be so, after a period from 740 to 750 years there would be nearly 19 million elephants descended from this first pair.

After about 1200 years, this hypothetical elephant population would cover the entire land area of the earth, Antarctica included, shoulder to shoulder and head to tail (Fig. 17.1). Clearly, then, not all offspring survive long enough to reproduce.

We examined the nature of the action and heritability of variant alleles (concepts 2 and 3) in the last chapter, with attention to a variety of allelic and genetic interactions. We saw in particular that most variants are recessive, which means they are unexpressed (thus hidden from selection) when present in only one copy. To begin this section we will look at various sources of variation and their consequences. We will then consider mechanisms of selection. The second major section of the chapter will explore several types of adaptation.

VARIATION FROM SEXUAL RECOMBINATION

A population is composed of many individuals. With rare exceptions, no two of these are exactly alike. In human beings we are well aware of the uniqueness of the individual, for we are accustomed to recognizing different people at sight, and we know from experience that each has distinctive anatomical and physiological characteristics, as well as distinctive abilities and behavioral traits. We are also fairly well aware of individual variation in such common domesticated animals as dogs, cats, and horses. But we tend to overlook the similar individual variation in less familiar species such as robins, squirrels, earthworms, sea stars, dandelions, and corn plants. Yet even though this variation may be less obvious to our unpracticed eye, it exists in all species.

The members of a population, then, share some important features, but differ from one another in numerous ways, some rather obvious, some very subtle. It follows that if there is selection against certain variants within a population and selection for other variants within it, the overall makeup of that population may change with time.

Variation arises from three main sources: crossing over in meiosis, the union of two unrelated haploid gametes (sexual recombination), and mutation (Fig. 17.2). The first two processes, which we examined earlier, do not lead to new alleles but to a recombination of existing ones. Although they

A

B

C

17.2 Sources of variation
One source of variation is spontaneous mutation. Most wisterias, like the one shown here, have lavender flowers (A). Several decades ago, however, the famous white-flowering Eno wisteria appeared behind a biology building on the Princeton campus (B). The white wisteria is almost certainly the result of a spontaneous mutation in a pigment gene of an offspring of one of the many lavender wisterias nearby. Another source of variation is recombination. Offspring of the same parents frequently do not resemble each other or either of their parents. Shown here, with their mother, are the various kittens of a single litter (C).

are in the long run beneficial to a species, crossing over and gamete fusion are part of the metabolically expensive and genetically chancy process of sexual reproduction, which is potentially costly to the mating pair. Since selection operates on individuals, the perpetuation of recombination suggests that sex persists because it allows a special kind of variation that serves evolutionary change. Mutation, which includes the point mutations mentioned in earlier chapters as well as large-scale exon recombinations, is a potential source of entirely new alleles; we will look at how new alleles arise in more detail in a later section. Taken together, these processes generate new alleles, and new combinations of alleles, and provide the genetic variability on which natural selection can act to produce evolutionary changes in populations.

Purely phenotypic variation Though we have been emphasizing the importance of variation in evolutionary change, some kinds of variation are immune or irrelevant to selection. Natural selection can act on genetic variation only when it is expressed in the phenotype. A completely recessive allele never occurring in the homozygous condition would be totally shielded from the action of natural selection; only a variation that affects the way an organism actually functions can be acted on.

Another potential complication is that any phenotypic variation within a population may give rise to reproductive differentials between individuals, whether or not the variation reflects corresponding genetic differences. Thus variations produced by exposure to different environmental conditions during development, or produced by disease or accidents, are subject to natural selection. In general, such selection has no effect on the overall genetic makeup of the population: there is no correlation between an individual's alleles and the chance events that lead to its lowered viability and reproductive potential. However, there are cases in which luck *can* cause

Exploring Further

THE LOGIC OF SEX

At first glance, sex makes no evolutionary sense. Half of a female's reproductive effort goes into sons, who will bear no offspring and so contribute little to her reproductive output. If males could be eliminated, each female could produce twice as many daughters, and far more grandchildren. This is easily demonstrated in groups of higher plants and animals that have both sexual and asexual morphs. In nearly all of these organisms, the asexual forms arose relatively recently from sexual precursors.

In addition to reducing reproductive output, sex tampers with success. A constellation of alleles that has worked well enough to allow an organism to reproduce is broken up, and half is assigned arbitrarily to each gamete, rather like taking cards randomly from a royal flush and a full house, say, and recombining them to form a new hand. The novel collection of cards is unlikely to be as good as either "parent" poker hand. Some asexual species are quite successful: the all-too-common dandelion is, despite its flowers, incapable of sexual reproduction.

The evolution of sex presumably began with haploid asexual species such as bacteria that normally reproduce asexually. *E. coli* can reproduce every 20 minutes if enough food is available; since it takes 18 minutes to replicate the chromosome, the contest to generate offspring while conditions are good is a race against the clock.

Diploidy is the next step toward sex, and we can see that a bacterium with two copies of its chromosome to replicate would be at an enormous reproductive disadvantage, leaving only eight offspring after two hours compared with 64 for its haploid counterpart. The cost of haploidy, however, is the immediate expression of any mutation, and since most genetic changes are deleterious, this penalty can be large. We will see in a later chapter that this is the most likely factor to explain the evolution of diploid dominance in plants, and the pattern is equally clear in animals, where the resources devoted to mutational repair activity increases with lifespan.

Diploidy by itself does not entail sexual recombination (dandelions are diploids) but even without it diploidy confers several advantages. Most obviously, deleterious mutations are hidden. In addition, the intact chromosome can be used as a template for the repair of double-stranded damage to the other. Diploids can also accumulate and harbor variants on the "spare" copy that may prove useful when conditions change.

Sexual reproduction adds crossing over and genetic recombination. Since crossing over involves the same enzymes that asexual diploids use for repair and that bacteria use for their rare bouts of conjugation, it seems likely that the first sexual species took advantage of crossing over as a way of correcting errors just prior to gamete formation. But changes bring new opportunities, and most biologists agree that crossing over creates more variation than it repairs.

Studies of the role of crossing over and recombination in evolution have led to two general hypotheses concerning the role of variation. The first, termed the *tangled-bank model,* is based on a reference to a diverse, multi-species environment near the end of *Origin of the Species*. Darwin realized that if conditions were everywhere the same, a few species would come to dominate any habitat. And yet, when he surveyed a square meter of his own lawn, he counted more than 20 species of plant. Darwin's description of an entangled bank emphasized the many different microhabitats found in even a small area. According to the tangled-bank model, asexual species are at a disadvantage because they are superior in only one microhabitat, whereas sexual species produce a range of offspring-microhabitat matches. Because of their adaptive narrowness, asexual siblings may even wind up competing primarily with each other, whereas the many unique sexual progeny of one pairing might be different enough to seek out their own most appropriate sites. Good evidence for this hypothesis is provided by at least some species.

The second hypothesis, the *red-queen model* takes its inspiration from the scene in *Through the Looking Glass* in which Alice and the Red Queen are running as quickly as they can and yet make no progress: however fast they move, they cannot outrun their surroundings (Fig. 1). The red-queen theory suggests that the need for variation arises from the need to keep up with a rapidly evolving "background" of predators, prey, and parasites, as well as changing conditions. The emphasis is on variation in time rather than variation in space.

The potential problem from parasites is espe-

Figure 1. Alice and the Red Queen, who were unable to outrun their environment, no matter how hard they tried.

cially easy to understand: most parasites are strain-specific, and can devastate an asexual or inbred population. New strains of barley in Britain, for example, have a useful life of only 3–5 years before fungal parasites adapt enough to ruin crops. Studies of thrips (juice-sucking insects) on long-lived hosts demonstrate that clones regularly develop specializations for that one plant. This degree of specialization is possible because parasites have a short generation time compared with their hosts; they can go through many rounds of selection before the host reproduces.

One of the predictions of the red-queen hypothesis is that crossing over should be most common among the longest-lived species; this would enable the host to compensate for rapid parasite evolution by making its offspring as different as possible. Crossover rates increasing with lifespan seem to be the pattern at least among mammals. What is hard to understand is why parasites have not wiped out dandelions and other asexual organisms.

With regard to habitat variation, both hypotheses are consistent with the pattern seen in many species that can reproduce either sexually or asexually. For instance, the first aphid to find a suitable new plant shoot reproduces asexually, creating a clone of daughters equally well adapted to this local habitat. But when conditions deteriorate, as when the shoot becomes overcrowded or suffers from drought, the foundress begins producing winged reproductive offspring that leave the shoot, mate with other aphids, and search out a suitable host (Fig. 2). Clearly the aphid life cycle can be interpreted as an adaptation to unpredictable spatial or temporal variation. There is every reason to believe that the degree and nature of the advantage of sexual reproduction can differ between species. The common thread, however, is that sex has been selected for because it creates variation. Without the capacity to change faster than mutations alone allow, long-lived species would be doomed.

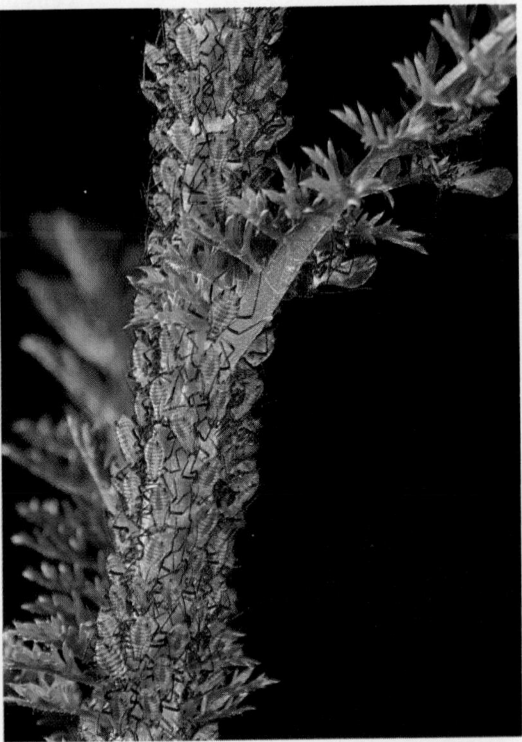

Figure 2. Sexual (winged) and asexual aphids on a yarrow stem

A

B

17.3 An okapi and a giraffe, two related African herbivores
The okapi (A) and giraffe (B) are thought to have had a common ancestor with a relatively short neck. The long-necked giraffe of today can reach food unavailable to shorter individuals.

evolution. If, for example, the population is small and the one carrier of a rare allele is killed by an accident, gene frequencies will change. We will return to the role of chance in small populations in both this chapter and the next.

An understanding of genetics makes it clear that development of athletic prowess by extensive practice, or development of intellectual powers by education, or maintenance of health by correct diet and prompt medical treatment of all ailments cannot alter the genes in the germ (gametic) cells. The gametes will carry the same genetic information regardless of postnatal experience. In short, selection that acts on variations produced exclusively by practice, education, diet, or medical treatment cannot bring about biological evolution.

Likewise, variation produced by somatic mutations is not raw material for evolutionary change. It would be possible, for example, for an important mutation to occur in an ectodermal cell of an early animal embryo. All the cells descended from the mutant cell would be of the mutant type. The result might be a major change in the animal's nervous system, but the change could not be passed on to the animal's offspring. The ectodermal cells are not the ones that give rise to gametes. Mutations in somatic cells cannot alter the genes in the germ cells, the cells that produce the gametes. Hence selection that acts on variations produced by somatic mutations cannot result in evolutionary change in sexually reproducing organisms.

Lacking genetic data, many prominent biologists of the last century and of the early part of this century assumed that exclusively phenotypic variation could serve as evolutionary raw material; that it cannot is far from obvious to many nonbiologists even today. As we saw in Chapter 1, the theory of ***evolution by natural selection*** proposed by Darwin and Wallace had a rival during the nineteenth century in the concept of ***evolution by the inheritance of acquired characteristics***—an idea often identified with Lamarck.

The Lamarckian hypothesis was that somatic characteristics acquired by an individual during its lifetime could be transmitted to its offspring. Thus the characteristics of each generation would be determined, in part at least, by all that happened to the members of the preceding generations—by all the modifications that occurred in them, including those caused by experience, use and disuse of body parts, and accidents. Evolutionary changes would be the gradual accumulation of such acquired modifications over many generations. The classic example is the evolution of the long neck of the giraffe (Fig. 17.3).

According to the Lamarckian view, ancestral giraffes with short necks tended to stretch their necks as much as they could to reach the tree foliage that served as a major part of their food. This frequent neck stretching caused their offspring to have slightly longer necks. Since these also stretched their necks, the next generation had still longer necks. And so, as a result of neck stretching to reach higher and higher foliage, each generation had slightly longer necks than the preceding generation.

The modern theory of natural selection, on the other hand, proposes that ancestral giraffes probably had short necks, but that the precise length of the neck varied from individual to individual because of their different genotypes. If the supply of food was limited, individuals with longer necks had a better chance of surviving and leaving progeny than those with shorter necks. This means, not that all individuals with shorter necks perished or that all with longer necks survived to reproduce, but simply that a slightly higher proportion of those with longer necks survived and left offspring. As a result, the proportion of individuals with genes for longer necks increased slightly in the succeeding generation. As the proportion of individuals with somewhat longer necks rose, the increased competition for food higher up on the trees resulted in a selective advantage for those with yet longer necks, and so evolution continued.

POPULATION GENETICS

The gene pool To understand evolution as change in the genetic makeup of populations in successive generations, it is necessary to know something about population genetics. Our study of the genetics of individuals was based on the concept of the genotype, which is the genetic constitution of an individual. Our study of the genetics of populations will be based in a similar manner on the concept of the gene pool, which is the genetic constitution of a population. The gene pool is the sum total of all the genes possessed by all the individuals in the population.

As we saw in the last chapter, the genome of a diploid individual contains, with a few exceptions, a maximum of only two alleles of any given gene.[1] But there is no such restriction on the gene pool of a population. It can contain any number of different allelic forms of a gene. The gene pool is characterized with regard to any given gene by the frequencies of the alleles of that gene in the population. Suppose, to use a simple example, that gene *A* occurs in only two allelic forms, *A* and *a*, in a particular sexually reproducing population. And suppose that allele *A* constitutes 90 percent of the total of both alleles while allele *a* constitutes 10 percent of the total. The frequencies of *A* and *a* in the gene pool of this population are therefore 0.9 and 0.1. If those frequencies were to change with time, the change would be evolution. Hence it is possible to determine what factors cause evolution by determining what factors can produce a shift in allelic frequencies.

Let us examine more carefully our hypothetical population in which alleles *A* and *a* have frequencies of 0.9 and 0.1 respectively. How can we calculate the genotypic frequencies that will be present in this population in the next generation? If we assume the population is large and that all genotypes have an equal chance of surviving, this calculation is not hard to make. If the frequency of *A* in the entire population is 0.9 and the frequency of *a* is 0.1, the alleles carried by sperm and eggs will also appear at these frequencies. Using this information, we can set up a Punnett square:

		Sperm	
		0.9*A*	0.1*a*
Eggs	0.9 *A*	0.81 *A*/*A*	0.09 *A*/*a*
	0.1 *a*	0.09 *a*/*A*	0.01 *a*/*a*

[1] Because the multiple copies of the rRNA genes are present in enormous numbers, it is almost inevitable that each individual will carry more than two alleles of these genes.

Notice that the only difference between this and a Punnett square for a cross between individuals is that here the sperm and eggs are not those produced by a single male and a single female, but those produced by all the males and females in the population, with the frequency of each type of sperm and egg shown on the horizontal and vertical axes respectively. Filling in the square (by combining the indicated alleles and multiplying their frequencies) tells us that the frequency of the homozygous dominant genotype (*A/A*) in the next generation of this population will be 0.81, the frequency of the heterozygous genotype (*A/a* and *a/A*) 0.18, and that of the homozygous recessive genotype (*a/a*) 0.01. Now we want to know whether the frequencies we have found will change in successive generations—in short, whether the population will evolve.

Evolution versus genetic equilibrium It is easy to believe that the more frequent allele (in our hypothetical case, *A*) will automatically increase in frequency while the less frequent allele (*a*) will automatically decrease and eventually be lost from the population, and thus that variation—the raw material for evolution—will inevitably vanish. A major problem for the early acceptance of Darwin's theory was the widespread belief in blending inheritance—that is, that offspring are wholesale mixtures of parental phenotypes, so that in the long run all individuals will tend toward some set of average species characteristics. This plausible assumption accords well with the common (but wholly incorrect) view that (except for humans and domesticated species) there is little difference between the members of a species. If there were little variation, evolution would be a minor phenomenon in nature. The discovery of particulate inheritance—of genes, alleles, and haploid gametes—disposed of this loss-of-variation problem. The rarity of a particular allele in a large population does not doom it to automatic disappearance, as we can show by using the known frequencies for one generation to compute the frequencies for the next generation.

We have said that the genotypic frequencies in the gene pool of the second generation of our hypothetical population will be 0.81, 0.18, and 0.01. We can use these figures to compute the allelic frequencies of the *A* and *a* in this generation. Since the frequency of the *A/A* individuals is 0.81, the frequency of their gametes in the gene pool will be 0.81. All these gametes will contain the *A* allele. Likewise, the frequency of the gametes of the *a/a* individuals will be 0.01, and each gamete will contain an *a* allele. The frequency of the heterozygous (*A/a* and *a/A*) individuals is 0.18 and the frequency of their gametes in the gene pool will be 0.18, but their gametes will be of two types, *A* and *a*, in equal numbers. Hence the frequency of the *A* and *a* alleles in the gametes of the population can be calculated as follows:

Frequency of genotypes	Frequency of *A* gametes	Frequency of *a* gametes
0.81 *A/A*	0.81	0
0.01 *a/a*	0	0.01
0.18 *A/a* + *a/A*	0.09	0.09
	0.9	0.1

We find that the allelic frequencies are 0.9 and 0.1, the same frequencies we started with in the preceding generation.

Since the allelic frequencies are unchanged, the genotypic frequencies in the succeeding generation will again be 0.81, 0.18, and 0.01; and in turn the allelic frequencies will be 0.9 and 0.1. We could perform the same calculation for generation after generation, always with the same results; neither the genotypic nor the allelic frequencies would change. In populations large enough to swamp chance effects, therefore, variation is retained. Moreover, evolution can occur only when something disturbs the genetic equilibrium. This was first recognized in 1908 by G. H. Hardy of Cambridge University and W. Weinberg, a German physician, working independently. According to the ***Hardy-Weinberg Law***, *under certain conditions of stability both phenotypic and allelic frequencies remain constant from generation to generation in sexually reproducing populations.*[2]

Let us examine the "certain conditions" that the Hardy-Weinberg Law says must be met if the gene pool of a population is to be in genetic equilibrium. They are as follows:

1. The population must be large enough to make it highly unlikely that chance alone could significantly alter allelic frequencies.
2. Mutations must not occur, or alternatively there must be mutational equilibrium.
3. There must be no immigration or emigration that alters allelic frequencies in the population in question.
4. Mating must be totally random with respect to genotype.
5. Reproductive success (that is, the number of offspring and the number of their eventual offspring) must be totally random with respect to genotype.

The Hardy-Weinberg Law demonstrates that variability and heritability, the two bases of natural selection, cannot alone cause evolution. Despite variability and heritability, if all five conditions of the Hardy-Weinberg Law are met, allelic frequencies will not change, and evolution cannot occur. But, in fact, these conditions are *never* completely met, and so evolution does occur. The present-day value of the Hardy-Weinberg Law is that it provides a baseline against which to judge data from actual populations. By defining the criteria for genetic equilibrium, it also indicates when a population is not in equilibrium, and helps to isolate the possible causative agents of evolution, some of which were by no means obvious to Darwin. The role of the investigator is then to discover the relative contribution of each factor to the evolution of a particular population. As we look at each factor in turn, we will see which ones are likely to be important.

With regard to the first condition, a population would have to be infinitely large for chance to be completely ruled out as a causal factor in the changing of allelic frequencies. In reality, of course, no population is infinitely large, but many natural populations are large enough so that chance alone is not likely to cause any appreciable alteration in the allelic frequencies in

[2] For this statement to hold true for phenotypes, the initial genotypic frequencies must be in equilibrium. If these frequencies are not in equilibrium, they will change in successive generations until the equilibrium is achieved. For example, if genotypes *A/A* and *a/a* were present in a population, but *A/a* were missing (let's suppose, because of human intervention), all three genotypes would appear in the next generation, and thereafter the genotypic and phenotypic frequencies would remain constant.

Exploring Further

THE HARDY-WEINBERG EQUILIBRIUM

On p. 453 we used a Punnett square to calculate the frequencies of the genotypes produced by alleles A and a, whose respective frequencies in a hypothetical population were given as 0.9 and 0.1. The same results can be obtained more rapidly by using an algebraic formula.

Expansion of the binomial expression $(p+q)^2$, where p is the frequency of one allele (in our case, A) and q the frequency of the other allele (a), yields the formula for the Hardy-Weinberg equilibrium:

$$p^2 + 2pq + q^2 = 1$$

Substituting the allelic frequencies 0.9 and 0.1 for p and q respectively, we obtain

$$\begin{array}{ccccccc} p^2 & + & 2pq & + & q^2 & = 1 \\ (0.9)(0.9) & + & 2(0.9)(0.1) & + & (0.1)(0.1) & = 1 \\ 0.81 & + & 0.18 & + & 0.01 & = 1 \end{array}$$

The three terms of the Hardy-Weinberg formula indicate the frequencies of the three genotypes:

$$\begin{array}{ll} p^2 = \text{frequency of } A/A & = 0.81 \\ 2pq = \text{frequency of } A/a + a/A & = 0.18 \\ q^2 = \text{frequency of } a/a & = 0.01 \end{array}$$

These are, of course, the same results we obtained using the Punnett square.

In this example we have assumed that we know the allelic frequencies and want to compute the corresponding genotypic frequencies. But the Hardy-Weinberg formula allows many other sorts of calculations as well. Suppose, for example, that we know a certain disease caused by a recessive allele d occurs in 4 percent of a certain population and that we want to find out what percent are heterozygous carriers of the disease. Since the disease occurs only in homozygous recessive individuals, the frequency of the d/d genotype is 0.04. Letting q^2 stand for the frequency of d/d in the formula, we can write

$$q^2 = 0.04$$

The frequency of allele d, then, is the square root of 0.04:

$$q = \sqrt{0.04} = 0.2$$

If the frequency of allele d is 0.2, the frequency of allele D must be 0.8, because the two frequencies must always add up to 1 (that is, $p + q = 1$). Substituting the frequencies of both alleles in the Hardy-Weinberg formula, we can compute the frequencies of the genotypes:

$$\begin{array}{ccccccc} p^2 & + & 2pq & + & q^2 & = 1 \\ (0.8)(0.8) & + & 2(0.8)(0.2) & + & (0.2)(0.2) & = 1 \\ 0.64 & + & 0.32 & + & 0.04 & = 1 \end{array}$$

Since the term $2pq$ stands for the frequency of the heterozygous genotype, which is what we wanted to know originally, our answer is that 0.32, or 32 percent, of the population are heterozygous carriers of the allele d that causes the disease we are studying. Powerful as this method is, however, it is important to remember that these equations apply only to populations in Hardy-Weinberg equilibrium.

Let us now see how this type of reasoning can be applied to calculate changes in allelic frequencies when only phenotypic frequencies can be measured directly. Suppose in a large population of freely interbreeding plants we find 59 percent with yellow blossoms (known to be a dominant phenotype) and 41 percent with white (a recessive phenotype). We return to the same place in the spring of the following year, after a very severe winter, and find 64 percent with yellow blossoms and 36 percent with white blossoms. Clearly, plants with the dominant allele survived better, but we want to know exactly how much the allelic frequencies changed.

Since white blossoms indicate a recessive phenotype, we know that the frequency of the genotype y/y was 0.41 initially. Setting $q^2 = 0.41$, we calculate that $q = 0.64$ (approximately), which means that the frequency of allele y was 0.64. The frequency of dominant allele Y was therefore 0.36 $(1 - 0.64 = 0.36)$. The next spring the frequency of white blossoms had fallen to 36 percent, which gives $q^2 = 0.36$ and $q = 0.60$. The frequency of allele y is therefore 0.60, and the frequency of Y must be 0.40. In summary, the frequency of allele y has changed from 0.64 to 0.60 and the frequency of Y from 0.36 to 0.40 in one year.

The preceding examples involved only two alleles. Similar procedures, even if considerably more complicated mathematically, can be used for situations involving multiple alleles. Thus the Hardy-Weinberg formula for a triallelic situation requires expansion of the trinomial $(p + q + r)^2$, where r is the frequency of the third allele. Similarly, a quadriallelic situation requires expansion of $(p + q + r + s)^2$.

their gene pools. A population with more than 10,000 members of breeding age is probably not significantly affected by random change. But allelic frequencies in small isolated populations of, say, fewer than 100 breeding-age members are highly susceptible to random fluctuations, which can easily lead to loss of an allele from the gene pool even when that allele is an adaptively superior one. In the absence of immigration or mutation, such an allele is lost forever. In such populations, in fact, there are relatively few alleles with intermediate frequencies; apparently the tendency is for most alleles either to be soon lost or to become fixed as the only allele present. In other words, small populations tend to have a high degree of homozygosity, while large populations tend to be more variable. Thus chance may cause evolutionary change in small populations (and even in large ones, given enough time and the absence of selection), but since this change, called ***genetic drift***, is not much influenced by the relative adaptiveness of the different alleles, it is essentially an indeterminate evolution, as likely to take one direction as another (Fig. 17.4). Because genetic drift can cause changes in allelic frequencies independent of natural selection, it is frequently called ***neutral selection***.

The second condition for genetic equilibrium—either no mutation or mutational equilibrium—is rarely met in populations. Mutations are always occurring. Most genes probably undergo mutation once every 1 million to 100 million replications; the rate of mutation for different genes varies greatly. As for mutational equilibrium, very rarely, if ever, are the mutations of alleles for the same character in exact equilibrium: the number of forward mutations per unit time is rarely exactly the same as the number of back mutations.[3] The result of this difference is a ***mutation pressure*** tending to cause a slow shift in the allelic frequencies in the population. The more stable allele will tend to increase in frequency, and the more mutable allele will tend to decrease in frequency, unless some other factor offsets the mutation pressure. Eventually, of course, the frequency of the more stable allele will become so high that it will undergo the same number of mutations per unit time as the more mutable allele, despite its lower mutation rate, and equilibrium will be achieved. This requires so much time, however, that other events almost always change allelic frequencies before mutational equilibrium is reached. But even though mutation pressure is almost always present, it is seldom a major factor in producing changes in allelic frequencies in a population in the short run. As we will see in a later section, however, gene duplication, exon recombination, and mutations in gene-control regions provide much of the basis for long-term genetic change.

According to the third condition for genetic equilibrium, a gene pool cannot accept immigrants from other populations that introduce new alleles or different allelic frequencies, and it cannot undergo changes in allelic frequencies by emigration. A high percentage of natural populations, however, probably experience at least a small amount of gene migration, generally called ***gene flow***, and this factor, which enhances variation, tends to upset the Hardy-Weinberg equilibrium and lead to evolution. Such evo-

17.4 Possible genetic drift in a cichlid fish *Pseudotropheus zebra*, one of hundreds of species of cichlid fish living in the rift lakes of Africa, is divided into numerous isolated populations, many of which have evolved their own distinctive morphology. As there is no known selective force that accounts for this diversity, the varied colors are thought by many researchers to be the result of genetic drift in each small population.

[3] By convention, the mutation from the more common allele to the less common one is called the forward mutation, and the reverse is called the back mutation.

A

B

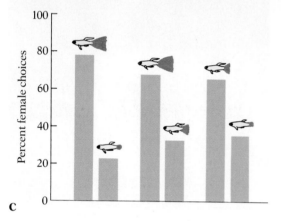

17.5 Female-choice sexual selection in guppies In most species of vertebrates, including guppies, the female chooses the mate. Male guppies exhibit great variety in spot size, color, and location, and in tail size and patterning (A, B). These features are largely heritable. Females choose males on the basis of their conspicuousness. Given a choice between males with tails of two different sizes—large and small, large and medium (the latter produced by surgical shortening), or medium and small—females prefer to be near males with larger tails (C). Subsequent findings indicate that they also preferentially mate with them. Females also prefer high display rates and greater coloration (particularly orange spots).

lution—that is, a change in allelic frequencies—need not, at least in the short run, involve natural selection. But there are doubtless populations—those on distant islands or in isolated patches of habitat, for instance—that experience no gene flow, and in many instances where flow does occur it is probably so slight as to be negligible as a factor causing shifts in allelic frequencies. We can conclude, therefore, that the third condition for genetic equilibrium is sometimes met in nature.

The final two conditions for genetic equilibrium in a population are that mating and reproductive success be totally random. Among the vast number of factors involved in mating and reproduction are choice of a mate, physical efficiency and frequency of the mating process, fertility, total number of zygotes produced at each mating, percentage of zygotes that lead to successful embryonic development and birth, survival of the young until they are of reproductive age, fertility of the young, and even survival of postreproductive adults when their survival affects either the chances of survival or the reproductive efficiency of the young. For mating and reproductive fitness to be totally random, all these factors must be random—that is, they must be independent of genotype, so that natural selection cannot operate. This condition is probably never met in any real population. An organism's genotype almost always influences each of these important factors. To take an obvious case, female guppies do not mate at random, but instead choose showy males with large tails (Fig. 17.5). In short, few aspects of reproduction are totally uncorrelated with genotype. Nonrandom reproduction is the usual rule. Aside from mate choice, which is considered part of sexual selection (to be discussed later in the chapter), nonrandom reproduction is a component of natural selection. Natural selection, then, is almost always operative in populations; there is always ***selection pressure*** acting to disturb the Hardy-Weinberg equilibrium and cause evolution, even if selection serves merely to limit the frequency of deleterious mutations.

In summary, the five conditions necessary to achieve the genetic equilibrium described by the Hardy-Weinberg Law correspond to five possible causative factors in evolution. The first, genetic drift, is likely to be important in small populations and need not be accompanied by natural selection. The second, mutation, is always at work, but in the short run is rarely significant. Like genetic drift, mutation is independent of natural selection, at least initially. The third, immigration and emigration, depends not only on the life history of the population, but also on the physical environment, which affects the ease of movement between populations; hence, immigration and emigration too can influence evolution apart from natural selection. The fourth and fifth, selection pressure from nonrandom mating and from variations in reproductive fitness, are almost always important in the evolutionary history of a species, though exactly which phenotypic traits are subject to the ongoing pressures of natural selection must be separately determined for each population.

We will concentrate on natural selection as the most influential mechanism in bringing about evolution in the natural world, but it is a useful exercise to think about Lamarck's long-necked giraffes again in the context of the Hardy-Weinberg factors. We've already seen that though the Lamarckian concept of inheritance of acquired characteristics has long been discredited, the Darwinian explanation of natural selection in favor of longer necks because they are more adaptive is by no means the only possibility. There are several alternative evolutionary scenarios that might conceivably have given rise to the present-day giraffe.

The most plausible alternative to natural selection is that giraffes have long necks as a result of genetic drift. Suppose that the ancestral population had a wide variety of neck lengths, or that a mutation caused a small subset of the population to have unusually long necks. If we assume that long necks are adaptively neutral—neither advantageous nor disadvantageous, or, more likely, that the benefits they confer are balanced by physiological costs—there is no reason according to the Hardy-Weinberg Law for selection pressure to cause neck length in the population as a whole to change. But suppose that the population suddenly declined because of some environmental factor like disease or bad weather, so that only a few individuals survived the crisis. If, by chance, a disproportionate number of the survivors were long-necked, the trait could become established without the intervention of natural selection (Fig. 17.6).

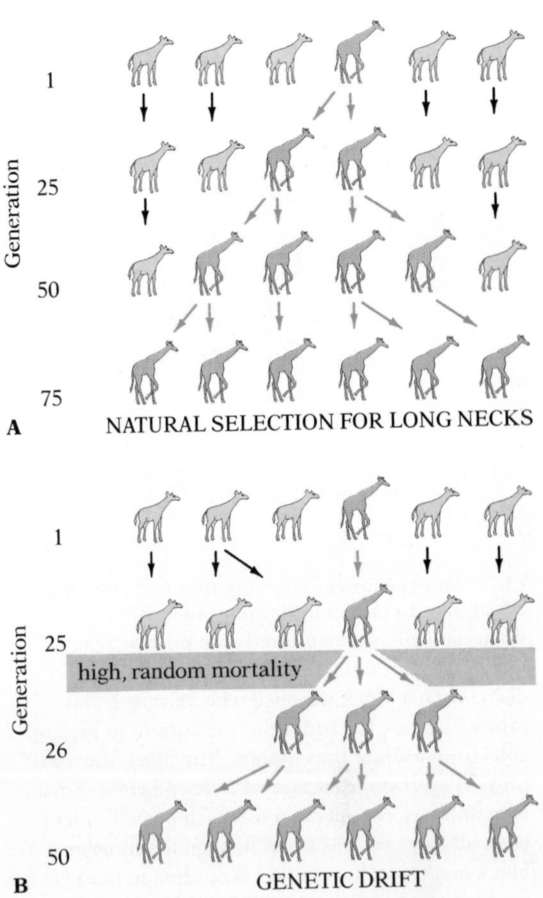

17.6 Natural selection versus genetic drift
In this hypothetical example, a small group of long-necked individuals arises in an otherwise stable population. (For simplicity, each animal in the drawing represents a fraction of the total population.) In evolution by natural selection (A), long-necked individuals become prevalent because they are able to reach more vegetation, and so survive to have proportionately more offspring in succeeding generations. In evolution by genetic drift (B), long-necked individuals become prevalent by chance: the frequency of long-necked individuals does not change until a chance catastrophe—fire, flood, heavy predation, or disease, for example—kills most members of the population, or the population size becomes small in some other way. Because only long-necked individuals happen to survive, their offspring multiply to fill the habitat even though their distinctive trait is of no net selective advantage.

As this hypothetical example indicates, the phenomenon of evolution does not depend exclusively on any single, particular mechanism, whether natural selection, genetic drift, mutation, or migration. It also underscores the potential error of assuming that all the traits of the living things around us are necessarily the adaptive result of natural selection. Wherever possible throughout this chapter and the next, therefore, we will consider the alternatives to natural selection. It is important to keep in mind, however, that most of the available evidence indicates that natural selection is the most important factor in evolution.

NATURAL SELECTION

Changes in individual allelic frequencies caused by natural selection

Let us now return for a moment to our hypothetical population in which the initial frequencies of the alleles A and a are 0.9 and 0.1 and the genotypic frequencies are 0.81, 0.18, and 0.01. According to the Hardy-Weinberg Law, in the absence of any of the five factors that cause evolution, these frequen-

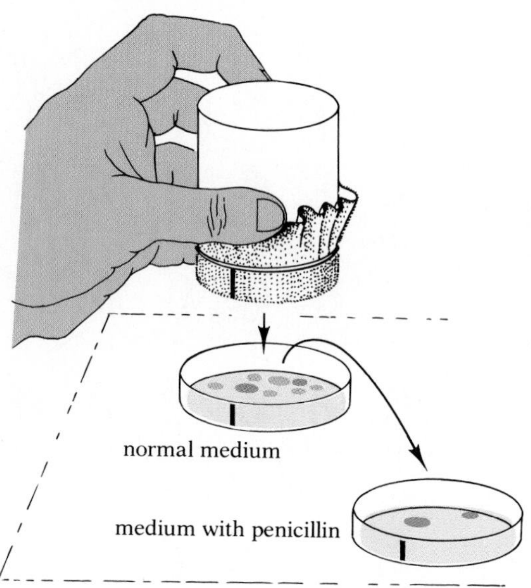

17.7 An experiment showing that mutation for resistance to penicillin is spontaneous Bacterial cells were cultured on a normal agar medium; many colonies developed (upper culture dish). Then a block wrapped with velveteen was pressed against the surface of the culture to pick up cells from each of the colonies. The block was next pressed against the surface of a second culture dish, containing sterile medium to which penicillin had been added; care was taken to align the transfer block and the culture dishes according to markers on the block and dishes (black lines). The cells from most of the colonies on the original dish failed to grow on the penicillin medium, but those of a few colonies (two are shown here) did grow. Had the cells of those two colonies spontaneously become penicillin-resistant before being transferred to the penicillin medium, or did exposure to the penicillin induce a mutation for resistance?

Because the transfer block had been aligned with each dish in the same way, according to the markers, it was possible to tell precisely which original colonies had given rise to the two colonies on the penicillin medium. Cells could therefore be taken from those original colonies, which had never been exposed to penicillin, and tested for resistance. They were found to be resistant. Hence the mutation for resistance must already have arisen; it was not induced by exposure to the drug.

cies will not change with the passage of time. Let's look now at the most important of these factors in guiding evolution: selection pressure.

Suppose that selection acts against the dominant phenotype in our example, and that this negative selection pressure is strong enough to reduce the frequency of A in the present generation from 0.9 to 0.8 before reproduction occurs. (Of course there will be a corresponding increase in the frequency of a from 0.1 to 0.2 among the survivors, since the two frequencies must total 1.) Now let us set up a Punnett square and calculate the genotypic frequencies that will be present in the zygotes of the second generation:

		Sperm	
		0.8 A	0.2 a
Eggs	0.8 A	0.64 A/A	0.16 A/a
	0.2 a	0.16 a/A	0.04 a/a

We find that the genotypic frequencies of the zygotes in the second generation are different from those in the parental generation; instead of 0.81, 0.18, and 0.01, the frequencies are 0.64, 0.32, and 0.04. If selection now acts against the dominant phenotype in this generation, and thereby again reduces the frequency of A, the genotypic frequencies in the third generation will be different from those of both preceding generations; the frequency of A/A will be lower and that of a/a higher. If this same selection pressure were to continue for many generations, the frequency of A/A would fall to a very low level and the frequency of a/a would rise to a very high one. Thus natural selection would have caused a change from a population in which 99 percent of the individuals showed the dominant phenotype and only 1 percent the recessive phenotype to a population in which very few showed the dominant phenotype and most showed the recessive phenotype. This evolutionary change from the prevalence of one phenotype to the prevalence of another would have occurred without the necessity of any new mutation, simply as a result of natural selection.

Instead of dealing with a hypothetical example, let us consider an actual situation in which selection has produced a radical shift in allelic frequencies. Soon after the discovery of the antibiotic activity of penicillin, *Staphylococcus aureus* (a bacterial species that can cause numerous infections, including boils and abscesses) quickly developed resistance to the drug. Higher and higher doses of penicillin were necessary to kill the bacteria, and the resistant bacteria became a serious problem in hospitals. Clearly, under the influence of the strong selection exerted by the penicillin, the bacterial population has evolved. But many studies have shown that the drug itself does not induce mutations for resistance; it simply selects against susceptible bacteria by killing them (Fig. 17.7). Apparently some genes determining metabolic pathways that confer resistance to penicillin are already present in low frequency in most populations, having arisen earlier as a result of random mutations. Individuals possessing these genes are thus ***preadapted*** to survive the antibiotic treatment, and, since it is they that reproduce and perpetuate the population (the susceptible individuals having been killed), the next generation shows a marked resistance to penicillin. If such genes were not already present in a population exposed to penicillin, no cells would survive and the population would be wiped out.

The primacy of selection pressure does not mean that new mutations cannot improve the resistance; in fact, continued selection with penicillin usually leads to gradually increased resistance, which is almost certainly in part a result of the differential survival of individuals with new mutations that enhance resistance. But it is purely a matter of chance that mutations beneficial in an environment containing penicillin should arise when this drug is administered; the same mutations arise at the same rate in the absence of penicillin, but are not selected for. They nevertheless persist if they have no adverse effect.[4]

Evolution of drug resistance in bacteria is not entirely comparable to evolution in biparental organisms, because intense selection can change gene frequencies much more rapidly in haploid asexual organisms than in biparental ones. The recombination that occurs at every generation in a biparental species often reestablishes genotypes eliminated in the previous generation; this does not happen in asexual organisms. Nevertheless, even very small selection pressures can produce major shifts in gene frequencies in biparental populations over an evolutionarily brief period. J. B. S. Haldane showed that if the individuals carrying a given dominant allele consistently benefit by as little as 0.001 in their capacity to survive (that is, if 1000 *A/A* or *A/a* individuals survive to reproduce for every 999 *a/a* individuals that survive to reproduce), then the frequency of the dominant allele could increase from 0.00001 to 1.0 in fewer than 24,000 generations. Now, 24,000 generations may sound like an incredibly large number, but remember that many plants and animals have at least one generation a year—*Drosophila*, for example, has more than 30. In very few species is the generation time more than 10 years (humans are among the few exceptions). Hence 24,000 generations often means fewer than 2400 years and rarely more than 240,000 years. Both of these are relatively short time spans when measured on the geologic time scale. Recent evidence suggests that many selection pressures in nature are much larger than 0.001; hence major changes in allelic frequencies sometimes probably take less than a century, perhaps even less than a decade.

Directional selection of polygenic characters So far, we have discussed idealized situations involving only two clearly distinct phenotypes determined by two alleles of a single gene. But in reality, as we saw in the last chapter, the vast majority of characters on which natural selection acts are influenced by many different genes, most of which have multiple alleles in the population; the expression of many characters, moreover, is influenced

[4] There is some evidence that in the face of adverse conditions, some species of bacteria can increase the rate at which their genomes mutate, either by neglecting to repair some errors, or by actively creating mistakes (as in the hypermutation process at work on the antibody genes, described in Chapter 15). In the face of declining reproductive potential, such a system would increase the variability in the population, and hence the chance of obtaining a favorable mutation that could "rescue" at least one individual, thus allowing it to found a new clone able to thrive despite the altered conditions. A more controversial observation suggests that hypermutation can be focused on a defective gene when the bacterium begins to need a working copy of it.

considerably by environmental conditions. Consequently such characters —height, for example—usually vary continuously over a wide range. When graphed, the frequency distribution often approximates the so-called normal, or bell-shaped, curve (Fig. 17.8).

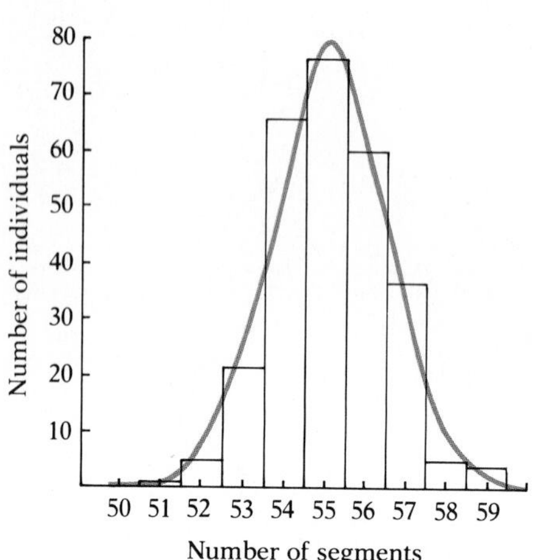

17.8 Frequency distribution of number of body segments in a population of the millipede *Narceus annularis*
The pattern of variation in number of segments (shown by the vertical bars) approximates, but does not exactly fit, the bell-shaped normal curve of probability.

If the environmental conditions should change, creating a consequent shift in the selection pressure, we would expect the curve of phenotypic variation to shift as a result of changing allelic frequencies. To illustrate this, let us assume that the conditions under which a certain plant grows best are genetically determined. The hypothetical case of such a plant is presented in Figure 17.9. The first curve (Fig. 17.9A) shows the annual rainfall at which the various plants in a particular population would grow best. The actual rainfall in the area where this population occurs averages 40 cm, as indicated by arrow 1, though it will vary around this mean from year to year. The population contains a very few plants (S) that would grow best if the annual rainfall were about 32 cm and a very few (W) that would grow best if the rainfall were about 48 cm. Plants that would grow best if the annual rainfall were about 36 cm (T) or 44 cm (V) are fairly common in the population. This phenotypic diversity is maintained because of the variability of the rainfall about the mean, so that in some unusually dry years, for example, the group S plants would have the advantage. In any given year, then, there is selection for an optimum phenotype, but the optimum varies. Since the average rainfall is 40 cm, the U plants are best adapted in the long run, and so are the most common.

Now let us suppose that the average annual rainfall in the area in question slowly increases over a period of years until it is 44 cm (arrow 2). Under these new environmental conditions, the V plants (which grow best when the annual rainfall is about 44 cm) will do better than before; a higher percentage of them can be expected to survive and reproduce, and their frequency should increase. Similarly, the W plants (which grow best when the annual rainfall is about 48 cm) will now grow better than formerly, and they too should increase in frequency. Conversely, the T plants and the U plants will not grow as well as formerly, and they should decrease in frequency. And the S plants, only a few of which managed to survive when the annual rainfall was 40 cm, would now be so poorly adapted to the prevalent conditions that none could survive. These changing frequencies, produced by the shift in the selection operating on the population, would give rise to the new curve shown in Figure 17.9B.

If the average annual rainfall continues to increase over a period of years until it reaches 48 cm (arrow 3), the W plants and X plants should increase in frequency, the U plants and V plants should decrease in frequency, and the T plants should disappear. These shifts would give rise to the curve shown in Figure 17.9C. If the average annual rainfall then slowly increases to 52 cm (arrow 4), it should cause further shifts in frequencies, producing the curve shown in Figure 17.9D.

The changing environmental conditions, then, have given rise to what is called ***directional selection***, which has caused the population to evolve along a particular functional line. If the population had not been sufficiently variable genetically to have the potential to change when the environment changed, it would have been much reduced, or it might even have become extinct. Before the average rainfall began to increase, there was selection each year in one direction or another, depending on whether the rainfall that year was greater or less than normal. This led to the distribution of plants centered on the 40-cm-per-year average.

Creation of novel phenotypes through natural selection Notice that in our hypothetical plant population the directional selection did not just shift the peak of the curve to the right, as might happen when the variation in the gene pool is small. Instead, it caused the entire curve—the extremes as well as the peak—to shift to the right. The shift was eventually so great, in fact, that a class of plants (X) not even present in the original population became the largest class. But, you might ask, if X plants, Y plants, and Z plants were not present initially, how did they arise in the descendant populations? One possibility is that, purely by chance, new genes or alleles arose that made their possessors grow better in wetter habitats; such novel genes or alleles would have been strongly selected for and would have spread rapidly through the population. But if moisture preference is influenced by many different genes, as is highly likely, new phenotypes such as X, Y, and Z could arise without the necessity of any new genetic variation, simply through the separate increase in frequency of particular alleles already present, which would then be more likely to occur together and produce a new phenotype.

Haldane calculated how long it would take for a new phenotype to be created in this way. He showed that if one particular allele of each of 15 independent genes is present in 1 percent of the individuals of a population, then all 15 alleles will occur together in only one of 10^{30} individuals. But there has never been a population of higher organisms containing anywhere near 10^{30} individuals. Hence the chances that all 15 alleles would occur together in even one individual in a real population are exceedingly small—zero for all practical purposes. But, according to Haldane, if there is moderate natural selection for each of the 15 alleles, it would take only about 10,000 generations (1000 to 100,000 years) for the frequency of each allele to increase from 1 percent to 99 percent. Once each allele is present in 99 percent of the population, 86 percent of the individuals in it will have all 15 alleles and hence will show the phenotype that was previously nonexistent in the population. Thus recombination and selection, even in the absence of new alleles, can produce new phenotypes by combining old genes in new ways and systematically favoring certain combinations.

An actual illustration of the sort of change outlined in the preceding hypothetical example is provided by a long-term selection experiment per-

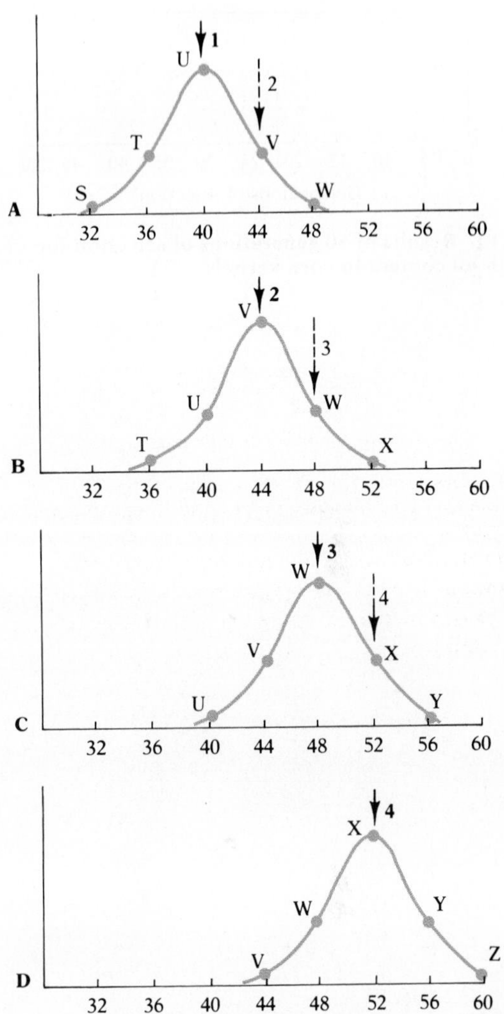

17.9 Evolutionary change of a hypothetical plant population in response to directional selection by changing rainfall The various phenotypes (S, T, U, etc.) in a hypothetical plant population reflect different genetically determined growth responses to annual rainfall. The four curves show the frequency distributions of the phenotypes at different times (A, B, C, D), under conditions of average annual rainfall indicated by the solid black arrows. As described in the text, a systematic change in average rainfall exerts directional selection on the plant population, shifting the curve of relative phenotypic frequencies more and more to the right.

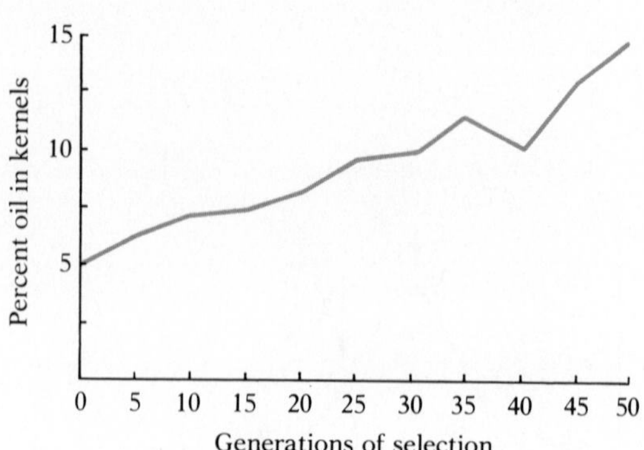

17.10 Results of 50 generations of selection for high oil content in corn kernels

formed on corn by agronomists at the University of Illinois. These investigators selected for high oil content of the corn kernels; the directional selection was continued for 50 generations. There was a steady increase in oil content throughout most of this period (Fig. 17.10). The kernels of the original stock of corn plants averaged about 5 percent oil; those of the plants in the 50th generation after selection averaged about 15 percent (higher than any individuals in the first generation), and there was no indication that a maximum had been reached.

That this steady change over the course of 50 generations must have resulted primarily from the formation of new genetic combinations through selection rather than from the occurrence of a series of new mutations can be seen from a few simple calculations. The agronomists raised between 200 and 300 corn plants in each generation. In 50 generations, then, they raised between 10,000 and 15,000. But the usual rate of mutation per gene in corn is never greater than one in 50,000 plants, and it is usually lower. Hence it is unlikely that even one mutation contributing to an increase in oil content occurred in any particular gene affecting this phenotype during the experiment. The gradual increase in oil content during the 50 generations of directional selection must have resulted largely or entirely from the formation of new genetic combinations.

To summarize: In biparental populations, selection—whether natural or artificial—determines the direction of change largely by altering the frequencies of alleles (and genes) that arose through duplication, transposition, and random mutation many generations before, thus establishing new genetic combinations and gene activities that produce new phenotypes. The processes that create new alleles and genes are not usually major *directing* forces in evolution; the principal evolutionary role of genetic changes consists in replenishing the store of variability in the gene pool and thereby providing the potential upon which future selection can act.

Disruptive selection It can sometimes happen that a polygenic character of a population is subject to two (or more) directional selection pressures favoring the two extremes of the distribution. Suppose, for example, that a certain population of birds shows much variation in bill length. Suppose, further, that as conditions change there are increasingly good feeding opportunities for the birds with the shortest bills and also for those with the longest bills, but decreasing opportunities for birds with bills of intermediate length. This might happen if the population of plants producing fruits suitable for intermediate-sized bills began to decline, or if a competing species more efficient at harvesting such fruits were to immigrate and become established. The effect of such selection, at least in the short run, would be to divide the population into two distinct types, one with short bills and one with long bills. The combined action of the opposing directional pressures would thus disrupt the smooth curve of phenotypes in a population (Fig. 17.11B); hence this sort of selection is called ***disruptive selection***.

The most important presumed example of disruptive selection is gamete dimorphism: large eggs and tiny sperm/pollen. If we assume that gametes were originally all the same size, as is still the case in many primitive species, any that were slightly larger would have enhanced the ultimate survival of the zygotes they contributed to by providing more food to fuel initial

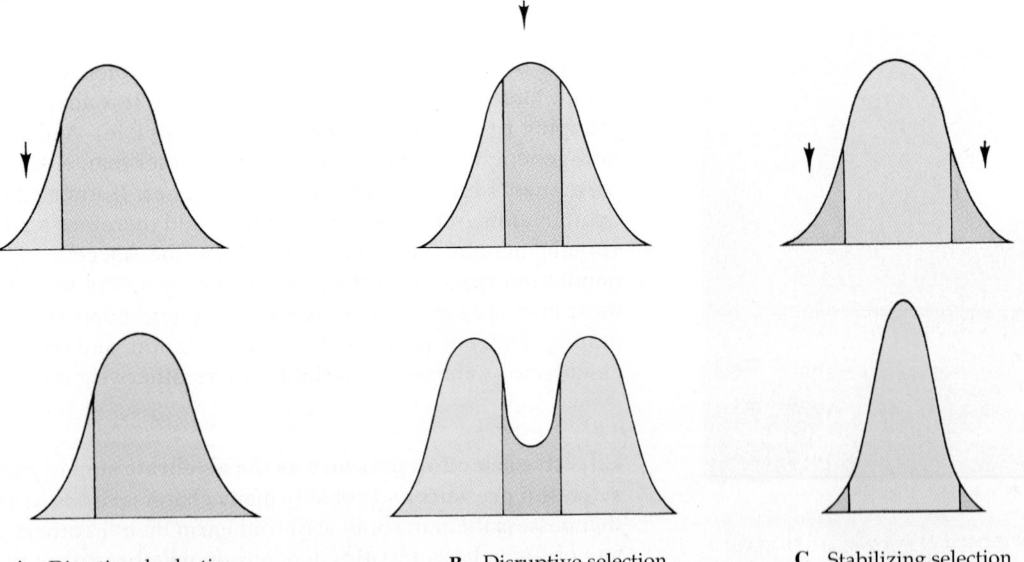

17.11 Directional, stabilizing, and disruptive selection compared
Each graph indicates the relative abundance of individuals of various heights in a population. In each instance the original condition of the population is shown above, and the later condition, after the specified selection, below. (A) Directional selection acts (arrow) against individuals exhibiting one extreme of a character (here the shortest individuals —represented by the blue area under the upper curve). The eventual result (bottom curve) is that the distribution of heights in the population has shifted to the right, indicating that the population has evolved in the direction of greater height. (B) By contrast, disruptive selection acts against individuals in the mid-part of a distribution, thereby favoring both extremes; in our example both the shortest and the tallest individuals would be favored, but individuals of medium height would be selected against. The result is a tendency for the population to split into two contrasting subpopulations. (C) Stabilizing selection acts against both extremes, culling individuals that deviate too far from the mean condition and thus decreasing diversity and preventing evolution away from the standard condition.

development. At the same time, organisms that produced small gametes would have gained a reproductive advantage by virtue of making more gametes with the same amount of food, thereby outnumbering the competition from other organisms. In addition, smaller gametes can swim faster because they encounter less resistance, so where speed is important they would have been at an advantage. Of course, selection must have worked to prevent small–small fusions, since the resulting zygote would have too little food to compete efficiently. If this widely accepted scenario is correct, then all the vast diversity of sex-specific morphology, physiology, and behavior could be a consequence of disruptive selection.

Stabilizing selection Most often, when a polygenic character in a population is subject to two or more opposing directional selection pressures operating simultaneously, the pressures select against the two extremes of the distribution—plants that are too tall to resist high winds, for example, and those of the same species that are so short that they are shaded by other plants, and so lose essential sunlight. Similarly, unusually severe winter storms frequently kill a disproportionate number of the largest and smallest birds in a population. When selection operates against individuals at the two ends of the distribution for a polygenic trait, the process is called ***stabilizing selection*** (Fig. 17.11C).

Stabilizing selection goes on all the time, operating on a much larger scale than that of the single characters we have been discussing, and as a result plays an extremely important conservative role. Each species, in the course of its evolution, comes to have a constellation of genes that interact in very precise ways in governing the developmental, physiological, and biochemical processes on which the continued existence of the species depends. Anything that disrupts the harmonious interaction of its genes is usually deleterious to the species. But in a sexually reproducing population,

17.12 Male and female guppies
The showy coloration of the male (top) contrasts with the drab gray of the female.

favorable groupings of alleles tend to be dispersed and new groupings formed by the recombination that occurs when each generation reproduces. Most of these new groupings will be less adaptive than the original grouping (though a few may be more adaptive). And the vast majority of new genetic variations tend to disrupt rather than enhance the established harmonious relationships among the genes. If unchecked, forces like recombination and random mutation would therefore tend to destroy the favorable genetic groupings on which the success of members of the population rests. Selection, by constantly acting to eliminate all but the most favorable genetic combinations, counteracts the disrupting, disintegrating tendency of recombination, mutation, and the like, and is thus the chief factor maintaining stability where otherwise there would be chaos.

Effective selection pressure as the algebraic sum of numerous separate selection pressures Probably many characteristics benefit the organisms that possess them in some ways and harm them in others. The evolutionary fate of such characteristics depends on whether or not the various positive selection pressures produced by their advantageous effects outweigh the negative selection pressures produced by their harmful effects. If the algebraic sum (an addition taking into account plus and minus signs) of all the separate selection pressures is positive, the trait will increase in frequency, but if the algebraic sum is negative, the trait will decrease in frequency.

As an example of the determination of a complex character having both beneficial and deleterious effects, let us consider the selection pressures on showiness in male guppies, a fish native to the freshwater streams of South America (Fig. 17.12). As we have already seen, males with larger tails and brighter spots are more likely to mate, evidently because females prefer such males. Competition for females should result in very showy males, since males with alleles for large tails and bright spots will leave more offspring; and, indeed, males in large aquarium populations have been found to become increasingly showy with succeeding generations. This process, which will be described in more detail shortly, is known as female-choice sexual selection.

No such selection pressure has operated on the females. These gray, nondescript fish closely resemble one another and are much less easily seen by predators than the males, whose bright markings and showy tails make them more subject to predation. This liability causes strong selection against such features; indeed, in laboratory situations, predators capture the showiest males first. As we might expect, males found in the wild are much less conspicuous than their counterparts in predator-free aquaria. The showiness of males, then, exemplifies stabilizing selection—a balance between selection for greater showiness (by the females, who confer reproductive success on the most "attractive" males) and selection for inconspicuousness (by the predators, who can terminate any prospect for further mating if a male is too visible).

Just as polygenic traits often have both advantageous and disadvantageous effects, the alleles of a single gene also usually have multiple effects (pleiotropy), and it is most unlikely that all of them will be advantageous. For example, *Drosophila* carrying alleles that disrupt the activity of specific systems involved in communication between nerve cells have altered patterns of wing venation; those with a muscle defect that causes them to hold their wings upright will not walk toward light, though this is a powerful response in normal fruit flies. Whether an allele increases or decreases in frequency is determined, as in the case of polygenic traits, by whether the sum of the various selection pressures favoring it is greater or smaller than the sum of the selection pressures acting against it.

Many instances are known in which the effects of a given allele are more advantageous in the heterozygous than in the homozygous condition. As we mentioned in the last chapter, for example, in some parts of Africa the allele for sickle-cell anemia occurs in humans much more often than we might expect in view of its highly deleterious effect when homozygous (Fig. 17.13). This is because the allele, when heterozygous, confers on the possessor a partial resistance to malaria. The equilibrium frequency of the sickle-cell allele is thus determined by at least four separate selection pressures: (1) the strong selection against the recessive homozygotes, who suffer the full debilitating effects of sickle-cell anemia; (2) the weaker selection against the heterozygotes as a result of their mild anemia; (3) the selection against the dominant homozygotes, who are more susceptible to malaria; and (4) the fairly strong selection favoring the heterozygotes as a result of their resistance to malaria.

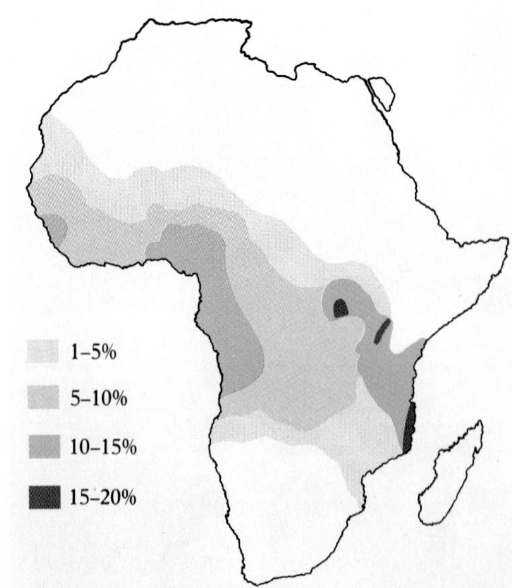

17.13 The distribution of sickle-cell anemia in Africa
The various colors indicate the percentage of the population in each area that has the disease.

Balanced polymorphism and the maintenance of genetic variability

Polymorphism is the occurrence in a population of two or more distinct forms, or morphs, of a genetically determined character. These phenotypes do not grade into one another like those represented by a bell-shaped curve for height, where "short" and "tall" designate not distinct morphs, but rather the extremes of a continuum. Instead, individuals fall into separate categories known as ***discontinuous phenotypes***, and intermediates are rare or absent. For example, Mendel's peas were polymorphic with regard to the color of their flowers: some plants had red flowers, others white, but none had pink. Human populations are polymorphic with regard to blood groups: the same population usually includes type A, type B, type AB, and type O individuals. Polymorphism is common in wild populations too: several species of snails, for instance, occur in banded and unbanded forms, and the red fox has both red and silver-colored morphs. In some instances the genetic basis of the polymorphism is known, as in Mendel's peas and human blood groups. But in many other instances, especially when the traits are polygenic, it is not known.

When the relative frequencies of the different morphs in a population are stably balanced over time, we speak of ***balanced polymorphism***. Sometimes the balance is maintained because the polymorphism itself is advantageous. Thus, if a polymorphic species lives in an environment subdivided into many local areas where different conditions prevail, one of the morphs may do better in one area, while others may do better in other areas. If the various morphs are all present among an individual's descendants, these can exploit more completely the subdivisions of the variable environment. Sometimes one morph is adaptively superior at one time of year or in one habitat, and another is superior at another time or in another place (Fig.

A

B

17.14 A diet-induced polymorphism
A moth common in northern Mexico and the southwestern United States lays its eggs on oak trees. The caterpillars that hatch in the early spring feed on oak catkins, and develop to resemble these reproductive organs of the host plant (A). After pupation during the summer, the emerging adults mate and then lay eggs on oaks long after the catkins have been dropped. Caterpillars hatching from these eggs consume oak leaves and develop to resemble twigs (B). Laboratory experiments demonstrate that what a newly hatched caterpillar begins eating determines which morphology will be produced.

17.14); hence an individual's descendants may have a better chance of survival if they are polymorphic than if all belong to a single form. Polymorphism of this type, accordingly, may be the result of a kind of disruptive selection: each of several contrasting forms is being selected for. This strategy will work best if the offspring can adopt the form that best suits the conditions in which they find themselves (as in the case of the caterpillars of Figure 17.14), but even the random production of alternative progeny morphs will be advantageous to a parental organism if selection against the wrong morph is high enough. In such a situation individuals that produce a pure culture of one form—thereby putting all their eggs in one basket—will run the risk of having none survive.

There are many instances of balanced polymorphism in which some of the contrasting morphs, instead of being themselves selected for, are the unavoidable by-products of selection for a heterozygous phenotype. As we saw with sickle-cell anemia in Africa, the heterozygotes are sometimes better adapted than the homozygotes—a condition known as ***heterozygote superiority*** (or sometimes heterosis or overdominance).

Heterozygote superiority favors balanced polymorphism, because it makes for retention in the population of both alleles of a given gene at frequencies higher than would be predicted from the selection acting on the homozygous phenotypes. Thus, if A/a individuals are adaptively superior to both A/A and a/a individuals, both allele A and allele a will be retained in fairly high frequency in the population. Neither allele will be eliminated, as might tend to happen if one of the homozygotes were far superior. Therefore all three possible genotypes—A/A, A/a, and a/a—will occur frequently in each generation, and if each of these produces a noticeably different phenotype the population will be polymorphic. The relative frequencies of A and a, and of the three morphs they produce, will be determined, as in sickle-cell anemia, by the balance between the several selection pressures acting on the system.

Frequency-dependent selection Sometimes the selection pressure on an individual displaying one morph (whether physiological, anatomical, or behavioral) depends on the frequency of that morph compared with the alternatives. For example, all female salmon and some males leave their home streams to feed and grow for several years at sea. They fight their way back to their natal streams and spawn. Some males, however, never leave; these "precocious parr" overwinter in the home stream, where they suffer high mortality. Still other males—"jacks"—return after only a year. When it comes time to mate, returning full-sized males defend territories and drive off the much smaller jacks; the jacks attempt to defend their own territories if the population of full-sized males is low, and otherwise compete for space near the large males. The surviving parr, which are tiny, are often able to hide on a territory; when a courting pair of salmon shed their gametes into the nest, the parr sneak in and deposit their own sperm.

Clearly the relative success of large males, jacks, and parr depends on how many others are "playing" each of these genetically determined strategies; the fewer that opt for one strategy, the better that strategy works. For example, the parr are basically parasites; they cannot attract and court females, but rather must depend on larger males. An individual parr can fertilize vast numbers of eggs in a season if there are many courting males and few competing parr. If, on the other hand, there are few territorial males and many parr, the individual success of these parasites plummets. The stable distribution of male morphs has been dramatically altered in the last century by fishing. In many areas, so many full-sized males have been caught that the genes for that option are disappearing, and 90 percent of the returning males are jacks.

17.15 A dominance ritual in mountain sheep Male mountain sheep work out a dominance hierarchy through a ritualized duel in which a run on hind legs culminates in a loud collision, followed by a head-turning display. Dominant males obtain unchallenged access to females.

SEXUAL SELECTION

Darwin distinguished between selection that affects physical survival and selection that operates on traits used exclusively to attract and keep mates. In evolutionary terms, the failure to reproduce is just as fatal as early death, but since selection for characters that enhance an individual's chance of mating are often very different for the two sexes, he considered *sexual selection* to be nearly independent of natural selection.

Male contests Darwin distinguished two basic varieties of sexual selection. The more common form involves contests between members of one sex (usually the males) for access to the other sex. These duels can take the form of dominance fights, which establish a male hierarchy and thus access to reproductively ready females (Fig. 17.15). More often the males fight for the best territories in the habitats females favor; the strongest males obtain the highest-quality territories, and therefore the largest number of matings (or, in monogamous species, the female that is best able to gain access to the territory). Male-specific features like large size, offensive weapons (including the horns of mountain sheep), and defensive structures (like the lion's mane, which protects his neck from bites inflicted by other males) are each the result of sexual selection.

17.16 A normal male widow bird in flight Experimental shortening or lengthening of tails demonstrates that females select males with the longest tails.

Female choice Darwin proposed that other uniquely male features exist only to attract females—the elaborate tail of the peacock, for instance, is not used to attack males, but rather is displayed whenever a peahen is nearby. A variety of experiments in which these dimorphisms have been altered demonstrate that females of some species do indeed select males on the basis of, for example, the length of their tails (Fig. 17.16; see also Fig. 17.5). Considerable controversy has surrounded the idea of female choice. Skeptics argue that such a system could not evolve if it did not enhance female reproduction, and a male's long feathers are unlikely to help females. But since many male dimorphisms exaggerate a species-specific recognition sign, the female benefits from the increased certainty of mating with a male of the right species, who also carries genes that will enhance the attractiveness (and therefore the reproductive potential) of her sons. She also benefits (perhaps) from genes that confer the sort of physiological superiority that have enabled the male to survive despite the burden of his morphological "handicap"—genes that might contribute to the physiological health of her daughters. In other cases the female's attraction to a male's dimorphism seems to have been favored by selection because the presence of bright colors and the like indicate that the male is free from the sorts of parasites that threaten the population, and the progeny would benefit from this immunity. The mating systems of many species appear to include elements of both male-contest and female-choice sexual selection.

HOW DO NEW ALLELES ARISE?

We have so far been concerned with the distribution of alleles in a population over time by sexual reproduction and selection. One of the troubling mysteries of evolution for Darwin and his early supporters, however, was how a complex organ or other structure could arise through selection on small variants of existing traits. How, for example, could the elaborate camera eye of vertebrates evolve? Wouldn't all the intermediate steps, as well as the first, be useless or even maladaptive, altering some important pre-existing structure with no assurance that anything worthwhile would ever emerge from the change? At the molecular level, the question can be phrased in terms of amino-acid sequences in proteins: wouldn't alterations destroy the specificity and activity of an existing enzyme, for instance, long before there would be much chance of mutations producing something new and better? Today we know that Darwin's general idea was correct—new genes, new alleles, new morphological structures generally arise out of previous ones by a combination of well-documented genetic mechanisms.

Our present understanding of point mutations—deletions, substitutions, and additions of one or a few bases—tells us that changes at this level can have two useful effects. First, they can fine-tune one allele of an existing gene, safe from negative selection during intermediate steps by virtue of the protection afforded by diploidy. (The orthodox version of the gene is on the other copy of the chromosome, performing its job.) Second, when small mutations occur in control regions—the binding sites for transcription factors, for example—they can dramatically alter the specificity, timing, and degree of responsiveness of gene activity in one or a group of alleles on that copy of a chromosome. These are potentially useful sources of variation, to be sure, but the origin of much major evolutionary change is based on two other mechanisms: (a) gene duplication followed by the independent evolution of the spare copies of the gene, and (b) novel recombinations of pre-existing exons.

Gene and exon duplication The discovery that eucaryotic chromosomes contain introns, exons, seemingly functionless pseudogenes, and repetitive DNA came as a great surprise. Most biologists expected that natural selection would have operated to weed out apparently unnecessary DNA, as it seems to have done in procaryotes. The bacterium *E. coli*, for example, has lost the ability to synthesize the few amino acids always found in abundance in its natural habitat; the genes for the corresponding biosynthetic pathways are missing, and essentially no space in the chromosome of this rapidly multiplying organism is wasted. By contrast, well over 90 percent of eucaryotic DNA lies apparently functionless and dormant. The subsequent discovery that the fetal immune system actively disposes of redundant gene segments was also unexpected, and made it immediately clear that the genome is not necessarily as stable as had been thought. The discovery of transposons underscores this point.

The unique structure and organization of antibody genes, MHC proteins, and T-cell receptors, which we discussed in Chapter 15, provides food for thought about the evolution of genes in general. One widely accepted hypothesis is that the exons of the variable and constant regions of both the light and heavy chains of antibodies, coding as they do for very similar polypeptides, arose from the duplication and arrangement of some prehistoric gene sequence—probably one of the cell-adhesion molecules. You may recall that cell-adhesion molecules are themselves part of a closely related family believed to be derived by duplication from a still more ancient gene. The duplication of exons (and the associated introns) or of entire genes can occur in several well-documented ways. Most involve chromosomal abnormalities. Two other mechanisms were described in Chapter 10: one is accidental reverse transcription of mRNA followed by incorporation of the resulting cDNA into a chromosome; the other is transposon-mediated duplication.

It is fairly easy to see that duplication of genes or exons followed by small-scale evolutionary change in the superfluous copies is more likely to lead to functional genes with novel properties than random changes alone would be. Imagine how rarely we could generate a meaningful sentence by randomly arranging letters and spaces, whereas if we began with a meaningful sentence (a gene for a functional protein) consisting of words and spaces (exons and introns) and changed a few existing letters, the odds of ending up with an intelligible sentence with a new meaning would be fairly high.

Duplication, then, is one possible source of the vast number of multiple copies of similar exons we see in antibody genes. Introns provide the most suitable points for insertion of new copies of exons. If exons were inserted into other exons, for instance, chances are that the insertion would occur within the codons for particular amino acids, so all subsequent codons would be out of phase and misread during translation. This problem does not arise if an exon is inserted into an intron. As you may recall, during mRNA processing introns are removed with the aid of snRNP, which binds to start and end signals that are part of the introns themselves. New exons successfully inserted into introns would probably bring with them enough of their flanking introns to provide the signals for transcript processing.

The consequence of this organization in the antibody gene is that the various exons, though they may have arisen from the same original sequence, can be modified independently through mutation and selection. That is, of course, particularly important in generating the many slightly different alternative exons of the variable region.

Evidence for the duplication of genes that code for non-antibody proteins The vertebrate immune system's ability to generate antibody diversity seems dependent upon a previous evolutionary bout of exon duplication. But has duplication been even partially responsible elsewhere for the evolution of genes for enzymes and structural proteins? The evidence for the role of duplication is particularly clear in the genes for myoglobin and hemoglobin. As you may remember, myoglobin, the oxygen-storage protein in muscles, consists of a single polypeptide chain, while hemoglobin, which carries oxygen in the blood, has two pairs of chains, for a total of four. The three-dimensional conformations of the α and β chains in hemo-

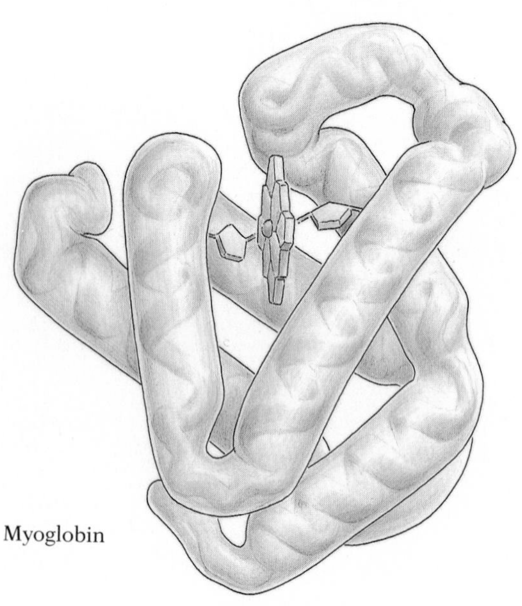

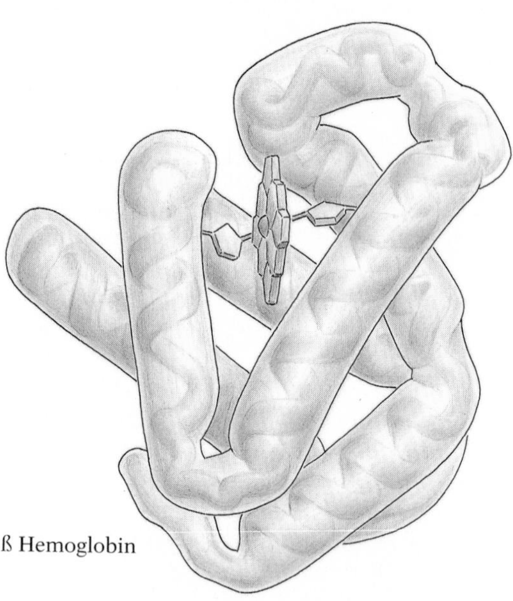

17.17 Myoglobin and the β chain of hemoglobin compared

The similarity in conformation between these peptide chains is evident from this representation. The genes for these two chains are thought to have arisen from a duplication event.

globin are nearly identical, and they also closely resemble the conformation of the single chain in myoglobin; the genes for all are thought to have evolved by duplication from a single ancestral gene (Fig. 17.17). This conclusion is reinforced by the discovery that introns are located in the same places in all these genes.

Hemoglobin itself is synthesized in slightly different forms at different times in an organism's life—during embryonic development and during adult life, for instance—and is thus specialized for differing conditions of pH and oxygen concentration. Again, the genes for these alternative forms of hemoglobin, which in human beings lie near one another on chromosome 7, are thought to have originated by duplication, and then to have followed independent evolutionary pathways. This seems all the more likely because the same region contains many pseudogenes with sequences very similar to those of the hemoglobin genes. Perhaps the pseudogenes are duplications whose subsequent evolution never led (or has yet to lead) to improved functional gene products.

The same pattern is seen in many groups of enzymes. For example, the digestive enzymes trypsin, chymotrypsin, and elastase, and the blood-clotting enzyme thrombin all have different functions, but the genes for them have base sequences and intron locations that are nearly identical. It is highly unlikely that each evolved independently into a near duplicate of the others. Most researchers believe that the genes for these enzymes began separately as duplications of the gene for some primordial enzyme, and then went their separate evolutionary ways.

There is now considerable evidence, then, that eucaryotic gene evolution depends in part on gene duplication, followed by changes in base sequence that give rise to functionally different products. But keep in mind that such changes occur slowly, and most random mutations result in genes with products of reduced function—or no function at all—rather than of altered specificity. For every new gene that produces a functional protein, there may be tens or hundreds of incomplete or failed "experiments" involving duplications. In short, unless some process is at work to edit out useless duplications, the eucaryotic chromosome should be full of nonfunctional base sequences with clear similarities to those of functional genes. And indeed, as we have seen, well over 90 percent of the mammalian genome does not code for functional products. It may be that the enormous number of nonfunctional pseudogenes in eucaryotic chromosomes is evidence of past duplications that never evolved into functional genes. Many of the examples of repetitive DNA discussed in Chapter 11 may also fit into this category.

Exon recombination We have seen that a gene can be duplicated and perhaps then evolve so that it codes for a functionally different product. We have also seen that the antibody genes, which appear to have arisen through repeated duplication of gene sequences, contain exons that are selected randomly and combined to generate diverse antibodies. These observations suggested to Walter Gilbert at Harvard and Colin Blake of Oxford another way in which new genes could evolve: perhaps individual exons from different genes could be brought together to produce new combinations.

This widely entertained proposal would make sense only if exons code for parts of the resulting protein called "domains"—distinct subunits that, like building blocks of various shapes, can form new structures when put together in new ways. Careful examination of the genes for dozens of proteins confirms that this is frequently the case. For instance, both introns in the gene for myoglobin occur between regions that code for sections joined at a major turn of this highly folded globular protein. Thus each intron defines a boundary of sorts between compact domains, and each myoglobin exon can be thought of as coding for one of these domains, or subunits. In genes for some other proteins, introns fall at the boundaries between regions coding for sections of α helix and regions coding for sections of the β pleated sheet. Alternatively, the introns sometimes flank regions coding for sections of the protein containing the active site. Because the regions encoded by exons form distinct subunits, a recombination of exons would have a real chance of generating a working enzyme with novel properties. A new gene could evolve by combining, say, exon 2 of gene *A*, exon 4 of gene *B*, exons 1 and 2 of gene *C*, and exon 2 of gene *D* at a new site (Fig. 17.18). Such a recombination could be effected by a simple movement of the exons from their original genes or from duplicates. It might involve movement of the exons themselves, or of copies of them. In either case, insertion of exons (with their flanking introns) within an intron region of the chromosome would improve the chances of generating a functional new gene.

The comparative study of DNA sequences in recent years has led to much more evidence of probable exon recombination in the past. Gilbert has estimated that all the genes that exist today (50,000 in humans alone, incorporating hundreds of thousands of exons) evolved from perhaps as few as 1000 unique exons. More striking in some ways is the use of this evolutionary scenario by molecular biologists to create novel enzymes to order by recombining existing exons to create hybrid, "designer" genes. It seems almost certain that the processes of exon duplication and recombination have been efficient mechanisms of major evolutionary innovation.

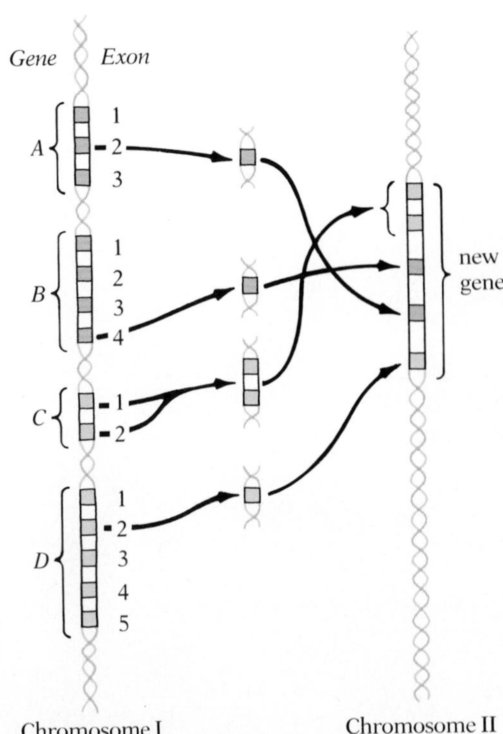

17.18 Hypothetical recombination of duplicate exons to create a new gene
Duplicates of five exons, from four different genes on a single chromosome (designated chromosome I), are imagined to be inserted near one another in a different chromosome (chromosome II). The intron regions flanking each exon being inserted provide the signals necessary for the new gene's transcript to be processed correctly. Since many exons code for protein subunits, the newly combined exons are more likely to encode a functional product than they would have been if they began and ended at random points. If their product is at least partially functional, the new gene will survive in the germ line (the gametes transmitted to offspring) and be subject to improvement through natural selection.

ADAPTATION

Every organism is, in a sense, a complex bundle of immense numbers of adaptations. We will examine a host of adaptations in later chapters—adaptations concerned with nutrient procurement, gas exchange, internal transport, regulation of body fluids, hormonal and nervous control, muscle activity, reproduction, and behavior. We should pause here to say more explicitly what is meant by adaptation.

In biology, an adaptation is any genetically controlled characteristic that increases an organism's fitness. ***Fitness***, as the term is used in evolutionary biology, is an individual's (or allele's or genotype's) probable genetic contribution to succeeding generations. An adaptation, then, is a characteristic that enhances an organism's chances of perpetuating its genes, usually by leaving descendants. Notice that we did not say adaptations increase the organism's chances of surviving, as is sometimes erroneously stated. While an adaptation, if it is to enhance the production of descendants, will ordinarily also enhance prereproductive survival, it will not necessarily enhance postreproductive survival. In many species it is, in fact, adaptive for the adults to die soon after they have reproduced.

Adaptations may be structural, physiological, or behavioral. They may be genetically simple or complex. They may involve individual cells or subcellular components, or whole organs or organ systems. They may be highly specific, of benefit only under very limited circumstances, or they may be general, of value under many and varied circumstances.

A population may become adapted to changed environmental conditions with extreme rapidity. A good example is provided by a study published in 1937 by W. B. Kemp of the Maryland Agricultural Experiment Station. The owner of a pasture in southern Maryland had seeded the pasture with a mixture of grasses and legumes. Then he divided the pasture into two parts, allowing one to be heavily grazed by cattle, while protecting the other from the livestock and letting it produce hay. Three years after this division, Kemp obtained specimens of blue grass, orchard grass, and white clover from each part of the pasture and planted them in an experimental garden where all the plants were exposed to the same environmental conditions. He found that the specimens of all three species from the heavily grazed half of the pasture exhibited dwarf, rambling growth, while specimens of the same three species from the ungrazed half exhibited vigorous, upright growth. In only three years the two populations of each species, known to have been identical initially because one batch of seed was used for the entire pasture, had become markedly different in their genetically determined growth pattern. Apparently the grazing cattle in the one half of the pasture had devoured most of the upright plants, and only plants low enough to be missed had survived and set seed. There had been, in short, intense selection against upright growth in this half of the pasture and correspondingly intense selection for the adaptively superior dwarf, rambling growth pattern. In the half of the pasture where there was no grazing, by contrast, upright growth was adaptively superior, and dwarf plants were unable to compete effectively.

Experimental tests of adaptiveness are usually not as easy to design as straightforward laboratory evaluations of cause and effect. There may be several alternative explanations for why a trait may be adaptive. It is important to remember, furthermore, that not all characters present in an organism need be adaptive in the first place. Some will be incidental pleiotropic effects of genes whose main effects—those that make them adaptive—are quite different. Other traits are essentially "historical": we have five fingers because the vertebrate developmental program produces five (or, rarely, six) digits; five may or may not be the optimal number for every kind of amphibian, reptile, bird, and mammal, but selection can only operate on existing variation. In the case of digits, we saw in Chapter 1 that selection has acted to "customize" digit size rather than number, enlarging or reducing the size of individual digits; the basic five-digit plan remains remarkably consistent. Apparently, some changes are simply not genetically feasible. This sort of evolutionary constraint is often called ***phylogenetic inertia***.

Devising tests to discover the selective pressures that have led to particular traits requires ingenuity and persistence. The Dutch biologist Niko Tinbergen was one of the first scientists to insist on putting his evolutionary theories to the test. When Tinbergen wondered, for instance, why ground-nesting gulls meticulously remove broken eggs from their nests (Fig. 17.19A), he formulated a variety of possible explanations: the damaged eggs might be a source of disease, which would infect the newly hatched young;

A

B

17.19 Eggshell removal by gulls
Ground-nesting gulls remove broken eggshells and other debris from their nests and carry them at least a meter away (A), while cliff-nesting species like the kittiwake do not remove eggshells (B). The nesting habit of the kittiwakes protects them from predation, while ground-nesting gulls must instead keep their nests inconspicuous.

the jagged edges of the broken shells might endanger the chicks; the unrelieved white of the exposed interiors of the broken shells might nullify the camouflage provided by the olive-drab exteriors, and so attract predators.

Tinbergen solved this problem (and many similar ones) in two steps: first, interspecific comparisons, then experimental tests. The species-comparison step allowed him to isolate the most likely hypothesis for testing. In the case of the damaged eggs, he observed that kittiwake gulls, which live on cliffs and are therefore exposed to virtually no predation, do not remove broken eggshells (Fig. 17.19B). Since disease and cuts, if they posed significant threats, ought to be as dangerous for the kittiwakes as for ground-nesting gulls, Tinbergen decided to test the predation hypothesis first. He did this by setting out an array of nests containing normal eggs, with broken eggs placed at varying distances from them. The results were rewardingly clear-cut: broken eggshells nearby called the attention of predators to an otherwise inconspicuous nest (Table 17.1). By such experiments Tinbergen provided new insight into the evolution of adaptive behavior. In the absence of direct experimental tests or compelling species comparisons, however, "explanations" of adaptiveness remain highly speculative, and should simply serve as working hypotheses to stimulate research.

Let us now look at some particularly clear examples of adaptations, which help clarify the processes of evolution.

TABLE 17.1 *Survival value of eggshell removal*[a]

Distance from egg to eggshell (cm)	*Percent eggs taken by predators*
5	65
15	42
100	32
200	21
No eggshell	22

[a] N. Tinbergen et al., Egg Shell Removal by the Black-headed Gull, *Behaviour*, vol. 19, 1963.

ADAPTATIONS FOR POLLINATION IN FLOWERING PLANTS

Flowering plants depend on external agents to carry pollen from the male parts in the flowers of one plant to the female parts in the flowers of another

A

B

C

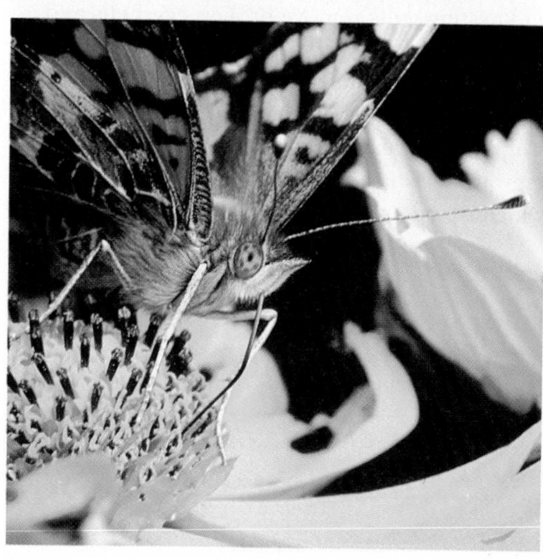
D

17.20 A variety of plant pollinators
(A) A bat at a flower, its face liberally dusted with pollen. (B) A honey possum feeding on nectar from flowers. (C) A hummingbird hovering over nectar-bearing flowers. (D) A butterfly inserting its long proboscis into a flower to obtain nectar. The proboscis is likely to be dusted with pollen, which the butterfly may carry to the next flower it visits.

plant (Fig. 17.20). The flowers of each species are adapted in shape, structure, color, and odor to the particular pollinating agents on which they depend, and they provide an especially clear illustration of the evolution of adaptedness. Evolving together, the plants and their pollinators became more finely tuned to each other's peculiarities—a process often termed ***coevolution***.

There are indeed striking correspondences between the pollinators and the species they pollinate. Bees are attracted innately to bright colors, ultraviolet bull's-eye patterns, and sweet, aromatic, or minty odors; they are active only during the day, and they usually alight on a petal before moving into the part of the flower containing the nectar and pollen. Bee flowers have showy, brightly colored petals that are usually blue or yellow but seldom red (bees can see blue or yellow light well, but they cannot see red at all); indeed, most bee-pollinated flowers have a UV bull's-eye, a sweet, aromatic, or minty fragrance, daytime opening or nectar production, and a special protruding lip or other suitable landing platform. But these observations tell us only that correspondences exist between certain flowers and the preference of bees. They do not indicate how these correspondences may have come about: whether the preferences of the pollinators provided the exclusive selective force on flower morphology and odor, or whether, instead, early flowers provided the selection pressures leading to the innate preferences of modern bees, or whether both factors have been at work. A look at other species of pollinators provides some clues.

Hummingbirds, for example, can see red well but blue only poorly; they have a weak sense of smell; and they ordinarily do not land on flowers, but hover in front of them while sucking the nectar. Flowers pollinated primarily by hummingbirds are usually red or yellow, are nearly odorless, and lack any protruding landing platform. Since flowers of the same genus can have very different morphologies to suit different pollinators (Fig. 17.21), it is probably the flowers that have done most of the adapting. However, pollinators have probably been adapting to flowers too, though to a lesser extent; different species of bees, for example, can have very different tongue lengths, suitable for different flower morphologies.

This pattern of coadaptation between pollinators and flowers extends to other nectar-feeding species as well. For example, in contrast to both bees and hummingbirds, moths and bats are generally most active at dusk and during the night, and the flowers they pollinate are mostly white and are open only during the late afternoon and night. These flowers often have a heavy fragrance that helps guide the moths and bats to them.

Moths play a role in a particularly interesting adaptation of plants to their pollinators. The flowers produced by scarlet gilia plants near Flagstaff, Arizona, range from red through pink to white. The dark-red flowers are most effective in attracting hummingbirds, but these pollinators emigrate a month after the season begins; the white flowers are most effective in attracting hawk moths, the pollinators available throughout the blooming season. The plants compensate for this shift in relative pollinator abundance by doubling the production of white flowers late in the season, while at the same time ceasing to produce any red blossoms (Fig. 17.22).

Unlike bees and moths, the short-tongued flies (which feed primarily on carrion, dung, humus, sap, and blood) are attracted by rank rather than sweet odors, and they rely very little on vision in locating food. The flowers of plants that depend on these flies for pollination are usually dull-colored and ill-smelling.

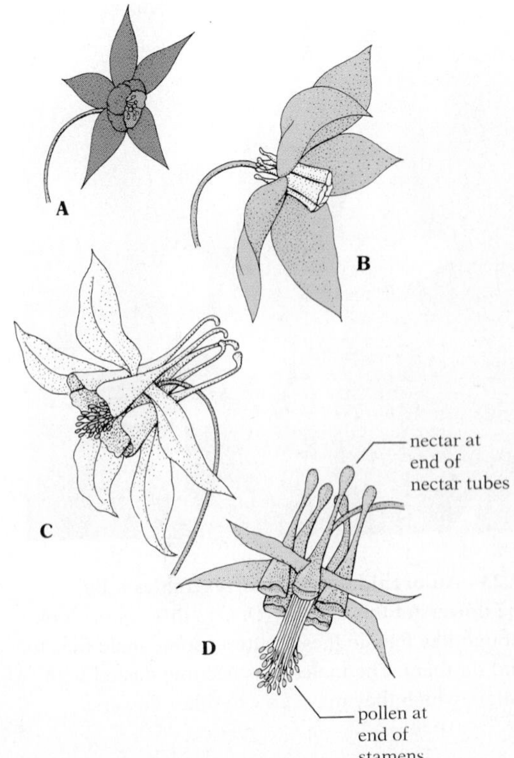

17.21 Characters of columbine flowers correlated with their pollinators

(A) *Aquilegia ecalcarata*, pollinated by bees.
(B) *A. nivalis*, pollinated by long-tongued bees.
(C) *A. vulgaris*, pollinated by long-tongued bumble bees.
(D) *A. formosa*, pollinated by hummingbirds.
The length and curvature of the nectar tubes of the flowers are correlated with the length and curvature of the bees' tongues and the hummingbirds' bills. The length of the pollen-bearing stamens is suited to the size of the pollinator, and the reduced petal width of *A. formosa* reflects the hummingbird's lack of any need for a perch.

17.22 Pollinator tracking in scarlet gilia (*Ipomopsis aggregata*)

During the early part of the summer, when these plants are pollinated by both hummingbirds and hawk moths, they produce about twice as many red flowers (preferred by hummingbirds) as white flowers (A). Later in the summer, as the hummingbird populations leave the area, scarlet gilia plants shift over to producing pink and—especially—white flowers, which are more attractive to the sole remaining pollinator, a local species of hawk moth (B).

17.23 An orchid flower that resembles a fly
The flowers of this species (*Ophrys insectifera*) look enough like female flies to attract some male flies to land on them. The males thus become dusted with pollen, which they may carry to other flowers.

A particularly dramatic example of adaptation for pollination is seen in some species of orchids, whose flowers resemble in shape, odor, and color the females of certain species of wasps, bees, or flies (Fig. 17.23). The male insect is stimulated to attempt to copulate with the flower and becomes covered with pollen in the process. When he later attempts to copulate with another flower, some of the pollen from the first flower is deposited on the second. So complete is the deception that sperm have actually been found inside the orchid flowers after a visit by the male insect.

Flowers pollinated by wind or water rather than animals characteristically lack bright colors, special odors, and nectar. In fact, most of them have no petals, and their sexual parts are freely exposed to the air currents. The pollen grains produced by these flowers are particularly small and light, and it is not unusual for them to be blown hundreds of miles.

We see from this species comparison, then, that the characteristics of flowers are not simply pleasing curiosities of nature that serve no practical function. They are important adaptations that have evolved in response to fundamental selection pressures.

17.24 Cryptic coloration of a crab
The sargassum crab, which lives in dense growths of the brown alga *Sargassum* off Bermuda, is the same color as the alga; its rounded body resembles the floats of the alga.

17.25 Flounders on two different backgrounds
The fish can change color to match the background, whether it is light-colored (top) or dark (bottom).

DEFENSIVE ADAPTATIONS OF ANIMALS

Cryptic appearance Many animals blend into their surroundings so well as to be nearly undetectable. Frequently their color matches the background almost perfectly (Fig. 17.24). In some cases animals even have the ability to alter the condition of their own pigment cells and change their appearance to harmonize with their background (Figs. 17.25 and 17.26). Often, rather than match the color of the general background, the animals may resemble inanimate objects commonly found in their habitat, such as leaves (Fig. 17.27) or twigs (Fig. 17.28). When the shape or color of an animal offers concealment against its background, it is said to have a ***cryptic*** appearance.

Careful studies have confirmed that cryptic appearance is an adaptive characteristic that helps animals escape predation. One such study was conducted by F. B. Sumner of the Scripps Institution of Oceanography in California. Sumner investigated predation by Galápagos penguins on mosquito fish (*Gambusia partuelis*), which can contract or expand their pigment cells to become lighter or darker, depending on their background. He established that the penguins caught 70 percent of the fish that contrasted with their background but only 34 percent of the fish that resembled their background. In a similar experiment F. B. Isely of Trinity University in San Antonio, Texas, studied predation by chickens, turkeys, and native birds on grasshoppers of various colors on differently colored backgrounds. He found that 88 percent of the nonprotected grasshoppers were eaten whereas only 40 percent of the cryptically colored ones were eaten.

17.26 Color change by the frog *Hyla versicolor*
Individuals of this species are able to change color to match either a tree trunk or vegetation.

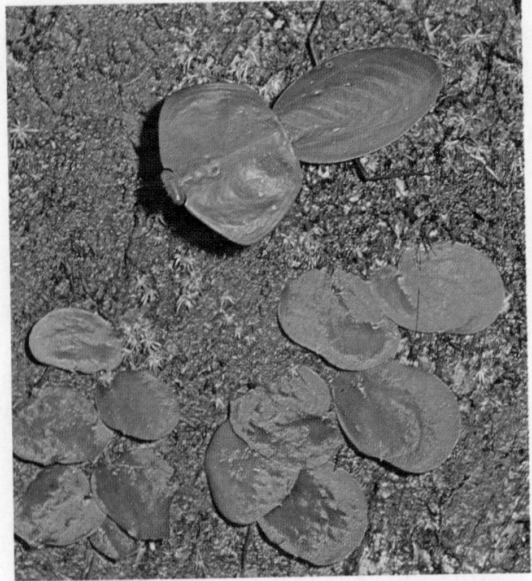

17.27 Leaflike mantis
The mantis (at top in picture) looks strikingly like the green leaves below it. Even the relationship of its thorax and abdomen reflects the way the leaves often occur in pairs.

17.28 A moth that looks like a broken twig
The perfection of the cryptic appearance of this moth (*Phalera bucephala*) is a triumph of evolutionary adaptation.

17.29 Cryptic coloration in a crab spider Matching the color of the flower on which it waits, a crab spider remains unseen long enough to trap an unwary honey bee.

Predators as well as prey may exhibit cryptic coloration. We are all familiar with the camouflaging stripes or spots of carnivores like tigers and leopards. Patterns and colors that match a particular background are found also in other predatory species, large and small (Fig. 17.29).

One of the most extensively studied cases of cryptic coloration is the so-called industrial melanism of moths. Since the mid-1880s many species of moths have become decidedly darker in industrial regions. This is actually a case of polymorphism in which the less frequent of two forms has become the more frequent; the originally predominant light form in these species of moths has given way in industrial areas to the dark (melanic) form. In the Manchester area of England, the first black specimens of the species *Biston betularia* were caught in 1848; by 1895 melanics constituted about 98 percent of the total population in the area. For such a remarkable shift in frequency to have occurred in so short a time, the melanic form must have had at least a 30 percent advantage over the light form.

The various species of moths exhibiting the rapid shift to melanism, though unrelated to one another, all habitually rest during the day in an exposed position on tree trunks or rocks, being protected from predation only by their close resemblance to their background. In former years the tree trunks and rocks were light-colored and often covered with light-colored lichens. Against this background the light forms of the moths were difficult to see, whereas the melanic forms were quite conspicuous (Fig. 17.30). Under these conditions it seems likely that predators captured melanics far more easily than the cryptically colored light moths. The light forms would thus have been strongly favored, and they would have occurred in much higher frequency than melanics. But with the advent of extensive industrialization tree trunks and rocks were blackened by soot, and the lichens, which are particularly sensitive to such pollution, disappeared. In this altered environment the melanic moths resembled the background more closely than the light moths. Thus selection should now have favored the melanics, which would explain why they have increased in frequency.

This scenario was put to a test in the mid-1950s, when H. B. D. Kettlewell of Oxford released approximately equal numbers of the light and melanic forms of *Biston betularia* onto trees in a rural area in the country of Dorset, England, where the tree trunks were light-colored, lichens were abundant, and the wild population of the moth was about 94.6 percent light-colored. A direct watch on the resting moths was maintained from blinds. Of 190 moths observed to be captured by birds, 164 were melanics and only 26 were light forms. Furthermore, of approximately 500 marked individuals of each color form released in another experiment, roughly twice as many light moths as melanic moths were recaptured in traps set up in the Dorset woods, an indication that more of the light moths had survived.

Taken alone, however, these experiments did not prove conclusively that the factor favoring the light moths over the melanics was their resemblance to their background. The results could be explained by assuming, for example, that the birds preferred the melanics because of some difference in flavor. Therefore Kettlewell duplicated the experiments under the reverse environmental conditions—in woods near Birmingham, England, where the tree trunks were blackened with soot, lichens were absent, and the wild population of the moth was about 85 percent melanic. The results of these experiments were the reverse of those in the Dorset experiments. Now birds were observed to capture nearly three times as many light moths as melanics, and roughly twice as many melanics were recaptured in traps. These experiments by Kettlewell prove that those moths that most closely resemble the background on which they rest have much the best chance of escaping predation. And, as we would predict, the imposition of pollution controls has led to a shift back toward the light morph.

17.30 Cryptic coloration of peppered moths
Top: Light and dark morphs of *Biston betularia* at rest on a tree trunk in unpolluted countryside. Bottom: Light and dark morphs on a soot-covered tree trunk. Here the light form is easier to see.

Warning coloration Whereas some animals have evolved cryptic coloration, others have evolved colors and patterns that contrast boldly with their background and thus render them clearly visible to potential predators (Fig. 17.31). Nearly all of these animals are in some way disagreeable to predators; they may taste bad, or smell bad, or sting, or secrete poisonous substances. In other words, they are animals that a predator will usually reject after one or two unpleasant encounters. Such animals benefit by being gaudily colored and conspicuous because predators that have experienced their unpleasant features learn to recognize and avoid them more easily in the future. Their flashy appearance is protective because it warns potential predators that they should stay away; such a warning appearance is said to be ***aposematic***.

The warning is sometimes so effective that, after unpleasant experiences with one or two warningly colored insects, some vertebrate predators simply avoid all flashily colored insects, whether or not they resemble the ones they encountered earlier. G. D. H. Carpenter demonstrated this by offering over 200 different species of insects to an insectivorous monkey. The monkey accepted 83 percent of the cryptically colored insects but only 16 percent of those with the warning coloration, even though many of the insects belonged to species the monkey had probably not previously encountered. It is possible, therefore, that avoidance of aposematic insects does not depend entirely on learning. Predators that have a genetic predisposition to avoid brightly colored prey would probably have an adaptive advantage over predators that waste time and energy pursuing inedible prey; hence there may be a tendency for predators to evolve an avoidance response to prey displaying warning coloration.

Mimicry Species not naturally protected by some unpleasant character of their own may closely resemble (mimic) in appearance and behavior some dangerous or unpalatable aposematic species. Such a resemblance can be

17.31 Aposematic coloration
The bright color of the poison-arrow frog (from which South American Indians obtain poison for their arrow tips) makes it easily recognizable by predators, which carefully avoid it.

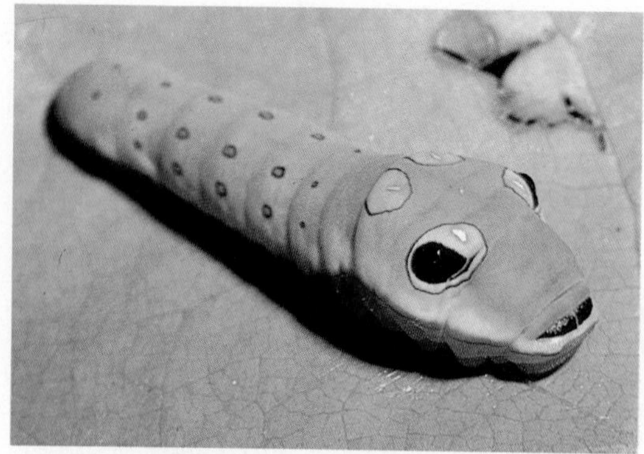

A

B

17.32 Examples of Batesian and aggressive mimicry

(A) The prominent imitation eyes and mouth of the spicebush swallowtail larva give it the appearance of a predator, to be avoided. (B) The markings of the harmless syrphid fly resemble those of a stinging bumble bee. (Upon close inspection, flies are readily distinguished from bees because their antennae are very short and they have only a single pair of wings.) (C) The warty frogfish is an aggressive mimic: its body, which resembles an algae-covered rock, is merely cryptic, but it displays a lure that resembles a small fish.

adaptive: the mimics may suffer little predation because predators cannot distinguish them from their models, which the predators have learned are unpleasant. This phenomenon is called ***Batesian mimicry*** (Fig. 17.32).

Convincing evidence for the potential effectiveness of this type of mimicry in protecting the mimic species comes from the elegant experiments of Jane van Z. Brower, then at Oxford. Brower produced an artificial model-mimic system with starlings as the predators and mealworms, which starlings ordinarily eat voraciously, as prey. She painted a tasteless color band on the mealworms, and dipped some in a distasteful solution. She then presented various ratios of noxious (dipped) models and mimics to different groups of birds. After a few unpleasant encounters with the models, the birds learned to recognize and avoid the painted worms, with the result that the mimics among them also escaped predation, particularly when the percentage of mimics presented to the starlings was 60 percent or less.

A few species pursue the opposite approach: aggressive mimics provide lures of apparently palatable prey to attract potential victims (Fig. 17.32C). In addition to Batesian mimicry, which is based on deception—mimicry of a distasteful or dangerous species by individuals of a species that is neither—there is a second kind of mimicry, called ***Müllerian mimicry***, which involves the evolution of a similar appearance by two or more distasteful or dangerous species. In this type of mimicry, individuals of each species act as both model and mimic. The members of each species have some defensive mechanism, but if each species had its own characteristic appearance, the predators would have to learn to avoid each of them separately; the learning process would thus be more demanding, and would involve the death of some individuals of each prey species. Selection favors evolution toward one appearance; the various protected species thus come to constitute a single prey group from the standpoint of the predators, which accordingly learn avoidance more easily.

One striking case of Müllerian mimicry involves the monarch butterfly *(Danaus plexippus)* and the unrelated viceroy *(Limenitis archippus)*. These two species look very much alike (Fig. 17.33) and are each distasteful to birds, though in different ways: monarchs sequester poisons from the milk-

C

weed plant upon which their caterpillars feed, while viceroys synthesize their own bad-tasting chemical. Birds experiencing one member of either species learn to avoid them both. (Until recently, the viceroy was thought to be a Batesian mimic of the monarch because blue jays, the predators used in early tests, are one of the few species of birds that find viceroys only slightly distasteful.)

The selective advantage of Müllerian mimicry may explain the similar markings of many unrelated species of wasps and bees, or of the group of poisonous reptiles known as coral snakes. If avoidance involves any genetic predisposition, then resemblances among the prey animals would facilitate more rapid selection for improved prey-recognition mechanisms in the predators. In fact, there is evidence that some predators have evolved the ability to recognize coral snakes innately; perhaps only Müllerian mimicry can provide a strong enough selection pressure to cause the evolution of such specialized recognition—a recognition that benefits individuals of both predator and prey species.

17.33 Müllerian mimicry in butterflies
The monarch butterfly (top) and the viceroy (bottom) have evolved strikingly similar color patterns.

SYMBIOTIC ADAPTATIONS

The term "symbiosis" is used in a variety of ways in the biological literature. Some authors apply it only to cases in which two species live together to their mutual benefit. We use the word in a broader sense.

Etymologically, symbiosis simply means "living together," without any implied value judgments. This is the meaning it was given when it was first introduced into biology, and this is the meaning it will have in this book. We will, however, recognize three categories of symbiosis. The first is ***commensalism***, a relationship in which one species benefits while the other receives little or no benefit or harm. The second is ***mutualism***, in which both species benefit. The third is ***parasitism***, in which one species benefits and individuals of the other species are harmed (Table 17.2).

TABLE 17.2 *Varieties of symbiotic relationships*

Relationship	*Species A*	*Species B*
Commensalism	+	**0**
Mutualism	+	+
Parasitism	+	−

+ = benefit   − = harm   **0** = no effect

Commensalism The advantage derived by the commensal species from its association with the host often involves shelter, support, transport, food, or several of these. For example, in tropical forests numerous small plants, called epiphytes, usually grow on the branches of the larger trees or in forks of their trunks (see Fig. 40.27, right p. 1176). These commensals, among which species of orchids and bromeliads are prominent, are not parasites. They use the host trees only as a base of attachment and do not obtain nourishment from them. They apparently do no harm to the host except very rarely when so many of them are on one tree that they stunt its growth or cause limbs to break. A similar type of commensalism is the use of trees as nesting places by birds.

Sometimes it is difficult to tell what benefit is involved in a commensal relationship. For example, certain species of barnacles occur nowhere except attached to the backs of whales, and other species of barnacles occur nowhere except attached to the barnacles that are attached to whales. Just what advantages either of these groups of barnacles enjoys is not clear. They do, of course, get a relatively unoccupied base for attachment, enhanced freedom from predation, and a form of transport that increases the dispersal of their offspring. But it is not certain that these benefits alone would have sufficed for the evolution of such specificity.

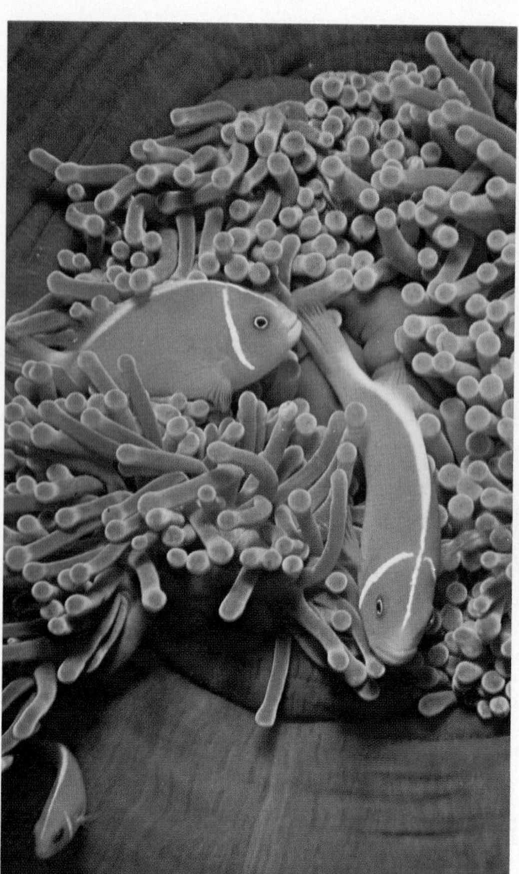

17.34 Three anemone fish living among the tentacles of a sea anemone from the Palau Islands

In some cases of commensalism, however, the benefit is dramatically obvious. For example, certain species of fish live in association with sea anemones, deriving protection and shelter from them and sometimes stealing some of their food (Fig. 17.34). These fish swim freely among the tentacles of the anemones, even though the tentacles quickly paralyze other fish that touch them. The anemones regularly feed on fish; yet the particular species that live as commensals with them sometimes actually enter the gastrovascular cavity of their host, emerging later with no apparent ill effects. The physiological and behavioral adaptations that make such a commensal relationship possible must be quite extensive. Another striking example is a small tropical fish that lives in the respiratory tree of a particular species of sea cucumber. The fish emerges at night to feed and then returns to its curious abode by first poking its host's rectal opening with its snout and then quickly turning so that it is drawn tail first through the rectal chamber into the respiratory tree. Still another example is a tiny crab that lives in the mantle cavity of oysters. The crab enters the cavity as a larva and eventually grows too big to escape through the narrow opening between the two valves of the oyster's shell. It is thus a prisoner of its host, but a well-sheltered prisoner. It steals a few particles of food from the oyster, but apparently does it no significant harm.

Mutualism Symbiotic relationships beneficial to both species are common. Figure 17.35 illustrates two instances of the widespread phenomenon called cleaning symbiosis, which is patently mutualistic. Other examples of mutualism include the relationship between a termite or a cow and the cellulose-digesting microorganisms in its digestive tract, or between a human being and the bacteria in the intestine that synthesize vitamin K. The plants we call lichens are actually formed of an alga or a cyanobacterium and a fungus united in such close mutualistic symbiosis that they give the appearance of being one plant (see Fig. 23.9, p. 652). Apparently the fungus benefits from the photosynthetic activity of its "guest," and the alga or bacterium benefits from the water-retaining properties of the fungal walls.

As is apparent from this discussion of commensalism and mutualism (particularly the former), the division of symbiosis into three subcategories is in many ways arbitrary. Commensalism, mutualism, and parasitism are all parts of a continuous spectrum of possible interactions. Fortunately, it really isn't very important which category we apply to most cases. The categories are only devices to help us organize what we know about nature and to form testable hypotheses. What is important is to keep in mind how commensalism, mutualism, and parasitism grade into each other, and to recognize that each case of symbiosis is different from all others and must be studied and analyzed on its own.

Parasitism Just as there are no sharp boundaries between parasitism and commensalism, or even between parasitism and mutualism, there is no strict line between parasitism and predation. Mosquitoes and lice both suck the blood of mammals, yet we usually call only the latter parasites. Foxes and tapeworms may both attack rabbits, but foxes are called predators and tapeworms are called parasites. The usual distinction is that a predator eats its prey quickly and then goes on its way, while a parasite passes much of its life on or in the body of a living host, from which it derives food in a manner harmful to the host. Parasites generally do not kill their hosts; those that eventually do so are called ***parasitoids*** (Fig. 17.36). Obviously the distinction between predator and parasite is not always clear. How long must one organism live on the body of another to be classed as a parasite? But though there will always be intermediate cases, it is profitable to distinguish between predation and parasitism, because each of these is a mode of existence followed by many kinds of organisms and each involves its own characteristic sorts of adaptations.

Parasites are customarily divided into two types: external parasites and internal parasites. The former live on the outer surface of their host, usually either feeding on the hair, feathers, scales, or skin of the host or sucking its blood. Internal parasites may live in the various tubes and ducts of the host's body, particularly the digestive tract, respiratory passages, or urinary ducts; or they may bore into and live embedded in tissues such as muscle or liver; or, in the case of viruses and some bacteria and protozoans, they may actually live inside the individual cells of their host.

Internal parasitism is usually marked by much more extreme specializations than external parasitism. The habitats available inside the body of another living organism are completely unlike those outside, and the unusual problems they pose have resulted in evolutionary adaptations entirely different from those seen in free-living forms. For example, internal parasites have often lost organs or whole organ systems that would be essential in a free-living species. Tapeworms, for instance, have no digestive system. They live in their host's intestine, where they are bathed by the products of the host's digestion, which they can absorb directly across their body wall without having to carry out any digestion themselves.

17.35 Cleaning symbiosis

Top: A giant seabass being cleaned by a cleanerfish. Bottom: Yellow-billed oxpeckers search for parasitic insects on a Black Rhinoceros. In both cases the symbiosis is mutualistic: the cleaner obtains food, and the host gets rid of parasites that could endanger its health.

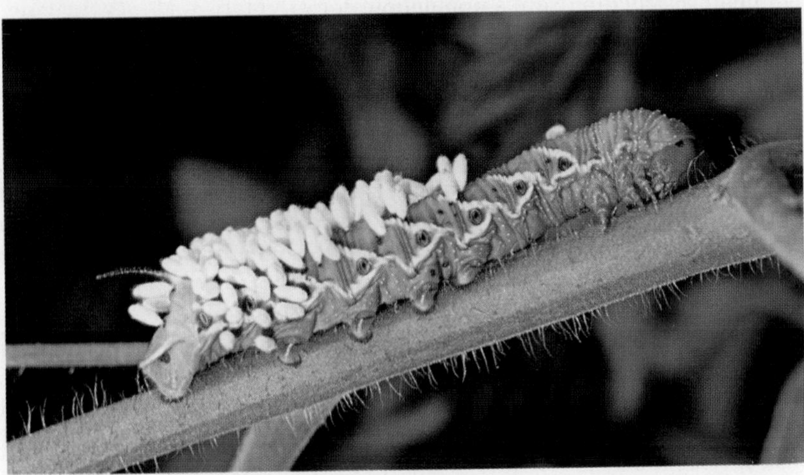

17.36 A caterpillar (tomato hornworm) with numerous pupae of a parasitoid wasp attached to its body

The wasp laid her eggs inside the body of the host caterpillar, and the larvae fed on it until they emerged and pupated.

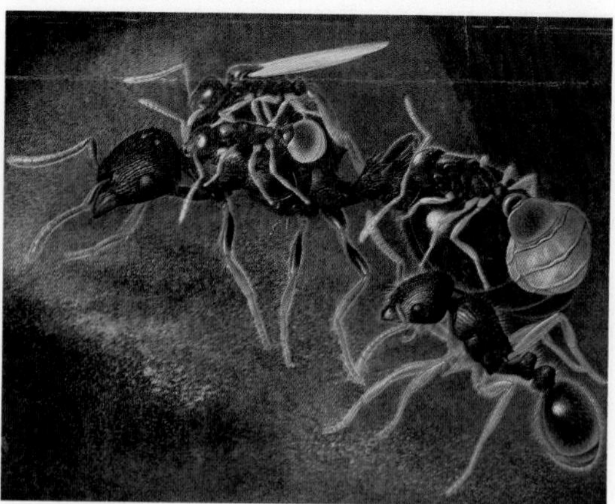

17.37 A parasitic ant
Queens of the "ultimate ant," *Teleutomyrmex,* ride on the host queen and are fed by a host worker (lower right); they dispense their eggs along with those of the host queen. Their offspring, all reproductives, mate in the colony.

Because of their frequent evolutionary loss of structures, certain parasites are often said to be degenerate. "Degenerate," of course, implies no value judgment, but simply refers to the lack, common in internal parasites, of many structures present in their free-living ancestors. The ant species *Teleutomyrmex* (Fig. 17.37) is a degenerate parasite living in the nests of another species of ant. Though not an internal parasite, it is degenerate to the extent that it cannot survive outside its hosts' nest: it has lost many of its glands, its sting, its pigmentation, its ability to digest anything but liquid fed to it by its hosts. Even its brain is degenerate. But from an evolutionary point of view, loss of structures useless in a new environment is an instance of specialized adaptation.

Specialization, then, does not necessarily mean increased structural complexity; it only means the evolution of characteristics particularly suited to some special situation or way of life. In internal parasites—or cave animals, which frequently lack eyes—the development and maintenance of structures that no longer serve a useful function would require energy that the organism might use to more advantage in some other way. And some useless structures, such as eyes in both internal parasites and cave animals, might well be a handicap in these special environments, because they would be a likely point of infection. It is readily understandable, therefore, that natural selection might favor those individuals in which such useless organs are either relatively small or lacking entirely. Alternatively, the unused structures could have been adaptively neutral and simply lost through genetic drift. In either case, the concept of evolutionary loss of unused structures has nothing to do with the erroneous Lamarckian notion that use and disuse can directly influence the size of a structure in an organism's progeny. Only the superior qualifications of a genotype lacking the structure, or the spontaneous loss of the genotype through genetic drift, could eliminate a structure from a population.

Structural degeneracy is far from being the only sort of special adaptation commonly seen in internal parasites. They often have body walls highly resistant to the destructive enzymes and antibodies of the host. Tapeworms, for example, are constantly bathed by the potent digestive juices of their host; yet their enzyme-resistant cuticle protects them from being digested. And tapeworms have very specialized heads, with hooks and suckers, that enable them to anchor themselves and avoid being expelled by the often vigorous peristaltic contractions that move the other contents of the host's intestine (see Fig. 24.27, p. 669). The equally elaborate specializations necessary to find a new host will be described in Chapter 24.

As the parasites evolve, the host species is also evolving, and there is strong selection pressure for the evolution of more effective defenses against the ravages of parasites. This constant interplay between host and parasite is at the heart of the red-queen hypothesis. Those individuals of the host species with superior defenses will be better able to survive and reproduce. Correspondingly, those individuals of the parasite species with the best ways of counteracting these defenses will be most likely to prosper. In turn, their counteractions will lead to pressure on the host to evolve still better defenses, against which the parasite may then evolve new means of surviving, and so forth.

Probably most long-established host-parasite relationships are balanced ones. Relationships that result in serious disease in the host are usually relatively new, or they are relationships in which a new and more virulent form of the parasite has recently arisen, or in which the host showing the serious disease symptoms is not the main host of the parasite. American Indians, for example, suffered severely when exposed to diseases first brought to North America by European colonists, even though some of those same diseases caused only mild symptoms in the Europeans, who had been exposed to the disease-causing organisms for many centuries and in whom the host-parasite relationship had nearly reached a balance. Many examples are known in which humans are only occasional hosts for a particular parasite and suffer severe disease symptoms, though the wild animal that is the major host shows few ill effects from its relationship with the same parasite.

It is worth keeping in mind, however, that relatively benign interactions between the host and a parasite need not ever evolve regardless of how much time passes. The reason is that the optimal strategy for a parasite depends critically on the life histories of itself and its host, and for some combinations there is no advantage to the parasite in achieving a balance and no way for the host to impose one. In some diseases—for example, rabies (which is always fatal to humans and many other mammals) and smallpox (which, untreated, kills about 30 percent of its victims)—the parasite apparently benefits from a massive attack on the host, which enables it to spread its offspring rapidly. A slower, longer-lasting release of progeny produces lower reproductive success. Much recent work has shown, in fact, that an evolutionary perspective on host-parasite dynamics—particularly life histories and transmission efficiencies—can lead to optimal designs for disease treatment and prevention, whereas many intuitively attractive alternatives that fail to consider these interactions turn out to be futile.

STUDY QUESTIONS

1. Imagine a relatively large, long-lived animal (say at least 5 kg, living a decade on average) which, because of its habitat and life style, does not benefit from sexual recombination. Derive a list of the rare conditions that would favor cloning in such an animal. (pp. 448–52)

2. Preadaptations can enable a species to expand into a novel niche. Many species appear to have occupied the niches created by humans—rats, mice, roaches, silverfish, sea gulls, pigeons, and so on. What sorts of adaptations, already useful or neutral in their normal habitat, enable these species to exploit new opportunities? Is there a common theme? Is it just a coincidence that many of these species are standard laboratory animals? (pp. 460–61)

3. What sorts of ecological factors might favor male contests over female choice? How might such factors tip the balance toward harems rather than monogamy or bigamy? (pp. 469–70)

4. Some researchers believe that human females are, on average, inherently better at some tasks (verbal expression, making subtle distinctions, and so on) than males, while males are, on average, better at mathematical and geometric problems. (Others, of course, believe the differences

that have been observed are the result of conditioning.) Is it possible for natural selection to work in different directions on the two sexes of one species? Cite examples to support your conclusion. If the answer is yes, what is the likely logic of differential specializations? (pp. 466–67, 469–70)

5. What, if anything, keeps parasites from driving their hosts to extinction? (There are several possibilities.) (pp. 484–87)

CONCEPTS FOR REVIEW

- Evolution versus natural selection
- Sources of variation
- Mutation
  - Point mutations versus larger events
  - Duplication
  - Exon recombination
- Sex
  - Costs and benefits of diploidy versus haploidy
  - Disadvantages of sexual recombination versus cloning
  - Evolution of sex
    - evolution of recombination
    - tangled-bank hypothesis
    - red-queen hypothesis
- Requirements for genetic equilibrium
- Genetic drift versus natural selection
- Preadaptation and evolution
- Types of selection
  - Directional
  - Disruptive
  - Stabilizing
  - Frequency-dependent
- Creation of novel phenotypes without mutation
- Balanced polymorphism
- Sexual selection versus natural selection
  - Male contests
  - Female choice
- Adaptation versus phylogenetic inertia
- Coevolution
- Mimicry
  - Cryptic versus aposematic appearance
  - Batesian versus Müllerian mimicry
- Symbiosis
  - Commensalism
  - Mutualism
  - Parasitism

SUGGESTED READING

BISHOP, J. A., and L. M. COOK, 1975. Moths, melanism and clean air, *Scientific American* 232 (1). (Offprint 1314) *On the lessening of air pollution in Britain and the diminishing frequency of melanics in some moth populations.*

CLARKE, B., 1975. The causes of biological diversity, *Scientific American* 233 (2). (Offprint 1326)

COLES, C. J., 1984. Unisexual lizards, *Scientific American* 250 (1). *On the evolution of asexual species from sexual ones.*

DAWKINS, R., 1976. *The Selfish Gene.* Oxford University Press, New York. *A well-written exposition of the controversial idea that the gene, not the individual organism, is the unit of selection, the organism being merely the robot vehicle of its selfish genes.*

FUTUYMA, D. J., 1986. *Evolutionary Biology,* 2nd ed. Sinauer Associates, Sunderland, Mass.

GOULD, J. L., and C. G. GOULD, 1989. *Sexual Selection.* Scientific American Books, New York. *Well-illustrated treatment of the evolution of sex and mate choice.*

GRANT, V., 1951. The fertilization of flowers, *Scientific American* 184 (6). (Offprint 12) *The special adaptations of flowers that help ensure their pollination.*

KETTLEWELL, H. B. D., 1959. Darwin's missing evidence, *Scientific American* 200 (3). (Offprint 842) *The story of the industrial melanism of the peppered moth in England.*

LEWONTIN, R. C., 1978. Adaptation. *Scientific American* 239 (3). (Offprint 1408) *An excellent discussion of the process of adaptation, emphasizing that most features are compromises between different selection pressures, and that chance plays a role in evolution when more than one solution to a problem is possible.*

LI, W. H., and D. GRAUR, 1991. *Fundamentals of Molecular Evolution.* Sinauer, Sunderland, Mass. *An excellent account of gene evolution through duplication and exon recombination.*

MAYR, E., 1978. Evolution, *Scientific American* 239 (3). (Offprint 1400) *A nice history of evolutionary thought.*

SPENCER, C. H. 1987. Mimicry in plants, *Scientific American* 257 (3).

STROBEL, G. A., 1991. Biological control of weeds, *Scientific American* 265 (1). *Illustrates how weeds, like diseases, can reproduce unchecked in new habitats, whereas in their normal range they are in balance with locally evolved parasites (usually insects or fungi).*

Chapter **18**

SPECIATION AND PHYLOGENY

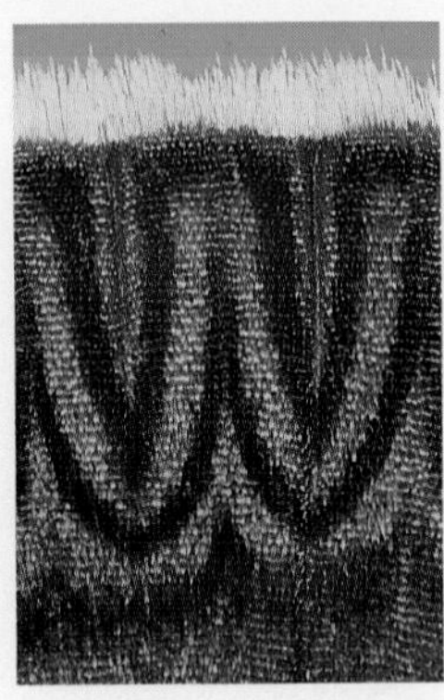

We have so far discussed only one major aspect of evolution, the change of a given population through time. Now we must turn to another of its major aspects, the processes by which a single population may split, giving rise to two or more different descendant populations. But before we can discuss this topic meaningfully, we must examine more carefully the populations that we have so far casually taken for granted. With reference to sexually reproducing organisms, we can define a ***population*** as a group of individuals that interbreed and so share a common gene pool.

UNITS OF POPULATION

Demes A ***deme*** is a small local population, such as all the deer mice or all the red oaks in a certain woodland, or all the perch or all the waterstriders in a given pond. Though no two individuals in a deme are exactly alike, the members of a deme do usually resemble one another more closely than they resemble the members of other demes. There are at least two reasons for this: (1) the individuals in a deme are more closely related genetically, because pairings occur more frequently between members of the same deme than between members of different demes; and (2) the individuals in a deme are exposed to more similar environmental influences and hence to more nearly the same selection pressures.

We can see that demes are not clear-cut permanent units of population. Though the deer mice in one farm's woodlot are more likely to mate among themselves than with deer mice in the next woodlot down the road, there will almost certainly be occasional matings between mice from different woodlots. Similarly, though the female parts of a particular red oak tree are more likely to receive pollen from another red oak tree in the same woodlot, they will sometimes receive pollen from a tree in another nearby woodlot. And the woodlots themselves are not permanent ecological features. They have only a transient existence as separate and distinct ecological units; neighboring woodlots may fuse after a few years, or a single woodlot may become divided into two or more separate smaller ones. Such changes in ecological features will produce corresponding changes in the demes of deer mice and red oak trees. Demes, then, are usually temporary units of population that intergrade with other similar units.

Species Notice that intergradation is between "similar" demes. We expect some interbreeding between deer mice from adjacent demes, but we do not expect deer mice to interbreed with house mice or black rats or gray squirrels. Nor do we expect to find crosses between red oaks and sugar maples or even between red oaks and pin oaks, even if they occur together in the same woodlot. In short, we recognize the existence of populations units larger than demes and both more distinct from each other and longer-lasting than demes. One such populations unit is that containing all the demes of deer mice. Another is that containing all the demes of red oaks. These larger units are known as species.

For centuries it has been recognized that plants and animals seem to fall naturally into many separate and distinct "kinds," or species. This does not mean that all the individuals of any one species are precisely alike—far from it. Any two individuals are probably distinguishable from each other in a variety of ways. But it does mean that all the members of a single species share certain biologically important attributes and that, as a group, they are genetically separated from other such groups. That such groups exist in nature has long been recognized by tribal peoples. Ernst Mayr of Harvard University cites a tribe in New Guinea that had 136 different names for what biologists later showed to be 137 species of local birds.

But though the existence of discrete clusters of living things that can be called species has long been recognized, the concept of what a species is has changed many times in the course of history. One idea widely held by nonbiologists, and once popular among biologists as well, is that each species is a static, immutable entity typified by some ideal form, of which all the real individuals belonging to that species are rough approximations; individual variation is supposed to result from the imperfection with which the individuals reflect the ideal characteristics. This static, typological concept contradicts all that we have learned about evolution. A ***species***, in the modern view, is a genetically distinctive group of natural populations (demes) that share a common gene pool and that are reproductively isolated from all other such groups. Or, to put it another way, a species is the largest unit of population within which effective gene flow (exchange of genetic material) occurs or can occur. The key word here is "effective"; we will see later why two species whose members mate but produce infertile hybrids are not classified as a single species.

Notice that the modern concept of species says nothing about how different from each other two populations must be to qualify as separate species. Admittedly, most species can be separated on the basis of fairly obvious anatomical, physiological, or behavioral characters, and biologists often rely on these in determining species. But the final criterion for living species is always reproduction—whether or not there is actual or potential gene flow.[1] If there is complete intrinsic reproductive isolation between two outwardly almost identical populations—that is, if there can be no gene flow between them—then those populations belong to different species despite their great similarity. On the other hand, if two populations show striking differences, but there is effective gene flow between them, those populations belong to the same species (Fig. 18.1). Anatomical, physiological, or behavioral characters simply serve as clues toward the identification of reproductively isolated populations; they do not in themselves determine whether a population constitutes a species.

Intraspecific variation We have already discussed the sorts of variation that may occur between individuals of a single deme as a result of mutation and recombination, particularly the latter. And we have seen that this variation is very important biologically, whether it involves almost imperceptible and intergrading differences or striking polymorphic discontinuities. But there is another sort of intraspecific variation that we have not yet discussed. This is variation between the demes of a single species, variation that is often correlated with geographical distribution.

There is usually so much gene flow between adjacent demes of the same species that differences between them are slight. Thus the frequencies of alleles *A* and *a* may be 0.90 and 0.10 in one deme and 0.89 and 0.11 in the adjacent deme. But the farther apart geographically two demes are, the smaller the chance of direct gene flow between them, and hence the greater

[1] Because paleontologists deal almost exclusively with fossils, they must rely to a great extent on morphological criteria in distinguishing between species.

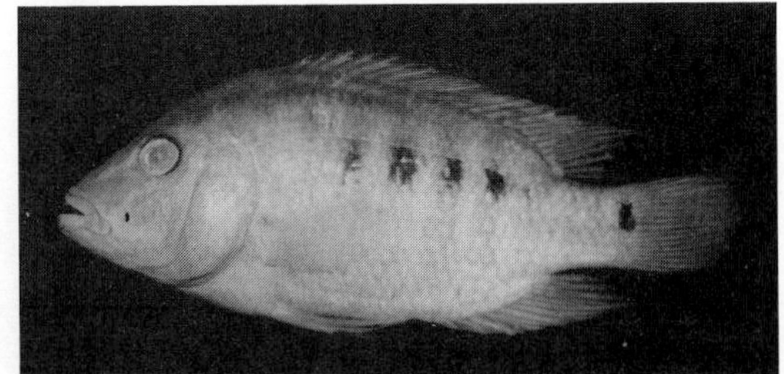

A

B

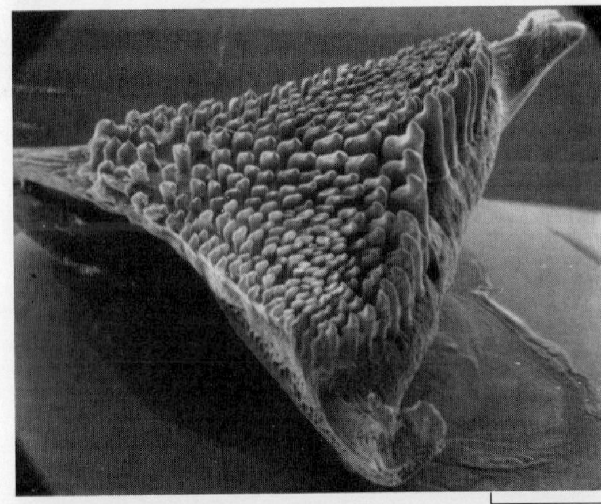

C

D

18.1 Variations in morphology between populations within a species

In some species, populations within a single area show easily observable morphological variations, as well as less obvious ones. A male of the cichlid species *Cichlasoma minckleyi*, found in the Cuatro Ciénegas basin in Mexico, exhibits the so-called deep-bodied form, or morph (A), while another male of the same species from the same basin exhibits the slender-bodied morph (B). An independent variation of the food-grinding structures in the lower pharyngeal jaw of males of this species also exists. Some males, regardless of body form, display the papilliform morph (C), while others display the molariform morph (D). Despite these major differences, individuals of the various morphs readily interbreed and produce fertile offspring; hence they are all considered members of the same species.

18.2 Clinal variation
(A) Map showing by means of isophene lines (lines connecting equal values) the geographic variation in the mean number of subcaudal scales of the snake *Coluber constrictor* (the racer)(B). Isophene map (C) showing geographic variation in the apical taper of leaves of the milkweed *Asclepias tuberosa* (D). The numbers represent degree of apical taper.

the likelihood that the differences between them will be more marked. If, for example, we collect samples of 500 deer mice each from Plymouth County, Massachusetts, Crawford County, Pennsylvania, and Roanoke County, Virginia, we will find numerous differences that enable us to distinguish between the three populations quite readily—much more readily than we could between populations from three adjacent counties in Massachusetts or Pennsylvania. Some of this geographic variation may reflect chance events such as genetic drift or the occurrence in one deme of a mutation that would be favorable in all demes but has not yet spread to them. But much of the geographic variation probably reflects differences in the selection pressures operating on the populations as a result of the differences between the environmental conditions in their respective ranges. (A species' range is the geographical area in which members of the species as a whole live or travel in the course of their normal activities, an area that may encompass an entire continent; individual animals generally occupy only a restricted part of this area—a home range. A species' habitat, by contrast, is where its members live—ponds, grasslands, forests in general, or only in or on oak trees, for example.) In other words, much geographic variation is adaptive. Each local population or deme tends to evolve adaptations to the specific environmental conditions in its own small portion of the species range. Such geographic variation is found in the vast majority of animal and plant species.

Environmental conditions often vary geographically in a more or less regular manner. There are changes in temperature with latitude or with altitude on mountain slopes, or changes in rainfall with longitude, as in many parts of the western United States, or changes in topography with latitude or

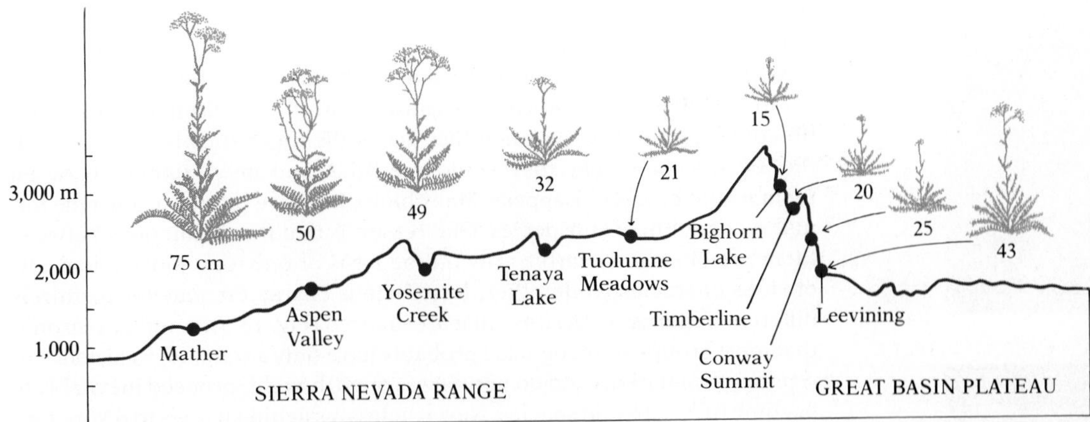

18.3 Altitudinal cline in height of the herb *Achillea lanulosa*
The higher the altitude, the shorter the plants. This variation was shown to be genetic (as opposed to merely environmental) by collecting seeds from the locations indicated and planting them in a test garden at Stanford where all were exposed to the same environmental conditions. The differences in height were still evident in the plants grown from these seeds.

longitude. Such environmental gradients are usually accompanied by genetic gradients—gradients of allelic frequency—in the species of animals and plants that inhabit the areas involved. Most species show north–south gradients in many characters; east–west gradients (Fig. 18.2C) and altitudinal gradients (Fig. 18.3) in various characters are not uncommon.

When a character of a species shows a gradual variation correlated with geography, we speak of that variation as forming a ***cline***. For example, many mammals and birds exhibit north–south clines in average body size, being larger in the colder climates toward the poles and smaller in the warmer climates toward the equator. Similarly, many mammalian species show north–south clines in the size of such extremities as the tails and ears: these exposed parts are smaller in the more polar demes.[2] A single widespread species often has many characters that vary clinally, but the several clines frequently do not coincide in direction, location, or intensity; one character may show clinal variation from north to south, another from east to west, and still another from northwest to southeast.

Sometimes geographically correlated genetic variation is not as gradual as in the clines discussed above. There may be a rather abrupt shift in some character in a particular part of the species range. When such an abrupt shift in a genetically determined character occurs in a geographically variable species, some biologists designate the populations involved as ***subspecies***. This term is also sometimes applied to more isolated populations —such as those on different islands or in separate mountain ranges or, as in fish, in separate rivers—when the populations are recognizably different genetically but are potentially capable of interbreeding freely. Subspecies may be defined, then, as groups of natural populations within a species that

[2] Increase in average body size with increasing cold is so common in warm-blooded animals that this tendency has been formally recognized as Bergmann's rule; the tendency toward decrease in the size of the extremities with increasing cold is called Allen's rule. The adaptive significance of these clines reflects the role of surface-to-volume ratios in heat exchange.

18.4 Two subspecies of Canada goose
These two subspecies of Canada goose have different breeding grounds and different ranges. *Branta canadensis maxima* (top) breeds in the central and south-central United States. *Branta canadensis moffitti* (bottom) breeds largely in central and western Canada.

differ genetically and that are partly isolated from each other because they have different ranges (Fig. 18.4).

Note that two subspecies of the same species cannot, by definition, long occur together geographically, because it is only the limitation on interbreeding imposed by distance that keeps them genetically distinctive. If they occurred together, they would interbreed and any distinction between them would quickly disappear. Many biologists have argued against the formal recognition of subspecies. One reason is that the distinctions between them are often made arbitrarily on the basis of only one morphologically obvious character while other, less obvious characters may form entirely different patterns of variation that are ignored (Fig. 18.5). Another reason is that most groups so recognized probably have only a transitory existence as separate populations and do not, as was once thought, proceed inevitably to become fully separate species. Nevertheless, assigning names to distinguish separate populations is frequently a great convenience.

As a result of intraspecific geographic variation, two populations belonging to the same species but occurring in two widely separated localities often show no more resemblance to each other than to populations belonging to other species. Such intraspecific dissimilarity serves to emphasize the point made earlier, that it is not the degree of morphological resemblance that determines whether or not two populations belong to the same species; it is whether they are reproductively isolated from each other. There are even instances where two widely separated populations are regarded as belonging to the same species even though the respective individuals, when brought together, are incapable of producing viable offspring. They are considered members of the same species because the populations are connected by an unbroken chain of intermediate populations that permit gene flow between them.

SPECIATION

In considering how species originate—the phenomenon of speciation—we will concentrate particularly on divergent speciation, the process by which one ancestral species gives rise to two or more descendant species, which grow increasingly unlike (diverge) as they evolve.

The role of geographic isolation Since species are defined in terms of reproductive isolation, the fundamental question of divergent speciation must be: How do two populations that initially share a common gene pool come to have completely separate gene pools? That is, how does the possibility of effective gene flow between the two populations disappear? How do barriers to the exchange of genes arise?

Most zoologists agree that in the majority of cases the initiating factor in speciation is geographic separation. As long as all the populations of a species are in direct or indirect contact, gene flow will normally continue throughout the system and splitting will not occur. Various populations within the system, however, may diverge in numerous characters and thus give rise to much intraspecific variation of the sorts discussed above. But if the initially continuous system of populations is divided by some geographic feature that constitutes a barrier to the dispersal of the species, then the separated population systems will no longer be able to exchange genes and further evolution will therefore be independent. Such populations are said to be ***allopatric***—literally "different groups"; ***sympatric*** populations, on the other hand, share a habitat. Given sufficient time, the two separate population systems may become quite unlike each other as each evolves in its own way. At first, the only reproductive isolation between them will be geographic—isolation by physical separation—and they will still be potentially capable of interbreeding; according to the modern concept of species, they will still belong to the same species. Eventually, however, they may become genetically so different that there would be no effective gene flow between them even if they should again come into contact. When this point in their divergence has been reached, the two population systems constitute two separate species.

There are at least three factors that can make geographically separated population systems diverge:

1. The chances are good that the two systems will have somewhat different initial gene frequencies. Because most species exhibit geographic variation, it is highly unlikely that a geographic barrier would divide a variable species into portions exactly alike genetically; the barrier would be much more likely to separate populations already genetically different, such as the terminal portions of a cline. Separation can occur in ways other than through the splitting of a once continuous distribution by a new geographic barrier. For example, small numbers of individuals frequently manage to cross an already existing barrier and found a new geographically isolated colony. Regardless of what causes the population to become divided, if one group is relatively small, its members will, of course, carry with them in their own genotypes a relatively small percentage of the total genetic variation present in the gene pool of the parental population, and the new colony will therefore have allelic frequencies very different from those of the parental population; this is a special form of genetic drift called the ***founder effect***. Obviously, if from the moment of their separation two populations have different genetic potentials, their future evolution may follow different paths. The present consensus is that most cases of geographic isolation probably involve small populations that exhibit the founder effect and genetic drift.

2. Separated population systems will probably experience different mutations. Mutations are random (though some are more probable than others), and the chances are good that some mutations will occur in one of the populations and not in the other, and vice versa. Since there is no gene flow between the populations, a new mutant gene arising in one of them cannot spread to the other.

3. Isolated populations will almost certainly be exposed to different environmental selection pressures, since they occupy different ranges. The chances that two separate ranges will be identical in every significant environmental factor are essentially nil.

The barriers that can cause the initial spatial separation leading to speciation are of many different types. A barrier is any physical or ecological feature that prevents the movement across it of the species in question. What is a barrier for one species may not be a barrier for another. Thus a prairie is a barrier for forest species but not, obviously, for prairie species. A mountain

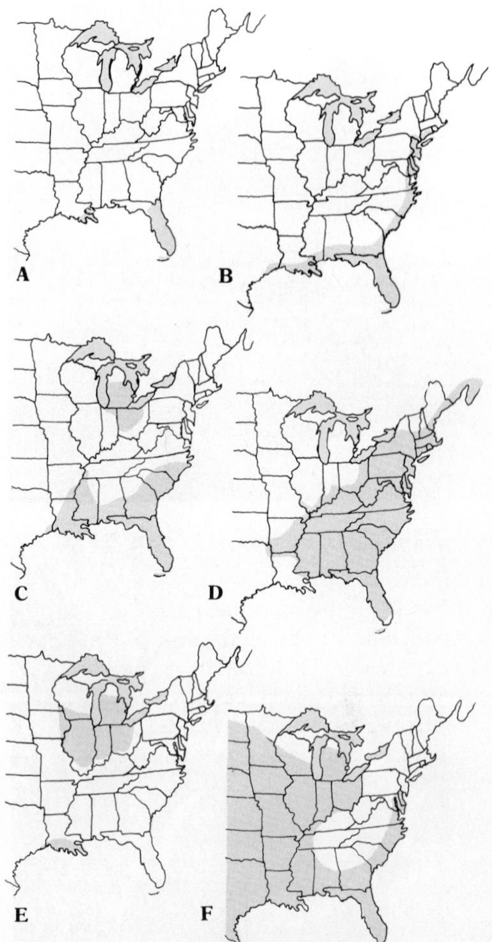

18.5 Discordant geographic variation in six characters of the snake *Coluber constrictor* in the eastern United States

Since no two of the characters vary together, selection of any one of them as a criterion for recognition of subspecies is largely arbitrary. (A) Areas where red eyes are found in juveniles. (B) Areas where red ventral spots are found on juveniles. (C) Areas where the loreal scale is in contact with the first supralabial scale in at least 10 percent of the specimens. (D) Areas where black adults are found. (E) Areas where dark postocular stripes are found. (F) Areas where full-grown adults have white chins.

18.6 Squirrels of the Grand Canyon
Two populations of squirrels that live in different ranges of the Grand Canyon in Arizona are morphologically distinct. The Kaibab squirrel (top), which lives on the Kaibab Plateau on the northern rim of the canyon, is darker than the Abert squirrel (bottom), of the related population that inhabits a range on the southern rim.

range is a barrier to species that can live only in lowlands, a desert is a barrier to species that require a moist environment, and a valley is a barrier to montane species. On a grander scale, oceans and glaciers have played a role in the speciation of many plants and animals. Let us look at a few actual examples of geographic isolation leading to speciation.

One of the most frequently cited examples is that of the Kaibab squirrel, which occurs on the north side of the Grand Canyon, and of the Abert squirrel, which occurs on the south side. The two are clearly very closely related and doubtless evolved from the same ancestor, but they almost never interbreed at present, because they do not cross the Grand Canyon. Biologists are not agreed whether these two morphologically distinct groups of squirrels have reached the level of full species or whether they should be considered well-marked geographic variants of a single species, but the fact remains that the Grand Canyon has acted as a barrier separating the two populations, and that those populations have, as a result, evolved divergently until they have at least approached the level of fully distinct species (Fig. 18.6). The Grand Canyon also separates the range of the gray-tailed antelope squirrel from that of the closely related white-tailed antelope squirrel, and it separates the range of the rock pocket mouse from that of the long-tailed pocket mouse.

On islands of the Pacific, in many instances, two closely related species of snails, clearly descended from the same ancestral population, live in valley woodlands separated by treeless ridges that the snails cannot cross. Blind cave beetles (genus *Pseudanophthalmus*) living in different caves in the eastern United States have often diverged to the level of full species.

Intrinsic reproductive isolation According to the model of divergent speciation just outlined, the initial factor preventing gene flow between two closely related population systems is ordinarily an extrinsic one—geography. Then, the model says, as the two populations diverge, they accumulate differences that will lead, given enough time, to the development of intrinsic isolating mechanisms—biological characteristics involving morphology, physiology, chromosomal compatibility, or behavior that prevent the two populations from occurring together or from interbreeding effectively when (or if) they again occur together. In other words, speciation is initiated when through external barriers the two population systems become entirely allopatric (come to have different ranges), but is not completed until the populations have evolved intrinsic mechanisms that will keep them allopatric or that will keep their gene pools separate even when they are sympatric (have the same range) (Fig. 18.7). Let us now examine the various kinds of intrinsic isolating mechanisms that may arise.

1. ***Ecogeographic isolation*** Two population systems, initially separated by some extrinsic barrier, may in time become so specialized for different environmental conditions that even if the original extrinsic barrier is removed they may never become sympatric, because neither can survive under the conditions where the other lives. In other words, they may evolve genetic differences that will maintain their geographic separation. An example is seen in two well-known tree species of the genus *Platanus: P. occidentalis* (the sycamore or buttonwood tree), which occurs in the eastern United States, and *P. orientalis* (the Oriental plane tree), which occurs in the

eastern part of the Mediterranean region. They can be artificially crossed and the hybrids are vigorous and fertile. But each species is adapted to the climate in its own native range, and the climates in the two ranges are so different that neither species will long survive in the range of the other. Thus there are genetic differences that under natural conditions would prevent gene flow between the two species. Their separation is not merely geographic; it is both geographic and genetic.

2. ***Habitat isolation*** When two sympatric populations occupy different habitats within their common range, the individuals of each population will be more likely to encounter and mate with members of their own population than with members of the other population. Their genetically determined preference for different habitats thus helps keep the two gene pools separate. There are numerous examples of such habitat isolation. *Bufo woodhousei* and *B. americanus* are two closely related toads that can cross and produce viable offspring. But in those areas where the ranges of the two toads overlap, *B. woodhousei* normally breeds in the quieter water of streams, while *B. americanus* breeds in shallow rainpools. The dragonfly *Progomphus obscurus* lives in northern Florida, while its close relative *P. alachuensis* lives in southern Florida. The ranges of the two species overlap in north-central Florida, but there the two species occupy different habitats, *P. obscurus* being restricted to rivers and streams and *P. alachuensis* to lakes. In California the ranges of *Ceanothus thyrsiflorus* and *C. dentatus*, two species of buckthorn shrubs, overlap broadly, but *C. thyrsiflorus* grows on moist hillsides with good soil, while *C. dentatus* grows on drier, more exposed sites with poor or shallow soil.

3. ***Seasonal isolation*** If two closely related species are sympatric, but breed during different seasons of the year, interbreeding between them will be effectively prevented. For example, *Pinus radiata* and *P. muricata*, two species of pine, are sympatric in some parts of California. They are capable of crossing, but rarely do so under natural conditions because *P. radiata* sheds its pollen early in February while *P. muricata* waits until April. *Reticulitermes hageni* and *R. virginicus*, two closely related species of termites, are sympatric in southern Florida, but the mating flights of the former occur from March through May while those of the latter occur in the fall and winter months. Five species of frogs belonging to the genus *Rana* are sympatric in much of eastern North America, but the period of most active mating is different for each species (Fig. 18.8).

4. ***Behavioral isolation*** In Chapter 38 we will discuss the immense importance of behavior in courtship and mating, particularly with respect to

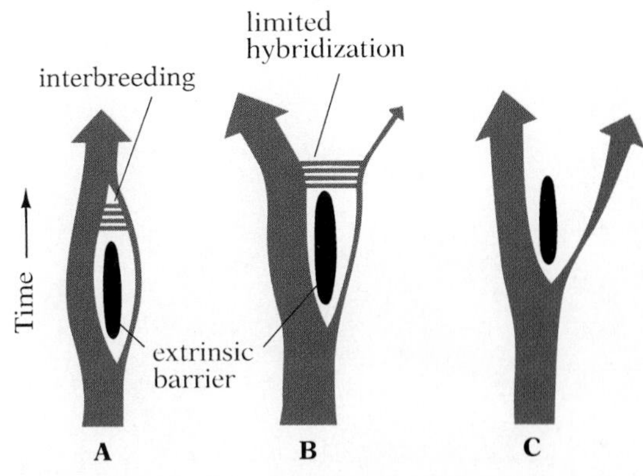

18.7 Model of geographic speciation
An ancestral population is split by an extrinsic (geographic) barrier for a time, and diverges with regard to various traits (only one of which is considered in these hypothetical examples). (A) If the barrier breaks down before the two subpopulations have been isolated long enough to have evolved intrinsic reproductive isolating mechanisms, the populations will interbreed and fuse back together. (B) Two populations are isolated by an extrinsic barrier long enough to have evolved incomplete intrinsic reproductive isolating mechanisms. When the extrinsic barrier breaks down, some hybridization occurs, but the hybrids are not as well adapted as the parental forms. Hence there is a strong selection pressure favoring forms of intrinsic isolation that prevent mating, and the two populations diverge more rapidly until mating between them is no longer possible. This rapid divergence is called character displacement. (C) Two populations are isolated by a geographic barrier so long that by the time the barrier breaks down they are too different to interbreed. In most cases, one population has many fewer members than the other, as indicated by the width of the "branches" where they separate. The smaller population—often quite small—usually diverges more from the common ancestor than does the larger population. This greater divergence is the result of the founder effect, a greater tendency for genetic drift, and a smaller and perhaps more specialized habitat.

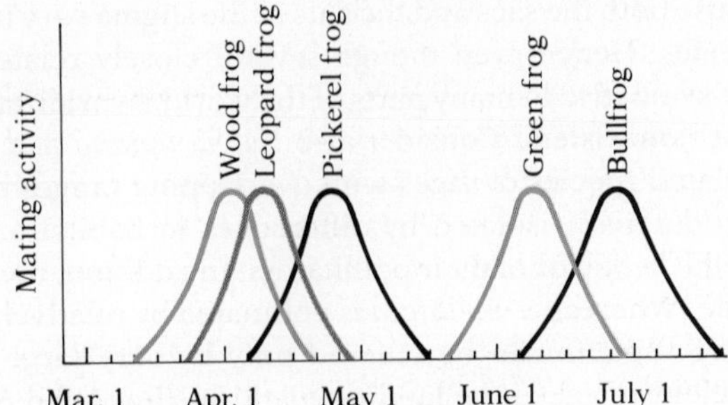

18.8 Mating seasons at Ithaca, New York, for five species of frogs of the genus *Rana*
The period of most active mating is different for each species. Where the mating seasons for two or more species overlap, different breeding sites are used.

18.9 A male fiddler crab (*Uca*)
The animal waves its large cheliped in the air as a courtship display. Critical details of the display differ among the various species of fiddler crabs.

species recognition. In many cases, species have unique courtship patterns, and where two species are sympatric, crosses rarely occur, because a courtship between members of different species involves so many wrong responses that it is unlikely to proceed all the way to copulation.

A particularly interesting example of the functioning of visual displays in species recognition has been reported by Jocelyn Crane of the New York Zoological Society. She found twelve species of fiddler crabs of the genus *Uca* actively courting on the same small beach (only about 56 m^2) in Panama. Each species had its own characteristic display, which included waving the large claw (cheliped), elevating the body, and moving around the burrow (Fig. 18.9). Crane found that the displays were so distinctive that she could recognize each species from a considerable distance merely by the form of its display; presumably the female crabs do likewise.

Auditory stimuli are important in species recognition among many animals, particularly birds and insects, and help prevent mating between related species. In several instances specialists have noticed that two or three very different calls were produced by what had been considered members of a single species of cricket. On investigation, they found that each call was, in fact, produced by a different species, but the species were so similar morphologically that no one had previously distinguished among them. Despite the morphological similarity of these closely related crickets, they do not hybridize in nature even when they are sympatric, because females do not respond to the sounds made by a male of a different species. ***Sibling species*** —species so closely related that humans can hardly distinguish them, at least until some diagnostic character, such as the call of the crickets, is found—may be a fairly common phenomenon.

5. ***Mechanical isolation*** If structural differences between two closely related species make it physically impossible for matings between males of one species and females of the other to occur, the two populations will obviously not exchange genes. If, for example, one species of animal is much larger than the other, matings between them may be very difficult, if not impossible. Or if the genital organs of the males of one species and the females of the other do not fit, mating will be prevented.

Mechanical isolation is probably more important in plants than in animals, particularly in plants that depend on insect pollinators. Consider milkweeds, for example. In these plants the pollen—which carries the male gametes—is contained in small sacs that stick to the legs of insects. The female part of each flower (stigma) has slits in it into which the sac must be inserted if pollination is to take place. Insertion of a pollen sac from one species of milkweed into the stigma of a different species is essentially impossible, because both the sacs and the slits in the stigma vary in shape from species to species. Hence even though several closely related species of milkweeds are sympatric in many parts of the world, hybridization between them is almost nonexistent. Consider also *Salvia apiana* and *S. mellifera*, two closely related species of sages with overlapping ranges in California. They are reproductively isolated by differences in habitat and flowering season and by the behavior of their pollinators. In addition, mechanical features play a role. Whereas *S. mellifera* is pollinated by relatively small bees, the flowers of *S. apiana* can be entered only by very large bees whose weight is sufficient to cause the landing platform (lower lip of corolla) to unfold and permit free entrance into the flower (Fig. 18.10).

6. ***Gametic isolation*** If individuals of two animal species mate or if the pollen from one plant species gets onto the stigma of another, fertilization still may not take place. Some 68 interspecific combinations are known in tobacco in which no cross-fertilization will occur even when the pollen is placed on the stigma, because the sperm nucleus from the pollen is unable to reach the egg cell in the ovary. In *Drosophila*, if cross-insemination occurs between *D. virilis* and *D. americana*, the sperm are rapidly immobilized by the unsuitable environment in the reproductive tract of the female and they never reach the egg cells. In other species of *Drosophila*, interspecific matings cause an antigenic reaction in the genital tract of the female, killing the sperm before they reach the eggs.

The mechanisms we have discussed so far exact little cost from the individual organisms: the machinery of isolation is either imposed from without, or generated internally from existing behavior, physiology, and morphology. As a result, no great loss of fitness is involved in maintaining isolation. The remaining four mechanisms, though effective, do exact a significant cost from the individual, and this cost acts as a selection pressure favoring evolution of the more efficient mechanisms of reproductive isolation we have already mentioned.

7. ***Developmental isolation*** Even when cross-fertilization occurs, the development of the embryo is often irregular and may cease before birth. The eggs of fish can often be fertilized by sperm from a great variety of other species, but development usually stops in the early stages. Crosses between sheep and goats produce embryos that die long before birth.

8. ***Hybrid inviability*** Hybrids are often weak and malformed and frequently die before they reproduce; hence there is no gene flow through them from the gene pool of the one parental species to the gene pool of the other parental species. An example of hybrid inviability is seen in certain tobacco hybrids, which develop tumors in their vegetative parts and die before they flower.

9. ***Hybrid sterility*** Some interspecific crosses produce vigorous but sterile hybrids. The best-known example is, of course, the cross between a female horse and a male donkey, which produces the mule. Mules have many characteristics superior to those of both parental species, but they are sterile. No matter how many mules are produced, the gene pools of horses and donkeys remain distinct, because there is no gene flow between them. The same is true of horses and zebras, which can hybridize to produce sterile zebroids.

10. ***Selective hybrid elimination*** The members of two closely related populations may be able to cross and produce fertile offspring. If those offspring and their progeny are as vigorous and well adapted as the parental forms, then the two original populations will not remain distinct for long if they are sympatric, and it will no longer be possible to regard them as full species. But if the fertile offspring and their progeny are less well adapted than the parental forms, then they will soon be eliminated. There will be some gene flow between the two parental gene pools through the hybrids, but not much. The parental populations are consequently regarded as separate species.

These last four mechanisms do effectively isolate species from one another, but their cost is significant: valuable metabolic resources are invested in doomed embryos or in frail or possibly sterile young, and the

A

B

C

18.10 Pollination of Scotch broom *(Cytisum scoparius)* by a bumble bee
The nectar and pollen are inaccessible to lighter pollinators like honey bees, whose weight is insufficient to trip release of the reproductive structures. (A) A bumble bee gathering nectar; (B) an untripped bloom; (C) a flower after tripping.

TABLE 18.1 *Intrinsic isolating mechanisms*

Effect	*Mechanism*	*Individuals affected*
Mating prevented	1. Ecogeographic isolation	**Parents:** fertilization prevented
	2. Habitat isolation	
	3. Seasonal isolation	
	4. Behavioral isolation	
	5. Mechanical isolation	
Production of hybrid young prevented	6. Gametic isolation	
	7. Developmental isolation	**Hybrids:** success prevented
Perpetuation of hybrids prevented	8. Hybrid inviability	
	9. Hybrid sterility	
	10. Selective hybrid elimination	

seasonal nature of many reproductive cycles may preclude a second chance for the individuals involved to mate and rear young. It is important to remember that success in finding a mate or in copulating does not necessarily lead to reproductive success—the production of fit offspring. Individuals that tend to mate with members of the wrong species will leave fewer descendants than those that mate with members of their own species. Wrong matings waste gametes, whether fertilization takes place or not, and whether the hybrids are viable or not. Selection, therefore, will strongly favor individuals whose behavior, morphology, or physiology reduces the chance of a mismatch in the first place, and if the parental populations are sympatric, there will be strong selection for the evolution of more effective intrinsic isolating mechanisms. Gene combinations that lead to correct mate selection will increase in frequency, and combinations that lead to incorrect selection will decrease, until eventually all hybridization ceases. The tendency of closely related sympatric species to diverge rapidly in characteristics that reduce the chances of hybridization and/or minimize competition between them is called ***character displacement*** (see Fig. 18.7). Table 18.1 summarizes the different intrinsic isolating mechanisms.

Situations in which only one of the ten isolating mechanisms is operative are extremely rare. Ordinarily several contribute to keeping two species apart. For example, closely related sympatric plant species often exhibit habitat and seasonal isolation in addition to some form of hybrid incapacity. And as we have seen, sympatric species, whether plant or animal, tend rapidly to evolve one or more of the forms of isolation that prevent mating (habitat, seasonal, behavioral, mechanical) rather than depend only on those forms that prevent the birth or perpetuation of hybrids.

Speciation by polyploidy The model of speciation discussed above involves the divergence of geographically separated populations. There are other ways in which new species may arise. Speciation that does not involve geographic isolation is called ***sympatric speciation***. One important example is speciation by polyploidy, the condition of having more than two sets of chromosomes. This condition can arise so quickly that it is entirely possible for a parent to belong to one species and its offspring to another. (Other forms of sympatric speciation will be discussed later.) Speciation by polyploidy has apparently been common in plants but rare in animals.

One type of polyploid speciation, called ***autopolyploidy***, involves a sudden increase in the number of chromosomes, usually as a result of the nondisjunction of chromosomes during meiosis. An example of this type of polyploidy was discovered by Hugo De Vries, one of the early geneticists, while he was making extensive studies of the evening primrose, *Oenothera lamarckiana*. This diploid species has 14 chromosomes. During De Vries' studies, a new form suddenly arose. This new form, to which he gave the species name *Oenothera gigas*, had 28 chromosomes. This tetraploid was reproductively isolated from the parental species because hybrids between *O. lamarckiana* and *O. gigas* were triploid (they received one of each type of chromosome from their *O. lamarckiana* parent and two of each type from their *O. gigas* parent), and triploid individuals, because of the highly irregular distribution of their chromosomes at meiosis, are sterile. It is characteristic of autopolyploidy that the polyploids are fertile and can breed with each other, but cannot cross with the diploid species from which they arose. Hence polyploid populations fulfill all the requirements of the modern definition of species—they are genetically distinctive, and they are reproductively isolated—though botanists do not always choose to give each polyploid population a formal species name.

The reproductive isolation of the polyploid daughter species from the ancestral stock sometimes permits adaptive divergence that would not otherwise be possible. For example, new polyploid species of certain plants have become adapted to mineral soils like mine tailings and serpentine outcrops. If the plants were not reproductively isolated, gene flow from the large surrounding population of normal diploids would probably prevent any local adaptation to these special soils. Polyploidy, then, substitutes for geographic isolation, and is one way in which speciation can occur sympatrically.

A second type of polyploid speciation, called ***allopolyploidy***, involves a multiplication (usually a doubling) of the number of chromosomes in a hybrid between two species. The hybrid has one set of chromosomes from each of the two species. Unless these two species are so closely related that they have homologous chromosomes capable of pairing (synapsing) in meiosis, the hybrid will almost certainly be sterile because of an inability to produce gametes. But if the hybrid undergoes chromosome doubling before meiosis, it will have a complete diploid set of chromosomes from each species, and viable gametes are far more likely to be produced. This type of polyploidy has probably been far more important in speciation than autopolyploidy. Allopolyploid individuals will be able to breed freely among themselves, but they will be unable to cross with either of the parental species. Consequently the allopolyploid population must be regarded as a distinct species.

Allopolyploid plants are only rarely more vigorous than the parental diploid plants, probably because each parental species is a delicately balanced mixture of genes, while the allopolyploid mixes together two sets of gene products, metabolic pathways, and control systems. But occasionally allopolyploids, because of the combination of genes from two different species, can grow in habitats which neither parental species can colonize. Allopolyploid speciation, therefore, has probably played an important role in the perpetuation of some plant groups during periods of widespread environmental change.

Allopolyploidy has also proven of great importance in the production of valuable new crop plants. As soon as plant breeders realized that many of our most useful plants, such as oats, wheat, cotton, tobacco, potatoes, bananas, coffee, and sugarcane are polyploids, they began trying to stimulate polyploidy, and obtained many new varieties. The chemical colchicine readily induces polyploidy. One of the first artificially produced allopolyploids came from a cross made in 1924 between the radish and the cabbage. Unfortunately it had the root of the cabbage and the shoot of the radish. Other crosses have yielded more desirable results (see Fig. 16.27, p. 441).

Chromosome doubling creates an obvious problem: if tetraploids cannot breed with diploids, they are restricted to breeding with each other. Hence, the rare diploid pollen grain must find an equally rare diploid egg cell to produce a fertile tetraploid. And, most likely, the resulting plant must then fertilize itself, since other tetraploids will be exceedingly rare. The same is

18.11 Sympatric speciation in tree hoppers Two sympatric populations of the tree hopper *Enchenopa binotata* have evolved adaptations to different host plants. The tree hopper at the top lives on bittersweet, while the one on the bottom lives on butternut. Host specificity may take the place of physical separation (allopatry) in preventing these two populations from interbreeding.

true of allopolyploids: the sole hybrid is extremely unlikely to find a genetically compatible plant, and so will probably have to fertilize itself. The result in both cases will be severe inbreeding of the new species, a strong founder effect, and rapid genetic drift. The new population will persist only if it has a strong competitive advantage in the habitat.

As we have seen, polyploid speciation generates new species almost instantaneously. Some researchers believe that the same principle operating on a smaller scale may also be a powerful but unrecognized mechanism of speciation. Chromosomal rearrangements, whether the result of a few major breakage-and-fusion events or of smaller but more numerous transpositions, duplications, and deletions, might give rise to individuals or small populations genetically incompatible with the rest of their species. If the variant organisms are at a competitive advantage, selection would favor further intrinsic isolation, and sympatric speciation might occur as a result. There is some evidence for this mechanism of speciation at least among the *Drosophila*.

Nonchromosomal sympatric speciation Though most speciation not involving gross chromosomal changes is certainly allopatric (requiring a period of geographic isolation), there is increasing evidence that sympatric speciation can occur without polyploidy or other major chromosomal rearrangements. Reproductive isolation is just as essential to these types of sympatric speciation, but is effected by other means. For example, relatively small changes in habitat preference may produce habitat isolation with effects sufficient to compensate for the absence of geographic separation in the development of isolating mechanisms. Thus a species of clover adapted to soil containing mine tailings is now reproductively isolated from the widespread and closely related species from which it evolved, not by polyploidy but by virtue of having a different flowering season. Another alternative to geographic isolation, strongly implicated in arthropods like tree hoppers, is related to host specificity. These insects often mate on the plants on which they feed, so subspecies that are adapted for a particular species of host plant will tend to inbreed. According to this scenario, selection would favor those individuals that breed strictly on the host species; consequently, intrinsic isolating mechanisms might develop that would serve to isolate tree hoppers with adaptations to different host plants. In effect, sympatric speciation in tree hoppers would become possible because host specificity could create reproductive isolation even in the absence of true geographic isolation (Fig. 18.11).

Yet another mechanism that may contribute to sympatric speciation is the behavioral phenomenon known as sexual imprinting. (Imprinting behavior will be discussed in Chapter 38.) The members of one sex of many species automatically memorize while young a particular feature or set of features displayed by their parents or siblings. The features may be visual, auditory, or olfactory. Memorization enables these individuals later to identify suitable mates with great precision. The lasting effect of sexual imprinting in birds was shown in experiments by Klaus Immelmann involving zebra finches and Bengalese finches. Birds of these two species are physiologically capable of interbreeding, but imprint so forcefully on their own species that they do not. In Immelmann's experiments, male zebra finches raised by Bengalese parents invariably courted female Bengalese finches when given a choice. Now, if the parent on which the young imprint displays a mutation in the feature being committed to memory (a brightly colored eye ring, for example, or novel elements in the courtship song), and if the young are later able to locate mates displaying the same mutation, the mutant individuals are likely to pair with each other. The result could be the founding of a population that does not interbreed with the rest of the species. In birds, then, instant reproductive isolation based on imprinting may play an important role in shaping populations.

One recently discovered example of at least incipient sympatric speciation illustrates how imprinting can create a kind of habitat preference that leads to reproductive isolation. The picture-wing fruit fly *Rhagoletis pomenella* is a well-known species that courts and deposits its eggs on hawthorn berries; the larvae feed on the fruit, pupate, and emerge to continue the cycle. At some stage in its life, the fly imprints on the odor of hawthorn, and uses that food-based memory to locate a suitable host plant. Jeff Feder at Princeton University has shown that the larvae are well adapted to hawthorn berries, growing and pupating about twice as well as on any other kind of tree fruit. However, there are two specialized species of parasitic wasps that seek out hawthorn berries and insert their eggs into the growing fruit fly larvae, ultimately killing about 90 percent of them.

About 150 years ago, land was cleared and large commercial apple orchards were established in the Hudson Valley of the northeastern United States; hawthorns became less common, and subsequently (probably by mistake) a few *Rhagoletis* began to lay their eggs on apples. Imprinting committed the next generation to apples, and reproductive isolation must have been nearly complete from the outset. Although *Rhagoletis* larvae grow only about half as well on apples, they suffer far less parasitism: not only are the wasps not looking for them on apples, but also the fruits appear earlier (so that the larvae may have entered the pupal stage before the height of the season for parasites) and apples are larger, allowing the larvae to burrow so far in that the wasps cannot reach them. The result, as measured by Feder, is that six times as many of the eggs laid on apples produce adult flies as compared with hawthorn. At about the same time another subgroup of *Rhagoletis* began to infest sour cherries, but this line is dying out as the cherry orchards in which they thrive are being abandoned. Whether either group is technically a separate species yet is not clear, but it is obvious that there is no inherent barrier in this pathway to sympatric speciation.

Adaptive radiation One of the most striking aspects of life is its extreme diversity. A bewildering array of species now occupies this globe. And the fossil record shows that of the species that have existed at one time or another those now living represent only a tiny fraction (probably less than one-tenth of 1 percent, all the other species being now extinct). Clearly, then, divergent evolution or ***radiation***—the evolutionary splitting of species into many separate descendant species—has been as frequent as extinction. How could opportunities for geographic isolation—thought to be by far the most common precursor to speciation—have been sufficient to lead to all the speciation not caused by sympatric speciation? After all, it is not unusual for a complex of four, five, or more closely related species to

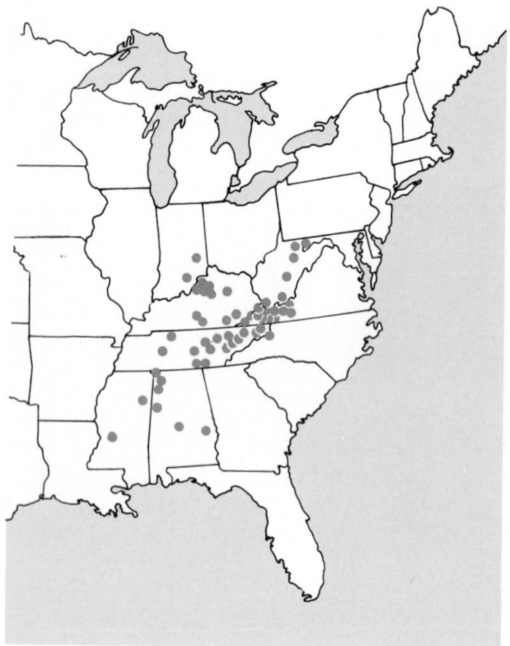

18.12 Distribution of 28 species of millipedes of the genus *Brachoria*

The color dots indicate all known localities for these millipedes, many of which are sympatric. All the speciation in the genus must have occurred within this very limited area in the eastern United States.

occur within a rather limited area. For instance, 28 species of the millipede genus *Brachoria*, many of them sympatric, are confined to a small portion of the deciduous forests of the eastern United States (Fig. 18.12). How could so much speciation have occurred in so small an area if geographic isolation was a necessary factor? In an attempt to answer such questions, let us turn to a particularly instructive and historically important example—the finches on the Galápagos Islands, which (along with tortoises and mockingbirds on those islands) played a major role in leading Charles Darwin to formulate his theory of evolution by natural selection.

The Galápagos Islands lie astride the equator in the Pacific Ocean roughly 950 km west of the coast of Ecuador, the country to which they now belong (Fig. 18.13). The islands have never been connected to one another. They apparently arose from the ocean floor as volcanoes approximately five million years ago. At first, of course, they were completely devoid of life, and were therefore open to exploitation by whatever species might chance to reach them from the mainland. Relatively few species did manage to reach the islands and become established. The only land vertebrates present on the islands before human beings got there were at least seven species of reptiles (one or more snakes, a species of huge tortoise, and at least five species of lizards, including two very large iguanas), seven species of mammals (five rats and two bats), and a limited number of birds (including two species of owls, one hawk, one dove, one cuckoo, one warbler, two flycatchers, one martin, four mockingbirds, and the famous Darwin's finches).

The 14 species of Darwin's finches constitute a separate subfamily found nowhere else in the world. Thirteen of them are believed to have evolved on the Galápagos Islands, and one on Cocos Island, from some unknown finch ancestor that colonized the islands from the South American or Central American mainland. It is readily understandable that the descendants of the geographically isolated colonizers should have undergone so much evolutionary change as to become, in time, very unlike their mainland ancestors. More perplexing at first glance is the manner in which the descendants of the original immigrants split into the separate populations that gave rise to today's 14 species.

The point to remember is that we are dealing not with a single island but with a cluster of more than 25 separate islands. The finches will not readily fly across wide stretches of water, and they show a strong tendency to remain near their home area. Hence a population on any one of the islands is effectively isolated from the populations on the other islands. We suppose that the initial colony was established on one of the islands where the colonizers, perhaps blown by high winds, chanced to land. Later, stragglers from this colony wandered or were blown to other islands and founded new colonies. The allelic frequencies in the new colonies differed from those in the original colonies from the moment they started, because of the founder effect. In time, the colonies on the different islands diverged even more, for the reasons already outlined in the model of geographic speciation (different mutations, different selection pressures, and, in such small populations as some of these must have been, genetic drift). What we might expect, therefore, is a different species, or at least a different subspecies, on each of the islands. But this is not what has actually been found; most of the islands have more than one species of finch, and the larger islands have 10 (Fig. 18.13, right). What is the explanation?

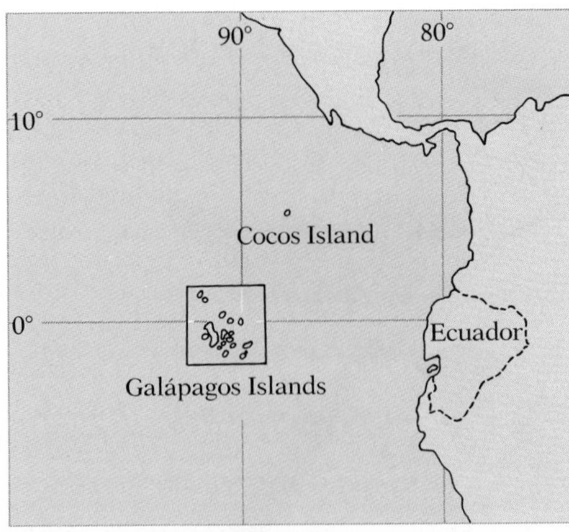

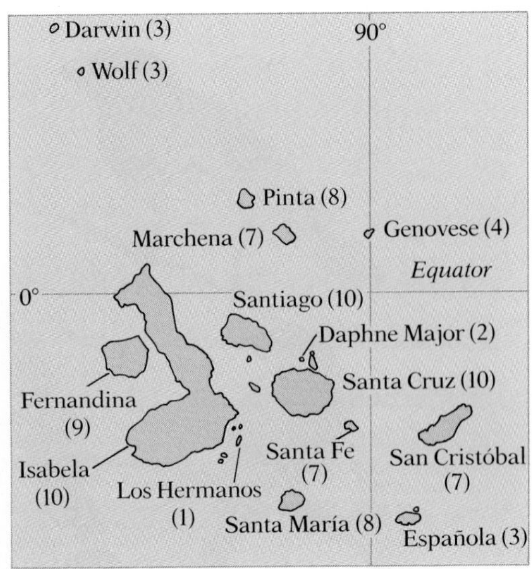

18.13 The Galápagos Islands
Left: The islands are located about 950 km off the coast of Ecuador. Cocos Island is about 700 km northeast of the Galápagos. Right: The islands shown in greater detail. The number in parentheses after each name indicates the number of species of Darwin's finches that occur on the island.

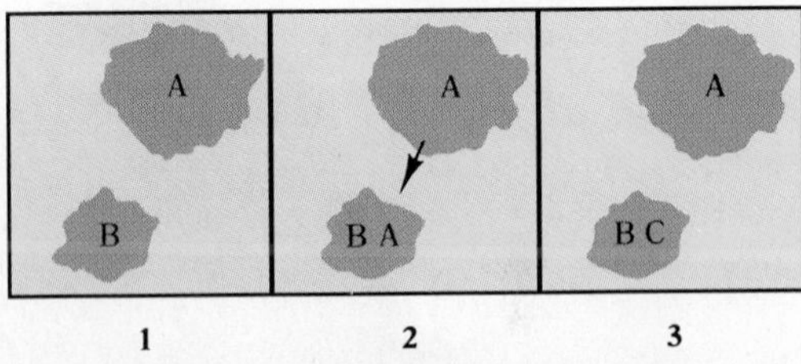

18.14 Model of speciation on the Galápagos Islands
An ancestral form colonized the larger of these two hypothetical islands. Later, part of the population dispersed to the smaller island. (1) Eventually the two populations, being isolated from each other, evolved into separate species A and B. (2) Some individuals of A dispersed to B's island. The two species coexisted, but intense competition between them led to rapid divergent evolution. (3) This rapid evolution of the population of A on B's island caused it to become more and more different from the original species A, until eventually it was sufficiently distinct to be considered a full species, C, in its own right. At the same time, the selection pressure imposed by the small invading population caused the large population of species B to evolve to a small degree as well.

Let us suppose that form A evolved originally on the island of Santa Cruz and that the closely related form B evolved on Santa María. If, later, form A had spread to Santa María before the two forms had been isolated long enough to evolve any but minor differences, the two forms might have interbred freely and merged with each other. But if A and B had been separated long enough to have evolved major differences before A invaded Santa María, then A and B might have been intrinsically isolated from each other, having developed into separate species, and they might have been able to coexist on the same island without interbreeding (Fig. 18.14). If they formed occasional hybrids, those hybrids might well have been less viable than the parental forms. Accordingly, natural selection would have favored individuals that mated only with their own kind, and this selection pressure would have led rapidly to more effective intrinsic isolating mechanisms, which would prevent the gamete wastage involved in cross-matings. In fact, Darwin's finches readily recognize members of their own species and show little interest in members of a different species.

We have now arrived at a point in our hypothetical example where Santa Cruz is occupied by species A and Santa María by both A and B. It would be highly unlikely that A and B could coexist indefinitely if they used the same food supply; the ensuing competition would be severe, and the less well-adapted species would tend to be eliminated by the other unless it evolved differences that minimized the competition. In short, wherever two or more very closely related species occur together, competition will lead either to the extinction of one, or to character displacement—in this instance the evolution of different feeding specializations.

18.15 Darwin's finches
Darwin's finches fall into four genera: birds 1, 3, 4, 5, 6, and 10 are the tree finches (*Camarhynchus*); birds 7, 8, 11, 12, 13, and 14 are the ground finches (*Geospiza*); bird 2 is the unfinchlike warbler finch; bird 9 is the one species inhabiting Cocos Island.

1. Vegetarian tree finch (*C. crassirostris*)
2. Warbler finch (*Certhidea olivacea*)
3. Large insectivorous tree finch (*C. psittacula*)
4. Medium insectivorous tree finch (*C. pauper*)
5. Mangrove finch (*C. heliobates*)
6. Small insectivorous tree finch (*C. parvulus*)
7. Large cactus ground finch (*G. conirostris*)
8. Cactus ground finch (*G. scandens*)
9. Cocos finch (*Pinaroloxias inornata*)
10. Woodpecker finch (*C. pallidus*)
11. Large ground finch (*G. magnirostris*)
12. Sharp-beaked ground finch (*G. difficilis*)
13. Medium ground finch (*G. fortis*)
14. Small ground finch (*G. fuliginosa*)

Character displacement is indeed what we find in Darwin's finches (Fig. 18.15). The 14 species form four groups (genera). One group contains six species that live primarily on the ground; of these, some feed primarily on seeds and others mostly on cactus flowers. Of the species that feed on seeds, some feed on large seeds, some on medium-sized seeds, and some on small seeds. These feeding preferences result from the morphological specialization of the beaks: small beaks are most efficient at handling small seeds, while larger beaks can crack large seeds. (The specialization is not symmetrical: large beaks can handle small seeds, though not very efficiently, but small beaks cannot crack large seeds at all.) From a series of careful beak measurements, David Lack of Oxford University was able to find clear evidence of character displacement. For example, when the small and medium ground finches coexist on the larger islands, their beak sizes are widely separated, with depths averaging about 8.4 and 13.2 mm respectively. But on small islands where only one of the two species exists, the birds of that species tend to have beaks of intermediate size, on the order of 9.7 mm (Fig. 18.16).

The second group of finches contains six species that live primarily in trees. Of these, one is vegetarian and the others eat insects, but the insect eaters differ from one another in the size of their prey and in where they catch them (Fig. 18.17). A third group contains only one species, which has become very unfinchlike and strongly resembles the warblers of the mainland. The fourth group also contains one species, restricted to Cocos Island, which is about 700 km northeast of the Galápagos Islands and about 500 km from Panama. Correlated with the differences in diet among the species are major differences in the size and shape of their beaks. (Noteworthy varia-

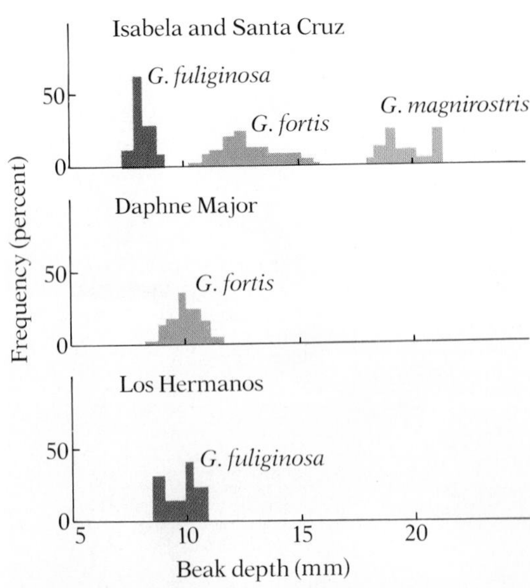

18.16 Beak sizes of ground finches
On large islands like Isabela and Santa Cruz, three species of ground finch coexist. Though the beak sizes of individuals within each species show the sort of variation essential for evolution by natural selection, they fall into three separate ranges, specialized respectively for efficient feeding on small, medium-sized, and large seeds. However, on small islands where only one species exists (either the small or the medium ground finch) the beak sizes fall into an intermediate range regardless of the size specialization found in the same species on the larger islands. In the face of competition for food on the larger islands, character displacement has taken place.

18.17 The tool-using finch
One of the insectivorous tree finches of the Galápagos has evolved most unusual feeding habits. Sometimes called the woodpecker finch, it chisels into wood after insects, but it lacks the long tongue that a woodpecker uses to probe insects out of a crack. Instead, it pokes into the crack with a cactus spine or twig that it holds in its beak. The mangrove finch does the same thing in a different habitat. These are two of the few known cases of tool use by a bird.

18.18 Beak differences in Hawaiian honeycreepers Differences in beak size and shape are apparent in related species of Hawaiian honeycreepers, which, like Darwin's finches, are thought to have evolved from a common finchlike ancestor.

tions in beak size and shape among birds thought to share a common ancestry can be found in other island habitats as well, as shown, for example, in Figure 18.18. These characteristics of the beak are an important means by which the birds recognize other members of their own species; song is another.)

Now, if selection on Santa María favored character displacement between species A and B, the population of species A on Santa María would become less and less like the population of species A on Santa Cruz (Fig. 18.14). Eventually these differences might become so great that the two populations would be intrinsically isolated from each other and would thus be separate species. We might now designate as species C the Santa María population derived from species A. The geographic separation of the two islands would thus have led to the evolution of three species (A, B, and C) from a single original species. The process of island hopping followed by divergence could continue indefinitely and produce many additional species. It was doubtless such a process that led to the formation of the 14 species of Darwin's finches.

Now let us apply the principles learned from Darwin's finches to the case of the 28 species of *Brachoria* millipedes confined to a small area in the eastern United States (see Fig. 18.12). These animals live in the humus layer on the floor of deciduous forests. They are rather sluggish and seldom move very far. It would have been easy for populations to become isolated in local forested areas separated by less hospitable regions. Such allopatric populations could have become sufficiently different so that, when conditions changed and they became sympatric, they would behave as full species. Clearly, the sorts of processes seen on islands can account also for radiation in a continental area. And on a somewhat larger geographic scale, the same processes can account for the observed adaptive radiation in insects, fish, reptiles, birds, mammals, and many plant groups. In short, adaptive radiation on islands, as in the case of Darwin's finches, is dramatic and lends itself particularly well to analysis, but it does not differ in principle from adaptive radiation under other circumstances. The model of speciation we have outlined can account for the great amount of divergence necessary to produce the immense diversity among living things.

It should be clear, as Darwin himself pointed out, that the rate of evolutionary divergence is not constant. When the first colonizing finches reached the Galápagos Islands, they would have encountered environmental conditions quite unlike those they had left behind in Central or South America. The selection pressures to which they were subjected would probably have been quite different from those in their former home; differences in the resources available, for example, may have led to selection for different morphology, physiology, and behavior. On the other hand, if there was initially little or no competition from other species, the result would have been a temporary relaxation of selection pressures. Only when the new habitat became saturated with finches would intraspecific competition for the available resources have become important. And so, sooner or later, selection pressures must have led to rapid divergence from the ancestral population. Later, as the finches became increasingly well adapted to conditions on the Galápagos, the rate of evolutionary change probably slowed down.

In general, when conditions change radically and organisms have new evolutionary opportunities for which they are at least modestly preadapted, they may undergo an evolutionary burst—a period of rapid adaptational change—which may then be followed by a more stable period during which any further evolutionary changes are merely fine-tuning of their characteristics. Such bursts of rapid evolutionary divergence probably characterized the tremendous radiation of amphibians when they moved onto land for the first time, and the explosive radiation of mammals when the demise of the dinosaurs left many biological niches—that is, ways of surviving—unoccupied. (We will examine the concept of niche in detail in Chapter 39.)

How important is competition? In the last chapter we contrasted the roles played by genetic drift (chance) and natural selection in the evolution of populations. Our discussion emphasized that evolution is a vital, ongoing phenomenon, while drift (potentially very important in small populations) and selection are two contributing—and clearly demonstrated—mechanisms by which it occurs. In our discussion of speciation we have touched on a similar theme: barriers that contribute to reproductive isolation can arise by chance (primarily genetic drift), by natural selection, or by a combination of the two. Throughout our discussion we have emphasized the role of competition in the formation of the species: even reproductively isolated populations may not be able to coexist indefinitely if they compete for precisely the same food, since even a slight but systematic superiority of one will tend to lead to the extinction of the other.

G. F. Gause of the University of Moscow, who first observed this phenomenon in the laboratory in the 1930s, formulated the competitive exclusion principle, often called the Niche Rule: no two species occupying the same niche can long coexist (Fig. 18.19). Only the character displacement that natural selection produces, resulting in changes in food preference, habitat choice, and the like, will allow closely related species to coexist. This Darwinian interpretation of how separate species form, with its emphasis on reproductive isolation and character displacement in the face of competition, probably accounts for much of what we can observe of speciation. During the last decade, however, the role of competition in speciation has come under increasingly close scrutiny. Today, competition is no longer thought to be as widespread and important as was formerly believed, while chance is now seen as clearly a greater force than Darwin and most scientists since him have thought.

There are at least two ways in which chance can take precedence over competition in causing species to diverge. The first we have already discussed: in small populations, genetic drift can be very powerful, driving alleles to extinction before the generally more gradual process of natural selection can produce stability. The other process, however, is more fundamental to our understanding of evolution. Most Darwinian analyses tacitly assume that the selection pressures organisms face change relatively slowly, and that large continuous populations therefore have ample time, generation by generation, to evolve specialized adaptations to their environments. But do conditions really remain sufficiently constant for a species to achieve equilibrium—a stable set of allelic frequencies—in the face of all the selection pressures affecting it?

Apparently, many populations are not always at equilibrium. In one par-

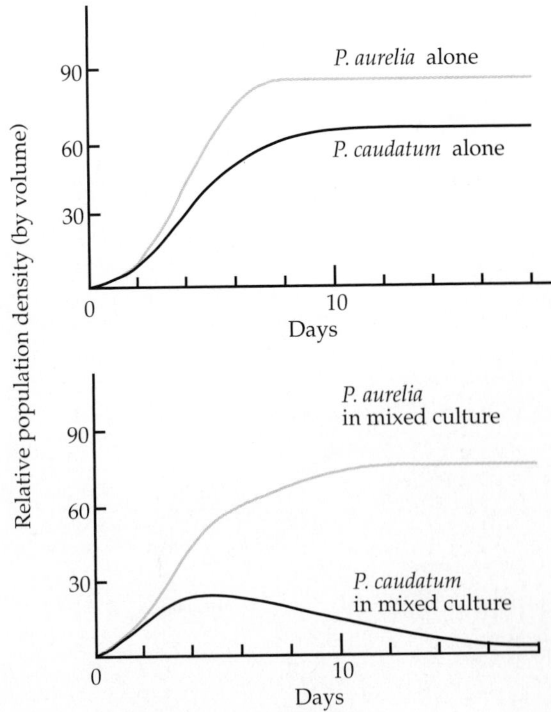

18.19 Competition and extinction
When a few individuals of either species of *Paramecium* are introduced into a tank alone (top), they multiply until they reach a limiting density. But when individuals of both species are introduced together (bottom), they multiply independently for the first three days, then begin to compete for resources. *P. aurelia* is in some way more efficient under these conditions, and drives *P. caudatum* to extinction in only three weeks.

ticularly thorough study Peter Grant of Princeton University and his colleagues tagged 1500 medium ground finches on one of the smaller Galápagos islands, Daphne Major. In 1977 only 20 percent of the normal amount of rain fell, and the plants on the island produced many fewer seeds than usual. Relatively dry or wet seasons, a surprisingly early or late hard frost, or especially warm or cold years are familiar conditions that organisms can generally take in stride, but the effects on plants and animals of extreme events can be drastic.

On Daphne Major, the consequences of the reduction in food were dramatic for the medium ground finch: with only 35 percent of the usual amount of food available, 387 of the 388 nestlings of 1976 died, as did the majority of adults. Furthermore, the adult survivors were not a representative cross section of the former population: roughly 180 of the 500 adult males survived, but only about 30 of the 500 females, and the birds that did survive the year of starvation were significantly larger than those that succumbed. Exceptionally intense natural selection in favor of large size—especially larger beak size—had occurred.

Thus one of several rare and unpredictable crises sharply altered the selection pressures for a year, and reduced the population to a level at which genetic drift suddenly became a real possibility. What had seemed to be an adaptive equilibrium was upset by a chance disturbance, and a population with altered characteristics emerged. The finch population had been forced through a period of environmental crisis. When a crisis is so severe as to cause major changes in allelic frequencies in a population, it is called an ***evolutionary bottleneck***.

The assumption that interspecific and intraspecific competition, leading gradually to character displacement and to new species, is the major mode of speciation is currently being challenged by the view that chance crises may occur often enough to upset whatever stability exists in the allelic frequencies within a species. Crises can be caused by any environmental factor—pestilence or extreme weather, for example—that severely affects an isolated population, or that itself serves to isolate one part of a population from the remaining body. Because of their small population size, endangered species are in this position. During such a crisis, as an isolated population passes through the evolutionary bottleneck produced by extraordinary environmental conditions, one character or another may gain ascendancy. Having been selected by a founder effect, such a character is still at risk if the surviving population is small, because it may decline in frequency or even become extinct through genetic drift. Alternatively, if a character that was not adaptive prior to the crisis proves adaptive thereafter, it may increase in frequency in the population. In the case of the finches, the effects of the drought of 1977 (and another in 1982) were canceled by another freak climatic event: record rainfall in 1983. Unprecedented numbers of seeds were produced, and selection strongly favored smaller beaks.

Though most researchers still believe that competition is the predominant force leading to speciation over time, many studies now suggest that the effects of rare boom-or-bust events have been important in the evolution of at least certain populations.

Punctuated equilibrium One consequence of the increasing evidence that natural catastrophes can affect the allelic frequencies of populations has been a growing interest in ***punctuated equilibrium***. This much-debated hypothesis concerning the mode and tempo of speciation was originally formulated by Niles Eldredge of the American Museum of Natural History and Stephen Jay Gould of Harvard. Based on careful study of certain fossil records, their hypothesis suggests that most allopatric speciation events are geologically "instantaneous," the result of crises or major genetic alterations that punctuate long periods of equilibrium, or ***stasis***, in which the morphology of the species remains relatively constant. Gould and Eldredge maintain that the fossil record does not support ***gradualism***—the view, which they believe was Darwin's, that speciation occurs as a gradual accumulation of morphological and physiological changes (Fig. 18.20). "Instantaneous" speciation, however, is by no means as rapid as it sounds: they believe that it takes thousands of generations, up to 100,000 years, to complete. Perhaps most important, because the population that survives and evolves after the crisis would seem to arise suddenly in the fossil record, Eldredge and Gould's theory provides an explanation of the gaps that exist in the record. According to the supporters of punctuated equilibrium, more gradual processes are of minor importance in speciation, and serve mostly to fine-tune a species to its environment.

The debate between punctuated equilibrium and the more conventional gradualistic perspective is fueled to a great extent by the many differences in the way the fossil record is interpreted. One difference arises because there is variation in rates of morphological change. Current evidence suggests that in some species certain morphological features may change rapidly in response to the selection pressures upon them, while others may remain relatively constant. A researcher who examines or is only able to discern the characters that remain constant may overlook this so-called ***mosaic evolution*** and assume that the species exhibits stasis—no change at all. On the other hand, supporters of gradualism often asssume that slow morphological or physiological change is occurring even when it is not obvious from the record; of course, neither physiological nor morphological changes in most soft body parts are preserved in the fossil record.

Another difference in interpretation concerns the great discontinuities in the record. The conventional view of these gaps is that they are, in most cases, just anomalies, and that the missing pieces in the picture of gradual evolutionary change might, if found, provide valuable information about transitions between the species preceding and following the gaps. The supporter of punctuated equilibrium, by contrast, regards the gaps as the norm, and as proof of the "instantaneous" speciation events the hypothesis postulates. But gaps in the record may frequently result from the kinds of environmental crises we have already discussed. Consider what happens when a lake, whose sediments have provided an excellent record of its organic inhabitants for hundreds of thousands of years, suddenly dries up as a result of a major climatic change. The formation of the fossil record in this location is terminated just at the time when the inhabitants of the lake are faced with extraordinary selection pressures and potential extinction.

With these reservations in mind, let us return to the hypothesis of punctuated equilibrium. The hypothesis holds that besides crises, punctuational change could be generated by major genetic alterations—polyploidy, hybridization, translocations, and deletions—particularly if they affect developmental-control genes, which can change the expression of entire constellations of other genes. One widely cited example is the idea that vertebrates

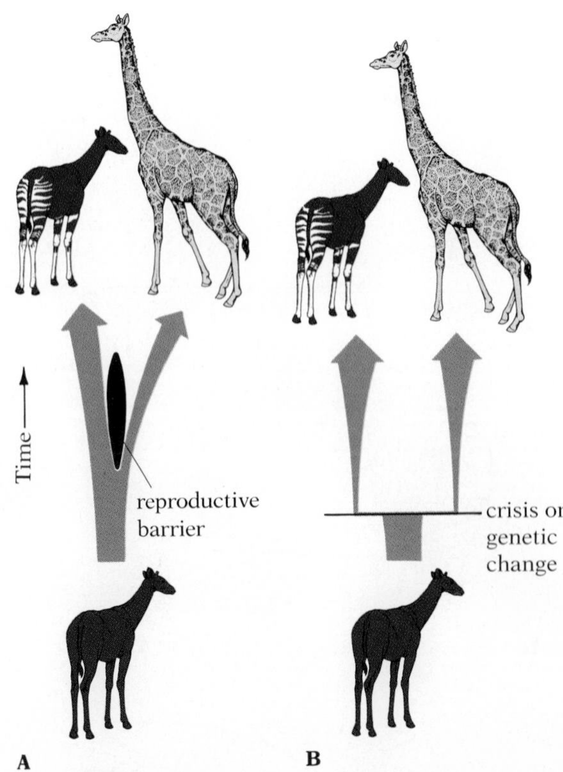

18.20 Gradualism versus punctuated equilibrium These drawings represent two hypotheses of how different species—here, the okapi and the giraffe—might have evolved from a common ancestor, the pre-okapi. Forms intermediate between the pre-okapi and the okapi and giraffe are not shown. (A) In the conventional interpretation of natural selection, a population is imagined to evolve gradually (as suggested by the gentle leftward movement of the lower part of the arrow, indicating a slow change in one trait) until a reproductive barrier (usually geographic isolation) separates two parts of the population. Generally one of the subgroups is very small, as indicated by the narrowness of the right-hand branch; it may diverge slowly from the larger subgroup through selection or genetic drift or both. (B) According to the punctuated-equilibrium model, there is little or no gradual change of a trait over time. Instead, a crisis or genetic event selects or gives rise by chance to one or more populations with traits very different from those of the ancestral population.

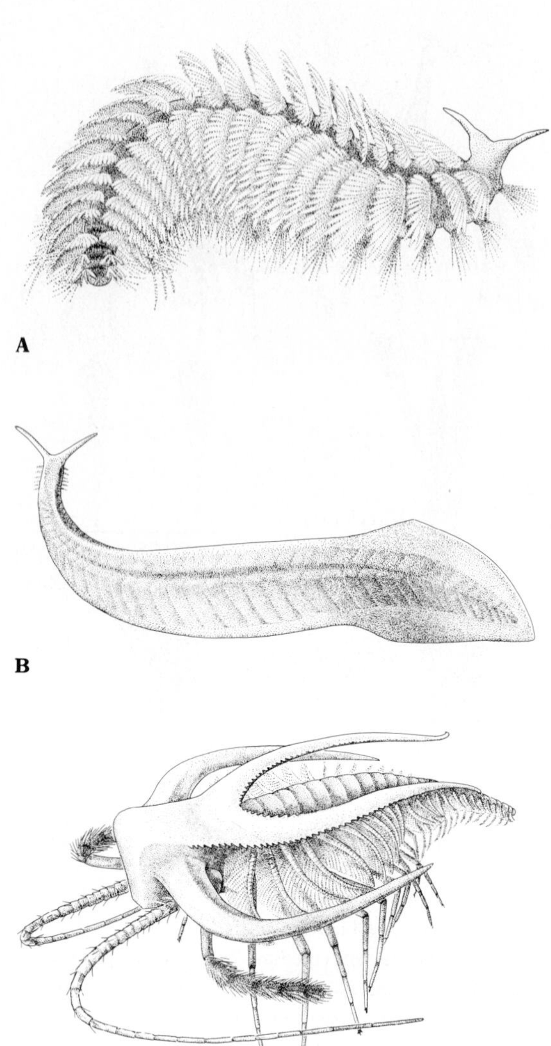

18.21 Some species from the Burgess shale The Burgess fauna includes annelids (segmented worms) like the polychaete worm shown here (A) and the oldest preserved chordate (B), whose notochord and somite-produced muscle bands are clearly visible. Among the many kinds of arthropods are curious versions of modern groups (not shown). The remaining species, of which only a small fraction are illustrated here (C–F), represent 20 or more extinct groups of arthropods.

evolved from the larval form of an aquatic invertebrate (p. 704). A change in developmental control could have blocked transformation into the adult morphology; this would have been followed by an elaboration of the larval phase and the evolution of the first fish. (Another possible example, involving the evolution of arthropods, is discussed in the next section.) Of course, genetic anomalies face the same problems as polyploids: with only siblings to mate with (at best), inbreeding and genetic drift become serious problems.

The Burgess fauna One of the most spectacular cases of rapid evolutionary radiation occurred in the geologic time division called the Cambrian period, about 500–570 million years ago. The "sudden" appearance of complex animals (older fossils are nearly all the remains of single-celled creatures) is a potentially compelling instance of punctuation. Two sites with well-preserved remains have been discovered, each about the size of a city block. The more extensively studied one is in the Burgess shale deposit in the mountains of western Canada; the other one is near Chengjiang in southern China. Because many of the fossils are similar, the animals probably had a world-wide distribution, and must have evolved over the course of a few million years. Until these discoveries, we had no clear picture of the kind of diversity that existed before selection "pruned" the tree of life.

The animals that typify a given place or, as in this case, a particular period are known collectively as a ***fauna***. The Burgess fauna contains a group of polychaetes (one kind of segmented worms, whose fascinating physiology will be examined in Part V) and the oldest known chordate (Fig. 18.21A), but mostly it consists of arthropods. These include some trilobites (an extinct group), a few uniramians (the group of invertebrates with mandibles —that is, chewing mouthparts—to which crustaceans and insects belong), and cheliceratans (the group that includes spiders, scorpions, and horseshoe crabs). However, none of these Burgess arthropods look like any modern species. In the same deposits there are at least 20 (and perhaps 30) other classes of arthropods, all now extinct (Fig. 18.21B–F).

The diversity of design is staggering, and poses two evolutionary problems: how did they come into being over such a relatively brief period, and why is it that most of these creatures—80–90 percent of the species represented in the fauna—have left no modern descendants, while other groups, with no obvious advantages, continue to thrive? We will deal with this second question in the next section. A potential answer to the first, favored by many biologists, is that the explosion in diversity was the natural consequence of a biological breakthrough: the invention of a novel developmental strategy. As we saw in Chapter 14, the embryonic development of animals involves the formation of a blastula, followed by gastrulation (and, in many species, neurulation). The breakthrough came with the appearance of compartmentalization: the embryo divides along the anterior-posterior axis into a number of nearly identical units, each of which can differentiate to produce specialized organs and appendages. It is likely that the evolution of the first multicellular creatures was followed quickly by the evolution of this more efficient multisegmental plan.

The remarkable diversity that followed would have been a consequence of the absence at that point of the highly redundant and hierarchical system of morphogens that now exist to assure precision and stability during development; as a result, variation would have been generated at an extraordinary rate, while competition in an environment with nothing occupying the many niches available to multicellular organisms must have been relatively lax initially. As the number of different species began to saturate the environment, however, natural selection would necessarily become more intense, and less efficient designs and less stable developmental programs probably went extinct. This scenario, therefore, postulates a kind of punctuation, followed by more gradual evolution and selective pruning. As we will soon see, there is another school of thought—an alternative to selection—to explain how many species may go extinct.

Whatever the fate of punctuation theory, the evolution of organisms can be less uniform and gradual than many biologists have supposed. Indeed, Darwin himself clearly pointed out that both local and global environmental phenomena must redefine and magnify selection pressures enormously and hence alter the direction and rate of evolution; contrary to the frequent misrepresentations of some modern writers, he rejected simple gradualism. The usual tempo of speciation probably lies somewhere between the gradual-change and the punctuated-equilibrium models.

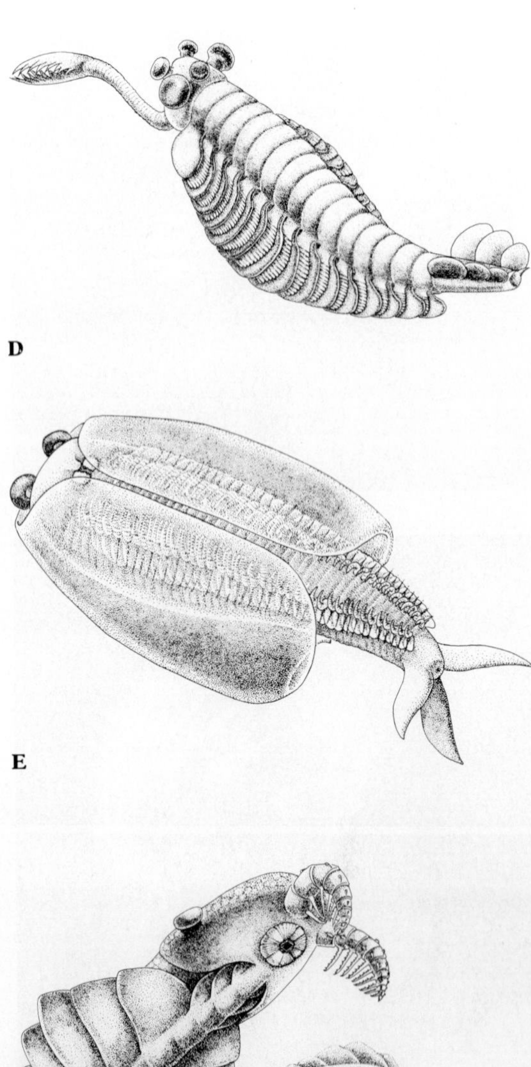

18.21 Some species from the Burgess shale (*continued*)

Chance and major patterns of evolution As we saw, in the Galápagos one species of finch underwent enormous evolutionary radiation into a variety of available lifestyles. The first group of birds to reach the islands might just as well have been some other species of finch, or some other sort of bird altogether. Chance events led to the creation of the new habitat and the immigration of the first birds. Later avian arrivals failed to thrive because the habitat was already being efficiently exploited by the (now) well-adapted first arrivals.

On a larger scale, there have been massive bouts of extinction, probably resulting from some chance worldwide calamity such as an asteroid impact or other catastrophic events. These random prunings of the tree of life have cleared away thousands of species at a time, leaving the survivors the opportunity to radiate and exploit novel lifestyles in the absence of competition. Perhaps this is what happened to most of the Burgess species. As we will see in Chapter 24, the enormous radiation of mammals after the Cretaceous-Tertiary catastrophe probably occurred only because this global disaster eliminated the dinosaurs, which had dominated the terrestrial large-animal opportunities. In some sense, therefore, the evolution of humans depended on this accident, and subsequent crises—warmings, ice ages, the continental movement that opened the Rift Valley in Africa (where the earliest human fossils are found), and so on—may also have been critical to the history of our species. Just as in day-to-day life, we may be where we are now because distant ancestors happened to be in the right place at the right time. According to the increasingly popular view that many large-scale evolutionary developments depend on chance, simple luck may often be more important than natural selection. Had the cards fallen differently, the argument goes, the dominant species of animals today might be something quite different.

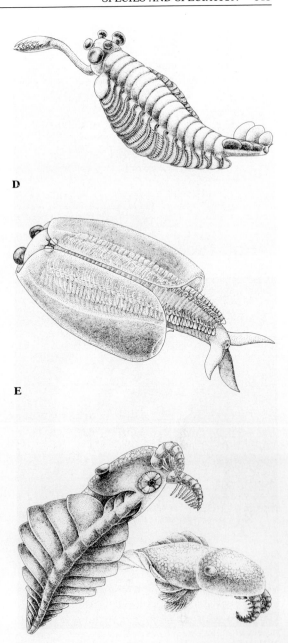

18.22 Flower and seeds of the dandelion, an asexual organism

Despite appearances to the contrary, the dandelion reproduces asexually at all times. The flower of the dandelion, which was inherited from a sexually reproducing precursor, now has infertile pollen and diploid eggs that go on to form seeds without fertilization.

THE SPECIES PROBLEM

We have been discussing speciation without worrying about any potential problems in identifying species as opposed to subspecies, populations in a cline, and so on. But it would be wrong to leave the impression that the modern definition of species can be applied without difficulty in all cases, or, indeed, that it is valid in all cases. The details of the definition itself are controversial and can provoke heated argument among biologists. Though most of them accept the major ideas on which the definition rests, and though most of them, if they were to study the same set of natural populations, would probably agree in the great majority of instances about which populations represent full species and which do not, there would be a small percentage of populations on which they could not agree. These are the cases where the modern definition of species is hard to apply or invalid. Let us examine a few such cases.

Asexual organisms Since the modern definition of species assumes interbreeding, it obviously does not apply to asexual organisms. Though most so-called asexual organisms actually do have provision for occasional sexual recombination, a few lack sexual mechanisms altogether. Dandelions, for instance, despite their flowers, are totally asexual (Fig. 18.22). Can such truly asexual organisms be said to form species in any sense? And would such species be comparable to sexual ones?

Asexual organisms do seem to form recognizable groups or kinds, even though the members of a group cannot exchange genes. Gaps, or discontinuities in the variation, occur between the various kinds just as they do between sexual species. One possible explanation for the groupings is that each group of asexual organisms evolved from a sexual species. The flowers of dandelions, which now have infertile pollen and diploid eggs that go on to form seeds without fertilization, originated in sexually reproducing ancestors. Since not all the variations that occur over the course of time are likely to be equally well adapted, asexual organisms like dandelions would continue to form recognizable groups: only those individuals whose genotypes produced well-adapted phenotypes would survive in significant numbers; hence there would be a limited number of superiorly adapted "types," and all individuals falling within the bounds of one such type would constitute a natural group that could be called a species, while all individuals falling within the bounds of another adaptive type would constitute a second species. The asexual species thus determined would resemble sexual species in that the latter, too, represent adaptive peaks. Sexual and asexual species therefore play comparable ecological roles, and both would be subject to natural selection. In the long run, however, the inability of asexual organisms to exploit the potential of genetic recombination to meet the challenges of changing conditions makes them less able to survive the vicissitudes of environmental change. And, indeed, nearly all multicellular asexual species are of recent evolutionary origin.

Fossil species The modern definition of species can be applied formally only to organisms that coexist, since the criterion of interbreeding cannot be used when comparing an organism with its likely ancestors of a million years earlier. Therefore, paleontologists can compare organisms from different periods in the earth's history only by using morphological criteria and geographic distribution. Two forms can be classified as separate species when they differ to about the same degree as related organisms from the present day that are known to constitute reproductively isolated species. For practical purposes, paleontologists usually regard gaps in the fossil record as breaks between species, even though they are fully aware that no gaps actually occurred in the lineages of the organisms.

Populations at an intermediate stage of divergence Our model of allopatric speciation assumes that geographically isolated populations will slowly diverge by essentially imperceptible stages until they have reached the level of full species. The intrinsic reproductive isolation that makes them full species itself evolves gradually. Hence there is no precise point at which the diverging populations suddenly reach the level of full species. There will be a period in the history of any two diverging lineages when the populations are in a hazy intermediate state between obviously belonging to the same species and obviously belonging to two separate species. But our definition of species makes no provision for such intermediate stages. Consequently intermediate stages, when they are encountered, must always pose a problem to any biologist intent on rigid categorization of what is in nature a fluid system. But the existence of intermediate stages does not invalidate the concept of speciation, because that very concept, in its modern form, predicts them.

Allopatric species One of the most obvious and frequently encountered problems in applying the modern definition of species arises when two populations are closely related and completely allopatric. Since they are allopatric, they are obviously not exchanging genes. But there must be neither actual nor *potential* effective gene flow if the two populations are to be regarded as separate species. How can potential gene flow be determined? One way that immediately comes to mind is to release a large sample of individuals from one population in the range of the other and then see whether free interbreeding takes place and, if it does, whether the hybrids are as viable as the parents. The wholesale introduction of foreign plants and animals is seldom desirable, however. In fact, in many cases it is illegal. An alternative would be to bring individuals from the two allopatric populations together in the laboratory and see if they will interbreed. Sometimes this procedure is useful. If the individuals will breed freely with other members of their own population but not with members of the other population, then we may reasonably conclude that the two populations are intrinsically isolated and should be considered separate species.

But what if interbreeding occurs freely between members of different populations in the laboratory? Are the two populations then to be regarded as belonging to the same species? No; the interbreeding simply demonstrates that certain types of intrinsic isolation do not exist between the populations. It says nothing about other types of intrinsic isolation. For example, under natural conditions ecogeographic or habitat isolation may exist, but these might very well be inoperative under laboratory conditions. Or behavioral isolation may be operative in nature but not in the laboratory.

Many species of animals that will have nothing to do with each other in the wild, because of important differences in their behavior patterns, will mate in the laboratory, where their normal behavior patterns break down. Lions and tigers, for example, are distinct allopatric species in the wild, and never mate in their small region of range overlap; in the unnatural environment of a zoo, however, they will mate, and produce living offspring. Clearly, when members of two different allopatric populations mate in the laboratory and produce viable offspring, the question of whether they belong to the same or to different species remains unanswered. The same ambiguity exists for the many organisms that will not breed at all under laboratory conditions: after all, males and females of the *same* species often refuse to mate outside of their natural habitats.

In many cases, then, there is no good test for determining whether two allopatric populations belong to the same or to different species. Although character displacement among sympatric species adds a level of complication, the usual practice in such cases is to determine the extent to which the two populations differ, and then to compare this degree of difference with that seen in related sympatric species. If the differences between the allopatric populations are of the same order of magnitude as (or greater than) those that distinguish sympatric species, the allopatric populations are considered fully separate species. If the differences are less than those that usually distinguish sympatric species, the two allopatric populations are likely to be regarded as belonging to the same species.

THE CONCEPT OF PHYLOGENY

Evolution implies that many unlike species have a common ancestor and that all forms of life probably stem from the same remote beginnings. Hence one of the tasks it sets for biologists is to discover the relationships among the species alive today and to trace the ancestors from which they descended (Fig. 18.23).

DETERMINING PHYLOGENETIC RELATIONSHIPS

When systematists,[3] also known as taxonomists, set out to reconstruct the evolutionary history—the ***phylogeny***—of a group of species that they think are related, they have before them the species living today and the fossil record. To reconstruct phylogenetic history as closely as possible, they must make inferences based on observational and experimental data. The difficulty is that what can be measured is *similarity*, whereas the goal is to determine *relatedness*. There are four major approaches to systematics—classical evolutionary taxonomy, phenetics, cladistics, and molecular taxonomy. Each uses different techniques to infer relatedness from similarity.

Classical evolutionary taxonomy Classical systematics depends more than any other approach on experience and subjective judgment. The usual procedure in reconstructing phylogenies by the classical method is to ex-

[3] Systematics, or taxonomy, in the words of G. G. Simpson, is "the scientific study of the kinds and diversity of organisms and of any and all relationships among them."

Electronic Review Guide

Essence of Biology for Macintosh® and IBM® with Windows®*

A powerful, interactive electronic review and study tool. Special features of *Essence of Biology* include:

* **1700 REVIEW "CARDS,"** carefully referenced to the text, give illustrated synopses of the most important points of each section.

* **A THOROUGH GLOSSARY** of all key terms is accessible from *anywhere* in the program.

* **INTERACTIVE QUESTIONS** for each chapter provide feedback for both correct and incorrect multiple-choice responses, helping you understand not just *which* answer is right, but *why* it is right.

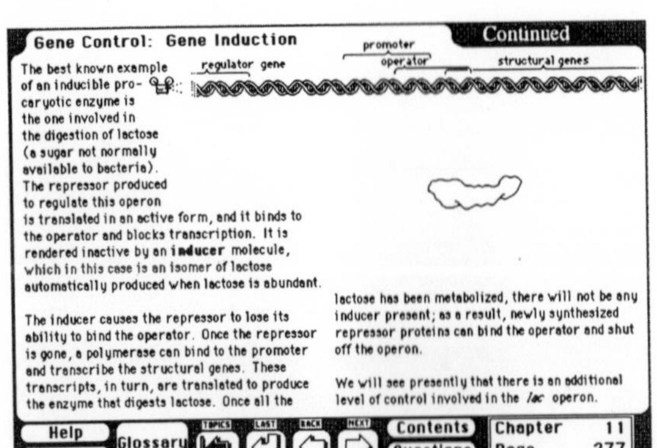

* **28 ONSCREEN ANIMATIONS** bring to life various complex biological processes, from the Jacob-Monod model of gene induction to the interplay of factors governing speciation.

* Note: The Macintosh version requires either HyperCard or Player and at least 1MB of memory; all Macintosh computers are shipped with one of these two programs. The IBM PC and compatible version requires Windows, a hard drive, a 3.5" high-density drive, a VGA- or super VGA-compatible screen, and 2MB of memory (4MB recommended).

* **SPECIAL PACKAGE**
Essence of Biology is available for only $15 with an original coupon (no copies accepted). Even without a coupon, the program is only $49.95. Call customer service at (800) 233-4830 with any questions.

Yes, please send me *Essence of Biology:*

Choose one: ☐ Windows® (IBM® PC 3.5") ☐ Macintosh®

Choose one: ☐ One-volume $15 ☐ Volume 1 $8 ☐ Volume 2 $8.

Add $1.50 for shipping (residents of CA, NY, PA, IL must add their local sales tax)

☐ My check is enclosed or

Please charge to my: ☐ Visa ☐ MasterCard ☐ American Express

Acct. # ____________ Exp. date ______ Telephone ______

Name ____________ Signature ____________

Ship to: Address ____________

City ____________ State/Prov. _____ Zip/PC _____

Billing Address (if different): Address ____________

City ____________ State/Prov. _____ Zip/PC _____

In the United States and Canada, simply return the attached coupon. Elsewhere, please write to the office that serves your area. Prices subject to change.

United Kingdom, Eire, Europe, the Middle East, Africa
W. W. Norton & Company Ltd.
10 Coptic Street
London WC1A 1PU

Australia
Jacaranda Wiley Ltd.
33 Park Road
Milton, Queensland 4064
Australia

New Zealand
Longman Paul Ltd.
Private Bag 102902
North Shore Mail Centre
Glenfield, Auckland 10
New Zealand

Japan
MK International Ltd.
1-50-7-203 Itabashi
Itabashi-ku
Tokyo 173
Japan

Taiwan & Korea
Bookman Books Ltd.
2F-5, #88 Hsin Sheng South Road
Sec. 3, Taipei
Taiwan

Hong Kong
Transglobal Publishers Service Ltd.
27/F Unit E Shield Industrial Centre
84/92 Chai Wan Kok Street
Tsuen Wan, N.T.
Hong Kong

Mexico, South & Central America, the Caribbean
EDIREP
5500 Ridge Oak Drive
Austin, Texas 78731
U.S.A

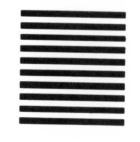

NO POSTAGE NECESSARY IF MAILED IN THE UNITED STATES

BUSINESS REPLY MAIL

FIRST CLASS PERMIT NO. 4008 NEW YORK, N.Y.

POSTAGE WILL BE PAID BY ADDRESSEE

W. W. NORTON & COMPANY, INC.
500 FIFTH AVENUE
NEW YORK, NEW YORK 10109-0145

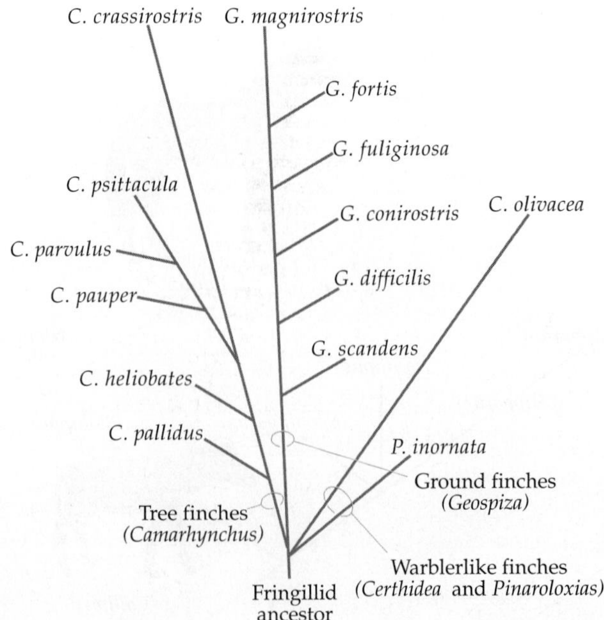

18.23 A phylogenetic tree for Darwin's finches This tree is based on both the degree and the nature of the morphological differences between species. The distances along branches from any one species to another reflect inferred degrees of relatedness.

amine as many independent characters of the species in question as possible, and to determine in which characters the species differ and in which they are alike. The assumption is that the differences and resemblances will reflect, at least in part, their true phylogenetic relationships. Ordinarily, as many different types of characters as possible are used in the hope that misleading data from any single character will be detected by a lack of agreement with the data from other characters.

The most easily studied and widely used characters pertain to morphology—including external morphology, internal anatomy and histology (tissue types), and the morphology of the chromosomes in cell nuclei. It is particularly helpful, of course, when morphological characters of living species can be compared with those of fossil forms. The characters chosen for comparison must be ones that vary within the group being analyzed; especially useful are characters that are unique to the group, and so have a common and relatively recent evolutionary origin. Among Darwin's finches (Fig. 18.15), for example, morphological characters that could also be checked in fossils include the beak depth, the ratio of beak depth to beak width, the angle of the beak relative to the head, the linear dimensions of the various bones that fuse to create the skull, the relative areas of the skull bones, the pattern of muscle insertions (visible as small indentations in bone) in the vocal apparatus, and so on. Characters that are usually lost in fossils but can be measured in live finches include tail length, markings of the young (degree of streaking, for instance), the changes in markings with age, and the like. Of these, the wide range of beak sizes, the depressed angle of the beaks, and the relatively short tails set Darwin's finches apart from their cousins on the South American mainland.

Characters preserved in fossils are of special importance because the fos-

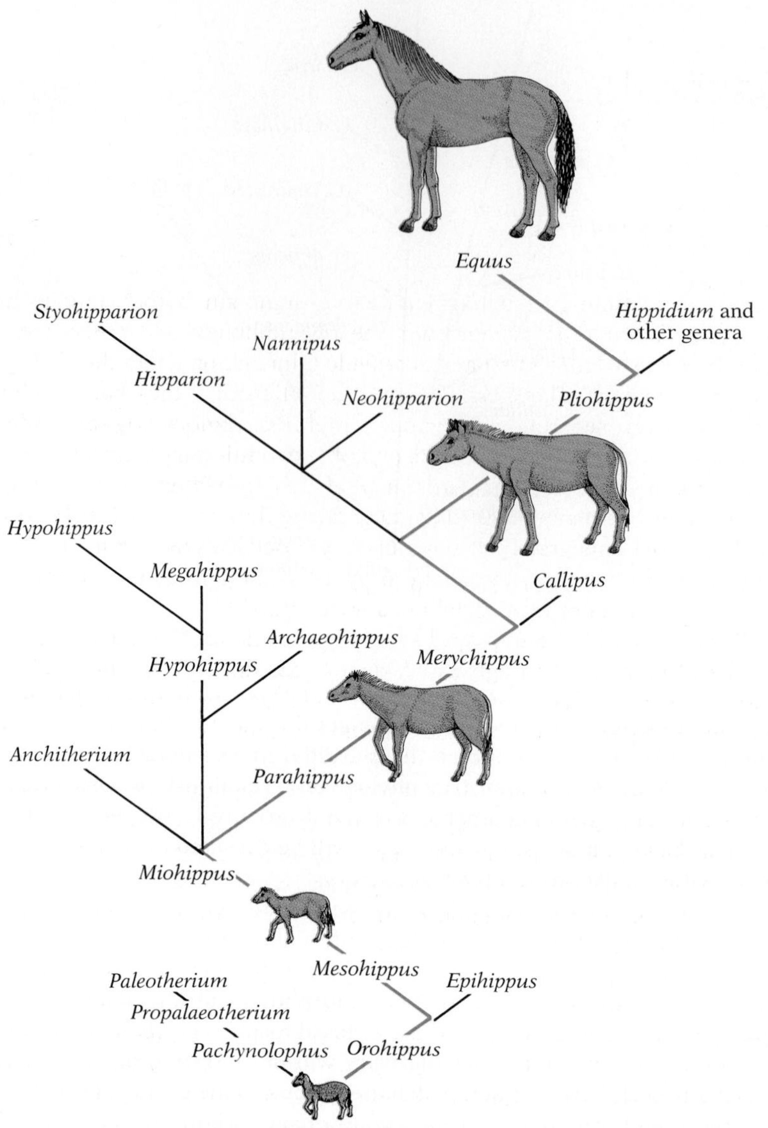

18.24 Presumed evolution of horses
The fairly complete fossil record of horses has enabled paleontologists to work out a reasonable picture of the evolutionary history of the group. The emphasis here is on the direct ancestors of the modern horse; many major branches left no modern descendants. *Hyracotherium* lived in the Eocene epoch about 55 million years ago. It was a small animal, weighing only a few kilograms. It had four toes on each front foot and three on each rear foot. It was a browser, feeding on trees and bushes. *Mesohippus*, which lived during the Oligocene epoch about 35 million years ago, was a bit larger, and its front feet, like the rear feet, had only three toes. *Merychippus*, a grazer, lived during the Miocene about 25 million years ago. It had three toes on each foot, the middle one much larger than the others, which were short and thin and did not reach the ground. *Pliohippus*, of the Pliocene, often had only one toe on each foot, though in some individuals tiny remnants of other toes persisted. *Equus*, the modern horse, is much larger than the ancestors shown here. It has only one toe on each foot.

sil record is the most direct source of evidence about the stages through which ancient forms of life passed. Unfortunately, that record is usually incomplete, and for many groups of organisms there is no suitable fossil record at all. At best, fossils may suggest the broad outlines of the evolution of major groups. In some groups, notably the horses, the fossil record has provided much phylogenetic information that could have been obtained from no other source (Fig. 18.24).

Another frequently used source of information is embryology. Morphological characters are often easier to interpret if the manner in which they develop is known. For example, if a particular structure in organism A and a

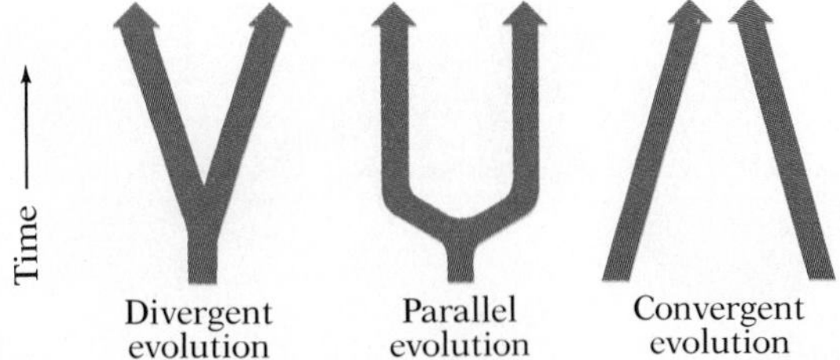

18.25 Patterns of evolution
In divergent evolution one stock splits into two, which become less and less like each other as time passes. In parallel evolution two related species evolve in much the same way for a long period of time, probably in response to similar environmental selection pressures. Convergent evolution occurs when two groups that are not closely related come to resemble each other more and more as time passes; this is usually the result of living in similar habitats and adopting similar lifestyles; as a result they experience similar selection pressures.

structure of quite different appearance in organism B both develop from the same embryonic primordium, the resemblances and differences between those structures in A and B provide information about the phylogeny of the organisms. The message would be different if they had developed from entirely separate embryonic structures. Embryological evidence often allows biologists to trace the probable evolutionary changes that have occurred in important structures, and helps them reconstruct the probable chain of evolutionary events that led to the modern forms of life. For example, the brief appearance of pharyngeal gill pouches during the early embryology of mammals, including humans, is thought to indicate that the distant ancestors of land vertebrates were aquatic.

Life histories have also played an important role in classical phylogenetic studies. The stages through which plants pass during their life cycles are particularly important sources of information, as we will see when we examine the algae and the vascular plants, for example.

The problem of convergence The classical approach, then, is based on evaluating similarities in a range of characters. But similarity, by itself, does not necessarily indicate common evolutionary descent. A particular similarity might simply reflect similar adaptation to the same environmental situation. The latter phenomenon is common in nature and is a serious source of confusion in phylogenetic studies.

When organisms that are not closely related become more similar in one or more characters because of independent adaptation to similar environmental situations, they are said to have undergone convergent evolution and the phenomenon is called ***convergence*** (Fig. 18.25). Whales, which are mammals descended from terrestrial ancestors, have evolved flippers from the legs of their ancestors; those flippers superficially resemble the fins of fish, but the resemblances result from convergence and they do not indicate a close relationship between whales and fish. Both arthropods and terrestrial vertebrates have evolved jointed legs and hinged jaws, but these similarities do not indicate that arthropods and vertebrates have evolved from a common ancestor that also had jointed legs and hinged jaws; these two groups of animals evolved their legs and jaws independently from a legless ancestor. The "moles" of Australia are not truly moles but marsupials (mammals whose young are born at an early stage of embryonic development and complete their development in a pouch in their mother's abdomen, rather than in a placenta inside the womb); they occupy the same habitat in Australia as do the true moles in other parts of the world and have, as a result, convergently evolved many startling similarities to the true

A

B

C

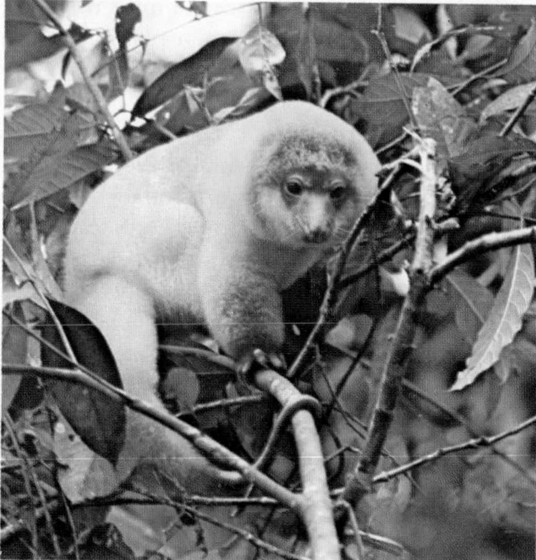

D

18.26 Marsupials that are convergent with placental mammals of other continents
(A) A marsupial mouse. (B) A marsupial glider, convergent with placental flying squirrels. (C) The tiger cat, a marsupial carnivore. (D) A cuscus, a marsupial monkey with a prehensile tail.

moles. The marsupial mole is but one of a vast array of Australian marsupials that are strikingly convergent with placental mammals of other continents (Fig. 18.26).

The preceding discussion makes it evident that when classical systematists find similarities between two species, they must try to determine whether the similarities are probably ***homologous*** (inherited from a common ancestor) or merely ***analogous*** (similar in function and often in superficial structure, but of different evolutionary origins). Thus the wings of robins and those of bluebirds are considered homologous, since the evidence indicates that both were derived from the wings of a common avian ancestor. But the wings of robins and those of butterflies are only analogous because, though they are functionally similar structures, they were not inherited from a common ancestor; they evolved independently and from different ancestral structures.

It is always important to indicate in what sense two structures are considered homologous or analogous. For example, the wings of birds and of bats are not homologous as wings, for they evolved independently, but they contain homologous bones: both types of wings evolved from the forelimbs of ancient terrestrial vertebrates that were ancestors to both birds and mammals. In short, the wings of birds and bats are analogous as wings and homologous as forelimbs. Similarly, the flippers of whales and seals evolved independently of each other, but both evolved from the front legs of land-mammal ancestors. Hence the flippers are homologous in the sense that both are forelimbs, with the same basic bone structure as that of other vertebrate forelimbs, but the modifications that make them flippers are analogous, not homologous.

Phenetics As we have seen, classical taxonomists must utilize personal judgment in deciding which characters should be considered and how they should be weighted. To be sure, intuition does play an important role in the scientific process, as was pointed out in Chapter 1, but the degree of subjectivity evident in classical taxonomy has motivated many systematists to attempt to develop more objective methods. One, less popular now than two decades ago, is phenetics. This approach to taxonomy uses as many morphological characters as possible, weights all characters equally, and ignores the issue of analogy versus homology. Since evidence from fossils, embryology, and behavior is difficult to quantify, it is generally not included. The expectation is that if enough characters are compared, the subjective judgments necessary for making relative weightings and identifying cases of analogy will be unnecessary; any errors will be canceled out, or will be swamped by the mass of other data.

In phenetics (as well as in the two other techniques to be described presently), each species' characters are compared to those of each other species, generating a value for the degree of difference for each pairing. A branch diagram is then constructed that places each species at a distance from each other species that corresponds to the calculated difference values (Fig. 18.27, top). The analysis is complicated by two problems: First, such a tree is "unrooted"—that is, there is no indication of which species are most like the ancestral form, or the order of branching points in the evolutionary tree. To determine each species' distance from the common

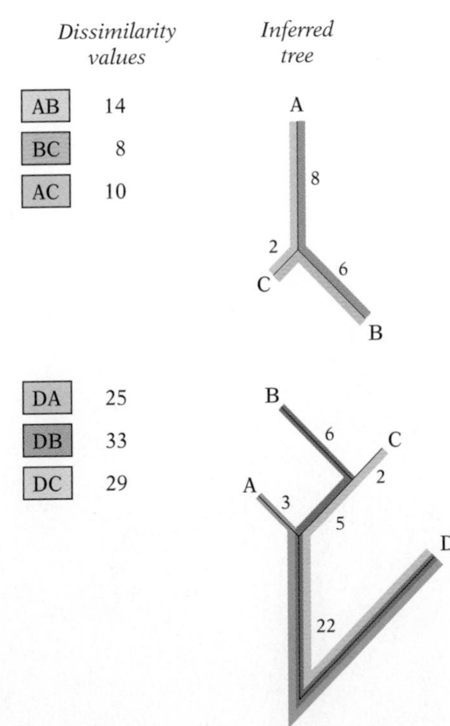

18.27 Construction of evolutionary trees from data on similarity
Species are compared in pairs and dissimilarity values are obtained (left). In the simplest case (top) these values are used to construct a branching diagram (right). In this example, since AB is four units greater than AC, the AB side of the BC branch must be four units longer than the AC side, which yields this unrooted tree. To locate the point of common ancestry of this group, an outgroup must be included (bottom). From the differences in path lengths to each species from the outgroup, the deepest branch among the species being analyzed can be determined; in this case, it falls between species A and the branch point between B and C. (The branch point between the ABC group and the outgroup cannot be calculated on the basis of these data; it is placed here arbitrarily as a reminder that the outgroup and the species being analyzed also have a common ancestor.) As the text points out, the measurements usually are slightly inconsistent; you can see how this complicates matters by attempting to recompute this tree after substituting, for instance, a value of 13 for AB.

ancestor, a suitable outgroup—a species not closely related to those being analyzed—must be added to the calculations to orient the tree correctly (Fig. 18.27, bottom). Second, since errors, uncertainties, and chance effects enter into the measurements and the evolutionary changes they attempt to quantify, the various branch lengths are never entirely consistent from pair to pair. Sometimes these inevitable anomalies are so small that they create no problem: only one branching diagram is consistent with the data, and discordant distances inferred from various pairings can simply be averaged. But when many species are included and branch points are close together, ambiguities are possible.

In theory, the branch lengths reflect differences that have accumulated as a result of drift and selection. If only drift were at work, the total distance from the deepest branch point to each present-day species would be the same, reflecting the time that has passed since the last common ancestor. Deviations—that is, longer paths—imply unusual degrees of selective pressure, and suggest something about the tempo and mode of evolution in the group.

Phenetics (often called numerical taxonomy) encounters serious problems when, as frequently happens, its two assumptions—that all characters are equally useful in determining phylogeny, and that little convergent evolution has taken place—are invalid. For example, phenetics, strictly applied, would classify the true mole and the marsupial mole as close relatives, a conclusion pheneticists themselves recognize cannot be correct. As we will see, molecular approaches use phenetic mathematical techniques with less risk of error.

Cladistics Cladists seek to avoid confusing analogy with homology by focusing on what they call ***shared derived characters***—traits that are common to the several species in question and that are of relatively recent rather than ancient origin.[4] Hence, in classifying two species of bats, the traits shared by mammals in general would not be considered; phenetics, on the other hand, would include any measurable trait. The cladistic approach is helpful in providing a rule for ignoring the absence (secondary loss) of traits—the hind limbs of aquatic mammals, for example—that causes many species to stand out as obvious exceptions to the general taxonomic patterns of their group. Cladists weight equally each of the traits they consider, but by ignoring features presumed to be shared at the point at which the speciation event in question occurred, they can be more selective than pheneticists: they have a criterion, or rule, for choosing the traits to be included. In the view of critics, however, deciding what traits are shared derived characters still calls for subjective judgment. And cladistic analysis, like phenetics, usually ignores fossils, embryology, and behavior. The result is sometimes controversial. For example, cladistic analysis places the crocodiles with the birds, and places mammals close to snakes and turtles (Fig. 18.28). Whether or not this view reflects the true phylogeny of higher verte-

[4] Cladistics takes its name from its term for a group of related organisms: the "clade." A clade is simply a monophyletic group—that is, a group with a single common ancestor. Thus, a clade can be a genus, a class, a kingdom (groups described at the end of this chapter), or all living things.

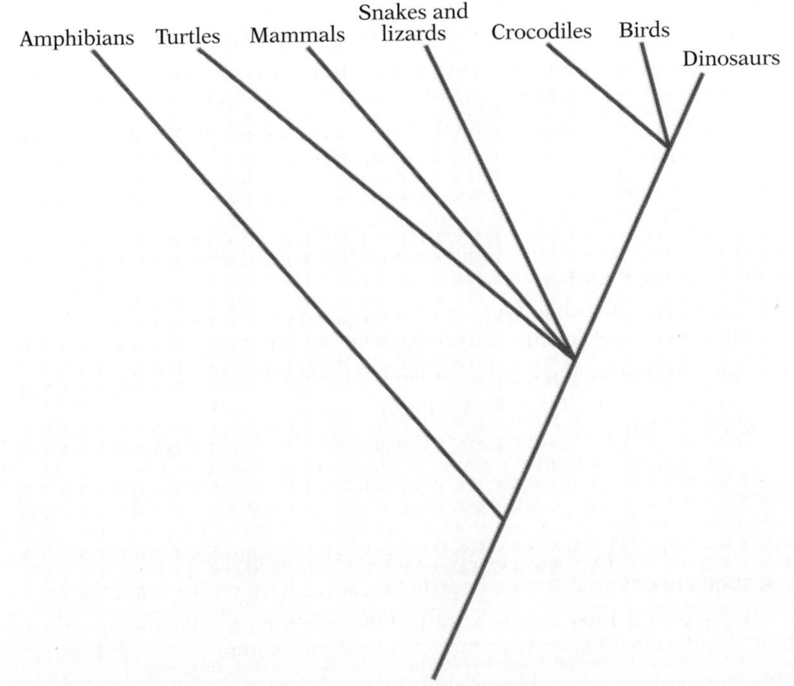

18.28 Cladistic analysis of the phylogeny of higher vertebrates
The major feature of this reconstruction is that reptiles do not belong to a single class. Depending on how a cladist chooses to group species, mammals, snakes, and turtles could form one class, with birds and crocodiles in the other; alternatively, there could be five different classes of reptiles, with birds and mammals in their own classes.

brates, there is widespread reluctance among biologists to accept this revision of the familiar picture of vertebrate evolution. And, like phenetics and classical taxonomy, cladistic analysis can be fooled by convergences, such as those exhibited by certain Australian marsupials. For analyzing the phylogeny of closely related genera or species, however, cladistics is now widely accepted. (Figure 18.23, for instance, is based on cladistic techniques.)

Molecular taxonomy The most popular new way to classify organisms is molecular taxonomy. This approach avoids the analogy/homology issue by focusing on the molecular level. The assumption is that convergent molecular evolution is very unlikely, and that analysis at this level is therefore probably accurate. Measurements can be made in several ways. One technique denatures the DNA of two species into single-stranded molecules, mixes them together, and allows them to form double-stranded hybrids. The greater the degree of hybridization, the more closely related the species are assumed to be. At least within groups no more diverse than the birds, this seems to work fairly well; rapidly evolving regions of repetitive DNA, however, can overwhelm the differences from drift and selection that this procedure seeks to measure, and some controversy exists over the best way to correct for this factor.

Another technique compares the amino acid sequences of proteins (see the *Exploring Further* box). The assumption here is that most amino acid substitutions are neutral—the exchange of one nonpolar peptide for another, for instance—and so differences should accumulate slowly with

Sequence of amino acids in cytochrome c for 28 organisms[a]

	Position	1 5 10 15 20 25 30 35 40 45 50
Mammals	Human, chimpanzee	G D V E K G K K I F I M K C S Q C H T V E K G G K H K T G P N L H G L F G R K T G Q A P G Y S Y T A
	Rhesus monkey	G D V E K G K K I F I M K C S Q C H T V E K G G K H K T G P N L H G L F G R K T G Q A P G Y S Y T A
	Horse	G D V E K G K K I F V Q K C A Q C H T V E K G G K H K T G P N L H G L F G R K T G Q A P G F T Y T D
	Donkey	G D V E K G K K I F V Q K C A Q C H T V E K G G K H K T G P N L H G L F G R K T G Q A P G F S Y T D
	Cow, pig, sheep	G D V E K G K K I F V Q K C A Q C H T V E K G G K H K T G P N L H G L F G R K T G Q A P G F S Y T D
	Dog	G D V E K G K K I F V Q K C A Q C H T V E K G G K H K T G P N L H G L F G R K T G Q A P G F S Y T D
	Rabbit	G D V E K G K K I F V Q K C A Q C H T V E K G G K H K T G P N L H G L F G R K T G Q A V G F S Y T D
	California gray whale	G D V E K G K K I F V Q K C A Q C H T V E K G G K H K T G P N L H G L F G R K T G Q A V G F S Y T D
	Great gray kangaroo	G D V E K G K K I F V Q K C A Q C H T V E K G G K H K T G P N L N G I F G R K T G Q A P G F T Y T D
Other vertebrates	Chicken, turkey	G D I E K G K K I F V Q K C S Q C H T V E K G G K H K T G P N L H G L F G R K T G Q A E G F S Y T D
	Pigeon	G D I E K G K K I F V Q K C S Q C H T V E K G G K H K T G P N L H G L F G R K T G Q A E G F S Y T D
	Pekin duck	G D V E K G K K I F V Q K C S Q C H T V E K G G K H K T G P N L H G L F G R K T G Q A E G F S Y T D
	Snapping turtle	G D V E K G K K I F V Q K C A Q C H T V E K G G K H K T G P N L N G L I G R K T G Q A E G F S Y T E
	Rattlesnake	G D V E K G K K I F T M K C S Q C H T V E K G G K H K T G P N L H G L F G R K T G Q A V G Y S Y T A
	Bullfrog	G D V E K G K K I F V Q K C A Q C H T C E K G G K H K V G P N L Y G L I G R K T G Q A A G F S Y T D
	Tuna	G D V A K G K K T F V Q K C A Q C H T V E N G G K H K V G P N L W G L F G R K T G Q A E G Y S Y T D
	Dogfish shark	G D V E K G K K V F V Q K C A Q C H T V E N G G K H K T G P N L S G L F G R K T G Q A Q G F S Y T D
Insects[b]	Tobacco hornworm moth	G N A D N G K K I F V Q R C A Q C H T V E A G G K H K V G P N L H G F F G R K T G Q A P G F S Y S N
	Fruit fly (*Drosophila*)	G D V E K G K K L F V Q R C A Q C H T V E A G G K H K V G P N L H G L I G R K T G Q A A G F A Y T N
Fungi[b]	Baker's yeast	G S A K K G A T L F K T R C E L C H T V E K G G P H K V G P N L H G I F G R H S G Q A Q G Y S Y T D
	Red bread mold	G D S K K G A N L F K T R C A E C H G E G G N L T Q K I G P A L H G L F G R K T G S V D G Y A Y T D
Plants[b]	Wheat	G N P D A G A K I F K T K C A Q C H T V D A G A G H K Q G P N L H G L F G R Q S G T T A G Y S Y S A
	Sunflower	G D P T T G A K I F K T K C A Q C H T V E K G A G H K Q G P N L N G L F G R Q S G T T A G Y S Y S A
	Castor bean	G D V K A G E K I F K T K C A Q C H T V E K G A G H K Q G P N L N G L F G R Q S G T T A G Y S Y S A
Number of different amino acids		1 3 5 5 4 1 3 3 4 1 4 3 2 1 3 3 1 1 2 3 3 4 2 3 4 2 1 4 1 1 2 1 5 1 3 2 1 1 3 2 1 3 3 6 1 2 3 1 2 4

[a] Adapted from M. O. Dayhoff, ed., *Atlas of Protein Sequence and Structure* (Washington, D.C.: National Biomedical Research Foundation, 1972), vol. 5; and R. E. Dickerson, The structure and history of an ancient protein, *Sci. Am.*, April 1972, copyright © 1972 by Scientific American, Inc.; all rights reserved.

[b] In cytochrome *c* from insects, fungi, and plants, a few (4–8) amino acids are usually ahead of what is here labeled Position 1; these are omitted from the table.

Exploring Further

NUCLEIC ACIDS AND PROTEINS AS TAXONOMIC CHARACTERS

A mutation that changes a single base in a gene may affect the gene product to varying degrees. At one extreme, it may not affect the gene product at all; if the change is in the third base of a codon, the new codon will often specify the same amino acid as the old one (see Table 9.1, p. 240). In most cases, however, the change may result in a codon for either a similar amino acid (one hydrophobic amino acid instead of another, for example) or a very different one. Very rarely, the change may create a termination codon, which will cause translation to end prematurely. Base changes that survive are of two major types: neutral mutations that do not significantly alter the activity of the gene product, and mutations that are fixed by selection because they improve the gene product, making it better suited to the needs of a particular group of organisms. In practice, distinguishing between neutral and selectively advantageous changes is not easy.

Mutations that do not alter the meaning of the codon are probably nearly neutral, and most of them probably accumulate randomly with time after two groups of organisms have diverged from a common ancestor. By comparing the extent of single-base changes of this type in the sequences of genes common to a wide range of species—genes for the enzymes of glycolysis, the citric-acid cycle, or the electron-transport chain, for example—we might get some measure of the time elapsed since two groups diverged. As a tool for dating evolutionary events, therefore, neutral single-base changes have great potential. At the moment, however, very few genes have been fully sequenced in a significant number of unrelated species.

Another approach is to compare the amino acid sequences of the gene products. Consider, for example, cytochrome *c*, an essential component of the respiratory chain in mitochondria. The complete amino acid sequence of this enzyme has been worked out for a variety of organisms, both plant and animal. The table above shows the sequence for some of the species so far examined, with the various functional groupings of amino acids (as determined by their R groups) indicated by a color code.

Perhaps the most obvious feature of this table is that cytochrome *c* is remarkably similar in all the species, even though some of them have probably

Sequence (positions 51–104; 55, 60, 65, 70, 75, 80, 85, 90, 95, 100, 104 marked)
A N K N K G I I W G E D T L M E Y L E N P K K Y I P G T K M I F V G I K K K E E R A D L I A Y L K K A T N E
A N K N K G I T W G E D T L M E Y L E N P K K Y I P G T K M I F V G I K K K E E R A D L I A Y L K K A T N E
A N K N K G I T W K E E T L M E Y L E N P K K Y I P G T K M I F A G I K K K T E R E D L I A Y L K K A T N E
A N K N K G I T W K E E T L M E Y L E N P K K Y I P G T K M I F A G I K K K T E R E D L I A Y L K K A T N E
A N K N K G I T W G E E T L M E Y L E N P K K Y I P G T K M I F A G I K K K G E R E D L I A Y L K K A T N E
A N K N K G I T W G E E T L M E Y L E N P K K Y I P G T K M I F A G I K K T G E R A D L I A Y L K K A T K E
A N K N K G I T W G E D T L M E Y L E N P K K Y I P G T K M I F A G I K K K D E R A D L I A Y L K K A T N E
A N K N K G I T W G E E T L M E Y L E N P K K Y I P G T K M I F A G I K K K G E R A D L I A Y L K K A T N E
A N K N K G I I W G E D T L M E Y L E N P K K Y I P G T K M I F A G I K K K G E R A D L I A Y L K K A T N E
A N K N K G I T W G E D T L M E Y L E N P K K Y I P G T K M I F A G I K K K S E R V D L I A Y L K D A T S K
A N K N K G I T W G E D T L M E Y L E N P K K Y I P G T K M I F A G I K K K A E R A D L I A Y L K Q A T A K
A N K N K G I T W G E D T L M E Y L E N P K K Y I P G T K M I F A G I K K K S E R A D L I A Y L K D A T A K
A N K N K G I T W G E E T L M E Y L E N P K K Y I P G T K M I F A G I K K K A E R A D L I A Y L K D A T S K
A N K N K G I I W G D D T L M E Y L E N P K K Y I P G T K M I F T G L S K K K E R T N L I A Y L K E K T A A
A N K N K G I T W G E D T L M E Y L E N P K K Y I P G T K M I F A G I K K K G E R Q D L I A Y L K S A C S K
A N K S K G I V W N N D T L M E Y L E N P K K Y I P G T K M I F A G I K K K G E R Q D L V A Y L K S A T S –
A N K S K G I T W Q Q E T L R I Y L E N P K K Y I P G T K M I F A G L K K K S E R Q D L I A Y L K K T A A S
A N K A K G I T W Q D D T L F E Y L E N P K K Y I P G T K M V F A G L K K A N E R A D L I A Y L K Q A T K –
A N K A K G I T W Q D D T L F E Y L E N P K K Y I P G T K M I F A G L K K P N E R G D L I A Y L K S A T K –
A N I K K N V L W D E N N M S E Y L T N P K K Y I P G T K M A F G G L K K E K D R N D L I T Y L K K A C E –
A N K Q K G I T W D E N T L F E Y L E N P K K Y I P G T K M A F G G L K K D K D R N D I I T F M K E A T A –
A N K N K A V E W E E N T L Y D Y L L N P K K Y I P G T K M V F P G L K K P Q D R A D L I A Y L K K A T S S
A N K N M A V I W E E N T L Y D Y L L N P K K Y I P G T K M V F P G L K K P Q E R A D L I A Y L K T S T A –
A N K N M A V Q W G E N T L Y D Y L L N P K K Y I P G T K M V F P G L K K P Q D R A D L I A Y L K E A T A –
1 1 2 5 2 3 2 6 1 6 4 3 2 2 5 3 1 1 3 1 1 1 1 1 1 1 1 1 1 1 3 1 5 1 2 2 1 6 9 2 1 7 2 2 2 2 2 2 1 6 4 3 5 4

Symbol	*Amino acid*
☐	NONPOLAR
G	glycine
A	Alanine
V	Valine
L	Leucine
I	Isoleucine
M	Methionine
F	Phenylalanine
W	Tryptophan
P	Proline
☐	POLAR
S	Serine
T	Threonine
C	Cysteine
Y	Tyrosine
N	Asparagine
Q	Glutamine
☐	ACIDIC
D	Aspartic acid
E	Glutamic acid
☐	BASIC
K	Lysine
R	Arginine
H	Histidine

G is printed in color, because glycine, despite its technically nonpolar R group, behaves like a polar amino acid.

not had a common ancestor for more than a billion years. For example, all the cytochromes have the same amino acid sequence from positions 70 through 80. In fact, cytochrome *c* is an evolutionarily conservative protein; its amino acid sequence has changed at a considerably slower average rate (about 20 million years for a 1 percent change) than, for example, the amino acid sequence of hemoglobin (5.8 million years) or that of fibrin (only one million years). The minimal change in cytochrome *c* suggests that only minor alterations can be tolerated if the enzyme is to continue functioning properly. Notice that even at points along the chain where there are differences, the amino acids are often functionally similar ones (with one polar amino acid substituted for another, one nonpolar amino acid for another, and so on). Some of the alterations may be neutral, having essentially no effect on the activity of the gene product, while others may represent minor but adaptive species-specific modifications of the protein. But still others almost certainly indicate major changes in the gene product. Among these are substitution of a polar for a nonpolar amino acid and alterations involving proline (which induces turns in polypeptide chains) or cysteine (which forms strong covalent bonds with other cysteines).

If we compare various species in the table with one another, we find that the number of differences in amino acids usually agrees reasonably well with the presumed evolutionary distances among the species. Thus the mammals differ less among themselves than any of them differ from the fish. Human beings and chimpanzees do not differ at all; both differ by one amino acid from the rhesus monkey, by an average of 10.4 amino acids from the other mammals, by an average of 14.5 from the reptiles, by 18 from the amphibian, and by an average of 22.5 from the fish. This is an accurate reflection of the generally accepted evolutionary sequence of fish ⟶ amphibian ⟶ reptile ⟶ mammal.

Because cytochrome *c* is evolutionarily so conservative, its value as a taxonomic character is limited to studies of the relationships among evolutionarily distant organisms. It cannot be used in comparing families or genera. On the other hand, more rapidly changing proteins, such as fibrin, may prove useful in cases of closely related species. In any comparison, reliable conclusions depend on using several different proteins; because of unusual selection pressures or an unusually large or small number of mutations in the gene, the rate of change in one particular protein may be too great or too small to be representative.

TABLE 18.2 *Classification hierarchies*

Biological	*Postal*
Kingdom	Country
Phylum/Division	State/Province
Class	City
Order	Street
Family	Number
Genus	Last name
Species	First name

time. Where there are strong divergent selection pressures (as, for instance, on the hemoglobin of mountain-dwelling or deep-diving aquatic mammals), differences are likely to be exaggerated far beyond what is expected from drift.

Finally, DNA sequences can be compared directly. This is very tedious, but sequencing of rRNA genes has revolutionized our understanding at the level of kingdoms and the next-lower classifications, called phyla; most of the branching diagrams that will be used to indicate the likely phylogenies of organisms in Chapters 19–25 are derived from this technique. Examination of changes in the third codon of any protein-encoding gene (most of which alterations are neutral because they code for the same peptide) permits detailed analysis within genera, and even within subspecies; the phylogeny of *Drosophila* is being worked out by this method. An early hope for sequence analysis was that the number of differences in whatever is being measured could be used as a molecular clock to date branch points precisely. It has become clear, however, that different genes within the same organism can change at very different rates; the same is true for the same gene in different species. Even dates inferred by averaging many different genes must at present be considered highly approximate.

PHYLOGENY AND CLASSIFICATION

Over a million species of animals and over 325,000 species of plants are known (and many more are yet to be discovered). To deal with this vast array of organic diversity, we clearly need a system by which species can be classified in a logical and meaningful manner. Many different kinds of classification are possible. We could, for example, classify flowering plants according to their color: all white-flowered species in one group, all red-flowered species in a second group, all yellow-flowered species in a third group, and so on. This sort of system has its usefulness, and indeed provided a general basis for classification before the goal of recognizing evolutionary affinities had become clear. But the information such categories convey—about flower color, for example—is of an incidental kind that fails to set apart fundamentally different organisms. The classification system used in biology today is strictly an attempt to encode the evolutionary history of the organisms; it is thus often a means of conveying information about many of their characteristics to those familiar with the various taxonomic groups (taxa) to which the organisms are assigned.

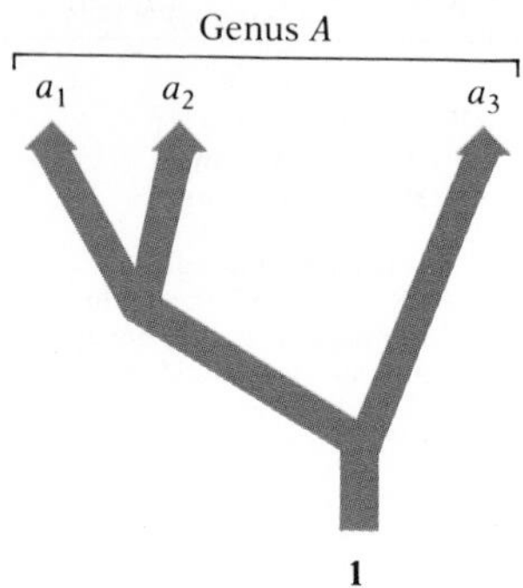

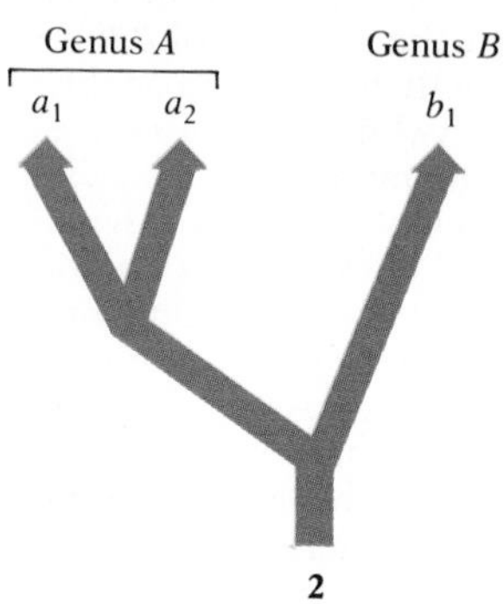

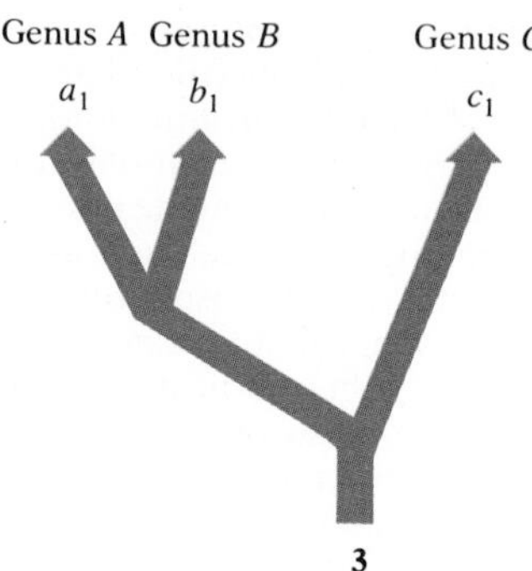

18.29 Alternative generic grouping for three related species Biologists try to group species in a way that will indicate their phylogenetic relationships. Thus a genus is a group of related species. But how closely related? There is no absolute answer to this question. Some biologists (the "lumpers") like large genera containing many subdivisions (subgenera or "species groups"); others (the "splitters") prefer small compact genera, containing only species that are very closely related. In the drawing, three alternative ways of grouping three related species are shown. The first recognizes only one genus, the second recognizes two, and the third recognizes three.

TABLE 18.3 *Classification of six species*

Category	*Haircap moss*	*Red oak*	*House fly*	*Herring gull*	*Wolf*	*Human*
Kingdom	Plantae	Plantae	Animalia	Animalia	Animalia	Animalia
Phylum or Division	Bryophyta	Tracheophyta	Arthropoda	Chordata	Chordata	Chordata
Class	Musci	Angiospermae	Insecta	Aves	Mammalia	Mammalia
Order	Bryales	Fagales	Diptera	Charadriiformes	Carnivora	Primata
Family	Polytrichaceae	Fagaceae	Muscidae	Laridae	Canidae	Hominidae
Genus[a]	*Polytrichum*	*Quercus*	*Musca*	*Larus*	*Canis*	*Homo*
Species[a]	*commune*	*rubra*	*domestica*	*argentatus*	*lupus*	*sapiens*

[a] The name of a particular species consists of the genus name and the species designation, both customarily in italics, as shown in this table.

The classification hierarchy Suppose you had to classify all the people on earth on the basis of where they live. You would probably begin by dividing the world population into groups based on country. This subdivision separates inhabitants of the United States from the inhabitants of France or Argentina, but it still leaves very large groups that must be further subdivided. Next, you would probably subdivide the population of the United States by states, then by counties, then by city or village or township, then by street, and finally by house number. You could do the same thing for Canada, Mexico, England, Australia, and all the other countries (using whatever political subdivisions in those countries correspond to states, cities, etc., in the United States). This procedure would enable you to place every individual in an orderly system of hierarchically arranged categories (Table 18.2). Note that each level in this hierarchy is contained within and is partly determined by all levels above it. Thus, once the country has been determined as the United States, a Mexican state or a Canadian province is excluded. Similarly, once the state has been determined as Pennsylvania, a city in New York or in California is excluded.

The same principles apply to the classification of living things on the basis of phylogenetic relationships. Again a hierarchy of categories is used (Table 18.2). Each category (taxon) in this hierarchy is a collective unit containing one or more groups from the next-lower level in the hierarchy. Thus a genus is a group of closely related species (Fig. 18.29); a family is a group of related genera; an order is a group of related families; a class is a group of related orders; and so on. The species in any one genus are believed to be more closely related to each other than to species in any other genus, the genera in any one family are believed to be more closely related to each other than to genera in any other family, and so on.

Table 18.3 gives the classification of six species. Notice that the table shows us immediately that the six species are not closely related, but that a human being and a wolf are more closely related to each other, both being mammals, than either is to a bird such as the herring gull. And it shows us that the mammals and the bird are more closely related to each other than to the house fly, which is in a different phylum, or to moss or red oaks, which are in a different kingdom. These relationships are correlated with similarities and differences in the morphology, physiology, and ecology of the six species.

The six species of Table 18.3 are all well known, and the relationships between them are probably intuitively clear to you. But many species are not so well known, and the relationships between them not so clear. Much research may be necessary before they can be fitted into the classification system with any degree of certainty, and their assignment to genus or family (or even order) may have to be changed as more is learned about them.

Hierarchical classification systems similar to the one in current use have been employed by naturalists for many centuries. The current system dates from the work of the great Swedish naturalist Carolus Linnaeus (1707–1778), who wrote extensively on the classification of both plants and animals. His system used kingdoms, classes, orders, genera, and species; the phylum and family categories were added later. The rationale on which Linnaeus based his system was necessarily very different from the phylogenetic one employed today. He worked a century before Darwin, and he had no conception of evolution, thinking of each species as an immutable entity, the product of a divine creation. He was simply grouping organisms according to similarities, primarily morphological. His results were similar to those obtained today because morphological characters, being products of evolution, tell us much about evolutionary relationships. The Linnaean system, however, produces results quite different from those of the modern phylogenetic system whenever it has to deal with convergence.

An outline of a modern classification of living things is given on pp. A1–A6, and will be explored in some detail in the next seven chapters.

Nomenclature The modern system of naming species also dates from Linnaeus. Before him, there had been little uniformity in the designation of species. Some species had a one-word name, others had two-word names, and still others had names consisting of long descriptive phrases. For example, the pre-Linnaean name for the common carnation was *dianthus floribus solitariis, squamis calycinis subovatis brevissimis, corollis crenatis,* and the name for the honey bee was *Apis pubescens, thorace subgriseo, abdomine fusco, pedibus posticis glabris utrinque margine ciliatis.* Linnaeus simplified things by giving each species a name consisting of two words: first, the name of the genus to which the species belongs and, second, a designation for that particular species. Thus the above-mentioned species of carnation became *Dianthus caryophyllus,* and the honey bee became *Apis mellifera.* Other species in the genus *Dianthus* have the same first word in their names, but each has its own specific designation (*Dianthus prolifer, Dianthus barbatus, Dianthus deltoides*). No two species can have the same name.[5] Notice that the names are always Latin (or Latinized) and that the

[5] More precisely, no two species of plants can have the same name, and no two species of animals can have the same name. Since the International Rules of Botanical Nomenclature and the International Code of Zoological Nomenclature are completely separate, it is possible for a plant and an animal to have the same name. There is also a separate International Bacteriological Code of Nomenclature.

genus name is capitalized while the species name is not.[6] Both names are customarily printed in italics (underlined if handwritten or typed). The correct name for any species, according to the present rules, is usually the oldest validly proposed name.[7]

The same Latin scientific names are used throughout the world. This uniformity of usage ensures that each scientist will know exactly which species another scientist is discussing. There would be no such assurance if common names were used, for not only does a given species have a different common name in each language, but it often has two or three names in a single language. For example, the plant *Bidens frondosa* is known by all the following English names: beggar-ticks, sticktight, bur marigold, devil's bootjack, pitchfork weed, and rayless marigold. To confuse matters further, a single common name is frequently applied to several species. For example, "gopher" is the name of a turtle in Florida and of a rodent in Kansas, and "raspberry" is the common name for more than a hundred species of plants.

STUDY QUESTIONS

1. One problem with molecular techniques is the issue of whether a given difference in base sequence between two species is neutral or adaptive. After all, there can be subtle differences created by the substitution of one nonpolar amino acid for another, and even third-codon changes can have an effect since they use different tRNAs, which may have different abundances, specificities, and activities. How might one go about attempting to resolve this issue for one specific gene product—an rRNA or a particular enzyme, for example? Feel free to imagine plausible advances in molecular techniques. (pp. 523–26)

2. The argument is sometimes made (almost always in the context of humans) that the concept of subspecies is useless because the variation within a subspecies for most characters exceeds the difference between the two subspecies. Is this line of reasoning correct? Does it work for species within a genus? Could it be used to prove that there are no "important" differences between males and females of a species? (pp. 493–94, 514–15)

3. How could you use one of the taxonomic techniques described in this chapter to determine which if any species in a genus were likely to have been produced through an evolutionary bottleneck? (pp. 510, 516–26)

4. For what sorts of species (including factors like size, generation time, reproductive strategy, habitat, and niche) is chance more likely to have been important than competition? What kinds of patterns would you expect to find through molecular-taxonomic analysis in such cases? (pp. 509–10, 523–26)

[6] This rule always holds for zoological names, but specific botanical names are sometimes capitalized when they are derived from the name of a person or from other proper nouns.

[7] For purposes of priority, botanical naming dates from the publication of Linnaeus' *Species Plantarum* in 1753, and zoological naming dates from the publication of the 10th edition of his *Systema Naturae* in 1758.

CONCEPTS FOR REVIEW

- Species
  - Deme versus species
  - Definition of species
  - Interspecific versus intraspecific variation
  - Subspecies
- Speciation
  - Founder effect, bottlenecks, and genetic drift
  - Sympatric versus allopatric speciation
  - Types of intrinsic isolation
  - Avenues of sympatric speciation
  - Character displacement
  - Competition versus chance
  - Punctuated equilibrium
- Phylogeny
  - Goal
  - Homology versus analogy
  - Advantages and limitation of techniques
    - classical taxonomy
    - phenetics
    - cladistics
    - molecular taxonomy

SUGGESTED READING

AYALA, F. J., 1978. The mechanisms of evolution, *Scientific American* 239 (3). (Offprint 1407) *On the large amount of hidden genetic variation in a species, and its consequences for speciation and taxonomy.*

CLARKE, B., 1975. The causes of biological diversity, *Scientific American* 233 (2). (Offprint 1326)

FUTUYMA, D. J., 1986. *Evolutionary Biology,* 2nd ed. Sinauer Associates, Sunderland, Mass.

GOULD, S. J., 1985. *The Flamingo's Smile: Reflections in Natural History.* W. W. Norton, New York. *A compellingly written collection of essays on evolutionary theory and other biological topics by the coauthor of the theory of punctuated equilibrium.*

GOULD, S. J., 1989. *Wonderful Life.* W. W. Norton, New York. *A punctuationist's view of the Burgess fauna.*

GRANT, P. R., 1991. Natural selection and Darwin's finches. *Scientific American* 265(4). *On the effects of a season of drought on the finch population in the Galápagos. A well-written and modern analysis of speciation.*

GRANT, V., 1985. *The Evolutionary Process.* Columbia University Press, New York. *Excellent treatment, with special emphasis on speciation.*

KIMURA, M., 1979. The neutral theory of molecular evolution, *Scientific American* 241 (5). *On the idea that most single-base mutations are neutral, and therefore provide an evolutionary "clock."*

LEVINTON, J. S., 1992. The big bang of animal evolution. *Scientific American* 267 (5). *On the apparent explosion of diversity at the beginning of the Cambrian.*

LI, W. H., and D. GRAUR, 1991. *Fundamentals of Molecular Evolution.* Sinauer, Sunderland, Mass. *Excellent treatment of how molecular techniques can be applied to phylogeny.*

SCHOENER, T. W., 1982. The controversy over interspecific competition, *American Scientist* 70, 586–95.

SIBLEY, C. G., and J. E. AHLQUIST, 1986. Reconstructing bird phylogeny by comparing DNAs, *Scientific American* 254 (2). *On the DNA-hybridization technique.*

SIMPSON, G. G., 1983. *Fossils and the History of Life.* Scientific American Books, New York. *Clearly written and well-illustrated exposition of evolution.*

STANLEY, S. M., 1987. *Extinction.* Scientific American Books, New York. *Another excellent summary in this series.*

STEBBINS, G. L., and F. J. AYALA, 1985. The evolution of Darwinism, *Scientific American* 253 (1). *A modern summary of evolutionary theory, with particular attention to the claims for punctuated equilibrium.*

WILSON, A. C., 1985. The molecular basis of evolution, *Scientific American* 253 (4). *On methods for tracing evolution through similarities in nucleotide or amino acid sequences.*

APPENDIX: A CLASSIFICATION OF LIVING THINGS

The classification given here is one of many in current use. Some other systems recognize more or fewer divisions and phyla, and combine or divide classes in a variety of other ways; but compared with the large areas of agreement, the differences between the various classifications are minor. Chapters 19–25 discuss certain of the points at issue between advocates of different systems.

Botanists have traditionally used the term "division" for the major groups that zoologists have called phyla. In classifications recognizing only two or three kingdoms, this difference in terminology causes little difficulty, because usage can be consistent within each kingdom. But when a kingdom Protista is recognized, as it is here, consistency is achieved only at the expense of violating well-established usage. The Protista contain some plantlike and funguslike groups traditionally called divisions and some animal-like groups traditionally called phyla. These usages we have respected.

Most classes within a division or phylum are listed here, but where there is only one class it is not named. For some classes—Insecta and Mammalia, for example—orders are given too. Except for a few extinct groups of particular evolutionary importance, like Placodermi, only groups with living representatives are included. A few of the better-known genera are mentioned as examples in each of the taxons.

Whenever possible, an estimate (a very rough one) of the number of living species is provided for higher taxons.

KINGDOM EUBACTERIA* (3,000)

DIVISION PURPLE BACTERIA

CLASS PURPLE SULFUR BACTERIA. *Pseudomonas, Escherichia*
CLASS RHODOPSEUDOMONAS. *Rhodopseudomonas, Rhizobium*
CLASS PURPLE NONSULFUR. *Sphaerotilus, Alcaligenes*
CLASS DESULFOVIBRIO. *Desulfovibrio*
CLASS RICKETTSIAE. *Rickettsia, Coxiella*
CLASS MYXOBACTERIA. *Myxococcus, Chondromycos*

DIVISION PROCHLOROPHYTA. *Chloroxybacteria, Prochloron*

DIVISION CYANOBACTERIA. *Oscillatoria, Nostoc, Gloeocapsa, Microcystis*

DIVISION GREEN SULFUR BACTERIA. *Chlorobium*

DIVISION SPIROCHETES. *Treponema, Spirochaeta*

DIVISION GRAM-POSITIVE BACTERIA

CLASS CLOSTRIDIA. *Bacillus, Clostridium, Mycoplasma, Staphylococcus, Streptococcus*
CLASS ACTINOMYCES. *Micrococcus, Actinomyces, Streptomyces*
CLASS MYCOPLASMA. *Mycoplasma, Acholeplasma*

KINGDOM ARCHAEBACTERIA (200)

DIVISION METHANOGENS. *Methanobacterium*

DIVISION HALOPHILES. *Halobacterium*

DIVISION THERMOACIDOPHILES. *Thermoplasma*

DIVISION SULFUR REDUCERS.

KINGDOM ARCHAEZOA (400)†

PHYLUM METAMONADA. *Giardia, Entermorias, Spironucleus*

PHYLUM MICROSPORIDIA. *Vairimorpha*

PHYLUM ARCHAEAMOEBA. *Pelomyxa*

* There is no generally accepted classification for bacteria at the higher taxon level. One recent classification divides the bacteria into 17 distinct divisions, some without formal names. Another important classification assigns them to 19 "parts," most without formal names. This one, admittedly incomplete, is based for the most part on similarities in the sequences of ribosomal RNAs, as described in G. E. Fox et al., *Science*, vol. 209, 1980 and M. L. Sogin et al., *Science*, vol. 243, 1989.

† There are many alternative classification systems for protists, algae, and plants. This one, based primarily on rRNAs, is derived largely from Sogin et al., and T. Cavalier-Smith, *Nature*, vol. 339, 1989.

KINGDOM PROTISTA†

SECTION PROTOPHYTA: Algal protists**

DIVISION EUGLENOIDEA. Euglenoids (800). *Euglena, Eutreptia, Phacus, Colacium*

DIVISION DINOFLAGELLATA. Dinoflagellates (1,000). *Gonyaulax, Gymnodinium, Ceratium, Gloeodinium*

SECTION PROTOMYCOTA: Fungal protists

PHYLUM SPOROZOA. Sporulation protozoans (3,600)

CLASS TELOSPOREA. *Monocystis, Gregarina, Eineria, Toxoplasma, Plasmodium*
CLASS PIROPLASMEA. *Babesia, Theileria*

PHYLUM MYXOZOA. Myxosporians (800). *Myxobolus, Myxidium, Ceratomyxa*

SECTION GYMNOMYCOTA: Slime molds

DIVISION PLASMODIOPHOROIDA. Plasmodiophores (or endoparasitic slime molds). *Plasmodiophora, Spongospora, Woronina*

DIVISION MYCETOZOA.

CLASS ACRASIOIDA. Cellular slime molds (26). *Dictyostelium, Polysphondylium*
CLASS MYXOIDA. True slime molds (400). *Physarum, Hemitrichia, Stemonitis*

SECTION PROTOZOA: Animal-like protists**

PHYLUM KINETOPLASTIDA (17,000)

Subphylum Mastigophora.

CLASS ZOOFLAGELLATA (or Zoomastigina). "Animal" flagellates. *Trypanosoma, Calonympha, Chilomonas, Trichonympha*
CLASS OPALINATA.‡ Opalinids. *Opalina, Zelleriella*

† There are many alternative classification systems for protists, algae, and plants. This one, based primarily on rRNAs, is derived largely from Sogin et al., and T. Cavalier-Smith, *Nature*, vol. 339, 1989.

** This traditional classification, though useful, has no phylogenetic validity.

‡ The opalinids are sometimes placed in the Ciliata, because they have cilia instead of flagella, but they lack the other diagnostic characters of Ciliata. It must be admitted, however, that they do not fit well in the Mastigophora either.

Subphylum Sarcodina.

CLASS RHIZOPODEA. Naked and shelled amoebae, foraminiferans. *Amoeba, Pelomyxa, Entamoeba, Arcella, Globigerina, Textularia*

CLASS ACTINOPODEA. Radiolarians, heliozoans, acantharians. *Aulacantha, Acanthometron, Actinosphaerium, Actinophrys*

PHYLUM CILIATA. Ciliates (6,000). *Paramecium, Stentor, Vorticella, Spirostomum*

KINGDOM CHROMISTA*

DIVISION PHAEOPHYTA*. Brown algae (1,500). *Sargassum, Ectocarpus, Fucus, Laminaria*

DIVISION CHRYSOPHYTA.

CLASS CHRYSOPHYCEAE. Golden-brown algae (650). *Chrysamoeba, Chromulina, Synura, Mallomonas*

Class Haptophyceae (or Prymnesiophyceae). Haptophytes and coccolithophores. *Isochrysis, Prymnesium, Phaeocystis, Coccolithus, Hymenomonas*

CLASS XANTHOPHYCEAE. Yellow-green algae (360). *Botrydiopsis, Halosphaera, Tribonema, Botrydium*

CLASS EUSTIGMATOPHYCEAE. Eustigmatophytes. *Pleurochloris, Visheria, Pseudocharaciopsis*

CLASS RAPHIDIOPHYCEAE. Chloromonads. *Gonyostomum*

CLASS BACILLARIOPHYCEAE. Diatoms (10,000). *Pinnularia, Arachnoidiscus, Triceratium, Pleurosigma*

DIVISION CRYPTOPHYTA. Cryptomonads. *Cryptomonas, Chroomonas, Chilomonas, Hemiselmis*

DIVISION LABYRINTHULEA. Net slime molds. *Labyrinthula*

DIVISION OOMYCOTA. Water molds, white rusts, downy mildews (400). *Saprolegnia, Phytophthora, Albugo*

KINGDOM PLANTAE

SECTION THALLOPHYTA†

DIVISION CHLOROPHYTA. Green algae (7,000). *Chlamydomonas, Volvox, Ulothrix, Spirogyra, Oedogonium, Ulva*

DIVISION CHAROPHYTA. Stoneworts (300). *Chara, Nitella, Tolypella*

* There are many alternative classification systems for protists, algae, and plants. This one, based primarily on rRNAs, is derived largely from Sogin et al., and T. Cavalier-Smith, *Nature*, vol. 339, 1989.

† This traditional classification, though useful, has no phylogenetic validity.

DIVISION RHODOPHYTA. Red algae (4,000). *Nemalion, Polysiphonia, Dasya, Chondrus, Batrachospermum*

SECTION EMBRYOPHYTA†

DIVISION BRYOPHYTA (23,600)

CLASS HEPATICAE. Liverworts. *Marchantia, Conocephalum, Riccia, Porella*

CLASS ANTHOCEROTAE. Hornworts. *Anthoceros*

CLASS MUSCI. Mosses. *Polytrichum, Sphagnum, Mnium*

DIVISION TRACHEOPHYTA. Vascular plants

Subdivision Psilopsida. *Psilotum, Tmesipteris*

Subdivision Lycopsida. Club mosses (1,500). *Lycopodium, Phylloglossum, Selaginella, Isoetes, Stylites*

Subdivision Sphenopsida. Horsetails (25). *Equisetum*

Subdivision Pteropsida. Ferns (10,000). *Polypodium, Osmunda, Dryopteris, Botrychium, Pteridium*

Subdivision Spermopsida. Seed plants

CLASS PTERIDOSPERMAE. Seed ferns. No living representatives

CLASS CYCADAE. Cycads (100). *Zamia*

CLASS GINKGOAE (1). *Gingko*

CLASS CONIFERAE. Conifers (500). *Pinus, Tsuga, Taxus, Sequoia*

CLASS GNETEAE (70). *Gnetum, Ephedra, Welwitschia*

CLASS ANGIOSPERMAE. Flowering plants

SUBCLASS DICOTYLEDONEAE. Dicots (225,000). *Magnolia, Quercus, Acer, Pisum, Taraxacum, Rosa, Chrysanthemum, Aster, Primula, Ligustrum, Ranunculus*

SUBCLASS MONOCOTYLEDONEAE. Monocots (50,000). *Lilium, Tulipa, Poa, Elymus, Triticum, Zea, Ophyrys, Yucca, Sabal*

KINGDOM FUNGI

DIVISION HYPHOCHYTRIDIOMYCOTA. Hyphochytrids (25). *Rhizidiomyces*

DIVISION CHYTRIDIOMYCOTA. Chytrids (1,000). *Olpidium, Rhizophydium, Diplophylctis, Cladochytrium*

DIVISION ZYGOMYCOTA. Conjugation fungi (250)

CLASS ZYGOMYCETES. *Rhizopus, Mucor, Phycomyces, Choanephora, Entomophthora*

CLASS TRICHOMYCETES. *Stachylina*

DIVISION ASCOMYCOTA. Sac fungi (12,000)

CLASS HEMIASCOMYCETES. Yeasts and their relatives. *Saccharomyces, Schizosaccharomyces, Endomyces, Eremascus, Taphrina*

† This traditional classification, though useful, has no phylogenetic validity.

CLASS PLECTOMYCETES. Powdery mildews, fruit molds, etc. *Erysiphe, Podosphaera, Aspergillus, Penicillium, Ceratocystis*

CLASS PYRENOMYCETES. *Sordaria, Neurospora, Chaetomium, Xylaria, Hypoxylon*

CLASS DISCOMYCETES. *Sclerotinia, Trichoscyphella, Rhytisma, Xanthoria, Pyronema*

CLASS LABOULBENIOMYCETES. *Herpomyces, Laboulbenia*

CLASS LOCULOASCOMYCETES. *Cochliobolus, Pyrenophora, Leptosphaeria, Pleospora*

DIVISION BASIDIOMYCOTA. Club fungi (15,000)

CLASS HETEROBASIDIOMYCETES. Rusts and smuts, *Ustilago, Urocystis, Puccinia, Phragmidium, Melampsora*

CLASS HOMOBASIDIOMYCETES. Toadstools, bracket fungi, mushrooms, puffballs, stinkhorns, etc. *Coprinus, Marasmius, Amanita, Agaricus, Lycoperdon, Phallus*

KINGDOM ANIMALIA*

SUBKINGDOM PARAZOA

PHYLUM PORIFERA. Sponges (5,000)

CLASS CALCAREA. Calcareous (Chalky) sponges. *Scypha, Leucosolenia, Sycon, Grantia*

CLASS HEXACTINELLIDA. Glass sponges. *Euplectella, Hyalonema, Monoraphis*

CLASS DEMOSPONGIAE. *Spongilla, Euspongia, Axinella*

CLASS SCLEROSPONGIAE. Coralline sponges. *Ceratoporella, Stromatospongia*

SUBKINGDOM PHAGOCYTELLOZOA

PHYLUM PLACOZOA (1). *Trichoplax*

SUBKINGDOM EUMETAZOA

SECTION RADIATA

PHYLUM CNIDARIA (or Coelenterata)

CLASS HYDROZOA. Hydrozoans (3,700). *Hydra, Obelia, Gonionemus, Physalia*

CLASS CUBOZOA. Sea wasps (20). *Tripedalia*

CLASS SCYPHOZOA. Jellyfish (200). *Aurelia, Pelagia, Cyanea*

CLASS ANTHOZOA. Sea anemones and corals (6,100). *Metridium, Pennatula, Gorgonia, Astrangia*

PHYLUM CTENOPHORA. Comb jellies (90). *Pleurobrachia, Haeckelia*

* The classification of Animalia has been adapted from *Synopsis and Classification of Living Organisms*, ed. S. P. Parker, McGraw-Hill, New York, 1982.

SECTION PROTOSTOMIA

PHYLUM PLATYHELMINTHES. Flatworms (10,000)

CLASS TURBELLARIA. Free-living flatworms. *Planaria, Dugesia, Leptoplana*

CLASS TREMATODA. Flukes. *Fasciola, Schistosoma, Prosthogonimus*

CLASS CESTODA. Tapeworms. *Taenia, Dipylidium, Mesocestoides*

PHYLUM GNATHOSTOMULIDA (100). *Gnathostomula, Haplognathia*

PHYLUM NEMERTEA (or Rhynchocoela). Proboscis or ribbon worms (650)

CLASS ANOPLA. *Tubulanus, Cerebratulus*

CLASS ENOPLA. *Amphiporus, Prostoma, Malacobdella*

PHYLUM ACANTHOCEPHALA. Spiny-headed worms (500). *Echinorhynchus, Gigantorhynchus*

PHYLUM MESOZOA (50)

CLASS RHOMBOZOA. *Dicyema, Pseudicyema, Conocyema*

CLASS ORTHONECTIDA. *Rhopalura*

PHYLUM ROTIFERA.* Rotifers (1,700). *Asplanchna, Hydatina, Rotaria*

PHYLUM GASTROTRICHA* (2,000). *Chaetonotus, Macrodasys*

PHYLUM KINORHYNCHA* (or Echinodera) (100). *Echinoderes, Semnoderes*

PHYLUM NEMATA.* Roundworms or nematodes (12,000). *Ascaris, Trichinella, Necator, Enterobius, Ancylostoma, Heterodera*

PHYLUM NEMATOMORPHA.* Horsehair worms (230). *Gordius, Paragordius, Nectonema*

PHYLUM ENTOPROCTA (150). *Urnatella, Loxosoma, Pedicellina*

PHYLUM LORICIFERA (3, newly discovered and probably very common)

PHYLUM PRIAPULIDA (8). *Priapulus, Halicryptus*

PHYLUM BRYOZOA† (or Ectoprocta). Bryozoans, moss animals (4,000)

CLASS GYMNOLAEMATA. *Paludicella, Bugula*

CLASS PHYLACTOLAEMATA. *Plumatella, Pectinatella*

CLASS STENOLAEMATA

PHYLUM PHORONIDA† (10). *Phoronis, Phoronopsis*

PHYLUM BRACHIOPODA.† Lamp shells (300)

CLASS INARTICULATA. *Lingula, Glottidia, Discina*

* Formerly considered a class of Phylum Aschelminthes.

† Bryozoa, Phoronida, and Brachiopoda are often referred to as the lophophorate phyla.

CLASS ARTICULATA. *Magellania, Neothyris, Terebratula*

PHYLUM MOLLUSCA. Molluscs

CLASS CAUDOFOVEATA (70). *Chaetoderma*

CLASS SOLENOGASTRES. Solenogasters (180). *Neomenia, Proneomenia*

CLASS POLYPLACOPHORA. Chitons (600). *Chaetopleura, Ischnochiton, Lepidochiton, Amicula*

CLASS MONOPLACOPHORA (8). *Neopilina*

CLASS GASTROPODA. Snails and their allies (univalve molluscs) (25,000). *Helix, Busycon, Crepidula, Haliotis, Littorina, Doris, Limax*

CLASS SCAPHOPODA. Tusk shells (350). *Dentalium, Cadulus*

CLASS BIVALVA. Bivalve molluscs (7,500). *Mytilus, Ostrea, Pecten, Mercenaria, Teredo, Tagelus, Unio, Anodonta*

CLASS CEPHALOPODA. Squids, octopuses, etc. (600). *Loligo, Octopus, Nautilus*

PHYLUM POGONOPHORA. Beard worms (100). *Siboglinum, Lamellisabella, Oligobrachia, Polybrachia*

PHYLUM SIPUNCULA (300). *Sipunculus, Phascolosoma, Dendrostomum*

PHYLUM ECHIURA (140)

CLASS ECHIUROINEA. *Echiurus, Ikedella*

CLASS XENOPNEUSTA. *Urechis*

CLASS HETEROMYOTA. *Crisia, Tubulipora*

PHYLUM ANNELIDA. Segmented worms

CLASS POLYCHAETA (including Archiannelida). Sandworms, tubeworms, etc. (8,000). *Nereis, Chaetopterus, Aphrodite, Diopatra, Arenicola, Hydroides, Sabella*

CLASS OLIGOCHAETA. Earthworms and many freshwater annelids (3,100). *Tubifex, Enchytraeus, Lumbricus, Dendrobaena*

CLASS HIRUDINOIDEA. Leeches (500). *Trachelobdella, Hirudo, Macrobdella, Haemadipsa*

PHYLUM ONYCHOPHORA (65). *Peripatus. Peripatopsis*

PHYLUM TARDIGRADA. Water bears (300). *Echiniscus, Macrobiotus*

PHYLUM ARTHROPODA (at least 2,000,000)

Subphylum Trilobita. No living representatives

Subphylum Chelicerata

CLASS EURYPTERIDA. No living representatives

CLASS MEROSTOMATA. Horseshoe crabs (4). *Limulus*

CLASS ARACHNIDA. Spiders, ticks, mites, scorpions, whipscorpions, daddy longlegs, etc. (55,000; at least 500,000 undiscovered species of mites are thought to exist). *Archaearanea, Latrodectus, Argiope, Centruroides, Chelifer, Mastigoproctus, Phalangium, Ixodes*

CLASS PYCNOGONIDA. Sea spiders (1,000). *Nymphon, Ascorhynchus*

Subphylum Crustacea (26,000). *Homarus, Cancer, Daphnia, Artemia, Cyclops, Balanus, Porcellio*

Subphylum Uniramia

CLASS CHILOPODA. Centipedes (2,500). *Scolopendra, Lithobius, Scutigera*

CLASS DIPLOPODA. Millipedes (10,000; another 50,000 species are thought to exist). *Narceus, Apheloria, Polydesmus, Julus, Glomeris*

CLASS PAUROPODA (500). *Pauropus*

CLASS SYMPHYLA (160). *Scutigerella*

CLASS INSECTA. Insects (900,000; another 1,000,000 species are thought to exist)

ORDER THYSANURA. Bristletails, silverfish, firebrats. *Machilis, Lepisma. Thermobia*

ORDER EPHEMERIDA Mayflies. *Hexagenia, Callibaetis, Ephemerella*

ORDER ODONATA. Dragonflies, damselflies. *Archilestes, Lestes, Aeshna, Gomphus*

ORDER ORTHOPTERA. Grasshoppers, crickets, etc. *Schistocerca, Romalea, Nemobius, Megaphasma*

ORDER PHASMATOPTERA. Walking sticks. *Phyllium*

ORDER BLATTARIA. Cockroaches. *Blatta, Periplaneta*

ORDER MANTODEA. Mantids. *Mantis*

ORDER GRYLLOBLATTARIA. *Grylloblatta*

ORDER ISOPTERA. Termites. *Reticulitermes, Kalotermes, Zootermopsis, Nasutitermes*

ORDER DERMAPTERA. Earwigs, *Labia, Forficula, Prolabia*

ORDER EMBIIDINA (or Embiaria or Embioptera). *Oligotoma, Anisembia, Gynembia*

ORDER PLECOPTERA. Stoneflies. *Isoperla, Taeniopteryx, Capnia, Perla*

ORDER ZORAPTERA. *Zorotypus*

ORDER PSOCOPTERA. Book lice. *Ectopsocus, Liposcelis, Trogium*

ORDER MALLOPHAGA. Chewing lice. *Cuclotogaster, Menacanthus, Menopon, Trichodectes*

ORDER ANOPLURA. Sucking lice. *Pediculus, Phthirius, Haematopinus*

ORDER THYSANOPTERA. Thrips. *Heliothrips, Frankliniella, Hercothrips*

ORDER HEMIPTERA. True bugs. *Belostoma, Lygaeus, Notonecta, Cimex, Lygus, Oncopeltus*

ORDER HOMOPTERA. Cicadas, aphids, leafhoppers, scale insects, etc. *Magicicada, Circulifer, Psylla, Aphis, Saissetia*

ORDER NEUROPTERA. Dobsonflies, alderflies, lacewings, mantispids, snakeflies, etc. *Corydalus, Hemerobius, Chrysopa, Mantispa, Agulla*

ORDER COLEOPTERA. Beetles, weevils. *Copris, Phyllophaga, Harpalus, Scolytus, Melanotus, Cicindela, Dermestes, Photinus, Coccinella, Tenebrio, Anthonomus, Conotrachelus*

ORDER HYMENOPTERA. Wasps, bees, ants, sawflies. *Cimbex, Vespa, Glypta, Scolia, Bembix, Formica Bombus, Apis*

ORDER STREPSIPTERA. Endoparasites

ORDER MECOPTERA. Scorpionflies. *Panorpa, Boreus, Bittacus*

ORDER SIPHONAPTERA. Fleas. *Pulex, Nosopsyllus, Xenopsylla, Ctenocephalides*

ORDER DIPTERA. True flies, mosquitoes. *Aedes, Asilus, Sarcophaga, Anthomyia, Musca, Chironomus, Tabanus, Tipula, Drosophila*

ORDER TRICHOPTERA. Caddisflies. *Limnephilus, Rhyacophilia, Hydropsyche*

ORDER LEPIDOPTERA. Moths, butterflies. *Tinea, Pyrausta, Malacosoma, Sphinx, Samia, Bombyx, Heliothis, Papilio, Lycaena*

Subphylum Pentastomida

CLASS PENTASTOMIDA. Parasites (60)

SECTION DEUTEROSTOMIA

PHYLUM CHAETOGNATHA. Arrow worms (70). *Sagitta, Spadella*

PHYLUM ECHINODERMATA

Subphylum Crinozoa

CLASS CRINOIDEA. Crinoids, sea lilies (630). *Antedon, Ptilocrinus, Comactinia*

Subphylum Asterozoa

CLASS STELLEROIDEA. Sea stars, brittle stars (2,600). *Asterias, Ctenodiscus, Luidia, Oreaster, Asteronyx, Amphioplus, Ophiothrix, Ophioderma, Ophiura*

Subphylum Echinozoa

CLASS ECHINOIDEA. Sea urchins, sand dollars, heart urchins (860). *Cidaris, Arbacia, Strongylocentrotus, Echinanthus, Echinarachnius, Moira*

CLASS HOLOTHUROIDEA. Sea cucumbers (900). *Cucumaria, Thyone, Caudina, Synapa*

PHYLUM HEMICHORDATA (90)

CLASS ENTEROPNEUSTA. Acorn worms. *Saccoglossus, Balanoglossus, Glossobalanus*

CLASS PTEROBRANCHIA. *Rhabdopleura, Cephalodiscus*

CLASS PLANCTOSPHAEROIDEA

PHYLUM CHORDATA. Chordates

Subphylum Tunicata (or Urochordata). Tunicates (2,000)

CLASS ASCIDIACEA. Ascidians or sea squirts. *Ciona, Clavelina, Molgula, Perophora*

CLASS THALIACEA. *Pyrosoma, Salpa, Doliolum*

CLASS APPENDICULARIA. *Appendicularia, Oikopleura, Fritillaria*

Subphylum Cephalochordata. Lancelets, amphioxus (30). *Branchiostoma, Asymmetron*

Subphylum Vertebrata. Vertebrates

CLASS AGNATHA. Jawless fish (50). *Cephalaspis,* Pteraspis,* Petromyzon, Entosphenus, Myxine, Eptatretus*

CLASS ACANTHODII. No living representatives

CLASS PLACODERMI. No living representatives

CLASS CHONDRICHTHYES. Cartilaginous fish, including sharks and rays (800). *Squalus, Hyporion, Raja, Chimaera*

CLASS OSTEICHTHYES. Bony fish (18,000)

SUBCLASS SARCOPTERYGII

ORDER CERATODIFORMES. Australian lungfish. *Neoceratodus*

* Extinct.

ORDER LEPIDOSIRENIFORMES. Lungfish. *Protopterus, Lepidosiren*

SUBCLASS ACTINOPTERYGII. Ray-finned fish. *Amia, Cyprinus, Gadus, Perca, Salmo*

CLASS AMPHIBIA (3,100)

ORDER ANURA. Frogs and toads. *Rana, Hyla, Bufo*

ORDER CAUDATA (or Urodela). Salamanders, *Necturus, Triturus, Plethodon, Ambystoma*

ORDER GYMNOPHIONA (or Apoda). *Ichthyophis, Typhlonectes*

CLASS REPTILIA (6,500)

ORDER TESTUDINES. Turtles. *Chelydra, Kinosternon, Clemmys, Terrapene*

ORDER RHYNCHOCEPHALIA. Tuatara, *Sphenodon*

ORDER CROCODYLIA. Crocodiles and alligators. *Crocodylus, Alligator*

ORDER LEPIDOSAURIA. Snakes and lizards, *Iguana, Anolis, Sceloporus, Phrynosoma, Natrix, Elaphe, Coluber, Thamnophis, Crotalus*

CLASS AVES. Birds (8,600). *Anas, Larus, Columba, Gallus, Turdus, Dendroica, Sturnus, Passer, Melospiza*

CLASS MAMMALIA. Mammals (4,100)

SUBCLASS PROTOTHERIA

ORDER MONOTREMATA. Egg-laying mammals. *Ornithorhynchus, Tachyglossus*

SUBCLASS THERIA. Marsupial and placental mammals

ORDER METATHERIA (or Marsupialia). Marsupials. *Didelphis, Sarcophilus, Notoryctes, Macropus*

ORDER INSECTIVORA. Insectivores (moles, shrews, etc.). *Scalopus, Sorex, Erinaceus*

ORDER DERMOPTERA. Flying lemurs. *Galeopithecus*

ORDER CHIROPTERA.* Bats, *Myotis, Eptesicus, Desmodus*

ORDER PRIMATA. Lemurs, monkeys, apes, humans, *Lemur, Tarsius, Cebus, Macacus, Cynocephalus, Pongo, Pan, Homo*

ORDER EDENTATA. Sloths, anteaters, armadillos, *Bradypus, Myrmecophagus, Dasypus*

ORDER PHOLIDOTA. Pangolin. *Manis*

ORDER LAGOMORPHA. Rabbits, hares, pikas. *Ochotona, Lepus, Sylvilagus, Oryctolagus*

ORDER RODENTIA. Rodents. *Sciurus, Marmota, Dipodomys, Microtus, Peromyscus, Rattus, Mus, Erethizon, Castor*

ORDER ODONTOCETA. Toothed whales, dolphins, porpoises. *Delphinus, Phocaena, Monodon*

ORDER MYSTICETA. Baleen whales. *Balaena*

ORDER CARNIVORA. Carnivores, *Canis, Procyon, Ursus, Mustela, Mephitis, Felis, Hyaena, Eumetopias*

ORDER TUBULIDENTATA. Aardvark. *Orycteropus*

ORDER PROBOSCIDEA. Elephants. *Elephas, Loxodonta*

ORDER HYRACOIEDEA. Hyraxes, conies. *Procavia*

ORDER SIRENIA. Manatees. *Trichechus, Halicore*

ORDER PERISSODACTYLA. Odd-toed ungulates. *Equus, Tapirella, Tapirus, Rhinoceros*

ORDER ARTIODACTYLA. Even-toed ungulates. *Pecari, Sus, Hippopotamus, Camelus, Cervus, Odocoileus, Giraffa, Bison, Ovis, Bos*

* Some authorities propose dividing the echo-locating bats and fruit-eating bats into separate orders.

CREDITS

TABLE OF CONTENTS

p. vi (left) © Copyright Boehringer Ingelheim International; (right) Courtesy New York Public Library, Rare Books Division. Arent Collection, Astor, Lenox and Tilden Foundations. **p. viii** (right) Jane Burton/Bruce Coleman, Inc. **p. xi** (left) Courtesy E. S. Ross; (right) Courtesy Irenäus Eibl-Eibesfeldt, Max Planck Institute for Behavioral Physiology. **p. xii** (left) Courtesy Carolina Biological Supply Company; (right) Courtesy Gregory S. Paul. **p. xiv** (left) © Ed Reschke; (right) Photography by L. Nilsson, from L. Nilsson, *Behold Man*, English translation © 1974, Albert Bonniers Förlag, Stockholm, and Little, Brown & Co. (Canada) Ltd. **p. xviii** (left) © Norman Myers/Bruce Coleman, Inc.; (right) © Howard Hall.

1.1 Courtesy NASA. **1.4** California Institute of Technology Archives. **1.5** SCALA/Art Resource, New York. **1.7** The Master and Fellows of Trinity College, Cambridge. **1.8** Courtesy Rare Book Division, The New York Public Library, Astor, Lenox and Tilden Foundations. **1.9** The Royal Society, London. **1.10** Warder Collection. **1.11** Rijksmuseum, Amsterdam. **1.12** Musée D'Orsay, © RMN. **1.13** Neg./Trans no. 32–666–2, Courtesy Department of Library Services, American Museum of Natural History, New York. **1.14** (A) Sovfoto; (B) F. Erize/Bruce Coleman, Inc. **1.15** Library of the Museum of Natural History, Paris. **1.17** The New York Public Library, Astor, Tilden, and Lenox Foundations. **1.18** Leonard Lee Rue III/Bruce Coleman, Inc. **1.19** Photographs by L. Nilsson; from L. Nilsson, *Behold Man*, English translation © 1974, Albert Bonniers Förlag, Stockholm, and Little, Brown and Co. (Canada) Ltd. **1.20** Modified from *The Illustrated Origin of Species*, by Charles Darwin, abridged and introduced by Richard E. Leakey, 1979; courtesy of Hill and Wang, a division of Farrar, Straus & Giroux, Inc. **1.21** By permission to the Syndics of Cambridge University Library. **1.22** Courtesy National Portrait Gallery, London. **1.23** (center) Kenneth W. Fink/Bruce Coleman, Inc.; (other photographs) Courtesy Louise B. Van der Meid. **1.24** Clockwise from left: (archaebacterium) H. W. Jannasch and C. O. Wirsen, *Bio Science*, vol. 29, 1979; copyright © 1979 by the American Institute of Biological Science; (archaezoan) Courtesy E. W. Daniels; (a protist) © M. I. Walker/Science Source—Photo Researchers, Inc.; (fly agaric) © G. R. Roberts; (great egret) © M. P. Kahl, 1972/Photo Researchers, Inc.; (dahlias) Gene Ahrens/Bruce Coleman, Inc.; (kelp) © Jeff Rotman; (a true bacterium) CNRI/Photo Science Library/Photo Researchers, Inc. **1.25** Courtesy Cold Spring Harbor Laboratory Archives Telescope Board, 1981.

PART I THE CHEMICAL AND CELLULAR BASIS OF LIFE

(1) Model of DNA © Will and Deni McIntyre/Photo Researchers, Inc.; (2) Mitochondria © Bill Longcore/Photo Researchers, Inc.; (3) Sunflowers © Gene Ahrens/Bruce Coleman, Inc.; (4) Endocytosis in a blood capillary © Secchi-Lecaque/Roussel-UCLAF/CNRI/Science Photo Library.

2.1 Anglo-Australian Telescope Board, 1981. **2.3** Science VU/Visuals Unlimited. **2.4** R. M. Feenstra and J. A. Stroscio, IBM Watson Res. Ctr., Yorktown Heights. **2.10** © David Newman/Visuals Unlimited (VU). **2.14** © Dwight Kuhn **2.22** © Carl Purcell, 1990. **2.27** Herman Eisenbeiss/Photo Researchers, Inc. **2.28** © Dwight Kuhn, 1986. **2.30** (B) John Shaw/Tom Stack & Associates. **2.31** Spenser Swanger/Tom Stack & Associates.

3.11 (A) Dwight Kuhn © 1980. (B) Courtesy W. Cheng, International Paper Company. **3.13** From R. G. Kessel and R. H. Kardon, *Tissues and Organs: A Text-Atlas of Scanning Electron Microscopy*, W. H. Freeman, San Francisco, copyright © 1979. ***Exploring Further*** p. 65, Courtesy V. Ingram, *Biochimica et Biophysica Acta*, vol. 28: 543, 1958. Copyright © 1958 Elsevier Science Publishers, Amsterdam. **3.23** Adapted by permission from *The Structure and Action of Proteins* by Richard E. Dickerson and Irving Geis, W. A. Benjamin, Inc., Menlo Park, Calif., Publisher; copyright © 1969 by Dickerson and Geis. **3.24** Tony Brain, Science Photo Library. **3.25** Adapted by permission from *The Structure and Action of Proteins* by Richard E. Dickerson and Irving Geis, W. A. Benjamin, Inc., Menlo Park, Calif., Publisher; copyright © 1969 by Dickerson and Geis. **3.26** Adapted by permission from *The Structure and Action of Proteins* by Richard E. Dickerson and Irving Geis, W. A. Benjamin, Inc., Menlo Park, Calif., Publisher; copyright © 1969 by Dickerson and Geis. **3.28** J. Kendrew, Cambridge University. **3.29** Adapted by permission from *The Structure and Action of Proteins* by Richard E. Dickerson and Irving Geis, W. A. Benjamin, Inc., Menlo Park, Calif., Publisher; copyright © 1969 by Dickerson and Geis. **3.37** John E. Swedborg/Bruce Coleman, Inc. **3.47** Modified from A. L. Lehninger, *Biochemistry*, 2nd ed., Worth Publishers, New York, 1975, p. 233. **3.50** © K. Talaro/VU.

4.1 (A) © K. Talaro/Visuals Unlimited. **4.4** Courtesy J. D. Pickett-Heaps, University of Melbourne. **4.5** (D) J. J. Cardamone and B. A. Phillips, University of Pittsburgh. **4.6** Photo courtesy B. Michel/IBM Research Division, Zurich.

From M. Amrein, *Science*, 240:515, 1988. Courtesy American Association for the Advancement of Science (A.A.A.S.). **4.7** Courtesy Jean-Paul Revel, California Institute of Technology. **4.12** © David M. Phillips/Visuals Unlimited. **4.14** The Liposome Company. **4.15** Courtesy J. David Robertson, Duke University. **4.18** Courtesy Daniel Branton, Harvard University. **4.26** (B) Courtesy Dorothy F. Bainton, University of California, San Francisco. **4.27** © R. L. Roberts, R. G. Kessel and H. N. Tung, Freeze Fracture Images of Cells and Tissues, Oxford University Press, 1991. **4.28** (A) J. Ross, J. Olmstead, and J. Rosenbaum, *Tissue and Cell*, vol. 7, 1975. **4.29** M. M. Perry and A. B. Gilbert, *J. Cell Sci.*, vol. 39, 1979, by copyright permission of the Rockefeller University Press. **4.30** N. Hirokawa and J. Heuser, *Cell*, vol. 30, 395–406, 1982. © 1982 by Massachusetts Institute of Technology (MIT). **4.32** (B) V. Herzog, H. Sies, and F. Miller, *J. Cell Biol.*, vol. 70, 1976, by copyright permission of the Rockefeller University Press. **4.33** © Biophoto Associates/Science Source—Photo Researchers, Inc. **4.34** Courtesy Eva Frei and R. D. Preston, University of Leeds. **4.35** (B) H. Latta, W. Johnson, and T. Stanley, *J. Ultrastruct. Res.*, vol. 51, 1975.

5.1 Biophoto Associates/Science Source—Photo Researchers, Inc. **5.2** Courtesy A. H. Sparrow and R. F. Smith, Brookhaven National Laboratory. **5.3** (A) Courtesy Barbara Hamkalo and J. B. Rattner, University of California, Irvine; (B) Courtesy Victoria Foe. **5.5** W. G. Whaley, H. H. Mollenhauer, and J. H. Leech, *Am J. Bot.*, vol. 47, 1960. **5.6** Courtesy Daniel Branton, Harvard University. **5.7** Courtesy K. R. Porter, University of Colorado. **5.9** Micrograph courtesy D. S. Friend, University of California, San Francisco. **5.11** Courtesy D. S. Friend, University of California, San Francisco. **5.13** Courtesy D. S. Friend, University of California, San Francisco. **5.15** Micrograph by S. E. Frederick and E. H. Newcomb, *J. Cell Biol.*, vol. 43, 1969. **5.16** Courtesy D. S. Friend, University of California, San Francisco. **5.17** (top) Micrograph by W. P. Wergin, courtesy E. H. Newcomb, University of Wisconsin; (bottom) M. C. Ledbetter, Photo Researchers, Inc. **5.18** Courtesy M. C. Ledbetter, Brookhaven National Laboratory. **5.21** Courtesy Elias Lazarides, California Institute of Technology. **5.22** Courtesy Susumu Ito, Harvard Medical School. **5.24** (C) Boehringer Ingelheim International; (D) Photograph by C. Lin, courtesy P. Forscher, Yale University. **5.25** (left) H. Kim, L. I. Binder, and J. L. Rosenbaum, *J. Cell Biol.*, vol. 80, 1979, by copyright permission of the Rockefeller University Press; (right) courtesy D. W. Fawcett, Harvard Medical School. **5.28** Photo from A. Ashkin, K. Schütze, J. M. Dziedzic, U. Euteneurt, and M. Schliwa, *Nature*, 348; 1990, pp. 346–48. **5.30** U. Aebi, University of Basel. **5.31** Micrograph from J. Heuser and S. R. Salpeter, *J. Cell Biol.*, vol. 82, 1979, by copyright permission of the Rockefeller University Press. **5.32** M. McGill, D. P. Highfield, T. M. Monahan, and B. R. Brinkley, *J. Ultrastruct Res.*, vol. 57, 1976. **5.33** (A) Courtesy R. W. Linck, Harvard Medical School, and D. T. Woodrum. **5.34** Courtesy E. R. Dirksen, University of California, Los Angeles. **5.35** Micrograph by K. Roberts, John Innes Institute, Norwich, England; from B. Alberts et al., *Molecular Biology of the Cell*, Garland Press, New York, 1983. **5.37** C. J. Brokaw, *Science*, vol. 178, 1972; copyright © 1972 by A.A.A.S. **5.39** Micrograph produced by J. W. Heuser, Washington University School of Medicine, St. Louis. From J. Heuser, et al., *J. Cell Biol.*, 95, p. 800, 1982. Reprinted by copyright permission of Rockefeller University Press. **5.40** (mitochondrion) Courtesy K. R. Porter, University of Colorado; (lysosome) Courtesy A. B. Novikoff, Albert Einstein College of Medicine; (glycocalyx) Courtesy A. Ryter, Institut Pasteur, Paris; (Golgi apparatus) D. W. Fawcett/VU; (centrioles) M.McGill, D.P. Highfield, T. M. Monahan, and B. R. Brinkley, *J. Ultrastruct Res*, vol. 57, 1976; (all other micrographs) Courtesy N. B. Gilula, Baylor College of Medicine. **5.41** (plasmodesma) Courtesy W. G. Whaley, et al., *J. Biophys. Biochem. Cytol.*, (now *J. Cell Biol.*), vol. 5, 1959; by copyright permission of the Rockefeller University Press; (nucleus, chloroplast, leucoplast, endoplasmic reticulum, mitochondrion and Golgi apparatus) Courtesy M. C. Ledbetter, Brookhaven National Laboratory. **5.42** Courtesy A. Ryter, Institut Pasteur, Paris. **5.43** Courtesy J. Griffith, School of Medicine, University of North Carolina, Chapel Hill.

Exploring Further p. 183, Efraim Racker, Cornell University. **6.10** (A) © K. R. Porter, D. W. Fawcett/Visuals Unlimited. (C) Photo by H. Fernández-Morán, courtesy E. Valdivia, University of Wisconsin. From Fernández-Morán, et al., *J. Cell Biol.* 22:63–100, 1964. Reproduced by copyright permission of the Rockefeller University Press.

7.11 (left) Micrograph by W. P. Wergin, courtesy E. H. Newcomb, University of Wisconsin. **7.15** (top left and right) Print Collection, Miriam and Ira D. Wallach Division of Art, Prints and Photographs; (middle left, and bottom left and right) General Research Division; (middle right) Rare books and Manuscripts Division, The New York Public Library Astor, Lenox and Tilden Foundations. **7.17** Courtesy Raymond Chollet, University of Nebraska. **7.20** Courtesy G. R. Roberts.

PART II THE PERPETUATION OF LIFE

(3) Human lymphocyte cell © CNRI/Science Photo Library/Photo Researchers, Inc.; (4) TEM of human cancer cells © Dr. Bryan Eyden/Science Photo Library/Photo Researchers, Inc.

8.2 Carolina Biological Supply Company. **8.3** M. McCarty, *Journal of Experimental Medicine*, vol. 79, 1944, 137–58. **8.5** Courtesy Lee D. Simon, Waksman Institute, Rutgers University. **8.12** From J. D. Watson, *The Double Helix*, Atheneum, New York, 1968. © J. D. Watson. Photo courtesy of Cold Spring Harbor Laboratory Archives. **8.13** Courtesy Cold Spring Harbor Laboratory Archives. **8.17** Courtesy A. C. Arnberg, Biochemical Laboratory, State University, Groningen, The Netherlands. **8.18** Redrawn from M. S. Meselson and F. W. Stahl, *Proc. Natl. Acad. Sci. U.S.A.*, vol. 44, 1958.

9.2 (B) Jack R. Griffith, University of North Carolina, Chapel Hill. **9.5** Photo by B. Tagawa, courtesy F. Perrin and P. Chambon. From P. Chambon, *Sci. Am.*, 244 (1981): 60–71. Used by permission. **9.7** Jack R. Griffith, University of North Carolina, Chapel Hill. **9.9** Adapted from R. Gupta, J. M. Lanter, and C. R. Woese, *Science*, vol. 221, 1983; copyright © 1983 by A.A.A.S. **9.11** Micrograph from O. L. Miller, Jr., B. A. Hamkalo, and C. A. Thomas, Jr., *Science*, vol. 169, 1970; copyright © 1970 by A.A.A.S. **9.12** Photograph courtesy Nigel Unwin, Stanford University School of Medicine. **9.14** Modified from B. Alberts et al., *Molecular Biology of the Cell*, Garland Press, New York, 1983. **9.15** Photo from M. A. Rould, J. J. Perona, D. Söll, and T. A. Steitz, *Science*, 246, 1989. Copyright © 1989 by A.A.A.S. **9.17** M. R. Hanson et al., *Mol. Gen. Genet.*, vol. 132, 1974. **9.19** Courtesy Bruce N. Ames, University of California, Berkeley.

10.1 (A,C) Jack R. Griffith, University of North Carolina, Chapel Hill; (B) Courtesy S. N. Cohen, Stanford University. **10.2** Courtesy L. G. Caro, University of Geneva, and R. Curtiss, University of Alabama. ***Exploring Further*** p. 278, Photo courtesy of Lark Sequencing. **10.13** John D. Cunningham/VU.

11.1 (A,B) Cold Spring Harbor Laboratory Archives. **11.2** (D) J. Griffith. **11.5** (F) Micrograph by W. Engler, courtesy of G. F. Bahr, *Fed. Proc., Fed. Am. Soc. Exp. Biol.*, vol. 34, 1975. **11.6** Courtesy Steven Henikoff, Fred Hutchinson Cancer Research Center, Seattle, Washington. **11.8** Photo courtesy J. G. Gall, Carnegie Institution, reproduced from M. B. Roth and J. G. Gall, *J. Cell Biol.*, 105, 1047–1054 (1987). Copyright © 1987 by Rockefeller University Press. **11.9** Michael Ashburner, Cambridge University. **11.11** Photograph courtesy O. L. Miller, Jr., and B. R. Beatty, *J. Cell Physiol.*, vol. 74, 1969. **11.13** Courtesy Gunter Albrecht-Buehler, Cold Spring Harbor Laboratory, and Frank Solomon, Massachusetts Institute of Technology. **11.14** K. Porter, G. Fonte, and Weiss, *Cancer Res.*, vol. 34, 1974. **11.16** Curve for retinablastoma based on data from H. W. Hethcote and A. G. Knudson, *Proc. Natl. Acad. Sci. U.S.A.*, vol. 75, 1978; curves for prostate and skin cancer based on data from Japanese Cancer Association, *Cancer Mortality and Morbidity Statistics*, Japanese Scientific Press, Tokyo, 1981.

12.2 Courtesy R.G.E. Murray, University of Western Ontario. **12.3** M. P. Marsden and U. K. Laemmli; *Cell*, 17:849–58, 1979, used by permission of MIT Press, Cambridge, Mass. **12.4** Courtesy M. W. Shaw, University of Michigan, Ann Arbor. **12.6** Courtesy A. S. Bajer, University of Oregon. **12.10** Courtesy A. S. Bajer, University of Oregon. **12.11** Modified from B. Alberts et al., *Molecular Biology of the Cell*, Garland Press, New York, 1983. **12.12** From M. S. Fuller, *Mycologia*, vol. 60, 1968. **12.13** From H. W. Beams and R. G. Kessel, *Am. Sci.*, vol. 64, 1976; reprinted by permission of American Scientist, *Journal of Sigma Xi*, the Scientific Research Society. **12.15** Courtesy A. S. Bajer, University of Oregon. **12.16** From W. G. Whaley et al., *Am. J. Bot.*, vol. 47, 1960. **12.18** From D. von Wettstein, *Proc. Natl. Acad. Sci. U.S.A.*, vol. 68, 1971. **12.20** Courtesy James Kezer, University of Oregon. **12.23** Courtesy A. S. Bajer, University of Oregon. **12.28** Ed Reschke. **12.30** Photograph by L. Nilsson, from L. Nilsson, *Behold Man*, English translation © 1974, Albert Bonniers Förlag, Stockholm, and Little, Brown and Co. (Canada) Ltd.

13.1 Micrograph by P. Sundstrom, Gamma-Liaison. **13.2** (F) D. M. Phillips, *J. Ultrastruct Res*, 72:1–12. 1980. By permission of Academic Press, Inc., Orlando. **13.7** Courtesy R. G. Kessel and C. Y. Shih, *Scanning Electron Microscopy in Biology*, Springer-Verlag, New York, 1974. **13.10** Oxford Scientific Films. **13.11** Photographs by L. Nilsson; from L. Nilsson, *Behold Man*, English translation © 1974, Albert Bonniers Förlag, Stockholm, and Little, Brown and Co. (Canada) Ltd. **13.13** Redrawn from G. J. Romanes, *Darwin and After Darwin*, Open Court Publishing Co., 1901. **13.14** Redrawn from D'A. W. Thompson, *On Growth and Form*, Cambridge, The University Press, 1942 (from G. Backman, after Stefanowska). **13.15** Redrawn from D'A. W. Thompson, *On Growth and Form*, Cambridge, The University Press, 1942, (after W. Ostwald). **13.16** Modified from V. B. Wigglesworth, *The Principles of Insect Physiology*, Methuen, 1947. **13.17** Redrawn from D'A. W. Thompson, *On Growth and Form*, Cambridge, The University Press, 1942 (from Quetelet's data). **13.19** (A–C,H) Photographs by D. Overcash, (D) photograph by L. West; (E–G) Photographs by E. R. Degginger/Bruce Coleman, Inc.

14.5 (B) Courtesy Douglas Melton, Harvard University. **14.6** From I. Mann, *The Development of the Human Eye*, copyright © 1964 by Grune and Stratton, New York. **14.7** From W. Krommenhoek, J. Sebus, and G. J. van Esch, *Biological Structures*, copyright © 1979 by L.C.G. Malmberg B. V., The Netherlands. **14.12** (B) Carolina Biological Supply Company. **14.15** (A,B) E. B. Lewis, California Institute of Technology, Pasadena. **14.16** Photograph by K. W. Tosney, as printed in N. K. Wessels, *Tissue Interactions and Development*, copyright © 1977 by Benjamin-Cummings, Menlo Park, California. **14.17** Photographs courtesy Dennis Summerbell, National Institute for Medical Research, London. **14.18** (D) Gregor Eichele, Baylor College of Medicine. **14.19** Adapted from M. Singer, *Sci. Am.*, October 1958, copyright © 1958 by Scientific American, Inc. All rights reserved. **14.21** Micrograph by Corey S. Goodman; from Corey S. Goodman and Michael J. Bastiani, *Sci. Am.*, December 1984, copyright © 1984 by Scientific American, Inc. All rights reserved. **14.23** Modified from J. E. Sulston and H. R. Horvitz, *Dev. Biol.*, vol. 56, 1977 copyright © 1977 by Academic Press, New York.

15.3 Photo from R. G. Kessel and R. H. Kardon, *Tissues and Organs: A Text-Atlas of Scanning Electron Microscopy*, © W.H. Freeman and Company, San Francisco, California, 1970. **15.4** Photo courtesy Peter Marks and Fredrick Maxfield, from P. Marks and F. R. Maxfield, *J. Cell Biol.*, 110:43–52 (1990). Reprinted by copyright permission of Rockefeller University Press. **15.5** (B) Computer graphic modelling and photography by A. J. Olson, The Scripps Research Institute. Copyright © 1992. **15.17** Reproduced from D. Lawson, C. Fewtrell, B. Gomperts, and M. Raff, *Journal of Experimental Medicine*, 142:391–402, 1975. Used by copyright permission of Rockefeller University Press. **15.19** Courtesy Center for Disease Control, Atlanta. ***Exploring Further*** p. 412, Courtesy Steven T. Brentano.

16.5 Jane Burton/Bruce Coleman, Inc. **16.9** Courtesy Ralph Somes, University of Connecticut. **16.11** Modified from Francisco H. Ayala and John A. Kiger, Jr., *Modern Genetics*, Benjamin-Cummings, Menlo Park, California, 1980. **16.14** Courtesy S. B. Moore. **16.15** Modified from A. M. Winchester, *Genetics*, 5th ed., Houghton Mifflin, Boston, 1977. **16.16** Larry LeFever/Grant Heilman Photography, Inc. **16.17** Courtesy Marion I. Barnhart, Wayne State University Medical School, Detroit, Michigan. **16.20** Courtesy M. L. Barr, *Can. Cancer Conf.*, vol. 2, Academic Press, New York, 1957. **16.21** H. Chaumeton/Nature. **16.22** K. R. Dronamraju, in E. J. Gardner, *Principles of Genetics*, copyright © 1975 by John Wiley & Sons, Inc.; reprinted with their permission. **16.25** Modified from *Biological Science: An Ecological Approach*, 4th ed., Houghton Mifflin, Boston, 1982. Used by permission. **16.27** E. J. Bingham, University of Wisconsin, Madison.

PART III EVOLUTIONARY BIOLOGY

(1) Courtesy Norbert Wu; (2) Peacock © Michael Giannechini, Photo Researchers, Inc; (3) Hawaiian honeycreepers drawing courtesy H. Douglas Pratt; (4) Beetles courtesy Bob Natalini.

17.2 (A,B) Courtesy J. L. Gould; (C) M. L. Estey, Photo Researchers, Inc. ***Exploring Further*** p. 451 (Fig. 2) © Kim Taylor/Bruce Coleman, Inc. **17.3** (A) George Holton, Photo Researchers, Inc.; (B) Masud Quraishy/Bruce Coleman, Inc. **17.4** A. J. Ribbink, *S. Afr. J. Zool.*, vol. 18, 1985. **17.5** (A,B) Hans Reinhard/Bruce Coleman, Inc. **17.10** Data from C. M. Woodworth et al., *Agron, J.*, vol. 44, 1952. **17.12** Hans Reinhard/Bruce Coleman, Inc. **17.14** (A,B) E. Greene, *Science*, vol. 243: 643–46, 1989. Copyright © A.A.A.S. **17.15** Stouffer Productions/Animals Animals. **17.16** John Wightman, Ardea London Ltd. **17.17** Adapted by permission from *The Structure and Action of Proteins* by Richard E. Dickerson and Irving Geis, W. A. Benjamin, Inc., Menlo Park, California, Publisher; copyright © 1969 by Dickerson and Geis. **17.19** (A) Colin Beer, Rutgers University, Newark; (B) Courtesy John Sparks, BBC (Natural History). **17.20** (A) Courtesy E. S. Ross; (B) Courtesy M. Morcombe; (C) © Michael Fogden/Bruce Coleman, Inc.; (D) Courtesy D. J. Howell, Purdue University. **17.22** Courtesy Ken N. Paige, Northern Arizona University. **17.23** Derek Washington/Bruce Coleman, Inc. **17.24** (left) Oxford Scientific Films/Bruce Coleman, Inc. **17.25** (top) Courtesy Jeff Rotman; (bottom) Jane Burton/Bruce Coleman, Inc. **17.26** (Top) David Overcash/Bruce Coleman, Inc.; (Bottom) John Shaw/Bruce Coleman, Inc. **17.27** (left) Peter Ward/Bruce Coleman, Inc. **17.28** Jane Burton/Bruce Coleman, Inc. **17.29** Courtesy E. S. Ross. **17.30** Breck P. Kent, Animals Animals. **17.31** Courtesy E. R. Degginger. **17.32** (A) James L. Castner, University of Florida; (B) Courtesy E. S. Ross; (C) David B. Grobecker, Pacific Ocean Research Foundation, Kailu-Kona. **17.33** (top) D. Overcash; (bottom) J. Shaw/Bruce Coleman, Inc. **17.34** Courtesy Douglas Faulkner. **17.35** (A) © Steinhardt Aquarium, Tom McHugh/Photo Researchers; (B) © Joe McDonald/Bruce Coleman, Inc. **17.36** R. P. Carr/Bruce Coleman, Inc. **17.37** Tr. #3058, courtesy Department Library Services, American Museum of Natural History, from H. Kutter, *Neujahrsblatt herausgegeben von der Naturforschenden Gesellschaft in Zurich*, vol. 171, 1969.

18.1 Irv Kornfield and Jeffrey N. Taylor, *Proc. Biol. Soc. Washington*, vol. 96. 1983. **18.2** (A) Modified from W. Auffenberg, *Tulane Stud. Zool. Bot.*, vol. 2, 1955; (B,D) Grant Heilman Photography; (C) modified from R. E. Woodson, *Ann. Mo. Bot. Gard.*, vol. 34, 1947. **18.3** Modified from J. Clausen et al., *Carnegie Inst. Washington Publ.*, no. 581, 1958. **18.4** (above) Lynn M. Stone; (below) Joseph Van Wormer/Bruce Coleman, Inc. **18.5** Redrawn from W. Auffenberg, *Tulane Stud. Zool. Bot.*, vol. 2, 1955. **18.6** Pat and Tom Leeson, Photo Researchers, Inc. **18.8** Modified from B. Wallace and A. M. Srb, *Adaptation*, copyright © 1964 by permission of Prentice-Hall, Inc., Englewood Cliffs, New Jersey. **18.9** J. Shaw/Bruce Coleman, Inc. **18.10** Courtesy E. S. Ross. **18.11** Courtesy T. K. Wood, University of Delaware. **18.13** Modified from D. Lack, *Darwin's Finches*, Cambridge, The University Press, 1947. **18.15** Drawing courtesy Sophie Webb. **18.16** From D. Lack, *Darwin's Finches*, Cambridge, The University Press, 1947. **18.17** Courtesy Irenäus Eibl-Eibesfeldt, Max Planck Institute for Behavioral Physiology. **18.18** Drawing courtesy H. Douglas Pratt. **18.19** Modified from G. F. Gause, *Science*, vol. 79, 1934; copyright © 1934 by A.A.A.S. **18.21** Illustrations by Marianne Collins are reproduced from *Wonderful Life: The Burgess Shale and the Nature of History*, by Stephen Jay Gould, with permission of W. W. Norton & Company, Inc.; copyright © 1989 Stephen Jay Gould. **18.22** Jane Burton/Bruce Coleman, Inc. **18.23** Redrawn from D. Lack, *Darwin's Finches*, Cambridge, The University Press, 1947. **18.26** (A) Chicago Zoological Park; photograph by Tom McHugh, Photo Researchers, Inc.; (B,C) Courtesy M. Morcombe; (D) Jack Fields, Photo Researchers, Inc.

PART IV THE GENESIS AND DIVERSITY OF ORGANISMS

(1) © William E. Ferguson; (2) © William E. Ferguson; (3) © David Scharf/Peter Arnold, Inc.; (4) © Jane Burton/Bruce Coleman, Inc.

19.1 Photographs courtesy NASA. **19.2** Courtesy D. W. Deamer, University of California, Davis. **19.3** Modified from R. E. Dickerson, *Sci. Am.*, September 1978; copyright © 1978 by Scientific American, Inc.; all rights reserved. **19.4** Courtesy Sigurgeir Jónasson. **19.6** Courtesy Sidney W. Fox, University of Miami, and Steven Brooke Studios, Coral Gables, Florida. **19.9** (A) TASS/Sovfoto; (B) courtesy of NASA; (C) Michael H. Carr, U. S. D. I. Branch of Astrogeology; (D) courtesy of NASA. **19.10** Courtesy J. W. Schopf, University of California, Los Angeles. **19.11** G. R. Roberts. **19.12** (A) Courtesy R. E. Lee, University of the Witwatersrand; (B) MMJP Plant Res. Lab., East Lansing. **19.13** D. A. Stetler and W. M. Laetsch, *Am. J. Bot*, vol. 56, 1969. **19.14** W. J. Larsen, *J. Cell Biol.*, vol. 47, 1970, by copyright permission of the Rockefeller University Press. **19.15** Courtesy I. B. Dawid, National Institutes of Health, Bethesda, Maryland, and D. R. Wolstenholme, University of Utah. **19.16** Modified from *Five Kingdoms* by Lynn Margulis and Karlene V. Schwartz, W. H. Freeman, New York; copyright © 1982. **19.19** (A, C) Courtesy E. V. Gravé; (B) courtesy of E. B. Daniels, from K. W. Jeon (ed.) *The Biology of Amoeba*, 1973, Academic Press, Orlando.

20.1 (A) Courtesy R. C. Williams, University of California, Berkeley; (B) courtesy M. Wurtz, University of Basel; (C) courtesy M. Gomersall, McGill University. **20.3** (bottom) From K. Corbett, *Virology*, vol. 22, 1964; reprinted by permission of Academic Press, Inc., New York. **20.4** Photo courtesy T. O. Diener, U. S. Department of Agriculture. **20.5** From T. O. Diener, *Am Sci.*, vol. 71, 1983. **20.8** (left) Courtesy David Scharf/Peter Arnold, Inc.; (right) Center for Disease Control, Atlanta, Georgia. **20.9** Turtox/Cambosco, Macmillan Science Co., Inc. **20.10** Courtesy Z. Skobe, Forsyth Dental Center/BPS. **20.11** Courtesy M. Gomersall, McGill University. **20.12** S. Kimoto and J. C. Russ, *Am. Sci.*, vol. 57, 1969. **20.14** G. B. Chapman, *J. Bacteriol.*, vol. 71, 1956. **20.15** Photograph by Ginny Fonte, from H. C. Berg, *Sci. Am.*, August 1975; copyright © 1975 by Scientific American, Inc.; all rights reserved. **20.16** From R. C. Johnson, M. P. Walsh, B. Ely, and L. Shapiro, *J. Bacteriol.*, vol. 138, 1979. **20.17** Courtesy W. Burgdorfer, S. F. Hayes, and D. Corwin, Rocky Mountain Laboratories. **20.18** Courtesy David Chase, Veterans Hospital, Sepulveda, California. **20.19** Courtesy Jack Griffith, University of North Carolina. **20.20** H. Reichenbach, Gesellschaft für Biotechnologische Forschung, mbH. **20.21** Courtesy Elliot Scientific Corp. **20.22** Courtesy Elliot Scientific Corp. **20.23** (A, B, C) T. E. Adams/Bruce Coleman, Inc. **20.24** Courtesy R. Malcolm Brown, Jr., University of Texas. **20.25** VU/© S. Thompson/Visuals Unlimited. **20.26** Courtesy Zell A. McGee and E. N. Robinson, Jr., University of Utah. **20.27** Photograph by Walther Stoeckenius, courtesy Carl Woese, University of Illinois. **20.28** From W. J. Jones, J. A. Leigh, F. Mayer, C. R. Woese, and R. S. Wolfe, *Arch. Microbiol.*, vol. 136, 1983; copyright © 1983 by Springer-Verlag. **20.29** H. W. Jannasch and C. O. Wirsen, *BioScience*, vol. 29, 1979; copyright © 1979 by the American Institute of Biological Sciences.

21.1 Photograph courtesy E. W. Daniels; art modified from K. W. Jeon (ed.), *The Biology of the Amoeba*, 1973, Academic Press, Orlando. **21.2** J. M. Jensen and S. R. Wellings, *J. Protozool.*, 19(2), 1972. **21.5** Courtesy of E. V. Gravé. **21.6** Photograph courtesy Ed Reschke. **21.7** Courtesy E. V. Gravé. **21.8** Courtesy E. V. Gravé. **21.9** Oxford Scientific Films/Bruce Coleman, Inc. **21.10** Courtesy E. V. Gravé. **21.12** (A) Courtesy E. V. Gravé; (B) H. Chaumeton/Nature; (C) Roman Vishniac Archives at the International Center of Photography New York. **21.13** Courtesy Thomas Eisner, Cornell University. **21.14** Manfred Kage/Peter Arnold, Inc. **21.17** Ray Simons/Photo Researchers, Inc. **21.20** (A, B) Courtesy K. B. Raper, *Proc. Am. Philos. Soc.*, vol. 104, 1960; (C–F) Carolina Biological Supply Company. **21.21** Photograph courtesy E. V. Gravé. **21.22** Courtesy E. V. Gravé.

22.2 B. S. C. Leadbeater, University of Birmingham. **22.4** Turtox/Cambosco, Macmillan Science Co., Inc. **22.5** E. R. Degginger/Bruce Coleman, Inc. **22.6** H. Chaumeton/Nature. **22.7** Anne Wertheim/Bruce Coleman, Inc. **22.8** Courtesy Douglas Faulkner. **22.10** (right) R. P. Carr/Bruce Coleman, Inc. **22.13** H. Chaumeton/Nature. **22.16** F. Sauer/Nature. **22.17** Courtesy Ed Reschke. **22.18** Roman Vishniac Archives at the International Center of Photography New York. **22.19** Courtesy Roman Vishniac. **22.21** Roman Vishniac Archives at the International Center of Photography New York. **22.26** E. R. Degginger/Bruce Coleman, Inc. **22.29** Modified from H. J. Fuller and O. Tippo, *College Botany*, Holt, Rinehart & Winston, Inc., New York, 1954. **22.31** Jane Burton/Bruce Coleman, Inc. **22.32** (A–C) From G. Shih and R. Kessel, *Living Images: Biological Microstructures Revealed by Scanning Electron Microscopy.* © 1982 by Science Books International; (D) Stephen Dalton/Photo Researchers, Inc. **22.33** Adrian Davies/Bruce Coleman, Inc. **22.35** Adrian Davies/Bruce Coleman, Inc. **22.37** Courtesy E. R. Degginger. **22.39** Portion of group in Carnegie Museum, Pittsburgh; used by permission. **22.40** Courtesy Field Museum of Natural History, Chicago. **22.41** Courtesy E. S. Ross. **22.42** John Shaw/Bruce Coleman, Inc. **22.43** Courtesy G. R. Roberts. **22.44** Ray Simons/Photo Researchers. Inc. **22.45** Modified from H. J. Fuller and O. Tippo, *College Botany*, Holt, Rinehart & Winston, Inc., New York, 1954. **22.48** Ed Reschke/Peter Arnold, Inc. **22.49** Courtesy Thomas Eisner, Cornell University. **22.50** Courtesy Nels R. Lersten, Iowa State University. **22.51** (left) Courtesy Victor B. Eichler; (right) courtesy Thomas Eisner, Cornell University. **22.54** Redrawn from H. N. Andrews, *Science*, vol. 142, 1963; copyright © 1963 by AAAS. **22.55** Courtesy E. S. Ross. **22.56** Roman Vishniac Archives at the International Center of Photography, New York. **22.57** From W. Krommenhoek, J. Sebus, and G. J. van Esch, *Biological Structures*, copyright © 1979 by L. C. G. Malmberg B. V., The Netherlands. **22.58** (A, B) Courtesy Thomas Eisner, Cornell University; (C) courtesy E. S. Ross. **22.59** Modified from H. J. Fuller and O. Tippo, *College Botany*, Holt, Rinehart & Winston, Inc., New York, 1954. **22.63** Courtesy E. R. Degginger **22.65** Courtesy E. S. Ross. **22.67** (right) Jane Burton, and (left) R. P. Carr/Bruce Coleman, Inc. **22.68** Modified from H. J. Fuller and O. Tippo, *College Botany*, Holt, Rinehart & Winston, Inc., New York, 1954.

23.1 Courtesy Pfizer Inc. **23.3** Photograph W. H. Amos/Bruce Coleman, Inc. **23.5** M. P. L. Fogden/Bruce Coleman, Inc. **23.6** From L. W. Sharp, *Fundamentals of Cytology*, McGraw-Hill Book Co., New York, copyright © 1943; used by permission. **23.9** (A) Jane Burton/Bruce Coleman Inc; (B) L. West/Bruce Coleman, Inc.; (C, D) V. Ahmadjian and J. B. Jacobs, *Nature*, vol. 289, 1981 © 1981, Macmillan Journals Ltd. **23.10** (left) Masana Izawa and (right) Satoshi Kuribyashi/Nature Production, Tokyo. **23.11** From L. W. Sharp, *Fundamentals of Cytology*, McGraw-Hill Book Co., New York, copyright © 1943; used by permission.

24.1 Courtesy Jeff Rotman. **24.4** Oxford Scientific Films. **24.5** H. Chaumeton/Nature. **24.8** Carolina Biological Supply Company. **24.11** R. N. Mariscal/Bruce Coleman, Inc. **24.12** Courtesy Howard Hall. **24.16** After K. G. Grell, University of Tübingen (1974). **24.17** Courtesy K. G. Grell, University of Tübingen. **24.19** After K. G. Grell, University of Tübingen. **24.21** H. Chaumeton/Nature. **24.26** CNRI/Science Photo Library–Photo Researchers, Inc. **24.28** (left) Courtesy Ed Rescke; (right) H. Chaumeton/Nature. **24.29** H. Chaumeton/Nature. **24.33** Based on a phylogenetic tree drawn by R. P. Higgins. **24.35** Adapted from *Life* by W. K. Purves and G. H. Orians, Sinauer Associates, Sunderland, Massachusetts; copyright © 1983. **24.36** Courtesy T. E. Adams. **24.38** H. Chaumeton/Nature. **24.39** H. Chaumeton/Nature. **24.41** Fred Bavendam/Peter Arnold, Inc. **24.42** Oxford Scientific Films. **24.44** H. Chaumeton/Nature. **24.45** (left) H. Chaumeton/Nature; (right) Rod Borland/Bruce Coleman, Inc. **24.46** Based in part on drawings by Louise G. Kingsbury. **24.47** F. Sauer/Nature. **24.48** H. Chaumeton/Nature. **24.49** Modified from W. Stempell, *Zoologie im Grundriss*, Borntraeger, 1926. **24.50** H. Chaumeton/Nature. **24.52** Photograph courtesy J. R. Pawlik, University of North Carolina, Wilmington. **24.53** E. R. Degginger/Bruce Coleman, Inc. **24.54** Oxford Scientific Films. **24.55** Courtesy E. S. Ross. **24.56** Jane Burton/Bruce Coleman, Inc. **24.58** Courtesy Ed Reschke. **24.59** © Gunter Ziesler/Peter Arnold, Inc. **24.60** E. R. Degginger/Bruce Coleman, Inc. **24.61** H. Chaumeton/Nature. **24.62** (top) H. Chaumeton/Nature; (bottom) Hans Reinhard/Bruce Coleman, Inc. **24.63** (A, D) H. Chaumeton/Nature; (B) Kim Taylor/Bruce Coleman, Inc.; (C) G. R. Roberts. **24.64** Oxford Scientific Films. **24.65** Courtesy G. R. Roberts. **24.66** A. Cosmos Blank, National Audubon Society/Photo Researchers, Inc. **24.68** Modified from T. I. Storer and R. L. Usinger, *General Zoology*, McGraw-Hill Book Co., New York, copyright © 1957; used by permission. **24.69** Courtesy Stephen Dalton/NHPA. **24.73** Modified from L. H. Hyman, *The Invertebrates*, McGraw-Hill Book Co., New York, copyright © 1955; used by permission. **24.74** (left) Robert Dunne, and (right) Charlie Ott/Photo Researchers, Inc. **24.75** © Andrew J. Martinez/Photo Researchers, Inc. **24.76** A. Kerstitch, Sea of Cortez Enterprises. **24.77** Courtesy Smithsonian Institution, photo #62419. **24.78** H. Chaumeton/Nature.

25.1 Photograph by Bill Wood/Bruce Coleman, Inc. **25.3** Photograph by H. Chaumeton/Nature. **25.4** Courtesy Ed Reschke. **25.6** Courtesy Ed Reschke. **25.7** Courtesy Ed Reschke. **25.8** (left) D. Claugher, by courtesy of the Trustees, The British Museum (Natural History); (left) courtesy Jerome Gross, Massachusetts General Hospital. **25.10** (right) Ed Reschke/Peter Arnold, Inc.; (left) Manfred P. Kage/Peter Arnold, Inc. **25.12** Photograph courtesy Ed Reschke. **25.14** Courtesy Heather Angel, Biofotos. **25.16** Modified from A. S. Romer, *The Vertebrate Body*, W. B. Saunders, 1949. **25.17** Chip Clark, courtesy Field Museum of Natural History, GEO 84533, Chicago. **25.18** Courtesy Howard Hall. **25.19** Courtesy The Royal Society, London. **25.20** Photograph by Chip Clark, courtesy National Museum of Natural History, Washington, D.C. **25.21** (left) S. C. Bisserot, and (right) Hans Reinhard/Bruce Coleman, Inc. **25.22** Courtesy E. S. Ross. **25.23** Hans Pfletschinger/Peter Arnold, Inc. **25.24** (A) G. R. Roberts; (B) Ferrero/Nature; (C,

D) Hans Reinhard/Bruce Coleman, Inc.; (E) John Moss/Photo Researchers, Inc. **25.26** From a mural by Maidi Wiebe, courtesy Field Museum of Natural History, Chicago. **25.27** (top) Royal Tyrrell Museum, Alberta Culture and Multiculturalism; (bottom) © Gregory S. Paul. **25.28** © Gregory S. Paul. **25.29** © Gregory S. Paul. **25.30** (B, C) Tektile glass from Mimbral, Mexico, courtesy Stanley V. Margolis, University of California, Davis. **25.31** Courtesy D. G. Allen. **25.32** Warren Garst/Tom Stack Associates. **25.33** Courtesy Australian Information Service. **25.34** Courtesy American Museum of Natural History. **25.35** Grospas/Nature. **25.37** M. P. L. Fogden/Bruce Coleman, Inc. **25.38** (left) Jan Lindblad/Photo Researchers, Inc.; (right) Peter Jackson/Bruce Coleman, Inc. **25.39** Jack Dermid/Bruce Coleman, Inc. **25.40** Monkey Jungle, Miami, Florida, R. P. Fontaine/Photo Researchers, Inc. **25.41** Brian Parker/Tom Stack and Associates. **25.42** Baron Hugo van Lawick, copyright © National Geographic Society. **25.45** From R. E. Leakey and R. Lewin, *Origins*, Dutton, New York, 1977; reproduced by permission of Rainbird Publishing Group. **25.46** Margo Crabtree, courtesy AAAS, used by permission of (A) University of the Witwatersrand Medical School, and (B–D) the National Museum of Kenya, Nairobi. **25.47** Neg. #333242, Courtesy Department of Library Services, American Museum of Natural History. **25.49** David Brill, © 1985 National Geographic Society; photographed at the Institut du Quarternaire, Université de Bordeaux, Talence, France.

PART V THE BIOLOGY OF ORGANISMS

(1) Carolina Biological Supply Company; (2) M. P. L. Fogden/Bruce Coleman, Inc.; (3) Courtesy Dr. D. W. Tank, AT&T Bell Laboratories; (4) Courtesy Philip Green.

26.4 G. R. Roberts. **26.5** Courtesy Ed Reschke. **26.8** Carolina Biological Supply Company. **26.12** Breck P. Kent/Earth Scenes. **26.13** U. N. Food and Agricultural Organization. **26.14** (A) Courtesy Hans W. Paerl, University of North Carolina; (B) Hans W. Paerl and Kathleen K. Gallucci, *Science*, vol. 227, 1985; copyright © 1985 by AAAS. **26.15** Courtesy E. R. Degginger. **26.16** (left) Oxford Scientific Films; (right) D. Lyons/Bruce Coleman, Inc. **26.17** Courtesy E. R. Degginger.

27.1 E. S. Ross. **27.2** Courtesy R. Farquharson, UNICEF. **27.4** From the *Vitamin Manual*, courtesy The Upjohn Company. **27.5** From the *Vitamin Manual*, courtesy The Upjohn Company. **27.7** (A–C) David Pramer, *Science*, vol. 144, 1964; copyright © 1964 by AAAS; (D) Courtesy G. Barron and N. Allin, University of Guelph. **27.8** Courtesy K. G. Grell, University of Tübingen. **27.12** Modified from W. D. Russell-Hunter, *A Biology of Lower Invertebrates*, Macmillan Publishing Co., New York, 1968. **27.13** Photograph courtesy E. V. Gravé. **27.14** Oxford Scientific Films/Bruce Coleman, Inc. **27.15** Turtox/Cambosco, Macmillan Science Co., Inc. **27.17** (A) Courtesy Howard Hall; (B) courtesy Thomas Eisner; (C) courtesy William A. Watkins. **27.20** Adapted from an original painting by Frank H. Netter, M.D., from *The CIBA Collection of Medical Illustrations*, copyright © 1959 by CIBA Pharmaceutical Company, division of CIBA-GEIGY Corp. **27.25** Adapted from Norman Kretckmer, *Sci. Am.*, October 1972; copyright © 1972 by Scientific American Inc.; all rights reserved. **27.26** From Warren Andrew, *Textbook of Comparative Histology*, Oxford, The University Press, 1959. **27.27** (top) Courtesy Susumu Ito, Harvard Medical School. **27.30** E. and P. Bauer, Wildstock.

28.3 Courtesy Thomas Eisner, Cornell University. **28.4** Courtesy J. H. Troughton, Department of Scientific and Industrial Research, Wellington, New Zealand. **28.6** Courtesy J. H. Troughton, Department of Scientific and Industrial Research, Wellington, New Zealand. **28.8** (top) Courtesy Thomas Eisner, Cornell University; (bottom) G. R. Roberts. **28.10** Modified from Ralph Buchsbaum, *Animals Without Backbones*, by permission of the University of Chicago Press, copyright © 1948 by the University of Chicago. **28.12** Courtesy J. H. Carmichael, Jr. **28.16** © F. Sauer/Nature. **28.17** Bob Gossington/Bruce Coleman, Inc. **28.18** Photograph from R. G. Kessel and R. H. Kardon, *Tissues and Organs: A Text-Atlas of Scanning Electron Microscopy*, © W. H. Freeman and Company, 1979. **28.23** Courtesy H-R. Duncker, Justus Liebig University, Geissen, Germany. **28.25** Modified from R. Margaria et al., *J. Appl. Physiol.*, vol. 18, 1963. **28.26** Roman Vishniac Archives at the International Center of Photography, New York. **28.28** (left) Photograph by D. Claugher, by courtesy of the Trustees, The British Museum (Natural History); (right) from Warren Andrew, *Textbook of Comparative Histology*, Oxford, The University Press, 1959. **28.29** Kim Taylor/Bruce Coleman, Inc.

29.3 Photograph courtesy Thomas Eisner, Cornell University. **29.4** Courtesy Nels R. Lersten, Iowa State University. **29.5** Courtesy Ed Reschke. **29.7** Photograph courtesy Ed Reschke. **29.9** B. Bracegirdle; from W. Krommenhoek, J. Sebus, and G. J. van Esch, *Biological Structures*, copyright © 1979 by L. C. G. Malmberg B. V., The Netherlands. **29.10** Courtesy Thomas Eisner, Cornell University. **29.11** Modified from V. A. Greulach and J. E. Adams, *Plants: An Introduction to Modern Botany*, Wiley, New York, 1962. **29.14** (A) Courtesy J. H. Troughton, Department of Scientific and Industrial Research, Wellington, New Zealand; (B, C) courtesy B. G. Butterfield, Canterbury University, and B. A. Meylan, Department of Scientific and Industrial Research, Wellington, New Zealand. **29.15** Courtesy Thomas Eisner, Cornell University. **29.16** Courtesy Thomas Eisner, Cornell University. **29.19** Laura Riley/Bruce Coleman, Inc. **29.21** S. and O. Biddulph et al., *Plant Physiol.*, vol. 33, no. 4, 1958. **29.23** M. H. Zimmermann, *Science*, vol. 133, 1961; copyright © 1961 by AAAS.

30.5 Micrograph from R. G. Kessel and R. H. Kardon, *Tissues and Organs: A Text-Atlas of Scanning Electron Microscopy*, © W. H. Freeman, San Francisco, 1979. **30.7** Modified from B. S. Guttman and J. W. Hopkins III, *Understanding Biology*; copyright © 1983 by Harcourt Brace Jovanovich, Inc.; used by permission of the publisher. ***Exploring Further*** p. 854, Roman Vishniac Archives at the International Center of Photography, New York; p. 855, courtesy Ed Reschke. **30.13** (left) Courtesy Thomas Eisner, Cornell University; (right) L. Nilsson; from L. Nilsson, *Behold Man*, English translation copyright © 1974 by Albert Bonniers Förlag, Stockholm, and Little, Brown and Co. (Canada) Ltd. **30.14** (left) Courtesy D. W. Fawcett, Harvard Medical School; (right) Micrograph from R. G. Kessel and R. H. Kardon, *Tissues and Organs: A Text-Atlas of Scanning Electron Microscopy*, W. H. Freeman, San Francisco, copyright © 1979. **30.18** Manfred Krage/Peter Arnold Inc. **30.20** Courtesy Turtox/Cambosco, Macmillan Science Co., Inc. **30.27** Courtesy Eila Kairinen, Gillette Research Institute. **30.28** Courtesy K. R. Porter and Ginny Fonte, University of Colorado.

31.1 Modified from E. G. Bollard, *J. Exp. Bot.*, vol. 4, 1953. **31.7** Modified from E. Baldwin, *An Introduction to Comparative Biochemistry*, Cambridge, The University Press, 1948. **31.9** After D. L. Hopkins, *Biol. Bull.*, vol. 90, 1946. **31.10** After E. H. Mercer, *Proc. Roy. Soc. Lond. B*, vol. 150, 1959. **31.11** Photographs courtesy Thomas Eisner, Cornell University. **31.12** Modified from Ralph Buchsbaum, *Animals Without Backbones*, by permission of the University of Chicago Press, copyright © 1948 by the University of Chicago. **31.17** Modified from H. W. Smith, *The Kidney*, Oxford, The University Press, 1951. **31.18** M. G. Farquhar, University of California, San Diego. **31.19** (A) Courtesy F. Spinelli, CIBA-GEIGY Research Laboratory, Basel, Switzerland.

32.2 (E) Redrawn from M. Schaffner, *The Ohio Naturalist*, 1906. **32.4** (B) Courtesy G. R. Roberts. **32.6** Photograph courtesy Donald Nevins, Iowa State University. **32.7** (left) Modified from K. Esau, *Plant Anatomy*, Wiley, New York, 1965; (right) courtesy E. R. Degginger. **32.8** Modified from Peter Albersheim, *Sci. Am.*, April 1975; copyright © 1975 by Scientific American, Inc.; all rights reserved. **32.11** Modified from V. A. Greulach and J. E. Adams, *Plants: An Introduction to Modern Botany*, Wiley, New York, 1962. **32.12** Redrawn from *Biologie: Ein Lehrbuch*, edited by G. Czihak et al., 2nd ed., Springer-Verlag, 1978. **32.13** © Ed Reschke. **32.14** (left) E. R. Degginger; (right) Used by permission from J. D. Dodd, *Course Book in General Botany*, copyright © 1977 by the Iowa State University Press. **32.15** Carolina Biological Supply Company. **32.18** Redrawn after Z. Schwartz-Sommer et al., *Science*, vol. 250, 1990; copyright © 1990 by AAAS. **32.27** Courtesy C. R. Hawes, Oxford Polytechnic. **32.31** Carolina Biological Supply Company. **32.33** Courtesy Robert Newman, University of Wisconsin. **32.35** Courtesy Sylvan Wittwer. **32.37** Redrawn from *Biologie: Ein Lehrbuch*, edited by G. Czihak et al., 2nd ed., Springer-Verlag, 1978. **32.40** Photograph by Stephen Gladfelter, Stanford University, from research by P. W. Oeller, Lu M.-W., L. P. Taylor, D. A. Pike, and A. Theologis at the Plant Gene Expression Center, University of California, Berkeley. **32.45** After W. A. Jensen and F. B. Salisbury, *Botany: An Ecological Approach*; copyright © 1972 by Wadsworth Publishing Co., Inc., Belmont, California; used by permission of the publisher. **32.46** Photo by M. E. Nuttall, DuPont Experimental Station, courtesy Ilya Raskin, Rutgers State University, New Brunswick.

33.1 Modified from H. A. Schneiderman and L. I. Gilbert, *Science*, vol. 143, 1964; copyright © 1964 by AAAS. **33.3** Painting by Frank H. Netter, M.D.; reprinted with permission from *The CIBA Collection of Medical Illustrations*, copyright © 1965 by CIBA Pharmaceutical Company, division of CIBA-GEIGY Corporation; all rights reserved. **33.5** Carolina Biological Supply. **33.10** © John P. Kay/Peter Arnold, Inc. **33.12** From A. J. Carlson et al., *The Machinery of the Body*, University of Chicago Press, 1961. **33.15** Modified from H. Curtis, *Biology*, 4th ed.; copyright © 1983 by Worth Publishers, Inc. **33.25** Adapted from an original model of a morphine molecule, courtesy Maitland Jones, Jr.

34.1 © Lionel Atwill/Peter Arnold, Inc. **34.2** Michael Fogden/Bruce Coleman, Inc. **34.3** © Ken M. Highfill/Photo Researchers, Inc. **34.7** (A) Courtesy D. M. Phillips, Population Council, New York; (B) from P. Motta, P. M. Andrews, and K. R. Porter, *Microanatomy of Cell and Tissue Surfaces*, Lea & Febiger, 1978; copyright © 1978 by Casa Editrice Dr. Francesco Vallardi, Societa Editrice Libraria, Milan, Italy. **34.10** (A/C) Courtesy Thomas Eisner, Cornell University; (B) C. Edelman—La Villette/Photo Researchers, Inc.

35.2 Courtesy S. L. Palay, Harvard Medical School. **35.4** (A) Courtesy H. deF. Webster, as printed in W. Bloom and D. W. Fawcett, *A Textbook of Histology*, 10th ed., W. B. Saunders Co., 1975; (B) R. L. Roberts, R. G. Kessel and H. N. Tung, *Freeze Fracture Images of Cells and Tissues*, Oxford University Press, 1991. **35.6** (A) H. Chaumeton/Nature; (B) adapted from E. R. Kandel, *Cellular Basis of Behavior*, W. H. Freeman, San Francisco, 1976. **35.7** Modified from E. R. Kandel, *Cellular Basis of Behavior*, W. H. Freeman, San Francisco, 1976. **35.10** Modified from T. H. Bullock and G. A. Horridge, *Structure and Function of the Nervous System of Invertebrates*, W. H. Freeman, San Francisco, 1965. **35.11** Modified from L. H. Hyman, *The Invertebrates*, vol. 2, McGraw-Hill Book Co., New York, copyright © 1951; and from Ralph Buchsbaum, *Animals Without Backbones*, University of Chicago Press, copyright © 1948 by the University of Chicago, used by permission of the publishers. **35.12** Modified from R. Goldschmidt, *Z. Wiss. Zool., Abt. A.*, vol. 92, 1909. **35.13** Photograph courtesy Thomas Eisner, Cornell University. **35.17** Courtesy A. L. Hodgkin, *J. Physiol. (London)*, vol. 131, 1956. **35.19** Adapted from W. A. Catterall, *Science*, vol. 242, 1988; copyright © 1988 by AAAS. **35.24** E. R. Lewis et al., *Science*, vol. 165, 1969; copyright © 1969 by AAAS. **35.25** R. L. Roberts, R. G. Kessel and H. N. Tung, *Freeze Fracture Images of Cells and Tissues*, Oxford University Press, 1991. **35.30** Photograph courtesy Ed Reschke. **35.31** Courtesy J. E. Heuser, Washington University Medical Center.

36.1 Redrawn from B. Katz, *J. Physiol. (London)*, vol. 111, 1950. **36.3** Graph modified from S. Skoglund, *Acta Physiol. Scand., Suppl. 124*, vol. 36, 1956. **36.6** Courtesy Ed Reschke. **36.7** After Murray, 1973. **36.8** D. Claugher, by courtesy of the Trustees, The British Museum (Natural History). **36.10** J. L. Lepore/Photo Researchers, Inc. **36.13** J. A. L. Cooke, Oxford Scientific Films. **36.14** Modified from R. E. Snodgrass, *Principles of Insect Morphology*, McGraw-Hill Book Co., New York, copyright © 1935; used with permission of the McGraw-Hill Book Co. **36.15** Courtesy J. L. Gould and C. G. Gould. **36.16** Douglas Faulkner/Photo Researchers, Inc. **36.19** Photograph by L. Nilsson; from L. Nilsson, *Behold Man*, English translation © 1974, Albert Bonniers Förlag, Stockholm, and Little, Brown and Co. (Canada) Ltd. **36.20** E. R. Lewis, F. S. Werblin, and Y. Y. Zeevi, University of California, Berkeley. **36.24** After A. F. MacNichol. **36.26** (A) Courtesy Thomas Eisner, Cornell University; (B) courtesy E. S. Ross. **36.28** Courtesy A. J. Hudspeth and R. A. Jacobs from A. J. Hudspeth, *Nature*, 341:397–404 (1989). **36.29** Modified from G. von Békésy, *Symp. Soc. Exp. Biol.*, vol. 16, 1962; G. von Békésy, *Experiments in Hearing*, McGraw-Hill Book Co., New York, copyright © 1960, used with the permission of the McGraw-Hill Book Co.; M. S. Gordon, G. Bartholomew, A. D. Grinnell, C. B. Jørgensen, and F. N. White, *Animal Function: Principles and Adaptations*, Macmillan, New York, 1968, copyright © 1968 by M. S. Gordon. **36.30** L. Nilsson; from L. Nilsson, *Behold Man*, English translation © 1974, Albert Bonniers Förlag, Stockholm, and Little, Brown and Co. (Canada) Ltd. **36.32** (A) © William E. Ferguson; (B) S. L. Craig/Bruce Coleman, Inc.; (C) E. R. Degginger/Bruce Coleman, Inc. **36.34** H. Vali, courtesy R. J. P. Williams, Oxford University. **36.36** Modified from W. M. Cornsweet, *Visual Perception*, Academic Press, New York, 1970; and S. Coren et al., *Sensation and Perception*, Academic Press, New York, 1978. **36.39** Modified from A. S. Romer and T. S. Parsons, *The Vertebrate Body*, 5th ed., copyright © 1977 by W. B. Saunders Company, reprinted by permission of CBS College Publishing. **36.40** Modified from A. S. Romer and T. S. Parsons, *The Vertebrate Body*, 5th ed., copyright © 1977 by W. B. Saunders Company, reprinted by permission of CBS College Publishing; and G. G. Simpson, C. S. Pittendrigh, and L. H. Tiffany, *Life: An Introduction to Biology*, copyright © 1957 by Harcourt Brace Jovanovich, Inc., used by permission of the publishers. **36.44** Modified from W. Penfield and T. Rasmussen, *The Cerebral Cortex of Man*, Macmillan, New York, 1950. **36.49** Based on data from John Allman, California Institute of Technology. **36.50** Institute of Medicine, National Academy of Sciences Press.

37.1 G. R. Roberts. **37.3** Photograph by D. Claugher, by courtesy of the Trustees, The British Museum (Natural History); drawing modified from J. Gray and H. W. Lissman from *J. Exp. Biol.*, vol. 15, 1938, by permission of Company of Biologists, Ltd. **37.4** H. Chaumeton/Nature. **37.7** © Biophoto Association/Photo Researchers, Inc. **37.9** © Ed Reschke. **37.14** Courtesy H. E. Huxley, Cambridge University. **37.15** Courtesy H. E. Huxley, Cambridge University. **37.18** Courtesy J. E. Heuser, Washington University Medical Center.

38.1 Ian Wyllie/Survival Anglia. **38.2** (A) Courtesy Paul Trötschel; (B) Stephen Dalton/Photo Researchers, Inc. **38.3** Courtesy John Sparks, BBC (Natural History). **38.4** Courtesy Thomas Eisner, Cornell University. **38.5** Modified from N. Tinbergen, *The Study of Instinct*, Oxford, The University Press, 1951. **38.6** Modified from N. Tinbergen, *The Study of Instinct*, Oxford, The University Press, 1951. **38.7** Courtesy E. R. Willis, Illinois State University. **38.8** Modified from N. Tinbergen and A. C. Perdeck, *Behaviour*, vol. 3, 1950. **38.9** Courtesy John Sparks, BBC (Natural History). **38.10** Modified from J-P. Ewert, *Sci. Am.*, March 1974; copyright © 1974 by Scientific American, Inc.; all rights reserved. **38.11** Adapted from K. Lorenz and N. Tinbergen, *Z. Tierpsychol.*, vol. 2, 1938. **38.12** (A, B, D, E) From E. H. Hess, *Sci. Am.*, July 1956; copyright © 1956 by Scientific American, Inc.; all rights reserved; (C) Wallace Kirkland, copyright © 1954, Time Inc. **38.13** Reprinted with permission of the author and the publishers from D. G. Freedman, *Human Infancy: An Evolutionary Perspective*, Lawrence Erlbaum Associates, Hillsdale, New Jersey, 1975. **38.14** Adapted from N. E. Collias and E. C. Collias, *Auk*, vol. 79, 1962. **38.15** After Drees. **38.16** Courtesy R. Silver, Barnard College. **38.17** From K. Lorenz, *Symp. Soc. Exp. Biol.*, vol. 4, 1950. **38.18** Based on J. E. Lloyd, *Misc. Publ. Mus. Zool. Univ. Mich.*, vol. 130, 1966; and A. D. Carlson and J. Copeland, *Am. Sci.*, vol. 66, 1978, reprinted by permission of *American Scientist*, Journal of Sigma Xi, the Scientific Research Society. **38.19** D. R. Bentley, *Science*, vol. 174, 1971; copyright © 1971 by AAAS. **38.20** From L. P. Brower, J. V. Z. Brower, and F. P. Cranston, *Zoologica*, vol. 50, 1965; used by permission of the New York Zoological Society. **38.21** Courtesy R. Thornhill, University of New Mexico. **38.22** (B) Redrawn from K. von Frisch, *Bees: Their Vision, Chemical Senses, and Language*, copyright © 1950 by Cornell University; used by permission of Cornell University Press; photograph courtesy Kenneth Lorenzen, University of California, Davis. **38.23** Based on data from R. Boch, *Z. Vergl. Physiol.*, vol. 40, 1957. **38.24** From K. Lorenz, *Zool. Anz.*, vol. 17, 1953. **38.25** TASS from Sovfoto. **38.26** From W. R. Miles, *J. Comp. Psychol.*, vol. 10, 1930; copyright © 1930 by the Williams and Wilkins Co., Baltimore, Maryland. **38.27** (A) Eric Hosking/Bruce Coleman, Inc.; (B) Nina Leen, *Life*, copyright © 1964 by Time, Inc. **38.28** From E. H. Hess, *Science*, vol. 130, 1959; copyright © 1959 by AAAS. **38.30** Warren Gurst and Genny Gurst, Tom Stack and Associates. **38.31** Adapted by permission from *The Honey Bee* by James L. Gould and Carol Grant Gould, W. H. Freeman, New York, copyright © 1988. **38.32** Adapted by permission from E. Gwinner and W. Wiltschko, *J. Comp. Physiol.*, vol. 125, 1978; copyright © 1978 by Springer-Verlag, New York. **38.33** (A) From B. Elsner, in *Animal Migration, Navigation, and Homing*, edited by K. Schmidt-Koenig and W. T. Keeton, Springer-Verlag, New York, 1978; (B) from M. Michener and C. Walcott, *J. Exp. Biol.*, vol. 47, 1967. **38.34** (right) Steve Johnson, courtesy Charles Walcott, Cornell University; (left) from K. Schmidt-Koenig and C. Walcott, *Animal Behav.*, vol. 26, 1978. **38.35** Courtesy E. R. Willis, Illinois State University. **38.36** Modified from N. Tinbergen and A. C. Perdeck, *Behaviour*, vol. 3, 1950. **38.37** Courtesy John Sparks, BBC (Natural History). **38.38** Photograph by Russ Charif, courtesy Charles Walcott, Cornell University.

PART VI ECOLOGY

(1) © Kjell Sandved; (2) © William E. Ferguson; (3) © Norbert Wu, 1991; (4) © Norbert Wu, 1992.

39.3 Courtesy George W. Barlow, *Anim. Beh.* vol. 22, 1974. **39.4** Courtesy John Sparks, BBC (Natural History). **39.6** Modified from T. Carlson, *Biochem. Z.*, vol. 57, 1913. **39.7** Modified from J. Davidson, *Trans. R. Soc. South Aust.*, vol. 62, 1938. **39.9** Modified from E. P. Odum, *Fundamentals of Ecology*, W. B. Saunders, 1959, after Deevey. **39.10** Population Reference Bureau, Inc. Updated by Charles F. Westoff, Office of Population Research, Princeton University. **39.12** Courtesy E. S. Ross. **39.13** Courtesy L. David Mech. **39.14** Redrawn from C. B. Huffaker, *Hilgardia*, vol. 27, 1958. **39.15** Photograph by Hans Reinhard/Bruce Coleman, Inc.; drawing based on H. N. Southern, *J. Zool.*, vol. 162, 1970. **39.16** Based on H. N. Southern, *J. Zool.*, vol. 162, 1970. **39.17** Redrawn after J. M. Scriber and F. Slansky, *Am. Rev. Entomology*, vol. 26, 1981. **39.18** Adapted from *Principles of Animal Ecology* by W. C. Allee, A. E. Emerson, O. Park, T. Park, and K. P. Schmidt, W. B. Saunders, Philadelphia; copyright © 1949. **39.19** Modified from J. H. Connell, *Ecology*, vol. 142, 1961; copyright © 1961 by the Ecological Society of America. **39.20** Drawings modified from R. H. MacArthur, *Ecology*, vol. 39, 1958; copyright © 1958 by the Ecological Society of America. Photographs by (top) Philip Boyer; (middle) R. Austing/Photo Researchers, Inc.; and (bottom) Edgar T. Jones/Bruce Coleman, Inc. **39.21** G. Tortoli/F. A. O. Photo. **39.22** Carol Hughes/Bruce Coleman, Inc. **39.23** Norman Myers/Bruce Coleman, Inc. **39.24** Redrawn by permission from E. O. Wilson, *Sociobiology*, Harvard University Press, Cambridge, Massachusetts, 1975. **39.25** © Hans Pfletschinger/Peter Arnold, Inc. **39.26** Adapted from R. E. Kenward, *J. Anim. Ecol.*, vol. 47. 1978. **39.27** Courtesy R. D. Estes, San Diego State University. **39.28** Courtesy L. T. Nash, Arizona State University. **39.30** George D. Lepp, Bio-Tec Images. **39.31** Roberto Bunge, Ardea London Ltd. **39.32** Redrawn with permission of Macmillan Publishing Co., Inc., from David R. Pilbeam, *The Ascent of Man: An Introduction to Human Evolution;* copyright © 1972 by David R. Pilbeam. **39.33** (A) J. Tanaka, Anthro-Photo; (B) M. Shostak, Anthro-Photo. **39.34** R. Lee, Anthro-Photo. **39.35** Redrawn from R. M. May, *Sci. Am.*, September, 1978; copyright © 1978 by Scientific American, Inc.; all rights reserved. **39.37** Courtesy E. J. Kormondy, *Smithsonian Magazine*, vol. 1, 1970. **39.38** Carolina Biological Supply Company. **39.40** Based on data in E. P. Odum, *Fundamentals of Ecology*, W. B. Saunders, 1959. **39.41** Redrawn from R. M. May, *Sci. Am.*, September 1978; copyright © 1978 by Scientific American, Inc.; all rights reserved. **39.42** Redrawn from R. H. Whittaker, *Communities and Ecosystems*, 2nd ed., Macmillan, New York, 1975; after B. Holt and G. M. Woodwell. **39.43** Redrawn from R. H. Whittaker, *Communities and Ecosystems*, 2nd ed., Macmillan, New York, 1975; after B. Holt and G. M. Woodwell. **39.44** Gene Ahrens/Bruce Coleman, Inc.

40.4 Modified from E. P. Odum, *Fundamentals of Ecology*, W. B. Saunders, 1959. **40.7** From T. E. Graedel and P. J. Crutzen, *Sci. Am.*, September, 1989; copyright © 1989 by Scientific American Inc.; all rights reserved. **40.8** From S. H. Schneider, *Sci. Am.*, September, 1989; copyright © 1989 by Scientific American Inc.; all rights reserved. **40.9** Modified from M. J. Pelczar et al., *Microbiology*, 4th ed., McGraw-Hill Book Co., New York, copyright © 1977; used by permission. **40.10** Courtesy U. S. Department of Agriculture. **40.12** Courtesy D. W. Schindler, *Science*, vol. 184, 1974; copyright © 1974 by AAAS. **40.14** Courtesy P. L. Ames. **40.15** Redrawn after K. P. Bowman, *Science*, vol. 239, 1988; copyright © 1988 by AAAS. **40.17** U. S. D. A. Forest Service. **40.18** Redrawn from G. E. Likens et al., *Ecol. Monogr.*, vol. 40, 1970; copyright © 1970 by the Ecological Society of America. **40.19** Oxford Scientific Films, Earth Scenes. **40.21** C. J. Tucker, J. R. G. Towshend, and T. E. Goff, *Science*, vol. 227, 1985. **40.22** Photographs courtesy E. S. Ross. **40.23** After Whittaker (1970). Reprinted with permission of Macmillan Publishing Co., Inc., from *Communities and Ecosystems* by Robert Whittaker. Copyright © Robert H. Whittaker. **40.24** (left) Leonard Lee Rue III/Bruce Coleman, Inc.; (right) courtesy E. R. Degginger. **40.25** (left) Paolo Koch/Photo Researchers, Inc.; (right) Wolfgang Bayer/Bruce Coleman, Inc. **40.26** Zig Leszczynski, Earth Scenes. **40.27** (left) Ferrero/Nature; (right) N. de Vore III/Bruce Coleman, Inc. **40.28** Courtesy E. R. Degginger. **40.29** (A) Courtesy E. R. Degginger; (B, C) courtesy E. S. Ross. **40.31** Bob and Clara Calhoun/Bruce Coleman, Inc. **40.33** After Schwarzbach, 1950. **40.34** (A) Courtesy Howard Hall; (B) courtesy James D. Jordan; (C) courtesy E. H. Newcomb and T. D. Pugh, University of Wisconsin/BPS. **40.36** Redrawn from John Napier, *The Roots of Mankind*, copyright © 1970, Smithsonian Institution Press, Washington, D.C.; used by permission. **40.38** Ferrero/Nature. **40.42** (A) Redrawn from R. H. MacArthur and E. O. Wilson, *The Theory of Island Biogeography*, copyright © 1967 by Princeton University Press; used by permission; (B) redrawn by C. J. Krebs, *Ecology*, 2nd ed., Harper & Row, New York, 1978; after Preston. **40.45** (top) C. C. Reijnvaan from W. M. Doctors van Leeuwen, *Ann. Jard. Bot. Buitenzorg*, 1936; (bottom) courtesy Stephen Self, University of Hawaii at Manoa.

GLOSSARY

The Glossary gives brief definitions of the most important terms that recur in the text, excluding taxonomic designations. For fuller definitions, consult the index, where italicized page numbers refer you to explanations of key terms in context.

Of the basic units of measurement, some are tabulated on p. A28, others have their own alphabetical entries.

Interspersed alphabetically with the vocabulary are the main prefixes and combining forms used in biology. You will notice that, while they are generally of Greek or Latin origin, many of them have acquired a new meaning in biology (examples: ***blasto-, -cyte, caryo-, -plasm***). Familiarity with these forms will make it easier for you to learn and remember the numerous terms in which they are incorporated.

TABLE 1 *Standard prefixes of the metric system*

kilo- (k)	1,000	10^3
deci- (d)	0.1	10^{-1}
centi- (c)	0.01	10^{-2}
milli- (m)	0.001	10^{-3}
micro- (μ)	0.000001	10^{-6}
nano- (n)	0.000000001	10^{-9}

TABLE 2 *Common units of length, weight, and liquid capacity*

kilometer (km)	1,000 m	0.62137 mile
meter (m)		39.37 inches
centimeter (cm)	0.01 m	0.39 inch
millimeter (mm)	0.001 m	0.039 inch
micrometer* (μm)	10^{-6} m	
nanometer (nm)	10^{-9} m	
angstrom† (Å)	10^{-10} m	
kilogram (kg)	1,000 g	2.2 pounds
gram (g)		0.035 ounce
milligram (mg)	0.001 g	
microgram (μg)	10^{-6} g	
liter (l)	1,000 cm^3	1.057 quarts
milliliter (ml)	0.001 l	

*Formerly called micron.
†No longer used; nanometer used instead.

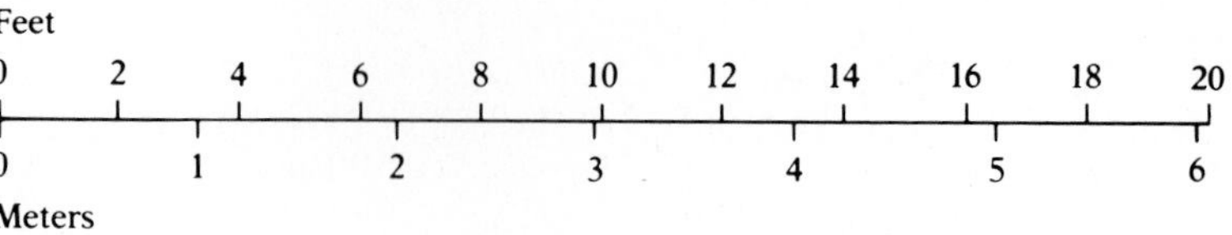

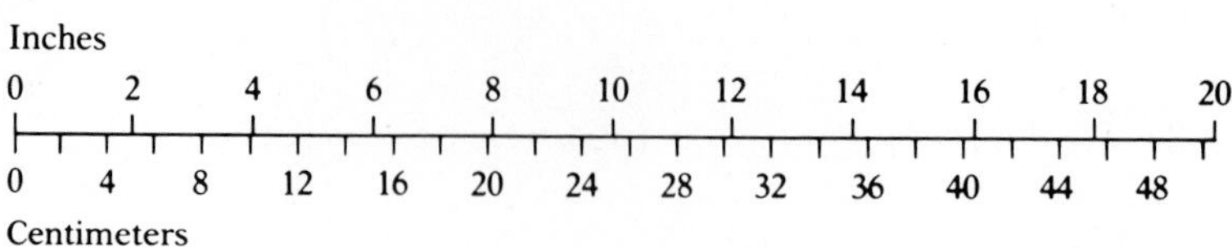

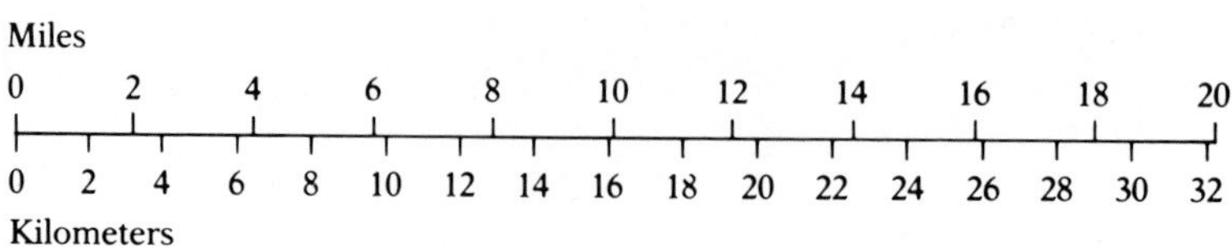

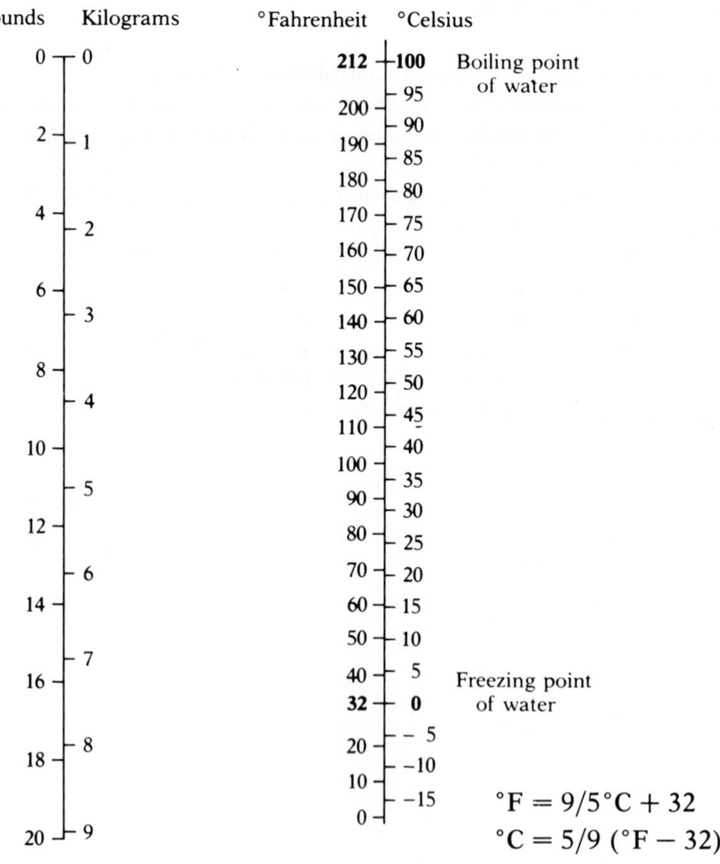

a- Without, lacking.

ab- Away from, off.

abdomen [L belly] In mammals, the portion of the trunk posterior to the thorax, containing most of the viscera except heart and lungs. In other animals, the posterior portion of the body.

absolute zero The temperature (—273°C) at which all thermal agitation ceases. The lowest possible temperature.

acellular Not constructed on a cellular basis.

acid [L *acidus* sour] A substance that increases the concentration of hydrogen ions when dissolved in water. It has a pH lower than 7.

acoelomate A body plan in which there is no cavity between the digestive tract and the body wall.

ACTH *See* adrenocorticotropic hormone.

action potential *See* potential.

active site In an enzyme, the portion of the molecule that reacts with a substrate molecule.

active transport Movement of a substance across a membrane by a process requiring expenditure of energy by the cell.

ad- Next to, at, toward.

adaptation In evolution, any genetically controlled characteristic that increases an organism's fitness, usually by helping the organism to survive and reproduce in the environment it inhabits. In neurobiology, the process that results in a short-lasting decline in responsiveness of a sensory neuron after repeated firing; *cf.* habituation, sensitization.

adenosine diphosphate (ADP) A doubly phosphorylated organic compound that can be further phosphorylated to form ATP.

adenosine monophosphate (AMP) A singly phosphorylated organic compound that can be further phosphorylated to form ADP.

adenosine triphosphate (ATP) A triply phosphorylated organic compound that functions as "energy currency" for organisms.

adipose [L *adeps* fat] Fatty.

ADP *See* adenosine diphosphate.

adrenal [L *renes* kidneys] An endocrine gland of vertebrates located near the kidney.

adrenalin A hormone produced by the adrenal medulla that stimulates "fight-or-flight" reactions.

adrenocorticotropic hormone (ACTH) A hormone produced by the pituitary that stimulates the adrenal cortex.

adsorb [L *sorbēre* to suck up] Hold on a surface.

advanced New, unlike the ancestral condition.

aerobic [L *aer* air] With oxygen.

alcohol Any of a class of organic compounds in which one or more —OH groups are attached to a carbon backbone.

alkaline Having a pH of more than 7. *See* base.

all-, allo- [Gk *allos* other] Other, different.

allele Any of several alternative gene forms at a given chromosomal locus.

allopatric [L *patria* homeland] Having different ranges.

allosteric Of an enzyme: one that can exist in two or more conformations. *Allosteric control:* control of the activity of an allosteric enzyme by determination of the particular conformation it will assume.

altruism The willingness of an individual to sacrifice its fitness for the benefit of another. *Reciprocal altruism:* the performance of a favor by one individual in the expectation of a favor in return, as when two animals groom each other.

alveolus [L little hollow] A small cavity, especially one of the microscopic cavities that are the functional units of lungs.

amino acid An organic acid carrying an amino group (—NH_2); the building-block compound of proteins.

amnion [Gk caul] An extraembryonic membrane that forms a fluid-filled sac containing the embryo in reptiles, birds, and mammals.

amoeboid [Gk *amoibē* change] Amoebalike in the tendency to change shape by protoplasmic flow.

AMP *See* adenosine monophosphate

amylase [L *amylum* starch] A starch-digesting enzyme.

an- Without.

anabolism [Gk *ana-* upward; *metabolē* change] The biosynthetic building-up aspects of metabolism.

anaerobic [L *aer* air] Without oxygen.

analogous Of characters in different organisms: similar in function and often in superficial structure but of different evolutionary origins.

anemia A condition in which the blood has lower than normal amounts of hemoglobin or red blood corpuscles.

angio-, -angium [Gk *angeion* vessel] Container, receptacle.

anion A negatively charged ion.

anisogamous Reproducing by the fusion of gametes that differ only in size, as opposed to gametes that are produced by oogamous species. Gametes of oogamous species, such as egg cells and sperm, are highly differentiated.

anterior Toward the front end.

antheridium [Gk *anthos* flower] Male reproductive organ of a plant; produces sperm cells.

antibody A protein, produced by the B lymphocytes of the immune system, that binds to a particular antigen.

antigen A substance, usually a protein or polysaccharide, that activates an organism's immune system.

anus [L ring] Opening at the posterior end of the digestive tract, through which indigestible wastes are expelled.

aorta The main artery of the systemic circulation.

apical At, toward, or near the apex, or tip, of a structure such as a plant shoot.

apo- Away from.

apoplast The network cell walls and intercellular spaces within a plant body; permits extensive extracellular movement of water within the plant.

aposematic [Gk *sēma* sign] Serving as a warning, with reference particularly to colors and structures that signal possession of defensive devices.

arch- [Gk *archein* to begin] Primitive, original.

archegonium [Gk *archegonos* the first of a race] Female reproductive organ of a higher plant; produces egg cells.

archenteron [Gk *enteron* intestine] The cavity in an early embryo that becomes the digestive cavity.

arteriole A small artery.

artery A blood vessel that carries blood away from the heart.

articulation A joint between bones. Articulating surfaces are those formed between bones and joints.

artifact A by-product of scientific manipulation rather than an inherent part of the thing observed.

ascus [Gk *askos* bag] The elongate spore sac of a fungus of the Ascomycota group.

asexual Without sex.

atmosphere (atm) (unit of pressure) The normal pressure of air at sea level: 101,325 newtons per square meter (approx. 14.7 pounds per square inch).

atom [Gk *atomos* indivisible] The smallest unit of an element, not divisible by ordinary chemical means.

atomic mass unit (amu) *See* dalton.

atomic weight The average weight of an atom of an element relative to ^{12}C, an isotope of carbon with six neutrons in the nucleus. The atomic weight of ^{12}C has arbitrarily been fixed as 12.

ATP *See* adenosine triphosphate.

auto- Self, same.

autonomic nervous system A portion of the vertebrate nervous system, comprising motor neurons that innervate internal organs and are not normally under direct voluntary control.

autosome [Gk *sōma* body] Any chromosome other than a sex chromosome.

autotrophic [Gk *trophē* food] Capable of manufacturing organic nutrients from inorganic raw materials.

auxin [Gk *auxein* to grow] Any of a class of plant hormones that promote cell elongation.

axon [Gk *axōn* axis] A fiber of a nerve cell that conducts impulses away from the cell body and can release transmitter substance.

bacteriophage [Gk *phagein* to eat] A virus that attacks bacteria, *abbrev.* phage.

basal At, near, or toward the base (the point of attachment) of a structure such as a limb.

basal body A structure, identical to the centriole, found at the base of cilia and eucaryotic flagella; consists of nine triplet microtubules arranged in a circle.

base (or alkali) A substance that increases the concentration of hydroxyl ions when dissolved in water. It has a pH higher than 7.

basidium The spore-bearing structure of Basidiomycota (club fungi).

bi- Two.

bilateral symmetry The property of having two similar sides, with definite upper and lower surfaces and anterior and posterior ends.

binary fission Reproduction by the division of a cell into two essentially equal parts by a nonmitotic process.

bio- [Gk *bios* life] Life, living.

biogenesis [Gk *genesis* source] Origin of living organisms from other living organisms.

biological magnification Increasing concentration of relatively stable chemicals as they are passed up a food chain from initial consumers to top predators.

biomass The total weight of all the organisms, or of a designated group of organisms, in a given area.

biome A large climatic region with characteristic sorts of plants and animals.

biotic Pertaining to life.

blasto- [Gk *blastos* bud] Embryo.

blastocoel [Gk *koilos* hollow] The cavity of a blastula.

blastopore [Gk *poros* passage] The opening from the cavity of the archenteron to the exterior in a gastrula.

blastula An early embryonic stage in animals, preceding the delimitation of the three principal tissue layers; frequently spherical and hollow.

B lymphocyte *See* lymphocyte.

buffer A substance that binds H^+ ions when their concentration rises and releases them when their concentration falls, thereby minimizing fluctuations in the pH of a solution.

C_3 plants Plants in which the Calvin cycle is the only pathway of CO_2 fixation. One of the products of photosynthesis is a three-carbon intermediary (C_3) of the Calvin cycle, from which the plants derive their name.

C_4 plants Also called **Kranz plants.** Plants in which one of the main early products of photosynthesis is a four-carbon compound (C_4). Kranz plants can carry out photosynthesis under conditions that are inhospitable to other plants.

caecum [L *caecus* blind] A blind diverticulum of the digestive tract.

calorie [L *calor* heat] The quantity of energy, in the form of heat, required to raise the temperature of one gram of pure water one degree from 14.5 to 15.5°C. The nutritionists' Calorie (capitalized) is 1,000 calories, or one kilocalorie.

cambium [L *cambiare* to exchange] The principal lateral meristem of vascular plants; gives rise to most secondary tissue.

cAMP *See* cyclic adenosine monophosphate.

capillarity [L *capillus* hair] The tendency of aqueous liquids to rise in narrow tubes with hydrophilic surfaces.

capillary [L *capillus*] A tiny blood vessel with walls one cell thick, across which exchange of materials between blood and the tissues takes place; receives blood from arteries and carries it to veins. Also, a similar vessel of the lymphatic system.

carbohydrate Any of a class of organic compounds composed of carbon, hydrogen, and oxygen in a ratio of about two hydrogens and one oxygen for each carbon; examples are sugar, starch, cellulose.

carbon fixation The process by which CO_2 is incorporated into organic compounds, primarily glucose; energy usually comes from the ATP and $NADP_{re}$ generated by photophosphorylation, and the metabolic pathway utilizing this energy is usually the Calvin cycle.

carboxyl group The —COOH group characteristic of organic acids.

cardiac [Gk *kardia* heart] Pertaining to the heart.

carnivore [L *carnis* of flesh; *vorare* to devour] An organism that feeds on animals.

carotenoid [L *carota* carrot] Any of a group of red, orange, and yellow accessory pigments of plants, found in plastids.

carrying capacity The maximum population that a given environment can support indefinitely.

cartilage A specialized type of dense fibrous connective tissue with a rubbery intercellular matrix.

caryo- [Gk *karyon* kernel] Nucleus.

Casparian strip A waterproof thickening in the radial and end walls of endodermal cells of plants.

cata- Down.

catabolism [Gk *katabolē* a throwing down] The degradational breaking-down aspects of metabolism, by which living things extract energy from food.

catalysis [Gk *katalyein* to dissolve] Acceleration of a chemical reaction by a substance that is not itself permanently changed by the reaction.

catalyst A substance that produces catalysis.

cation A positively charged ion.

caudal [L *cauda* tail] Pertaining to the tail.

cell cycle The cycle of cellular events from one mitosis through the next. Four stages are recognized, of which the last—distribution of genetic material to the two daughter nuclei—is mitosis proper.

cell sap *See* sap.

cellulose [L *cellula* cell] A complex polysaccharide that is a major constituent of most plant cell walls.

centi- [L *centum* hundred] One hundredth.

central nervous system A portion of the nervous system that contains interneurons and exerts some control over the rest of the nervous system. In vertebrates, the brain and the spinal cord.

centri- [L *centrum* center] Center.

centrifugation [L *fugere* to flee] The spinning of a mixture at very high speeds to separate substances of different densities.

centriole A cylindrical cytoplasmic organelle located just outside the nucleus of animal cells and the cells of some lower plants; associated with the spindle during mitosis and meiosis.

centromere [Gk *meros* part] A special region on a chromosome from which kinetochore microtubules radiate during mitosis or meiosis.

cephalization [Gk *kephalē* head] Localization of neural coordinating centers and sense organs at the anterior end of the body.

cerebellum (L small brain] A part of the hindbrain of vertebrates that controls muscular coordination.

cerebrum [L brain] Part of the forebrain of vertebrates, the chief coordination center of the nervous system.

channel *See* membrane channel.

character Any structure, functional attribute, behavioral trait, or other characteristic of an organism.

character displacement The rapid divergent evolution in sympatric species of characters that minimize competition and/or hybridization between them.

chemiosmotic gradient The combined electrostatic and osmotic-concentration gradient generated by the electron-transport chains of mitochondria and chloroplasts; the energy in this gradient is used, for the most part, to synthesize ATP.

chemosynthesis Autotrophic synthesis of organic materials, energy for which is derived from inorganic molecules.

chitin [Gk *chitōn* tunic] Polysaccharide that forms part of the hard exoskeleton of insects, crustaceans, and other invertebrates; also occurs in the cell walls of fungi.

chlorophyll [Gk *chlōros* greenish yellow; *phyllon* leaf] The green pigment of plants necessary for photosynthesis.

chloroplast A plastid containing chlorophyll.

chrom-, -chrome [Gk *chrōma* color] Colored; pigment.

chromatid A single chromosomal strand.

chromatin The mixture of DNA and protein (mostly histones in the form of nucleosome cores) that comprises eucaryotic nuclear chromosomes.

chromatography Process of separating substances by adsorption on media for which they have different affinities.

chromosome [Gk *sōma* body] A filamentous structure in the cell nucleus (or nucleoid), mitochondria, and chloroplasts, along which the genes are located.

cilium [L eyelid] A short hairlike locomotory organelle on the surface of a cell (*pl* cilia).

cisterna [L cistern] A cavity, sac, or other enclosed space serving as a reservoir.

classical conditioning *See* conditioning.

cleavage Division of a zygote or of the cells of an early embryo.

climax (ecological) A relatively stable stage reached in some ecological successions.

cline [Gk *klinein* to lean] Gradual variation, correlated with geography, in a character of a species.

cloaca [L sewer] Common chamber that receives materials from the digestive, excretory, and reproductive systems.

clone [Gk *klōn* twig] A group of cells or organisms derived asexually from a single ancestor and hence genetically identical.

co- With, together.

codon The unit of genetic coding, three nucleotides long, specifying an amino acid or an instruction to terminate translation.

coel-, -coel [Gk *koilos* hollow] Hollow, cavity; chamber.

coelom A body cavity bounded entirely by mesoderm.

coenocytic [Gk *koinos* common] Having more than one nucleus in a single mass of cytoplasm.

coenzyme A nonproteinaceous organic molecule that plays an accessory role, but a necessary one, in the catalytic action of an enzyme.

coevolution Two or more organisms evolving, each in response to the other.

coleoptile [Gk *koleon* sheath, *ptilon* feather] A sheath around the young shoot of grasses.

collagen A fibrous protein; the most abundant protein in mammals.

collenchyma [Gk *kolla* glue] A supportive tissue in plants in which the cells usually have thickenings at the angles of the walls.

colloid [Gk *kolla*] A stable suspension of particles that, though larger than in a true solution, do not settle out.

colon The large intestine.

com- Together.

commensalism [L *mensa* table] A symbiosis in which one party is benefited and the other party receives neither benefit nor harm.

community In ecology, a unit composed of all the populations living in a given area.

competition In ecology, utilization by two or more individuals, or by two or more populations, of the same limited resource; an interaction in which both parties are harmed.

condensation reaction A reaction joining two compounds with resultant formation of water.

conditioning Associative learning. *Classical conditioning:* the association of a novel stimulus with an innately recognized stimulus. *Operant conditioning:* learning of a novel behavior as a result of reward or punishment; trial-and-error learning.

conformation (of a protein) [L *conformatio* symmetrical forming] The three-dimensional pattern according to which the polypeptide chains of a protein coil (secondary structure), fold (tertiary structure), and—if there is more than one chain—fit together (quarternary structure).

conjugation [L *jugare* to join, marry] Process of genetic recombination between two organisms (e.g., bacteria, algae) through a cytoplasmic bridge between them.

connective tissue A type of animal tissue whose cells are embedded in an extensive intercellular matrix; connects, supports, or surrounds other tissues and organs.

contractile vacuole An excretory and/or osmoregulatory vacuole in some cells, which, by contracting, ejects fluids from the cell.

cooperativity The phenomenon of enhanced reactivity of the remaining binding sites of a protein as a result of the binding of substrate at one site.

cork [L *cortex* bark] A waterproof tissue, derived from the cork cambium, that forms at the outer surfaces of the older stems and roots of woody plants; the outer bark or periderm.

corpus luteum [L yellow body] A yellowish structure in the ovary, formed from the follicle after ovulation, that secretes estrogen and progesterone (*pl.* corpora lutea).

cortex [L bark] In plants, tissue between the epidermis and the vascular cylinder of stems and roots. In animals, the outer barklike tissue of some organs, as *cerebral cortex, adrenal cortex,* etc.

cotyledon [Gk *kotylē* cup] A "seed leaf," a food-digesting and -storing part of a plant embryo.

countercurrent exchange A strategy in which two streams move past each other in opposite directions, facilitating the exchange of substances between them across a membrane. The gills in gas exchange and kidneys in the production of concentrated urine are two sites of countercurrent exchange.

covalent bond A chemical bond resulting from the sharing of a pair of electrons.

Crassulacean acid metabolism (CAM) A variation of photosynthesis found in some plants that grow in hot, dry environments. Plants such as succulents avoid water loss by closing their stomata during the day and opening them at night.

crossing over Exchange of parts between two homologous chromosomes.

cross section *See* section.

cryptic [Gk *kryptos* hidden] Concealing.

cuticle [L *cutis* skin] A waxy layer on the outer surface of leaves, insects, etc.

cyclic adenosine monophosphate (cyclic AMP or **cAMP)** Compound, synthesized in living cells from ATP, that functions as an intracellular mediator of hormonal action; also plays a part in neural transmission and some other kinds of cellular control systems.

cyst [Gk *kystis* bladder, bag] (1) A saclike abnormal growth. (2) Capsule that certain organisms secrete around themselves and that protects them during resting stages.

-cyte, cyto- [Gk *kytos* container] Cell.

cytochrome Any of a group of iron-containing enzymes important in electron transport during respiration or photophosphorylation.

cytokinesis [Gk *kinēsis* motion] Division of the cytoplasm of a cell.

cytoplasm All of a cell except the nucleus.

cytosol The relatively fluid, less structured part of the cytoplasm of a cell, excluding organelles and membranous structures.

dalton A unit of mass equal to one twelfth the atomic weight of ^{12}C, or 1.66024×10^{-24} gram. Formerly called atomic mass unit (amu).

deamination Removal of an amino group.

deciduous [L *decidere* to fall off] Shedding leaves each year.

dehydration reaction A condensation reaction.

deme [Gk *dēmos* population] A local unit of population of any one species.

dendr-, dendro- [Gk *dendron* tree] Tree; branching.

dendrite A short unsheathed fiber of a nerve cell—often spiny, usually branched and tapering—that receives many synapses and carries excitation and inhibition toward the cell body.

deoxyribonucleic acid (DNA) A nucleic acid found in most viruses, all bacteria, chloroplasts, mitochondria, and the nuclei of eucaryotic cells, characterized by the presence of a deoxyribose sugar in each nucleotide; the genetic material of all organisms except the RNA viruses.

-derm [Gk *derma* skin] Skin, covering; tissue layer.

di- Two.

dicot A member of a subclass of the angiosperms, or flowering plants, characterized by the presence of two cotyledons in the embryo, a netlike system of veins in the leaves, and flower petals in fours or fives; *cf.* monocot. *Herbaceous dicot:* a perennial whose aboveground parts die annually. *Woody dicot:* a perennial whose aboveground parts—trunk and branches—remain alive and grow annually.

differentiation The process of developmental change from an immature to a mature form, especially in a cell.

diffusion The movement of dissolved or suspended particles from place to place as a result of their heat energy (thermal agitation).

digestion Hydrolysis of complex nutrient compounds into their building-block units.

diploid [Gk *diploos* double] Having two of each type of chromosome.

disaccharide A double sugar, one composed of two simple sugars.

distal [L *distare* to stand apart] Situated away from some reference point (usually the main part of the body).

diverticulum [L *devertere* to turn aside] A blind sac branching off a cavity or canal.

DNA *See* deoxyribonucleic acid.

dominant (1) Of an allele: exerting its full phenotypic effect despite the presence of another allele of the same gene, whose phenotypic expression it blocks or masks. *Dominant phenotype, dominant character:* one caused by a dominant allele. (2) Of an individual: occupying a high position in the social hierarchy.

dormancy [L *dormire* to sleep] The state of being inactive, quiescent. In plants, particularly seeds and buds, a period in which growth is arrested until environmental conditions become more favorable.

dorsal [L *dorsum* back] Pertaining to the back.

drive *See* motivation.

duodenum [From a Latin phrase meaning 12 *(duodecin)* finger's-breadths long] The first portion of the small intestine of vertebrates, into which ducts from the pancreas and gallbladder empty.

ecosystem [Gk *oikos* habitation] The sum of physical features and organisms occurring in a given area.

ecto- Outside, external.

ectoderm The outermost tissue layer of an animal embryo. Also, tissue derived from the embryonic ectoderm.

ectothermic *See* poikilothermic.

effector The part of an organism that produces a response, e.g., muscle, cilium, flagellum.

egg An egg cell or female gamete. Also a structure in which embryonic development takes place, especially in birds and reptiles; consists of an egg cell, various membranes, and often a shell.

electrochemical gradient Combined electrostatic and osmotic-concentration gradient, such as the chemiosmotic gradient of mitochondria and chloroplasts.

electron A negatively charged primary subatomic particle.

electronegativity The formal measure of an atom's attraction for free electrons. Atoms with few electron vacancies in their outer shell tend to be more electronegative than those with more. In covalent bonds, the shared electrons are, on average, nearer the more electronegative atom; this asymmetry, in part, gives rise to the polarity of certain molecules.

electronic charge unit The charge of one electron, or 1.6021×10^{-19} coulomb.

electron-transport chain A series of enzymes found in the inner membrane of mitochondria and (with somewhat different components) in the thylakoid membrane of chloroplasts. The chain accepts high-energy electrons and uses their energy to create a chemiosmotic gradient across the membrane in which it is located.

electrostatic force The attraction (also called *electrostatic attraction*) between particles with opposite charges, as between a proton and an electron, or between H^+ and OH^-; and the repulsion between particles with like charges, as between two H^+ ions.

electrostatic gradient The free-energy gradient created by a difference in charge between two points, generally the two sides of a membrane.

elimination (or defecation) The release of unabsorbed wastes from the digestive tract. *Cf.* excretion.

embryo A plant or animal in an early stage of development; generally still contained within the seed, egg, or uterus.

emulsion [L *emulsus* milked out] Suspension, usually as fine droplets, of one liquid in another.

-enchyma [Gk *parenchein* to pour in beside] Tissue.

end-, endo- Within, inside; requiring.

endergonic [Gk *ergon* work] Energy-absorbing; endothermic.

endocrine [Gk *krinein* to separate] Pertaining to ductless glands that produce hormones.

endocytosis The process by which the cell membrane forms an invagination which becomes a vesicle, trapping extracellular material that is then transported within the cell; in general, the invagination is triggered by the binding of membrane receptors to specific substances used by the cell.

endoderm The innermost tissue layer of an animal embryo.

endodermis A plant tissue, especially prominent in roots, that surrounds the vascular cylinder; all endodermal cells have Casparian strips.

endonuclease An enzyme that breaks bonds within nucleic acids, as opposed to an exonuclease, which can digest only a terminal group. *Restriction endonuclease:* an enzyme that breaks bonds only within a specific sequence of bases.

endoplasmic reticulum [L *reticulum* network] A system of membrane-bounded channels in the cytoplasm.

endoskeleton An internal skeleton.

endosperm [Gk *sperma* seed] A nutritive material in seeds.

endosymbiotic hypothesis Hypothesis that certain eucaryotic organelles—in particular mitochondria and chloroplasts—originated as free-living procaryotes that took up mutalistic residence in the ancestors of modern eucaryotes.

endothermic In thermodynamics, energy-absorbing (endergonic). In physiology, warm-blooded (homeothermic).

entropy Measure of the disorder of a system.

enzyme [Gk *zymē* leaven] A compound, usually a protein, that acts as a catalyst.

epi- Upon, outer.

epicotyl A portion of the axis of a plant embryo above the point of attachment of the cotyledons; forms most of the shoot.

epidermis [Gk *derma* skin] The outermost portion of the skin or body wall of an animal.

episome [Gk *sōma* body] Genetic element at times free in the cytoplasm, at other times integrated into a chromosome.

epithelium An animal tissue that forms the covering or lining of all free body surfaces, both external and internal.

equilibrium constant The ratio of products of a reaction to the reactants after the reaction has been allowed to proceed until there is no further change in these concentrations.

erythrocyte [Gk *erythros* red] A red blood corpuscle, i.e., a blood corpuscle containing hemoglobin.

esophagus [Gk *phagein* to eat] An anterior part of the digestive tract; in mammals it leads from the pharynx to the stomach.

essential fatty acid A fatty acid an organism needs but cannot synthesize, and so must obtain preformed (or in a precursor form) from its diet.

estrogen [L *oestrus* frenzy] Any of a group of vertebrate female sex hormones.

estrous cycles [L *oestrus*] In female mammals, the higher primates excepted, a recurrent series of physiological and behavioral changes connected with reproduction.

estuary That portion of a river that is close enough to the sea to be influenced by marine tides.

eu- [Gk *eus* good] Most typical, true.

eucaryotic cell A cell containing a distinct membrane-bounded nucleus, characteristic of all organisms except bacteria.

evaginated [L *vagina* sheath] Folded or protruded outward.

eversible [L *evertere* to turn out] Capable of being turned inside out.

evolution [L *evolutio* unrolling] Change in the genetic makeup of a population with time.

ex-, exo- Out of, outside; producing.

excretion Release of metabolic wastes and excess water. *Cf.* elimination.

exergonic [Gk *ergon* work] Energy-releasing; exothermic.

exocytosis The process by which an intracellular vesicle fuses with the cell membrane, expelling its contents into its surroundings.

exon A part of a primary transcript (and the corresponding part of a gene) that is ultimately either translated (in the case of mRNA) or utilized in a final product, such as tRNA.

exoskeleton An external skeleton.

extrinsic External to, not a basic part of; as in *extrinsic isolating mechanism.*

fauna The animals of a given area or period.

feature detector A circuit in the nervous system that responds to a specific type of feature, such as a vertically moving spot or a particular auditory time delay.

feces [L *faeces* dregs] Indigestible wastes discharged from the digestive tract.

feedback The process by which a control mechanism is regulated through the very effects it brings about. *Positive feedback:* the process by which a small effect is amplified, as when a depolarization triggers an action potential. *Negative feedback* (or feedback inhibition): the process by which a control mechanism is activated to restore conditions to their original state.

fermentation Anaerobic production of alcohol, lactic acid, or similar compounds from carbohydrate via the glycolytic pathway.

fertilization Fusion of nuclei of egg and sperm.

fetus [L *fetus* pregnant] An embryo in its later development, still in the egg or uterus.

fitness The probable genetic contribution of an individual (or allele or genotype) to succeeding generations. *Inclusive fitness:* the sum of an individual's personal fitness plus the fitness of that individual's relatives devalued in proportion to their genetic distance from the individual.

fixation (1) Conversion of a substance into a biologically more usable form, as the conversion of CO_2 into carbohydrate by photosynthetic plants or the incorporation of N_2 into more complex molecules by nitrogen-fixing bacteria. (2) Process of treating living tissue for microscopic examination.

flagellum [L whip] A long hairlike locomotory organelle on the surface of a cell.

flora The plants of a given area or period.

follicle [L *follis* bag] A jacket of cells around an egg cell in an ovary.

follicle-stimulating hormone (FSH) A gonadotropic hormone of the anterior pituitary that stimulates growth of follicles in the ovaries of females and function of the seminiferous tubules in males.

food chain Sequence of organisms, including producers, consumers, and decomposers, through which energy and materials may move in a community.

foot-candle Unit of illumination; the illumination of a surface produced by one standard candle at a distance of one foot; *cf.* lambert.

founder effect The difference between the gene pool of a population as a whole and that of a newly isolated population of the same species.

free energy Usable energy in a chemical system; energy available for producing change.

fruit A mature ovary or cluster of ovaries (sometimes with additional structures associated with the ovary).

fruiting body A spore-bearing structure (e.g., the aboveground portion of a mushroom).

FSH *See* follicle-stimulating hormone.

gamete [Gk *gametē(s)* wife, husband] A sexual reproductive cell that must usually fuse with another such cell before development begins; an egg or sperm.

gametophyte [Gk *phyton* plant] A haploid plant that can produce gametes.

ganglion [Gk tumor] A structure containing a group of cell bodies of neurons (*pl.* ganglia).

gastr-, gastro- [Gk *gastēr* belly] Stomach; ventral; resembling the stomach.

gastrovascular cavity An often branched digestive cavity, with only one opening to the outside, that conveys nutrients throughout the body; found only in animals without circulatory system.

gastrula A two-layered, later three-layered, animal embryonic stage.

gastrulation The process by which a blastula develops into a gastrula, usually by an involution of cells.

gated channel A membrane channel that can open or close in response to a signal, generally a change in the electrostatic gradient or the binding of a hormone, transmitter, or other molecular signal.

gel Colloid in which the suspended particles form a relatively orderly arrangement; *cf.* sol.

-gen; -geny [Gk *genos* birth, race] Producing; production, generation.

gene [Gk *genos*] The unit of inheritance; usually a portion of a DNA molecule that codes for some product such as a protein, tRNA, or rRNA.

gene amplification Any of the strategies that give rise to multiple copies of certain genes, thus facilitating the rapid synthesis of a product (such as rRNA for ribosomes) for which the demand is great.

gene flow The movement of genes from one part of a population to another, or from one population to another, via gametes.

gene pool The sum total of all the genes of all the individuals in a population.

gene regulation Any of the strategies by which the rate of expression of a gene can be regulated, as by controlling the rate of transcription.

generator potential *See* potential.

genetic drift Change in the gene pool as a result of chance and not as a result of selection, mutation, or migration.

genome The cell's total complement of DNA: in eucaryotes, the nuclear and organelle chromosomes; in procaryotes, the major chromosome, episomes, and plasmids. In viruses and viroids, the total complement of DNA or RNA.

genotype The particular combination of genes present in the cells of an individual.

germ cell A sexual reproductive cell; an egg or sperm.

gibberellin A plant hormone—one of its effects is stem elongation in some dwarf plants.

gill An evaginated area of the body wall of an animal, specialized for gas exchange.

gizzard A chamber of an animal's digestive tract specialized for grinding food.

glucose [Gk *glykys* sweet] A six-carbon sugar; plays a central role in cellular metabolism.

glycocalyx The layer of protein and carbohydrates just outside the plasma membrane of an animal cell; in general, the proteins are anchored in the membrane, and the carbohydrates are bound to the proteins.

glycogen [Gk *glykys*] A polysaccharide that serves as the principal storage form of carbohydrate in animals.

glycolysis [Gk *glykys*] Anaerobic catabolism of carbohydrates to pyruvic acid.

Golgi apparatus Membranous subcellular structure that plays a role in storage and modification particularly of secretory products.

gonadotropic Stimulatory to the gonads.

gonadotropin A hormone stimulatory to the gonads, a gonadotropic hormone.

gonads [Gk *gonos* seed] The testes or ovaries.

gram molecule *See* mole.

granum [L grain] A stacklike grouping of photosynthetic membranes in a chloroplast (*pl.* grana).

guard cell A specialized epidermal cell that regulates the size of stoma of a leaf.

habit [L *habitus* disposition] In biology, the characteristic form or mode of growth of an organism.

habitat [L it lives] The kind of place where a given organism normally lives.

habituation The process that results in a long-lasting decline in the receptiveness of interneurons (primarily) to the input from sensory neurons or other interneurons; *cf.* sensitization, adaptation.

haploid [Gk *haploos* single] Having only one of each type of chromosome.

hem-, hemat-, hemo- [Gk *haima* blood] Blood.

hematopoiesis [Gk *poiēsis* making] The formation of blood.

hemoglobin A red iron-containing pigment in the blood that functions in oxygen transport.

hepatic [Gk *hēpar* liver] Pertaining to the liver.

herbaceous [L *herbaceus* grassy] Having a stem that remains soft and succulent; not woody.

herbaceous dicot *See* dicot.

herbivore [L *herba* grass; *vorare* to devour] An animal that eats plants.

Hertz A unit of frequency (as of sound waves) equal to one cycle per second.

hetero- [Gk *heteros* other] Other, different.

heterogamy [Gk *gamos* marriage] The condition of producing gametes of two or more different types.

heterotrophic [Gk *trophē* food] Incapable of manufacturing organic compounds from inorganic raw materials, therefore requiring organic nutrients from the environment.

heterozygous [Gk *zygōtos* yoked] Having two different alleles of a given gene.

Hg [L *hydrargyrum* mercury] The symbol for mercury. Pressure is often expressed in *mm Hg*—the pressure exerted by a column of mercury whose height is measured in millimeters (at 0° C, 1 mm Hg = 133.3 newtons per square meter).

hilum Region where blood vessels, nerves, ducts, enter an organ.

hist- [Gk *histos* web] Tissue.

histology The structure and arrangement of the tissues of organisms; the study of these.

histone One of a class of basic proteins serving as structural elements of eucaryotic chromosomes.

homeo-, homo- [Gk *homoios* like] Like, similar.

homeostasis The tendency in an organism toward maintenance of physiological and psychological stability.

homeothermic [Gk *thermē* heat] Capable of self-regulation of body temperature; warm-blooded, endothermic.

home range An area within which an animal tends to confine all or nearly all its activities for a long period of time.

homologous Of chromosomes: bearing genes for the same characters. Of characters in different organisms: inherited from a common ancestor.

homozygous [Gk *zygōtos* yoked] Having two copies of the same allele of a given gene.

hormone [Gk *horman* to set in motion] A control chemical secreted in one part of the body that affects other parts of the body.

hybrid In evolutionary biology, a cross between two species. In genetics, a cross between two genetic types.

hydr-, hydro- [Gk *hydōr* water] Water; fluid; hydrogen.

hydration Formation of a sphere of water around an electrically charged particle.

hydrocarbon Any compound made of only carbon and hydrogen.

hydrogen bond A weak chemical bond formed when two polar molecules, at least one of which usually consists of a hydrogen bonded to a more electronegative atom (usually oxygen or nitrogen), are attracted electrostatically.

hydrolysis [Gk *lysis* loosing] Breaking apart of a molecule by addition of water.

hydrophilic Readily entering into solution by forming hydrogen bonds with water or other polar molecules.

hydrophobic Incapable of entering into solution by molecules that are neither ionic nor polar, and therefore cannot dissolve in water.

hydrostatic [Gk *statikos* causing to stand] Pertaining to the pressure and equilibrium of fluids.

hydroxyl ion The OH^- ion.

hyper- Over, overmuch; more.

hypertonic Of a solution (or colloidal suspension): tending to gain water from some reference solution (or colloidal suspension) separated from it by a selectively permeable membrane—usually because it has a higher osmotic concentration than the reference solution.

hypertrophy [Gk *trophē* food] Abnormal enlargement, excessive growth.

hypha [Gk *hyphē* web] A fungal filament.

hypo- Under, lower; less.

hypocotyl The portion of the axis of a plant embryo below the point of attachment of the cotyledons; forms the base of the shoot and the root.

hypothalamus [Gk *thalamos* inner chamber] Part of the posterior portion of the vertebrate forebrain, containing important centers of the autonomic nervous system and centers of emotion.

hypotonic Of a solution (or colloidal suspension): tending to lose water to some reference solution (or colloidal suspension) separated from it by a selectively permeable membrane—usually because it has a lower osmotic concentration than the reference solution.

imprinting A kind of associative learning in which an animal rapidly learns during a particular critical period to recognize an object, individual, or location in the absence of overt reward; distinguished from most other associative learning in that it is retained indefinitely, being difficult or impossible to reverse.

independent assortment The Principle of Independent Assortment is frequently referred to as Mendel's second law. Genes found on different chromosomes, so-called unlinked genes, assort independently in meiosis unless they are recombined by crossing over.

inducer In embryology, a substance that stimulates differentiation of cells or development of a particular structure. In genetics, a substance that activates particular genes.

inorganic compound A chemical compound not based on carbon.

in situ [L in place] In its natural or original position.

instinct Heritable, genetically specified neural circuitry that guides and directs behavior.

insulin [L *insula* island] A hormone produced by the β islet cells in the pancreas that helps regulate carbohydrate metabolism, especially conversion of glucose into glycogen.

integument [L *integere* to cover] A coat, skin, shell, rind, or other protective surface structure.

inter- Between (e.g., *interspecific*, between two or more different species).

interneuron A neuron that receives input from and synapses on other neurons, as distinguished from a sensory neuron (which receives sensory information) and a motor neuron (which synapses on a muscle).

intra- Within (e.g., *intraspecific*, within a single species).

intrinsic Inherent in, a basic part of; as in *intrinsic isolating mechanism*.

intron A part of a primary transcript (and the corresponding part of a gene) that lies between exons, and is removed before the RNA becomes functional.

invaginated [L *vagina* sheath] Folded or protruded inward.

invertebrate [L *vertebra* joint] Lacking a backbone, hence an animal without bones.

in vitro [L in glass] Not in the living organism, in the laboratory.

in vivo [L in the living] In the living organism.

ion An electrically charged atom.

ionic bond A chemical bond formed by the electrostatic attraction between two oppositely charged ions.

iso- Equal, uniform.

isogamy [Gk *gamos* marriage] The condition of producing gametes of only one type, with no distinction existing between male and female.

isolating mechanism An obstacle to interbreeding, either extrinsic, such as a geographical barrier, or intrinsic, such as structural or behavioral incompatibility.

isotonic Of a solution (or colloidal suspension): tending neither to gain nor to lose water when separated from some reference solution (or colloidal suspension) by a selectively permeable membrane—usually because it has the same osmotic concentration as the reference solution.

isotope [Gk *topos* place] An atom differing from another atom of the same element in the number of neutrons in its nucleus.

kilo- A thousand.

kin-, kino- [Gk *kinēma* motion] Motion, action.

kinase An enzyme that catalyzes the phosphorylation of a substrate by ATP.

lactic acid A three-carbon organic acid produced in animals and some microorganisms by fermentation.

lambert In metric system, unit of brightness of a light source; approximately equivalent to 929 foot-candles.

lamella [L thin plate] A thin platelike structure; a fairly straight intracellular membrane.

larva [L ghost, mask] Immature form of some animals that undergo radical transformation to attain the adult form.

lateral Pertaining to the side.

lateral inhibition Process by which adjacent sensory cells or their targets interact to inhibit one another when excited, the result being an exaggerated contrast; neural basis of feature-detector circuits.

lenticel [L *lenticella* small lentil] A porous region in the periderm of a woody stem through which gases can move.

leukocyte [Gk *leukos* white] A white blood cell; *cf.* lymphocyte, macrophage.

LH *See* luteinizing hormone.

ligament [L *ligare* to bind] A type of connective tissue linking two bones in a joint.

ligase An enzyme that catalyzes the bonding between adjacent nucleotides in DNA and RNA.

lignin [L *lignum* wood] An organic compound in wood that makes cellulose harder and more brittle.

linkage The presence of two or more genes on the same chromosome, which, in the absence of crossing over, causes the characters they control to be inherited together.

lip- [Gk *lipos* fat] Fat or fatlike.

lipase A fat-digesting enzyme.

lipid Any of a variety of compounds insoluble in water but soluble in ethers and alcohols; includes fats, oils, waxes, phospholipids, and steroids.

locus [L place] In genetics, a particular location on a chromosome, hence often used synonymously with gene (*pl.* loci).

lumen [L light, opening] The space or cavity within a tube or sac (*pl.* lumina).

lung An internal chamber specialized for gas exchange in an animal.

luteinizing hormone (LH) A gonadotropic hormone of the pituitary that stimulates conversion of a follicle into corpus luteum and secretion of progesterone by the corpus luteum; also stimulates secretion of sex hormone by the testes.

lymph [L *lympha* water] A fluid derived from tissue fluid and transported in special lymph vessels to the blood.

lymphocyte A white blood cell that responds to the presence of a foreign antigen. *B lymphocyte:* a cell that upon stimulation by an antigen secretes antibodies. *T lymphocyte:* a cell that attacks infected cells and modulates the activity of B lymphocytes.

-lysis, lyso- [Gk *lysis* loosing] Loosening, decomposition.

lysogenic Of bacteria: carrying bacteriophage capable of lysing, i.e., destroying, other bacterial cells.

lysosome A subcellular organelle that stores digestive enzymes.

macro- Large.

macrophage A phagocytic white blood cell that ingests material—particularly viruses, bacteria, and clumped toxins—bound by circulating antibodies.

Malpighian tubule An excretory diverticulum of the digestive tract in insects and some other arthropods.

mast cell Cells that are specialized for the secretion of histamine and other local chemical mediators as part of the immune response.

matrix [L *mater* mother] A mass in which something is embedded, e.g., the intercellular substance of a tissue.

medulla [L marrow, innermost part] (1) The inner portion of an organ, e.g., *adrenal medulla.* (2) The *medulla oblongata,* a portion of the vertebrate hindbrain that connects with the spinal cord.

medusa [*after* Medusa, mythological monster with snaky locks] The free-swimming stage in the life cycle of a coelenterate.

mega- Large.

megaspore A spore that will germinate into a female plant.

meiosis [Gk *meiōsis* diminution] A process of nuclear division in which the number of chromosomes is reduced by half.

membrane A structure, formed mainly by a double layer of phospholipids, which surrounds cells and organelles.

membrane channel A pore in a membrane through which certain molecules may pass.

membrane pump A permease that uses energy, usually from ATP, to move substances across the membrane against their osmotic-concentration or electrostatic gradients.

meristematic tissue [Gk *meristos* divisible] A plant tissue that functions primarily in production of new cells by mitosis.

meso- Middle.

mesoderm The middle tissue layer of an animal embryo.

mesophyll [Gk *phyllon* leaf] The parenchymatous middle tissue layers of a leaf.

meta- Posterior, later; change in.

metabolism [Gk *metabolē* change] The sum of the chemical reactions within a cell (or a whole organism), including the energy-releasing breakdown of molecules (catabolism) and the synthesis of complex molecules and new protoplasm (anabolism).

metamorphosis [Gk *morphē* form] Transformation of an immature animal into an adult. More generally, change in the form of an organ or structure.

micro- Small. Male. In units of measurement, one millionth.

microfilament A long, thin structure, usually formed from the protein actin; when associated with myosin filaments, as in muscles, microfilaments are involved in movement.

microorganism A microscopic organism, especially a bacterium, virus, or protozoan.

microspore A spore that will germinate into a male plant.

microtubule A long, hollow structure formed from the protein tubulin; found in cilia, eucaryotic flagella, basal bodies/centrioles, and the cytoplasm.

middle lamella A layer of substance deposited between the walls of adjacent plant cells.

milli- One thousandth.

mineral In biology, any naturally occurring inorganic substance, excluding water.

mitochondrion [Gk *mitos* thread; *chondrion* small grain] Subcellular organelle in which aerobic respiration takes place.

mitosis [Gk *mitos*] Process of nuclear division in which complex movements of chromosomes along a spindle result in two new nuclei with the same number of chromosomes as the original nucleus.

modulator A control chemical that stabilizes an allosteric enzyme in one of its alternative conformations.

mold Any of many fungi that produce a cottony or furry growth.

mole The amount of a substance that has a weight in grams numerically equal to the molecular weight of the substance. One mole of a substance contains 6.023×10^{23} molecules of that substance; hence one mole of a substance will always contain the same number of molecules as a mole of any other substance.

molecular weight The weight of a molecule calculated as the sum of the atomic weights of its constituent atoms.

molecule A chemical unit consisting of two or more atoms bonded together.

mono- One.

monocot A member of a subclass of angiosperms, or flowering plants, characterized by the presence of a single cotyledon in the embryo, parallel veins in the leaves, and flower petals in threes; *cf.* dicot.

-morph, morpho- [Gk *morphē* form] Form, structure.

morphogenesis The establishment of shape and pattern in an organism.

morphology The form and structure of organisms or parts of organisms; the study of these.

motivation The internal state of an animal that is the immediate cause of its behavior; drive.

motor neuron A neuron, leading away from the central nervous system, that synapses on and controls an effector.

motor program A coordinated, relatively stereotyped series of muscle movements performed as a unit, either innate (as the movements of swallowing) or learned (as in speech); also, the neural circuitry underlying such behavior; *cf.* reflex.

mouthparts Structures or appendages near the mouth used in manipulating food.

mucosa Any membrane that secretes mucus (a slimy protective substance), e.g., the membrane lining the stomach and intestine.

muscle [L *musculus* small mouse, muscle] A contractile tissue of animals.

mutation [L *mutatio* change] Any relatively stable heritable change in the genetic material.

mutualism A symbiosis in which both parties benefit.

mycelium [Gk *mykēs* fungus] A mass of hyphae forming the body of a fungus.

myo- [Gk *mys* mouse, muscle] Muscle.

NAD *See* nicotinamide adenine dinucleotide.

NADP *See* nicotinamide adenine dinucleotide phosphate.

nano- [L *nanus* dwarf] One billionth.

natural selection Differential reproduction in nature, leading to an increase in the frequency of some genes or gene combinations and to a decrease in the frequency of others.

navigation The initiation and/or maintenance of movement toward a goal.

negative feedback *See* feedback.

nematocyst [Gk *nēma* thread; *kystis* bag] A specialized stinging capsule in coelenterates; contains a hairlike structure that can be ejected.

neo- New.

neocortex Portion of the cerebral cortex in mammals, of relatively recent evolutionary origin; often greatly expanded in the higher primates and dominant over other parts of the brain.

nephr- [Gk *nephros* kidney] Kidney.

nephridium An excretory organ consisting of an open bulb and a tubule leading to the exterior; found in many invertebrates, such as segmented worms.

nephron The functional unit of a vertebrate kidney, consisting of Bowman's capsule, convoluted tubule, and loop of Henle.

nerve [L *nervus* sinew, nerve] A bundle of neuron fibers (axons).

nerve net A nervous system without any central control, as in coelenterates.

neuron [Gk nerve, sinew] A nerve cell.

neutron An electrically neutral subatomic particle with approximately the same mass as a proton.

niche The functional role and position of an organism in the ecosystem; the way an organism makes its living, including, for an animal, not only what it eats, but when, where, and how it obtains food, where it lives, etc.

nicotinamide adenine dinucleotide (NAD) An organic compound that functions as an electron acceptor, e.g., in respiration.

nicotinamide adenine dinucleotide phosphate (NADP) An organic compound that functions as an electron acceptor, e.g., in biosynthesis.

nitrogen fixation Incorporation of nitrogen from the atmosphere into substances more generally usable by organisms.

node (of plant) [L *nodus* knot] Point on a stem where a leaf or bud is (or was) attached.

nonhomologous Of chromosomes: two chromosomes that do not share the same genes and thus do not pair during meiosis.

notochord [Gk *nōtos* back; *chordē* string] In the lower chordates and in the embryos of the higher vertebrates, a flexible supportive rod running longitudinally through the back just ventral to the nerve cord.

nucleic acid Any of several organic acids that are polymers of nucleotides and function in transmission of hereditary traits, in protein synthesis, and in control of cellular activities.

nucleoid A region, not bounded by a membrane, where the chromosome is located in a procaryotic cell.

nucleolus A dense body within the nucleus, usually attached to one of the chromosomes; consists of multiple copies of the genes for certain kinds of rRNA.

nucleosome A complex consisting of several histone proteins, which together form a "spool," and chromosomal DNA, which is wrapped around the spool.

nucleotide A chemical entity consisting of a five-carbon sugar with a phosphate group and a purine or pyrimidine attached; building-block unit of nucleic acids.

nucleus (of cell) [L kernel] A large membrane-bounded organelle containing the chromosomes.

nutrient [L *nutrire* to nourish] A food substance usable in metabolism as a source of energy or of building material.

nymph [Gk *nymphē* bride, nymph] Immature stage of insect that undergoes gradual metamorphosis.

olfaction [L *olfacere* to smell] The sense of smell.

omnivorous [L *omnis* all; *vorare* to devour] Eating a variety of foods, including both plants and animals.

oncogene A gene that causes one of the biochemical changes that lead to cancer.

ontogeny [Gk *ōn* being] The course of development of an individual organism.

oo- [Gk *ōion* egg] Egg.

oogamy A type of heterogamy in which the female gametes are large nonmotile egg cells.

oogonium Unjacketed female reproductive organ of a thallophyte plant.

operant conditioning *See* conditioning.

operator A region on the DNA to which a control substance can bind, thereby altering the rate of transcription.

oral [L *oris* of the mouth] Relating to the mouth.

organ [Gk *organon* tool] A body part usually composed of several tissues grouped together into a structural and functional unit.

organelle A well-defined subcellular structure.

organic compound A chemical compound containing carbon.

organism An individual living thing.

orientation The act of turning or moving in relation to some external feature, such as a source of light.

osmol Measure of osmotic concentration; the total number of moles of osmotically active particles per liter of solvent.

osmoregulation Regulation of the osmotic concentration of body fluids in such a manner as to keep them relatively constant despite changes in the external medium.

osmosis [Gk *ōsmos* thrust] Movement of a solvent (usually water in biology) through a selectively permeable membrane.

osmotic potential The free energy of water molecules in a solution or colloid under conditions of constant temperature and pressure; since this free energy decreases as the proportion of osmotically active particles rises, a measure of the tendency of the solution or colloid to lose water.

osmotic pressure The pressure that must be exerted on a solution or colloid to keep it in equilibrium with pure water when it is separated from the water by a selectively permeable membrane; hence a measure of the tendency of the solution or colloid to take in water.

ov-, ovi- [L *ovum* egg] Egg.

ovary Female reproductive organ in which egg cells are produced.

ovulation Release of an egg from the ovary.

ovule A plant structure, composed of an integument, sporangium, and megagametophyte, that develops into a seed after fertilization.

ovum A mature egg cell (*pl.* ova).

oxidation Energy-releasing process involving removal of electrons from a substance; in biological systems, generally by the removal of hydrogen (or sometimes the addition of oxygen).

pancreas In vertebrates, a large glandular organ located near the stomach that secretes digestive enzymes into the duodenum and also produces hormones.

papilla [L nipple] A small nipplelike protuberance.

para- Alongside of.

parapodium [Gk *podion* little foot] One of the paired segmentally arranged lateral flaplike protuberances of polychaete worms.

parasitism [Gk *parasitos* eating with another] A symbiosis in which one party benefits at the expense of the other.

parasympathetic nervous system One of the two parts of the autonomic nervous system.

parathyroids Small endocrine glands of vertebrates located near the thyroid.

parenchyma A plant tissue composed of thin-walled, loosely packed, relatively unspecialized cells.

parthenogenesis [Gk *parthenos* virgin] Production of offspring without fertilization.

pathogen [Gk *pathos* suffering] A disease-causing organism.

pectin A complex polysaccharide that cross-links the cellulose fibrils in a plant cell wall and is a major constituent of the middle lamella.

pellicle [L *pellis* skin] A thin skin or membrane.

pepsin [Gk *pepsis* digestion] A protein-digesting enzyme of the stomach.

peptide bond A bond between two amino acids resulting from a condensation reaction between the amino group of one acid and the acidic group of the other.

perennial A plant that lives for several years, as compared to annuals and biennials, which live for one and two years respectively.

peri- Surrounding.

pericycle A layer of cells inside the endodermis but outside the phloem of roots and stems.

periderm The corky outer bark of older stems and roots.

peristalsis [Gk *stalsis* contraction] Alternating waves of contraction and relaxation passing along a tubular structure such as the digestive tract.

permeable [L *permeare* to go through] Of a membrane: permitting other substances to pass through.

permease [L *permeare*] A protein that allows molecules to move across a membrane; *cf.* gated channel, membrane channel, membrane pump.

petiole [L *pediculus* small foot] The stalk of a leaf.

PGAL *See* phosphoglyceraldehyde.

pH Symbol for the logarithm of the reciprocal of the hydrogen ion concentration; hence a measure of acidity. A pH of 7 is neutral; lower values are acidic, higher values alkaline (basic).

phage *See* bacteriophage.

phagocytosis [Gk *phagein* to eat] The active engulfing of particles by a cell.

pharynx Part of the digestive tract between the oral cavity and the esophagus; in vertebrates, also part of the respiratory passage.

phenotype [Gk *phainein* to show] The physical manifestation of a genetic trait.

pheromone [Gk *pherein* to carry + hormone] A substance that, secreted by one organism, influences the behavior or physiology of other organisms of the same species when they sense its odor.

phloem [Gk *phloios* bark] A plant vascular tissue that transports organic materials; the inner bark.

-phore [Gk *pherein* to carry] Carrier.

phosphoglyceraldehyde (PGAL) A three-carbon phosphorylated carbohydrate, important in both photosynthesis and glycolysis.

phospholipid A compound composed of glycerol, fatty acids, a phosphate group, and often a nitrogenous group.

phosphorylation Addition of a phosphate group.

photo- [Gk *phōs* light] Light.

photon A discrete unit of radiant energy.

photoperiodism A response by an organism to the duration and timing of the light and dark conditions.

photophosphorylation The process by which energy from light is used to convert ADP into ATP.

photosynthesis Autotrophic synthesis of organic materials in which the source of energy is light; *cf.* photophosphorylation.

-phyll [Gk *phyllon* leaf] Leaf.

phylogeny [Gk *phylē* tribe] Evolutionary history of an organism.

physiology [Gk *physis* nature] The life processes and functions of organisms; the study of these.

-phyte, phyto- [Gk *phyton* plant] Plant.

phytochrome A protein pigment of plants sensitive to red and far-red light.

pinocytosis [Gk *pinein* to drink] The active engulfing by cells of liquid or of very small particles.

pistil The female reproductive organ of a flower, composed of one or more megasporophylls.

pith A tissue (usually parenchyma) located in the center of a stem (rarely a root), internal to the xylem.

pituitary An endocrine gland located near the brain of vertebrates; known as the master gland because it secretes hormones that regulate the action of other endocrine glands.

placenta [Gk *plax* flat surface] An organ in mammals, made up of fetal and maternal components, that aids in exchange of materials between the fetus and the mother.

plasm-, plasmo-, -plasm [Gk *plasma* something formed or molded] Formed material; plasma; cytoplasm.

plasma Blood minus the cells and platelets.

plasma membrane The outer membrane of a cell.

plasmid A small circular piece of DNA free in the cytoplasm of a bacterial or yeast cell and replicated independently of the cell's chromosome.

plasmodesma [Gk *desma* bond] A connection between adjacent plant cells through tiny openings in the cell walls (*pl.* plasmodesmata).

plasmolysis Shrinkage of a plant cell away from its wall when in a hypertonic medium.

plastid Relatively large organelle in plant cells that functions in photosynthesis and/or nutrient storage.

pleiotropic [Gk *pleiōn* more] Of a gene: having more than one phenotypic effect.

poikilothermic [Gk *poikilos* various; *thermē* heat] Incapable of precise self-regulation of body temperature, dependent on environmental temperature; cold-blooded, ectothermic.

polar molecule A molecule with oppositely charged sections; the charges, which are far weaker than the charges on ions, arise from differences in electronegativity between the constituent atoms.

pollen grain [L *pollen* flour dust] A microgametophyte of a seed plant.

poly- Many.

polycistronic Pertaining to the transcription of two or more adjacent cistrons (structural genes) into a single messenger RNA molecule.

polymer [Gk *meros* part] A large molecule consisting of a chain of small molecules bonded together by condensation reactions or similar reactions.

polymerase An enzyme complex that catalyzes the polymerization of nucleotides; examples are DNA polymerase, which is involved in replication, and RNA polymerase, which is involved in transcription.

polymorphism [Gk *morphē* form] The simultaneous occurrence of several discontinuous phenotypes in a population.

polyp [Gk *polypous* many-footed] The sedentary stage in the life cycle of a coelenterate.

polypeptide chain A chain of amino acids linked together by peptide bonds.

polyploid Having more than two complete sets of chromosomes.

polysaccharide Any carbohydrate that is a polymer of simple sugars.

population In ecology, a group of individuals belonging to the same species.

portal system [L *porta* gate] A blood circuit in which two beds of capillaries are connected by a vein (e.g., *hepatic portal system).*

positive feedback *See* feedback.

posterior Toward the hind end.

potential Short for *potential difference:* the difference in electrical charge between two points. *Resting p.:* a relatively steady potential difference across a cell membrane, particularly of a nonfiring nerve cell or a relaxed muscle cell. *Action p.:* a sharp change in the potential difference across the membrane of a nerve or muscle cell that is propagated along the cell; in nerves, identified with the nerve impulse. *Generator p.:* a change in the potential difference across the membrane of a sensory cell that, if it reaches a threshold level, may trigger an action potential along the associated neural pathway.

predation [L *praedatio* plundering] The feeding of free-living organisms on other organisms.

presumptive Describing the developmental fate of a tissue that is not yet differentiated. Presumptive neural tissue, for example, is destined to become part of the nervous system once it has differentiated.

primary transcript Newly synthesized RNA—generally mRNA—before the introns are removed.

primitive [L *primus* first] Old, like the ancestral condition.

primordium [L *primus; ordiri* to begin] Rudiment, earliest stage of development.

pro- Before.

proboscis [Gk *boskein* to feed] A long snout; an elephant's trunk. In invertebrates, an elongate, sometimes eversible process originating in or near the mouth that often serves in feeding.

procaryotic cell A type of cell that lacks a membrane-bounded nucleus; found only in bacteria.

progesterone [L *gestare* to carry] One of the principal female sex hormones of vertebrates.

promoter The region of DNA to which the transcription complex binds.

prot-, proto- First, primary.

protease A protein-digesting enzyme.

protein A long polypeptide chain.

proteolytic Protein-digesting.

proton A positively charged primary subatomic particle.

proto-oncogene A gene that can, after certain sorts of mutation or translocation, or after mutation or translocation in associated control regions, become an oncogene and cause one of the changes leading to cancer.

protoplasm Living substance, the material of cells.

provirus Viral nucleic acid integrated into the genetic material of a host cell.

proximal Near some reference point (often the main part of the body).

pseudo- False; temporary.

pseudocoelom A functional body cavity not entirely enclosed by mesoderm.

pseudogene An untranscribed region of the DNA that closely resembles a gene.

pseudopod, pseudopodium [L *podium* foot] A transitory cytoplasmic protrusion of an amoeba or an amoeboid cell.

pulmonary [L *pulmones* lungs] Relating to the lungs.

purine Any of several double-ringed nitrogenous bases important in nucleotides.

pyloric [Gk *pylōros* gatekeeper] Referring to the junction between the stomach and the intestine.

pyrimidine Any of several single-ringed nitrogenous bases important in nucleotides.

pyruvic acid A three-carbon compound produced by glycolysis.

race A subspecies.

radial symmetry A type of symmetry in which the body parts are arranged regularly around a central line (in animals, running through the oral-anal axis) rather than on the two sides of a plane.

radiation As an evolutionary phenomenon, divergence of members of a single lineage into different niches or adaptive zones.

receptor In cell biology, a region, often the exposed part of a membrane protein, that binds a substance but does not catalyze a reaction in the chemical it binds; the membrane protein frequently has another region that, as a result of the binding, undergoes an allosteric change and so becomes catalytically active.

recessive Of an allele: not expressing its phenotype in the presence of another allele of the same gene, therefore expressing it only in homozygous individuals. *Recessive character, recessive phenotype:* one caused by a recessive allele.

reciprocal altruism *See* altruism.

recombination In genetics, a novel arrangement of alleles resulting from sexual reproduction and from crossing over (or, in procaryotes and eucaryotic organelles, from conjugation). In gene evolution, a novel arrangement of exons resulting from a variety of processes that duplicate and transport segments of the chromosomes within the genome; these processes include transposition, unequal crossing over, and chromosomal breakage and fusion.

rectum [L *rectus* straight] The terminal portion of the intestine.

redox reaction [*from red*uction-*ox*idation] A reaction involving reduction and oxidation, which inevitably occur together; *cf.* reduction, oxidation.

reduction Energy-storing process involving addition of electrons to a substance; in biological systems, generally by the addition of hydrogen (or sometimes the removal of oxygen).

reflex [L *reflexus* bent back] An automatic act consisting, in its pure form, of a single simple response to a single stimulus, as when a tap on the knee elicits a knee jerk. Distinguished from a motor program, which involves a coordinated response of several muscles.

reflex arc A functional unit of the nervous system, involving the entire pathway from receptor cell to effector.

reinforcement (psychological) Reward for a particular behavior.

releaser *See* sign stimulus.

renal [L *renes* kidneys] Pertaining to the kidney.

respiration [L *respiratio* breathing out] (1) The release of energy by oxidation of fuel molecules. (2) The taking in of O_2 and release of CO_2; breathing.

resting potential *See* potential.

restriction endonuclease *See* endonuclease.

reticulum [L little net] A network.

retina The tissue in the rear of the eye that contains the sensory cells of vision.

retrovirus An RNA virus that, by means of a special enzyme (reverse transcriptase), makes a DNA copy of its genome which is then incorporated into the host's genome.

rhizoid [Gk *rhiza* root] Rootlike structure.

ribonucleic acid (RNA) Nucleic acid characterized by the presence of a ribose sugar in each nucleotide. The primary classes of RNA are mRNA (messenger RNA, which carries the instructions specifying the order of amino acids in new proteins from the genes to the ribosomes where protein synthesis takes place), rRNA (ribosomal RNA, which is incorporated into ribosomes), and tRNA (transfer RNA, which carries amino acids to the ribosomes as part of protein synthesis).

ribosome A small cytoplasmic organelle that functions in protein synthesis.

RNA *See* ribonucleic acid.

salt Any of a class of generally ionic compounds that may be formed by reaction of an acid and a base, e.g., table salt, NaCl.

sap Water and dissolved materials moving in the xylem; less commonly, solutions moving in the phloem. *Cell sap:* the fluid content of a plant-cell vacuole

saprophyte [Gk *sapros* rotten] A heterotrophic plant or bacterium that lives on dead organic material.

sarcomere [Gk *sarx* flesh; *meros* part] The region of a skeletal-muscle myofibril extending from one Z line to the next; the functional unit of skeletal-muscle contraction.

sclerenchyma [Gk *sclēros* hard] A plant supportive tissue composed of cells with thick secondary walls.

section *Cross* or *transverse s.:* section at right angles to the longest axis. *Longitudinal s.:* section parallel to the longest axis. *Radial s.:* longitudinal section along a radius. *Sagittal s.:* vertical longitudinal section along the midline of a bilaterally symmetrical animal.

seed A plant reproductive entity consisting of an embryo and stored food enclosed in a protective coat.

segmentation The subdivision of an organism into more or less equivalent serially arranged units.

selection pressure In a population, the force for genetic change resulting from natural selection.

semipermeable Permeable only to solvent (usually water); less strictly: selectively permeable, i.e., permeable to some substances but not to others.

sensitization The process by which an unexpected stimulus alerts an animal, reducing or eliminating any preexisting habituation; *cf.* adaptation, habituation.

sensory neuron A neuron, leading toward the central nervous system, that receives input from a receptor cell or is itself responsive to sensory stimulation.

septum [L barrier] A partition or wall (*pl.* septa).

sessile [L *sessilis* of sitting, low] Of animals, sedentary. Of plants, without a stalk.

sex-linked Of genes: located on the X chromosome.

sexual dimorphism Morphological differences between the two sexes of a species, as in the size of tails of peacocks as compared to peahens.

sexual selection Selection for morphology or behavior directly related to attracting or winning mates. *Male-contest sexual selection:* selection for morphology or behavior that enables a male to win fights or contests for access to females, gaining a high position in a dominance hierarchy, for example, or possession of a territory. *Female-choice sexual selection:* selection for morphology or behavior that enables a male to attract females directly.

shoot A stem with its leaves, flowers, etc.

sieve element A conductile cell of the phloem.

sign stimulus (or releaser) A simple cue that orients or triggers specific innate behavior.

sinus [L curve, hollow] (1) A channel for the passage of blood lacking the characteristics of a true blood vessel. (2) A hollow within bone or another tissue (e.g. the air-filled sinuses of some of the facial bones).

sol Colloid in which the suspended particles are dispersed at random; *cf.* gel.

solute Substance dissolved in another (the solvent).

solution [L *solutio* loosening] A homogeneous molecular mixture of two or more substances.

solvent Medium in which one or more substances (the solute) are dissolved.

-soma, somat-, -some [Gk *sōma* body] Body, entity.

somatic Pertaining to the body; to all cells except the germ cells; to the body wall. *Somatic nervous system:* a portion of the nervous system that is at least potentially under control of the will; *cf.* autonomic nervous system.

specialized Adapted to a special, usually rather narrow, function or way of life.

speciation The process of formation of new species.

species [L kind] The largest unit of population within which effective gene flow occurs or could occur.

sperm [Gk *sperma* seed] A male gamete.

sphincter [Gk *sphinktēr* band] A ring-shaped muscle that can close a tubular structure by contracting.

spindle A microtubular structure with which the chromosomes are associated in mitosis and meiosis.

sporangium A plant structure that produces spores.

spore [Gk *spora* seed] An asexual reproductive cell, often a resting stage adapted to resist unfavorable environmental conditions.

sporophyll [Gk *phyllon* leaf] A modified leaf that bears spores.

sporophyte [Gk *phyton* plant] A diploid plant that produces spores.

stamen [L thread] A male sexual part of a flower; a microsporophyll of a flowering plant.

starch A glucose polymer, the principal polysaccharide storage product of vascular plants.

stele [Gk *stēlē* upright slab] The vascular cylinder in the center of a root or stem, bounded externally by the endodermis.

stereo- [Gk *stereos* solid] Solid; three-dimensional.

steroid Any of a number of complex, often biologically important compounds (e.g., some hormones and vitamins), composed of four interlocking rings of carbon atoms.

stimulus Any environmental factor that is detected by a receptor.

stoma [Gk mouth] An opening, regulated by guard cells, in the epidermis of a leaf or other plant part (*pl.* stomata).

stroma [Gk *strōma* bed, mattress] The ground substance within such organelles as chloroplasts and mitochondria.

subspecies A genetically distinct geographic subunit of a species.

substrate (1) The base on which an organism lives, e.g., soil. (2) In chemical reactions, a substance acted upon, as by an enzyme.

succession In ecology, progressive change in the plant and animal life of an area.

sucrose A double sugar composed of a unit of glucose and a unit of fructose; table sugar.

suspension A heterogeneous mixture in which the particles of one substance are kept dispersed by agitation.

sym-, syn- Together.

symbiosis [Gk *bios* life] The living together of two organisms in an intimate relationship.

sympathetic nervous system One of the two parts of the autonomic nervous system.

sympatric [L *patria* homeland] Having the same range.

symplast In a plant, the system constituted by the cytoplasm of cells interconnected by plasmodesmata.

synapse [Gk *haptein* to fasten] A juncture between two neurons.

synapsis The pairing of homologous chromosomes during meiosis.

synergistic [Gk *ergon* work] Acting together with another substance or organ to achieve or enhance a given effect.

systemic circulation The part of the circulatory system supplying body parts other than the gas-exchange surfaces.

-tactic Referring to a taxis.

taxis A simple continuously oriented movement in animals (e.g., phototaxis, geotaxis) (*pl.* taxes).

taxonomy [Gk *taxis* arrangement] The classification of organisms on the basis of their evolutionary relationships.

tendon [L *tendere* to stretch] A type of connective tissue attaching muscle to bone.

territory A particular area defended by an individual against intrusion by other individuals, particularly of the same species.

testis Primary male sex organ in which sperm are produced (*pl.* testes).

thalamus [Gk *thalamos* inner chamber] Part of the rear portion of the vertebrate forebrain, a center for integration of sensory impulses.

thallus [Gk *thallos* young shoot] A plant body exhibiting relatively little tissue differentiation and lacking true roots, stems, and leaves.

thorax [Gk *thōrax* breastplate] In mammals, the part of the trunk anterior to the diaphragm, which partitions it from the abdomen. In insects, the body region between the head and the abdomen, bearing the walking legs and wings.

thymus [Gk *thymos* warty excrescence] Glandular organ that plays an important role in the development of immunologic capabilities in vertebrates.

thyroid [Gk *thyreoeidēs* shield-shaped] An endocrine gland of vertebrates located in the neck region.

thyroxin A hormone, produced by the thyroid, that stimulates a speedup of metabolism.

tissue [L *texere* to weave] An aggregate of cells, usually similar in both structure and function, that are bound together by intercellular material.

T lymphocyte *See* lymphocyte.

toxin A proteinaceous substance produced by one organism that is poisonous to another.

trachea In vertebrates, the part of the respiratory system running from the pharynx into the thorax; the "windpipe." In land arthropods, an air duct running from an opening in the body wall to the tissues.

tracheid An elongate thick-walled tapering conductile cell of the xylem.

trans- Across; beyond.

transcription In genetics, the synthesis of RNA from a DNA template.

transduction [L *ducere* to lead] In genetics, the transfer of genetic material from one host cell to another by a virus. In neurobiology, the translation of a stimulus like light or sound into an electrical change in a receptor cell.

transformation The incorporation by bacteria of fragments of DNA released into the medium from dead cells.

translation In genetics, the synthesis of a polypeptide from an mRNA template.

translocation In botany, the movement of organic materials from one place to another within the plant body, primarily through the phloem. In genetics, the exchange of parts between nonhomologous chromosomes.

transpiration Release of water vapor from the aerial parts of a plant, primarily through the stomata.

transposition The movement of DNA from one position in the genome to another. *Transposon:* a mobile segment of DNA, usually encoding the enzymes necessary to effect its own movement.

-trophic [Gk *trophē* food] Nourishing; stimulatory.

tropic hormone A hormone produced by one endocrine gland that stimulates another endocrine gland.

tropism [Gk *tropos* turn] A turning response to a stimulus, primarily by differential growth patterns in plants.

turgid [L *turgidus* swollen] Swollen with fluid.

turgor pressure [L *turgēre* to be swollen] The pressure exerted by the contents of a cell against the cell membrane or cell wall.

tympanic membrane [Gk *tympanon* drum] A membrane of the ear that picks up vibrations from the air and transmits them to other parts of the ear; the eardrum.

urea The nitrogenous waste product of mammals and some other vertebrates, formed in the liver by combination of ammonia and carbon dioxide.

ureter The duct carrying urine from the kidney to the bladder in higher vertebrates.

urethra The duct leading from the bladder to the exterior in higher vertebrates.

uric acid An insoluble nitrogenous waste product of most land arthropods, reptiles, and birds.

uterus In mammals, the chamber of the female reproductive tract in which the embryo undergoes much of its development; the womb.

vaccine [L *vacca* cow] Drug containing an antigen, administered to induce active immunity in the patient.

vacuole [L *vacuus* empty] A membrane-bounded vesicle or chamber in a cell.

valence A measure of the bonding capacity of an atom, which is determined by the number of electrons in the outer shell.

vascular tissue [L *vasculum* small vessel] Tissue concerned with internal transport, such as xylem and phloem in plants and blood and lymph in animals.

vaso- [L *vas* vessel] Blood vessel.

vector [L *vectus* carried] Transmitter of pathogens.

vegetative Of plant cells and organs: not specialized for reproduction. Of reproduction: asexual. Of bodily functions: involuntary.

vein [L *vena* blood vessel] A blood vessel that transports blood toward the heart.

vena cava [L hollow vein] One of the two large veins that return blood to the heart from the systemic circulation of vertebrates.

ventral [L *venter* belly] Pertaining to the belly or underparts.

vessel element A highly specialized cell of the xylem, with thick secondary walls and extensively perforated end walls.

villus [L shaggy hair] A highly vascularized fingerlike process from the intestinal lining or from the surface of some other structure (e.g., a chorionic villus of the placenta) (*pl.* villi).

virus [L slime, poison] A submicroscopic noncellular, obligatorily parasitic entity, composed of a protein shell and a nucleic acid core, that exhibits some properties normally associated with living organisms, including the ability to mutate and to evolve.

viscera [L] The internal organs, especially those of the great central body cavity.

vitamin [L *vita* life] An organic compound, necessary in small quantities, that a given organism cannot synthesize for itself and must obtain prefabricated in the diet.

woody dicot *See* dicot.

X chromosome The female sex chromosome.

xylem [Gk *xylon* wood] A vascular tissue that transports water and dissolved minerals upward through the plant body.

Y chromosome The male sex chromosome.

yolk Stored food material in an egg.

zoo- [Gk *zōion* animal] Animal; motile.

zoospore A ciliated or flagellated plant spore.

zygote [Gk *zygōtos* yoked] A fertilized egg cell.

zymogen [Gk *zymē* leaven] An inactive precursor of an enzyme.

INDEX

Page numbers in **boldface** refer to illustrations; those in *italics* identify definitions or main treatment of subjects mentioned in several parts of the book.